Lecture Notes in Computer Science 2803

Edited by G. Goos, J. Hartmanis, and J. van Leeuwen

Springer
Berlin
Heidelberg
New York
Hong Kong
London
Milan
Paris
Tokyo

Matthias Baaz Johann A. Makowsky (Eds.)

Computer Science Logic

17th International Workshop CSL 2003
12th Annual Conference of the EACSL
8th Kurt Gödel Colloquium, KGC 2003
Vienna, Austria, August 25-30, 2003
Proceedings

Volume Editors

Matthias Baaz
Vienna University of Technology
Institute of Algebra and Computational Mathematics
Wiedner Hauptstr. 8-10, 1040 Wien, Austria
E-mail: baaz@logic.at

Johann A. Makowsky
Technion - Israel Institute of Technology
Faculty of Computer Science
Haifa 32000, Israel
E-mail: janos@cs.technion.ac.il

Cataloging-in-Publication Data applied for

A catalog record for this book is available from the Library of Congress

Bibliographic information published by Die Deutsche Bibliothek
Die Deutsche Bibliothek lists this publication in the Deutsche Nationalbibliografie;
detailed bibliographic data is available in the Internet at <http://dnb.ddb.de>.

CR Subject Classification (1998): F.4.1, F.4, I.2.3-4, F.3

ISSN 0302-9743
ISBN 3-540-40801-0 Springer-Verlag Berlin Heidelberg New York

Springer-Verlag Berlin Heidelberg New York
a member of BertelsmannSpringer Science+Business Media GmbH

http://www.springer.de

Printed in Germany

Typesetting: Camera-ready by author, data conversion by DA-TeX Gerd Blumenstein
Printed on acid-free paper SPIN 10949757 06/3142 5 4 3 2 1 0

Preface

The annual conference of the European Association for Computer Science Logic, CSL 2003, was held jointly with the 8th Kurt Gödel Colloquium, 8. KGC, at the Vienna University of Technology on 25–30 August 2003.

The conference series CSL started as a program of international workshops on Computer Science Logic, and then in its sixth meeting became the annual conference of the EACSL. This conference was the 17th meeting and 12th EACSL conference.

The KGC is the biennial conference of the Kurt Gödel Society. It has taken place in various formats and has been devoted to special topics in logic, such as computational logic, set theory, algebraic logic, and history of logic.

The CSL 2003 and 8. KGC were organized by the Kurt Gödel Society jointly with the Institut für Algebra und Computermathematik and the Institut für Computersprachen, Vienna University of Technology.

The CSL 2003 and 8. KGC joint program committee had the difficult task of choosing among 112 mostly high-quality submissions. Each paper was refereed by at least two reviewers and then discussed by the whole program committee. In the final stage, during a three day electronic discussion, the program committee selected 39 papers for presentation at the conference and publication in these proceedings. Unfortunately, many high-quality papers had to be rejected due to the limitations set by the conference duration.

The program committee chose as invited speakers Bruno Buchberger, Dov Gabbay, Helmut Veith, Nikolai Vorobjov, and Andrei Voronkov. Additionally, Sergei Artemov was invited jointly by CSL 2003 and 8. KGC and the European Summer School in Logic Language and Information (ESSLLI 2003) for the key note address of both events.

In addition to the main conference program, CSL 2003 and 8. KGC featured four tutorials given by Ahmed Bouajjani (*Verification of Infinite State Systems*), Georg Moser and Richard Zach (*The Epsilon Calculus*), Nikolai Vorobjov (*Effective Quantifier Elimination over Real Closed Fields*), Igor Walukiewicz (*Winning Strategies and Synthesis of Controllers*).

The last day of CSL 2003 and 8. KGC was jointly held with the 2nd annual workshop of the European Research Training Network GAMES (Games and Automata for Synthesis and Validation), organized by Erich Grädel.

We thank the program committee and all the referees for the work in reviewing the papers. We are grateful to Norbert Preining for taking care of the main organizational tasks and we thank the remaining members of the local organizing team, Arnold Beckmann, Agata Ciabattoni, Christian Fermüller, Rosalie Iemhoff, and Sebastiaan Terwijn.

We would like to thank the following institutions for supporting the meeting: the European Association for Computer Science Logic (EACSL), the Austrian Federal Ministry for Education, Science and Culture, the City Council of Vienna, the Vienna University of Technology, and the companies IBM and Siemens.

June 2003 Matthias Baaz and Johann Makowsky

Program Committee

Matthias Baaz *Vienna University of Technology (Chair)*
Arnold Beckmann *Vienna University of Technology*
Lev Beklemishev *Steklov Institute, Moscow/Utrecht Univ.*
Agata Ciabattoni *Vienna University of Technology*
Kousha Etessami *University of Edinburgh*
Chris Fermüller *Vienna University of Technology*
Didier Galmiche *LORIA - UHP Nancy*
Harald Ganzinger *Max Plank Institut Saarbrücken*
Erich Grädel *Aachen University*
Petr Hajek *Czech Academy of Science*
Martin Hyland *University of Cambridge*
Reinhard Kahle *Universidade Nova de Lisboa*
Helene Kirchner *LORIA CNRS, Nancy*
Daniel Leivant *Indiana University Bloomington*
Johann Makowsky *Technion-IIT Haifa (Co-chair)*
Jerzy Marcinkowski *Wroclaw University*
Franco Montagna *University of Siena*
Robert Nieuwenhuis *Tech. Univ. Catalonia Barcelona*
Michel Parigot *CNRS University of Paris 7*
Jeff Paris *Manchester University*
Helmut Schwichtenberg *Munich University*
Jerzy Tiuryn *Warsaw University*

Additional Referees

Klaus Aehlig
José Alferes
Rajeev Alur
Carlos Areces
Albert Atserias
Ralph Back
Ulrich Berger
Dietmar Berwanger
Lars Birkedal
Bruno Blanchet
Steve Bloch
Stefano Borgo
Glen Bruns
Marco Cadoli
Witold Charatonik
Petr Cintula
Horatiu Cirstea
Veronique Cortier
Victor Dalmau
Rod Downey
Jacques Duparc
Roy Dyckhoff
Mauro Ferrari
N. Francez
Carsten Führmann
Brunella Gerla
Silvio Ghilardi
Patrice Godefroid
Rajeev Gore
Bernhard Gramlich
Martin Grohe
O. Grumberg
Therese Hardin
Jerry den Hartog
Hugo Herbelin
M. Hermann
William Hesse
Wim Hesselinx
Pascal Hitzler
Michael Huth
Rosalie Iemhoff
Neill Immerman
Mark Jerrum
Felix Joachimski
Marcin Jurdzinski
Hans Jürgen Ohlbach
Lukasz Kaiser
Michael Kohlhase
Boris Konev
Jan Krajicek
Lars Kristiansen

Jean-Louis Krivine
Orna Kupferman
Dietrich Kuske
Oliver Kutz
Olivier Laurent
Salvatore La Torre
Martin Lange
Dominique Larchey
Alexander Leitsch
Pierre Lescanne
Luigi Liquori
Christof Loeding
Markus Lohrey
Byron Long
Carsten Lutz
P. Madhusudan
Savi Maharaj
Yevginiy Makarov
Jean-Yves Marion
Narciso Marti-Oliet
Maarten Marx
K. Meer
Daniel Mery
Stephan Merz
George Metcalfe
Dale Miller
Oskar Mis
Virgile Mogbil
Tobias Niopkow
Hans de Nivelle
Damian Niwinski
Gethin Norman
Jean-Marc Notin
Pawel Olszta
V. van Oostrom
Fernando Orejas
Geoffrey Ostrin
Martin Otto
J. van de Pol
Chris Pollett
Christophe Raffalli
Alexander Rabinovich
E. Ravve
Maarten de Rijke
Christophe Ringeissen
Philipp Rohde
Marie-Christine Rousset
Paul Roziere
V. Rybakov
Luigi Santocanale
Renate Schmidt
Manfred Schmidt-Schauß
Peter Selinger
Kurt Sieber
Harold Simmons
Alex Simpson
V. Sofronie-Stokkermans
V. Stoltenberg-Hansen
Martin Strecker
Thomas Streicher
Wieslaw Szwast
Lidia Tendera
Wolfgang Thomas
Alwen Tiu
Jacobo Toran
Christian Urban
Alvaro del Val
Helmut Veith
Laurent Vigneron
Albert Visser
Peter Vojtas
Benjamin Werner
Uwe Waldmann
I. Walukiewicz
Heinrich Wansing
Andreas Weiermann
ToMasz Wierzbicki
Lucian Wischik
Alberto Zanardo
Jeffrey Zucker

Local Organizing Committee

Matthias Baaz, Chair
Arnold Beckmann
Agata Ciabattoni
Christian Fermüller
Rosalie Iemhoff
Norbert Preining
Sebastiaan Terwijn

Table of Contents

Deciding Monotonic Games

Parosh Aziz Abdulla[1], Ahmed Bouajjani[2], and Julien d'Orso[1]

[1] Uppsala University, Sweden
[2] University of Paris 7, France

Abstract. In an earlier work [AČJYK00] we presented a general framework for verification of infinite-state transition systems, where the transition relation is monotonic with respect to a *well quasi-ordering* on the set of states. In this paper, we investigate extending the framework from the context of transition systems to that of *games*. We show that monotonic games are in general undecidable. We identify a subclass of monotonic games, called *downward closed* games. We provide algorithms for analyzing downward closed games subject to winning conditions which are formulated as safety properties.

1 Introduction

One of the main challenges undertaken by the model checking community has been to develop algorithms which can deal with infinite state spaces. In a previous work [AČJYK00] we presented a general framework for verification of infinite-state *transition systems*. The framework is based on the assumption that the transition relation is monotonic with respect to a *well quasi-ordering* on the set of states (configurations). The framework has been used both to give uniform explanations of existing results for infinite-state systems such as Petri nets, Timed automata [AD90], lossy channel systems [AJ96b], and relational automata [BBK77,Čer94]; and to derive novel algorithms for model checking of Broadcast protocols [EFM99,DEP99], timed Petri nets [AN01], and cache coherence protocols [Del00], etc.

A related approach to model checking is that of control [AHK97]. Behaviours of reactive systems can naturally be described as *games* [dAHM01,Tho02], where control problems can be reduced to the problem of providing winning strategies. Since the state spaces of reactive systems are usually infinite, it is relevant to try to design algorithms for solving games over infinite state spaces.

In this paper, we consider extending the framework of [AČJYK00] from the context of transition systems to that of games. This turns out to be non-trivial. In fact, for one of the simplest classes of monotonic transition systems, namely *Petri nets*, we show that the game problem is undecidable. The negative result holds for games with the simplest possible winning condition, namely that of *safety*. Such a game is played between two players A and B, where player A tries to avoid a given set of *bad* configurations, while player B tries to force the play into such a configuration. On the other hand, we show decidability of the safety game problem for a subclass of monotonic games, namely *downward*

M. Baaz and J.A. Makowsky (Eds.): CSL 2003, LNCS 2803, pp. 1–14, 2003.

closed games: if a player can make a move from a configuration c_1 to another configuration c_2, then all configurations which are larger than c_1 (with respect to the ordering on the state space) can also make a move to c_2. Typical examples of downward closed systems are those with *lossy behaviours* such as lossy channel systems [AJ96b] and lossy VASS [BM99].

We summarize our (un)decidability results as follows:

- Decidability of the safety problem for games where player B has a downward closed behaviour (a *B-downward closed game*). Considering the case where only one player is downward closed is relevant, since it allows, for instance, modelling behaviours of systems where one player (representing the environment) may lose messages in a lossy channel system (a so called B-LCS game). In case player A has a deterministic behaviour (has no choices), our algorithm for B-downward closed games degenerates to the symbolic backward algorithm presented in [AJ96b,AČJYK00] for checking safety properties. In fact, we give a characterization of the set of winning (and losing) configurations in such a game. Observe that this result implies decidability of the case when both players have downward closed behaviours.
- Decidability of the safety problem for A-downward closed games. In case player B has a deterministic behaviour, our algorithm for A-downward closed games degenerates to the forward algorithms described in [AJ96b,AČJYK00] and [Fin94,FS98] for checking eventuality properties (of the form $\forall \diamond p$). However, in contrast to B-downward closed games, we show it is not possible to give a characterization of the set of winning (or losing) configurations.
- Decidability results for downward closed games do not extend to monotonic games. In particular we show that deciding safety properties for games based on VASS (Vector Addition Systems with States). is undecidable. VASS is a variant of Petri nets. The undecidability result holds even if both players are assumed to have monotonic behaviours.
- Undecidability of *parity games* for both A- and B-downward closed games. In a parity game, each configuration is equipped with a *rank* chosen from a finite set of natural numbers. The winning condition is defined by the parity of the lowest rank of a configuration appearing in the play. In particular, we show undecidability of parity games for both A-LCS and B-LCS games. On the other hand, if both players can lose messages, then the problem is decidable.

Outline. In the next Section, we recall some basic definitions for games. In Section 3, we introduce monotonic and downward closed games. We present a symbolic algorithm for solving the safety problem for *B-downward closed* games in Section 4; and apply the algorithm to B-LCS in Section 5. In Section 6, we consider *A-downward closed* games. In Section 7, we show that the safety problem is undecidable for monotonic games. In Section 8, we study decidability of parity games for the above models.

2 Preliminaries

In this section, we recall some standard definitions for games.

A game G is a tuple $(C, C_A, C_B, \longrightarrow, C_F)$, where C is a (possibly infinite) set of *configurations*, C_A, C_B is a partitioning of C, $\longrightarrow \subseteq (C_A \times C_B) \cup (C_B \times C_A)$ is a set of *transitions*, and $C_F \subseteq C_A$ is a finite set of *final configurations*. We write $c_1 \longrightarrow c_2$ to denote that $(c_1, c_2) \in \longrightarrow$. For a configuration c, we define $Pre(c) = \{c' | \, c' \longrightarrow c\}$, and define $Post(c) = \{c' | \, c \longrightarrow c'\}$. We extend Pre to sets of configurations such that $Pre(D) = \cup_{c \in D} Pre(c)$. The function $Post$ can be extended in a similar manner. Without loss of generality, we assume that there is no deadlock, i.e., $Post(c) \neq \emptyset$ for each configuration c. For a set $D \subseteq C_A$ of configurations, we define $\overset{A}{\sim} D$ to be the set $C_A \setminus D$. The operator $\overset{B}{\sim}$ is defined in a similar manner. For a set $D \subseteq C_A$, we use $\overline{Pre}(D)$ to denote $\overset{B}{\sim} \left(Pre \left(\overset{A}{\sim} D \right) \right)$. For $E \subseteq C_B$, we define $\overline{Pre}(E)$ in a similar manner.

A *play* P (of G) from a configuration c is an infinite sequence $c_0, c_1, c_2, \ldots$ of configurations such that $c_0 = c$, and $c_i \longrightarrow c_{i+1}$, for each $i \geq 0$. A play $c_0, c_1, c_2, \ldots$ is *winning* (for A) if there is no $j \geq 0$ with $c_j \in C_F$.

A *strategy* for player A (or simply an A-*strategy*) is a partial function $\sigma_A : C_A \mapsto C_B$ such that $c \longrightarrow \sigma_A(c)$. A B-*strategy* is a partial function $\sigma_B : C_B \mapsto C_A$ and is defined in a similar manner to σ_A. A configuration $c \in C_A$ together with strategies σ_A and σ_B (for players A and B respectively) define a play $P(c, \sigma_A, \sigma_B) = c_0, c_1, c_2, \ldots$ from c where $c_{2i+1} = \sigma_A(c_{2i})$, and $c_{2i+2} = \sigma_B(c_{2i+1})$, for $i \geq 0$. A similar definition is used in case $c \in C_B$ (interchanging the order of applications of σ_A and σ_B to the configurations in the sequence).

An A-strategy σ_A is said to be *winning* from a configuration c, if for all B-strategies σ_B, it is the case that $P(c, \sigma_A, \sigma_B)$ is winning. A configuration c is said to be *winning* if there is a winning A-strategy from c.

We shall consider the *safety problem* for games:

The safety problem
Instance. A game G and a configuration c.
Question. Is c winning?

3 Ordered Games

In this section, we introduce *monotonic* and *downward closed* games.

Orderings. Let A be a set and let $\preceq$ be a quasi-order (i.e. a reflexive and transitive binary relation) on A. We say that $\preceq$ is a *well quasi-ordering (wqo)* on A if there is no infinite sequence $a_0, a_1, a_2, \ldots$ with $a_i \not\preceq a_j$ for $i < j$. For $B \subseteq A$, we say that B is *canonical* if there are no $a, b \in B$ with $a \neq b$ and $a \preceq b$. We use *min* to denote a function where, for $B \subseteq A$, the value of $min(B)$ is a canonical subset of B such that for each $b \in B$ there is $a \in min(B)$ with $a \preceq b$. We say that $\preceq$ is *decidable* if, given $a, b \in A$ we can decide whether $a \preceq b$. A set $B \subseteq A$

is said to be *upward closed* if $a \in B$ and $a \preceq b$ imply $b \in B$. A *downward closed* set is defined in a similar manner.

Monotonic Games. An *ordered game* G is a tuple $(C, C_A, C_B, \longrightarrow, C_F, \preceq)$, where $(C, C_A, C_B, \longrightarrow, C_F)$ is a game and $\preceq \subseteq (C_A \times C_A) \cup (C_B \times C_B)$ is a decidable wqo on the sets C_A and C_B. The ordered game G is said to be *monotonic* with respect to player A (or simply *A-monotonic*) if, for each $c_1, c_2 \in C_A$ and $c_3 \in C_B$, whenever $c_1 \preceq c_2$ and $c_1 \longrightarrow c_3$, there is a c_4 with $c_3 \preceq c_4$ and $c_2 \longrightarrow c_4$. A *B-monotonic game* is defined in a similar manner. A *monotonic game* is both A-monotonic and B-monotonic.

Downward Closed Games. An ordered game $G = (C, C_A, C_B, \longrightarrow, C_F, \preceq)$ is said to be *A-downward closed* if, for each $c_1, c_2 \in C_A$ and $c_3 \in C_B$, whenever $c_1 \longrightarrow c_3$ and $c_1 \preceq c_2$, then $c_2 \longrightarrow c_3$. A *B-downward closed game* is defined in a similar manner. A game is *downward closed* if it is both A- and B-downward closed. Notice that each class of downward closed games is included in the corresponding class of monotonic games. For instance, each A-downward closed game is A-monotonic. From the definitions we get the following property.

Lemma 1. *For an A-downward closed game G and any set $E \subseteq C_B$, the set $Pre(E)$ is upward closed. A similar result holds for B-downward closed games.*

4 B-Downward Closed Games

We present a symbolic algorithm for solving the safety problem for B-downward closed games. In the rest of this Section, we assume an B-downward closed game $G = (C, C_A, C_B, \longrightarrow, C_F, \preceq)$.

Scheme. Given a configuration c in G, we want to decide whether c is winning or not. To do that, we introduce a scheme by considering a sequence of sets of configurations of the form:

$$s: \quad D_0 \ , \ E_0 \ , \ D_1 \ , \ E_1 \ , \ D_2 \ , \ E_2 \ , \ \ldots$$

where $D_i \subseteq C_A$ and $E_i \subseteq C_B$. Intuitively, the sets D_i and E_i characterize the configurations (in C_A and C_B respectively) which are not winning. The elements of the sequence are defined by

$$D_0 = C_F \qquad E_0 = Pre(D_0)$$

$$D_{i+1} = D_i \cup \overline{Pre}(E_i) \qquad E_{i+1} = E_i \cup Pre(D_{i+1}) \qquad i = 0, 1, 2, \ldots$$

We say that s *converges (at ℓ)* if $D_{\ell+1} \subseteq D_\ell$ or $E_{\ell+1} \subseteq E_\ell$. In such a case, the set $D_\ell \cup E_\ell$ characterizes exactly the set of configurations which are not winning. The question of whether a given configuration c is winning amounts therefore to whether $c \notin (D_\ell \cup E_\ell)$. To show that our characterization is correct, we show the following two Lemmas. The first Lemma shows that if c appears in one of the generated sets then it is not a winning configuration. The second Lemma states that if the sequence converges, then the generated sets contain all non-winning configurations.

Lemma 2. *If $c \in D_i \cup E_i$, for some $i \geq 0$, then c is not winning.*

Lemma 3. *If s converges and $c \notin D_i \cup E_i$ for each $i \geq 0$, then c is winning.*

Below, we present a symbolic algorithm based on the scheme above. We shall work with *constraints* which we use as symbolic representations of sets of configurations.

Constraints. An *A-constraint* denotes a (potentially infinite) set $[\![\phi]\!] \subseteq C_A$ of configurations. A *B-constraint* is defined in a similar manner. For constraints ϕ_1 and ϕ_2, we use $\phi_1 \sqsubseteq \phi_2$ to denote that $[\![\phi_2]\!] \subseteq [\![\phi_1]\!]$. For a set ψ of constraints, we use $[\![\psi]\!]$ to denote $\bigcup_{\phi \in \psi} [\![\phi]\!]$. For sets of constrains ψ_1 and ψ_2, we use $\psi_1 \sqsubseteq \psi_2$ to denote that for each $\phi_2 \in \psi_2$ there is a $\phi_1 \in \psi_1$ with $\phi_1 \sqsubseteq \phi_2$. Notice that $\psi_1 \sqsubseteq \psi_2$ implies $[\![\psi_2]\!] \subseteq [\![\psi_1]\!]$. Sometimes, we identify constraints with their interpretations, so we write $c \in \phi$, $\phi_1 \subseteq \phi_2$, $\phi_1 \cap \phi_2$, $\neg\phi$, etc. We consider a particular class of B-constraints which we call *upward closed constraints.* An *upward closed constraint* is of the form $c \uparrow$, where $c \in C_B$, and has an interpretation $[\![c \uparrow]\!] = \{c' |\ c \preceq c'\}$.

A set Ψ of A-constraints is said to be *effective* with respect to the game G if

- The set C_F is characterized by a finite set $\psi_F \subseteq \Psi$ (i.e. $[\![\psi_F]\!] = C_F$).
- For a configuration $c \in C_A$ and a constraint $\phi \in \Psi$, we can decide whether $c \in [\![\phi]\!]$.
- For each $\phi \in \Psi$, we can compute a finite set ψ' of upward closed constraints such that $[\![\psi']\!] = Pre\,([\![\phi]\!])$. In such a case we use $Pre(\phi)$ to denote the set ψ'. Notice that $Pre\,([\![\phi]\!])$ is upward closed by Lemma 1. Also, observe that computability of $Pre(\phi)$ implies that, for a finite set $\psi \subseteq \Psi$, we can compute a finite set ψ' of upward closed constraints such that $[\![\psi']\!] = Pre\,([\![\psi]\!])$.
- For each finite set ψ of upward closed constraints, we can compute a finite set $\psi' \subseteq \Psi$ such that $[\![\psi']\!] = \overline{Pre}\,([\![\psi]\!])$. In such a case we use $\overline{Pre}(\psi)$ to denote the set ψ'.

The game G is said to be *effective* if there is a set Ψ of constraints which is effective with respect to G.

Symbolic Algorithm. Given a constraint system Ψ which is effective with respect to the game G, we can solve the safety game problem by deriving a symbolic algorithm from the scheme described above. Each D_i will be characterized by a finite set of constraints $\psi_i \in \Psi$, and each E_i will be represented by a finite set of upward closed constraints ψ_i'. More precisely:

$$\psi_0 = \psi_F \qquad \psi_0' = Pre(\psi_0)$$

$$\psi_{i+1} = \psi_i \cup \overline{Pre}(\psi_i') \qquad \psi_{i+1}' = \psi_i' \cup Pre(\psi_{i+1}) \qquad i = 0, 1, 2, \ldots$$

The algorithm terminates in case $\psi_j' \sqsubseteq \psi_{j+1}'$. In such a case, a configuration c is not winning if and only if $c \in [\![\psi_j]\!] \cup [\![\psi_j']\!]$. This gives an effective procedure for deciding the safety game problem according to the following

- Each step can be performed due to effectiveness of Ψ with respect to G.
- For a configuration $c \in C_A$ and a constraint $\phi \in \psi_i$, we can check $c \in [\![\phi]\!]$ due to effectiveness of Ψ with respect to G. For a configuration $c \in C_B$ and a constraint $\phi \in \psi'_i$, we can check $c \in [\![\phi]\!]$ due to decidability of $\preceq$.
- For a configuration c and an upward closed constraint $\phi = c' \uparrow$, we can check $c \in [\![\phi]\!]$, since $\preceq$ is decidable and since $c \in [\![\phi]\!]$ if and only if $c' \preceq c$.
- The termination condition can be checked due to decidability of $\preceq$ (which implies decidability of $\sqsubseteq$).
- Termination is guaranteed due to well quasi-ordering of $\preceq$ (which implies well quasi-ordering of $\sqsubseteq$).

From this we get the following

Theorem 1. *The safety problem is decidable for the class of effective B-downward closed games.*

5 B-LCS

In this section, we apply the symbolic algorithm presented in Section 4 to solve the safety game problem for *B-LCS* games: games between two players operating on a finite set of channels (unbounded FIFO buffers), where player B is allowed to lose any number of messages after each move.

A *B-lossy channel system (B-LCS)* is a tuple $(S, S_A, S_B, L, M, T, S_F)$, where S is a finite set of *(control) states*, S_A, S_B is a partitioning of S, L is a finite set of *channels*, M is a finite *message alphabet*, T is a finite set of *transitions*, and $S_F \subseteq S_A$ is the set of *final states.* Each transition in T is a triple (s_1, op, s_2), where

- either $s_1 \in S_A$ and $s_2 \in S_B$, or $s_1 \in S_B$ and $s_2 \in S_A$.
- op is of one of the forms: $\ell!m$ (sending message m to channel ℓ), or $\ell?m$ (receiving message m from channel ℓ), or *nop* (not affecting the contents of the channels).

A B-LCS $\mathcal{L} = (S, S_A, S_B, L, M, T, S_F)$ induces a B-downward closed game $G = (C, C_A, C_B, \longrightarrow, C_F, \preceq)$ as follows:

- *Configurations:* Each configuration $c \in C$ is a pair (s, w), where $s \in S$, and w, called a *channel state*, is a mapping from L to M^*. In other words, a configuration is defined by the control state and the contents of the channels. We partition the set C into $C_A = \{(s, w) \,|\, s \in S_A\}$ and $C_B = \{(s, w) \mid s \in S_B\}$.
- *Final Configurations:* The set C_F is defined to be $\{(s, w) \mid s \in S_F\}$.
- *Ordering:* For $x_1, x_2 \in M^*$, we use $x_1 \preceq x_2$ to denote that x_1 is a (not necessarily contiguous) substring of x_2. For channel states w_1, w_2, we use $w_1 \preceq w_2$ to denote that $w_1(\ell) \preceq w_2(\ell)$ for each $\ell \in L$. We use $(s_1, w_1) \preceq (s_2, w_2)$ to denote that both $s_1 = s_2$ and $w_1 \preceq w_2$. The ordering $\preceq$ is decidable and wqo (by Higamn's Lemma [Hig52]).

- *Non-loss transitions*: $(s_1, w_1) \longrightarrow (s_2, w_2)$ if one of the following conditions is satisfied
  - There is a transition in T of the form $(s_1, \ell!m, s_2)$, and w_2 is the result of appending m to the end of $w_1(\ell)$.
  - There is a transition in T of the form $(s_1, \ell?m, s_2)$, and w_1 is the result of appending m to the head of $w_2(\ell)$.
  - There is a transition in T of the form (s_1, nop, s_2), and $w_2 = w_1$.
- *Loss transitions:* If $s_1 \in S_B$ and $(s_1, w_1) \longrightarrow (s_2, w_2)$ according to one of the previous two rules then $(s_1, w_1) \longrightarrow (s_2', w_2')$ for each $(s_2', w_2') \preceq (s_2, w_2)$.

Remark. To satisfy the condition that there are no deadlock states in games induced by B-LCS, we can always add two "winning" states $s_1^* \in S_A$, $s_2^* \in S_B$, and two "losing" states $s_3^* \in S_A$, $s_4^* \in S_B$, where $s_3^* \in S_F$, and $s_1^* \notin S_F$. We add four transitions (s_1^*, nop, s_2^*), (s_2^*, nop, s_1^*), (s_3^*, nop, s_4^*), and (s_4^*, nop, s_3^*). Furthermore, we add transitions (s, nop, s_4^*) for each $s \in S_A$, and (s, nop, s_1^*) for each $s \in S_B$. Intuitively, if player A enters a configuration, where he has no other options, then he is forced to move to s_4^* losing the game. A similar reasoning holds for player B.

We show decidability of the safety problem for B-LCS using Theorem 1. To do that we first describe upward closed constraints for B-LCS, and then introduce constraints which are effective with respect to B-LCS. We introduce upward closed constraints in several steps. First, we define upward closed constraints on words, and then generalize them to channel states and configurations. An *upward-closed constraint* over M is of the form $X \uparrow$ where $X \subseteq M^*$, and has an interpretation $[\![X \uparrow]\!] = \{x | \exists x' \in X.\ x' \preceq x\}$. An upward closed constraint ϕ over channel states is a mapping from L to upward closed constraints over M, with an interpretation $[\![\phi]\!] = \{w | \forall \ell \in L.\ w(\ell) \in [\![\phi(\ell)]\!]\}$. We use $w \uparrow$ to denote the upward closed constraint ϕ over channel states where $\phi(\ell) = w(\ell) \uparrow$ for each $\ell \in L$. An upward closed constraint ϕ (over configurations) is of the form (s, ϕ'), where $s \in S$ and ϕ' is an upward closed constraint over channel states, with an interpretation $[\![\phi]\!] = \{(s, w) \mid w \in [\![\phi']\!]\}$. We use $(s, w) \uparrow$ to denote the upward closed constraint $(s, w \uparrow)$.

We introduce *extended upward-closed constraints* which we show to be effective with respect to B-LCS games. An *extended upward-closed constraint* over M is of the form $x \bullet \phi$, where $x \in M^*$ and ϕ is an upward closed constraint over M, and has an interpretation $[\![x \bullet \phi]\!] = \{x \bullet x' | x' \in [\![\phi]\!]\}$. Extended upward closed constraints are generalized to channel states and configurations in a similar manner to above.

In the rest of this section we prove the following lemma

Lemma 4. *Extended upward closed constraints are effective for B-LCS games.*

From Theorem 1 and Lemma 4 we get the following

Theorem 2. *The safety problem is decidable for B-LCS games.*

We devote the rest of this section to the proof of Lemma 4. This is achieved as follows:

- The set C_F is characterized by the (finite) set of constraints of the form (s, ϕ) where $s \in S_F$ and ϕ is an extended upward closed constraint over channels, where $\phi(\ell) = \epsilon \bullet \epsilon \uparrow$ for each $\ell \in L$ (notice that $[\![\epsilon \bullet \epsilon \uparrow]\!] = M^*$).
- For a configuration $c \in C_A$ and an extended constraint ϕ we can check whether $c \in [\![\phi]\!]$. (Lemma 5).
- For each extended upward closed constraint ϕ, we can compute a finite set ψ of upward closed constraints, such that $\psi = Pre(\phi)$ (Lemma 7).
- For each finite set ψ_1 of upward closed constraints, we can compute a finite set ψ_2 of extended upward closed constraints, such that $\psi_2 = \overline{Pre}(\psi_1)$. (Lemma 6).

For words x_1, x_2, we use $x_1 \sqcap x_2$ to denote the (finite) set of minimal (with respect to $\preceq$) words x_3 such that $x_1 \preceq x_3$ and $x_2 \preceq x_3$.

Lemma 5. *For a configuration $c \in C_A$ and an extended constraint ϕ we can check whether $c \in [\![\phi]\!]$.*

Lemma 6. *For each finite set ψ_1 of upward closed constraints, we can compute a finite set ψ_2 of extended upward closed constraints, such that $\psi_2 = \overline{Pre}(\psi_1)$.*

Lemma 7. *For each extended upward closed constraint ϕ, we can compute a finite set ψ of upward closed constraints, such that $\psi = Pre(\phi)$.*

Remark. Theorem 1 holds also in the case where both players can lose messages. In fact, we can show that negation constraints are effective with respect to such games.

6 *A*-Downward Closed Games

We present an algorithm for solving the safety problem for A-downward closed games. We use the algorithm to prove decidability of the safety problem for a variant of lossy channel games, namely A-LCS.

An A-downward closed game is said to be *effective* if for each configuration c we can compute the set $Post(c)$. Observe that this implies that the game is finitely branching.

Suppose that we want to check whether a configuration $c_{init} \in C_A$ is winning. The algorithm builds an AND-OR tree, where each node of the tree is labelled with a configuration. OR-nodes are labelled with configurations in C_A, while AND-nodes are labelled with configurations in C_B.

We build the tree successively, starting from the root, which is labelled with c_{init} (the root is therefore an OR-node). At each step we pick a leaf with label c and perform one of the following operations:

- If $c \in C_F$ then we declare the node *unsuccessful* and close the node (we will not expand the tree further from the node).
- If $c \in C_A$, $c \notin C_F$, and there is a predecessor of the node in the tree with label c' where $c' \preceq c$ then we declare the node *successful* and close the node.
- Otherwise, we add a set of successors, each labelled with an element in $Post(c)$. This step is possible by the assumption that the game is effective.

The procedure terminates by Köning's Lemma and by well quasi-ordering of $\preceq$. The resulting tree is evaluated interpreting AND-nodes as conjunction, OR-nodes as disjunction, successful leaves as the constant true and unsuccessful leaves as the constant false. The algorithm answers "yes" if and only if the resulting tree evaluates positively.

Theorem 3. *The safety problem is decidable for effective A-downward closed games.*

A-LCS An *A-LCS* has the same syntax as a B-LCS. The game induced by an A-LCS has a similar behaviour to that induced by a B-LCS. The difference is that in the definition of the *loss transitions*:

- If $s_1 \in S_A$ and $(s_1, w_1) \longrightarrow (s_2, w_2)$ according to a non-loss transition then $(s_1, w_1) \longrightarrow (s_2', w_2')$ for each $(s_2', w_2') \preceq (s_2, w_2)$.

It is straightforward to check that a game induced by an A-LCS is A-downward closed and effective. This gives the following.

Theorem 4. *The safety problem is decidable for A-LCS games.*

Although the safety problem is decidable for A-LCS games, it is not possible to give a characterization of the set of winning configurations as we did for B-LCS. By a similar reasoning to Lemma 1, the set $Pre(\overset{B}{\sim} E_i)$ is upward closed and therefore can be characterized by a finite set of upward closed constraints for each $i \geq 0$. In turn, the set $\bigcup_{i \geq 0} D_i$ can be characterized by a finite set of negation constraints. We show that we cannot compute a finite set of negation constraints ψ such that $[\![\psi]\!] = \bigcup_{i \geq 0} D_i$, as follows.

We reduce an uncomputability result reported in [BM99] for transition systems induced by lossy channel systems. The results in [BM99] imply that we cannot characterize the set of configurations c satisfying the property $c \models \exists_\infty \Box \neg S_F$, i.e., we cannot characterize the set of configurations from which there is an infinite computation which never visits a given set S_F of control states. Given a lossy channel system $\mathcal{L}$ (inducing a transition system) and a set S_F of states, we derive an A-LCS $\mathcal{L}'$ (inducing an A-downward closed game). For each configuration c in $\mathcal{L}$, it is the case that $c \models \exists_\infty \Box \neg S_F$ if and only if the configuration corresponding to c is winning in the game induced by $\mathcal{L}'$. Intuitively, player A simulates the transitions of $\mathcal{L}$, while player B follows passively. More precisely, each state s in $\mathcal{L}$ has a copy $s \in C_A$ in $\mathcal{L}'$. For each transition $t = (s_1, op, s_2)$ in $\mathcal{L}$, there is a corresponding "intermediate state" $s_t \in C_B$ and two corresponding transitions (s_1, op, s_t) and (s_t, nop, s_2) in $\mathcal{L}'$. Furthermore, we have two state $s_1^* \in C_A$ and $s_2^* \in C_B$ which are losing (defined in a similar manner to Section 5). Each configuration in C_A can perform a transition labelled with *nop* to s_2^*. It is straightforward to check that a configuration c is winning in $\mathcal{L}'$ if and only if $c \models \exists_\infty \Box \neg F$.

From this, we get the following:

Theorem 5. *We cannot compute a finite set of negation constraints characterizing the set of non-winning configurations in an A-LCS (although such a set always exists).*

7 Undecidability of Monotonic Games

We show that the decidability of the safety problem does not extend from downward closed games to monotonic games. We show undecidability of the problem for a particular class of monotonic games, namely *VASS games.* In the definition of VASS games below, both players are assumed to have monotonic behaviours. Obviously, this implies undecidability for A- and B-monotonic games.

In fact, it is sufficient to consider VASS with two dimensions (two variables). Let $\mathcal{N}$ and $\mathcal{I}$ denote the set of natural numbers and integers respectively.

VASS Games A *(2-dimensional) VASS (Vector Addition System with States) game* $\mathcal{V}$ is a tuple (S, S_A, S_B, T, S_F), where S is a finite set of *(control) states*, S_A, S_B is a partitioning of S, T is a finite set of *transitions*, and $S_F \subseteq S$ is the set of *final states.* Each transition is a triple $(s_1, (a, b), s_2)$, where

- either $s_1 \in S_A$ and $s_2 \in S_B$, or $s_1 \in S_B$ and $s_2 \in S_A$.
- $a, b \in \mathcal{I}$. The pair (a, b) represents the change made to values of the variables during the transition.

A VASS $\mathcal{V} = (S, S_A, S_B, T, S_F)$ induces a monotonic game $G = (C, C_A, C_B, \longrightarrow, C_F, \preceq)$ as follows:

- Each configuration $c \in C$ is a triple (s, x, y), where $s \in S$ and $x, y \in \mathcal{N}$. In other words, a configuration is defined by the state and the values assigned to the variables.
- $C_A = \{(s, x, y) \mid s \in S_A\}$.
- $C_B = \{(s, x, y) \mid s \in S_B\}$.
- $(s_1, x_1, y_1) \longrightarrow (s_2, x_2, y_2)$ iff $(s_1, (a, b), s_2) \in T$, and $x_2 = x_1 + a$, and $y_2 = y_1 + b$. Observe that since $x_2, y_2 \in \mathcal{N}$, we implicitly require $x_2 \geq 0$ and $y_2 \geq 0$; otherwise the transition is blocked.
- $C_F = \{(s, x, y) \mid s \in S_F\}$.
- $(s_1, x_1, y_1) \preceq (s_2, x_2, y_2)$ iff $s_1 = s_2$, $x_1 \leq x_2$, and $y_1 \leq y_2$.

We can avoid deadlock in VASS games in a similar manner to Section 5.

Theorem 6. *The safety problem is undecidable for VASS games.*

Undecidability is shown through a reduction from an undecidable problem for *2-counter machines.*

2-Counter Machines. A *2-counter machine* M is a tuple (S_M, T_M), where S_M is a finite set of *states*, and T_M is a finite set of *transitions.* Each transition is a triple of the form $(s_1, (a, b), s_2)$, or $(s_1, x = 0?, s_2)$, or $(s_1, y = 0?, s_2)$, where $s_1, s_2 \in S_M$.

A *configuration* of M is a triple (s, x, y) where $s \in S_M$ and $x, y \in \mathcal{N}$. We define a transition relation $\longrightarrow$ on configurations such that $(s_1, x_1, y_1) \longrightarrow (s_2, x_2, y_2)$ iff either

- $(s_1, (a, b), s_2) \in T_M$, and $x_2 = x_1 + a$, and $y_2 = y_1 + b$; or

- $(s_1, x = 0?, s_2) \in T_M$ and $x_2 = x_1 = 0$, and $y_2 = y_1$; or
- $(s_1, y = 0?, s_2) \in T_M$ and $x_2 = x_1$, and $y_2 = y_1 = 0$.

The *2-counter reachability problem* is defined as follows

2-counter reachability problem
Instance. A 2-counter machine $M = (S_M, T_M)$ and two states $s_{init}, s_f \in S_M$.
Question. Is there a sequence

$$(s_0, x_0, y_0) \longrightarrow (s_1, x_1, y_1) \longrightarrow (s_2, x_2, y_2) \longrightarrow \cdots \longrightarrow (s_n, x_n, y_n)$$

of transitions such that $s_0 = s_{init}$, $x_0 = 0$, $y_0 = 0$, and $s_n = s_f$?

It is well-known that the 2-counter reachability problem is undecidable. In the following, we show how to reduce the 2-counter reachability problem to the safety problem for VASS games. Given a 2-counter machine $M = (S_M, T_M)$ and two states $s_{init}, s_f \in S_M$, we construct a corresponding VASS game, such that the reachability problem has a positive answer if and only if the game problem has a negative answer. Intuitively, player B emulates the moves of the 2-counter machine, while player A is passive. Tests for equality with 0 cannot be emulated directly by a VASS system. This means that player B could try to make moves not corresponding to an actual move of the 2-counter machine. However, if player B tries to "cheat", i.e. make a forbidden move, then we allow player A to go into a winning escape loop. This means that player B always chooses to make legal moves. Furthermore, we add an escape loop accessible when the system has reached the final state. This loop is winning for player B. Thus, player B wins whenever the final state is reachable. More formally, we define the VASS game $\mathcal{V} = (S, S_A, S_B, T, S_F)$ as follows:

- $S_A = \{s_t^A |\, t \in T_M\} \cup \{s_*^A, s_{reached}^A, s_{init}^A\}$. In other words, for each transition $t \in T_M$ there is a state $s_t^A \in S_A$. We also add three special states s_*^A, $s_{reached}^A$ and s_{init}^A to S_A.
- $S_B = \{s^B |\, s \in S_M\} \cup \{s_*^B\}$. In other words, for each state in $s \in S_M$ there is a corresponding state $s^B \in S_B$. We also add a special state s_*^B to S_B.
- For each transition t of the form $(s_1, (a, b), s_2) \in T_M$, there are two transitions in T, namely $(s_1^B, (a, b), s_t^A)$ and $(s_t^A, (0, 0), s_2^B)$. Player B chooses a move, and player A follows passively.
- For each transition t of the form $(s_1, x = 0?, s_2) \in T_M$, there are three transitions in T, namely $(s_1^B, (0, 0), s_t^A)$, $(s_t^A, (0, 0), s_2^B)$, and $(s_t^A, (-1, 0), s_*^B)$. Player B may cheat here. However, if this is the case, player A will be allowed to move to s_*^B, which is winning.
- Transitions of the form $(s_1, y = 0?, s_2) \in T_M$ are handled in a similar manner to the previous case.
- There are five additional transitions in T, namely an initializing transition $(s_{init}^A, (0, 0), s_{init}^B)$; an escape loop to detect that the final state has been reached $\left(s_f^B, (0, 0), s_{reached}^A\right)$ and $\left(s_{reached}^A, (0, 0), s_f^B\right)$; a loop to detect illegal moves $(s_*^B, (0, 0), s_*^A)$ and $(s_*^A, (0, 0), s_*^B)$.
- $S_F = \{s_{reached}^A\}$.

Let $G = (C, C_A, C_B, \longrightarrow, C_F, \preceq)$ be the monotonic game induced by $\mathcal{V}$. We show that there is a sequence
$(s_0, x_0, y_0) \longrightarrow (s_1, x_1, y_1) \longrightarrow (s_2, x_2, y_2) \longrightarrow \cdots \longrightarrow (s_n, x_n, y_n)$ of transitions in M with $s_0 = s_{init}$, $x_0 = 0$, $y_0 = 0$, and $s_n = s_f$ iff the configuration $(s^A_{init}, 0, 0)$ is not winning in G.

8 Parity Games

A *parity game G of degree n* is a tuple $(C, C_A, C_B, \longrightarrow, r)$ where $C, C_A, C_B, \longrightarrow$ are defined as in games (Section 2), and r is a mapping from C to the set $\{0, \ldots, n\}$ of natural numbers. We use C^k to denote $\{c | \, r(c) = k\}$. The sets C^k_A and C^k_B are defined in a similar manner. We call $r(c)$ the *rank* of c. Abusing notation, we define the *rank* $r(P)$ of a play $P = c_0, c_1, c_2, \ldots$ to be $\min\{r(c_0), r(c_1), r(c_2) \ldots\}$. We say that P is *parity winning* if $r(P)$ is even. We say that c is *parity winning* if there is an A-strategy σ_A such that, for each B-strategy σ_B, it is the case that $P(c, \sigma_A, \sigma_B)$ is parity winning.

The parity problem
Instance. A parity game G and a configuration c in G.
Question. Is c (parity) winning?

Remark. Notice that our definition of parity games considers parity of configurations which *appear* in the play, rather than the configurations which appear *infinitely often* (which is the standard definition). Our undecidability result can be extended for the latter case, too.

We show below that the parity problem is undecidable for A-downward closed games. In particular, we show undecidability of the problem for A-LCS games. The proof for B-downward closed games is similar.

Theorem 7. *The parity problem is undecidable for A-LCS games.*

In [AJ96a] we show undecidability of the *recurrent state problem*, for transition systems based on lossy channel systems.

Recurrent State Problem
Instance. A lossy channel systems $\mathcal{L}$ and a control states s_{init}.
Question. Is there a channel state w such that there is an infinite computation starting from (s_{init}, w)?

We reduce the recurrent state problem for LCS to the parity problem for A-LCS. We construct a new $\mathcal{L}'$ to simulate $\mathcal{L}$. Intuitively, we let player A choose the moves of the original system, while player B follows passively. An additional loop at the beginning of $\mathcal{L}'$ allows us to guess the initial contents w of the channels. If the system deadlocks, then player B wins. So the only way for player A to win is to make the system follow an infinite sequence of moves. More formally, $\mathcal{L}' = (S, S_A, S_B, L, M, T, S_F)$ is defined as follows. For each control state s in $\mathcal{L}$, we create a control state $s^A \in S_A$. For each transition t in $\mathcal{L}$, we create a control state $s^B_t \in S_B$. For each transition $t = (s_1, op, s_2)$ in $\mathcal{L}$ there are two transitions

(s_1^A, op, s_t^B) and (s_t^B, nop, s_2^A) in $\mathcal{L}'$. Furthermore, there are five additional states $s_1^*, s_4^* \in S_A$, $s_2^*, , s_3^*, , s_5^* \in S_B$, together with the following transitions:

- Two transitions $(s_1^*, \ell!m, s_2^*)$ and (s_2^*, nop, s_1^*) for each $m \in M$ and $\ell \in L$. These two allow to build up the initial channel contents.
- Two transitions (s_1^*, nop, s_3^*) and (s_3^*, nop, s_{init}^A). This is to get to the initial state of $\mathcal{L}$ when the channel content is ready.
- A transition (s^A, nop, s_5^*) for each control state s in $\mathcal{L}$. This transition is only taken when $\mathcal{L}$ is deadlocked.
- Two transitions (s_4^*, nop, s_5^*), and (s_5^*, nop, s_4^*). This loop indicates a deadlock in $\mathcal{L}$.

The ranks of the configurations are defined as follows:

- $r((s_1^*, w)) = r((s_2^*, w)) = r((s_3^*, w)) = 3$, for each w.
- $r((s^A, w)) = r((s_t^B, w)) = 2$, for each w, each transition t in $\mathcal{L}$, and each control state s in $\mathcal{L}$.
- $r((s_4^*, w)) = r((s_5^*, w)) = 1$, for each w.

We show that (s_1^*, ϵ) is parity-winning if and only if there exists a w and an infinite sequence starting from (s_{init}, w).

Remarks.

- In case both players can lose messages, we can show that the parity problem is decidable. The reason is that the best strategy for each player is to empty the channels after the next move. The problem can therefore be reduced into an equivalent problem over finite-state graphs.
- Using results in [May00], we can strengthen Theorem 7, showing undecidability for A-VASS (and B-VASS) games. Such games are special cases of the ones reported here where the message alphabet is of size one (each channel behaves as a *lossy counter*).

References

[AČJYK00] Parosh Aziz Abdulla, Karlis Čerāns, Bengt Jonsson, and Tsay Yih-Kuen. Algorithmic analysis of programs with well quasi-ordered domains. *Information and Computation*, 160:109–127, 2000.

[AD90] R. Alur and D. Dill. Automata for modelling real-time systems. In *Proc. ICALP '90*, volume 443 of *Lecture Notes in Computer Science*, pages 322–335, 1990.

[AHK97] R. Alur, T. Henzinger, and O. Kupferman. Alternating-time temporal logic. In *Proc. 38th Annual Symp. Foundations of Computer Science*, pages 100–109, 1997.

[AJ96a] Parosh Aziz Abdulla and Bengt Jonsson. Undecidable verification problems for programs with unreliable channels. *Information and Computation*, 130(1):71–90, 1996.

[AJ96b] Parosh Aziz Abdulla and Bengt Jonsson. Verifying programs with unreliable channels. *Information and Computation*, 127(2):91–101, 1996.

[AN01] Parosh Aziz Abdulla and Aletta Nylén. Timed Petri nets and BQOs. In *Proc. ICATPN'2001: 22nd Int. Conf. on application and theory of Petri nets*, volume 2075 of *Lecture Notes in Computer Science*, pages 53 -70, 2001.

[BBK77] J. M. Barzdin, J. J. Bicevskis, and A. A. Kalninsh. Automatic construction of complete sample systems for program testing. In *IFIP Congress, 1977*, 1977.

[BM99] A. Bouajjani and R. Mayr. Model checking lossy vector addition systems. In *Symp. on Theoretical Aspects of Computer Science*, volume 1563 of *Lecture Notes in Computer Science*, pages 323–333, 1999.

[Čer94] K. Čerāns. Deciding properties of integral relational automata. In Abiteboul and Shamir, editors, *Proc. ICALP '94*, 21st *International Colloquium on Automata, Lnaguages, and Programming*, volume 820 of *Lecture Notes in Computer Science*, pages 35–46. Springer Verlag, 1994.

[dA98] L. de Alfaro. How to specify and verify the long-run average behavior of probabilistic systems. In *Proc. LICS' 98* 13th *IEEE Int. Symp. on Logic in Computer Science*, 1998.

[dAHM01] L. de Alfaro, T. Henzinger, and R. Majumdar. Symbolic algorithms for infinite state games. In *Proc. CONCUR 2001*, 12th *Int. Conf. on Concurrency Theory*, 2001.

[Del00] G. Delzanno. Automatic verification of cache coherence protocols. In Emerson and Sistla, editors, *Proc.* 12th *Int. Conf. on Computer Aided Verification*, volume 1855 of *Lecture Notes in Computer Science*, pages 53–68. Springer Verlag, 2000.

[DEP99] G. Delzanno, J. Esparza, and A. Podelski. Constraint-based analysis of broadcast protocols. In *Proc. CSL'99*, 1999.

[EFM99] J. Esparza, A. Finkel, and R. Mayr. On the verification of broadcast protocols. In *Proc. LICS' 99* 14th *IEEE Int. Symp. on Logic in Computer Science*, 1999.

[Fin94] A. Finkel. Decidability of the termination problem for completely specified protocols. *Distributed Computing*, 7(3), 1994.

[FS98] A. Finkel and Ph. Schnoebelen. Well-structured transition systems everywhere. Technical Report LSV-98-4, Ecole Normale Supérieure de Cachan, April 1998.

[Hig52] G. Higman. Ordering by divisibility in abstract algebras. *Proc. London Math. Soc.*, 2:326–336, 1952.

[May00] R. Mayr. Undecidable problems in unreliable computations. In *Theoretical Informatics (LATIN'2000)*, number 1776 in Lecture Notes in Computer Science, 2000.

[Tho02] W. Thomas. Infinite games and verification. In *Proc.* 14th *Int. Conf. on Computer Aided Verification*, volume 2404 of *Lecture Notes in Computer Science*, pages 58–64, 2002.

The Commuting V-Diagram
On the Relation of Refinement and Testing

Bernhard K. Aichernig

United Nations University
International Institute for Software Technology (UNU/IIST)
P.O. Box 3058, Macau
bka@iist.unu.edu

Abstract. This article discusses the relations between the step-wise development through refinement and the design of test-cases. It turns out that a commuting diagram inspired by the V-process model is able to clarify the issues involved. This V-diagram defines the dependencies of specifications, implementations and test-cases in the category of contracts. The objects in this category are contracts defined in the formalism of the refinement calculus. The maps are the refinement steps between these objects. Our framework is able to define the correctness notion of test-cases, testing strategies as refinement rules, and which test-cases should be added under refinement.

Keywords: formal methods, specification-based testing, refinement, refinement calculus, contracts.

1 Introduction

The synergy of formal methods and testing has become a popular area of research. In the last years, test-generation tools have been invented for almost every popular specification language. One reason is the industry's demand to cut the efforts of software testing. Another is the academics' insight that testing is complementary to proving the correctness of a program (see e.g. Hoare's comments on testing in [8]).

However, only little research has been put into the question how the related development techniques such as refinement contribute to testing. Our current research addresses this open issue. In our previous work we have demonstrated that test design can be viewed as a reverse program synthesis problem of finding adequate abstractions [1]. The consequence of this insight is that we are able to define test-case synthesis rules in order to calculate correct test-cases from specifications. The mathematical framework in our work is the refinement calculus of Back and von Wright [6] including a simple but powerful contract language.

Our general approach is able to cover rather different test-selection techniques like domain partitioning [3], interactive scenario selection [2], and mutation testing [4]. These previous work demonstrated that a test-case selection strategy can be represented by means of formal synthesis rules that define how specifications

M. Baaz and J.A. Makowsky (Eds.): CSL 2003, LNCS 2803, pp. 15–28, 2003.

should be changed into test-cases. This approach in inspired by the refinement calculus where refinement rules define correctness preserving development steps. The difference is that our rules represent the derivation of test-cases by means of abstraction steps.

In this paper we focus on the question: "Which new test-cases are needed if we refine a specification or implementation?". In order to give a scientific answer, the role of testing and step-wise development has to be clarified. This can be done using a simple diagram (Figure 1) to which we give a precise mathematical semantics.

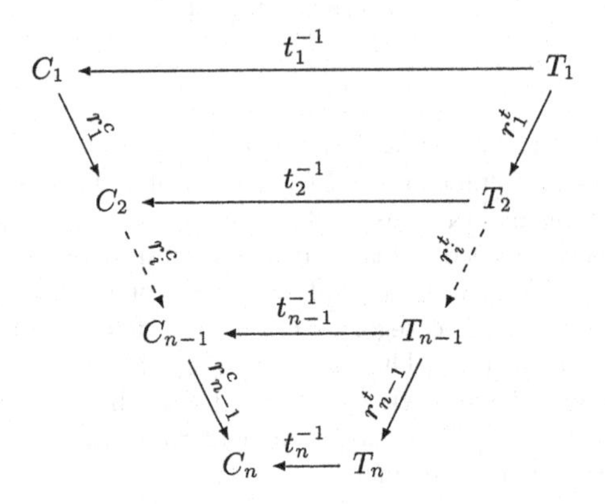

Fig. 1. The V-Diagram.

The V-diagram is inspired by the V process model, a derivative of the waterfall model where the development phases are explicitly linked to testing. Both visualizations stress the importance of testing. The left-hand side of the V in Figure 1 represents the step-wise development of a specification C_1 into an implementation C_n. In the following C_1 to C_n are called contracts representing commitments on different levels of abstractions. All the arrows in the diagram denote refinement. On the right-hand side of the V, the corresponding test-cases are shown. The test-cases are refined in accordance with the contracts. In our view of test-cases, contract C_i must be a refinement of its test-cases T_i. Consequently, test-cases can be viewed as a special form of specification. During discussions we have found that not too many colleagues are aware of this fact.

The formal refinement relation between test-cases T_i and a contract C_i can be interpreted in two directions:

- *test-cases as specifications*: if test-cases are given, an implementation or formal specification must be a correct refinement of the intended test-cases.
- *test-synthesis as an abstraction problem*: As a refinement calculus is a technique to derive correct implementations from a specification by following correctness preserving refinement rules, dual abstraction rules can be used to calculate test-cases. The names of the refinement arrows t_i^{-1} in Figure 1 should indicate this reverse process of abstraction.

In the following sections we will discuss this diagram and its consequences in more detail. Section 2 presents the mathematical interpretation of the V-diagram as a commuting diagram in the category of contracts. Section 3 briefly introduces the notions of contracts and refinements in the refinement calculus. This section covers the left-hand side of the V. Next, Section 4 shows that different kinds of test-cases are in fact abstractions of formal specifications. Section 5 presents a main contribution of this paper: properties for new test-cases under refinement. In Section 6 an example serves to illustrate these findings. Finally, we draw our conclusions in Section 7.

2 A Formal Interpretation of the V-Diagram

The explanations of the V-Diagram have been informal so far. We neither gave a definition of contracts, nor a definition of refinement. These definitions will be provided in Section 3. However, taking a category theoretic view on Figure 1 reveals interesting properties without a detailed knowledge of the refinement calculus.

A category **C** is an algebraic structure consisting of a class of *objects* and a class of *maps*[1]. Each map f, has one object A as *domain* and one object B as *codomain*, denoted $A \xrightarrow{f} B$. For each object A an *identity map* $A \xrightarrow{1_A} A$ exists. Furthermore, *composition* $A \xrightarrow{g \circ f} C$ is defined on maps. In a category the identity laws $1_B \circ f = f$ and $f \circ 1_A = f$ hold and composition is associative: $(h \circ g) \circ f = h \circ (g \circ f)$.

It can be easily seen that contracts form such a category, here called **Con**. The objects in **Con** are contracts formulated in the contract language of the refinement calculus (see Section 3). A map $A \xrightarrow{r} B$ in **Con** is a refinement step of A into B. We write $A \sqsubseteq B$ for B being a refinement of A. The identity refinement $A = B$ exists as well as composition of refinements which is associative.

In category theory equalities can be easily described by means of commuting diagrams. For example, the identity law $1_B \circ f = f$ can be expressed via the following commuting diagram:

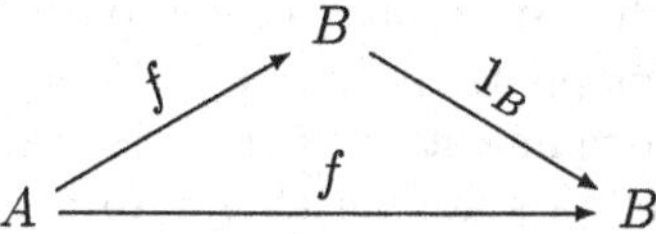

Consequently, by interpreting Figure 1 as a diagram in the category **Con**, several properties can be derived.

The commutations in the V-diagram represent $n-1$ equations for $1 \leq i < n$:

$$r_i^c \circ t_i^{-1} = t_{i+1}^{-1} \circ r_i^t$$

This equation implies that given two sets of test-cases, T_i for a contract C_i and T_{i+1} for a refined contract C_{i+1} such that $C_i \sqsubseteq C_{i+1}$, then test-cases T_{i+1} must

[1] We use the terminology of [9].

be a refinement of T_i or $T_i \sqsubseteq T_{i+1}$. This means that under refinement the test-cases have to be refined as well. What this refinement looks like will be discussed in Section 5.

Since composition exists we know that step-wise refinement preserves refinement. For example, if we extend $\sqsubseteq$ point-wise to maps, then $C_1 \sqsubseteq C_n$ can be expressed as

$$1_{C_1} \sqsubseteq r^c_{n-1} \circ \cdots \circ r^c_2 \circ r^c_1$$

The same holds for refining test-cases, e.g.:

$$1_{T_1} \sqsubseteq r^t_{n-1} \circ \cdots \circ r^t_2 \circ r^t_1$$

As a consequence even more similar equations relating the several refinements on different levels of abstraction can be derived. All these properties are captured in the V-diagram.

3 Refining Contracts: $C \xrightarrow{r^c} C'$

3.1 Contracts

The prerequisite for testing is some form of contract between the user and the provider of a system that specifies what it is supposed to do. In case of system-level testing usually user and software requirement documents define the contract. Formal methods propose mathematics to define such a contract unambiguously and soundly. In the following the formal contract language of Back and von Wright [6] is used. It is a generalization of the conventional pre- and post-condition style of formal specifications known from VDM, B and Z. The foundation of this refinement calculus is based on lattice-theory and classical higher-order logic (HOL).

A system is modeled by a global state space Σ. A single state x in this state space is denoted by $x : \Sigma$. Functionality is either expressed by functional state transformers f or relational updates R. A state transformer is a function $f : \Sigma \to \Gamma$ mapping a state space Σ to the same or another state space Γ.

A relational update $R : \Sigma \to (\Gamma \to \mathsf{Bool})$ specifies a state change by relating a state before $(\sigma : \Sigma)$ with a state after execution $(\gamma : \Gamma)$. In HOL, relations are modeled by functions mapping the states to Boolean valued predicates. For convenience, a relational assignment $(x := x'|b)$ is available and generalizes assignment statements. It sets a state variable x to a new state x' such that b, relating x and x', holds.

The language further distinguishes between the responsibilities of communicating agents in a contract. Here, the contract models the viewpoint of one agent called the *angel* who interacts with the rest of the system called the *demon*. In our work following [6, 5], the user is considered the angel and the system under test the demon. Relational contract statements denoted by $\{R\}$ express relational updates under control of the angel (user). Relational updates of the demon are denoted by $[R]$ and express updates that are non-deterministic from the angel's point of view. Usually, we take the viewpoint of the angel.

The contract statement $\langle f \rangle$ denotes a functional update of the state determined by a state transformer f. There is no choice involved here, neither for the angel nor the demon agent, since there is only one possible next state for a given state.

Two contracts can be combined by sequential composition $C_1; C_2$ or choice operators. The angelic choice $C_1 \sqcup C_2$ and the demonic choice $C_1 \sqcap C_2$ define non-deterministic choice of the angel or demon between two contracts C_1 and C_2. Furthermore, predicate assertions $\{p\}$ and assumptions $[p]$ define conditions the angel, respectively the demon, must satisfy. In this language of contract statements $\{p\}; \langle f \rangle$ denotes partial functions and $\{p\}; [R]$ pre-postcondition specifications. Furthermore, recursive contracts defined by means of least (μ) and greatest fix-point operators (ν) may express several patterns of iteration.

The core contract language used in this work can be summarized by the following BNF grammar, where p is a condition and R a relation over states.

$$C := \{p\} \mid [p] \mid \{R\} \mid [R] \mid C; C \mid C \sqcup C \mid C \sqcap C \mid \mu X \cdot C$$

As needed, we will extend this core language by our own contract statements. However, all new statements will be defined by means of the above core language. Thus, our language extensions are conservative. This means that no inconsistencies into the theory of the refinement calculus are introduced by our new definitions.

3.2 Example Contracts

A few simple examples should illustrate the contract language. The following contract is a pre- postcondition specification of a square root algorithm:

$$\{x \geq 0 \wedge e > 0\}; [x := x' \mid -e \leq x \; - \; x'^2 \leq e]$$

The precondition is an assertion about an input variable x and a precision e. A relational assignment expresses the demonic update of the variable x to its new value x'. Thus, the contract is breached unless $x \geq 0 \wedge e > 0$ holds in the state initially. If this condition is true, then x is assigned some value x' for which $-e \leq x - x'^2 \leq e$ holds.

Consider the following version of the square root contract that uses both kinds of non-determinism:

$$\{x, e := x', e' \mid x' \geq 0 \wedge e' > 0\}; [x := x' \mid -e \leq x \; - \; x'^2 \leq e]$$

In this contract the interaction of two agents is specified explicitly. This contract requires that our agent, called the angel, first chooses new values for x and e. Then the other agent, the demon, is given the task of computing the square-root in the variable x.

The following example should demonstrate that programming constructs can be defined by means of the basic contract statements. A conditional statement can be defined by an angelic choice as follows:

$$\textsf{if } P \textsf{ then } S_1 \textsf{ else } S_2 \textsf{ fi} \triangleq \{P\}; S_1 \sqcup \{\neg P\}; S_2$$

Thus, the angel agent can choose between two alternatives. The agent will, however, always choose only one of these, the one for which the assertion is true, because choosing the alternative where the guard is false would breach the contract. Hence, the agent does not have a real choice if he wants to satisfy the contract[2].

Iteration can be specified by recursive contracts $(\mu X \cdot C)$. Here X is a variable that ranges over contract statements, while $(\mu X \cdot C)$ is the contract statement C, where each occurrence of X in C is interpreted as a recursive invocation of the contract C. For example, the standard while loop is defined as follows:

$$\mathsf{while}\ g\ \mathsf{do}\ S\ \mathsf{od}\ \triangleq\ (\mu X \cdot \mathsf{if}\ g\ \mathsf{then}\ S; X\ \mathsf{else\ skip\ fi})$$

We write $\mathsf{skip} \triangleq \langle id \rangle$ for the action that applies the identity function to the present state.

3.3 Semantics

The semantics of the contract statements is defined by weakest precondition predicate transformers. A predicate transformer $C : (\Gamma \rightarrow \mathsf{Bool}) \rightarrow (\Sigma \rightarrow \mathsf{Bool})$ is a function mapping postcondition predicates to precondition predicates. The set of all predicate transformers from Σ to Γ is denoted by $\Sigma \mapsto \Gamma \triangleq (\Gamma \rightarrow \mathsf{Bool}) \rightarrow (\Sigma \rightarrow \mathsf{Bool})$.

The different roles of the angel and the demon are reflected in the following weakest-precondition semantics. Here q denotes a postcondition predicate and σ a particular state, p is an arbitrary predicate, and R a relation. The weakest-precondition predicate transfmore is denoted by wp. The notation $f.x$ is used for function application instead of the more common form $f(x)$.

$$\begin{aligned}
\mathsf{wp}.\{p\}.q &\triangleq p \cap q && \textit{(assertion)}\\
\mathsf{wp}.[p].q &\triangleq \neg p \cup q && \textit{(assumption)}\\
\mathsf{wp}.\{R\}.q.\sigma &\triangleq (\exists\, \gamma \in \Gamma \,.\, R.\sigma.\gamma \wedge q.\gamma) && \textit{(angelic update)}\\
\mathsf{wp}.[R].q.\sigma &\triangleq (\forall\, \gamma \in \Gamma \,.\, R.\sigma.\gamma \Rightarrow q.\gamma) && \textit{(demonic update)}\\
\mathsf{wp}.(C_1; C_2).q &\triangleq C_1.(C_2.q) && \textit{(sequential composition)}\\
\mathsf{wp}.(C_1 \sqcup C_2).q &\triangleq C_1.q \cup C_2.q && \textit{(angelic choice)}\\
\mathsf{wp}.(C_1 \sqcap C_2).q &\triangleq C_1.q \cap C_2.q && \textit{(demonic choice)}
\end{aligned}$$

In this semantics, the breaching of a contract by our angel agent, means that the weakest-precondition is false. If a demon agent breaches a contract, the weakest-precondition is trivially true. The semantics of the specification constructs above can be interpreted as follows:

- The weakest precondition semantics of an *assertion* contract reflects the fact that, if the final state of the contract should satisfy the post-condition q, then in addition the assertion predicate p must hold. It can be seen that the global state is not changed by an assertion statement. Consequently, the angel breaches this contract if $p \cap q$ evaluates to false.

[2] An alternative definition can be given by demonic choice.

- The semantics of an *assumption* shows that the demon is responsible for satisfying an assumption predicate p. If the assumption does not hold, the demon breaches the contract and the angel is released from the contract. In this case, the weakest-precondition trivially evaluates to true.
- The *angelic update* definition says that a final state γ must exist in the relation R, such that the postcondition q holds. The existential quantifier in the weakest-precondition shows that the angel has control of this update. The angel can satisfy the contract, as long as one update exists that satisfies the postcondition. In the set notation this update is defined as $\mathsf{wp}.\{R\}.q.\sigma \triangleq R.\sigma \cap q \neq \emptyset$.
- This is in contrast to the definition of the *demonic update*. Here, all possible final states γ have to satisfy the postcondition. The reason is that the demonic update is out of our control. It is not known, to which of the possible states, described by the relation R, the state variables will be set. In the set notation this update is defined as $\mathsf{wp}.[R].q.\sigma \triangleq R.\sigma \subseteq q$.
- The weakest-precondition of two sequentially combined contracts is defined by the *composition* of the two weakest-preconditions.
- The *angelic choice* definition shows that the weakest-precondition is the union of the weakest-precondition of the two contracts. Thus, a further choice of the angel further weakens the weakest-preconditions.
- The *demonic choice* is defined as the intersection of the weakest-preconditions of the two contracts. Thus, demonic choice represents a strengthening of the weakest-preconditions.

For further details of the predicate transformer semantics, we refer the reader to [6].

3.4 Refinement and Abstraction

The notion of contracts includes specification statements as well as programming statements. More complicated specification statements as well as programming statements can be defined by the basic contract statements presented above. The refinement calculus provides a synthesis method for refining specification statements into programming statements that can be executed by the target system. The refinement rules of the calculus ensure by construction that a program is correct with respect to its specification.

Formally, refinement of a contract C by C', written $C \sqsubseteq C'$, is defined by the point-wise extension of the subset ordering on predicates: For Γ being the after state space of the contracts, we have

$$C \sqsubseteq C' \triangleq \forall q \in (\Gamma \to \mathsf{Bool}) \cdot C.q \subseteq C'.q$$

This ordering relation defines a lattice of predicate transformers (contracts) with the lattice operators meet $\sqcap$ and join $\sqcup$. The top element $\top$ is $\mathsf{magic}.q \triangleq \mathsf{true}$, a statement that is not implementable since it can magically establish every postcondition. The bottom element $\bot$ of the lattice is $\mathsf{abort}.q \triangleq \mathsf{false}$ defining the notion of abortion. The choice operators and negation of contracts are defined by point-wise extension of the corresponding operations on predicates. A large collection of refinement rules can be found in [6, 10].

Abstraction is dual to refinement. If $C \sqsubseteq C'$, we can interchangeably say C is an abstraction of C'. In order to emphasize rather the search for abstractions than for refinements, we write $C \sqsupseteq C'$ to express C' is an abstraction of C. Trivially, abstraction can be defined as:

$$C \sqsupseteq C' \triangleq C' \sqsubseteq C$$

Hence, abstraction is defined as the reverse of refinement. The reader should keep this technical definition of abstraction in mind when we identify test-cases being abstractions in the following section.

4 Test-Cases through Abstraction: $T \xrightarrow{t^{-1}} C$

In the following we will demonstrate that test-cases are in fact contracts — highly abstract contracts. To keep our discussion simple, we do not consider parameterized procedures, but only global state manipulations. In [6] it is shown how procedures can be defined in the contract language. Consequently, our approach scales up to procedure calls.

4.1 Input-Output Tests

The simplest form of test-cases are pairs of input i and output o data. We can define such an input-output test-case TC as a contract between the user and the unit under test:

$$\mathsf{TC}\ i\ o \triangleq \{x = i\}; [y := y' | y' = o]$$

Intuitively, the contract states that if the user provides input i, the state will be updated such that it equals o. Here, x is the input variable and y the output variable.

In fact, such a TC is a formal pre-postcondition specification solely defined for a single input i. This demonstrates that a collection of n input-output test-cases TCs are indeed point-wise defined formal specifications:

$$\mathsf{TCs} \triangleq \mathsf{TC}\ i_1\ o_1 \sqcup \ldots \sqcup \mathsf{TC}\ i_n\ o_n$$

Moreover, such test-cases are abstractions of general specifications, if the specification is deterministic for the input-value of the test-case, as the following theorem shows.

Theorem 1. *Let $p : \Sigma \to \mathsf{Bool}$ be a predicate, $Q : \Sigma \to \Gamma \to \mathsf{Bool}$ a relation on states, and $TC\ i\ o$ a test-case, where all state variables (the whole state) are* *observable and not just a designated input variable x and an output variable y* [3]. *Thus, input $i : \Sigma$ and output $o : \Gamma$. Then*

$$\{p\}; [Q] \sqsupseteq \mathsf{TC}\ i\ o\ \equiv\ p.i \wedge (Q.i = o)$$

□

The intuition behind Theorem 1 is that input-output test-cases can be viewed as pre-postcondition specifications with a special pre-condition restricting the input to a single test-value. If the specification is deterministic for the given input, a derived test-case is an abstraction of this specification. Note that in the case of non-deterministic specifications a form of non-deterministic test-cases are needed, as has been shown in [2].

Furthermore, the selection of certain test-cases out of a collection of test-cases can be considered an abstraction:

Corollary 1.

$$\mathsf{TC}\ i_1\ o_1 \sqcup \ldots \sqcup \mathsf{TC}\ i_n\ o_n \sqsupseteq TC\ i_k\ o_k$$

for all k, $1 \leq k \leq n$.

Proof. The theorem is valid by definition of the join operator $a \sqcup b \sqsupseteq a$ or $a \sqcup b \sqsupseteq b$, respectively. □

The fact that test-cases are indeed formal specifications and, as Theorem 1 shows, abstractions of more general contracts shows an aspect of why test-cases are so popular: First, they are abstract in the technical sense, and thus easy to understand. Second, they are formal and thus unambiguous.

Here, only the commonly used pre-postcondition contracts have been considered. They are a normal form for all contracts not involving angelic actions. This means that arbitrary contracts excluding $\sqcup$ and $\{R\}$ can be formulated in a pre-postcondition style (see Theorem 26.4 in [6]). However, our result that test-cases are abstractions holds for general contract statements involving user interaction. We refer to [2, 1] where we have demonstrated that even sequences of interactive test-cases, so called scenarios, are abstractions of interactive system specifications. In the same work we have shown how test-synthesis rules can be formulated for deriving test-partitions or scenarios.

5 Refining Test-Cases: $T \xrightarrow{r^t} T'$

5.1 Defining the Problem

In the previous sections it has been demonstrated that test-cases are abstractions of specification or implementation contracts. If contracts are refined towards an implementation, the test-cases can be refined accordingly in order to test the correctness of the chosen refinement steps.

[3] The slightly more complex case of partly observable states including proofs can be found in [2]

However, the refinement relation $\sqsubseteq$ alone is a too weak criterion in order to design new test-cases. For example, the relation $\sqsubseteq$ includes the equality of contracts $=$. This means that the same test-cases could be used for the refined contract. Note that in this case the V-diagram still commutes, but that $T \xrightarrow{r^t} T' = T \xrightarrow{1_T} T$. More useful criteria for designing new test-cases are needed.

The problem of finding new test-cases can be formulated as follows:

Theorem 2. *Given a contract C and its refinement C'. Furthermore, it is assumed that test-cases T for C have been correctly designed, thus $C \sqsupseteq T$. Then the refined test-cases T' for testing C' have the general form*

$$T' = (T \sqcup T'_{new})$$

for arbitrary new test-cases T'_{new} such that $C' \sqsupseteq T'_{new}$.

Proof. It must be shown that $(T \sqcup T'_{new})$ is a refinement of T and that they are test-cases of C', thus $T \sqsubseteq (T \sqcup T'_{new})$ and $C' \sqsupseteq (T \sqcup T'_{new})$ must hold.

The first property $T \sqsubseteq (T \sqcup T'_{new})$ holds for any T'_{new} by definition of the join operator $\sqcup$ in the lattice of contracts.

Similarly, by the definition of $\sqcup$, the second property $C' \sqsupseteq (T \sqcup T'_{new})$ holds iff $C' \sqsupseteq T$ and $C' \sqsupseteq T'_{new}$. Trivially, $C' \sqsupseteq T'_{new}$ is a premise. $C' \sqsupseteq T$ follows from the facts that C' is a refinement of C, thus $C' \sqsupseteq C$, and that T are correct test-cases designed for C, thus $C \sqsupseteq T$ holds.

This proves that the diagram commutes for refined test-cases $(T \sqcup T'_{new})$. □

Theorem 2 shows that the problem of finding test-cases for a refined specification can be reduced to the question of finding additional new test-cases. However, not all new test-cases are useful. In order to be economical, they should cover the newly added parts or properties of the refined contract C'.

5.2 Abstract Contracts as Mutations

A main contribution of this paper is the insight that the problem of finding useful test-cases T'_{new} can be mapped to the author's approach to mutation testing [4].

In analogy to program mutation in the traditional approaches of mutation testing, in contract-based mutation testing introduced in [4] we produce a mutant contract by introducing small changes to the formal contract definition. Then we select test-cases that are able to detect the introduced mutations. What kind of changes to be introduced is heuristically defined by a set of mutation operators that take a contract and produce its mutated version. However, not all mutants are useful:

Definition 1. ***(useful mutant)*** *A useful mutant of a contract C is a mutated contract C_i such that there exists a test-case $TC_i \sqsubseteq C$ that is able to distinguish an implementation of C_i from an implementation of C.* □

Theorem 3. *A mutant C_i of a contract C is a useful mutant iff $C \not\sqsubseteq C_i$.* □

In [4] we pointed out that *abstract mutants* $C_i \sqsubset C$ are useful. Furthermore we defined adequate test-cases for mutation testing:

Definition 2. ***(adequate test-case)*** *A test-case for a contract C is called adequate with respect to a set of mutation operators M if it is able to distinguish at least one useful mutant M_i, with $m.C = M_i$ and $m \in M$.* □

Here, in our quest for useful new test-cases T'_{new} for a contract C' and $C \sqsubseteq C'$, we consider C as a useful mutant. Then T'_{new} should be adequate test-cases that are able to distinguish the abstract mutant from its refinement.

Thus the criterion for useful new test-cases is an adaptation of the abstraction rule for deriving mutation test-cases (Theorem 12 in [4]):

Theorem 4. *Given a contract C and its refinement C'. Furthermore, test-cases T for C have been previously designed following some selection strategy. Then the refined test-cases T' for testing C' have the general form*

$$T' = (T \sqcup T'_{new})$$

and

$$C' \sqsupseteq T'_{new} \ \wedge \ C \not\sqsupseteq T'_{new}$$

Alternatively an abstraction rule for producing useful new test-cases can be formulated as follows:

$$\frac{\begin{array}{c} C \sqsubseteq C', C \sqsupseteq T \\ C' \sqsupseteq T'_{new}, \ C \not\sqsupseteq T'_{new} \end{array}}{C' \sqsupseteq_t T \sqcup T'_{new}}$$

□

It can be seen that $C \not\sqsupseteq T'_{new}$ is the central property for adding new test-cases that are able to distinguish between contract C and its refinement C' or between their implementations, respectively. An example serves to illustrate this.

6 Example

Assume that we want to find an index i pointing to the smallest element in an array $A[1..n]$, where $n = len.A$ is the length of the array and $1 \leq n$ (so the array is nonempty).

We define the predicate $minat.i.A$ to hold when the minimum value in A is found at $A[i]$.

$$minat.i.A \triangleq \ 1 \leq i \leq len.A \wedge (\forall j \mid 1 \leq j \leq len.A \,\bullet\, A[i] \leq A[j])$$

Then the contract

$$MIN \triangleq \ \{1 \leq len.A\}; [i := i' \mid minat.i'.A]$$

constitutes the problem of finding an index i with the minimum $A[i]$ in A.

In this example only input-output test-cases are considered. We define $\mathsf{TC}\ i\ o \triangleq \{A = i\}; [i := i' | i' = o]$ to be the corresponding deterministic test-case contracts with the input array A and the output index i containing the minimum of the array.

Assume the following test-cases for MIN have been designed:

$$T_1 \triangleq \mathsf{TC}\ [2,1,3]\ 2$$

$$T_2 \triangleq \mathsf{TC}\ [2,3,1]\ 3$$

$$T_3 \triangleq \mathsf{TC}\ [1,2,3]\ 1$$

In [4] it is shown how these test-cases are derived by following a mutation testing strategy. Note that only deterministic test-cases are used, although the specification is non-deterministic. We have to add a non-deterministic test-case with two possipble outputs

$$\mathsf{T}_4 \triangleq \mathsf{TC}\ [1,1,3]\ (i = 1 \vee i = 2)$$

in order to obtain sufficient coverage with respect to mutation testing.

Next, consider the following implementation (refinement) of MIN using a guarded iteration statement:

$$\begin{aligned} MIN \sqsubseteq MIN' \triangleq\ & \mathsf{begin\ var}\ k := 2;\ i := 1; \\ & \quad \mathsf{do}\ k \leq n \wedge A[k] < A[i] \rightarrow i, k\ := k, k+1 \\ & \quad []\ k \leq n \wedge A[k] \geq A[i] \rightarrow k\ := k+1 \\ & \quad \mathsf{od} \\ & \mathsf{end} \end{aligned}$$

What test-cases should be added in order to test the refinement? The rule in Theorem 4 points us to the answer, since a correct test-case of MIN' should be found that is not an abstraction of MIN. It is straightforward to reconsider the non-deterministic test-case, since the implementation MIN' is deterministic. Hence, new test-cases should be able to distinguish the deterministic from the non-deterministic version (mutant):

$$\mathsf{T}_{new} \triangleq \mathsf{TC}\ [1,1,3]\ 1$$

is such a useful test-case. Our framework actually allows us to prove this fact by showing that $MIN' \sqsupseteq \mathsf{T}_{new}$ and $MIN \not\sqsupseteq \mathsf{T}_{new}$.

7 Conclusion

Summary. In this article we have presented a unified view on testing and refinement. First, category theory has been used to analyze the relations of testing and refinement on a high-level of abstraction. A commuting diagram was able to capture the essential properties. Then, the refinement calculus was used to take a closer look at this diagram. Refinement and abstraction have been defined formally. It has been shown that test-cases are in fact abstract contracts of specifications or implementations. Finally, the novel framework was able to answer the question "Which new test-cases are needed if we refine a specification or implementation?" A new formal inference rule for assessing new test-cases under refinement has been provided.

Related Work. To our current knowledge no other work has been published that uses a refinement calculus for deriving test-cases. Stepney was the first who made the abstraction relation between test-cases and object-oriented Z specifications explicitly [12]. Her group developed a tool for interactively calculating partition abstractions and to structure them for reuse. Our work can be seen as an extension of this view on testing. We are happy to see that recently other colleagues seem to pick up our ideas and transfer the results to other frameworks and application domains like modal logics and security [11].

Derrick and Boiten looked at testing refinements in state-based Z-specifications [7]. We assume that our framework can express the properties they have collected for refining this kind of specifications. This is part of future work .

Discussion. Our approach to testing is a novel unification of the area of testing and formal methods in several senses: First, test-cases are considered as a special form of specifications or contracts. Next, program synthesis techniques, like the refinement calculus can be applied to test-synthesis — it is just abstraction. Finally, the notion of refinement is sufficient to clarify the relations of testing and step-wise development — it forms a category of contracts. The work presented demonstrates that such a unification leads to simpler theories and to a better understanding of the different test-approaches available. In this paper, for example, we have found a scientifically defensible formula for which test-case to be added under refinement.

Here, we did not address the important issue of test coverage explicitly, but have given a rule for calculating test-cases for refinements. In our previous papers we have given abstraction rules for different selection strategies [2–4]. The new rule for refinement test-cases can be easily combined with these rules such that the new test-cases follow a certain test-strategy. One has just to assure that the new test-cases T'_{new} in Theorem 4 are calculated according to a strategy rule and that T'_{new} is not an abstraction of C.

Future Work. We hope that our newly gained insight that testing refinements can be mapped to mutation testing will stimulate further research. Our approach can be easily applied to other frameworks that form a similar category. In category theory we would say that a functor must exists that maps our category of contracts into other frameworks.

Finally, automation is of major concern. Future work must involve the research that analyzes to which extent our approach can be automated and which mathematical framework is best suited to support this automation.

References

1. Bernhard K. Aichernig. *Systematic Black-Box Testing of Computer-Based Systems through Formal Abstraction Techniques.* PhD thesis, Institute for Software Technology, TU Graz, Austria, January 2001. Supervisor: Peter Lucas.
2. Bernhard K. Aichernig. Test-case calculation through abstraction. In J. N. Oliveira and Pamela Zave, editors, *Proceedings of FME 2001: Formal Methods for Increasing Software Productivity, International Symposium of Formal Methods Europe, March 2001, Berlin, Germany*, volume 2021 of *Lecture Notes in Computer Science*, pages 571–589. Springer-Verlag, 2001.
3. Bernhard K. Aichernig. Test-Design through Abstraction: a Systematic Approach Based on the Refinement Calculus. *Journal of Universal Computer Science*, 7(8):710 – 735, August 2001.
4. Bernhard K. Aichernig. Contract-based mutation testing in the refinement calculus. In *REFINE'02, the British Computer Society - Formal Aspects of Computing refinement workshop, Copenhagen, Denmark, July 20-21, 2002, affiliated with FME 2002*, 2002.
5. Ralph Back, Anna Mikhajlova, and Joakim von Wright. Reasoning about interactive systems. In Jeannette M. Wing, Jim Woodcock, and Jim Davies, editors, *FM'99 — Formal Methods, World Congress on Formal Methods in the Development of Computing Systems, Toulouse, France, September 1999, Proceedings, Volume II*, volume 1709 of *Lecture Notes in Computer Science*, pages 1460–1476. Springer-Verlag, 1999.
6. Ralph-Johan Back and Joakim von Wright. *Refinement Calculus, a Systematic Introduction.* Graduate Texts in Computer Science. Springer-Verlag, 1998.
7. John Derrick and Eerke Boiten. Testing refinements of state-based formal specifications. *Software Testing, Verification and Reliability*, 9:27–50, July 1999.
8. Tony Hoare. Towards the Verifying Compiler. In *The United Nations University/ International Institute for Software Technology 10th Anniversary Colloquium: Formal Methods at the Crossroads, from Panacea to Foundational Support, Lisbon, March 18–21, 2002.* Springer-Verlag, 2003. To be published.
9. F. William Lawvere and Stephen H. Schanuel. *Conceptual Mathematics: a First Introduction to Categories.* Cambridge University Press, 1997.
10. Carroll Morgan. *Programming from Specifications.* Series in Computer Science. Prentice-Hall International, 2nd. edition, 1994.
11. Claus Pahl. Interference Analysis for Dependable Systems using Refinement and Abstraction. In *Formal Methods Europe 2002, Copenhagen, Denmark, July 22-24, 2002.* Springer-Verlag, 2002.
12. Susan Stepney. Testing as abstraction. In J. P. Bowen and M. G. Hinchey, editors, *Proceedings of ZUM '95, the 9th International Conference of Z Users, September 1997, Limerick, Ireland*, volume 967 of *Lecture Notes in Computer Science*, pages 137–151. Springer-Verlag, 1995.

Concurrent Construction of Proof-Nets

Jean-Marc Andreoli[1,2] and Laurent Mazaré[1]

[1] Xerox Research Centre Europe, Grenoble, France
[2] Institut de Mathématiques de Luminy, France
Jean-Marc.Andreoli@xrce.xerox.com

Abstract. The functional paradigm of computation has been widely investigated and given a solid mathematical foundation, initiated with the Curry-Howard isomorphism, then elaborated and extended in multiple ways. However, this paradigm is inadequate to capture many useful programming intuitions, arising in particular in the development of applications integrating distributed, autonomous components. Indeed, in this context, non-determinism and true concurrency are the rule, whereas functional programming stresses determinism, and, although it allows some degree of concurrency, it is more as a "nice feature to have" rather than a primary assumption.

This paper is part of a program the ambition of which is to provide a logical foundation to a set of programming intuitions which, until now, have not been adequately accounted for. In particular, we are interested in the intuitions which lie behind the concept of *transaction*, a powerful and essential concept in distributed component-based application development. This concept is independent of the application domain and usually captured in an abstract form in middleware architectural layers.

We claim here that proof-construction, and more precisely proof-net construction in Linear Logic, offers the adequate basis for our purpose. We outline the relation, which is of the same nature as the Curry-Howard isomorphism, between transactional concepts and mechanisms on one hand, and proof-net construction on the other. Finally, we describe an algorithm which performs concurrent proof-net construction, where each expansion step is viewed as a transaction. Conflicts between such transactions are minimised using general topological properties of proof-nets, based on a variant of the notion of "domination tree", introduced and proved here.

Keywords: Logical foundations of programming paradigms, Linear Logic, Proof-nets, Concurrency, Transactions.

1 Introduction

It has been recognised early on in the development of computer science that proof theory offers adequate concepts and tools to capture abstract computational mechanisms. Thus, the so-called Curry-Howard isomorphism establishes a direct correspondence between proof-theoretic notions such as Cut elimination and fundamental mechanisms of the functional programming paradigm, such as parameter passing and type-preserving evaluation. However, this paradigm is essentially characterised by its determinism: a well-typed computation may evolve in multiple ways, but always terminates and there

M. Baaz and J.A. Makowsky (Eds.): CSL 2003, LNCS 2803, pp. 29–42, 2003.

is a notion of "result" of the computation which is independent of the specific strategy chosen in the reductions (this is essentially the convergence property of typed lambda-calculus). This form of determinism, which is extremely intuitive when computation is viewed as pure (eg. arithmetic) calculation, becomes a hindrance when computation is viewed as coordinated interactions between a set of autonomous components through their environment. And many computer programs are more naturally understood in this way rather than as pure calculation: operating systems, electronic brokers, workflow engines, monitoring tools, etc. These programs may involve deterministic calculations at some points in their execution, but, overall, they do not produce any final "result", and constantly interact instead with their environment in a non fully deterministic way.

There is another computation paradigm, improperly called logic programming, which, especially in its concurrent flavour, stresses non-determinism and interaction rather than calculation. In proof-theoretic terms, it corresponds to a different intuition, namely proof-construction rather than proof-reduction. The mapping between computational and proof-theoretic concepts in the two paradigms can thus be summarised as follows:

Computation	**Proof theory**	
Paradigm	"functional"	"logic"
State	Proof – possibly with Cuts – without Proper axioms	Proof – without Cuts – possibly with Proper axioms
Transformation	Proof reduction ie. Cut elimination	Proof construction ie. Proper axiom elimination
Final state	Proof without Cut	Proof without Proper axiom
Type	Formula	Formula / Sequent

In this paper, we study the problem of the construction of a proof by a set of concurrent agents. The inference system considered here is the focussing bipolar sequent calculus of Linear Logic [1], which has been shown to be equivalent to that of full Linear Logic. However, to express concurrency in proof construction, a sequent system is not the adequate representation tool. Proof-nets, which offer a desequentialised representation of proofs, are more appropriate. The paradigm of proof-net construction by concurrent agents in the multiplicative, transitory fragment of bipolar Linear Logic has been presented in [2]. It is recalled in Section 2. The agents participating in the construction are bipoles from the "Universal program", labelling the inferences of the sequent system.

Whereas it is quite easy to ensure that the structure collaboratively built by the bipole agents remains a proof-structure, it is not so easy to guarantee that it remains a proof-net. This requires checking a correctness criterion, which is a topological property of the structure that ensures that it can be sequentialised, and hence is a proof-net. A priori, checking the correctness criterion may lead each bipole agent to freeze an arbitrary large portion of the topology, resulting in conflicts between the agents, forcing them to take turn and, in the end, restoring the artificial sequentialisations of the sequent system that we had precisely tried to get rid of by turning to proof-nets. The challenge, addressed here, is therefore to identify precisely the portion of the topology that needs to be frozen by a candidate inference to ensure that it does not violate the correctness criterion, whatever the structure outside this portion. Only those inferences for which these portions overlap (conflict) need to be sequentialised. In [2], a simplification procedure was presented, which, when applied to a restricted fragment of transitory multiplicative Linear Logic (basically modelling multiset rewriting), completely solves the concurrency problem: in fact, in this fragment, no correctness conflict ever occurs. Here, we address the problem in the whole fragment of transitory multiplicative Linear Logic, where conflicts may occur, but we show that they can be minimised, based on topological properties of proof-nets which are introduced and proved in Section 3. Note that we still restrict to transitory inferences (with at least one premiss), meaning that we ignore the problem of terminating agents (which, in fact, poses the same problem as the multiplicative unit in correctness checking). Section 4 defines an abstract implementation of a concurrent, collaborative proof-net construction mechanism using these properties.

The bipole agents act as typical infrastructure software (a.k.a. middleware). Indeed, they do not capture any application-specific behaviour (justifying their name of "Universal program"), but their role is to guarantee that some invariant is preserved, namely that the overall structure they build is a proof-net, ie. could be sequentialised into a sequent proof. In that sense, they are closely analogous to transaction schedulers in distributed applications. Indeed, like bipole agents, transaction schedulers are not concerned with the application-level semantics of the actions they involve but only with their interdependences, and, while bipole agents ensure that the inferences they perform could be sequentialised, transaction schedulers ensure that the transactions they enact could be serialised. Thus, proof-net construction by concurrent bipole agents provides a generic logical model of the behaviour of transaction schedulers, which are essential components of middleware infrastructures. The correspondence between transaction concepts, in particular the Atomicity and Isolation properties of "ACID" transactions, and proof-theoretic concepts is at the core of the present paper. It is claimed here that this correspondence is not just a coincidence, or a by-product of logic's powerful expressivity, but reveals the deep computational nature of proof-construction just as the Curry-Howard correspondence does to proof-reduction. In particular, the ability to sequentialise, which characterises proof-nets, has an operational meaning in the paradigm of proof-net construction as a set of isolation constraints (the "I" of ACID transactions) among the agents that perform the construction. Not surprisingly, the correctness criterion for proof-nets is remarkably close to the traditional serialisability criterion of transaction schedulers (absence of cycle in the dependency graph of transactions). Hence, we have the following correspondence

Transactions	Proof-Net construction
Serialisability	Sequentialisability
Isolation	Correctness

2 Focussing Proof-Net Construction

We assume given an infinite set $\mathcal{P}$ of *places*, and we define a *link* by a set of top places and a set of bottom places, together with a polarity (positive or negative). The sets of top and bottom places of a link must be disjoint; furthermore, a negative link must have *exactly* one bottom place, while a positive link must have *at least* one bottom place. A

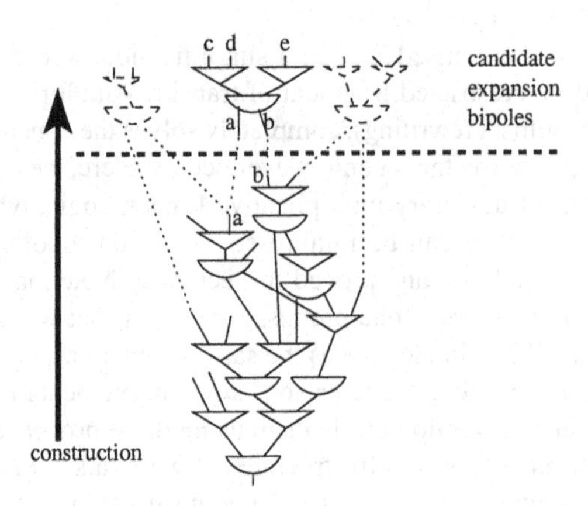

Fig. 1. Example of a focussing proof-structure construction

link L_1 is said to be *just-below* a link L_2, notation $L_1 \nearrow L_2$, if there exists a place which is both at the top of L_1 and at the bottom of L_2. The relation *just-above*, notation $\searrow$, is the converse of $\nearrow$, and can be defined in the same way inverting top and bottom. Two links are said to be *adjacent* if one is just-below (or just-above) the other.

Negative links represent connex combinations of the traditional "par" links of Linear Logic (modulo associativity-commutativity). Positive links represent associative-commutative connex combinations of the traditional "tensor" links, together with identity axioms connected on one side to the input of a tensor link and pending on the other side (hence the possibly multiple bottom places). Note that a positive link does not distinguish between the main output of the tensor combination and the pending places of the identity links. Graphically, the polarities of links are distinguished by their shape: triangular for the negative links and round for the positive links, as follows.

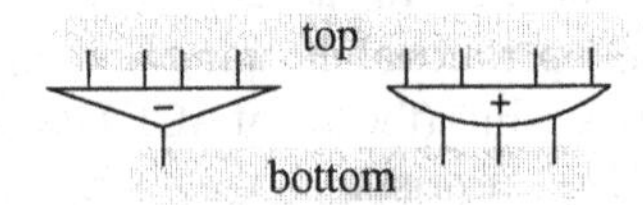

Definition 1. *A (multiplicative)* focussing *proof-structure is a set* π *of links satisfying the following conditions:*

1. *The sets of top (resp. bottom) places of any two links in* π *are disjoint.*
2. *If two links in* π *are adjacent, their polarities are opposite.*

A place which is at the top of a link but not at the bottom of any link in a focussing proof-structure is called a top place of the structure. The bottom places of a structure are defined similarly, by permuting top/bottom. An example of focussing proof-structure is given in Figure 1. In the sequel, except when mentioned otherwise, we take proof-structure to mean focussing proof-structure.

Definition 2. *A proof-structure* π *is said to be* bipolar *if any place occurring at the top of some positive link in* π *also occurs at the bottom of some negative link in* π *and vice-versa. Furthermore,* π *is said to be* elementary *if it is bipolar and contains exactly one positive link.*

The example of Figure 1 is bipolar. Note that any bipolar proof-structure is the union of disjoint elementary proof-structures, and this decomposition is unique. Furthermore, each elementary proof-structure corresponds to a bipole.

It has been shown in [2] how each (possibly open) proof in the focussing bipolar sequent calculus can be mapped into a (focussing) proof-structure. This process is called desequentialisation, and is strictly analogous to the desequentialisation of sequent proofs in Linear Logic [3].

Definition 3. *A* proof-net *is a proof-structure obtained by desequentialisation of a (possibly open) proof in the focussing bipolar sequent calculus.*

A correctness criterion, characterising proof-nets among arbitrary proof-structures, is provided in [2] for transitory focussing proof-structures. It is an adaptation of the Danos-Regnier criterion for proof-modules [4], the focussing proof-structures manipulated here being, properly speaking, proof-modules.

Definition 4. *A proof-structure is said to be* transitory *if all its positive links have at least one top place.*

The criterion for a (transitory) proof-structure π is expressed as a property of the graph $\nearrow_\pi$ (ie. the restriction to π of the relation "just-below" on links). We use the following notations. Let $\mathcal{L}$ be the set of all links. For any $x \in \mathcal{L}$, the expression x^+ (resp. x^-) means that x has a positive (resp. negative) polarity. For any binary graph $\mathcal{R}$ over $\mathcal{L}$, $|\mathcal{R}|$ denotes the support set of $\mathcal{R}$, ie. the set $\bigcup_{x\mathcal{R}y}\{x,y\}$; also, $\mathcal{R}^{op}$ denotes the reverse of $\mathcal{R}$ and $\mathcal{R}^*$ its transitive closure.

Definition 5. *A* trip α *is a non-empty binary relation on $\mathcal{L}$ which is finite, connex and such that any element x in $\mathcal{L}$ has at most one successor (written $\overset{\cdot\cdot\bullet}{\alpha}(x)$ when it exists) and at most one predecessor (written $\overset{\bullet\cdot\cdot}{\alpha}(x)$ when it exists) by α. An element of $\mathcal{L}$ is called a* start-point, stop-point, middle-point *of α if it has, respectively, a successor but no predecessor, a predecessor but no successor, both a predecessor and a successor.*

It is easy to show that for a given trip α, one and only one of the following conditions is true: (i) either α has no start-point nor stop-point, in which case, α is called a *loop*; (ii) or α has a unique start-point and a unique stop-point, and they are distinct. Note that a loop can never be strictly contained in another loop.

Definition 6. *Let $\mathcal{R}$ be a binary graph over $\mathcal{L}$.*

- *$\mathcal{R}$ is said to be* polarised *if adjacent elements in $\mathcal{R}$ are of opposite polarity:*
  $\forall x, y \in \mathcal{L}\ \ x\mathcal{R}y\ \Rightarrow\ (x^- \wedge y^+) \vee (x^+ \wedge y^-)$
- *$\mathcal{R}$ is said to be* bipolarised *if it is finite, polarised, and negative elements have a unique predecessor:*
  $\forall x \in |\mathcal{R}|\ \ (x^-\ \Rightarrow\ \exists! y \in \mathcal{L}\ y\mathcal{R}x)$
- *A trip α is said to be* over $\mathcal{R}$ *if $\alpha \subset \mathcal{R} \cup \mathcal{R}^{op}$. A* singularity *for $\mathcal{R}$ of a trip α is a negative middle-point x of α such that:*
  $\neg(\overset{\bullet\cdot\cdot}{\alpha}(x)\mathcal{R}x\mathcal{R}\overset{\cdot\cdot\bullet}{\alpha}(x) \vee \overset{\cdot\cdot\bullet}{\alpha}(x)\mathcal{R}x\mathcal{R}\overset{\bullet\cdot\cdot}{\alpha}(x))$
- *$\mathcal{R}$ is said to be* correct *if any loop over $\mathcal{R}$ has at least one singularity for $\mathcal{R}$.*

The correctness criterion is defined for proof-structures as follows, and leads to the following result shown in [2].

Definition 7. *A proof-structure π is said to be* correct *if the relation $\upharpoonleft_\pi$ (the restriction to π of the "just-below" relation on links) is correct.*

Theorem 1. *Any proof-net is a correct, bipolar proof-structure. Any correct transitory bipolar proof-structure is a proof-net.*

3 Properties of the Focussing Proof-Nets

In this section, we analyse some properties of transitory focussing proof-nets which are used, in the next section, to ensure correctness preservation during the bottom-up construction of a shared proof-structure by multiple, concurrent agents performing only (transitory) bipole expansions. The demonstration of the main results (provided in the submitted version) are available from the authors.

3.1 Notations and Conventions

$\mathcal{R}$ denotes a bipolarised, correct, binary relation over $\mathcal{L}$. Being correct, $\mathcal{R}$ is obviously acyclic, and it is useful to visualise $\mathcal{R}$ as being oriented bottom-up. Unless specified otherwise, trips and singularities are all taken to be relative to $\mathcal{R}$. The elements of $|\mathcal{R}|$ are called *points*. A trip is said to be *proper* if it is not reduced to a loop with two points. A *path* is a trip that always go upward (while trips in general may go any direction). A point x is said to be *below* a point y if there is a path from x to y. A point is called a *root* if there is no point below it. Note that, since $\mathcal{R}$ is bipolarised, roots are always positive. Let α be a trip and x a point of $|\alpha|$.

- If x is not a stop-point of α then α is said to *exit* x either *upward* when $x\mathcal{R}\overset{\cdot\cdot\bullet}{\alpha}(x)$, or *downward* when $\overset{\cdot\cdot\bullet}{\alpha}(x)\mathcal{R}x$.
- If x is not a start-point of α, then α is said to *enter* x either *downward* when $x\mathcal{R}\overset{\bullet\cdot\cdot}{\alpha}(x)$, or *upward* when $\overset{\bullet\cdot\cdot}{\alpha}(x)\mathcal{R}x$.

3.2 The Domination Forest

We make use below of a notion of domination order which is different from the standard one [5]: whereas the standard notion is applicable to any kind of flow graph, our definition applies only to bipolarised and correct graphs. Our definition is formally similar to the standard one, but considers only singularity-free trips in the graphs, and the proofs of the main results are essentially different, as they exploit the specific properties of bipolarised correct graphs. Moreover, properly speaking, the domination order here is not a tree (as in the standard case) but a forest.

Definition 8. *Let x, y be negative points. x is said to* dominate *y (notation $x \leq y$) if any singularity-free trip starting at a root and stopping upward at y visits x upward.*

Note that for any negative point x, there exists at least one singularity-free trip starting at a root and stopping upward at x: any path from any root below x to x will do (a path is obviously a singularity-free trip); such paths always exist, since $\mathcal{R}$ is acyclic and x cannot itself be a root since it is negative while the roots are positive.

Proposition 1. *The relation $\leq$ on negative points is a forest order.*

We write $x < y$ for $x \leq y \wedge x \neq y$. Since $\leq$ is a forest order, we have:

- Any set of negative points X which has a lower bound by $\leq$ has a greatest lower bound, written $\sqcap(X)$ and called the *joint dominator* of X; as usual, $\sqcap(\{x, y\})$ is written $x \sqcap y$.
- The set of predecessors by $<$ of any negative point x, if not empty, has a greatest element by $\leq$, written $\delta(x)$ and called the *immediate dominator* of x.

Note that the knowledge of any one of $\leq, <, \sqcap, \delta$ is sufficient to recover all the others.

Theorem 2. *Let x, y be negative points, and let α be a singularity-free trip starting downward at x and stopping upward at y. Then any negative point visited by α is strictly dominated by $x \sqcap y$ (when defined), ie.*

$$\forall z \in |\alpha| \;\; z^- \;\Rightarrow\; x \sqcap y < z$$

3.3 Domination under Expansion

In a bipolar expansion step of the proof-net construction process, $\mathcal{R}$ is turned into $\mathcal{R}' = \mathcal{R} \cup (N \times \{p\})$ where p is the (unique) positive link introduced by the bipole and N the set of already present negative links at the top places of which p is connected. In fact, $\mathcal{R}'$ also contains the set of pairs (p, n) where n ranges over the negative links introduced by the bipole, but we will not consider them here as they do not affect the correctness of $\mathcal{R}'$.

Proposition 2. *Let $N \subset |\mathcal{R}|$ be a set of negative elements, and $p \in \mathcal{L} \setminus |\mathcal{R}|$ be a positive element. The graph $\mathcal{R}' = \mathcal{R} \cup (N \times \{p\})$ is bi-polarised. Furthermore, it is correct if and only if there is no singularity-free trip in $\mathcal{R}$ starting downward at some $x \in N$ and stopping upward at some $y \in N$.*

We now assume that the graph $\mathcal{R}' = \mathcal{R} \cup (N \times \{p\})$ is correct. The following result, expliciting the relation between $\leq, \delta$ induced by $\mathcal{R}$ and their counterparts $\leq', \delta'$, induced by $\mathcal{R}'$, underlies our concurrent proof-net construction algorithm.

Theorem 3. *Let x be a negative point.*
Either $\delta(x) \leq' x$ and $\delta'(x) = \delta(x)$, or $\neg\delta(x) \leq' x$ and $\sqcap(N) \leq \delta'(x) < \delta(x)$.

4 A Proof-Net Construction Algorithm

4.1 Information Containers

The proof-net construction mechanism considered here proceeds by expansion steps performed by concurrent agents, each applying a bipole to a set of places that match its trigger. To ensure that the construction remains a proof-net, each agent must check, for each candidate expansion, that its application would not violate the criterion given by Theorem 1 (we assume here that the agents apply only transitory bipoles, so the condition of the Theorem is satisfied). Performing such a check amounts to some kind of traversal of the proof-net built so far: it is useful to visualise it in terms of a token moving from each link to adjacent links in the proof-net, gathering and checking information on the way. Hence, an expansion step is itself a transaction composed of a set of micro-steps, including the traversal of the net and, eventually, the installation of the successful candidate expansion. In each micro-step, the transaction retrieves and/or updates information about the proof-net. For example, it may retrieve information like: what are the links adjacent to link x in the proof-net ? It may also update information, like: link x has a new adjacent link y in the proof-net (this happens when the expansion is actually performed, after a successful correctness check). Adjacency information is obviously crucial to correctness checking. We will see that other pieces of information are also relevant.

To be useful, the information retrieved in each micro-step of a transaction t needs to be protected against updates of that same information by concurrent transactions, until t is completed. For this purpose, we assume that the information is encapsulated into containers that keep track of which transactions access them. Whenever a transaction t accesses a container, it locks it, thus denying access to its content to any other transactions, until t either commits or aborts. We assume here a simple locking model which does not distinguish between access modes (read or write) of the operations. We assume that (i) whenever a transaction commits, it releases all the locks it has acquired; (ii) whenever a transaction aborts, all the updates made to the containers it has locked are cancelled before the locks are released; (iii) in order to avoid deadlocks, whenever a transaction is denied access to a container, the whole transaction is aborted (thus freeing all its locks) and retried from the start (this is a rather crude way of sequentialising conflicting transactions, used here for simplicity; any of the other traditional strategies, eg. based on arbitrary prioritisation of the transactions, is applicable here).

There is one container for each link in the proof-net, holding the adjacency information, ie. pointers to the containers of the links which are just-above and just-below in the proof-net. Note that, due to the way the construction proceeds, the information concerning the links just-below is never changed after creation of a link. The information concerning the links just-above is also never changed in the case of a positive link, whereas it may be changed by some further expansion of the proof-net in the case of a negative link. Hence, locking is only needed for negative links, and only when the information concerning the links just-above is accessed.

As an optimisation, two links need not be put in separate containers if it can be guaranteed that whenever a transaction locks one, it also eventually locks the other. Hence, containers may group more than one link. Grouping may be done by *sessions* as proposed in [2]. This optimisation could straightforwardly be incorporated in the sequel, but will be omitted here for simplification purpose.

Finally, the information attached to each place is also encapsulated into containers. Typically, a place container holds the type attached to that place (a negative atom, never changed) as well as pointers to the containers of the links of the proof-net at the top or

Table 1. A naive implementation of expansion checking

Procedure CHECK(T: set of places)
  Let $N = \emptyset$
  For-each $a \in T$ **Do**:
    Lock a; **If** a is consumed **Then**: **Abort**
    Add to N the (negative) link at the top of which a appears
  For-each $x \in N$ **Do**: **Call** CHECK-TRIP$^{\downarrow}$(x,N,$\emptyset$)

Procedure CHECK-TRIP$^{\downarrow}$(x: negative link, N, A: set of negative links)
  If $x \notin A$ **Then**:
    Let y be the (positive) link just-below x
    For-each (negative) link $x' \neq x$ just-above y **Do**: **Call** CHECK-TRIP$^{\uparrow}$(x',N,A)
    For-each (negative) link x' just-below y **Do**: **Call** CHECK-TRIP$^{\downarrow}$(x',N,A)

Procedure CHECK-TRIP$^{\uparrow}$(x: negative link, N, A: set of negative links)
  If $x \in N$ **Then**: **Abort Else**: **Lock** x
  For-each (positive) link y just-above x **Do**:
    For-each (negative) link x' just-above y **Do**: **Call** CHECK-TRIP$^{\uparrow}$(x',N,$A \cup \{x\}$)
    For-each (negative) link $x' \neq x$ just-below y **Do**: **Call** CHECK-TRIP$^{\downarrow}$(x',N,$A \cup \{x\}$)

bottom of which that place appears. The link at the top of which a place appears is never changed after creation of that place. On the other hand, the link at the bottom of which a place appears may be added dynamically, by an expansion step, but once it is set, it is never changed (the place is then said to be consumed). Place containers are used to ensure that the construction remains a proof-structure (a pre-requisite to correctness), simply by preventing that a place be consumed by several concurrent transactions.

To perform an expansion of the proof-net built so far from a given set of places, a bipole agent proceeds in two steps: (i) it first checks that the candidate expansion would not break the proof-structure nor violate the correctness criterion; (ii) it then installs the candidate expansion, creating and/or updating place and link containers with the appropriate information. The first phase reads information from the already existing containers, while the second phase updates the information in these containers, and creates new containers. These two phases are executed in a single transaction.

4.2 A Naive Procedure

A naive procedure for the first phase (preservation of proof-structure and correctness) is given in Table 1. It is a direct application Proposition 2 reformulated here as follows:

> An expansion of a proof-net π from a set of places T (matching the trigger of a bipole) yields a proof-net if and only if (i) no place of T is at the bottom of a link of π and (ii) there exists no singularity-free trip over $\nearrow_{\pi}$ starting downward at some link x and stopping upward at some link y such that both x, y have a top place in T.

Table 2. Computation of the joint dominator of a set of points

Procedure JOINT-DOM(N: set of negative links) **Returns** negative link (or undefined)
  Sort N by decreasing height
  While N has more than 1 element:
    Remove the first point x from N
    If $\delta(x)$ is undefined **Then**: **Return** undefined **Else**: **Lock** x
    If $\delta(x) \notin N$ **Then**: Insert $\delta(x)$ in N, so that it remains sorted
  Return the single remaining element of N

Thus, procedure CHECK of Table 1 first computes the set N of points having a top place in T, checking at the same time condition (i) above. It then follows all the possible singularity-free trips starting downward at a link of N. If one of them stops upward at another link of N, then it means that the candidate expansion induces a violation of the correctness criterion and the transaction should be aborted. Otherwise, correctness is preserved and the transaction can proceed with the second phase (installation of the expansion) and be committed.

Following the singularity-free trips is achieved by the two procedures CHECK-TRIP. Their first parameter x denotes the current *negative* link in the trip being built, and the direction of the arrow ($\uparrow$ or $\downarrow$) denotes the direction of traversal of that link (upward or downward). The second parameter N is the set of links which must not be reached upward. The third parameter A denotes the set of negative links already visited upward by the trip, and is used to ensure that the trip does not visit them again downward (this is sufficient to ensure that no link is ever visited twice). Locking is only needed in the upward visit phases, since these are the only phases which use information which may be modified by concurrent transactions (namely the information concerning the links just-above).

However, the systematic exploration of all singularity-free trips starting at N amounts to locking a large portion of the proof-net, and decreases the potential concurrency of the transactions. We use below the results of Section 3 to reduce the amount of exploration required by each candidate expansion and thus improve the potential for concurrency in the construction. Basically, we enrich the containers with additional information about the proof-net (beyond bare adjacency), and use it to limit the scope of the exploration of singularity-free trips. The constraint here is two-fold: (i) the additional information should enable to compute a practical limit to the exploration (and locking); (ii) the update of the additional information in case of success should not itself require new explorations (and locking).

4.3 An Improved Procedure

Given a candidate expansion at a set T of places, the procedure of Table 1 computes all the singularity-free trips starting downward at a point x of N, to check that none of them stops upward at another point y of N (where N is the set of links of π with a top place in T). Now, by Theorem 2, any negative point z of such a trip satisfies $x \sqcap y < z$, hence $\sqcap(N) < z$. Thus, in the procedure of Table 1, the exploration of the singularity-free trips

Table 3. An improved implementation of expansion checking

Procedure CHECK(T: set of places):
  Let $N = \emptyset$
  For-each $a \in T$ **Do**:
    Lock a; **If** a is consumed **Then**: **Abort**
    Add to N the (negative) link at the top of which a appears
  Let $\xi =$ JOINT-DOM(N)
  If ξ is defined **Then**: Let $A = \{\xi\}$ **Else**: Let $A = \emptyset$
  Let $U = \emptyset$
  For-each $x \in N$ **Do**: Add CHECK-TRIP$^{\downarrow}$(x,N,A) to U
  For-each $x \in U$ **Do**: Set $\delta(x) =$ JOINT-DOM($\{\delta(x), \xi\}$)

Procedure CHECK-TRIP$^{\downarrow}$(x: negative link, N, A: set of negative links)
Returns set of negative links
  Let $U = \emptyset$
  If $x \notin A$ **Then**:
    Let y be the (positive) link just-below x
    For-each (negative) link $x' \neq x$ just-above y **Do**: Add CHECK-TRIP$^{\uparrow}$(x',N,A) to U
    For-each (negative) link x' just-below y **Do**: Add CHECK-TRIP$^{\downarrow}$(x',N,A) to U
  Return U

Procedure CHECK-TRIP$^{\uparrow}$(x: negative link, N, A: set of negative links)
Returns set of negative links
  If $x \in N$ **Then**: **Abort Else**: **Lock** x
  Let $U = \{x\}$
  For-each (positive) link y just-above x **Do**:
    For-each (negative) link x' just-above y **Do**: Add CHECK-TRIP$^{\uparrow}$(x',N,$A \cup \{x\}$) to U
    For-each (negative) link $x' \neq x$ just-below y **Do**: Add CHECK-TRIP$^{\downarrow}$(x',N,$A \cup \{x\}$) to U
  Return U

in the current proof-net can be limited by $\sqcap(N)$ (when it exists). The computation of $\sqcap(N)$ can itself be obtained using only mapping δ (the immediate dominator mapping). Thus, the additional information we propose to store in each container for a (negative) link x is a pointer to the container of $\delta(x)$ (when it exists, otherwise, some null pointer).

Note that each expansion step may update the immediate dominator information in several containers, and it needs to be locked when used in a transaction to compute $\sqcap(N)$, as shown in the procedure of Table 2. This procedure also makes use of a height information, included in the containers. This information is a positive integer computed each time a container is created (and never modified afterwards) so as to respect the following condition: the height of a container is strictly greater than the height of any of the containers which are just-below it (the set of which is known at the time of the creation of the container and never changes).

Now, the joint dominator $\sqcap(N)$ sets a good limit on the exploration of the singularity-free trips starting downward at N. However, there is a price to pay: the immediate dominator information from which the joint dominator is computed has to be maintained. In particular, when a candidate expansion is successfully installed, the immediate dominator information stored in an a priori unknown number of containers may have to be updated. Fortunately, Theorem 3 shows that the area which has to be considered for the update is itself limited by $\sqcap(N)$. More precisely, there is no need to consider for update the containers other than those which have been visited, and locked. The procedure of Table 3 modifies the naive one to compute the set of containers which need to be considered for update (U in procedure CHECK).

It is not obvious to compute exactly the updated value of $\delta(x)$ at each container x in U. Instead, we use Theorem 3 to assign a lower bound of this value: $\delta(x) \sqcap \sqcap(N)$. This means that the actual algorithm does not maintain the exact value of the immediate dominator at each negative point, but a lower bound. Hence, the computation of the joint dominator may return a lower bound of the true value. This may result in wider explorations than strictly needed, but does not compromise the validity of the algorithm. The figure below illustrates the difference between the naive and improved procedure on a simple example (black/white nodes are positive/negative links):

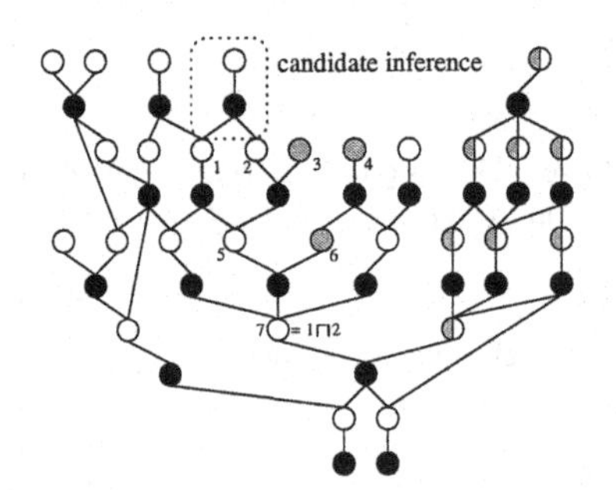

● locked by both procedures
◐ locked by naive procedure only

	immediate dominator		
node	before	after	inference
1	7	7	
2	5	5	
3	5	7	
4			
5	7	7	
6			
7			

5 Related Work

The standard notion of domination tree has been used in the past for correctness checking of proof-nets [5]. It is not clear whether this work, which is not concerned with proof-net construction, has any relationship with the work presented here, since our notion of domination does not coincide with the standard one. The actual properties proved in each case are different and support entirely different uses of proof-nets.

Proof-search in the sequent system of Linear Logic has been widely investigated. Proof-net search by a sequential process has also been studied in [6]. Our approach addresses concurrency in proof-net construction as an essential feature.

Petri-nets [7], introduced to study concurrency of processes competing for bounded resources, are the closest formal tool to our proposal. It was remarked early in the development of Linear Logic that proof-nets have strong connections with Petri-nets [8]. There are major differences though.

- In the proof-net construction paradigm, proof-nets are used as a mean of representing a trace of past actions (the construction so-far) in order to constrain future actions (by imposing transactional isolation constraints). Petri-nets, on the other hand, are used as a specification tool for concurrent programs. In particular, as a trace, a proof-net (or even a proof-structure) never contains cycles, while, as a specification, a Petri-net naturally allows cycles. In a middleware paradigm, the purpose is not to model what should take place in an application (specification) but rather to capture what actually takes place (trace). Proof-nets are therefore more appropriate for our purpose. Note that, of course, proof-nets as used here can naturally represent the trace of Petri-net executions, with a straightforward mapping between the bipoles used in the proof-nets and the actual execution of transitions in the Petri-net. However, usual Petri-net transitions (multiset rewriting) lead to simple proof-nets which do not need the kind of correctness checking procedure presented here: in fact they fall in the fragment considered in [2] for which no correctness problem can occur.

- The difference in purpose between specification and trace, which, by the way, corresponds to a deep cultural gap in the computing community between application programmers looking at how to implement things and system managers trying rather to understand what is going on in a system, has important consequences on the subsequent use of the tools. Petri-nets have been extended in many ways, in order to enhance their expressivity, according to needs arising from applications or intuition. And indeed, extensions of Petri-nets have been proposed to capture various form of transactional and contextual features [9–11] at the specification level. On the other hand, proof-nets need to be kept as bare as possible and avoid ad-hoc extensions which would not fit in the abstract proof machinery (the multiplication of concepts would make them useless as a tool to understand what is going on). The present paper tries to provide an understanding of concurrency and transaction concepts only in terms of the basic operations provided by Linear Logic.

The same discussion applies to other tools used to specify concurrent systems (or analyse them based on such specifications), such as the many process calculi [12–14] (the join calculus, in particular, is directly related to the work presented here), or the various coordination models based on shared tuple spaces (which also have many relations, including historical ones, with the work presented here).

6 Conclusion

Proof-nets are a powerful tool to represent and understand concurrency in computations. In particular, proof-net construction by a set of concurrent, decentralised agents provides an abstract model of infrastructure software in distributed applications (a.k.a. middleware). More precisely, proof-net construction captures in a natural way the behaviour of an essential component of middleware infrastructures, namely transaction schedulers. The concept of transaction has been widely investigated in the literature, and addresses basic intuitions such as atomicity and isolation in concurrent actions.

In this paper, we claim that the true computational content of proof-construction is given in terms of atomicity and isolation constraints between a set of agents performing expansion steps in a proof-net. We give a naive algorithm ensuring that the shared structure being built remains a proof-net. The proposed algorithm is at the same time incremental and truly concurrent. It does not rely on any kind of "behind the scene" interaction between the agents in order to synchronise: the proof-net being built is the only piece of data shared between the agents.

However, the naive algorithm may result in unbounded explorations of the net at each expansion step. A fine-grain analysis of proof-net properties allows us to improve the algorithm, by setting a computationally tractable bound to the required exploration. One of the main tool used in this analysis is the concept of "domination" order attached to a graph. This concept is not new, and has been used in various contexts involving the study of flow-graphs in general. We use a different version of this concept here, to derive important properties of proof-nets used in the concurrent, incremental construction algorithm.

References

1. Andreoli, J.M.: Focussing and proof construction. Annals of Pure and Applied Logic **107** (2001) 131–163
2. Andreoli, J.M.: Focussing proof-net construction as a middleware paradigm. In Voronkov, A., ed.: Proc. of 18th Int'l Conference on Automated Deduction, Copenhagen, Denmark. Volume 2392 of Lecture Notes in Computer Science. Springer Verlag (2002) 501–516
3. Girard, J.Y.: Linear logic. Theoretical Computer Science **50** (1987) 1–102
4. Danos, V., Regnier, L.: The structure of multiplicatives. Archive for Mathematical Logic **28** (1989) 181–203
5. Murawski, A., Ong, C.: Dominator trees and fast verification of proof nets. In: Proc. of 15th Conference on Logic in Computer Science, Santa Barbara, Ca, U.S.A., IEEE Computer Society Press (2000)
6. Habert, L., Notin, J.M., Galmiche, D.: Link: A proof environment based on proof nets. In Egly, U., Fermüller, C., eds.: Proc. of Tableaux 02, Copenhagen, Denmark. Volume 2381 of Lecture Notes in Computer Science. Springer Verlag (2002) 330–334
7. Reisig, W.: Petri Nets. Springer Verlag (1985)
8. Marti-Olliet, N., Meseguer, J.: From petri-nets to linear logic. Mathematical Structures in Computer Science **1** (1991) 69–101
9. Bruni, R., Montanari, U.: Transactions and zero-safe nets. In Ehrig, H., Juhás, G., Padberg, J., Rozenberg, G., eds.: Unifying Petri Nets, Advances in Petri Nets. Volume 2128 of Lecture Notes in Computer Science. Springer Verlag (2001) 380–426
10. Lomazova, I., Schnoebelen, P.: Some decidability results for nested petri nets. In Bjørner, D., Broy, M., Zamulin, A., eds.: Proc. of 3rd Perspectives of System Informatics, Novosibirsk, Russia. Volume 1755 of Lecture Notes in Computer Science. Springer Verlag (1999) 198–207
11. Ciardo, G.: Petri nets with marking-dependent arc cardinality. In Valette, R., ed.: Proc. of Application and Theory of Petri-Nets, Zaragoza, Spain. Volume 815 of Lecture Notes in Computer Science. Springer Verlag (1994) 179–198
12. Milner, R., Parrow, J., Walker, D.: A calculus of mobile processes. Information and Computation **100** (1992) 1–77
13. Fournet, C., Gonthier, G.: The reflexive cham and the join-calculus. In: Proc. of 23rd Symposium on Principles of Programming Languages, St. Petersburg Beach, Fl, U.S.A., ACM Press (1996) 372–385
14. Bruni, R., Laneve, C., Montanari, U.: Orchestrating transactions in join calculus. In Brim, L., Jancar, P., Kretìnsky, M., Kucera, A., eds.: Proc. of 13th Int'l Conference on Concurrency Theory, Brno, Czech Republic. Volume 2421 of Lecture Notes in Computer Science. Springer Verlag (2002) 321–337

Back to the Future: Explicit Logic for Computer Science

Sergei Artemov[1,2]

[1] The Graduate Center of the City University of New York
365 Fifth Avenue, New York
NY, 10016, USA
sartemov@gc.cuny.edu
[2] Moscow University

We will speak about three traditions in Logic:

- *Classical*, usually associated with Frege, Hilbert, Gödel, Tarski, and others;
- *Intuitionistic*, founded by Brouwer, Heyting, Kolmogorov, Gödel, Kleene, and others;
- *Explicit*, which we trace back to Skolem, Curry, Gödel, Church, and others.

The classical tradition in logic based on quantifiers $\forall$ and $\exists$ essentially reflected the 19th century mathematician's way of representing dependencies between entities. A sentence $\forall x \exists y A(x, y)$, though specifying a certain relation between x and y, did not mean that the latter is a function of the former, let alone a computable one. The Intuitionistic approach provided a principal shift toward the effective functional reading of the mathematician's quantifiers. A new, non-Tarskian semantics had been suggested by Kleene: realizability that revealed a computational content of logical derivations. In a decent intuitionstic system, a proof of $\forall x \exists y A(x, y)$ yields a program f that computes $y = f(x)$.

Explicit tradition makes the ultimate step by using representative systems of functions instead of quantifiers from the very beginning. Since the work of Skolem, 1920, it has been known that the classical logic can be adequately recast in this way. Church in 1936 showed that even the very basic system of function definition and function application is capable of emulating any computable procedure. However, despite this impressive start, the explicit tradition remained a Cinderella of the mathematical logic for decades. Now things have changed: due to its very explicitness, this third tradition became the one most closely connected with Computer Science.

In this talk we will show how switching from quantifiers to explicit functional language helps problem solving in both theoretical logic and its applications. A discovery of a natural system of self-referential proof terms, *proof polynomials*, was essential in the solution to an open problem of Gödel concerning formalization of provability. Proof polynomials considerably extend the Curry-Howard isomorphism and lead to a joint calculus of propositions and proofs which unifies several previously unrelated areas. It changes our conception of the appropriate syntax and semantics for reasoning about knowledge, functional programming languages, formalized deduction and verification.

M. Baaz and J.A. Makowsky (Eds.): CSL 2003, LNCS 2803, p. 43, 2003.

Constraint Satisfaction with Countable Homogeneous Templates

Manuel Bodirsky[1,*] and Jaroslav Nešetřil[2,**]

[1] Humboldt Universität zu Berlin, Germany
bodirsky@informatik.hu-berlin.de
[2] Institute for Theoretical Computer Science (ITI)
and Department of Applied Mathematics (KAM),
Charles University, Prague, Czech Republic
nesetril@kam.mff.cuni.cz

Abstract. For a fixed countable homogeneous relational structure Γ we study the computational problem whether a given finite structure of the same signature homomorphically maps to Γ. This problem is known as the constraint satisfaction problem CSP(Γ) for Γ and was intensively studied for finite Γ. We show that - as in the case of finite Γ - the computational complexity of CSP(Γ) for countable homogeneous Γ is determinded by the clone of polymorphisms of Γ. To this end we prove the following theorem which is of independent interest: The primitive positive definable relations over an ω-categorical structure Γ are precisely the relations that are invariant under the polymorphisms of Γ.
Constraint satisfaction with countable homogeneous templates is a proper generalization of constraint satisfaction with finite templates. If the age of Γ is finitely axiomatizable, then CSP(Γ) is in NP. If Γ is a digraph we can use the classification of homogeneous digraphs by Cherlin to determine the complexity of CSP(Γ).

1 Introduction

For a fixed relational structure Γ (called the *template*), the constraint satisfaction problem CSP(Γ) is the following computational problem: Given a finite structure S of the same signature as Γ, is there a homomorphism from S to Γ?

Constraint satisfaction problems frequently occur in theoretical computer science, and have attracted much attention for finite templates Γ. It is conjectured that CSP(Γ) has a *dichotomy* in the sense that every constraint satisfaction problem CSP(Γ) for finite structure Γ is either tractable or NP-complete. This is true for templates that are undirected graphs [20], for two element templates [35] or three element templates [6]. It is known that every constraint satisfaction problem is polynomial time equivalent to a digraph-homomorphism problem [17]. There are powerful classes of algorithms solving the known tractable

* Member of the European Graduate Program "Combinatorics, Geometry and Computation", supported by DFG grant GRK 588/2 and supported by the COMBSTRU network of the European Union.
** Supported by grant LN00A056 of the Czech Ministry of Education and supported by the COMBSTRU network of the European Union.

M. Baaz and J.A. Makowsky (Eds.): CSL 2003, LNCS 2803, pp. 44–57, 2003.

constraint satisfaction problems [17], namely group theoretic algorithms and local-consistency based algorithms [14, 21].

But many constraint satisfaction problems in the literature can not be formulated as a constraint satisfaction problem with a finite template. One example is Allen's interval algebra [1] that has applications in temporal reasoning in artificial intelligence. The classification of the tractable and hard subalgebras of Allen's algebra was completed only recently [24, 30], and they also exhibit a complexity dichotomy. Other examples are tree description languages that were introduced in computational linguistics [3, 4, 11]. Even digraph-acyclicity can not be formulated as a constraint satisfaction problem with finite template Γ. However, arbitrary infinite templates Γ might have undecidable constraint satisfaction problems.

We propose to study constraint satisfaction with *countable homogeneous templates.* This can be seen as a strict generalization of constraint satisfaction with finite templates, since every constraint satisfaction problem with a finite template is polynomial-time equivalent to a constraint satisfaction problem with a homogeneous template (see Section 3). Moreover, the constraint satisfaction problems mentioned above can be formulated naturally in this new framework. To prove tractability or hardness of constraint satisfaction problems with homogeneous templates reductions to different hard problems and new algorithms are used, which have not yet been considered for CSP(Γ) with finite Γ. Countable homogeneous structures are intensively studied by model theorists, and they have many remarkable properties. For finite signatures they allow quantifier elimination and are ω-categorical, i.e. their first-order theories have only one countable model up to isomorphism. Countable homogeneous structures have been classified for all digraphs [10].

Adding relations to a template Γ that are *primitive positive* definable over Γ does not change the computational complexity of CSP(Γ). The central theorem here [5] is that a relation is primitive positive definable over a finite relational structure Γ if and only if it is left invariant under the polymorphisms of Γ. This was first used in the context of constraint satisfaction by Jeavons et al. [23], and initiated the algebraic approach to constraint satisfaction, which has successfully been carried further e.g. in [13].We generalize this result to ω-categorical structures Γ: A relation is p.p.-definable in Γ if and only if it is invariant under the polymorphisms of Γ.

We can determine the complexity of CSP(Γ) and prove a dichotomy if Γ is a homogeneous graph or a tournament. Since there are uncountably many countable homogeneous digraphs Γ and uncountably many corresponding constraint satisfaction problems, the class of problems CSP(Γ) contains undecidable problems. However, if we assume that the class of finite induced substructures of a countable homogeneous digraph Γ is finitely axiomatized, one can determine the complexity of CSP(Γ) with a classification result of Cherlin [10].

The paper is organized as follows. We first give some background on relational homogeneous structures. In the next section on combinatorial constraint satisfaction problems we explain the rôle of *primitive positive definability* in constraint satisfaction. We give a characterization of primitive positive definability on homogeneous structures in Section 5 after introducing the necessary tools from universal algebra in Section 4. We end with a catalog of homogeneous relational structures and a discussion of their constraint satisfaction problems.

2 Background

A *relational signature* τ is a (in this paper always at most countable) set of *relation symbols* R_i, each associated with an *arity* k_i. A *(relational) structure* Γ *over relational signature* τ (also called τ*-structure*) is a set D_Γ (the *domain*) together with a relation $R_i \subseteq D_\Gamma^{k_i}$ for each relation symbol of arity k_i. For simplicity we denote both a relation symbol and its corresponding relation with the same symbol. For a τ-structure Γ and $R \in \tau$ it will also be convenient to say that $R(u_1, \ldots, u_k)$ *holds in* Γ if $(u_1, \ldots, u_k) \in R$. We sometimes use the shortened notation $\overline{x}$ for a vector $x_1, \ldots, x_n$ of any length.

A first-order formula φ over the signature τ is said to be *primitive positive* (we say φ is a *p.p.-formula*, for short) if it is of the form

$$\exists \overline{x}(\varphi_1(\overline{x}) \wedge \cdots \wedge \varphi_k(\overline{x})) .$$

where $\varphi_1, \ldots, \varphi_k$ are atomic formulas. (For an introduction to first order logic and model theory see [22].) Let Γ be a relational structure of signature τ. Then a p.p.-formula φ over τ with k free variables defines a k-ary relation $R \subseteq D_\Gamma^k$: the relation R is the set of all tuples satisfying the formula φ in Γ. Equivalently, R is contained in $\langle \Gamma \rangle_{pp}$ if and only if there exists a finite relational τ-structure S containing k designated vertices $x_1, \ldots, x_k$ such that

$$R = \{ (f(x_1), \ldots, f(x_k)) \mid f \colon S \to \Gamma \text{ homomorphism} \} .$$

We call these relations *p.p.-definable*, and denote the relational structure that contains all such relations for a given Γ by $\langle \Gamma \rangle_{pp}$. Likewise, the larger set of all first order definable relations is denoted by $\langle \Gamma \rangle_{fo}$.

A relational structure Γ is called *homogeneous* (in the literature also *ultrahomogeneous*) if every partial isomorphism between two finite substructures can be extended to an automorphism of Γ. Prominent examples of countable homogeneous structures are the *Rado graph* $\mathbf{R}$ and the dense linear order $(\mathbb{Q}, <)$. The Rado graph can be defined as the unique (up to isomorphism) model of the almost-sure theory of finite random graphs. Homogeneous structures have been classified for graphs [27], for tournaments, for posets [36], and finally digraphs [10] (there are continuum many homogeneous digraphs). For homogeneous structures with arbitrary relational signatures a classification is not yet known.

The *age* $\mathrm{Sub}(\Gamma)$ of a relational structure Γ over τ is the set of all finite structures over τ that (isomorphically) embed in Γ. An important property of countable homogeneous structures is their characterization by *amalgamation classes*. A class of finite structures $\mathcal{C}$ is an *amalgamation class* if $\mathcal{C}$ is nonempty, closed under isomorphism and taking induced substructures, and has the *amalgamation property*. The amalgamation property says that for all $A, B_1, B_2 \in \mathcal{C}$ and embeddings $e : A \to B_1$ and $f : A \to B_2$ there exists $C \in \mathcal{C}$ and embeddings $g : B_1 \to C$ and $h : B_2 \to C$ such that $ge = hf$.

Theorem 1 (Fraïssé [18]). *A countable class $\mathcal{C}$ of finite relational structures with countable signature is the age of a unique (up to isomorphism) countable homogeneous structure if and only if $\mathcal{C}$ is an amalgamation class.*

If $\mathcal{C}$ is an amalgamation class, we call the corresponding countable homogeneous structure the *Fraïssé-limit* of $\mathcal{C}$. By definition amalgamation classes can be defined by a set of forbidden induced finite substructures. For a set of finite structures $\mathcal{N}$ over τ we denote by $Forb(\mathcal{N})$ the set of finite structures S over τ such that no structure in $\mathcal{N}$ is embeddable in S. We say that a class of finite structures $\mathcal{C}$ over τ is finitely axiomatizable if there exists a first order formula φ over τ such that for all τ-structures A we have $A \in \mathcal{C}$ if and only if A is a model of φ. By compactness it follows that an amalgamation class $\mathcal{C}$ is finitely axiomatizable if and only if $\mathcal{C} = Forb(\mathcal{N})$ for some *finite* set of forbidden induced substructures $\mathcal{N}$.

A homogeneous structure Γ over a finite signature is ω-categorical, i.e. every countable structure satisfying the same first order formulas as Γ is isomorphic to Γ. A relational structure is called *oligomorphic* iff the automorphism group $Aut(G)$ of the structure Γ has only a finite number of orbits on the set of n-tuples of elements of Γ. The following theorem is essential (see [22]):

Theorem 2 (Engeler, Ryll-Nardzewski, Svenonius). *A countable structure Γ is ω-categorical if and only if $Aut(\Gamma)$ is oligomorphic.*

3 Combinatorial Constraint Satisfaction

Let Γ be an arbitrary structure with relational signature τ - also called the *template*. Then the constraint satisfaction problem CSP(Γ) is the following computational problem:

Given: A finite τ-structure S.

Question: Is there some homomorphism from S to Γ?

Formally, we denote by CSP(Γ) the set of all finite τ-structures that homomorphically map to Γ. For finite Γ we can assume without loss of generality that Γ is a *core*, i.e. all endomorphisms of Γ are automorphisms. If Γ is a core, adding all the singleton relations to Γ does not change the complexity of CSP(Γ) (as stated in [7]). In this case Γ becomes a homogeneous relational structure. Therefore constraint satisfaction with homogeneous templates can be seen as a generalization of constraint satisfaction with finite templates.

All constraint satisfaction problems with finite Γ are clearly contained in NP. If the age of a relational homogeneous structure Γ of finite signature is finitely axiomatizable then CSP(Γ) is also contained in NP. To see this, suppose we are given an instance S of CSP(Γ). An algorithm can then guess the image of S under a homomorphism, and verify that the image belongs to the age of Γ in polynomial time using the finite axiomatization. Thus we have

Proposition 1. *Let Γ be a countable homogeneous relational structure of finite signature τ with a finitely axiomatizable age. Then CSP(Γ) is in NP.*

Note that we need the axiomatizability assumption in Proposition 1 as there exist homogeneous Γ such that CSP(Γ) is undecidable, see Section 6. In analogy with the dichotomy conjecture of Feder and Vardi [17], we can make the following conjecture.

Conjecture 1 (Dichotomy). Let Γ be a countable homogeneous relational structures with a finitely axiomatizable age. Then the class of constraint satisfaction problems CSP(Γ) has a dichotomy.

For both finite and infinite Γ, the following simple lemma explains the relevance of p.p.-definable relations in constraint satisfaction. Suppose we extend a relational structure Γ by a p.p.-definable relation R. This does not change the computational complexity of the corresponding constraint satisfaction problem, since we can replace every occurence of R in an instance of CSP(Γ) by the τ-structure that defines R.

Lemma 1. *Let Γ be a τ-structure and let Γ' be the extension of this structure by a relation R that is p.p.-definable over Γ. Then* CSP(Γ) *is polynomial-time equivalent to* CSP(Γ').

In the next section we introduce the algebraic notions that will be needed to characterize p.p.-definability.

4 The Clone of Polymorphisms

In this section, D will stand for a countable set and O for the set of *finitary operations* on D, i.e., functions from D^k to D for finite k. We say that $f \in O$ *preserves* a k-ary relation $R \subseteq D^k$ if R is a subalgebra of $(D, f)^k$. An operation that preserves all relations of a relational structure Γ is called a *polymorphism* of Γ. The set of all k-ary polymorphisms of Γ is denoted by $Pol^{(k)}(\Gamma)$, and we write $Pol(\Gamma)$ for the set of all finitary polymorphisms $Pol(\Gamma) = \bigcup_{i=1} Pol^{(i)}(\Gamma)$.

The notion of a *product* of relational structures allows an equivalent definition of polymorphisms, relating polymorphisms to homomorphisms. The *(categorical- or cross-) product* $\Gamma_1 \times \Gamma_2$ of two relational τ-structures Γ_1 and Γ_2 is a τ-structure on the domain $D_{\Gamma_1} \times D_{\Gamma_2}$. For all relations $R \in \tau$ the relation $R((x_1, y_2), \ldots, (x_k, y_k))$ holds in $\Gamma_1 \times \Gamma_2$ iff $R(x_1, \ldots, x_k)$ holds in Γ_1 and $R(y_1, \ldots, y_k)$ holds in Γ_2. Comparing the corresponding definitions we see that a k-ary polymorphism f of a relational structure is a homomorphism from $\Gamma^k = \Gamma \times \ldots \times \Gamma$ to Γ, i.e., for an m-ary relation R in τ, if $R(x_1, \ldots, x_m)$ holds in Γ^k then $R(f(x_1), \ldots, f(x_m))$ holds in Γ.

An operation π is a *projection* (or a *trivial polymorphism*) if for all n-tuples, $\pi(x_1, \ldots, x_n) = x_i$ for some fixed $i \in \{1, \ldots, n\}$. The *composition* of a k-ary operation f and k operations $g_1, \ldots, g_k$ of arity n is an n-ary operation defined by

$$f(g_1, \ldots, g_k)(x_1, \ldots, x_n) = f(g_1(x_1, \ldots, x_n), \ldots, g_k(x_1, \ldots, x_n)) \ .$$

A *clone* F is a set of operations from O that is closed under composition and that contains all projections. We write D_F for the *domain* D of the clone F. For a set of operations F from O we write $\langle F \rangle$ for the smallest clone containing all operations in F (the clone *generated* by F). Observe that $Pol(\Gamma)$ is a clone with the domain D_Γ.

Moreover, $Pol(\Gamma)$ is also closed under interpolation: We say that an operation $f \in O$ is an *interpolation* of a subset F of O if for every finite subset B of D there is some operation $g \in \langle F \rangle$ such that $f|_B = g|_B$ (f restricted to B equals g restricted to B, i.e., $f(a) = g(a)$ for every $a \in B^k$). The set of interpolations of F is called the *local closure* of F. If the maximal arity of Γ is bounded, $Pol(\Gamma)$ is also locally closed.

The converse was proved by Rosenberg and Schweigert [34]. Together with a theorem of Larose and Tardif [28] on infinite graphs this is one of the few known results on infinite structures and their polymorphisms. Many results on clones in general can be found in [38].

Proposition 2 (Rosenberg and Schweigert [34]). *A set $F \subseteq O$ of operations is locally closed if and only if $X = Pol(\Gamma)$ for some relational structure Γ of bounded maximal arity.*

Important properties of operations in a clone: a k-ary operation f is *idempotent* iff $f(x, \ldots, x) = x$ for all $x \in D$. An operation f is called *essentially unary* iff there is a unary operation f_0 such that $f(x_1, \ldots, x_k) = f_0(x_i)$ for some $i \in \{1, \ldots, k\}$. A relational structure Γ is called *projective*, iff all idempotent polymorphisms of Γ are projections, and *strongly projective*, iff all polymorphisms of Γ are projections [33].

Let F be a (local) clone with domain D. Then $R \subseteq D^m$ *is invariant under* F, if every $f \in F$ preserves R. We denote by $Inv(F)$ the relational structure containing the set of all relations left invariant under F. A fundamental result of Bodnarčuk et al. [5] (other presentations can be found in [12, 32]) says that for arbitrary finite relational structures Γ the p.p.-definable relations can be characterized as the invariants of the polymorphisms of Γ.

Theorem 3 (Bodnarčuk et al. [5]). *Let Γ be a finite relational structure. Then*

$$\langle \Gamma \rangle_{pp} = Inv(Pol(\Gamma)) \ .$$

The proof of Theorem 3 also shows that it is decidable whether for a given finite relational structure Γ a given relation R is p.p.-definable or not. Generalizations of Theorem 3 and the related Galois correspondences were also studied for infinite domains. For arbitrary relational structures Γ the set of relations $Inv(Pol(\Gamma))$ was characterized with local closure operators on relational algebras in [37] (see also [31], page 32).

In the next section we will show that for countable homogeneous structures Γ any first-order definable relation is in $\langle\Gamma\rangle_{pp}$ if and only if it is left invariant under all polymorphisms of finite arity. But first we note that the following is well-known for arbitrary cardinalities of the domain.

Proposition 3 (see e.g. [32]). *Let Γ be a relational structure. Then*

$$\langle\Gamma\rangle_{pp} \subseteq Inv(Pol(\Gamma)) .$$

Proof. Let R be a relation in $\langle\Gamma\rangle_{pp}$. We prove that $R \in Inv(Pol(\Gamma))$ by induction on the length of a defining p.p.-formula φ. The claim is true for $\varphi = R(x_1, \ldots, x_n)$. For $\varphi = \exists x.\varphi'$ we observe that every polymorphism that is left invariant by φ' also leaves φ invariant. The same holds for $\varphi = \varphi_1 \wedge \varphi_2$. □

For a structure with a countable domain the inclusion of Proposition 3 might be strict. Consider for instance the following relational structure $\Gamma = (\mathbb{N}; R_1, R_2, R_3)$ on the natural numbers communicated to the authors by Ferdinand Börner. We show that $Inv(Pol(\Gamma))$ contains relations that are not p.p.-definable.

$$\begin{aligned} R_1 &= \{(a,b,c,d) \mid a=b \text{ or } c=d,\ a,b,c,d \in \mathbb{N}\} \\ R_2 &= \{(0)\} \\ R_3 &= \{(a,a+1) \mid a \in \mathbb{N}\} \end{aligned}$$

Every function preserving R_1 is essentially unary. If f is unary and preserves R_2 then $f(0) = 0$. Furthermore, if f preserves R_3 we have $f(a+1) = f(a)+1$ for all a, and inductively follows $f(a) = a$. Therefore $Pol(\Gamma)$ is the set of all projections. Every projection preserves all relations, but even the unary first-order definable relation $\{x \mid x = 1 \vee x = 3\}$ is not p.p.-definable.

For ω-categorical structures Γ the situation looks better: It is known that the first-order definable relations are precisely the relations that are invariant under the automorphisms of Γ, i.e. $\langle\Gamma\rangle_{fo} = Inv(Aut(\Gamma))$ (see e.g. [8, 22]). The structure $Inv(Aut(\Gamma)$ is homogeneous and called the *canonical structure* of the permutation group $Aut(\Gamma)$. We prove a corresponding theorem for primitive positive definability in the next section.

5 A Characterization of Primitive Positive Definability

We characterize the primitive positive first-order definable relations over an ω-categorical structure Γ by the polymorphisms of Γ of finite arity.

Theorem 4. *Let Γ be an ω-categorical structure with relational signature τ. Then a relation R on Γ is invariant under the polymophisms of Γ if and only if R is p.p.-definable, i.e.,*

$$\langle\Gamma\rangle_{pp} = Inv(Pol(\Gamma)).$$

Proof. We already stated in Proposition 3 that the p.p.-definable relations over Γ are invariant under the polymophisms of Γ.

For the converse, let R be a k-ary relation from $Inv(Pol(\Gamma))$. Note that R is first-order definable in Γ: By ω-categoricity and Ryll-Nardzewski, and since Γ and $Inv(Pol(\Gamma))$ have the same automorphism group, the relation R is a union of *finitely* many orbits of the automorphism group of Γ, and it can be defined by a disjunction φ of τ-formulas that define these orbits. Let $M_1, \ldots, M_w$ be the satisfiable monomials in this disjunction, and let $x_1, \ldots, x_k$ be the variables of the monomials.

We have to construct a finite τ-structure Q with designated vertices $v_1, \ldots, v_k$ such that

$$R = \{(f(v_1), \ldots, f(v_k)) \mid f\colon Q \to \Gamma \text{ homomorphism}\}.$$

The idea is to first consider an *infinite* τ-structure, namely the categorical product Γ^w, and then to apply König's Lemma to prove the existence of a suitable finite substructure.

For each monomial $M_j \in M_1, \ldots, M_w$ of φ we find a substructure $a_1^j, \ldots, a_k^j$ of Γ, such that $a_1^j, \ldots, a_k^j$ satisfies M_j in Γ. Let $b_1, b_2, \ldots$ be an enumeration of the w-tuples in D_Γ^w, starting with $b_i = (a_1^i, \ldots, a_w^i)$ for $1 \leq i \leq k$. Let us call a partial mapping from Γ^w to Γ a *bad* mapping if it maps $b_1, \ldots, b_k$ to a tuple not satisfying φ. Since R is invariant under all polymorphisms, no homomorphism from Γ^w to Γ is bad.

We now claim that there is a finite substructure Q of Γ^w such that no homomorphism from Q to Γ is bad. Assume for contradiction that all finite substructures of Γ^w containing $b_1, \ldots, b_k$ have a homomorphism to Γ mapping $b_1, \ldots, b_k$ to a tuple not satisfying φ. We now construct a bad homomorphism from Γ^w to Γ, i.e. the images of $b_1, \ldots, b_k$ do not satisfy φ. This contradicts the fact that R is invariant under all polymorphisms.

To this end, consider the following infinite but finitely branching tree. The nodes on level n in the tree are the equivalence classes of the bad homomorphisms from $\Gamma^w|_{b_1,\ldots,b_n}$ to Γ, where two homomorphisms f_1 and f_2 are equivalent if $f_1 = gf_2$ for some $g \in Aut(\Gamma)$. Adjacency between nodes on consecutive levels is defined by restriction. By our assumption, for each finite substructure of Γ^w there is a bad homomorphism, and thus the tree contains a node on each level. By the Ryll-Nardzewski, there are only finitely many nodes at each level. By König's Lemma the tree contains an infinite path. We use this path to define a bad homomorphism from Γ^w to Γ.

We proved by contradiction that there must be a finite substructure Q containing the vertices $b_1, \ldots, b_k$ of Γ^w such that all homomorphisms from Q to Γ map $b_1, \ldots, b_k$ to a tuple satisfying φ. Conversely, every mapping $f : Q \to \Gamma$ such that the tuple $(f(b_1), \ldots, f(b_k))$ satisfies in Γ the monomial M_j can be extended to a homomorphism $f : \Gamma^w \to \Gamma$. To see this note that both $a_1^j, \ldots, a_k^j$ and $(f(b_1), \ldots, f(b_k))$ satisfy M_j and thus both lie in the same orbit of $Aut(\Gamma)$. Thus we can choose f to be the jth projection combined with the automorphism sending $(a_1^j, \ldots, a_k^j)$ to $(f(b_1), \ldots, f(b_k))$. This completes the proof. □

The clone of polymorphisms of an infinite structure is usually a very complicated object. However for homogeneous structures of finite signature we have the following:

Proposition 4. *Let Γ be a homogeneous structure with finite relational signature. Then the polymorphisms of Γ are locally generated by countably many polymorphisms of Γ and the automorphism group $Aut(\Gamma)$.*

Proof. Let $u_1, u_2, \ldots, u_k$ be m-tuples from Γ, and let $v_i = g(u_i)$ for some m-ary polymorphism g of Γ. Depending on m and k there are only finitely many isomorphism types of the substructure of Γ induced by the elements of $u_1, u_2, \ldots, u_k$ and the elements $v_1, \ldots, v_k$. Let F be a set of polymorphisms of arity k containing a polymorphism g for each of these isomorphism types in Γ.

Now let f be an m-ary polymorphism of Γ. We show that f is locally generated by $F \cup Aut(\Gamma)$. Let $B \subseteq D_\Gamma$ a set of finite cardinality. By the definition of F, the restriction $f|_B$ is isomorphic to the restriction of one of the operations $g \in F$. Let π be the isomorphism. By homogeneity of Γ, π can be extended to an automorphism π' of Γ. The identity $f|_B = (\pi' g)|_B$ implies that f is in the local closure of $Aut(\Gamma) \cup F$. □

6 A Catalog of Homogeneous Templates

We consider various homogeneous and ω-categorical structures, some of their polymorphisms and their corresponding constraint satisfaction problems. In particular we look at the binary structures from the classification project for countable homogeneous structures.

The Countable Homogeneous Tournaments. We start with the homogeneous tournaments, which have been classified by Lachlan [26]. There are a few types only: The oriented cycle C_3, the dense linear order $(\mathbb{Q}, <)$, the dense local order $S(2)$, and the generic tournament for the set of all finite tournaments.

The problem CSP(C_3) is known to be tractable. The constraint satisfaction problem of the dense linear order $(\mathbb{Q}, <)$ is computationally equivalent to the problem whether a given digraph D is acyclic. This tractable problem can not be formulated as a constraint satisfaction problem with a finite template. Note that the relational structure $(\mathbb{Q}, <)$ is not projective, e.g. $x, y \mapsto \max(x, y)$ is a polymorphism. The homogeneous tournament which is the Fraïssé-limit of all finite tournaments has a trivial constraint-satisfaction problem: Every finite tournament homomorphically maps to it. Thus the only interesting remaining case is the dense linear order $S(2)$ (see [10]). The problem CSP($S(2)$) is NP-hard, since it can simulate the hard problem *Betweenness* [19].

To define the *dense local order* $S(2)$, consider two disjoint dense subsets X and Y of the rational numbers $\mathbb{Q}$ (i.e., for every rational number we will find sequences in X and in Y that converge against this number). Then the relation $<$ of $S(2)$ equals the dense linear order of $\mathbb{Q}$ on $X \cup Y$, but we reverse the edges between the sets X and Y. It is easy to see that $S(2)$ is the up to isomorphism unique countable tournament that satisfies the following property: for every vertex, both in-neighbourhood and out-neighbourhood are isomorphic to $(\mathbb{Q}, <)$.

The problem *Betweenness* can be stated as a constraint satisfaction problem CSP$(\mathbb{Q}, R)$ where $R = \{(x_1, x_2, x_3) \subseteq \mathbb{Q}^3 \mid x_2 < x_1 \text{ or } x_2 > x_3\}$. This relation can be simulated by the following p.p.-formula in $S(2)$:

$$\exists u, v : u < x_1 \wedge u < x_2 \wedge u < x_3 \wedge x_1 < x_3 \wedge x_2 < v \wedge v < x_1 \wedge v < x_3$$

The Countable Homogeneous Graphs. Lachlan and Woodrow [27] showed that every infinite such graph is either the Rado-graph, the Fraïssé-limit of all K_n-free graphs, the complete n-partite graphs, or a complement of these.

The Rado graph $\mathbf{R}$ is the Fraïssé-limit of the class of all graphs, therefore the constraint satisfaction problem for the Rado graph is trivial: Every graph can be homomorphically mapped to it. The automorphism group of $\mathbf{R}$ has a rich structure (see e.g. [9] for an overview). Łuczak and Nešetřil [29] showed that the Rado graph as well as the generic K_n-free graphs are projective. This is in interesting opposition to the finite case, where projectivity of a core of cardinality at least three implies NP-hardness for the corresponding constraint satisfaction problem, which can be seen using Theorem 3 by reduction of k-colorability. Again the constraint satisfaction problem is easy, since every graph which does not contain the K_n as a subgraph embeds into the generic K_n-free graph. The only interesting remaining cases are the complete n-partite graphs. Each such graph has a homomorphism to the finite graph K_n, which has an NP-complete constraint satisfaction problem for $n \geq 3$, and is tractable for $n \leq 2$.

Remark: It is perhaps interesting to note that we can use the same results to give a new and short proof that almost all constraint satisfaction problems are NP-complete, if the template is a finite undirected graph. The fact that almost all graphs are strongly projective (Nešetřil and Łuczak [29]) combined with Theorem 3 shows that almost all graphs can simulate the inequality-relation. This implies NP-hardness of the constraint satisfaction problem on a domain of size at least three. Note that we did not use the involved proof of the dichotomy for graphs in [20].

Countable Homogeneous Digraphs. The countable homogeneous digraphs have been classified by Cherlin [10], and there are uncountably many. But the classification shows that the age of all but a countable well-understood class of homogeneous digraphs has the strong *free amalgamation property.* Therefore it is easy to see that CSP(Γ) is the set of all (weak) subgraphs of Γ, and that the set of constraint satisfaction problems is also uncountable. Thus there are homogeneous digraphs with an undecidable constraint satisfaction problem. However, if we consider the homogeneous structures Γ that have a finitely axiomatizable age and free amalgamation, it is easy to see that CSP(Γ) is tractable. The remaining homogeneous digraphs we described by Cherlin and its straightforward to determine the complexity of their constraint satisfaction problems ([10]; note that $CSP(S(2)) = CSP(S(3)) = CSP(P(3))$).

To summarize: Every homogeneous digraph problem is either undecidable, NP-complete or tractable. Note that this does not say anything about the mentioned constraint satisfaction problems for finite digraphs, since the corresponding homogeneous templates have a larger signature.

Tree Descriptions. The following constraint satisfaction problem was studied in [11]. Given: a finite structure S over the signature $\tau = \{\longrightarrow, \perp\}$ containing two binary relation symbols. Question: can we find a rooted forest F on the vertices of S such that every edge from $\longrightarrow$ lies in the transitive closure of F, and every edge $\perp$ does not? Let us call such τ-structures S *solvable.*

Using the notion of *dense trees* (see [15]) we can formulate this problem (and related problems) as a constraint satisfaction problem CSP(Γ) for appropriate Γ. This means, it is possible to find an ω-categorical structure Γ such that CSP(Γ) contains precisely the solvable τ-structures. The problem can be decided by a polynomial time algorithm [4]. The graph algorithm presented there can be generalized to various constraint satisfaction problems for tree-like structures. We already mentioned that every ω-categorical structure Γ can be made homogeneous by expanding the signature and Γ by some first-order definable relations. In the case of countable dense trees this is possible by an additional ternary relation [16].

Allen's Interval Algebra and Its Fragments. Consider as a base set D the closed intervals on the rational numbers, and the following binary relations on these intervals: Let $x = (x^-, x^+)$ and $y = (y^-, y^+)$ be closed intervals. We define

- The interval x *precedes* y, $x \mathsf{p} y$, iff $x^+ < y^-$.
- The interval x *overlaps* y, $x \mathsf{o} y$, iff $x^- < y^- < x^+$ and $x^+ < y^+$.
- The interval x is *during* y, $x \mathsf{d} y$, iff $y^- < x^-$ and $x^+ < y^+$.
- The interval x *starts* y, $x \mathsf{s} y$, iff $x^- = y^-$ and $x^+ > y^-$.
- The interval x *finishes* y, $x \mathsf{f} y$, iff $x^+ = y^+$ and $x^- > y^-$.
- The interval x *meets* y, $x \mathsf{m} y$, iff $x^+ = y^-$.
- The interval x *equals* y, $x \equiv y$, iff $x^- = y^-$ and $x^+ = y^+$.

For any set of relations derived from $\mathsf{p}, \mathsf{o}, \mathsf{d}, \mathsf{s}, \mathsf{m}, \mathsf{f}$ and $\equiv$ by union and complementation the corresponding countable relational structure is ω-categorical. The constraint satisfaction problems for these structures have a dichotomy [24, 30]. Whereas for finite templates all known hard constraint satisfaction problems are hard because they can express the relation one-in-three-sat, here again the problem *Betweennness* [19] is used to prove hardness.

7 Related Work

We want to relate our work to previous unifying approaches to constraint satisfaction. The literature on *combining constraint solving* [2, 25] has a broader view on constraint satisfaction, and also uses tools from universal algebra. However they are concerned mainly with decidability questions of more expressive constraint languages.

Various logical formalisms have been proposed to formulate constraint satisfaction problems as the model-checking problems of certain higher-order logics [17]. On of them is the class SNP, the class of existential second-order formulas Φ with a universal first-order part, which might use the relation symbols of the given signature τ and existentially quantified relation symbols. One of the results in [17] says that every problem in NP is equivalent to a problem in SNP, even if the relation symbols from τ occur only negatively in Φ (in which case the class is called *monotone* SNP). To answer the question whether the model checking problem of a given monotone SNP formula can be described as a constraint satisfaction problem with a countable homogeneous structure, the following problem posed by Cherlin [10] is of importance:

Problem 1. Let τ be a relational signature and $\mathcal{N}$ a finite set of finite τ-structures. Give a good criterion for $Forb(\mathcal{N})$ to be an amalgamation class.

8 Conclusion

Constraint satisfaction problems with countable homogeneous templates cover several classes of constraint satisfaction problems that were investigated in the literature. Examples are tree description languages and subalgebras of Allen's interval algebra. Using the classification of homogeneous digraphs we can determine the complexity of the constraint satisfaction problems for all homogeneous digraphs. For larger signatures the classification of homogeneous structures is a difficult task. To study the complexity of a constraint satisfaction problem CSP(Γ) it is useful to know whether a given first-order relation is p.p.-definable over Γ. In Section 5 we show that p.p.-definability of a first-order definable relation is characterized by a countable set of polymorphisms. We ask the following question:

Problem 2. Let $\mathcal{N}$ be a finite set of relational structures over signature τ such that Forb($\mathcal{N}$) is the age of a homogeneous structure Γ. Given a first-order formula φ over τ, is it decidable whether φ is on Γ equivalent to a primitive positive formula?

The techniques to describe p.p.-definability might be applied to simplify the technical and intricate proofs in the classification of the tractable fragments of Allen's interval algebra.

References

1. J. F. Allen. Maintaining knowledge about temporal intervals. *Communications of the ACM*, 26(11):832–843, 1983.
2. F. Baader and K. Schulz. Combining constraint solving. *H. Comon, C. March, and R. Treinen, editors, Constraints in Computational Logics*, 2001.
3. M. Bodirsky, D. Duchier, J. Niehren, and S. Miele. A new algorithm for normal dominance constraints. 2003.

4. M. Bodirsky and M. Kutz. Pure dominance constraints. In *Proceedings of the 19th Annual Symposium on Theoretical Aspects of Computer Science (STACS'02)*, pages 287–298, Antibes - Juan le Pins, 2002.
5. V. G. Bodnarčuk, L. A. Kalužnin, V. N. Kotov, and B. A. Romov. Galois theory for post algebras, part I and II. *Cybernetics*, 5:243–539, 1969.
6. A. Bulatov. Tractable constraint satisfaction problems on a 3-element set. *Research Report*, 2002.
7. A. Bulatov, A. Krokhin, and P. G. Jeavons. Classifying the complexity of constraints using finite algebras. *submitted*, 2003.
8. P. J. Cameron. *Oligomorphic Permutation Groups.* Cambridge University Press, 1990.
9. P. J. Cameron. The random graph. *R.L.Graham and J. Nesetril, editors, The Mathematics of Paul Erdős*, 1996.
10. G. Cherlin. The classification of countable homogeneous directed graphs and countable homogeneous n-tournaments. *AMS Memoir*, 131(621), January 1998.
11. T. Cornell. On determining the consistency of partial descriptions of trees. In *32nd Annual Meeting of the Association for Computational Linguistics (ACL'94)*, pages 163–170, 1994.
12. V. Dalmau. Computational complexity of problems over generalized formulas. PhD-thesis at the Departament de Llenguatges I Sistemes Informatics at the Universitat Politecnica de Catalunya, 2000.
13. V. Dalmau. A new tractable class of constraint satisfaction problems. In *6th International Symposium on Mathematics and Artificial Intelligence*, 2000.
14. R. Dechter. Local and global relational consistency. *Journal of Theoretical Computer Science*, 1996.
15. M. Droste. Structure of partially ordered sets with transitive automorphism groups. *AMS Memoir*, 57(334), September 1985.
16. M. Droste, W. Holland, and D. Macpherson. Automorphism groups of infinite semilinear orders (i). *Proc. London Math. Soc.*, 58:454–478, 1989.
17. T. Feder and M. Vardi. The computational structure of monotone monadic SNP and constraint satisfaction: A study through datalog and group theory. *SIAM J. Comput.*, 28:57–104, 1999.
18. R. Fraïssé. *Theory of Relations.* North-Holland, 1986.
19. Garey and Johnson. *A Guide to NP-completeness.* CSLI Press, Stanford, 1978.
20. P. Hell and J. Nešetřil. On the complexity of H-coloring. *Journal of Combinatorial Theory, Series B*, 48:92–110, 1990.
21. P. Hell, J. Nešetřil, and X. Zhu. Duality and polynomial testing of tree homomorphisms. *Trans. Amer. Math. Soc.*, 348(4):1281–1297, 1996.
22. W. Hodges. *A shorter model theory.* Cambridge University Press, 1997.
23. P. Jeavons, D. Cohen, and M. Gyssens. Closure properties of constraints. *Journal of the ACM*, 44(4):527–548, 1997.
24. P. Jeavons, P. Jonsson, and A. A. Krokhin. Reasoning about temporal relations: The tractable subalgebras of allen's interval algebra. *JACM*, To appear.
25. H. Kirchner and C. Ringeissen. Combining symbolic constraint solvers on algebraic domains. *Journal of Symbolic Computation*, 18(2):113–155, 1994.
26. A. H. Lachlan. Countable homogeneous tournaments. *Trans. Amer. Math. Soc.*, 284:431–461, 1984.
27. A. H. Lachlan and R. Woodrow. Countable ultrahomogeneous undirected graphs. *Trans. Amer. Math. Soc.*, 262(1):51–94, 1980.
28. B. Larose and C. Tardif. Strongly rigid graphs and projectivity. *Multiple-Valued Logic 7*, pages 339–361, 2001.

29. T. Łuczak and J. Nešetřil. Projective graphs. *Submitted*, 2003.
30. B. Nebel and H.-J. Bürckert. Reasoning about temporal relations: A maximal tractable subclass of Allen's interval algebra. *Journal of the ACM*, 42(1):43–66, 1995.
31. R. Pöschel. A general galois theory for operations and relations and concrete characterization of related algebraic structures. *Technical Report of Akademie der Wissenschaften der DDR (Berlin)*, 1980.
32. R. Pöschel and L. A. Kalužnin. *Funktionen- und Relationenalgebren.* DVW, 1979.
33. I. G. Rosenberg. Strongly rigid relations. *Rocky Mountain Journal of Mathematics*, 3(4):631–639, 1973.
34. I. G. Rosenberg and D. Schweigert. Locally maximal clones. *J. Algorithms, Languages and Combinatorics*, 5(4):421–455, 2000.
35. T. J. Schaeffer. The complexity of satisfiability problems. *In Proc. 10th ACM Symp. on Theory of Computing*, pages 216–226, 1978.
36. J. H. Schmerl. Countable homogeneous partially ordered sets. *Algebra Universalis*, 9:317–321, 1979.
37. L. Szabó. Concrete representation of relatied structures of universal algebras. *Acta Sci. Math. (Szeged)*, 40:175–184, 1978.
38. A. Szendrei. *Clones in universal Algebra.* Seminaire de mathematiques superieures. Les Presses de L'Universite de Montreal, 1986.

Quantified Constraints: Algorithms and Complexity

Ferdinand Börner[1], Andrei Bulatov[2,*],
Peter Jeavons[2,*], and Andrei Krokhin[3,**]

[1] Institut für Informatik, University of Potsdam
Potsdam, D-14482, Germany
fboerner@cs.uni-potsdam.de
[2] Oxford University Computing Laboratory
Oxford, OX1 3QD, UK
{andrei.bulatov,peter.jeavons}@comlab.ox.ac.uk
[3] Department of Computer Science, University of Warwick
Coventry, CV4 7AL, UK
andrei.krokhin@dcs.warwick.ac.uk

Abstract. The standard constraint satisfaction problem over an arbitrary finite domain can be expressed as follows: given a first-order sentence consisting of a conjunction of predicates, where all of the variables are existentially quantified, determine whether the sentence is true. This problem can be parameterized by the set of allowed constraint predicates. With each predicate, one can associate certain predicate-preserving operations, called polymorphisms, and the complexity of the parameterized problem is known to be determined by the polymorphisms of the allowed predicates. In this paper we consider a more general framework for constraint satisfaction problems which allows arbitrary quantifiers over constrained variables, rather than just existential quantifiers. We show that the complexity of such extended problems is determined by the surjective polymorphisms of the constraint predicates. We give examples to illustrate how this result can be used to identify tractable and intractable cases for the quantified constraint satisfaction problem over arbitrary finite domains.

1 Introduction

The constraint satisfaction problem (CSP) provides a general framework in which a wide variety of combinatorial problems can be expressed in a natural way [12, 28, 37]. A constraint satisfaction problem instance can be viewed as a collection of predicates on overlapping sets of variables. The aim is to determine whether there exist values for all of the variables such that all of the specified predicates hold simultaneously. The standard constraint satisfaction problem can

* Partially supported by the UK EPSRC grant GR/R29598.
** Partially supported by the UK EPSRC grant GR/R29598 and by the IST Programme of the EU under contract number IST-1999-14186 (ALCOM-FT).

M. Baaz and J.A. Makowsky (Eds.): CSL 2003, LNCS 2803, pp. 58–70, 2003.

be parameterized by restricting the set of allowed predicates which can be used as constraints. The problem of determining (up to complete classification) the complexity of the CSP (and its many variants) for all possible parameter sets has attracted much attention, partly because constraint satisfaction problems play an important role in Artificial Intelligence [28, 37], and partly because they "present a reasonably accurate bird's-eye view on complexity theory" [12]. One important outcome of research in this direction has been the design of sophisticated new polynomial-time algorithms for solving a wide variety of problems (see, e.g., [5, 6, 15]). Another outcome has been progress with some important issues in complexity theory, such as the discovery of large subclasses of complexity classes that avoid intermediate complexity (e.g., dichotomy results in the case of **NP**, see [4, 6, 17, 35]).

For the Boolean (i.e., two-valued) case, the complexity of the standard constraint satisfaction problem has been studied from the above perspective [35], as well as a number of related problems, including quantified and counting problems, maximum (and minimum) satisfiability, generating all solutions, optimizing the number of positive truth values in a solution (all of these are in [12]), minimal satisfiability [25], deciding equivalence and isomorphism of instances [2], maximizing Hamming distance between solutions [13], finding lexicographically minimal (or maximal) solutions [34], finding a second solution [22], inverse satisfiability [23], and solving random instances [11].

Analysing the complexity of non-Boolean CSPs is a significantly more difficult task: these problems usually withstand a direct combinatorial approach, and so require more involved techniques. A far-reaching approach via graph theory, logic and games has been developed in [16, 17, 27]. However, the most successful approach so far has been the algebraic approach developed in [8, 20, 21]. This approach has led to a number of new results (see, e.g., [5, 7, 9, 15, 16]), and has culminated (so far) in a complete classification of the complexity of parameterized CSPs for the three-valued case [4] and for the case when all unary predicates are available [6].

The standard CSP can be expressed as follows: given a first-order sentence consisting of a conjunction of predicates, where all of the variables are existentially quantified, determine whether the sentence is true. One of the most natural generalisations of this framework is to consider the *quantified constraint satisfaction problem* (QCSP), in which universal quantifiers are allowed in the sentence, as well as existential quantifiers [12, 14]. This generalisation greatly increases the expressive power of the framework, but also increases the complexity of deciding whether an arbitrary instance is true — from **NP**-complete to **PSPACE**-complete.

Boolean QCSP (also known as QSAT or QBF) and some of its restrictions (such as Q3SAT) have always been standard examples of **PSPACE**-complete problems [19, 29, 35]. For some parameter sets, Boolean QCSP has been shown to be tractable: for all binary predicates in [1], and for Horn predicates in [26]. Finally, a complete classification for the Boolean case was obtained in [12, 14]. For the non-Boolean case, some superpolynomial algorithms were given in [38].

However, to the best of our knowledge, only trivial results are known about the complexity of non-Boolean QCSPs.

In this paper we extend the algebraic approach for the first time to the more general framework of the quantified constraint satisfaction problem over an arbitrary finite set of values. We show that certain algebraic objects (surjective polymorphisms) determine the complexity of the QCSP for any given choice of parameter set. We then give examples to show how this result can be used to identify quantified constraint satisfaction problems lying in more restricted complexity classes such as **NL** and **PTIME**. Finally, we obtain the first complete classification result for a class of general (non-Boolean) quantified constraint satisfaction problems.

2 Preliminaries

Throughout the paper we use the standard correspondence between predicates and relations: a relation consists of all tuples of values for which the corresponding predicate holds. We will use the same symbol for a predicate and its corresponding relation, since the meaning will always be clear from the context. We will use $R_D^{(m)}$ to denote the set of all m-ary relations (or predicates) over a set D, and R_D to denote the set $\bigcup_{m=1}^{\infty} R_D^{(m)}$.

Definition 1. *Let $\Gamma \subseteq R_D$. An* instance *of* CSP(Γ) *is a first-order sentence $\exists x_1 \ldots \exists x_l\, (\varrho_1 \wedge \ldots \wedge \varrho_q)$, where each ϱ_i is an atomic formula involving a predicate from Γ, and $x_1, \ldots, x_l$ are the variables appearing in the ϱ_i. The* question *is whether the sentence is true.*

The predicates ϱ_i appearing in an instance will be referred to as *constraints*, since each of them restricts the possible models for the instance in some way.

In addition to predicates and relations we will also consider arbitrary *operations* on the set of values. We will use $O_D^{(n)}$ to denote the set of all n-ary operations on a set D (that is, the set of mappings $f\colon D^n \to D$), and O_D to denote the set $\bigcup_{n=1}^{\infty} O_D^{(n)}$.

Any operation on D can be extended in a standard way to an operation on tuples over D, as follows. For any operation $f \in O_D^{(n)}$, and any collection of tuples $\boldsymbol{a}_1, \boldsymbol{a}_2, \ldots, \boldsymbol{a}_n \in D^m$, where $\boldsymbol{a}_i = (\boldsymbol{a}_i(1), \ldots, \boldsymbol{a}_i(m))$ $(i = 1 \ldots n)$, define $f(\boldsymbol{a}_1, \ldots, \boldsymbol{a}_n)$ to be $(\, f(\boldsymbol{a}_1(1), \ldots, \boldsymbol{a}_n(1)), \ldots, f(\boldsymbol{a}_1(m), \ldots, \boldsymbol{a}_n(m))\,)$.

Definition 2. *For any relation $\varrho \in R_D^{(m)}$, and any operation $f \in O_D^{(n)}$, if $f(\boldsymbol{a}_1, \ldots, \boldsymbol{a}_n) \in \varrho$ for all choices of $\boldsymbol{a}_1, \ldots, \boldsymbol{a}_n \in \varrho$, then ϱ is said to be* invariant *under f, and f is called a* polymorphism *of ϱ.*

The set of all relations that are invariant under each operation from some set $C \subseteq O_D$ will be denoted $\mathsf{Inv}(C)$. The set of all operations that are polymorphisms of every relation from some set $\Gamma \subseteq R_D$ will be denoted $\mathsf{Pol}(\Gamma)$. We remark that the operators $\mathsf{Inv}()$ and $\mathsf{Pol}()$ form a Galois correspondence between R_D and O_D (see Proposition 1.1.14 of [32]). A basic introduction to this correspondence can be found in [30], and a comprehensive study in [32].

By considering certain properties of the set $\mathsf{Inv}(\mathsf{Pol}(\Gamma))$, the following result was obtained in [20].

Theorem 1. *Let Γ_1 and Γ_2 be sets of predicates over a finite set, such that Γ_1 is finite. If $\mathsf{Pol}(\Gamma_2) \subseteq \mathsf{Pol}(\Gamma_1)$ then $\mathrm{CSP}(\Gamma_1)$ is polynomial-time reducible to $\mathrm{CSP}(\Gamma_2)$.*

This result shows that, when the set of values is finite, finite sets of predicatess with the same polymorphisms give rise to constraint satisfaction problems which are mutually reducible. In other words, the complexity of $\mathrm{CSP}(\Gamma)$ is determined by the polymorphisms of Γ.

A number of results on the complexity of constraint satisfaction problems have been obtained via this approach (e.g., [4–9, 15, 16, 20, 21]). For example, Schaefer's Dichotomy Theorem [35], when appropriately re-stated, easily follows from Theorem 1 and well-known algebraic results [33] (see [20]).

Theorem 2 ([35]). *For any $\Gamma \subseteq R_{\{0,1\}}$, $\mathrm{CSP}(\Gamma)$ is in* **PTIME** *when $\mathsf{Pol}(\Gamma)$ contains at least one of the following:*

- *the constant 0 or constant 1 operations,*
- *the conjunction or disjunction operations,*
- *the affine operation $x - y + z \pmod 2$,*
- *the majority operation $(x \vee y) \wedge (x \vee z) \wedge (y \vee z)$.*

In all other cases $\mathrm{CSP}(\Gamma)$ is **NP***-complete.*

In this paper we consider the more general framework of the *quantified constraint satisfaction problem*, which is defined as follows.

Definition 3. *Let $\Gamma \subseteq R_D$. An* instance *of $\mathrm{QCSP}(\Gamma)$ is a first-order sentence $\mathcal{Q}_1 x_1 \dots \mathcal{Q}_l x_l \ (\varrho_1 \wedge \ldots \wedge \varrho_q)$, where each ϱ_i is an atomic formula involving a predicate from Γ, $x_1, \ldots, x_l$ are the variables appearing in the ϱ_i, and $\mathcal{Q}_1, \ldots, \mathcal{Q}_l$ are arbitrary quantifiers. The* question *is whether the sentence is true.*

Clearly, an instance of $\mathrm{CSP}(\Gamma)$ corresponds to an instance of $\mathrm{QCSP}(\Gamma)$ in which all the quantifiers happen to be existential.

For any finite set D, and any set of relations $\Gamma \subseteq R_D$, one can use an exhaustive search algorithm to show that $\mathrm{QCSP}(\Gamma)$ is in **PSPACE**. The complexity of $\mathrm{QCSP}(\Gamma)$ has been completely characterized in the Boolean case [12, 14].

Theorem 3 ([12, 14]). *For any $\Gamma \subseteq R_{\{0,1\}}$, $\mathrm{QCSP}(\Gamma)$ is in* **PTIME** *when $\mathsf{Pol}(\Gamma)$ contains at least one of the following:*

- *the conjunction or disjunction operations,*
- *the affine operation $x - y + z \pmod 2$,*
- *the majority operation $(x \vee y) \wedge (x \vee z) \wedge (y \vee z)$.*

In all other cases $\mathrm{QCSP}(\Gamma)$ is **PSPACE***-complete.*

Corollary 1. *$\mathrm{QCSP}(\{N\})$ is* **PSPACE***-complete, where N is the ternary "not-all-equal" relation on $\{0,1\}$, defined by $N = \{0,1\}^3 \setminus \{(0,0,0),(1,1,1)\}$*

Even though Theorem 3 is stated here using polymorphisms, it was proved using a combinatorial rather than an algebraic approach, and that method of proof does not easily generalize to larger sets of values. In the remaining sections we introduce an algebraic approach which enables us to systematically analyse the complexity of quantified constraint satisfaction problems over an arbitrary finite set D.

3 Reduction in QCSP

The next result is the main result of this paper. It shows that, for quantified constraint satisfaction problems, *surjective* polymorphisms play a similar role to that played by arbitrary polymorphisms for ordinary CSPs (cf. Theorem 1). Let $\mathsf{s\text{-}Pol}(\Gamma)$ denote the set of all surjective operations from $\mathsf{Pol}(\Gamma)$.

Theorem 4. *Let Γ_1 and Γ_2 be sets of predicates over a finite set, such that Γ_1 is finite. If $\mathsf{s\text{-}Pol}(\Gamma_2) \subseteq \mathsf{s\text{-}Pol}(\Gamma_1)$, then $\mathrm{QCSP}(\Gamma_1)$ is polynomial-time reducible to $\mathrm{QCSP}(\Gamma_2)$.*

This theorem follows immediately from the next two propositions whose proofs can be found in [3].

Definition 4. *For any set $\Gamma \subseteq R_D$, the set $[\Gamma]$ consists of all predicates that can be expressed using*

1. *predicates from Γ, together with the binary equality predicate $=_D$ on D,*
2. *conjunction,*
3. *existential quantification,*
4. *universal quantification.*

Proposition 1. *Let Γ_1 and Γ_2 be sets of predicates over a finite set, such that Γ_1 is finite. If $[\Gamma_1] \subseteq [\Gamma_2]$, then $\mathrm{QCSP}(\Gamma_1)$ is polynomial-time reducible to $\mathrm{QCSP}(\Gamma_2)$.*

Proposition 2. *For any set of predicates Γ over a finite set, $[\Gamma] = \mathsf{Inv}(\mathsf{s\text{-}Pol}(\Gamma))$.*

Note that this proposition intuitively means that the expressive power of constraints in QCSP is determined by their surjective polymorphisms. Hence, in order to show that some relation ϱ belongs to $[\Gamma]$, one does not have give an explicit construction, but instead one can show that ϱ is invariant under all surjective polymorphisms of Γ, which often turns out to be significantly easier.

We remark that the operators $\mathsf{Inv}()$ and $\mathsf{s\text{-}Pol}()$ used in Proposition 2 form a Galois connection between R_D and the set of all surjective members of O_D which has not previously been investigated (see, e.g., survey [31]).

4 Intractable Cases

In this section we will use Theorem 4 to show that certain small sets of predicates give rise to quantified constraint satisfaction problems that are intractable. In particular, we will show that QCSP(Γ) can be **PSPACE**-complete even in some cases where CSP(Γ) is trivial.

We first establish that a particular QCSP problem is **PSPACE**-complete. This problem corresponds to a generalized form of the standard GRAPH-$|D|$-COLORABILITY problem [19, 29] (which can be expressed as CSP($\{\neq_D\}$) where $\neq_D$ is the binary disequality predicate on D).

Proposition 3. QCSP($\{\neq_D\}$) *is* **PSPACE**-*complete when* $|D| \geq 3$.

Proof. By reduction from QCSP($\{N\}$), where N is the ternary not-all-equal predicate on a 2-element set, as defined in Corollary 1. For details see [3].

Theorem 5. *For any finite set D with $|D| \geq 3$, and any $\Gamma \subseteq R_D$, if every $f \in$ s-Pol(Γ) is of the form $f(x_1, \ldots, x_n) = g(x_i)$ for some $1 \leq i \leq n$ and some permutation g on D, then* QCSP(Γ) *is* **PSPACE**-*complete.*

Proof. By Lemma 1.3.1 (b) of [32], Pol($\{\neq_D\}$), for $|D| \geq 3$, consists of all operations of the form described in the Theorem. Hence Pol($\{\neq_D\}$) = s-Pol($\{\neq_D\}$), and we can apply Theorem 4 and Proposition 3.

The next example uses this result to show that even predicates that give rise to trivial constraint satisfaction problems can give rise to intractable quantified constraint satisfaction problems. This can happen because non-surjective operations, which may guarantee the tractability of the CSP, do not affect the complexity of the QCSP.

Example 1. Let τ_s be the s-ary "not-all-distinct" predicate holding on a tuple $(a_1, \ldots, a_s)$ if and only if $|\{a_1, \ldots, a_s\}| < s$. Note that $\tau_s \supseteq \{(a, \ldots, a) \mid a \in D\}$, so every instance of CSP($\{\tau_s\}$) is trivially satisfiable by assigning the same value to all variables.

However, by Lemma 2.2.4 of [32], the set Pol($\{\tau_{|D|}\}$) consists of all non-surjective operations on D, together with all operations of the form given in Theorem 5. Hence, $\{\tau_{|D|}\}$ satisfies the conditions of Theorem 5, and QCSP($\{\tau_{|D|}\}$) is **PSPACE**-complete. Similar arguments can be used to show that QCSP($\{\tau_s\}$) is **PSPACE**-complete, for any s in the range $3 \leq s \leq |D|$.

5 Tractable Cases

In spite of the results of the previous section, it is possible to identify sets of predicates which give rise to tractable QCSP problems. In this section we identify and describe two families of predicates of this kind.

5.1 Mal'tsev Predicates

An operation $m(x, y, z)$ on D is said to be *Mal'tsev* if it satisfies the identities $m(x, y, y) = m(y, y, x) = x$ for all x, y. For example, for an Abelian group G, the operation $f(x, y, z) = x - y + z$, called the *affine operation* of G, is a Mal'tsev operation. Relations invariant under the affine operation of a finite Abelian group play a significant role in the study of the complexity of the standard constraint satisfaction problem [5, 17, 20, 21].

Throughout this subsection, let $\Gamma = \mathsf{Inv}(\{m\})$. By developing and using a deep algebraic structural theory, a sophisticated polynomial-time algorithm for deciding CSP(Γ) was given in [5]. Moreover, it was proved there that a satisfying assignment to any CSP(Γ) instance can also be found in polynomial time. We will show now that QCSP(Γ) can be solved using these algorithms.

Theorem 6. *Let m be an arbitrary Mal'tsev operation on D. The problem class* QCSP($\mathsf{Inv}(\{m\})$) *is in* **PTIME**.

Proof. The proof is based on the following lemma.

Lemma 1. *Let $\mathcal{P} = \mathcal{Q}_1 x_1 \ldots \mathcal{Q}_n x_n\ \Phi(x_1, \ldots, x_n)$ be an instance of* QCSP(Γ), *and j the maximal index such that $\mathcal{Q}_j$ is the universal quantifier.*

(1) If $\Phi'(x_1, \ldots, x_{j-1}) = \forall x_j \exists x_{j+1} \ldots \exists x_n\ \Phi(x_1, \ldots, x_n)$ is satisfiable then, for any model $(c_1, \ldots, c_n)$ of Φ, the tuple $(c_1, \ldots, c_{j-1})$ is a model of Φ'.

(2) $\mathcal{P}$ is true if and only if so is $\mathcal{P}' = \mathcal{P}_1 \wedge \mathcal{P}_2$ where

$$\begin{aligned}\mathcal{P}_1 &= \mathcal{Q}_1 x_1 \ldots \mathcal{Q}_{j-1} x_{j-1} \exists x_j \exists x_{j+1} \ldots \exists x_n\ \Phi(x_1, \ldots, x_n),\\ \mathcal{P}_2 &= \exists x_1 \ldots \exists x_{j-1} \forall x_j \exists x_{j+1} \ldots \exists x_n\ \Phi(x_1, \ldots, x_n).\end{aligned}$$

Proof. (1) Let $(a_1, \ldots, a_{j-1})$ be a model for Φ', and $(a_j^b, \ldots, a_n^b)$, $b \in D$, its extensions such that $\boldsymbol{a}^b = (a_1, \ldots, a_{j-1}, a_j^b, \ldots, a_n^b)$ is a model of Φ and $a_j^b = b$.

Take an arbitrary model $\boldsymbol{c} = (c_1, \ldots, c_{j-1}, c_j, \ldots, c_n)$ of Φ. We need to show that $(c_1, \ldots, c_{j-1})$ is a model of Φ'. Fix an arbitrary $b \in D$ and let $\boldsymbol{d} = (d_1, \ldots, d_n)$ be equal to $m(\boldsymbol{a}^b, \boldsymbol{a}^{c_j}, \boldsymbol{c})$. Proposition 2 implies that the predicate defined by Φ is invariant under m, so $\boldsymbol{d}$ is a model of Φ, too. Moreover, we have $d_i = m(a_i, a_i, c_i) = c_i$ for $i \in \{1, \ldots, j-1\}$ and $d_j = m(a_j^b, a_j^{c_j}, c_j) = m(b, c_j, c_j) = b$. Thus, $(c_1, \ldots, c_{j-1})$ is a model of Φ'.

(2) Obviously, if $\mathcal{P}$ is true then $\mathcal{P}'$ is also true. The inverse implication easily follows from part (1). Indeed, since $\mathcal{P}_2$ is true, we can apply (1); then, (1) implies that every tuple $(c_1, \ldots, c_{j-1})$ that can be extended to a model of Φ can be extended so with c_j being any given element. Thus, since $\mathcal{P}_1$ is true, so is $\mathcal{P}$.

Repeatedly applying Lemma 1(2), one can show that every instance of QCSP(Γ) can be decomposed into a conjunction of instances which have the same quantifier-free part and each contain at most one universal quantifier. Moreover, if we can find a model of Φ then part (1) of the lemma implies that initial segments of this model can be used in deciding whether each of the instances is true. It remains to notice that, as is easy to check, fixing a value for any variable in a predicate from Γ gives another predicate invariant under m which implies that $\exists x_{l+1} \ldots \exists x_n \Phi(c_1, \ldots, c_{l-1}, b, x_{l+1}, \ldots, x_n)$ is also an instance of CSP($\mathsf{Inv}(\{m\})$). Now it follows that the algorithm shown in Fig. 1 is correct.

Input $\mathcal{P} = \mathcal{Q}_1 x_1 \ldots \mathcal{Q}_n x_n \, \Phi(x_1, \ldots, x_n)$ where $\Phi = \varrho_1 \wedge \ldots \wedge \varrho_q$, and $\varrho_1, \ldots, \varrho_q \in \Gamma$.
Output 'YES' if $\mathcal{P}$ is true, 'NO' otherwise.

Step 1 **Solve** the instance $\exists x_1 \ldots \exists x_n \, \Phi$
Step 2 **If** Φ has a model **then find** one, $(c_1, \ldots, c_n)$,
else OUTPUT('NO') and **STOP**.
Step 3 **For** $l = n, \ldots, 1$ **do**
If $\mathcal{Q}_l$ is the universal quatifier **then**
For each $b \in D$ **do**
Solve the instance $\exists x_{l+1} \ldots \exists x_n \Phi(c_1, \ldots, c_{l-1}, b, x_{l+1}, \ldots, x_n)$.
If the instance has no solution **then OUTPUT**('NO') and **STOP**.
enddo
enddo
Step 4 **OUTPUT**('YES').

Fig. 1. Algorithm for deciding QCSP(Γ) for Mal'tsev Γ

This algorithm uses $k|D| + 1$ applications of an algorithm solving CSP(Γ), where k is the number of universal quantifiers in an instance, and one application of an algorithm finding a model. Now we can use the general polynomial-time algorithms developed in [5]. This finishes the proof of Theorem 6.

Note that if the operation m has a special form then the method described above may lead to better algorithms. For example, let G be a finite Abelian group, with affine operation f, and unit element 0, and let Γ be a set of relations over G which are invariant under f. Note that, by straightforward algebraic manipulation, it can be shown that any (n-ary) relation invariant under f is a coset of a subgroup of the group G^n.

In the simplest case, when the order of G is prime, G can be considered as a prime field, and hence G^n as a vector space over G. In this case, each coset of a subgroup of G^n is a linear manifold, and it is well-known that such manifolds can be defined by systems of linear equations, whose coefficients are elements of the field G. Therefore, in this case, QCSP(Γ) can be considered as the problem of solving quantified linear systems over G, which can be done by applying standard techniques from linear algebra, or by using them in the above algorithm.

5.2 Implicational Predicates

Our second example of predicates which give rise to tractable quantified constraint satisfaction problems concerns predicates that are invariant under an operation known as the *dual discriminator*. These problems can be viewed as generalized Q2SAT, and our algorithm generalizes and extends the algorithm for Q2SAT given in [1].

Definition 5. *For any finite set D, the* dual discriminator *operation d on D, is given by*

$$d(x, y, z) = \begin{cases} y \text{ if } y = z \\ x \text{ otherwise} \end{cases}$$

Note that, in the special case when $D = \{0, 1\}$, the dual discriminator is exactly the majority operation mentioned in Theorems 2 and 3.

Theorem 7. *Let D be a finite set, and let d be the dual discriminator operation on D. The problem class* $\mathrm{QCSP}(\mathsf{Inv}(\{d\}))$ *is in* **NL**.

To establish this result, we consider a graph structure associated with any instance of a constraint satisfaction problem, which is sometimes known as the "microstructure graph" [18]. For any instance $\mathcal{P}$ of QCSP, with variables V and set of values D, the *microstructure graph* of $\mathcal{P}$ is the graph (W, E), where the set of nodes W, is the subset of pairs $V \times D$ representing all possible assignments of values to individual variables which are compatible with the constraints on those variables, and the set of edges, E, is the set of all assignments to pairs of variables which are compatible with the constraints on those pairs. In other words, E is the set of all (ordered) pairs of nodes $((v_i, a), (v_j, b)) \in W \times W$ such that $v_i \neq v_j$ and the partial assignment $v_i = a, v_j = b$ is compatible with each individual predicate in $\mathcal{P}$.

We will call an arc $((v_i, a), (v_j, b))$ in the microstructure graph of $\mathcal{P}$ *implicative* if it is the unique arc in E with first component (v_i, a) and second component (v_j, b'), for some value b'. If we remove all arcs from the microstructure graph of $\mathcal{P}$ which are *not* implicative, then we obtain a directed graph which will be called the *implication graph* of $\mathcal{P}$, and denoted $\mathrm{I}(\mathcal{P})$.

Note that the arcs in the implication graph $\mathrm{I}(\mathcal{P})$ represent logical implications which can be made about the possible models for the conjunction of predicates in $\mathcal{P}$. If any model gives value a to variable v_i, and $\mathrm{I}(\mathcal{P})$ contains the arc $((v_i, a), (v_j, b))$, then that model must also give the value b to the variable v_j. By the transitivity of implication, we obtain the same conclusion if there is any path from (v_i, a) to (v_j, b) in $\mathrm{I}(\mathcal{P})$.

For some sets of predicates the implication graph has certain additional properties, which can be used to obtain further restrictions on the possible models. We define one such property as follows.

Definition 6. *For any QCSP instance $\mathcal{P}$, the implication graph $\mathrm{I}(\mathcal{P})$ will be called* invertible *if whenever there is a path from a node (v_i, a) to a node (v_j, b), then for each node (v_j, b') with $b' \neq b$ there is a path from (v_j, b') to (v_i, a') for some $a' \neq a$ (which may depend on b').*

The second property we define holds when the implication graph captures all the restrictions on the possible models. In other words, it says that a given assignment is a model for the conjunction of predicates in $\mathcal{P}$ whenever it satisfies all of the implications encoded in the implication graph of $\mathcal{P}$.

Definition 7. *Let $\mathcal{P}$ be a QCSP instance of the form $\mathcal{Q}_1 x_1 \ldots \mathcal{Q}_l x_l \ \Phi$. The implication graph $\mathrm{I}(\mathcal{P})$ will be called* sufficient *if a given assignment σ is a model of Φ whenever $\mathrm{I}(\mathcal{P})$ contains the node $(v, \sigma(v))$ for each variable v, and does not contain an arc $((v_i, \sigma(v_i)), (v_j, b))$, with $b \neq \sigma(v_j)$.*

We will now show that, when the implication graph for an instance is both invertible and sufficient, then there is a non-deterministic logarithmic-space algorithm for that instance.

Proposition 4. *There is a non-deterministic logarithmic-space algorithm that decides any QCSP instance $\mathcal{P}$ whose implication graph* $\mathrm{I}(\mathcal{P})$ *is both invertible and sufficient.*

Proof. See [3].

Finally, we show that both of these properties hold for all predicates that are invariant under the dual discriminator.

Proposition 5. *Let D be a finite set, and let d be the dual discriminator operation on D. For any instance $\mathcal{P}$ in* $\mathrm{QCSP}(\mathsf{Inv}(\{d\}))$*, the implication graph* $\mathrm{I}(\mathcal{P})$ *is invertible and sufficient.*

Proof. It was shown in [21] that the predicates invariant under d are precisely the predicates which can be expressed as conjunctions of binary predicates, each of which is of the form of a "0/1/all" binary relation, as described in [10]. Such predicates are also described in [24], where they are referred to as "implicational".

The defining characteristic of a "0/1/all" relation on a pair of variables is that for both variables each value is either disallowed, or else allowed with either precisely one value, or all possible values, for the other variable [10]. Hence, it is straightforward to show that they give rise to implication graphs which are invertible. (In fact, an equivalent statement appears as Proposition 2.3 in [24].)

Furthermore, it is straightforward to verify that, in any problem instance whose constraints are specified by such relations, the restrictions imposed by these constraints on possible assignments are precisely those captured by the implication graph of the instance. Hence this implication graph is sufficient.

6 A Trichotomy Result

In this section we apply results from the previous sections to obtain a complete classification of complexity of $\mathrm{QCSP}(\Gamma)$ in those cases where Γ contains the set Δ of all graphs of permutations. Recall that the graph of a permutation π is the binary relation $\{(x, y) \mid y = \pi(x)\}$ (or the binary predicate $\pi(x) = y$). The complexity of $\mathrm{CSP}(\Gamma)$ for such sets Γ is completely classified in [15].

We will need two new surjective operations:

- The k-ary *near projection* operation,

$$l_k(x_1, \ldots, x_k) = \begin{cases} x_1 \text{ if } x_1, \ldots, x_k \text{ are all different,} \\ x_k \text{ otherwise.} \end{cases}$$

- The ternary *switching* operation,

$$s(x, y, z) = \begin{cases} x \text{ if } y = z, \\ y \text{ if } x = z, \\ z \text{ otherwise.} \end{cases}$$

Proposition 6. *If $\Gamma \subseteq R_D$, $|D| \geq 3$, and $l_{|D|} \in$ s-Pol(Γ) then* QCSP(Γ) *is polynomial-time reducible to* CSP(Γ') *where* $\Gamma' =$ Inv(Pol(Γ)). *In particular,* QCSP(Γ) *is in* **NP**.

Proof. See [3].

Theorem 8. *Let $\Delta \subseteq \Gamma \subseteq R_D$, and $|D| \geq 3$.*
- *If* s-Pol(Γ) *contains the dual discriminator d, or the switching operation s, or an affine operation, then* QCSP(Γ) *is in* **PTIME**;
- *else, if* s-Pol(Γ) *contains $l_{|D|}$, then* QCSP(Γ) *is* **NP**-*complete;*
- *else* QCSP(Γ) *is* **PSPACE**-*complete.*

Proof. Chapter 5 of [36] shows that, either s-Pol(Γ) consists of all projections (that is, all functions of the form $f(x_1, \ldots, x_n) = x_i$ for some $1 \leq i \leq n$), or else s-Pol(Γ) contains the dual discriminator operation, d, or the near-projection operation, $l_{|D|}$, or (when $|D| \in \{3, 4\}$) an affine operation. If s-Pol(Γ) consists of all projections then, by Theorem 5, QCSP(Γ) is **PSPACE**-complete. If s-Pol(Γ) contains d or an affine operation then, by Theorem 6 or Theorem 7, QCSP(Γ) is tractable.

Suppose that s-Pol(Γ) contains $l_{|D|}$. Then, by Proposition 6, this problem is polynomial-time reducible to CSP(Γ'), where $\Gamma' =$ Inv(Pol(Γ)). If s-Pol(Γ) contains s or d then clearly so does Pol(Γ'). It follows from the results of [15] that in this case CSP(Γ') is tractable, and hence so is QCSP(Γ). If s-Pol(Γ) contains neither s nor d then, by [15], CSP(Γ) is **NP**-complete. Since, obviously, CSP(Γ) is polynomial-time reducible to QCSP(Γ), and QCSP(Γ) is in **NP**, the result follows.

Note that, for any fixed finite set D, the conditions in Theorem 8 can be efficiently checked.

7 Conclusion

We have shown that the algebraic theory relating complexity and polymorphisms, which was originally developed for the standard constraint satisfaction problem allowing only existential quantifiers, can be extended to deal with the more general framework of the quantified constraint satisfaction problem.

In this extension of the theory it turns out that it is the *surjective* polymorphisms of the predicates used in problem instances which determine the complexity of the corresponding problems. Using this information we have been able to identify subproblems of the quantified constraint satisfaction problem lying in (or complete for) some standard complexity classes, and to obtain a complete classification of complexity for certain special cases. We expect that by developing the results and ideas from this paper one will be able to successfully build a far-reaching theory linking algebraic properties of relations with the computational complexity of certain associated problems, just as it was in the case of ordinary constraint satisfaction problems.

Acknowledgement The authors thank Reinhard Pöschel who helped launch this project.

References

1. Aspvall, B., Plass, M., Tarjan, R.: A linear time algorithm for testing the truth of certain quantified Boolean formulas. Information Processing Letters **8** (1979) 121–123
2. Böhler, E., Hemaspaandra, E., Reith, S., Vollmer, H.: Equivalence and isomorphism for Boolean constraint satisfaction. In: Proceedings 16th International Workshop on Computer Science Logic, CSL'02. Volume 2471 of Lecture Notes in Computer Science., Springer-Verlag (2002) 412–426
3. Börner, F., Krokhin, A., Bulatov, A., Jeavons, P.: Quantified constraints and surjective polymorphisms. Technical Report PRG-RR-02-11, Computing Laboratory, University of Oxford, UK (2002)
4. Bulatov, A.: A dichotomy theorem for constraints on a three-element set. In: Proceedings of 43rd IEEE Symposium on Foundations of Computer Science, FOCS'02. (2002) 649–658
5. Bulatov, A.: Mal'tsev constraints are tractable. Technical Report RR-02-05, Computing Laboratory, Oxford University (2002)
6. Bulatov, A.: Tractable conservative constraint satisfaction problems. In: Proceedings 18th IEEE Symposium on Logic in Computer Science, LICS'03. (2003) to appear.
7. Bulatov, A., Jeavons, P.: Algebraic structures in combinatorial problems. Technical Report MATH-AL-4-2001, Technische Universität Dresden, Germany (2001)
8. Bulatov, A., Krokhin, A., Jeavons, P.: Constraint satisfaction problems and finite algebras. In: Proceedings 27th International Colloquium on Automata, Languages and Programming, ICALP'00. Volume 1853 of Lecture Notes in Computer Science., Springer-Verlag (2000) 272–282
9. Bulatov, A., Krokhin, A., Jeavons, P.: The complexity of maximal constraint languages. In: Proceedings 33rd ACM Symposium on Theory of Computing, STOC'01. (2001) 667–674
10. Cooper, M., Cohen, D., Jeavons, P.: Characterising tractable constraints. Artificial Intelligence **65** (1994) 347–361
11. Creignou, N., Daudé, H.: Random generalized satisfiability problems. In: Proceedings of 5th International Symposium on Theory and Applications of Satisfiability Testing - SAT'02. (2002) 17–26
12. Creignou, N., Khanna, S., Sudan, M.: Complexity Classifications of Boolean Constraint Satisfaction Problems. Volume 7 of SIAM Monographs on Discrete Mathematics and Applications. (2001)
13. Crescenzi, P., Rossi, G.: On the Hamming distance of constraint satisfaction problems. Theoretical Computer Science **288** (2002) 85–100
14. Dalmau, V.: Some dichotomy theorems on constant-free quantified Boolean formulas. Technical Report TR LSI-97-43-R, Department LSI, Universitat Politecnica de Catalunya (1997)
15. Dalmau, V.: A new tractable class of constraint satisfaction problems. In: Proceedings 6th International Symposium on Artificial Intelligence and Mathematics. (2000)
16. Dalmau, V.: Constraint satisfaction problems in non-deterministic logarithmic space. In: Proceedings 29th International Colloquium on Automata, Languages and Programming, ICALP'02. Volume 2380 of Lecture Notes in Computer Science., Springer-Verlag (2002) 414–425

17. Feder, T., Vardi, M.: The computational structure of monotone monadic SNP and constraint satisfaction: A study through Datalog and group theory. SIAM Journal on Computing **28** (1998) 57–104
18. Freuder, E.: Exploiting structure in constraint satisfaction problems. In Mayoh, M., Tyugum, E., Penjam, J., eds.: Constraint Programming. Volume 131 of NATO ASI Series., Springer-Verlag (1993)
19. Garey, M., Johnson, D.: Computers and Intractability: A Guide to the Theory of NP-Completeness. Freeman, San Francisco, CA. (1979)
20. Jeavons, P.: On the algebraic structure of combinatorial problems. Theoretical Computer Science **200** (1998) 185–204
21. Jeavons, P., Cohen, D., Gyssens, M.: Closure properties of constraints. Journal of the ACM **44** (1997) 527–548
22. Juban, L.: Dichotomy theorem for generalized unique satisfiability problem. In: Proceedings 12th Conference on Fundamentals of Computation Theory, FCT'99. Volume 1684 of Lecture Notes in Computer Science., Springer-Verlag (1999) 327–337
23. Kavvadias, D., Sireni, M.: The inverse satisfiability problem. SIAM Journal on Computing **28** (1998) 152–163
24. Kirousis, L.: Fast parallel constraint satisfaction. Artificial Intelligence **64** (1993) 147–160
25. Kirousis, L., Kolaitis, P.: The complexity of minimal satisfiability problems. In: Proceedings 18th International Symposium on Theoretical Aspects of Computer Science, STACS'01. Volume 2010 of Lecture Notes in Computer Science., Springer-Verlag (2001) 407–418
26. Kleine Büning, H., Karpinski, M., Flögel, A.: Resolution for quantified Boolean formulas. Information and Computation **117** (1995) 12–18
27. Kolaitis, P., Vardi, M.: Conjunctive-query containment and constraint satisfaction. Journal of Computer and System Sciences **61** (2000) 302–332
28. Marriott, K., Stuckey, P.: Programming with Constraints: an Introduction. MIT Press (1998)
29. Papadimitriou, C.: Computational Complexity. Addison-Wesley (1994)
30. Pippenger, N.: Theories of Computability. Cambridge University Press, Cambridge (1997)
31. Pöschel, R.: Galois connections for operations and relations. Technical Report MATH-AL-8-2001, Technische Universität Dresden, Germany (2001)
32. Pöschel, R., Kalužnin, L.: Funktionen- und Relationenalgebren. DVW, Berlin (1979)
33. Post, E.: The two-valued iterative systems of mathematical logic. Volume 5 of Annals Mathematical Studies. Princeton University Press (1941)
34. Reith, S., Vollmer, H.: Optimal satisfiability for propositional calculi and constraint satisfaction problems. In: Proceedings 25th International Symposium on Mathematical Foundations of Computer Science, MFCS'00. Volume 1893 of Lecture Notes in Computer Science., Springer (2000) 640–649
35. Schaefer, T.: The complexity of satisfiability problems. In: Proceedings 10th ACM Symposium on Theory of Computing, STOC'78. (1978) 216–226
36. Szendrei, A.: Clones in Universal Algebra. Volume 99 of Seminaires de Mathematiques Superieures. University of Montreal (1986)
37. Tsang, E.: Foundations of Constraint Satisfaction. Academic Press, London (1993)
38. Williams, R.: Algorithms for quantified Boolean formulas. In: Proceedings ACM Symposium on Discrete Algorithms, SODA'02. (2002) 299–307

Verification of Infinite State Systems

Ahmed Bouajjani

LIAFA, University of Paris 7

The aim of this tutorial is to give an overview on the state-of-the-art in infinite-state model checking and its applications.

We present a unified modeling framework based on word/term rewrite systems and show its relevance in reasoning about several important classes of systems (communication protocols, parametrized distributed algorithms, multi-threaded programs, etc).

Then, we address the verification problem of various classes of such models. We consider especially the basic problem of reachability analysis which consists in computing a (finite) representation of the (potentially infinite) set of reachable configurations.

We show the main existing approaches to tackle this problem:

- Specialized constructions for several significant classes of models for which this problem is shown to be decidable,
- General principles to prove the termination and the completeness of the iterative computation of the reachability sets for classes of models,
- Generic constructions and fixpoint acceleration techniques, leading to powerful semi-algorithms applicable to general classes of models in order to compute exact/approximate reachability sets.

M. Baaz and J.A. Makowsky (Eds.): CSL 2003, LNCS 2803, p. 71, 2003.

Parity of Imperfection *or* Fixing Independence

Julian C. Bradfield

Laboratory for Foundations of Computer Science, School of Informatics,
King's Bldgs, University of Edinburgh, Edinburgh EH9 3JZ, UK
jcb@inf.ed.ac.uk

Abstract. We introduce a fixpoint extension of Hintikka and Sandu's IF (independence-friendly) logic. We obtain some results on its complexity and expressive power. We relate it to parity games of imperfect information, and show its application to defining independence-friendly modal mu-calculi.

1 Introduction

Independence-friendly logic [7] is a logic introduced by Hintikka and Sandu which gives an alternative account of branching quantifiers (Henkin quantifiers) in terms of games of imperfect information. It allows the expression of quantifiers where the choice must be independent of specified earlier choices; it has existential second-order power. In the last few years, it has attracted study from both philosophical logicians and mathematical logicians. Also, in earlier work, we have argued that its modal analogues have a role to play in concurrency theory. (See [2] and [3] for discussions of this role and its relation to other work in concurrency theory.)

Given a first-order logic, or a logic like IF that is supposed to *look* first-order (even though it isn't), it is natural to want to add fixpoint operators. One motivation is just the mathematical interest of studying inductive definability in many contexts; a more computer-science-based motivation is the desire to be able to produce an IF analogue of the modal mu-calculus, a popular and interesting temporal logic.

In [2], we asserted that using the semantics given to IF by Hodges [8], it was possible to define an IF fixpoint logic. In this article, we give a detailed definition of IF least fixpoint logic (which, typically of IF logics, is a little more subtle than one first thinks), and then study it.

In section 2, we deal with the preliminaries, the existing syntax and semantics of IF logic. Sections 3, 4 and 5 are the main part of the paper; in section 3 we give the detailed definitions of IF fixpoint logic and its semantics; in section 4 we give a couple of interesting examples; and in section 5 we establish some partial results on complexity and expressive power. Then in section 6 we return to the game-theoretic roots of IF by giving a suitable notion of parity game of imperfect information, which gives an alternative semantics for IF fixpoint logic. Finally, in section 7 we briefly sketch the application to IF modal mu-calculus that was one of the original motivations for looking at IF with fixpoints.

M. Baaz and J.A. Makowsky (Eds.): CSL 2003, LNCS 2803, pp. 72–85, 2003.

2 IF-FOL Syntax and Semantics

First of all, we state one important **notational convention**: we take the scope of all quantifiers and fixpoint operators to extend as far to the right as possible.

For the purposes of this article, we will use only a sublanguage of IF-FOL (sometimes just IF for short). The full languages advocated by Hintikka and analysed by Hodges and others include the possibility of conjunctions and disjunctions that are independent of previous quantifiers. These operators do not introduce inherently new problems, but they do introduce some additional complexity (and space) in defining the semantics. We will therefore ignore them, and consider only the independent quantifiers; the interested reader can use [8] to put back the independent junctions.

One of the more tedious features of IF-FOL is the need to be more pedantic than usual in keeping track of free variables etc., as not all the things one takes for granted in usual logic are true in IF-FOL. When introducing fixpoint operators, even more care is needed, and we shall therefore give the semantics even more pedantically than Hodges did.

Definition 1. *Assume the usual FOL set of proposition (P, Q etc.), relation (R, S etc.), function (f, g etc.) and constant (a, b etc.) symbols, with given arities. Assume also the usual* variables *v, x etc. We write $\vec{x}, \vec{v}$ etc. for tuples of variables, and similarly for tuples of other objects; we use concatenation of symbols to denote concatenation of tuples with tuples or objects.*

For formulae ϕ and terms t, the (meta-level) notations $\phi[\vec{x}]$ and $t[\vec{x}]$ mean that the free variables of ϕ or t are included in the variables $\vec{x}$, without repetition[1].

The terms of IF-FOL are as usual constructed from variables, constants and function symbols. The free variables of a term are as usual; the free variables of a tuple of terms are the union of the free variables of the terms.

We assume equality $=$ is in the language, and atomic formulae are defined as usual. The free variables of the formula $R(\vec{t})$ are those of $\vec{t}$.

The compound formulae are given as follows:

Conjunction and disjunction. *If $\phi[\vec{x}]$ and $\psi[\vec{y}]$ are formulae, then $(\phi \vee \psi)[\vec{z}]$ and $(\phi \wedge \psi)[\vec{z}]$ are formulae, where $\vec{z}$ is the union of $\vec{x}$ and $\vec{y}$.*

Quantifiers. *If $\phi[\vec{y}, x]$ is a formula, x a variable, and W a finite set of variables, then $(\forall x/W.\, \phi)[\vec{y}]$ and $(\exists x/W.\, \phi)[\vec{y}]$ are formulae. If W is empty, we write just $\forall x.\, \phi$ and $\exists x.\, \phi$.*

Game negation. *If $\phi[\vec{x}]$ is a formula, so is $({\sim}\phi)[\vec{x}]$.*

Flattening. *If $\phi[\vec{x}]$ is a formula, so is $(\downarrow \phi)[\vec{x}]$.*

(Negation. *$\neg\phi$ is an abbreviation for ${\sim} \downarrow \phi$.)*

Definition 2. *IF-FOL$^+$ is the logic in which ${\sim}$, $\downarrow$ and $\neg$ are applied only to atomic formulae.*

[1] [8] writes $\phi(\vec{x})$, but we wish to distinguish the meta-notation for free variables from the object-level syntax for atomic formulae and the meta-notation for assigning values to variables.

In the independent quantifiers the intention is that W is the set of independent variables, whose values the player is not allowed to know at this choice point: the Henkin quantifier $\begin{smallmatrix}\forall x & \exists y \\ \forall u & \exists v\end{smallmatrix}$ can be written as $\forall x/\varnothing.\,\exists y/\varnothing.\,\forall u/\{x,y\}.\,\exists v/\{x,y\}$. If one then plays the usual model-checking game with this additional condition, which can be formalized by requiring strategies to be uniform in the 'unknown' variables, one gets a game semantics which characterizes the Skolem function semantics in the sense that Eloise has a winning strategy iff the formula is true. However, these games are not determined, so it is *not* true that Abelard has a winning strategy iff the formula is untrue. For example, $\begin{smallmatrix}\forall x \\ \exists y\end{smallmatrix}.x = y$ (or $\forall x.\,\exists y/\{x\}.\,x = y$) is untrue in any structure with more than one element, but Abelard has no winning strategy.

The trump semantics of Hodges [8], with variants by others, gives a Tarski-style semantics for this logic, equivalent to the imperfect information game semantics given by Hintikka and Sandu. The semantics is as follows:

Definition 3. *Let a structure A be given, with constants, propositions and relations interpreted in the usual way. A* deal $\vec{a}$ *for $\phi[\vec{x}]$ or $\vec{t}[\vec{x}]$ is an assignment of an element of A to each variable in $\vec{x}$. Given a deal $\vec{a}$ for a tuple of terms $\vec{t}[\vec{x}]$, let $\vec{t}(\vec{a})$ denote the tuple of elements obtained by evaluating the terms under the deal $\vec{a}$.*

If $\phi[\vec{x}]$ is a formula and W is a subset of the variables in $\vec{x}$, two deals $\vec{a}$ and $\vec{b}$ for ϕ are $\simeq_W$-equivalent *($\vec{a} \simeq_W \vec{b}$) iff they agree on the variables not in W. A* $\simeq_W$-set *is a non-empty set of pairwise $\simeq_W$-equivalent deals.*

The interpretation $[\![\phi]\!]$ of a formula is a pair (T, C) where T is the set of trumps, *and C is the set of* cotrumps.

- *If $(R(\vec{t}))[\vec{x}]$ is atomic, then a non-empty set D of deals is a trump iff $\vec{t}(\vec{a}) \in R$ for every $\vec{a} \in D$; D is a cotrump iff it is non-empty and $\vec{t}(\vec{a}) \notin R$ for every $\vec{a} \in D$.*
- *D is a trump for $(\phi \wedge \psi)[\vec{x}]$ iff D is a trump for $\phi[\vec{x}]$ and D is a trump for $\psi[\vec{x}]$; D is a cotrump iff there are cotrumps E, F for ϕ, ψ such that every deal in D is an element of either E or F.*
- *D is a trump for $(\phi \vee \psi)[\vec{x}]$ iff it is non-empty and there are trumps E of ϕ and F of ψ such that every deal in D belongs either to E or F; D is a cotrump iff it is a cotrump for both ϕ and ψ.*
- *D is a trump for $(\forall y/W.\,\psi)[\vec{x}]$ iff the set $\{\,\vec{a}b \mid \vec{a} \in D, b \in A\,\}$ is a trump for $\psi[\vec{x}, y]$. D is a cotrump iff it is non-empty and there is a cotrump E for $\psi[\vec{x}, y]$ such that for every $\simeq_W$-set $F \subseteq D$ there is a b such that $\{\,\vec{a}b \mid \vec{a} \in F\,\} \subseteq E$.*
- *D is a trump for $(\exists y/W.\,\psi)[\vec{x}]$ trump iff there is a trump E for $\psi[\vec{x}, y]$ such that for every $\simeq_W$-set $F \subseteq D$ there is a b such that $\{\,\vec{a}b \mid \vec{a} \in F\,\} \subseteq E$; D is a cotrump iff the set $\{\,\vec{a}b \mid \vec{a} \in D, b \in A\,\}$ is a cotrump for $\psi[\vec{x}, y]$.*
- *D is a trump for $\sim\phi$ iff D is a cotrump for ϕ; D is a cotrump for $\sim\phi$ iff it is a trump for ϕ.*
- *D is a trump (cotrump) for $\downarrow \phi$ iff D is a non-empty set of members (non-members) of trumps of ϕ.*

A sentence is true in the usual sense if $\{\langle\rangle\} \in T$ (the empty deal is a trump set), and false in the usual sense if $\{\langle\rangle\} \in C$; this corresponds to Eloise or Abelard having a uniform winning strategy. Otherwise, it is undetermined.

Note that the game negation $\sim$ provides the usual de Morgan dualities.

A trump for ϕ is essentially a set of winning positions for the model-checking game for ϕ, for a given *uniform* strategy, that is, a strategy where choices are uniform in the 'hidden' variables. The most intricate part of the above definition is the clause for $\exists y/W.\,\psi$: it says that a trump for $\exists y/W.\,\psi$ is got by adding a witness for y, uniform in the W-variables, to trumps for ψ.

It is easy to see that any subset of a trump is a trump. In the case of an ordinary first-order $\phi(\vec{x})$, the set of trumps of ϕ is just the power set of the set of tuples satisfying ϕ. To see how a more complex set of trumps emerges, consider the following formula, which has x free: $\exists y/\{x\}.\, x = y$. Any singleton set of deals is a trump, but no other set of deals is a trump. Thus we obtain that $\forall x.\, \exists y/\{x\}.\, x = y$ has no trumps (unless the domain has only one element).

The following definition is for later convenience: a set T of sets of deals is *well-dealt* if for every $D \in T$, D is non-empty and $D' \in T$ for every non-empty $D' \subseteq D$. A formula has well-dealt semantics (T, C) if T and C are well-dealt; the above semantics ensures that all IF-FOL formulae have well-dealt semantics.

[8] shows that every well-dealt set is the semantics of some IF formula (given suitable atomic relations), giving us

Proposition 4. *On a structure A with n elements, IF formulae of length m require space exponential in n^m to represent their semantics.*

Proof. The set of tuples for m free variables has n^m elements; Given a k element set, there are 2^k subsets, but not all sets of subsets are well-dealt; however, there are about $2^k/\sqrt{k}$ sets of size $k/2$, and hence at least $2^{2^k/\sqrt{k}}$ well-dealt sets of subsets. (Cameron and Hodges [4] look in more detail at the combinatorics of trumps.) □

We can record the easy loose upper bounds on the time complexity of IF-FOL operations:

Proposition 5. *In a structure A of size n, the trump components of the IF operators can be calculated in the following times on formulae with m free variables, where $k = n^m$: $\vee$ and $\wedge$ in $2^{k+1} \cdot k^2$; $\forall x$ in $2^k \cdot k^3 n$; $\exists x/W$ in $2^{k+k \lg n}$.*

Proof. A crude analysis of the cost of computing the trump semantics more or less directly from the definitions. Note that the computation for $\exists$ has further exponential factors above the 2^k from the number of possible trumps, effectively due to the computation of choice functions. □

In the case of IF, these exponential upper bounds are much worse than is really required for determining whether a deal satisfies (i.e. is a singleton trump for) an IF formula, since IF expressible properties are in NP (because we can guess values for choice functions).

3 Adding Fixpoint Operators

The prime motivation for considering fixpoint extensions is in the modal setting, where it is a standard way to produce temporal logics from modal logics. However, fixpoint extensions to IF logics raise a number of issues, and it is useful to recall briefly the first-order case.

In the classical settings, fixpoint operators are added to allow sets or relations to be inductively defined by formulae: $\mu(x, X).\phi(x, X)$, where X is a set variable, is the least set A such that $A = \{ x \mid \phi(x, A) \}$, and the syntax of formulae is extended to allow terms of the form $t \in X$ or $t \in \mu(x, X).\phi(x, X)$ (among set theorists) or $X(t)$ and $(\mu(x, X).\phi(x, X))(t)$ (among finite model theorists).

In applying this directly to IF-FOL, there is the obvious problem that we no longer have a simple notion of an element satisfying a formula, so the usual definition no longer type-checks. There are two possible approaches, depending on how one views the use of fixpoint terms. If one takes the view that their purpose is to define sets, and the logic is a means to this end, then it is natural to retain the use of set variables, and work out how to make $\phi(x, X)$ reduce to a boolean. On the other hand, if one views fixpoint operators as a means of introducing recursion into the logical formulae, it is more natural to decide that fixpoint terms should have the same semantics as other formulae, namely sets of trumps, and that therefore the variables X range over trump sets rather than sets. We then have to decide the meaning of $X(t)$. This is the approach we suggested in [2], and will now pursue.

Definition 6. *IF-LFP extends the syntax of IF-FOL as follows:*

- *There is a set* $\mathrm{Var} = \{X, Y, \ldots\}$ *of fixpoint variables. Each variable* X *has an arity* $(\mathrm{ar}_1(X), \mathrm{ar}_2(X))$*;* $\mathrm{ar}_1(X)$ *is the arity of the fixpoint, and* $\mathrm{ar}_2(X)$ *is the number of free parameters of the fixpoint.*
- *If* X *is a fixpoint variable, and* $\vec{t}$ *an* $\mathrm{ar}_1(X)$*-vector of terms then* $X(\vec{t})$ *is a formula.*
- *The notation* $\phi(X)$ *indicates that* X *is among the free fixpoint variables of* ϕ*. If* $\phi(X)[\vec{x}, \vec{z}]$ *is a formula with* $\mathrm{ar}_1(X)$ *free individual variables* $\vec{x}$ *and* $\mathrm{ar}_2(X)$ *free individual variables* $\vec{z}$*, and* $\vec{t}$ *is a sequence of* $\mathrm{ar}_1(X)$ *terms with free variables* $\vec{y}$*, then* $(\mu(X, \vec{x}).\phi)(\vec{t})[\vec{z}, \vec{y}]$ *is a formula;* **provided that** ϕ *is IF-FOL*$^+$*.*
- *similarly for* $\nu(X, \vec{x}).\phi$*.*

The process of extending the trump semantics to fixpoint formulae is not entirely straightforward. First we define valuations for free fixpoint variables.

Definition 7. *A fixpoint valuation* $\mathcal{V}$ *maps each fixpoint variable* X *to a pair* $(\mathcal{V}_T(X), \mathcal{V}_C(X)) \in (\wp(\wp(A^{\mathrm{ar}_1(X)+\mathrm{ar}_2(X)})))^2$*.*

Let D *be a non-empty set of deals for* $X(\vec{t})[\vec{x}, \vec{z}, \vec{y}]$*, where* $\vec{y}$ *are the free variables of* $\vec{t}$ *not already among* $\vec{x}, \vec{z}$*. A deal* $d = \vec{a}\vec{c}\vec{b} \in D$*, where* $\vec{a}, \vec{c}, \vec{b}$ *are the deals for* $\vec{x}, \vec{z}, \vec{y}$ *respectively, determines a deal* $d' = \vec{t}(d)\vec{c}$ *for* $X[\vec{x}, \vec{z}]$*. Let* $D' = \{ d' \mid d \in D \}$*.* D *is a trump for* $X(\vec{t})$ *iff* $D' \in \mathcal{V}_T(X)$*; it is a cotrump iff* $D' \in \mathcal{V}_C X$*.*

Then we define a suitable complete partial order on denotations:

Definition 8. *If (T_1, C_1) and (T_2, C_2) are elements of $(\wp(\wp(A^n)))^2$, define $(T_1, C_1) \preceq (T_2, C_2)$ iff $T_1 \subseteq T_2$ and $C_1 \supseteq C_2$.*

Lemma 9. *If $\phi(X)[\vec{x}, \vec{z}]$ is an IF-FOL$^+$ formula and $\mathscr{V}$ is a fixpoint valuation, the map on $(\wp(\wp(A^{\mathrm{ar}_1(X)+\mathrm{ar}_2(X)}))^2$ given by*

$$(T, C) \mapsto [\![\phi]\!]_{\mathscr{V}[X:=(T,C)]}$$

is monotone with respect to $\preceq$; hence it has least and greatest fixpoints, with ordinal approximants defined in the usual way.

Definition 10. *$[\![\mu(X, x).\phi(X)[\vec{x}, \vec{z}]]\!]$ is the least fixpoint of the map just defined; $[\![\nu(X, x).\phi(x)[\vec{x}, \vec{z}]]\!]$ is the greatest fixpoint. $\mu^\zeta(X, x).\phi$ means the ζth approximant of $\mu(X, x)$.*

The following lemma records the usual basic properties (which have to be checked again in this setting), and one new basic property, particular to the IF case.

Lemma 11. *1. The trump and cotrump components of $[\![\mu(X, x).\phi]\!]$ are well-dealt.*
2. If Y is free in ϕ, then $[\![\mu(X, x).\phi]\!]$ is monotone in Y; hence the definition extends to further fixpoints in the usual way, as does this lemma.
3. μ and ν are dual: T is a trump for $\mu(X, x).\phi(X)$ iff it is a cotrump for $\nu(X, x).{\sim}\phi({\sim}X)$ (with the outer negation pushed in by duality).

Proof. (1) by induction on approximants; (2) as usual; (3) from definitions.

A distinctive feature of the definition, compared to the normal LFP definition, is the way that free variables are explicitly mentioned. Normally, one can fix values for the free variables, and then compute the fixpoint, but because of independent quantification this is not possible in the IF setting. For example, consider the formula fragment

$$\forall z. \ldots \mu(X, x). \ldots \vee \exists y/\{z\}. X(y)$$

The independent choice of y means that the trumps for the fixpoint depend on the possible deals for z, not just a single deal.

Another point is that the trump set of a least fixpoint is the union of the trump sets of its approximants; but the interpretation of logical disjunction is not union of trump sets, but union of trumps (applied pointwise to the trump sets). Thus the usual view of a least fixpoint as a transfinite disjunction is not valid in general. The following explains why, despite this, the IF-LFP semantics is consistent with classical LFP semantics.

Proposition 12. *Call a set T of trumps or cotrumps* full *iff it is the set of non-empty subsets of $\bigcup T$. Call a formula ϕ of IF-LFP* classical *iff it is in IF-FOL$^+$ and it contains no independent quantification (i.e. all quantifiers are $\exists x/\varnothing$ and $\forall x/\varnothing$). Then*

1. *if* ϕ *is fixpoint free, then* $[\![\phi]\!]_{\mathrm{IF}} = (T, C)$ *is full,* $\bigcup T = [\![\phi]\!]_{\mathrm{FO}}$*, and* $\bigcup C = [\![\neg\phi]\!]_{\mathrm{FO}}$ *;*
2. *if* T_ζ *is a (transfinite) sequence of full well-dealt deal sets, then* $\bigcup_\zeta T_\zeta$ *and* $\bigcap_\zeta T_\zeta$ *are full well-dealt sets;*
3. *hence (1) is true for any classical IF-LFP formula.*

4 Examples of IF-LFP

IF logic is not entirely easy to understand and mu-calculi are also traditionally hard to understand, so we now consider some examples that demonstrate interesting features of the combination. For convenience, we introduce the abbrevation $\phi \Rightarrow \psi$ for $\psi \vee {\sim}\phi$ provided that ϕ is atomic.

Let $G = (V, E)$ be a directed graph. The usual LFP formula $R(y, z) \stackrel{\text{def}}{=} (\mu(X, x).z = x \vee \exists w.\, E(x, w) \wedge X(w))(y)$ asserts that the vertex z is reachable from y. Hence the formula $\forall y.\, \forall z.\, R(y, z)$ asserts that G is strongly connected. Now consider the IF-LFP formula

$$\forall y.\, \forall z.\, (\mu(X, x).z = x \vee \exists w/\{y, z\}.\, E(x, w) \wedge X(w))(y).$$

At first sight, one might think this asserts not only that every z is reachable from every y, but that the path taken is independent of the choice of y and z. This is true exactly if G has a directed Hamiltonian cycle, a much harder property than being strongly connected.

Of course, the formula does not mean this, because the variable w is fresh each time the fixpoint is unfolded. In the trump semantics, the denotation of the fixpoint will include all the possible choice functions at each step, and hence all possible combinations of choice functions. Thus the formula reduces to strong connectivity.

It may be useful to look at the approximants of this formula in a little more detail, to get some intuitions about the trump semantics. Considering just

$$H \stackrel{\text{def}}{=} (\mu(X, x).z = x \vee \exists w/\{y, z\}.\, E(x, w) \wedge X(w))[x, y, z],$$

we see that in computing each approximant, the calculation of $[\![\exists w/\{y, z\}.\, \ldots]\!]$ involves generating a trump for every possible value of a choice function $f\colon x \mapsto w$. This is a feature of the original trump semantics, and can be understood by viewing it as a second-order semantics: just as the compositional Tarskian semantics of $\exists x.\, \phi(x)$ involves computing all the witnesses for $\phi(x)$, so computing the trumps of $\exists x/\{y\}.\, \phi$ involves computing all the Skolem functions; and unlike the first-order case, it is necessary to work with functions (as IF can express existential second-order logic). Consequently, the nth approximant includes all states such that $x \to f_1(x) \to f_2 f_1(x) \to \ldots \to f_n \ldots f_1(x) = z$ for any sequence of successor-choosing functions f_i. Thus we see that the cumulative effect is the same as for a normal $\exists w$, and the independent choice has indeed not bought us anything.

It is, however, possible[2] to produce a slightly more involved formula expressing the Hamiltonian cycle property in this inductively defined way, by using the standard trick for expressing functions in Henkin quantifier logics. We replace the formula H by

$$\begin{aligned}&\forall s.\, \exists t/\{y,z\}.\, E(s,t)\\&\quad \wedge \mu(X,x).x = z \vee \forall u.\, \exists v/\{x,z,s,t\}.\,(s = u \Rightarrow t = v) \wedge (x = u \Rightarrow X(v)).\end{aligned}$$

This works because the actual function f selecting a successor for every node is made outside the fixpoint by $\forall s.\, \exists t/\{y,z\}.\, E(s,t) \wedge \ldots$; then inside the fixpoint, a new choice function g is made so that $X(g(x))$, and g is constrained to be the same as f by the clause $(s = u \Rightarrow t = v)$. (The reader who is not familiar with the IF/Henkin to existential second-order translation might wish to ponder why $\forall s.\, \exists t/\{y,z\}.\, E(s,t) \wedge \mu(X,x).x = z \vee (x = s \Rightarrow X(t))$ does not work.)

5 Complexity and Expressive Power of IF-LFP

The above examples have shown IF-LFP being used to express relatively simple NP properties. Since, as remarked, it is well known that Henkin quantifiers and IF logic express just the NP properties, and since it is also known [6] that LFP plus Henkin quantifiers express P^{NP}, one might imagine that IF-LFP (which is not closed under classical negation) also expresses only NP properties, or at worst some subset of P^{NP}. This is not the case; adding fixpoints to the IF formulation gives a more significant increase in expressive power.

Firstly, we note that the approximant semantics of fixpoints gives the usual behaviour in simple upper bounds:

Proposition 13. *If $\phi(X)[x, z_1, \ldots, z_m]$ is an IF-FOL$^+$ formula, then in a structure of size n, the approximants of $\mu(X,x).\phi$ close after at most 2^{n^m} steps. Hence in an IF-LFP formula with d alternating fixpoints and m variables, 2^{dn^m} evaluations of IF formulae are required. If the formula size is l, this gives a total cost of $2^{dn^m} \cdot l \cdot 2^{n^m(1+\lg n)} = l \cdot 2^{n^m(1+d+\lg n)}$.*

Observe, however, that the contribution from fixpoint alternation is small compared to the cost of computing independent existential quantifiers.

Despite the relative weakness of adding fixpoints, they do in some sense release the power of independent quantification. This is shown by the following theorem.

Theorem 14. *There is an IF-LFP sentence (with one least fixpoint) which is EXPTIME-hard to evaluate.*

[2] Since IF logic is equi-expressive with Henkin quantified logic, it is also equi-expressive with existential second-order logic, and so can express 'Hamiltonian cycle' without using fixpoints. Thus we are not, in this example, adding technical expressive power. However, the pure IF definition is quite complex, as it involves defining a binary relation coded via functions; so we are adding expressive convenience.

Proof. We give a reduction from the EXPTIME-complete problem of determining whether Player 1 has a winning strategy for the game of generalized chess.

A structure for a generalized chess game between 1 and 2 of order n comprises a board R with n^2 (or any other fixed polynomial) squares r and a set P of n (or any other fixed polynomial) pieces p. A position of the game is a function $\pi : P \to R$. There may be some relations on P and R in the signature. The game is defined by three first-order formulae with parameter π: a formula $\phi_I(\pi)$ true only of the initial position, a formula $\phi_W(\pi)$ which is true if player 1 has won at π, and a formula $\phi_M(\pi, i, p, r)$ which is true if moving piece p to square r is a legal move for player i from position π. (Without loss of generality, we assume that a move consists of moving exactly one piece.)

Given a position π and a move p, r, the 'next position' formula $N(\pi')$ is defined to be $\forall p'. (p' = p \wedge \pi'(p) = r) \vee \pi'(p') = \pi(p')$.

The set X of winning positions (i.e. from which 1 can force a win) for 1 can then be inductively defined by the type 3 functional

$$F(X, \pi) \Leftrightarrow \Phi \stackrel{\text{def}}{=} \phi_W(\pi) \vee ((\forall p, r. \phi_M(\pi, 2, p, r) \Rightarrow \exists \pi'. N(\pi') \wedge X(\pi')) \\ \wedge (\exists p, r. \phi_M(\pi, 1, p, r) \wedge \exists \pi'. N(\pi') \wedge X(\pi'))).$$

We now show how to express this inductive definition in IF-LFP. Part of the coding is the well-known expression of existential second-order logic in IF or Henkin logic, which we have already seen in the Hamiltonian cycle example. The general technique is thus: assume given an ESO formula $\exists f. \psi$. Let $Q_1(f(\tau_1)), \ldots, Q_n(f(\tau_n))$ be the instances in ψ of applications of f occurring in atoms Q_i. Then the translation is

$$\forall x. \exists y. \forall x_1. \exists y_1/\{x\}. \ldots \forall x_n. \exists y_n/\{x, x_1, \ldots, x_{n-1}\}. \bigwedge_i (x_i = x \Rightarrow y_i = y) \wedge \hat{\psi},$$

where $\hat{\psi}$ is obtained from ψ by replacing $Q_i(f_i(\tau_i))$ with $x_i = \tau_i \Rightarrow Q_i(y_i)$.

The second part is passing a function through a fixpoint. This is fairly simple to do: one just passes the domain and codomain as normal parameters, and relies on the quantification outside forcing them to represent a function. In this case, the classical type 2 relation $X(\pi)$ is replaced by a binary IF type 1 relation $Y(s, t)$, so that the classical $\exists \pi. (\mu X.\Phi)(\pi)$ becomes $\forall x. \exists y. (\mu Y.\hat{\Phi})(x, y)$, where $\hat{\Phi}$ is obtained from Φ by applying the ESO–IF translation using s, t for π and replacing $X(\pi')$ with $Y(p'', r'')$, where p'', r'' are the variables bound by the translation $\forall p''. \exists r''/\ldots.$ of $\exists \pi'$.

One then shows by an inductive argument on ζ that $\exists \pi. (\mu^\zeta X.\Phi)(\pi)$ holds iff $\forall x. \exists y. (\mu^\zeta Y.\hat{\Phi})(x, y)$ holds. Finally, if we wish to determine whether the initial position is winning for 1, we evaluate $\exists \pi. \phi_I(\pi) \wedge (\mu X.\Phi)(\pi)$

(We should note that we have extended the IF abbrevation $\phi \Rightarrow \psi$ to the case where ϕ is classical, not just atomic. This is acceptable because game negation coincides with classical negation for classical formula.) □

The above argument was applied to the case of finite structures, but there is nothing in it that depends on finiteness. We can therefore obtain the following theorem, which refutes our conjecture in [2] that a fixpoint extension of IF would be within Δ_2^1.

Theorem 15. *Let $F(X,\alpha)$ be a positive Σ_1^1 type 3 functional in the language of arithmetic. Then a set of integers definable from the set of reals inductively defined by F can be expressed in IF-LFP. It follows that that IF-LFP (even with just one fixpoint) over the natural numbers can express Σ_2^1 properties.*

Proof. F is defined by a Σ_1^1 formula $\phi(X,\alpha)$. Use the technique of the previous proof to express $\exists\alpha.\,(\mu X.\phi)(\alpha)\wedge\psi(\alpha,n)$, where ψ is first-order. Cenzer [5] showed that any Σ_2^1 set of reals is the closure of a Σ_1^1 positive inductive definition over the reals. Since if α is a Σ_2^1 real, the set $\{\alpha\}$ is also Σ_2^1, we also have the stated consequence. □

Cenzer's results also allow us to obtain an improvement (for those who don't believe CH) on the closure ordinal for a single IF fixpoint over ω. The usual cardinality argument for fixpoints tells us merely that an IF fixpoint over ω must close by $2^{\aleph_0}$. The improvement is

Theorem 16. *If $\phi(X)$ is an IF-FOL$^+$ formula (i.e. with $\sim$ and $\downarrow$ applied only to atoms), then $\mu X.\phi$ has closure ordinal $\leq \aleph_1$.*

Proof. Seen as operations on $\wp(2^\omega)$, the semantics of the IF boolean operators and quantifiers are Σ_1^1. (This is not immediately apparent from the definitions as presented above, but a small amount of rearrangement reveals it.) Cenzer showed that the closure ordinal of a Σ_1^1 monotone inductive definition over the reals is $\leq \aleph_1$. □

It remains to investigate lower bounds on the complexity of multiple IF fixpoints. We remark only that the absence of classical negation makes this less easy than it otherwise would be.

6 IF Parity Games

We briefly recall the game semantics of first-order logic and of IF logic.

Given a FO formula ψ (in positive form) and a structure A, a position is a subformula $\phi(\vec{x})$ of ψ together with a *deal* for ϕ, that is, an assignment of values $\vec{v}$ to its free variables $\vec{x}$. At a position $(\forall x.\,\phi_1,\vec{v})$, Abelard chooses a value v for x, and play moves to the position $(\phi_1,\vec{v}\cdot v)$; similarly Eloise moves at $\exists x.\,\phi$. At $(\phi_1\wedge\phi_2,\vec{v})$, Abelard chooses a conjunct ϕ_i, and play moves to $(\phi_i(\vec{x}'),\vec{v}')$, where $\vec{x}',\vec{v}'$ are $\vec{x},\vec{v}$ restricted to the free variables of ϕ_i; and at $(\phi_1\vee\phi_2,\vec{v})$, Eloise similarly chooses a disjunct. A play of the game terminates at (negated) atoms $(P(\vec{x}),\vec{v})$ (resp. $(\neg P(\vec{x}),\vec{v})$), and is won by Eloise (resp. Abelard) iff $P(\vec{v})$ is true. Then it is standard that $M\vDash\phi$ exactly if Eloise has a winning strategy in this game, where a strategy is a function from sequences of legal positions to moves.

These games have *perfect information*; both players know everything that has happened, and in particular when one player makes a choice, they know the other player's previous choices. Game semantics for IF logic [7] use games of imperfect information: at the position $\exists x/W.\,\phi$, when Eloise chooses a value v for x, she does not know what Abelard chose for the values of the independent variables W. A *uniform* Eloise strategy for the game is one in which her choice of v is indeed uniform in the values of W, and we say a formula is true if Eloise has a uniform winning strategy.

Now recall that in a parity game the positions are assigned ranks $0, \ldots, r$, and if a run of the game is infinite, Eloise wins if the highest rank appearing infinitely often is even. The model-checking game for FOL extends to a model-checking game for LFP by assigning even ranks to maximal fixpoints and odd to minimal, such that the rank of an inner fixpoint is less than the rank of its enclosing fixpoints. Then the formula is true iff Eloise has a winning strategy for the defined parity game.

Combining these two concepts, a general parity game of imperfect information is given by a usual parity game together with imperfect information requirements at each position, requiring a player to move uniformly in some part of the game history. The winning runs are those given by the usual parity winning conditions; a player wins the game if she has winning strategy for the parity game that is uniform as required by the imperfect information requirements.

In general, infinite imperfect information games are undecidable even on finite structures, since they require players to keep arbitrary knowledge (and lack of knowledge) of the history of the game. To obtain a class of decidable imperfect parity games, we will first give a parity game semantics for IF-LFP, and then define a class of imperfect parity games characterized by IF-LFP.

Definition 17. *The model-checking game for an IF-LFP formula is defined by adding the following clauses to the Hintikka–Sandu game for IF. The moves are extended by the usual fixpoint unfolding rule: at a position* $((\mu(X, x).\phi)(t), \vec{u})$, *play moves to* $(\phi, \vec{u}v)$, *where* v *is the value of* t; *at a position* $(X(t), \vec{u}v\vec{w})$, *where* $\vec{u}$ *is the deal for the free variables of* X, v *for* x, *and* $\vec{w}$ *for the variables bound inside* ϕ, *play moves to* $(\phi, \vec{u}v')$ *where* v' *is the value of* t. *Parities are assigned to positions in the usual way, and the usual infinite parity winning condition is added.*

The independence requirements are that at a quantifier $\exists x/W.$ *(and dually), Eloise must choose* x *without knowing the values of the* W *variables* and without knowing the values of any variables bound in some currently enclosing fixpoint but chosen before the most recent unfolding of that fixpoint. *(In other words, she does not remember choices that have gone out of scope and have no value in the current deal.)*

Correspondingly, a uniform strategy in the parity game is a strategy where the choice function is uniform in the independent variables and the out-of-scope variables.

Theorem 18. *If* ϕ *is an IF-LFP sentence, then Eloise has a uniform winning strategy for the model-checking game if and only if* $\{\langle\rangle\}$ *is a trump for* ϕ. *Moreover, the strategy can be history-free.*

Proof. (Sketch) The argument relating parity conditions to alternating fixpoints applies to any set of monotone operators with fixpoints added, not just to FOL or modal logic. Thus by combining the usual proof of equivalence of parity games and fixpoints with the proof of equivalence of IF games and trump semantics, we get the result.

History-freeness follows from the inductive construction of the set of winning positions, but is intuitively obvious, as all the information a player is supposed to be able to remember is included in the game position.

This game account of the IF-LFP semantics brings out the key factor, which may have been less obvious in the trump semantics, that keeps model-checking decidable. This is that passing through a fixpoint variable throws away all information about choices made within the body of the fixpoint, unless they are explicitly passed as parameters. Of course, this is also true in usual LFP, but in the IF case knowledge of previous choices is explicitly part of the semantics.

This suggests the following definition:

Definition 19. *An imperfect information parity game on a structure A is* finite-memory *if each player is equipped with a finite memory in which they can remember previous moves. A player's choice at a move is required to depend only on the current position and memory, with additional imperfect information requirements imposed by the game on the memory (i.e. a player may have to temporarily forget things).*

A player wins the game if they have a uniform history-free winning strategy.

The expected theorem is

Theorem 20. *Given a finite-memory imperfect parity game on A, the statement 'Eloise wins the game' is expressible by an IF-LFP formula whose fixpoint alternation depth is the parity rank of the game.*

Proof. (Sketch) The finite memory is modelled by parameters of fixpoints. We will use fixpoints X which carry one parameter p for the position in the game, and parameters m_i for the memory 'cells'. The inner loop of an inductive definition of winning positions is the usual expression of 'it is Eloise's move and there exists a move such that the next position is in X, or it is Abelard's move and all next moves are in X', as in the formula we used early for generalized chess. The quantifiers are made explicitly independent of the memory items required to be unknown (which may require a case analysis of the moves of the game).

To deal with the parities, we use the first-order version of the usual 'parity game formula' from parity automata and modal mu-calculus (see [1] for a detailed explanation of the parity game formula): for each rank $j = 0, \dots, r$, there is a fixpoint variable X_j. Then the inner loop is enclosed by $\nu X_0.\mu X_1.\dots\mu/\nu.X_r.$, and the formula $X(p, \vec{m})$, where p and $\vec{m}$ are the position and memory after the next move, is conjoined with

$$\bigwedge_{0 \le j \le r} (R_j \Rightarrow X_j(p, \vec{m}))$$

where R_j is the formula expressing that the next position has rank j.

The usual proof now applies to give the result. □

Corollary 21. *Finite-memory imperfect parity games on finite structures are decidable*[3].

7 Application to IF Modal Mu-Calculus

Our original motivation for looking at fixpoint extensions of IF logic was the desire to combine two threads of work. Firstly, modal mu-calculus is a well studied and widely used temporal logic. Secondly, we have argued in [2] and [3] that modal versions of Henkin quantifiers and independence logics provide a natural expression of some properties of concurrent systems. Given a concurrent modal logic, it is natural to extend it to a concurrent temporal logic by adding fixpoint operators. In [2] we looked at modal analogues of Henkin quantifiers acting on systems composed of several concurrent components; since a single Henkin quantifier gives an operator on the powerset of states, there was no difficulty in adding such modalities to mu-calculus. In [3], we designed a modal analogue of IF logic, defined on certain structures appropriate for true concurrency. The full definition of the structures and the logic is, for a number of technical reasons, rather long and complex. We refer the reader to [3] for details, and here just give the idea.

IFML extends the syntax of usual modal logic as follows. Instead of the simple 'next step' modality $\langle a\rangle\Phi$, each modality carries a *tag* α, and may be declared to be independent of previous tags β by the Hintikka slash, giving a syntax $\langle a\rangle_{\alpha/\beta}\Phi$. The intended interpretation is that the choice of a action must be independent of the action chosen in the modality tagged by β; for this to make sense, the action at β should be concurrent (in the technical sense of event structures etc.) with the action at α. The semantics is given in terms of runs (sequences of states) of the system, directly via an imperfect information model-checking game.

As we have not the space to import the full definition of IFML here, we shall not give detailed propositions in this section, but remarks that can be refined into theorems with the material in [3].

Remark 22. The game semantics of IFML given in [3] can be equivalently expressed by translating to IF as a meta-language (modulo the introduction of some fairly messy defined functions and relations on runs of the system) such that the main variable holding the state ranges over runs (as in the game), and auxiliary variables range over actions. Consequently, IFML has a trump semantics. The evaluation of a formula on a finite system is decidable, since the maximum length of runs that must be considered is bounded by the modal depth of the formula.

Remark 23. We can define an IF modal mu-calculus by adding fixpoint formulae of the form $\mu(X,\chi).\Phi$ and $X(\alpha)$, where the fixpoint variable X has not only an implicit parameter for the current 'state', but also explicit parameters χ for tags to be passed through the fixpoint.

[3] This result is probably long known in some form, but I do not know the reference.

This can be given a semantics via IF-LFP. However, since a 'state' in the semantics is a run, not a system state, it is not obvious that decidability of model-checking is maintained for IF mu-calculus. (We conjecture that it is, but some results from concurrency theory, such as the undecidability of hereditary history-preserving bisimulation, give some cause for doubt.)

The IF modal mu-calculus has a model-checking game that is an IF version of the usual parity games for modal logic, as done above for IF-LFP.

8 Conclusion

We have defined a suitable fixpoint extension of independence-friendly logic, and established some results. We have related it to parity games of imperfect information, and we have shown how it may be applied to the construction of independence-friendly modal mu-calculi.

For IF-LFP itself, there are still many questions remaining. Chief among these are better upper and lower bounds on the complexity of model-checking (in the finite case) and descriptive complexity (in the infinite case). We have shown that IF-LFP is more complex than we surmised in earlier work, and it is not unlikely that it will turn out to be much more complex.

References

1. J. C. Bradfield, Fixpoints in arithmetic, transition systems and trees. *Theoretical Informatics and Applications*, **33** 341–356 (1999).
2. J. C. Bradfield, Independence: logics and concurrency, *Proc. CSL 2000*, LNCS **1862** 247–261 (2000).
3. J. C. Bradfield and S. B. Fröschle, Independence-friendly modal logic and true concurrency, *Nordic J. Computing* **9** 102–117 (2002).
4. P. Cameron and W. Hodges, Some combinatorics of imperfect information, *J. Symbolic Logic* **66** 673–684 (2001).
5. D. Cenzer, Monotone inductive definitions over the continuum, *J. Symbolic Logic* **41**:1 188–198 (1976).
6. G. Gottlob, Relativized logspace and generalized quantifiers over finite ordered structures. *J. Symbolic Logic* **62**:2, 545–574 (1997).
7. J. Hintikka and G. Sandu, A revolution in logic?, *Nordic J. Philos. Logic* **1**(2) 169–183 (1996).
8. W. Hodges, Compositional semantics for a language of imperfect information, *Int. J. IGPL* **5**(4), 539–563.

Atomic Cut Elimination for Classical Logic

Kai Brünnler

Technische Universität Dresden, Fakultät Informatik
D-01062 Dresden, Germany
kai.bruennler@inf.tu-dresden.de

Abstract. System SKS is a set of rules for classical propositional logic presented in the calculus of structures. Like sequent systems and unlike natural deduction systems, it has an explicit cut rule, which is admissible. In contrast to sequent systems, the cut rule can easily be reduced to atomic form. This allows for a very simple cut elimination procedure based on plugging in parts of a proof, like normalisation in natural deduction and unlike cut elimination in the sequent calculus. It should thus be a good common starting point for investigations into both proof search as computation and proof normalisation as computation.

Keywords: sequent calculus, natural deduction, cut elimination, classical logic, atomic cut.

1 Introduction

The two well-known connections between proof theory and language design, *proof search as computation* and *proof normalisation as computation*, have mainly used different proof-theoretic formalisms. While designers of functional programming languages prefer natural deduction, because of the close correspondence between proof normalisation and reduction in related term calculi [4,8], designers of logic programming languages prefer the sequent calculus [7], because infinite choice and much of the unwanted non-determinism is limited to the cut rule, which can be eliminated.

System SKS [2] is a set of inference rules for classical propositional logic presented in a new formalism, the *calculus of structures* [5]. This system admits the good properties usually found in sequent systems: in particular, all rules that induce infinite choice in proof search are admissible. Thus, in principle, it is as suitable for proof search as systems in the sequent calculus. In this paper I will present a cut elimination procedure for SKS that is very similar to normalisation in natural deduction. It thus allows us to develop, at least for the case of classical logic, both the proof search and the proof normalisation paradigm of computation in the same formalism and starting from the same system of rules.

Cut elimination in the sequent calculus and normalisation in natural deduction, widely perceived as 'morally the same', differ quite a bit, technically. Compared to cut elimination, (weak) normalisation is simpler, involving neither

M. Baaz and J.A. Makowsky (Eds.): CSL 2003, LNCS 2803, pp. 86–97, 2003.

permutation of a multicut rule, nor induction on the cut-rank. The equivalent of a cut in natural deduction, for example,

$$\supset_E \dfrac{\supset_I \dfrac{\begin{array}{c}\Delta_1\\ \Gamma, A \vdash B\end{array}}{\Gamma \vdash A \supset B} \qquad \begin{array}{c}\Delta_2\\ \Gamma \vdash A\end{array}}{\Gamma \vdash B} \quad ,$$

is eliminated as follows: first, assumption A and all its copies are removed from Δ_1. Second, the derivation Δ_2, with the context strengthened accordingly, is plugged into all the leaves of Δ_1 where assumption A was used.

This method relies on the fact that no rule inside Δ_1 can change the premise A, which is why it does not work for the sequent calculus. To eliminate a cut in the sequent calculus, one has to cope with the fact that logical rules may be applied to both eigenformulas of the cut. This is usually done by permuting up the cut rule step-by-step. However, given a cut with an *atomic* cut formula a inside a sequent calculus proof, we can trace the occurrence of a and its copies produced by contraction, identify all the leaves where they are used in identity axioms, and plug in subproofs in very much the same way as in natural deduction. The problem for the sequent calculus is that cuts are not atomic, in general.

The calculus of structures generalises the one-sided sequent calculus. It has led not only to inference systems with interesting new properties for classical and linear logic [2,9,10], but also to inference systems for new logics that are problematic for the sequent calculus [5,6,3].

Derivations in the calculus of structures enjoy a top-down symmetry that is not available in the sequent calculus: they are chains of one-premise inference rules. 'Meta-level conjunction' (the branching of the proof tree) and 'object-level conjunction' (the connective in a formula) are identified. The two notions of formula and sequent are also identified, they merge into the notion of *structure*, which is a formula subject to equivalences that are usually imposed on sequents. This simplification makes explicit the duality between the identity axiom and the cut rule [5]:

$$identity\ \frac{S\{true\}}{S\{R \vee \bar{R}\}} \qquad\qquad cut\ \frac{S\{R \wedge \bar{R}\}}{S\{false\}}$$

The identity rule is read bottom-up as: if inside a structure there occurs a disjunction of a structure R and its negation, then it can be replaced by the constant *true*. The notion of duality between cut and identity is precisely the one that is known as *contrapositive*.

Just like in the sequent calculus, the identity axiom can easily be reduced to atomic form. The symmetry of the calculus of structures allows to reduce the cut to atomic form in the same way as the identity axiom, i.e. without having to go through cut elimination. Atomicity of the cut then admits a very simple cut elimination procedure that is similar to normalisation in natural deduction.

Associativity

$$[\boldsymbol{R},[\boldsymbol{T}]]=[\boldsymbol{R},\boldsymbol{T}]$$
$$(\boldsymbol{R},(\boldsymbol{T}))=(\boldsymbol{R},\boldsymbol{T})$$

Commutativity

$$[\boldsymbol{R},\boldsymbol{T}]=[\boldsymbol{T},\boldsymbol{R}]$$
$$(\boldsymbol{R},\boldsymbol{T})=(\boldsymbol{T},\boldsymbol{R})$$

Singleton

$$[R]=R=(R)$$

Units

$$[\mathsf{f},\boldsymbol{R}]=[\boldsymbol{R}]$$
$$(\mathsf{t},\boldsymbol{R})=(\boldsymbol{R})$$
$$[\mathsf{t},\mathsf{t}]=\mathsf{t}$$
$$(\mathsf{f},\mathsf{f})=\mathsf{f}$$

Negation

$$\bar{\mathsf{t}}=\mathsf{f}$$
$$\bar{\mathsf{f}}=\mathsf{t}$$
$$\overline{[R_1,\ldots,R_h]}=(\bar{R}_1,\ldots,\bar{R}_h)$$
$$\overline{(R_1,\ldots,R_h)}=[\bar{R}_1,\ldots,\bar{R}_h]$$
$$\bar{\bar{R}}=R$$

Fig. 1. Equations on structures

After introducing basic notions of the calculus of structures, I show system SKS with atomic contraction, weakening, identity and, most significantly, atomic cut. Then, after establishing some lemmas, I present the cut elimination procedure.

2 The Calculus of Structures

Definition 1. Propositional variables p and their negations $\bar{p}$ are *atoms*, with the negation of $\bar{p}$ defined to be p. Atoms are denoted by a, b, ... The *structures* of the language KS are generated by

$$S ::= \mathsf{t} \mid \mathsf{f} \mid a \mid [\underbrace{S,\ldots,S}_{>0}] \mid (\underbrace{S,\ldots,S}_{>0}) \mid \bar{S} \quad ,$$

where t and f are the units *true* and *false*, $[S_1,\ldots,S_h]$ is a *disjunction* and $(S_1,\ldots,S_h)$ is a *conjunction*. $\bar{S}$ is the *negation* of the structure S. The units are not atoms. Structures are denoted by S, R, T, U and V. *Structure contexts*, denoted by $S\{\ \}$, are structures with one occurrence of $\{\ \}$, the *empty context* or *hole*, that does not appear in the scope of a negation. $S\{R\}$ denotes the structure obtained by filling the hole in $S\{\ \}$ with R. We drop the curly braces when they are redundant: for example, $S[R,T]$ stands for $S\{[R,T]\}$. Structures are *equivalent* modulo the smallest congruence relation induced by the equations shown in Fig. 1, where $\boldsymbol{R}$ and $\boldsymbol{T}$ are finite, non-empty sequences of structures. In general we do not distinguish between equivalent structures.

Definition 2. An *inference rule* is a scheme of the kind $\rho\,\dfrac{S\{T\}}{S\{R\}}$, where ρ is the *name* of the rule, $S\{T\}$ is its *premise* and $S\{R\}$ is its *conclusion*. In an instance of ρ, the structure taking the place of R is called *redex* and the structure taking the place of T is called *contractum*. A *(formal) system* $\mathscr{S}$ is a set of inference rules. To clarify the use of the equational theory where it is not obvious, I will use the rule $= \frac{T}{R}$ where R and T are equivalent structures.

Definition 3. The *dual* of a rule is obtained by exchanging premise and conclusion and replacing each connective by its De Morgan dual.

Definition 4. A *derivation* Δ in a certain formal system is a finite chain of instances of inference rules in the system:

$$\begin{array}{rc} \pi' & \dfrac{T}{V} \\ \pi & \overline{\quad\vdots\quad} \\ \rho' & \overline{\quad U \quad} \\ \rho & \overline{\quad R \quad} \end{array} .$$

A derivation can consist of just one structure. The topmost structure in a derivation is called the *premise* of the derivation, and the structure at the bottom is called its *conclusion*. A derivation Δ whose premise is T, whose conclusion is R, and whose inference rules are in $\mathscr{S}$ will be indicated with $\begin{array}{c} T \\ \Delta \Vert \mathscr{S} \\ R \end{array}$. A *proof* Π in the calculus of structures is a derivation whose premise is the unit true. It will be denoted by $\begin{array}{c} \Pi \Vert \mathscr{S} \\ R \end{array}$. A rule ρ is *derivable* for a system $\mathscr{S}$ if for every instance of $\rho \frac{T}{R}$ there is a derivation $\begin{array}{c} T \\ \Vert \mathscr{S} \\ R \end{array}$. A rule ρ is *admissible* for a system $\mathscr{S}$ if for every proof $\begin{array}{c} \Vert \mathscr{S} \cup \{\rho\} \\ S \end{array}$ there is a proof $\begin{array}{c} \Vert \mathscr{S} \\ S \end{array}$.

3 System SKS

System SKS, shown in Fig. 2, has been introduced and shown to be sound and complete for classical propositional logic in [2]. The first S stands for "symmetric" or "self-dual", meaning that for each rule, its dual (or contrapositive) is also in the system. The K stands for "klassisch" as in Gentzen's LK and the last S says that it is a system on structures.

The rules $\mathsf{ai}\downarrow, \mathsf{s}, \mathsf{m}, \mathsf{aw}\downarrow, \mathsf{ac}\downarrow$ are called respectively *atomic identity*, *switch*, *medial*, *atomic weakening* and *atomic contraction*. Their dual rules carry the same name prefixed with a "co-", so e.g. $\mathsf{aw}\uparrow$ is called *atomic co-weakening*. The rules s and m are their own duals. The rule $\mathsf{ai}\uparrow$ is special, it is called *atomic*

$$\mathsf{ai}\!\downarrow \frac{S\{\mathsf{t}\}}{S[a,\bar{a}]} \qquad \mathsf{ai}\!\uparrow \frac{S(a,\bar{a})}{S\{\mathsf{f}\}}$$

$$\mathsf{s}\ \frac{S([R,U],T)}{S[(R,T),U]}$$

$$\mathsf{m}\ \frac{S[(R,U),(T,V)]}{S([R,T],[U,V])}$$

$$\mathsf{aw}\!\downarrow \frac{S\{\mathsf{f}\}}{S\{a\}} \qquad \mathsf{aw}\!\uparrow \frac{S\{a\}}{S\{\mathsf{t}\}}$$

$$\mathsf{ac}\!\downarrow \frac{S[a,a]}{S\{a\}} \qquad \mathsf{ac}\!\uparrow \frac{S\{a\}}{S(a,a)}$$

Fig. 2. System SKS

$$\mathsf{ai}\!\downarrow \frac{S\{\mathsf{t}\}}{S[a,\bar{a}]} \qquad \mathsf{aw}\!\downarrow \frac{S\{\mathsf{f}\}}{S\{a\}} \qquad \mathsf{ac}\!\downarrow \frac{S[a,a]}{S\{a\}}$$

$$\mathsf{s}\ \frac{S([R,T],U)}{S[(R,U),T]} \qquad \mathsf{m}\ \frac{S[(R,T),(U,V)]}{S([R,U],[T,V])}$$

Fig. 3. System KS

cut. Rules ai↓, aw↓, ac↓ are called *down-rules* and their duals are called *up-rules.* In [2], by a semantic argument, all up-rules were shown to be admissible. By removing them we obtain system KS, shown in Fig. 3, which is complete.

Cut-free sequent systems fulfill the subformula property. Our case is different, because the notions of formula and sequent are merged. System KS does not fulfill a "substructure property" just as sequent systems do not fulfill a "subsequent property". However, when seen bottom-up, no rule in system KS introduces new atoms. It thus satisfies the main aspect of the subformula property: when given a conclusion of a rule there is only a finite number of premises to choose from. In proof search, for example, the branching of the search tree is finite.

Identity, cut, weakening and contraction are restricted to atoms in system SKS. The general versions of those rules are shown in Fig. 4.

Theorem 5. The rules i↓, w↓ and c↓ are derivable in {ai↓, s}, {aw↓, s} and {ac↓, m}, respectively. Dually, the rules i↑, w↑ and c↑ are derivable in {ai↑, s}, {aw↑, s} and {ac↑, m}, respectively.

$$\mathsf{i}\!\downarrow \frac{S\{\mathsf{t}\}}{S[R,\bar{R}]} \qquad \mathsf{i}\!\uparrow \frac{S(R,\bar{R})}{S\{\mathsf{f}\}}$$

$$\mathsf{w}\!\downarrow \frac{S\{\mathsf{f}\}}{S\{R\}} \qquad \mathsf{w}\!\uparrow \frac{S\{R\}}{S\{\mathsf{t}\}}$$

$$\mathsf{c}\!\downarrow \frac{S[R,R]}{S\{R\}} \qquad \mathsf{c}\!\uparrow \frac{S\{R\}}{S(R,R)}$$

Fig. 4. General identity, weakening, contraction and their duals

Proof. By an easy structural induction on the structure that is cut, weakened or contracted. Details are in [2]. The case for the cut is shown here. A cut introducing the structure (R,T) together with its dual structure $[\bar{R},\bar{T}]$ is replaced by two cuts on smaller structures:

$$\mathsf{i}\!\uparrow \frac{S(R,T,[\bar{R},\bar{T}])}{S\{\mathsf{f}\}} \quad \rightsquigarrow \quad \mathsf{i}\!\uparrow \frac{\mathsf{i}\!\uparrow \dfrac{\mathsf{s}\dfrac{\mathsf{s}\dfrac{S(R,T,[\bar{R},\bar{T}])}{S(R,[\bar{R},(T,\bar{T})])}}{S[(R,\bar{R}),(T,\bar{T})]}}{S(R,\bar{R})}}{S\{\mathsf{f}\}} \quad .$$

□

So, while general identity, weakening, contraction and their duals do not belong to SKS, they will be freely used in derivations in SKS to denote multiple instances of the corresponding rules in SKS according to Theorem 5.

Remark 6. Sequent calculus derivations easily correspond to derivations in system SKS. For instance, the cut of sequent systems in Gentzen-Schütte form [11]:

$$\mathsf{Cut}\ \frac{\vdash \Phi, A \qquad \vdash \Psi, \bar{A}}{\vdash \Phi, \Psi} \qquad \text{corresponds to} \qquad \mathsf{i}\!\uparrow \frac{\mathsf{s}\dfrac{\mathsf{s}\dfrac{([\Phi,A],[\Psi,\bar{A}])}{[\Phi,(A,[\Psi,\bar{A}])]}}{[\Phi,\Psi,(A,\bar{A})]}}{[\Phi,\Psi]} \quad .$$

4 Cut Elimination

In the calculus of structures, there is more freedom in applying inference rules than in the sequent calculus. While this allows for a richer combinatorial analysis of proofs, it is a significant challenge for cut elimination. During cut elimination,

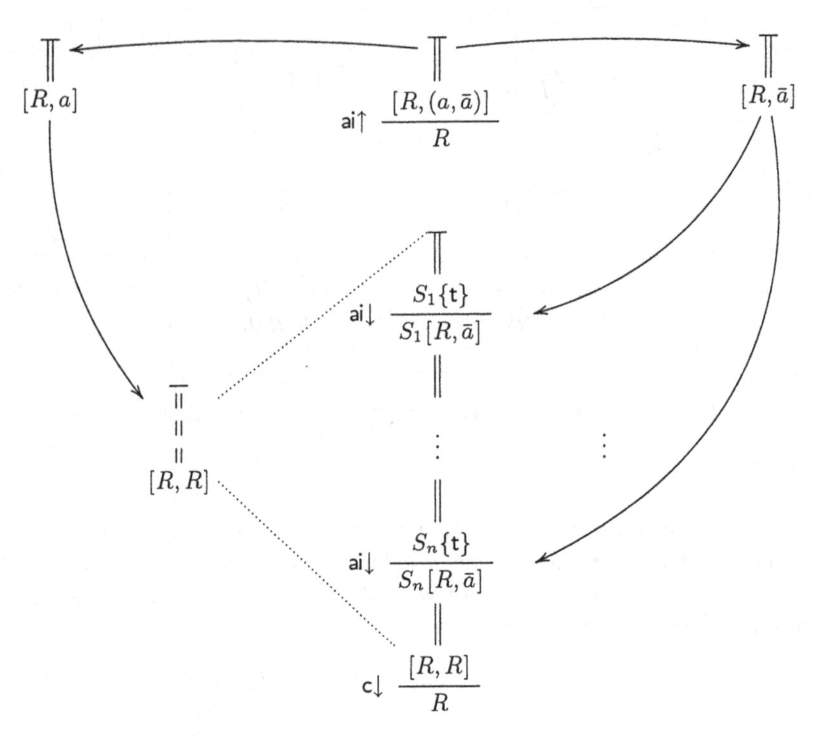

Fig. 5. Elimination of one atomic cut

the sequent calculus allows to get into the crucial situation where on one branch a logical rule applies to the main connective of the eigenformula and on the other branch the corresponding rule applies to the dual connective of the dual eigenformula. In the calculus of structures, rules apply deep inside a context, they are not restricted to main connectives. The methodology of the sequent calculus thus does not apply. For example, one cannot permute the cut over the switch rule. One can generalise the cut in order to permute it over switch, but this requires a case analysis that is far more complicated than in the sequent calculus. Contraction is an even bigger problem. Despite many efforts, no cut elimination procedure along these lines has been found for system SKS.

Two new techniques were developed to eliminate cuts in the calculus of structures. The first is called *decomposition*, and has been used in [6,9] for some systems related to linear logic. Proving termination of decomposition is rather involved [9]. It makes essential use of the exponentials of linear logic which restrict the use of contraction. So far, this technique could not be used for classical logic with its unrestricted contraction. The second technique is called *splitting* [5], and essentially makes available a situation corresponding to the one described above for the sequent calculus. Splitting covers the broadest range of systems in the calculus of structures, it not only applies to the systems mentioned above, but has recently also been applied to system SKS (but the proof is not published yet). Compared to splitting, the procedure given here is much simpler. In fact, I do not know of any other system with such a simple cut elimination procedure.

In the sequent calculus as well as in sequent-style natural deduction, a derivation is a tree. Seen bottom-up, a cut splits the tree into two branches. To apply a cut, one is forced to split the context among the two branches (in the case of multiplicative context treatment) or to duplicate the context (in the case of additive context treatment). In the calculus of structures, the cut rule does not split the proof.

The crucial idea, illustrated in Fig. 5, is that we can do that during cut elimination. This allows us to plug-in proofs just like in natural deduction: we duplicate the proof above a cut and remove atom a from the copy shown on the left and the atom $\bar{a}$ from the copy shown on the right. We choose one copy, the one on the left in this case, and replace a by R throughout the proof, breaking some instances of identity. They are fixed by substituting the proof on the right. A contraction is applied to obtain a cut-free proof of R.

In contrast to the sequent calculus, the cut is not the only problematic rule in system SKS. The rule $\mathsf{aw}\!\uparrow$ also induces infinite choice in proof-search. Fortunately, we can not only eliminate the cut rule, but also the other up-rules. Each up-rule individually can be shown to be admissible for system KS. However, since we are going to eliminate the cut anyway, to eliminate rules $\mathsf{aw}\!\uparrow$ and $\mathsf{ac}\!\uparrow$ the following lemma is sufficient.

Lemma 7. Each rule in SKS is derivable for identity, cut, switch and its dual rule.

Proof. An instance of $\rho\!\uparrow \dfrac{S\{T\}}{S\{R\}}$ can be replaced by

$$\mathsf{i}\!\downarrow \dfrac{\mathsf{s}\,\dfrac{\rho\!\downarrow\dfrac{\mathsf{i}\!\uparrow\dfrac{S\{T\}}{S(T,[R,\bar{R}])}}{S[R,(T,\bar{R})]}}{S[R,(T,\bar{T})]}}{S\{R\}}$$

.

The same holds for down-rules. □

When plugging in a derivation in natural deduction, its context has to be strengthened, to fit into the leaf into which it is plugged. Adding to a context in natural deduction is easy, since it is a flat object, a set or a multiset. In the calculus of structures, contexts are more general, nested objects. The following definition is used to strengthen contexts.

Definition 8. Given a derivation Δ, the derivation $S\{\Delta\}$ is obtained as follows:

$$\Delta = \begin{array}{rc} \pi' & \dfrac{T}{V} \\ \pi & \overline{\quad\vdots\quad} \\ \rho' & \overline{\quad U \quad} \\ \rho & \overline{\quad R \quad} \end{array} \qquad S\{\Delta\} = \begin{array}{rc} \pi' & \dfrac{S\{T\}}{S\{V\}} \\ \pi & \overline{\quad\vdots\quad} \\ \rho' & \overline{\quad S\{U\} \quad} \\ \rho & \overline{\quad S\{R\} \quad} \end{array} .$$

Definition 9. An instance of atomic cut is called *shallow* if it is of the following form:

$$\mathsf{ai}\!\uparrow \frac{[S,(a,\bar{a})]}{S} \quad .$$

Lemma 10. The atomic cut is derivable for shallow atomic cut and switch.

Proof. An easy induction locally replaces an instance of atomic cut by a shallow atomic cut followed by instances of switch. Details are in [1]. □

Lemma 11. Each proof $\overset{}{\underset{T\{a\}}{\Vert}}{\scriptstyle \mathsf{KS}}$ can be transformed into a proof $\underset{T\{\mathsf{t}\}}{\Vert}{\scriptstyle \mathsf{KS}}$.

Proof. Starting with the conclusion, going up in the proof, in each structure we replace the occurrence of a and its copies, that are produced by contractions, by the unit t. Replacements inside the context of any rule instance do not affect the validity of this rule instance. Instances of the rules m and s remain valid, also in the case that atom occurrences are replaced inside redex and contractum. Instances of the other rules are replaced by the following derivations:

$$\mathsf{ac}\!\downarrow \frac{S[a,a]}{S\{a\}} \quad \leadsto \quad = \frac{S[\mathsf{t},\mathsf{t}]}{S\{\mathsf{t}\}}$$

$$\mathsf{aw}\!\downarrow \frac{S\{\mathsf{f}\}}{S\{a\}} \quad \leadsto \quad = \frac{\mathsf{s}\,\dfrac{=\dfrac{S\{\mathsf{f}\}}{S([\mathsf{t},\mathsf{t}],\mathsf{f})}}{S[\mathsf{t},(\mathsf{t},\mathsf{f})]}}{S\{\mathsf{t}\}}$$

$$\mathsf{ai}\!\downarrow \frac{S\{\mathsf{t}\}}{S[a,\bar{a}]} \quad \leadsto \quad \mathsf{aw}\!\downarrow \frac{=\dfrac{S\{\mathsf{t}\}}{S[\mathsf{t},\mathsf{f}]}}{S[\mathsf{t},\bar{a}]} \quad .$$

□

Properly equipped, we now turn to cut elimination.

Theorem 12. Each proof $\underset{T}{\Vert}{\scriptstyle \mathsf{SKS}}$ can be transformed into a proof $\underset{T}{\Vert}{\scriptstyle \mathsf{KS}}$.

Proof. By Lemma 7, the only rule left to eliminate is the cut. By Lemma 10, we replace all cuts by shallow cuts. The topmost instance of cut, together with the proof above it, is singled out:

$$\underset{T}{\Vert}{\scriptstyle \mathsf{KS}\cup\{\mathsf{ai}\uparrow\}} \quad = \quad \begin{array}{c} \Pi \Vert {\scriptstyle \mathsf{KS}} \\ \mathsf{ai}\!\uparrow \dfrac{[R,(a,\bar{a})]}{R} \\ \Delta \Vert {\scriptstyle \mathsf{KS}\cup\{\mathsf{ai}\uparrow\}} \\ T \end{array} \quad .$$

Lemma 11 is applied twice on Π to obtain

$$\Pi_1 \left\| \begin{array}{c} \mathsf{KS} \\ {[R, a]} \end{array} \right. \quad \text{and} \quad \Pi_2 \left\| \begin{array}{c} \mathsf{KS} \\ {[R, \bar{a}]} \end{array} \right. .$$

Starting with the conclusion, going up in proof Π_1, in each structure we replace the occurrence of a and its copies, that are produced by contractions, by the structure R.

Replacements inside the context of any rule instance do not affect the validity of this rule instance. Instances of the rules m and s remain valid, also in the case that atom occurrences are replaced inside redex and contractum. Instances of ac↓ and aw↓ are replaced by their general versions:

$$\mathsf{ac}\!\downarrow \frac{S[a,a]}{S\{a\}} \quad \leadsto \quad \mathsf{c}\!\downarrow \frac{S[R,R]}{S\{R\}}$$

$$\mathsf{aw}\!\downarrow \frac{S\{\mathsf{f}\}}{S\{a\}} \quad \leadsto \quad \mathsf{w}\!\downarrow \frac{S\{\mathsf{f}\}}{S\{R\}} .$$

Instances of ai↓ are replaced by $S\{\Pi_2\}$:

$$\mathsf{ai}\!\downarrow \frac{S\{\mathsf{t}\}}{S[a,\bar{a}]} \quad \leadsto \quad \begin{array}{c} S\{\mathsf{t}\} \\ S\{\Pi_2\} \Big\| \mathsf{KS} \\ S[R,\bar{a}] \end{array} .$$

The result of this process of substituting Π_2 into Π_1 is a proof Π_3, from which we build

$$\begin{array}{c} \Pi_3 \Big\| \mathsf{KS} \\ \mathsf{c}\!\downarrow \dfrac{[R,R]}{R} \\ \Delta \Big\| \mathsf{KS} \cup \{\mathsf{ai}\!\downarrow\} \\ T \end{array}$$

Proceed inductively downward with the remaining instances of cut. □

5 Conclusion

System SKS seems a good starting point for developing both the proof search as well as the proof normalisation paradigm in one system. Since all up-rules are admissible, it is suitable for proof search as computation. The cut elimination procedure given is simpler than those for sequent calculi. The way in which proofs are substituted resembles normalisation in natural deduction. This hopefully allows for a computational interpretation in the proof normalisation as computation paradigm.

Of course, a lot of work remains to done. In the proof search as computation realm, given the admissibility of cut, a suitable notion of *uniform proof* as in [7] should be obtainable. For proof normalisation as computation, natural questions to be considered are strong normalisation and confluence of the cut elimination procedure when imposing as little strategy as possible. Similarly to [8], a term calculus should be developed and its computational meaning be made precise. Intuitionistic logic is a more familiar setting for this, so the possibility of treating intuitionistic logic should be explored.

A natural question is whether this procedure scales to more expressive cases, for example to predicate logic. System SKSq extends system SKS by first-order quantifiers [1]. There, cut elimination is proved via a translation to the sequent calculus. The procedure presented here does not appear to easily scale to system SKSq. The problem, which does not occur in shallow inference systems like sequent calculus or natural deduction, are existential quantifiers in the context of a cut which bind variables both in a and $\bar{a}$. The procedure easily extends to *closed* atomic cuts, that is, cuts where the eigenformula is an atom prefixed by quantifiers that bind all its variables. The question then is how to reduce general cuts to closed atomic cuts. If this problem were solved, then the procedure would scale to predicate logic. Hopefully this will lead to a cut elimination procedure for predicate logic, which is simpler than other cut elimination procedures, as happened for propositional logic.

Acknowledgements

This work has been supported by the DFG Graduiertenkolleg 334. I would like to thank Alessio Guglielmi, for inspiration, and Alwen Fernanto Tiu and Michel Parigot, for comments which helped simplifying an earlier version of the cut elimination procedure. Alessio Guglielmi, Charles Stewart and Lutz Straßburger carefully read previous drafts of this work and prompted improvements.

Web Site

Information about the calculus of structures is available from:

http://www.wv.inf.tu-dresden.de/~guglielm/Research/ .

References

1. Kai Brünnler. Locality for classical logic. Technical Report WV-02-15, Dresden University of Technology, 2002.
Available at http://www.wv.inf.tu-dresden.de/~kai/LocalityClassical.pdf.
2. Kai Brünnler and Alwen Fernanto Tiu. A local system for classical logic. In R. Nieuwenhuis and A. Voronkov, editors, *LPAR 2001*, volume 2250 of *Lecture Notes in Artificial Intelligence*, pages 347–361. Springer-Verlag, 2001.

3. Paola Bruscoli. A purely logical account of sequentiality in proof search. In Peter J. Stuckey, editor, *Logic Programming, 18th International Conference*, volume 2401 of *Lecture Notes in Artificial Intelligence*, pages 302–316. Springer-Verlag, 2002.
4. Jean Gallier. Constructive logics. Part I: A tutorial on proof systems and typed λ-calculi. *Theoretical Computer Science*, 110:249–339, 1993.
5. Alessio Guglielmi. A system of interaction and structure. Technical Report WV-02-10, 2002. Available at http://www.wv.inf.tu-dresden.de/~guglielm/Research/Gug/Gug.pdf.
6. Alessio Guglielmi and Lutz Straßburger. Non-commutativity and MELL in the calculus of structures. In L. Fribourg, editor, *CSL 2001*, volume 2142 of *Lecture Notes in Computer Science*, pages 54–68. Springer-Verlag, 2001.
7. Dale Miller, Gopalan Nadathur, Frank Pfenning, and Andre Scedrov. Uniform proofs as a foundation for logic programming. *Annals of Pure and Applied Logic*, 51:125–157, 1991.
8. M. Parigot. $\lambda\mu$-calculus: an algorithmic interpretation of classical natural deduction. In *LPAR 1992*, volume 624 of *Lecture Notes in Computer Science*, pages 190–201. Springer-Verlag, 1992.
9. Lutz Straßburger. MELL in the calculus of structures. Technical Report WV-2001-03, Dresden University of Technology, 2001. Accepted by TCS. Available at http://www.ki.inf.tu-dresden.de/~lutz/els.pdf.
10. Lutz Straßburger. A local system for linear logic. In Matthias Baaz and Andrei Voronkov, editors, *Logic for Programming, Artificial Intelligence, and Reasoning, LPAR 2002*, volume 2514 of *LNAI*, pages 388–402. Springer-Verlag, 2002.
11. Anne Sjerp Troelstra and Helmut Schwichtenberg. *Basic Proof Theory*. Cambridge University Press, 1996.

Computational Mathematics, Computational Logic, and Symbolic Computation

Bruno Buchberger

Research Institute for Symbolic Computation
Johannes Kepler University
4040 Linz, Austria

"Computational mathematics" (algorithmic mathematics) is the part of mathematics that strives at the solution of mathematical problems by algorithms. In a superficial view, some people might believe that computational mathematics is the easy part of mathematics in which trivial mathematics is made useful by repeating trivial steps sufficiently many times on current powerful machines in order to achieve marginally interesting results. The opposite is true: Many times, computational mathematics needs and stimulates deeper mathematics, i.e. deeper mathematical theorems with more difficult proofs than "pure" mathematics. This is so because, in order to establish an algorithmic method for a given mathematical problem, i.e. in order to reduce the solution of a given problem to the few operations that can be executed on machines, deeper insight on the given problem domain is necessary than the insight necessary for establishing the reduction of the given problem to powerful nonalgorithmic abstract mathematical operations as, for example, choice functions and the basic quantifiers of set theory.

Computational mathematics comes in two flavors: "numerical mathematics", in which the original problems are replaced by approximate versions and one is satisfied with approximate solutions to the approximate problems, and "exact algorithmic mathematics" in which the original problems are solved by algorithms in the original domains or isomorphic representations of these domains. Exact algorithmic mathematics can be divided into "discrete mathematics", in which the objects in the underlying mathematical domains are finitary, and "computer algebra", in which the objects in the underlying mathematical domains according to their original defintion are infinite and the possibility of an isomorphic finitary representation in itself is a non-trivial mathematical question.

For many mathematical problems it can be mathematically proved that exact algorithmic solutions are not possible or are possible only by algorithms with a certain complexity. Even in these cases, algorithmic mathematics can and should go on by considering either approximate versions or special cases of the problem.

Mathematical logic is the mathematical meta-theory of mathematics. The characteristic feature of mathematics is its method of gaining knowledge from given knowledge by reasoning. Hence, the meta-theory of mathematics is essentially the theory of reasoning. As any other mathematical theory, one can and should ask the question of how much of reasoning can be made algorithmic. The

M. Baaz and J.A. Makowsky (Eds.): CSL 2003, LNCS 2803, pp. 98–99, 2003.

part of mathematical logic that deals with algorithmic methods for reasoning is called "computational logic". Although it is well known that the algorithmization of mathematical reasoning in its most general form in a certain sense is not possible, the algorithmization of reasoning under certain restrictions or for certain limited - but still extremely broad - areas of mathematics is possible and, in fact, is one of the most challenging mathematical endeavors with enormous practical significance.

In fact, as a result of analyzing mathematical invention in the various areas of mathematics, it turns out that the transition from the object level to the meta-level is not limited to mathematical logic but is one of the main - but mostly hidden - instruments of mathematical progress in every field of mathematics and at the core of mathematical intelligence and invention. We therefore advocate that future mathematical systems must provide a frame for considering both the object and the meta-level of mathematical theories and must provide a means for the transition from the object level to the metalevel. In fact, "symbolic computation" is a term that more and more is used as a common term for both computer algebra flavored computational mathematics on the object level and computational logic on the meta-level. In other words, "symbolic computation" grows into the most general frame for all aspects of algorithmization. More concretely, in the recent research efforts of the symbolic computation community, the interaction of computer algebra and computational logic and the applications of the results of this interaction for the future automation of "mathematical knowledge management" moves into the center of interest.

In the talk, we will illustrate the above general outline by examples of symbolic computation algorithms and its underlying theories and by some demos in the Theorema software system.

Simple Stochastic Parity Games*

Krishnendu Chatterjee, Marcin Jurdziński, and Thomas A. Henzinger

Department of Electrical Engineering and Computer Sciences
University of California, Berkeley, USA

Abstract. Many verification, planning, and control problems can be modeled as games played on state-transition graphs by one or two players whose conflicting goals are to form a path in the graph. The focus here is on simple stochastic parity games, that is, two-player games with turn-based probabilistic transitions and ω-regular objectives formalized as parity (Rabin chain) winning conditions. An efficient translation from simple stochastic parity games to nonstochastic parity games is given. As many algorithms are known for solving the latter, the translation yields efficient algorithms for computing the states of a simple stochastic parity game from which a player can win with probability 1.
An important special case of simple stochastic parity games are the Markov decision processes with Büchi objectives. For this special case a first provably subquadratic algorithm is given for computing the states from which the single player has a strategy to achieve a Büchi objective with probability 1. For game graphs with m edges the algorithm works in time $O(m\sqrt{m})$. Interestingly, a similar technique sheds light on the question of the computational complexity of solving simple Büchi games and yields the first provably subquadratic algorithm, with a running time of $O(n^2/\log n)$ for game graphs with n vertices and $O(n)$ edges.

1 Introduction

Many verification, AI planning, and control problems can be formalized as state-transition graphs, and solved by finding paths in these graphs that meet certain criteria. Uncertainty about a process evolution is often modeled by probabilistic transitions, and then instead of searching for paths we are interested in measuring the probability that a path satisfies a given criterion, or finding controllers that maximize this probability. For decades there have been several separated communities studying such problems in the context of stochastic games [13], Markov decision processes (MDP's) [9], AI planning, and model checking. Only recently some unification has been attempted. MDP's can be naturally viewed as 1-player stochastic games and the book of Filar and Vrieze [8] provides a unified rigorous treatment of the theories of MDP's and stochastic games. They coin the term Competitive MDP's to encompass both 1- and 2-player stochastic

* This research was supported in part by the DARPA grant F33615-C-98-3614, the ONR grant N00014-02-1-0671, the NSF grants CCR-9988172 and CCR-0225610, and the Polish KBN grant 7-T11C-027-20.

M. Baaz and J.A. Makowsky (Eds.): CSL 2003, LNCS 2803, pp. 100–113, 2003.

games. We also suggest to cast various games based on state-transition models into a unified framework. For this purpose we use the following parameters.

- Number of players: "$1/2$": Markov chains; 1: nondeterministic state-transition systems; "$1\,1/2$": MDP's; 2: game graphs; "$2\,1/2$": stochastic games.
- The players' knowledge about the course of the game: *simple* (or turn-based) games: the state determines who plays next; *concurrent* games: the players choose moves simultaneously and independently, without knowing each other's choices [13, 1, 5].
- Winning objectives: *qualitative* (ω-regular) objectives [14]: finite objectives (reachability and safety), or infinite objectives (liveness, such as Büchi or general parity conditions), *quantitative* (reward) objectives [8]: discounted reward, or limiting average reward, or total reward.
- Winning criteria: qualitative criteria [5]: sure winning, almost-sure winning (with probability 1), or limit-sure winning (with probability arbitrarily close to 1), quantitative criteria: exact probability of winning, or expected reward.

We mention a few notable examples of models and problems studied in various communities that fit into the above categorization.

- Summable MDP's [9, 8]: simple $1\,1/2$-player games with maximum expected discounted reward.
- Mean-payoff games [17]: simple 2-player games with maximum limiting average reward.
- Parity games [12, 14]: simple 2-player games with parity objectives.
- Quantitative simple stochastic games [2]: simple $2\,1/2$-player games with reachability objectives and exact probability of winning.
- Qualitative concurrent ω-regular games [4]: concurrent $2\,1/2$-player games with parity objectives and various qualitative winning criteria.
- Quantitative concurrent ω-regular games [6]: concurrent $2\,1/2$-player games with parity objectives and exact probability of winning.

In earlier work [4, 11] we studied the complexity of algorithms for solving concurrent parity games. In particular, we have given efficient reductions from the problem of solving concurrent Büchi and co-Büchi games (under the almost-sure winning criterion) to the extensively studied problem of solving simple (nonconcurrent) parity games [15, 10, 16]. In this paper we focus on the following three types of games:

- *Qualitative simple stochastic parity games*: simple $2\,1/2$-player games with parity objectives and almost-sure winning criterion.
- *Qualitative Büchi MDP's*: simple $1\,1/2$-player games with Büchi winning objectives and almost-sure winning criterion.
- *Simple Büchi games*: simple 2-player games with Büchi winning objectives.

We use n to denote the number of vertices and m to denote the number of edges of a game graph. Our main results can be summarized as follows.

Theorem 1. *Every qualitative simple stochastic parity game with priorities in the set $\{0, 1, 2, \ldots, d-1\}$ can be translated to a simple parity game with the same set of priorities, with $O(d \cdot n)$ vertices and $O(d \cdot (m+n))$ edges, and hence it can be solved in time $O(d \cdot (m+n) \cdot (nd)^{\lceil d/2 \rceil})$.*

Theorem 2 (Pure memoryless determinacy). *From every vertex of a simple stochastic parity game either one player has a pure memoryless strategy to win with probability 1, or there is a $\delta > 0$ such that the other player has a pure memoryless strategy to win with probability at least δ. Hence the almost-sure and limit-sure winning criteria coincide.*

Theorem 3. *Qualitative Büchi MDPs can be solved in time $O(m\sqrt{m})$.*

This implies also a complexity improvement for solving MDP's with reachability objectives under the almost-sure winning criterion, for which the best algorithm so far had $O(mn)$ running time [3]. Interestingly, the novel technique we use for Büchi MDP's allows us to shed some light on the important problem of finding subquadratic algorithms for simple Büchi games.

Theorem 4. *Simple Büchi games with $O(n)$ edges can be solved in time $O(n^2/\log n)$.*

This result and reductions in [11] prove that concurrent games with a constant number of actions and with reachability and Büchi objectives under the almost-sure and limit-sure winning criteria can also be solved in subquadratic time (the best algorithms so far had $O(n^2)$ running time [4]).

2 Simple Stochastic Parity Games

Given $n \in \mathbb{N}$, we write $[n]$ for the set $\{0, 1, 2, \ldots, n\}$ and $[n]_+$ for the set $\{1, 2, \ldots, n\}$. A *2½-player game* (or *simple stochastic game*, or SSG) $G = (V, E, (V_\Box, V_\Diamond, V_\bigcirc))$ consists of a directed graph (V, E) and a partition $(V_\Box, V_\Diamond, V_\bigcirc)$ of the vertex set V. For technical convenience we assume that every vertex has at least one outgoing edge. For simplicity we only consider the case when G is binary. An *infinite path* in G is a infinite sequence $\langle v_0, v_1, v_2, \ldots \rangle$ of vertices such that $(v_k, v_{k+1}) \in E$ for all $k \in \mathbb{N}$. We write Ω for the set of all infinite paths. The game is played with three players that move a token from vertex to vertex so that an infinite path is formed: from vertices in $V_\Box$, player Even ($\Box$) moves the token along an outgoing edge to a successor vertex; from vertices in $V_\Diamond$, player Odd ($\Diamond$) moves the token; and from vertices in $V_\bigcirc$, player Random ($\bigcirc$) moves the token. If there are two outgoing edges, then player Random always moves the token to one of the two successor vertices with probability $1/2$. Since player Random does not have a proper choice of moves, as the other two players do, we use the ½-player terminology for player Random. The *2-player games* are the special case of the 2½-player games with $V_\bigcirc = \emptyset$. The *1½-player games* (or MDP's) are the special case of the 2½-player games with $V_\Diamond = \emptyset$. In other words, in 2-player games, the only players are Even and Odd; and in $1\frac{1}{2}$-player games, the only players are Even and Random.

Strategies. For a finite set A, a probability distribution on A is a function $f : A \to [0,1]$ such that $\sum_{a \in A} f(a) = 1$. We denote the set of probability distributions on A by $\mathcal{D}(A)$. A *(mixed) strategy* for player Even is a function $\sigma : V^* \cdot V_\Box \to \mathcal{D}(V)$ such that for every finite and nonempty sequence $\overline{v} \in V^* \cdot V_\Box$ of vertices, which represents the history of the play so far, $\sigma(\overline{v})$ is the next move to be chosen by player Even. A strategy must prescribe only available moves, i.e., if $(v, u) \notin E$, then $\sigma(\overline{w} \cdot v)(u) = 0$. The strategy σ is *pure* if for all $\overline{w} \in V^*$ and $v \in V_\Box$, there is a vertex u such that $\sigma(\overline{w} \cdot v)(u) = 1$. The strategies for player Odd are defined analogously. We write Σ and Π for the sets of all strategies for players Even and Odd, respectively. A *memoryless strategy* is a strategy which does not depend on the history of the play but only on the current vertex. A pure memoryless strategy for player Even can be represented as a function $\sigma : V_\Box \to V$ such that $(v, \sigma(v)) \in E$ for all $v \in V_\Box$.

For an initial vertex v, and two strategies $\sigma \in \Sigma$ and $\pi \in \Pi$ for players Even and Odd, respectively, we define $\mathit{Outcome}(v, \sigma, \pi) \subseteq \Omega$ to be the set of paths that can be followed when a play starts from vertex v and the players use the strategies σ and π. Formally, $\langle v_0, v_1, v_2, \ldots \rangle \in \mathit{Outcome}(v, \sigma, \pi)$ if $v_0 = v$, and for all $k \geq 0$, we have that $v_k \in V_\bigcirc$ implies $(v_k, v_{k+1}) \in E$, $v_k \in V_\Box$ implies $\sigma(v_0, v_1, \ldots, v_k)(v_{k+1}) > 0$, and $v_k \in V_\Diamond$ implies $\pi(v_0, v_1, \ldots, v_k)(v_{k+1}) > 0$. Once a starting vertex v and strategies $\sigma \in \Sigma$ and $\pi \in \Pi$ for the two players have been chosen, the probabilities of events are uniquely defined, where an *event* $\mathcal{A} \subseteq \Omega$ is a measurable set of paths. For a vertex v and an event $\mathcal{A} \subseteq \Omega$, we write $\Pr_v^{\sigma,\pi}[\mathcal{A}]$ for the probability that a path belongs to $\mathcal{A}$ if the game starts from v and the players use the strategies σ and π.

Winning objectives. A *winning objective* for a SSG G is a set $\mathcal{W} \subseteq \Omega$ of infinite paths. We consider the following winning objectives.

- *Büchi objective.* For a set $T \subseteq V$ of *target* vertices, the Büchi objective is defined as $\text{Büchi}(T) = \{\langle v_0, v_1, v_2 \ldots \rangle \in \Omega : v_k \in T \text{ for infinitely many } k \geq 0\}$.
- *Parity objective.* Let $p : V \to [d]$ be a function that assigns a *priority* $p(v)$ to every vertex $v \in V$, where $d \in \mathbb{N}$. For an infinite path $\overline{v} = \langle v_0, v_1, \ldots \rangle \in \Omega$, we define $\text{Inf}(\overline{v}) = \{ i \in [d] : p(v_k) = i \text{ for infinitely many } k \geq 0 \}$. The *Even parity objective* is defined as $\text{Parity}(p) = \{ \overline{v} \in \Omega : \min\big(\text{Inf}(\overline{v})\big) \text{ is even} \}$, and the *Odd parity objective* as $\text{co-Parity}(p) = \{ \overline{v} \in \Omega : \min\big(\text{Inf}(\overline{v})\big) \text{ is odd} \}$.

Note that for a priority function $p : V \to [1]$ with only two priorities (0 and 1), an even parity objective $\text{Parity}(p)$ is equivalent to the Büchi objective $\text{Büchi}(p^{-1}(0))$, i.e., the target set consists of the vertices with priority 0. A $2\frac{1}{2}$*-player parity game* (or parity SSG) is a pair (G, p), where G is a $2\frac{1}{2}$-player game and p is a priority function. If G is a 2-player (resp. $1\frac{1}{2}$-player) game, then (G, p) is a *2-player parity game* (resp. $1\frac{1}{2}$*-player parity game*, or parity MDP). A *2-player Büchi game* is a pair (G, T), where G is a 2-player game and T is a set of target vertices. If G is a $1\frac{1}{2}$-player game, then (G, T) is a $1\frac{1}{2}$*-player Büchi game* (or Büchi MDP).

Winning criteria. Consider an SSG G with winning objective $\mathcal{W}$. We say that a strategy $\sigma \in \Sigma$ for player Even is

- *sure winning* from vertex v if for all strategies $\pi \in \Pi$ of player Odd, we have $Outcome(v, \sigma, \pi) \subseteq \mathcal{W}$;
- *almost-sure winning* from vertex v if for all strategies $\pi \in \Pi$ of player Odd, we have $\Pr_v^{\sigma,\pi}[\mathcal{W}] = 1$;
- *positive-probability winning* from vertex v if there is a $\delta > 0$, such that for all strategies $\pi \in \Pi$ of player Odd, we have $\Pr_v^{\sigma,\pi}[\mathcal{W}] \geq \delta$.

The definitions for player Odd are similar. We shall see that player Even has an almost-sure winning strategy for $\mathcal{W}$ from v if and only if player Odd does not have a positive-probability winning strategy for $\Omega \setminus \mathcal{W}$ from v. For 2-player games all three of the above winning criteria coincide, i.e., the existence of a positive-probability winning strategy for a player implies the existence of a sure winning strategy. In this paper we consider the dual criteria of almost-sure winning (i.e., winning with probability 1) and positive-probability winning, for $2^1/_2$- and $1^1/_2$-player games, and the criterion of sure winning for 2-player games:

- The problem of *solving a* $2^1/_2$*-player* (resp. $1^1/_2$*-player*) *parity game* (G, p) is to compute the set of vertices of G from which the player Even has an almost-sure winnning strategy for the objective $\mathcal{W} = \text{Parity}(p)$.
- The problem of *solving a 2-player parity game* (G, p) is to compute the set of vertices of G from which the player Even has a sure winnning strategy for the objective $\mathcal{W} = \text{Parity}(p)$.

3 Solving $2^1/_2$-Player Parity Games

The main result of this section is an algorithm for solving $2^1/_2$-player parity games, which is obtained by an efficient reduction to 2-player parity games, i.e., a proof of Theorem 1. As in our earlier work [11], the key technical tool for the correctness proof of the reduction is the notion of ranking functions, which witness the existence of winning strategies for the players. Our ranking functions are closely related to the semantics of the μ-calculus formulas that express the winning sets of *concurrent* stochastic parity games [4], but due to the lack of concurrency in our games, the defining conditions for our ranking functions are considerably simpler. Two corollaries of our proof are of independent interest. First, we establish the existence of pure memoryless winning strategies for both players in simple stochastic parity games (Theorem 2). This is in contrast to concurrent games, where players need mixed strategies with infinite memory [4]. Second, in simple stochastic parity games the almost-sure and limit-sure winning criteria coincide (Theorem 2), which is not the case for concurrent games [4].

3.1 Ranking Functions

In this subsection we provide a characterization of the "universal" parity MDP problem. We define certain sufficient conditions for establishing that for *all* strategies π, the Markov chain M_π satisfies the parity condition with probability 1, or with probability at least $\delta > 0$. These sufficient conditions are then used in the next subsection to prove correctness of our solution for $2^1/_2$-player parity games: a strategy σ is winning for a player if and only if the parity MDP G_σ is a solution to the universal parity MDP problem.

Consider a parity MDP $M = (V, E, (V_\diamond, V_\bigcirc), p : V \to [d])$. Without loss of generality assume that d is even. A *ranking function* for player Even labels vertices with $(d/2)$-tuples of natural numbers: $\varphi = (\varphi^1, \varphi^3, \ldots, \varphi^{d-1}) : V \to [n]^{d/2} \cup \{\infty\}$, for some $n \in \mathbb{N}$. For succinctness, for all odd $k \in [d]$ we write $\overrightarrow{\varphi}^k(v)$ to denote the tuple $(\varphi^1(v), \varphi^3(v), \ldots, \varphi^k(v))$. We often call $\varphi(v)$ the *rank* of vertex v, and we call $\overrightarrow{\varphi}^k(v)$ the k-th rank of vertex v. A ranking function for player Odd is a function $\psi = (\psi^0, \psi^2, \ldots, \psi^d) : V \to [n]^{d/2+1} \cup \{\infty\}$; we use similar notational conventions as with ranking functions for player Even. For all $v \in V_\bigcirc$, we write $\mathrm{Pr}_v[\overrightarrow{\varphi}^k_<]$ (resp. $\mathrm{Pr}_v[\overrightarrow{\varphi}^k_\leq]$) for the one-step probability of reaching from vertex v a successor u of v such that $\overrightarrow{\varphi}^k(u) <_{\mathrm{lex}} \overrightarrow{\varphi}^k(v)$ (resp. $\overrightarrow{\varphi}^k(u) \leq_{\mathrm{lex}} \overrightarrow{\varphi}^k(v)$). In other words, $\mathrm{Pr}_v[\overrightarrow{\varphi}^k_<]$ is the probability in vertex v of strictly decreasing the k-th rank in one step, and $\mathrm{Pr}_v[\overrightarrow{\varphi}^k_\leq]$ is the probability of not increasing the k-th rank. Moreover, we write $\mathrm{Pr}_v[\varphi_{<\infty}]$ (or for notational convenience $\mathrm{Pr}_v[\overrightarrow{\varphi}^{-1}_\leq]$) for the one-step probability of reaching from v a successor u of v such that $\varphi(u) \neq \infty$. We always use these notations in the context of expressions such as $\mathrm{Pr}_v[\overrightarrow{\varphi}^k_\leq] = 1$ or $\mathrm{Pr}_v[\overrightarrow{\varphi}^k_<] \geq \varepsilon$. By slight abuse of notation, for vertices $v \in V_\Box$ we also write $\mathrm{Pr}_v[\overrightarrow{\varphi}^k_<]$ and $\mathrm{Pr}_v[\overrightarrow{\varphi}^k_\leq]$ in such expressions, and then we mean those expressions to hold if and only if they hold for *all* mixed one-step strategies in vertex v. It is easy to verify that if either of the two expressions above holds for all pure one-step strategies in v, then it also holds for all mixed one-step strategies.

Definition 1 (Almost-sure ranking). *A ranking function $\varphi : V \to [n]^{d/2} \cup \{\infty\}$ for player Even is an almost-sure ranking if there is an $\varepsilon \geq 0$ such that for every vertex v with $\varphi(v) \neq \infty$, the following condition C_v holds:*

- *$p(v)$ even:* $\bigvee_{\mathrm{odd}\ i \in [p(v)]} (\mathrm{Pr}_v[\overrightarrow{\varphi}^{i-2}_\leq] = 1 \wedge \mathrm{Pr}_v[\overrightarrow{\varphi}^i_<] \geq \varepsilon) \vee (\mathrm{Pr}_v[\overrightarrow{\varphi}^{p(v)-1}_\leq] = 1)$,
- *$p(v)$ odd:* $\bigvee_{\mathrm{odd}\ i \in [p(v)]} (\mathrm{Pr}_v[\overrightarrow{\varphi}^{i-2}_\leq] = 1 \wedge \mathrm{Pr}_v[\overrightarrow{\varphi}^i_<] \geq \varepsilon)$.

Proposition 1. *Let $k \in [d]$ be an odd priority. Then for every vertex v with $\varphi(v) \neq \infty$ the following conditions hold.*

(a) If $p(v) = k$, then in one step from vertex v the k-th rank decreases with probability at least ε.
(b) If $p(v) > k$, then in one step from vertex v either the k-th rank decreases with probability at least ε, or the k-th rank does not increase (with probability 1).

Lemma 1. *Let φ be an almost-sure ranking for a parity MDP. Then for every (mixed) strategy of player Odd, the Even parity objective is satisfied with probability 1 from every vertex v with $\varphi(v) \neq \infty$.*

Proof. Once the strategy for player Odd is fixed, a play in the parity MDP is an infinite random walk. We argue that this random walk satisfies the Even parity objective with probability 1. From our discussion above it follows that the conditions expressed in the definition of an almost-sure ranking hold for all mixed one-step strategies of player Odd, and since our reasoning below is carried out using only these conditions, it applies to all mixed strategies for player Odd.

In order to prove that with probability 1, the lowest priority occurring infinitely often is even, it suffices to show that for every odd priority $k \in [d]$, if vertices of priority k keep occurring in the random walk, then with probability 1 eventually a vertex of lower priority occurs. First note that from the definition of an almost-sure ranking, it follows that all successors of a vertex with finite rank have finite rank: one of the conditions $\Pr_v[\overrightarrow{\psi}^i_{\leq}] = 1$ must hold so the i-th rank cannot increase in any step and thus the rank stays finite. Let $k \in [d]$ be odd. For the sake of contradiction assume that from some point on vertices of priority k keep occurring, but no vertex of a lower priority ever occurs. In this case Proposition 1 implies that in every step either the k-th rank does not increase with probability 1, or it decreases with probability at least ε, and moreover, in every step from a vertex of priority k the k-th rank decreases with probability at least ε. As there are $N = (n+1)^{(k+1)/2}$ different values of a k-th rank, within at most N visits to a vertex of priority k the k-th rank must decrease to $(0, \dots, 0)$ with probability at least ε^N. Thus with probability 1 a vertex with k-th rank $(0, \dots, 0)$ is eventually reached. This contradicts the assumption that priority k occurs infinitely often: no vertex of priority k can have its k-th rank equal to $(0, \dots, 0)$, because by Proposition 1(a) a step from such a vertex has to decrease the k-th rank with positive probability and $(0, \dots, 0)$ is the smallest rank. ∎

Definition 2 (Positive-probability ranking). *A ranking function $\psi : V \to [n]^{d/2+1} \cup \{\infty\}$ for player Odd is a positive-probability ranking if there is an $\varepsilon \geq 0$ such that for every vertex v with $\psi(v) \neq \infty$, the following condition D_v holds:*

- $p(v)$ *even:* $(\Pr_v[\overrightarrow{\psi}^0_{<}] \geq \varepsilon) \vee \bigvee_{\text{even } i \in [p(v)]_+} (\Pr_v[\overrightarrow{\psi}^{i-2}_{\leq}] = 1 \wedge \Pr_v[\overrightarrow{\psi}^i_{<}] \geq \varepsilon)$,
- $p(v)$ *odd:* $(\Pr_v[\overrightarrow{\psi}^0_{<}] \geq \varepsilon) \vee \bigvee_{\text{even } i \in [p(v)]_+} (\Pr_v[\overrightarrow{\psi}^{i-2}_{\leq}] = 1 \wedge \Pr_v[\overrightarrow{\psi}^i_{<}] \geq \varepsilon) \vee (\Pr_v[\overrightarrow{\psi}^{p(v)}_{\leq}] = 1)$.

A proof similar to that of Lemma 1 can be used to prove the following lemma.

Lemma 2. *Let ψ be a positive-probability ranking for a parity MDP. Then there is a $\delta > 0$ such that for every (mixed) strategy of player Even, the Odd parity objective is satisfied with probability at least δ from every vertex v with $\psi(v) \neq \infty$.*

3.2 The Reduction

Given a $2^1\!/\!_2$-player parity game $G = (V, E, (V_\Box, V_\Diamond, V_\bigcirc), p : V \to [d])$, we construct a 2-player parity game $\overline{G}$ with the same set $[d]$ of priorities. For every vertex $v \in V_\Box \cup V_\Diamond$, there is a vertex $\overline{v} \in \overline{V}$ with "the same" outgoing edges, i.e., $(v, u) \in E$ if and only if $(\overline{v}, \overline{u}) \in \overline{E}$. Each random vertex $v \in V_\bigcirc$ is substituted by the gadget presented in Figure 1. More formally, the players play the following 3-step game in $\overline{G}$ from vertex $\overline{v}$ of priority $p(v)$. First, in vertex $\overline{v}$ player Odd

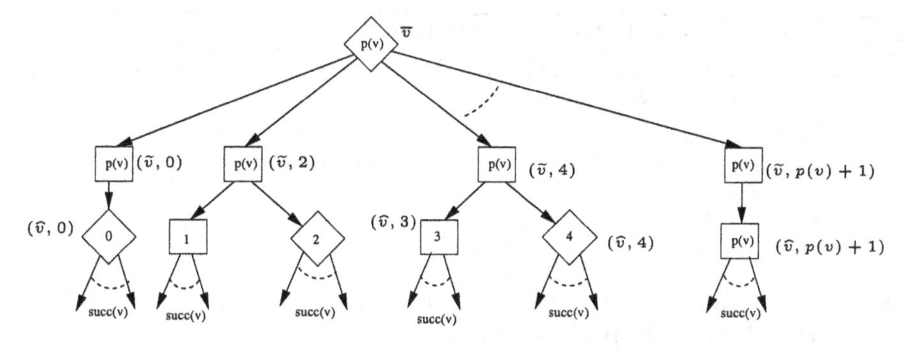

Fig. 1. Gadget for reducing 2½-player parity games to 2-player parity games.

chooses a successor $(\widetilde{v}, k)$, each of priority $p(v)$, where $k \in [p(v)+1]$ is even. Then in vertex $(\widetilde{v}, k)$ player Even chooses from at most two successors: vertex $(\widehat{v}, k-1)$ of priority $k-1$ if $k > 0$, or vertex $(\widehat{v}, k)$ of priority k if $k \leq p(v)$. Finally, in a vertex $(\widehat{v}, k)$ the choice is between all vertices $\overline{u}$ such that $(v, u) \in E$, and it belongs to player Even if k is odd, and to player Odd if k is even.

Lemma 3. *For every vertex v in G, if player Even (resp. Odd) has a sure winning strategy from vertex $\overline{v}$ in $\overline{G}$, then player Even (resp. Odd) has an almost-sure (resp. positive-probability) winning strategy from v.*

Proof. We prove the claim for player Even. The case of player Odd is similar and is omitted here. If $\overline{W_\Box}$ is the set of vertices from which player Even has a winning strategy in $\overline{G}$, then there is a ranking function (also called a *progress measure* [10]) $\overline{\varphi} : \overline{V} \to [n]^{d/2} \cup \{\infty\}$ (where $n \leq |\overline{V}|$) such that $\overline{\varphi}(w) \neq \infty$ for all $w \in W_\Box$. This ranking function induces a memoryless winning strategy $\overline{\sigma}$ for player Even in $\overline{G}$ [10]. We define a ranking function φ and a memoryless strategy for player Even σ for G by setting $\varphi(v) = \overline{\varphi}(\overline{v})$ and $\sigma(v) = \overline{\sigma}(\overline{v})$ for every $v \in V$. Taking the strategy subgraph of σ in G we obtain a parity MDP M. In order to prove the claim, by Lemma 1 it suffices to argue that φ is an almost-sure ranking for M. It is easy to verify that the ranking condition C_v holds for all vertices $v \notin V_\bigcirc$ with $\varphi(v) \neq \infty$. We prove that the ranking condition C_v holds for all vertices $v \in V_\bigcirc$ with $\varphi(v) \neq \infty$. Let $k \in [p(v)]_+$. Since vertex $\overline{v}$ belongs to player Odd, the edge leading to vertex $(\widetilde{v}, k)$ must be in M. Vertex $(\widetilde{v}, k)$ belongs to player Even, so either the edge leading to vertex $(\widehat{v}, k-1)$ or the one leading to vertex $(\widehat{v}, k)$ belongs to M. In the former case, by analyzing the inequalities between $\overline{\varphi}$-ranks of vertices on the path from $\overline{v}$ to the successor of $(\widehat{v}, k-1)$ in M which hold by the definition of a ranking function [10], we can deduce that $\Pr_v[\overrightarrow{\varphi}^{k-1}_<] \geq 1/2$ holds. In the latter case, we get that $\Pr_v[\overrightarrow{\varphi}^{k-1}_\leq] = 1$ holds. Considering all edges that lead out of vertex $\overline{v}$ in M, we conclude that the following condition holds: $(\Pr_v[\overrightarrow{\varphi}_{<\infty}] = 1) \wedge \bigwedge_{\text{odd } i \in [p(v)]} (\Pr_v[\overrightarrow{\varphi}^i_\leq] = 1 \vee \Pr_v[\overrightarrow{\varphi}^i_<] \geq 1/2)$. This condition can be shown to imply the ranking condition C_v using the two simple properties that if $i, j \in [p(v)]$ are odd and $i < j$, then $\Pr_v[\overrightarrow{\varphi}^j_\leq] = 1$ implies $\Pr_v[\overrightarrow{\varphi}^i_\leq] = 1$, and $\Pr_v[\overrightarrow{\varphi}^i_<] \geq \varepsilon$ implies $\Pr_v[\overrightarrow{\varphi}^j_<] \geq \varepsilon$. ∎

Algorithm 1 **Classical algorithm for Büchi MDP's**

Input : $1^1\!/_2$-player Büchi game (G,T). **Output:** $W_{\bigcirc}$ and $W_{\Box} = V \setminus W_{\bigcirc}$.
1. $G_0 := G$; 2. $W_0 := \emptyset$; 3. $i := 0$
4. **repeat**
 4.1 $W_{i+1} :=$ *One-Iteration-Of-The-Classical-Algorithm*(V_i)
 4.2 $V_{i+1} := V_i \setminus W_{i+1}$; $i := i+1$
 until $W_i = \emptyset$
5. $W_{\bigcirc} := \bigcup_{k=1}^{i} W_k$

Procedure *One-Iteration-Of-The-Classical-Algorithm*
Input: set $V_i \subseteq V$. **Output:** set $W_{i+1} \subseteq V_i$.
1. $R_i := \mathit{Reach}(T \cap V_i, V_i)$; 2. $\mathit{Tr}_i := V_i \setminus R_i$; 3. $W_{i+1} := \mathit{Attr}_{\bigcirc}(\mathit{Tr}, V_i)$

4 An $O(m\sqrt{m})$ Algorithm for $1^1\!/_2$-Player Büchi Games

In this section we consider $1^1\!/_2$-player games with Büchi winning objectives, i.e., Büchi MDP's. There are two players, Even and Random. We write T for the set of target vertices, which player Even attempts to visit infinitely often. By $\mathcal{W}_{\Box}$ we denote the set of vertices from which player Even has an almost-sure winning strategy, and by $\mathcal{W}_{\bigcirc}$ the set from which the Büchi objective is violated with positive probability for all strategies of player Even. We call these sets the winning sets for player Even and Random, respectively. The main result of this section is an $O(n\sqrt{n})$ algorithm for computing $\mathcal{W}_{\Box}$ and $\mathcal{W}_{\bigcirc}$ for a $1^1\!/_2$-player Büchi game with n vertices and $O(n)$ edges. This proves Theorem 3, because a game graph with m edges can be easily converted in $O(m+n)$ time to an equivalent game graph with $O(m)$ vertices and $O(m)$ edges.

In the rest of the paper we use the following notations for a graph $G = (V, E)$ and a set $S \subseteq V$ of vertices. We write $\mathit{succ}(v, G) = \{ u \in V : (v, u) \in E \}$ for the set of immediate successors of vertex v. We define $\mathit{In}(S, G) = \{ (v, u) \in E : v \notin S \text{ and } u \in S \}$ to be the set of edges that enter set S, and $\mathit{Source}(S, G) = \{ v \in V : (v, u) \in \mathit{In}(S, G) \text{ for some } u \}$ is the set of sources of edges that enter S. We write $\mathit{Reach}(S, G)$ for the set of vertices from which there is a path in graph G to a vertex in S. Let $(V_{\Box}, V_*)$ be a partition of the set V of vertices (player $\Box$ moves from the vertices in $V_{\Box}$ and player $*$ moves from the vertices in V_*, where $* \in \{\Diamond, \bigcirc\}$). We inductively define the set $\mathit{Attr}_{\Box}(S, G)$ of vertices from which player $\Box$ has a strategy to reach the set S in the following way. Set $R_0 = S$, and for $k \geq 0$, set $R_{k+1} = R_k \cup \{ v \in V_{\Box} : (v, u) \in E \text{ for some } u \in R_k \} \cup \{ v \in V_* : u \in R_k \text{ for all } (v, u) \in E \}$. Let $\mathit{Attr}_{\Box}(S, G) = \bigcup_k R_k$. We define the set $\mathit{Attr}_*(S, G)$ in a similar way. We fix a graph G until the end of the paper and by a slight abuse of notation instead of putting graphs as the second parameters of all the above definitions, we will write a subset S of the vertices to stand for the subgraph of G that is induced by the set S of vertices.

The *classical algorithm* (Algorithm 1 [3]) works as follows. First it finds the set of vertices R from which the target set T is reachable. The rest of the vertices $\mathit{Tr} = V \setminus R$ (a "trap" for player Even) are identified as winning for player Random. Then the set of vertices W, from which player Random has a strategy to reach its winning set Tr, is computed. Set W is identified as a subset of the winning set for player Random and it is removed from the vertex set. The algorithm then iterates on the reduced game graph. In every iteration it performs a *backward* search from the current target set to find vertices which can reach the target set. Each iteration takes $O(n)$ time.

Our *improved algorithm* (Algorithm 2) differs from the classical algorithm by selectively performing a backward search as the classical algorithm does, or a cheap *forward* exploration of edges from vertices that are good candidates to be included in the winning set of player Random. In Step 4.1 of the improved algorithm, if the number of edges entering the set of vertices included in the previous iteration into the winning set of player Random is at least as big as a constant k, then we run an iteration of the classical algorithm. Otherwise, i.e., if this number is smaller than k, then let S be the set of sources of edges that enter the set of vertices winning for player Random in the previous iteration. The vertices in S are considered as candidates to be included into the winning set of player Random in the current iteration. In Step 4.2.2.1 (procedure *Dovetail-Explore*) a dovetailing exploration of edges is performed from all vertices in S. From all vertices $v \in S$, up to ℓ edges are explored in a round-robin fashion. If the forward exploration of edges from a vertex $v \in S$ terminates before ℓ edges are explored, and none of the explored vertices is in the target set, then the explored subgraph is included in the winning set of player Random. If Step 4.2.2.1 fails to identify a winning set of player Random, then in Step 4.2.4 an iteration of the classical algorithm is executed. The winning set discovered by the iteration of the classical algorithm in Step 4.2.4 must contain at least ℓ edges as otherwise it would have been discovered in Step 4.2.2.1.

Lemma 4. *Algorithm 2 correctly computes the sets* $\mathcal{W}_{\bigcirc}$ *and* $\mathcal{W}_{\Box}$.

Proof. We prove by induction that W_i computed in any iteration of the improved algorithm satisfies $W_i \subseteq \mathcal{W}_{\bigcirc}$. Base case: $W_0 = \emptyset \subseteq \mathcal{W}_{\bigcirc}$.
Inductive case: we argue that $W_i \subseteq \mathcal{W}_{\bigcirc}$ implies $W_{i+1} \subseteq \mathcal{W}_{\bigcirc}$.

1. If Step 4.1 is executed, then $W_{i+1} \subseteq \mathcal{W}_{\bigcirc}$ by correctness of the classical algorithm.
2. If Step 4.2.2 is executed, then a nonempty set R_v is included in Tr_{i+1} in Step 3.3 of procedure *Dovetail-Explore*. By the condition in Step 3.3 of *Dovetail-Explore*, no vertex in R_v can reach a vertex outside of R_v, and since $R_v \cap T \cap V_i = \emptyset$, we conclude that T_i cannot be reached from any vertex in R_v. Therefore $R_v \subseteq \mathcal{W}_{\bigcirc}$ and $Tr_{i+1} \subseteq \mathcal{W}_{\bigcirc}$. Hence $U \subseteq \mathcal{W}_{\bigcirc}$ and $W_{i+1} \subseteq \mathcal{W}_{\bigcirc}$.
3. If Steps 4.2.2 and 4.2.4 are executed, then $U \subseteq \mathcal{W}_{\bigcirc}$ and the correctness of an iteration of the classical algorithm imply $W_{i+1} \subseteq \mathcal{W}_{\bigcirc}$.

The other inclusion $\mathcal{W}_{\bigcirc} \subseteq W_{\bigcirc}$ follows from the correctness of the classical algorithm, because the termination condition of the loop in Step 4 implies that $Tr_i = \emptyset$ holds so the last iteration was an iteration of the classical algorithm. ∎

Lemma 5. *The total work in Step 4.2.2.1 of Algorithm 2 is* $O(kn)$.

Algorithm 2 Improved Algorithm for Büchi MDP's

Input: 1½-player Büchi game (G,T). **Output:** $W_\bigcirc$ and $W_\Box = V \setminus W_\bigcirc$.
1. $G_0 := G$; 2. $W_0 := \emptyset$; 3. $i := 0$
4. **repeat**
  4.1 **if** $|In(W_i, V_i \cup W_i)| \geq k$ **then**
    4.1.1 $W_{i+1} :=$ *One-Iteration-Of-The-Classical-Algorithm*(V_i)
  4.2 **else** ($|In(W_i, V_i \cup W_i)| < k$)
    4.2.1 $U := \emptyset$
    4.2.2 **repeat**
      4.2.2.1 $Tr_i =$ *Dovetail-Explore*$(V_i \setminus U, Source(W_i \cup U, V_i \cup W_i))$
      4.2.2.2 $U =: U \cup Attr_\bigcirc(Tr_i, V_i \setminus U)$
      until $|In(W_i \cup U, V_i \cup W_i)| \leq k$ **and** $Tr_i \neq \emptyset$
    4.2.3 **if** $Tr_i \neq \emptyset$ **then** $W_{i+1} := W_i \cup U$
    4.2.4 **else** $W_{i+1} := U \cup ($*One-Iteration-Of-The-Classical-Algorithm*$(V_i \setminus U))$
  4.3 $V_{i+1} := V_i \setminus W_{i+1}$
  4.4 $i := i + 1$
  until $W_i = \emptyset$
5. $W_\bigcirc := \bigcup_{k=1}^{i} W_k$

Procedure *Dovetail-Explore*
Input: set $S \subseteq V_i$ and graph $G_i = (V_i, E_i)$. **Output:** set $Tr_i \subseteq V_i$.
1. $Tr_i := \emptyset$; 2. $ec := \ell$
3. **repeat**
  3.1 $ec := ec - 1$
  3.2 **for each** vertex $v \in S$
    extend the sub-graph $R_v \subseteq V_i$ by exploring a new edge
  3.3 **if** there is $v \in S$ s.t. for all $u \in R_v$, $succ(u, V_i) \subseteq R_v$ **and** $R_v \cap T \cap V_i = \emptyset$
    then return $Tr_i := R_v$
  until $ec = 0$
4. **return** Tr_i

Proof. Consider the following two cases.

1. If a nonempty set of vertices R_v is included in the set Tr_i in Step 3.3 of *Dovetail-Explore*, and the number of edges in the induced subgraph R_v is e_i, then the total work done in Step 3 is $O(ke_i)$, because $|Source(W_i \cup U, V_i \cup W_i)| \leq k$. Since e_i edges are removed from the graph and the number of all edges in the graph is $O(n)$, the total work over all iterations of *Dovetail-Explore* when a nonempty set R_v is included in a set Tr_i is $O(kn)$.
2. If $Tr_i = \emptyset$ after executing the procedure *Dovetail-Explore*, then the work done there is $O(k\ell)$. Whenever this happens, the subgraph induced by the set of vertices Tr_i discovered by the following iteration of the classical algorithm must have more than ℓ edges. This can happen at most $O(n/\ell)$ times, because the number of edges in the graph is $O(n)$. Hence the total work over all iterations is $O((n/\ell)k\ell) = O(kn)$. ∎

Lemma 6. *The total work in Step 4.2.2.2 of Algorithm 2 is $O(n)$, in Step 4.2.2 it is $O(kn)$, in Step 4.1 it is $O(n^2/k)$, and in Step 4.2.4 it is $O(n^2/\ell)$.*

Lemma 7. *Algorithm 2 solves* $1\,{}^1\!/_2$*-player Büchi games with* n *vertices and* $O(n)$ *edges in time* $O(n\sqrt{n})$.

Proof. Correctness follows from Lemma 4. By Lemma 6 the work in Steps 4.1, 4.2.2, and 4.2.4 is $O(n^2/k + kn + n^2/\ell)$. Take $k = \ell = \sqrt{n}$ to get the $O(n\sqrt{n})$ bound for the total work. ∎

5 An $O(n^2/\log n)$ Algorithm for 2-Player Büchi Games

In this section we consider 2-player games of the form (G, T), where T for the set of target vertices. By $\mathcal{W}_\Box$ we denote the set of vertices from which player Even has a strategy to visit a state in T infinitely often, and by $\mathcal{W}_\Diamond$ the set of vertices from which player Odd can avoid visiting T infinitely often. These are the winning sets for player Even and Odd, respectively. By determinacy of parity games [7] we have $\mathcal{W}_\Diamond = V \setminus \mathcal{W}_\Box$. Inspired by the algorithm of the previous section, we provide an algorithm for computing the set $\mathcal{W}_\Diamond$ in time $O(n^2/\log n)$ if G has n vertices and $O(n)$ edges, a proof of Theorem 4. A graph is *binary* if every vertex has at most two successors. For simplicity we present the algorithm for the case when the game graph is binary.

The *classical algorithm* for solving 2-player Büchi games (Algorithm 3 [15]) is very similar to the classical algorithm of the previous section. The only difference is that in step 1 of each iteration i we compute set R_i to be the set of vertices from which player Even has a strategy to reach set T, i.e., $R_i = Attr_\Box(T, V_i)$; and in step 3 the set W_{i+1} is the set of vertices from which player Odd has a strategy to reach set Tr_i. Then the winning set of player Even is obtained as the union of the sets W_i over all iterations.

Note that in step 1 of every iteration i an $O(n)$ *backward* alternating search is performed to compute the set R_i. The key idea of our *improved algorithm* (Algorithm 4) is to perform a cheap *forward* exploration of edges in some iterations in order to discover subsets of the winning set for player Odd. The improved algorithm for 2-player Büchi games differs from the improved algorithm of the previous section in the way the forward exploration is performed. In order to detect a trap for player Even in which player Odd has a winning strategy, we need to consider all successors of every vertex in the forward exploration. Let S be the set of sources of edges entering the winning set of player Odd discovered in the previous iteration, and let $|S| \le k$. The vertices in set S are new candidates to be included in the winning set of player Odd. From these vertices a BFS of depth $\log \ell$ is performed in Step 4.2.2.1 of Algorithm 4. In step 4.2.2.4 we check if the explored subgraph contains a trap for player Even in which player Odd has a winning strategy. If no such trap is detected then one iteration of the classical algorithm is executed. The key for the subquadratic bound of our algorithm is the observation that if step 4.2.2 fails to identify a non-empty winning subset for player Odd, then the set discovered by the following iteration of the classical algorithm has at least $\log \ell$ vertices.

We say that a set of vertices S is an *Even trap* in a graph G if for all $v \in S$, we have $succ(v, G) \subseteq S$ if $v \in V_\Box$, and $succ(v, G) \cap S \neq \emptyset$ if $v \in V_\Diamond$. It is easy

Algorithm 3 Classical Algorithm for 2-player Büchi Games

Input : 2-player Büchi game (G, T). **Output:** $W_\diamond$ and $W_\Box = V \setminus W_\diamond$.
[Steps 1.–4. are the same as in Algorithm 1]
5. $W_\diamond := \bigcup_{k=1}^{i} W_k$

Procedure *One-Iteration-Of-The-Classical-Algorithm*
Input: set $V_i \subseteq V$. **Output:** set $W_{i+1} \subseteq V_i$.
1. $R_i := Attr_\Box(T \cap V_i, V_i)$; 2. $Tr_i := V_i \setminus R_i$; 3. $W_{i+1} := Attr_\diamond(Tr_i, V_i)$

Algorithm 4 Improved Algorithm for 2-player Büchi Games

Input : 2-player Büchi game (G, T). **Output:** $W_\diamond$ and $W_\Box = V \setminus W_\diamond$.
[Steps 1.–3. and 4.1 are the same as in Algorithm 2]
4. **repeat**
  4.2 **else** $(|In(W_i, V_i \cup W_i)| < k)$
    4.2.1 $Tr_i := \emptyset$
    4.2.2 **for each** vertex $v \in Source(W_i, V_i \cup W_i)$
      4.2.2.1 Find the reachable subgraph R_v by a BFS of depth $\log \ell$
      4.2.2.2 Let F_v denote the set of vertices at depth $\log \ell$
      4.2.2.3 $T'_v := \{v \in V_\diamond \cap F_v : succ(v, G_i) \cap R_v = \emptyset\} \cup (V_\Box \cap F_v)$
      4.2.2.4 $R'_v := Attr_\Box((R_v \cap T \cap V_i) \cup T'_v, R_v)$
      4.2.2.5 $Tr_i := Tr_i \cup (R_v \setminus R'_v)$
    4.2.3 **if** $Tr_i \neq \emptyset$ **then** $W_{i+1} := Attr_\diamond(Tr_i, V_i)$
    4.2.4 **else** $W_{i+1} :=$ *One-Iteration-Of-The-Classical-Algorithm*(V_i)
 until $W_i = \emptyset$
5. $W_\diamond := \bigcup_{k=1}^{i} W_k$

to verify that if $P \subseteq V$ then the set $V \setminus Attr_\Box(P, V)$, i.e., the complement of an *Even attractor*, is always an Even trap in the graph induced by vertices in set V. Intuitively, player Odd can prevent player Even from leaving an Even trap, and hence if $S \cap T = \emptyset$ then a trap is included in the winning set of player Odd in the Büchi game (G, T); we call such a set an *Even trap winning for player Odd.*

Lemma 8. *Algorithm 4 correctly computes the sets $\mathcal{W}_\diamond$ and $\mathcal{W}_\Box$.*

Proof. We prove by induction that W_i computed in any iteration of the improved algorithm satisfies $W_i \subseteq \mathcal{W}_\diamond$. The proof is similar to Lemma 4 and differs only for case 2, which we prove now. If Steps 4.2 and 4.2.3 get executed in iteration i, then every nonempty set $R_v \setminus R'_v$ included in the set Tr_{i+1} is an Even trap winning for player Odd in the subgraph of G induced by the set of vertices V_i. It is an Even trap because for every vertex $u \in V_\Box \cap (R_v \setminus F_v)$, we have $succ(u, V_i) \subseteq R_v$, and as a complement of an Even attractor it is an Even trap in the subgraph induced by set R_v. The set $R_v \setminus R'_v$ is moreover an Even trap *winning for player Odd*, because by step 4.2.2.4 all target vertices in the set R_v are included in the set R'_v. ∎

Lemma 9. *Let R_v be a set computed in Step 4.2.2.1. Let S be an Even trap winning for player Odd such that $v \in S$, all vertices of S are reachable from v, and $|S| < \log \ell$. Then $S \subseteq R_v \setminus R'_v$, and hence S is discovered in Step 4.2.2.*

Lemma 10. *The total work in Step 4.1 of Algorithm 4 is $O(n^2/k)$, in Step 4.2.2 it is $O(k\ell n)$, and in Step 4.2.3 it is $O(n)$.*

Lemma 11. *Algorithm 4 solves 2-player binary Büchi games in time $O(n^2/\log n)$.*

Proof. Correctness follows from Lemma 8. The work of Step 4.2.4 is $O(n^2/\log \ell)$ by Lemma 9. The work of Steps 4.1, 4.2.2, and 4.2.4 is $O(n^2/k + k\ell n + n^2/\log \ell)$ by Lemma 10. Take $\ell = n^\varepsilon$ with $0 < \varepsilon < 1$ and $k = \log n$ to get the $O(n^2/\log n)$ upper bound for the total work. By Lemma 10 the work in Step 4.2.3 is $O(n)$, hence the time complexity of Algorithm 4 is $O(n^2/\log n)$. ∎

References

1. R. Alur, T.A. Henzinger, and O. Kupferman. Alternating-time temporal logic. *Journal of the ACM*, 49:672–713, 2002.
2. A. Condon. The complexity of stochastic games. *Information and Computation*, 96:203–224, 1992.
3. L. de Alfaro. Computing minimum and maximum reachability times in probabilistic systems. In *CONCUR'99*, vol. 1664 of *LNCS*, pages 66–81. Springer, 1999.
4. L. de Alfaro and T.A. Henzinger. Concurrent ω-regular games. In *LICS'00*, pages 141–154. IEEE, 2000.
5. L. de Alfaro, T.A. Henzinger, and O. Kupferman. Concurrent reachability games. In *FOCS'98*, pages 564–575. IEEE, 1998.
6. L. de Alfaro and R. Majumdar. Quantitative solution of ω-regular games. In *STOC'01*, pages 675–683. ACM, 2001.
7. E.A. Emerson and C. Jutla. Tree automata, μ-calculus, and determinacy. In *FOCS'91*, pages 368–377. IEEE Computer Society Press, 1991.
8. J. Filar and K. Vrieze. *Competitive Markov Decision Processes*. Springer, 1997.
9. R.A. Howard. *Dynamic Programming and Markov Processes*. Wiley, 1960.
10. M. Jurdziński. Small progress measures for solving parity games. In *STACS'00*, vol. 1770 of *LNCS*, pages 290–301. Springer, 2000.
11. M. Jurdziński, O. Kupferman, and T.A. Henzinger. Trading probability for fairness. In *CSL'02*, vol. 2471 of *LNCS*, pages 292–305. Springer, 2002.
12. R. McNaughton. Infinite games played on finite graphs. *Annals of Pure and Applied Logic*, 65:149–184, 1993.
13. L.S. Shapley. Stochastic games. *Proc. Nat. Acad. Science*, 39:1095–1100, 1953.
14. W. Thomas. On the synthesis of strategies in infinite games. In *STACS'95*, vol. 900 of *LNCS*, pages 1–13. Springer, 1995.
15. W. Thomas. Languages, automata, and logic. In G. Rozenberg and A. Salomaa, eds., *Handbook of Formal Languages*, vol. 3, pages 389–455. Springer, 1997.
16. J. Vöge and M. Jurdziński. A discrete strategy-improvement algorithm for solving parity games. In *CAV'00*, vol. 1855 of *LNCS*, pages 202–215. Springer, 2000.
17. U. Zwick and M. Paterson. The complexity of mean-payoff games on graphs. *Theoretical Computer Science*, 158:343–359, 1996.

Machine Characterizations of the Classes of the W-Hierarchy

Yijia Chen and Jörg Flum

Abteilung für Mathematische Logik, Universität Freiburg,
Eckerstr. 1, 79104 Freiburg, Germany
chen@zermelo.mathematik.uni-freiburg.de, Joerg.Flum@math.uni-freiburg.de

Abstract. We give machine characterizations of the complexity classes of the W-hierarchy. Moreover, for every class of this hierarchy, we present a parameterized halting problem complete for this class.

1 Introduction

Parameterized complexity theory provides a framework for a refined complexity analysis of algorithmic problems that are intractable in general. Central to the theory is the notion of *fixed-parameter tractability*, which relaxes the classical notion of tractability, polynomial time computability, by admitting algorithms whose runtime is exponential, but only in terms of some *parameter* that is usually expected to be small. As a complexity theoretic counterpart, a theory of *parameterized intractability* has been developed. In classical complexity theory, the notion of NP-completeness is central to a nice and simple theory for intractable problems. Unfortunately, the world of parameterized intractability is more complex: there is a big variety of seemingly different classes of parameterized intractability. Nevertheless, it can be argued that the classes W[1], W[2], ... of the W-hierarchy together with some other classes like W[P] correspond to NP in classical complexity theory. In particular, all these classes are defined by parameterized variants of the NP-complete satisfiability problem. Unfortunately, the definition of these classes by means of complete problems makes it not easy to understand them. The authors of [4] tried to remedy this situation by presenting machine characterizations for some of the classes. But as they remarked "it remains an interesting open problem to find natural machine characterizations for the classes W[t] for $t \geq 2$". In this paper we obtain such characterizations.

By definition, a parameterized problem is fixed-parameter tractable, if it is decidable by a (deterministic) algorithm in at most $f(k) \cdot p(n)$ steps for some computable function f and some polynomial p. Here k denotes the size of the parameter and n the size of the input. Problems in any of the "intractable" parameterized complexity classes mentioned so far, also are decidable in at most $f(k) \cdot p(n)$ steps but by a *nondeterministic* algorithm. Thus, a first attempt to characterize one of these intractable classes by machines consists in considering nondeterministic algorithms that perform at most $f(k) \cdot p(n)$ steps but restricting further the number of nondeterministic steps. This approach led to the following

M. Baaz and J.A. Makowsky (Eds.): CSL 2003, LNCS 2803, pp. 114–127, 2003.

results in [4]: A problem is in W[P] if and only if it is decidable in at most $f(k) \cdot p(n)$ steps by an algorithm whose number of nondeterministic steps is bounded in terms of the parameter. Similarly, a problem is in W[1], if in addition the nondeterministic steps are performed at the end of the computation.

Already in [4], nondeterministic random access machines turned out to be the appropriate machine model in order to get clear formulations of the characterizations. In their nondeterministic steps these machines are able to guess natural numbers ($\leq f(k) \cdot p(n)$); these numbers are considered as names of objects. To obtain machine characterizations of the classes of the W-hierarchy we have to consider the corresponding alternating random access machines, but at the same time we have to ensure that the programs only have access to (the properties of) the elements named by the guessed numbers and not to the numbers themselves.

In [3], Cesati and Di Ianni prove that the halting problem p-HPNMT for nondeterministic multitape Turing machines, parameterized by the number of steps, is W[2]-hard by reducing the parameterized dominating set problem to it: the machine first guesses the elements of a dominating set and then checks that they really constitute a dominating set. An analysis of the use of and the access to the guessed elements in this algorithm helped the authors to find the machine characterizations of the classes of the W-hierarchy. In Section 4 we present a (short) proof of membership in W[2] of p-HPNMT by reducing it to a model-checking problem in W[2], a result obtained in [2] by different means.

As the corresponding proof in [4] shows, the machine characterization of W[1] is closely related to the W[1]-completeness of the halting problem for nondeterministic Turing machines. For $t \geq 2$, the W[t]-complete halting problem we present in the last section refers to alternating Turing machines with oracles and it is not so closely related to the corresponding machine characterization but to a logical model-checking problem complete for W[t].

2 Preliminaries

In this section we recall some definitions and results and fix our notations.

2.1. Relational Structures and First-Order Logic. A *vocabulary* τ is a finite set of relation symbols. Each relation symbol has an *arity*. A *(relational) structure* $\mathcal{A}$ of vocabulary τ, or τ-structure, consists of a set A called the *universe*, and an interpretation $R^{\mathcal{A}} \subseteq A^r$ of each r-ary relation symbol $R \in \tau$. We synonymously write $\bar{a} \in R^{\mathcal{A}}$ or $R^{\mathcal{A}}\bar{a}$ to denote that the tuple $\bar{a} \in A^r$ belongs to the relation $R^{\mathcal{A}}$. For example, we view a *directed graph* as a structure $\mathcal{G} = (G, E^{\mathcal{G}})$, whose vocabulary consists of one binary relation symbol E. $\mathcal{G}$ is an (undirected) *graph*, if $E^{\mathcal{G}}$ is irreflexive and symmetric.

The class of all first-order formulas is denoted by FO. They are built up from atomic formulas using the usual boolean connectives and existential and universal quantification. Recall that *atomic formulas* are formulas of the form $x = y$ or $Rx_1 \dots x_r$, where $x, y, x_1, \dots, x_r$ are variables and R is an r-ary relation symbol. For $t \geq 1$, Σ_t denotes the class of all first-order formulas of the form

$$\exists x_{11} \dots \exists x_{1k_1} \forall x_{21} \dots \forall x_{2k_2} \; \dots \; Qx_{t1} \dots Qx_{tk_t} \; \psi,$$

where $Q = \forall$ if t is even and $Q = \exists$ otherwise, and where ψ is quantifier-free. Π_t-formulas are defined analogously starting with a block of universal quantifiers. Let $t, u \geq 1$. A formula φ is $\Sigma_{t,u}$, if it is Σ_t and all quantifier blocks after the leading existential block have length $\leq u$.

If Φ is a class of formulas, then $\Phi[\tau]$ denotes the class of all formulas of vocabulary τ in Φ. If $\mathcal{A}$ is a τ-structure and $\varphi \in \mathrm{FO}[\tau]$ a sentence, i.e., a formula without free variables, then we write $\mathcal{A} \models \varphi$ to denote that $\mathcal{A}$ satisfies φ.

The proof of the following lemma is easy.

Lemma 1. *Let $\varphi_1(\bar{x}), \ldots, \varphi_m(\bar{x})$ and $\psi_1(\bar{x}, \bar{y}), \ldots, \psi_m(\bar{x}, \bar{y})$ be formulas in $\mathrm{FO}[\tau]$, where $\bar{x} = x_1 \ldots x_r$ and $\bar{y} = y_1 \ldots y_s$ are sequences of variables that have no variable in common. Assume $Q_1, \ldots, Q_r, Q'_1, \ldots, Q'_s \in \{\forall, \exists\}$. If $\mathcal{A}$ is a τ-structure with $\mathcal{A} \models \forall \bar{x} \neg(\varphi_i \wedge \varphi_j)$ for $i \neq j$, then $\mathcal{A}$ satisfies both or none of the formulas*

$$Q_1 x_1 \ldots Q_r x_r \bigwedge_{i=1}^{m} (\varphi_i \rightarrow Q'_1 y_1 \ldots Q'_s y_s \psi_i),$$

$$Q_1 x_1 \ldots Q_r x_r Q'_1 y_1 \ldots Q'_s y_s \bigwedge_{i=1}^{m} (\varphi_i \rightarrow \psi_i).$$

2.2. Parameterized Complexity. A *parameterized problem* is a set $Q \subseteq \Sigma^* \times \Pi^*$, where Σ and Π are finite alphabets. If $(x, y) \in \Sigma^* \times \Pi^*$ is an instance of a parameterized problem, we refer to x as the *input* and to y as the *parameter.* We usually denote the length of the input string x by n and the length of the parameter string y by k.

For example, for a class S of structures and a class L of first-order formulas,

$$\text{p-MC(S, L)} := \{(\mathcal{A}, \varphi) \mid \mathcal{A} \in \mathrm{S},\ \varphi \text{ a sentence in L, and } \mathcal{A} \models \varphi\}$$

is the *parameterized model-checking problem for* S *and* L; mostly, for easier readability, we present parameterized problems in the following form:

p-MC(S,L)	*Input:* $\mathcal{A} \in \mathrm{S}$.
	Parameter: φ, a sentence in L.
	Problem: $\mathcal{A} \models \varphi$?

Definition 1. *A parameterized problem $Q \subseteq \Sigma^* \times \Pi^*$ is* fixed-parameter tractable, *if there are a computable function $f : \mathbb{N} \rightarrow \mathbb{N}$, a polynomial p, and an algorithm that, given a pair $(x, y) \in \Sigma^* \times \Pi^*$, decides if $(x, y) \in Q$ in at most $f(k) \cdot p(n)$ steps.*

FPT *denotes the complexity class consisting of all fixed-parameter tractable parameterized problems.*

Complementing the notion of fixed-parameter tractability, there is a theory of parameterized intractability. It is based on the following notion of reduction:

Definition 2. *An* FPT-reduction *from the parameterized problem $Q \subseteq \Sigma^* \times \Pi^*$ to the parameterized problem $Q' \subseteq (\Sigma')^* \times (\Pi')^*$ is a mapping $R : \Sigma^* \times \Pi^* \rightarrow (\Sigma')^* \times (\Pi')^*$ such that:*

1. *For all* $(x, y) \in \Sigma^* \times \Pi^*$*:* $(x, y) \in Q \iff R(x, y) \in Q'$.
2. *There exists a computable function* $g : \mathbb{N} \to \mathbb{N}$ *such that for all* $(x, y) \in \Sigma^* \times \Pi^*$*, say with* $R(x, y) = (x', y')$*, we have* $k' \leq g(k)$ *(where* $k = |y|$ *and* $k' = |y'|$*).*
3. *There exist a computable function* $f : \mathbb{N} \to \mathbb{N}$ *and a polynomial* p *such that* R *is computable in time* $f(k) \cdot p(n)$*.*

We write $Q \leq^{\text{FPT}} Q'$ if there is an FPT-reduction from Q to Q' and set $[Q]^{\text{FPT}} := \{Q' \mid Q' \leq^{\text{FPT}} Q\}$. For a class C of parameterized problems, we let $[\mathrm{C}]^{\text{FPT}} := \bigcup_{Q \in \mathrm{C}} [Q]^{\text{FPT}}$.

Denote by GRAPH the class of all finite graphs and by STR the class of all finite structures. The parameterized complexity classes W[1], W[2], ... of the W-hierarchy are defined as the closure of a family of parameterized problems under FPT-reductions. For W[t], the defining family of problems consists of parameterized versions of the satisfiability problems for circuits of *weft* t (and varying depth).

For the purposes of this paper, the most appropriate way to introduce the complexity classes of the W-hierarchy is the following (cf. [7], [8]):

Definition 3. *For* $t \geq 1$*,* W[t] *is the class of all parameterized problems that, for some* $u \geq 1$*, are* FPT*-reducible to* p-MC(GRAPH, $\Sigma_{t,u}$)*, that is,*

$$\mathrm{W}[t] = [\{\text{p-MC}(\mathrm{GRAPH}, \Sigma_{t,u}) \mid u \geq 1\}]^{\text{FPT}}.$$

The following equivalent characterization of W[t] is well-known:

Proposition 1. $\mathrm{W}[t] = [\{\text{p-MC}(\mathrm{STR}, \Sigma_{t,u}[\tau]) \mid u \geq 1, \tau \textit{ vocabulary}\}]^{\text{FPT}}$.

At various places of this paper we tacitly make use of the following remark:

Remark 1. Sometimes, in order to show that a given parameterized problem is in W[t], we will present a reduction to p-MC(STR, $\Sigma_{t,u}[\tau]$) for some vocabulary τ also containing a fixed finite number of constant symbols or even, we will consider a reduction where the number of constants depends on the parameter. This will allow to express properties in a more readable fashion. If we have a fixed finite number of constant symbols, they can be eliminated by using appropriate unary relations that are singletons; then, in the $\Sigma_{t,u}$-formula to be defined, the constants are replaced by variables that are existentially quantified (in the first block) and get their right value using the corresponding relations. If the number of constants symbols depends on the parameter, e.g., we use constants for $0, 1, \ldots, k$, in order to stay within a fixed vocabulary (independent of k) these constants can be eliminated by a singleton relation for 0 and the binary successor relation on $\{0, 1, \ldots, k\}$ and again by existentially quantified variables.

Sometimes we refer to the complexity classes of the A-hierarchy (cf. [8]):

Definition 4. *For* $t \geq 1$*,* $\mathrm{A}[t] = [\text{p-MC}(\mathrm{GRAPH}, \Sigma_t)]^{\text{FPT}}$.

3 Machine Characterization

In [4], machine characterizations of the classes W[1], A[t] for $t \geq 1$, and W[P] using nondeterministic and alternating random access machines (RAMs) were presented. The nondeterministic RAMs are based on the standard random access machines (cf. [10]). The model was non-standard when it came to nondeterminism. Instead of allowing the machines to nondeterministically choose one bit, or an instruction of the program to be executed next, the authors allowed them to nondeterministically choose a natural number, more precisely, there was an additional instruction "GUESS i j" whose semantics was: Guess a natural number less than or equal to the number stored in register i and store it in register j. In [4] it was remarked: "While this form of nondeterminism may seem unnatural at first sight, we would like to argue that it is very natural in many typical 'applications' of nondeterminism. For example, a nondeterministic algorithm for finding a clique in a graph guesses a sequence of vertices of the graph and then verifies that these vertices indeed form a clique. Such an algorithm is much easier described on a machine that can guess the numbers representing the vertices of a graph at once, rather than guessing their bits."

The alternating RAMs, in addition to the instructions of the form "GUESS i j" (denoted by "EXISTS i j" in the context of alternating machines) also had "FORALL i j" instructions. The semantics was defined as usually for alternating machines. The computations of alternating machines suitable for A[t] have a parameter-bounded final part which contains all the EXISTS- and FORALL-instructions and have at most $t-1$ alternations (see [4] for the precise statement). For W[t] it seems not to suffice (see Section 6) to bound the length of the blocks without alternation, but we have to restrict the access to the numbers guessed in the EXISTS- and FORALL-instructions: as just remarked in the algorithm for finding a clique, we view these numbers as labels of certain objects and the type of machine we are going to introduce only has access to the properties of these objects and not directly to the labels.

We turn to the precise definition of the random access machines we are going to use and that we call W-RAMs. A W-RAM has the

- the *standard registers* $0, 1, \ldots$, their contents are denoted by $r_0, r_1, \ldots$, respectively.
- the *guess registers* $0, 1, \ldots$, their contents are denoted by $g_0, g_1, \ldots$, respectively.

All registers have initial value 0. Often we denote g_{r_i}, i.e., the contents of the guess register whose index is the content of the ith standard register, by $g(r_i)$.

The W-RAM has all the standard instructions for the standard registers (cf. Section 2.6 of [10]; e.g., the arithmetic operations are addition, subtraction, and division by two (rounded off)). Moreover, it has four additional instructions:

Instruction	Semantics
EXISTS $\uparrow j$	guess a natural number $\leq r_0$; store it in the r_jth guess register
FORALL $\uparrow j$	guess a natural number $\leq r_0$; store it in the r_jth guess register
JG= i j c	if $g(r_i) = g(r_j)$, then jump to the instruction with label c
JG0 i j c	if $r_{\langle g(r_i), g(r_j) \rangle} = 0$, then jump to the instruction with label c.

Here, $\langle\ ,\ \rangle : \mathbb{N} \times \mathbb{N} \to \mathbb{N}$ is any simple coding of ordered pairs of natural numbers by natural numbers such that $\langle i, j\rangle \leq (1+\max\{i, j\})^2$ and $\langle 0, 0\rangle = 0$. Of course, the semantics of the EXISTS- and FORALL-instructions are the same, but we view them as existential and universal instructions, respectively. All other instructions are said to be deterministic. If the machine stops, it *accepts* its input, if $r_0 = 0$, otherwise it rejects it.

The following lemma, whose proof is immediate, is crucial for the main theorem of this section; it shows that the contents of the standard registers depend only on the sequence of executed instructions:

Lemma 2. *Assume that, for a given input, we have two (partial) computations on a W-RAM. If the same sequence of instructions is carried out in both computations, then the contents of the standard registers will be the same.*

Definition 5. *A program $\mathbb{P}$ for a W-RAM is an* AW-program, *if there is a computable function f and a polynomial p such that for every input (x, y) with $|x| = n$ and $|y| = k$ the program $\mathbb{P}$ on every run*

1. *performs at most $f(k) \cdot p(n)$ steps;*
2. *at most $f(k)$ steps are existential or universal;*
3. *at most the first $f(k) \cdot p(n)$ standard registers are used;*
4. *at every point of the computation the registers contain numbers $\leq f(k) \cdot p(n)$.*

The promised machine characterization of W[t] reads as follows:

Theorem 1. *Let Q be a parameterized problem and $t \geq 1$. Then Q is in* W[t] *if and only if there is a $u \geq 1$ and there are a computable function h and an AW-program $\mathbb{P}$ for a W-RAM such that $\mathbb{P}$ decides Q and such that for every run of $\mathbb{P}$ on an instance (x, y) of Q as input (with $|y| = k$)*

- *all existential and universal steps are among the last $h(k)$ steps of the computation,*
- *there are at most $t - 1$ alternations between existential and universal states, and the first guess step is existential,*
- *every block without alternations, besides the first one, contains at most u guess steps.*

Proof. Assume first that $Q \in$ W[t], then $Q \leq^{\mathrm{FPT}}$ p-MC(GRAPH, $\Sigma_{t,u}$) for some $u \geq 1$. Hence there are computable functions f and g, a polynomial $p \in \mathbb{N}[x]$, and an algorithm $\mathbb{A}$ assigning to every (x, y), in time $\leq f(k) \cdot p(n)$, a graph $\mathcal{G} = \mathcal{G}_{x,y}$ and a sentence $\varphi = \varphi_{x,y} \in \Sigma_{t,u}$, say,

$$\varphi = \exists x_{11} \ldots \exists x_{1k_1} \forall x_{21} \ldots \forall x_{2k_2} \ldots Q x_{t1} \ldots Q x_{tk_t} \psi,$$

with $k_2, \ldots, k_t \leq u$, with $|\varphi| \leq g(k)$, and with a quantifier-free ψ, such that

$$Qxy \iff \mathcal{G} \models \varphi.$$

The claimed AW-program $\mathbb{P}$ for a W-RAM, on input (x, y), proceeds as follows:

1. It computes the graph $\mathcal{G} = (G, E^{\mathcal{G}})$, say with $G = \{1, \ldots, m\}$, and stores its adjacency matrix in the standard registers: $r_{\langle i,j \rangle} = 0 \iff E^{\mathcal{G}}ij$.
2. It computes φ.
3. It checks whether $\mathcal{G} \models \varphi$.

To carry out point 3, the program $\mathbb{P}$, using the EXISTS- and FORALL-instructions guesses the values of the quantified variables. Then, it checks the quantifier-free part using the JG=- and JG0-instructions. The number of steps needed for point 3 can be bounded by $h(k)$ for some computable h. Hence, all existential and universal steps are among the last $h(k)$ steps of the computation.

Now assume that $\mathbb{P} = (\pi_1, \ldots, \pi_m)$ is an AW-program deciding Q and that $u \geq 1$, h, and $\mathbb{P}$ have the properties stated at the right side of the equivalence claimed in the theorem. For the program $\mathbb{P}$ choose the function f and the polynomial p according to Definition 5. We claim that $Q \in \mathrm{W}[t]$. By Proposition 1, it suffices to show that $Q \leq^{\mathrm{FPT}}$ p-MC(STR, $\Sigma_{t,2\cdot u}[\tau]$) for some τ. The set of instruction numbers of $\mathbb{P}$ is $\{1, \ldots, m\}$, more precisely, π_i is the instruction of $\mathbb{P}$ with instruction number i. We denote instruction numbers (= potential contents of the program counter) by $c, c_1, \ldots$ and finite sequences of instruction numbers by $\bar{c}$. $\ell(\bar{c})$ denotes the last instruction number of the sequence $\bar{c}$, and $[\bar{c})$ the sequence obtained from $\bar{c}$ by omitting its last member. Fix an instance (x, y) of Q. Let

$$C := \bigcup_{0 \leq r \leq h(k)-1} \{1, \ldots, m\}^r \quad \text{and} \quad N := \{0, 1, \ldots, f(k) \cdot p(n)\}.$$

with $k = |y|$. We look for a structure $\mathcal{A}$ and a $\Sigma_{t,2\cdot u}$-sentence φ such that

$$\mathbb{P} \text{ accepts } (x, y) \iff \mathcal{A} \models \varphi.$$

Let $\bar{c}_0$ be the sequence of instruction numbers of the deterministic part of $\mathbb{P}$ on input (x, y) ending with the instruction number of the first existential instruction. As universe A of $\mathcal{A}$ we take $A := C \cup N$. Moreover, in $\mathcal{A}$ there are the binary relation $\leq^{\mathcal{A}}$, the natural ordering on N, and ternary relations $R^{\mathcal{A}}$ and $T^{\mathcal{A}}$ defined by

$$\begin{aligned} R^{\mathcal{A}}\bar{c}ij \iff & \bar{c} \in C,\ i, j \in N, \text{and if } \mathbb{P}, \text{ on input } (x, y), \text{ carries out} \\ & \text{the sequence of instructions } [\bar{c}_0\, \bar{c}), \text{ then } r_i = j. \\ T^{\mathcal{A}}\bar{c}ij \iff & \bar{c} \in C,\ i, j, \langle i, j \rangle \in N, \text{and if } \mathbb{P}, \text{ on input } (x, y), \text{ carries out} \\ & \text{the sequence of instructions } [\bar{c}_0\, \bar{c}), \text{ then } r_{\langle i,j \rangle} = 0. \end{aligned}$$

Moreover, we have a constant for 0. This finishes the definition of $\mathcal{A}$, which can be constructed within the time allowed by an FPT-reduction.

We turn to the definition of φ. First, we fix $\bar{c} \in C$: Let $i = i(\bar{c})$ be the number of blocks without alternations of the sequence of instructions determined by $[\bar{c}_0\bar{c})$;

let $j = j(\bar{c})$ be the number of guesses in the ith block. If $i \leq t$ and if each block, besides the first one, has length $\leq u$, we introduce a formula

$$\varphi_{\bar{c}}(\bar{x}_1, \ldots, \bar{x}_{i-1}, x_{i,1} \ldots x_{i,j}) \tag{1}$$

where $\bar{x}_1 := x_{1,1}, \ldots, x_{1,h(k)}$ and $\bar{x}_s := x_{s,1} \ldots x_{s,u}$ for $s = 2, \ldots, i-1$ with the intuitive meaning

> if a partial run of $\mathbb{P}$ has $\bar{c}_0\,\bar{c}$ as sequence of instructions numbers and if every variable $x_{i',j'}$ displayed in (1) has, as value, the j'th guess of the i'th block (and $x_{i',j'} = 0$, if there was no such guess), then there is an accepting continuation of this run.

Then, for the empty sequence $\emptyset$ of instruction numbers and $\varphi := \varphi_\emptyset$, we have

$$\mathbb{P} \text{ accepts } (x, y) \iff \mathcal{A} \models \varphi.$$

For $\bar{c} \in C$ of maximal length, $|\bar{c}| = h(k) - 1$, we set (recall that, by definition, a computation accepts its input, if $r_0 = 0$ at the end of the computation)

$$\varphi_{\bar{c}} := \begin{cases} \text{TRUE}, & \text{if } \pi_{\ell(\bar{c})} = \text{STOP (i.e., } \pi_{\ell(\bar{c})} \text{ is the STOP-instruction) and } R\bar{c}\,0\,0 \\ \text{FALSE}, & \text{otherwise.} \end{cases}$$

If $\bar{c} \in C$ and $|\bar{c}| < h(k) - 1$, we assume that $\varphi_{\bar{c}'}$ has already been defined for all $\bar{c}'$ with $|\bar{c}| < |\bar{c}'|$. The definition depends on the type of the instruction of $\mathbb{P}$ with instruction number $\ell(\bar{c})$.

If $\pi_{\ell(\bar{c})} = \text{STOP}$, then again

$$\varphi_{\bar{c}} := \begin{cases} \text{TRUE}, & \text{if } R\bar{c}\,0\,0 \\ \text{FALSE}, & \text{otherwise.} \end{cases}$$

The definition of $\varphi_{\bar{c}}$ is simple for the standard instructions, e.g., if $\pi_{\ell(\bar{c})} = \text{STORE}\uparrow u$ (i.e., "$\pi_{\ell(\bar{c})} = r_{r_u} := r_0$"), then $\varphi_{\bar{c}} := \varphi_{\bar{c}\,\ell(\bar{c})+1}$.

We give the definitions for the new instructions: If $\pi_{\ell(\bar{c})} = \text{EXISTS}\uparrow v$, then (for $i = i(\bar{c})$ and $j = j(\bar{c})$)

$$\varphi_{\bar{c}} := \begin{cases} \exists x_{i,j+1} \exists y (R\bar{c}\,0\,y \wedge x_{i,j+1} \leq y \wedge \varphi_{\bar{c}\,\ell(\bar{c})+1}), & \text{if } i = 1 \text{ or } (i \text{ is odd and } j < u) \\ \exists x_{i+1,1} \exists y (R\bar{c}\,0\,y \wedge x_{i+1,1} \leq y \wedge \varphi_{\bar{c}\,\ell(\bar{c})+1} & \\ \qquad \wedge\, x_{i,j+1} = 0 \wedge \ldots \wedge x_{i,u} = 0), & \text{if } i \text{ is even, } i < t, \text{ and } j \leq u \\ \text{FALSE}, & \text{otherwise.} \end{cases}$$

The definition is similar for instructions of the type FORALL $\uparrow v$, but then the variables are quantified universally.

Assume $\pi_{\ell(\bar{c})} = \text{JG=}\ v\ w\ c$. We need $g(r_v)$ and $g(r_w)$. Determine the actual contents v_0 and w_0 of the vth and the wth standard register, i.e., v_0 and w_0 with $R^{\mathcal{A}}\bar{c}vv_0$ and $R^{\mathcal{A}}\bar{c}ww_0$. Consider the sequence of instructions given by $\bar{c}$ and determine the last instructions in it of the form FORALL $\uparrow z$ or EXISTS $\uparrow z$ such that at that time $r_z = v_0$, say, it is the j_0th guess in the i_0th block. Similarly, let the j_1th guess in the i_1th block be the last instruction of the form FORALL $\uparrow z$ or EXISTS $\uparrow z$ such that at that time $r_z = w_0$ (the case that such instructions do no exist is treated in the obvious way). Then set

$$\varphi_{\bar{c}} := (x_{i_0,j_0} = x_{i_1,j_1} \to \varphi_{\bar{c}\,c}) \wedge (\neg x_{i_0,j_0} = x_{i_1,j_1} \to \varphi_{\bar{c}\,\ell(\bar{c})+1}).$$

Assume $\pi_{\ell(\bar{c})} =$ JG0 u v c. As in the preceding case, let x_{i_0,j_0} and x_{i_1,j_1} denote the actual values of the r_uth and the r_vth guess register, respectively. Then set

$$\varphi_{\bar{c}} := (T\bar{c}x_{i_0,j_0}x_{i_1,j_1} \to \varphi_{\bar{c}\,c}) \wedge (\neg T\bar{c}x_{i_0,j_0}x_{i_1,j_1} \to \varphi_{\bar{c}\,\ell(\bar{c})+1}).$$

As already mentioned above, we set $\varphi := \varphi_{\emptyset}$. By Lemma 1, one easily verifies that φ is equivalent to a $\Sigma_{t,2\cdot u}$-formula. Clearly, the size of φ can be bounded in terms of $h(k)$ and

$$\begin{aligned} Qxy &\iff \mathbb{P} \text{ accepts } (x, y) \\ &\iff \mathcal{A} \models \varphi, \end{aligned}$$

which gives the desired reduction showing that $Q \in$ W$[t]$. □

4 W[2] and Multitape Machines

Among the many known W[1]-complete problems, of course, the most generic one is the halting problem p-HPN for nondeterministic Turing machines (cf. [1]):

> p-HPN *Input:* A nondeterministic Turing machine M.
> *Parameter:* $k \in \mathbb{N}$.
> *Problem:* Does M accept the empty word in at most k steps?

Surprisingly, the same problem for nondeterministic Turing machines with several tapes is W[2]-complete (cf. [3] and [2]). Mike Fellows pointed out this result to the second author and encouraged him to look for a machine characterization of W[2]. In this section we present a simple proof that the halting problem for multitape machines is in W[2]; it avoids the equality W[2] = W*[2] (cf. [5]), used in [2] and for which no simple proof is known.

Proposition 2. *For some vocabulary* τ, p-HPNMT $\leq^{\text{FPT}}$ p-MC(STR, $\Sigma_{2,3}[\tau]$).

Here p-HPNMT denotes the halting problem for nondeterministic multitape Turing machines:

> p-HPNMT *Input:* A nondeterministic Turing machine M with an arbitrary finite number of tapes.
> *Parameter:* $k \in \mathbb{N}$.
> *Problem:* Does M accept the empty word in at most k steps?

Proof. Let M be a nondeterministic Turing machine with w_0 (work) tapes. We aim at a structure $\mathcal{A} = \mathcal{A}_{M,k}$ and a $\Sigma_{2,3}$-sentence $\varphi = \varphi_{M,k}$ such that

$$(M, k) \in \text{p-HPNMT} \iff \mathcal{A} \models \varphi.$$

Let Σ and Q be the alphabet and the set of states of M, respectively. The instructions of M have the form

$$q\,(a_1, \ldots, a_{w_0}) \to q'(a'_1, \ldots, a'_{w_0})\,(h_1, \ldots, h_{w_0})$$

where $q, q' \in Q$, $a_1, \ldots, a_{w_0}, a'_1, \ldots, a'_{w_0} \in \Sigma \cup \{*\}$ ($*$ is the blank symbol) and $h_1, \ldots, h_{w_0} \in \{-1, 0, 1\}$. Let T be the set of tuples $(b_1, \ldots, b_{w_0}) \in (\Sigma \cup \{*\})^{w_0}$ and H the set of tuples $(h_1, \ldots, h_{w_0}) \in \{-1, 0, 1\}^{w_0}$ occurring in instructions.

The structure $\mathcal{A}$ has the universe

$$A := Q \cup (\Sigma \cup \{*\}) \cup \{-1, 0, 1, \ldots, \max\{w_0, k\}\} \cup T \cup H.$$

We need the natural ordering relation $\leq^{\mathcal{A}}$ on $\{-1, 0, 1, \ldots, \max\{w_0, k\}\}$, the 5-ary relation $D^{\mathcal{A}}$ (the "transition relation") and the ternary relation $P^{\mathcal{A}}$ (the "projection relation") defined by

$$\begin{aligned} D^{\mathcal{A}}qtq't'h &\iff qt \to q't'h \text{ is an instruction of } M \\ P^{\mathcal{A}}wba &\iff 1 \leq w \leq w_0,\ b \in T \cup H,\ b = (b_1, \ldots, b_{w_0}), \text{ and } b_w = a. \end{aligned}$$

Moreover, we have constant symbols for the initial state q_0, the accepting state q_{acc}, for $*$, and for $-1, w_0, 0, 1, \ldots, k$ (cf. Remark 1).

The formula φ we aim at, will express that there is an accepting run of length $\leq k$ (w.l.o.g. of length $= k$); among others, it will contain the variables $q_i, t_i, q'_i, t'_i, h_i$ for $i = 1, \ldots, k$, in fact, $q_i\, t_i \to q'_i\, t'_i\, h_i$ represents the ith instruction applied in the run; φ is obtained by existentially quantifying all variables in

$$(\varphi_{\text{init}}(q_1, t_1) \wedge \bigwedge_{i=1}^{k} Dq_i\, t_i\, q'_i\, t'_i\, h_i \wedge \bigwedge_{i=1}^{k-1} q'_i = q_{i+1} \wedge q_k = q_{\text{acc}} \wedge \psi),$$

where $\varphi_{\text{init}}(q_1, t_1) := (q_1 = q_0 \wedge \forall w(1 \leq w \leq w_0 \to Pwt_1*))$ and where ψ is a universal formula expressing that the sequence of instructions can be applied: For this purpose, for $i = 1, \ldots, k$, we introduce quantifier-free formulas

$$\varphi_i^L(w, p, x, \bar{v}_i) \text{ and } \varphi_i^S(w, p, \bar{v}_i)$$

with $\bar{v}_i := q_1, t_1, q'_1, t'_1, h_1, \ldots, q_{i-1}, t_{i-1}, q'_{i-1}, t'_{i-1}, h_{i-1}$ and with the meaning

> if, starting with the empty tape, the sequence of instructions $\bar{v}_i$ has been carried out, then the pth cell of the wth tape contains the letter x,

and

> if, starting with the empty tape, the sequence of instructions $\bar{v}_i$ has been carried out, then the head of the wth tape scans the pth cell,

respectively. Then, as ψ, we can take

$$\psi := \forall w \forall p \forall x \bigwedge_{i=1}^{k} \Big((\varphi_i^L(w,p,x,\bar{v}_i) \wedge \varphi_i^S(w,p,\bar{v}_i)) \to Pwt_i x \Big).$$

The simultaneous definition of φ_i^L and φ_i^P by induction on i is routine, e.g.,

$$\varphi_1^L(w,p,x) := (1 \leq w \leq w_0 \wedge 1 \leq p \leq k \wedge x = *);$$

$$\varphi_{i+1}^L(w,p,x,\bar{v}_{i+1}) := (1 \leq w \leq w_0 \wedge 1 \leq p \leq k) \wedge$$
$$((\neg\varphi_i^S(w,p,\bar{v}_i) \wedge \varphi_i^L(w,p,x,\bar{v}_i)) \vee (\varphi_i^S(w,p,\bar{v}_i) \wedge Pwt_i'x)).$$

One easily verifies that φ is (logically equivalent to) a $\Sigma_{2,3}$-sentence. □

Corollary 1. *The halting problem* p-HPNMT *for multitape machines is* W[2]*-complete.*

Proof. By Proposition 1 and the preceding proposition, p-HPNMT is in W[2]. To show that p-HPNMT is W[2]-hard we recall the proof of [3] that the parameterized dominating set problem p-DS can be reduced to p-HPNMT. The essential problem is to obtain a corresponding machine in the time allowed by an FPT-reduction.

Suppose $(\mathcal{G}, k)$ is an instance of p-DS with $G = \{a_1, \ldots, a_n\}$. Let the multitape machine M have $n+1$ tapes, numbered by 0 to n, and let $\Sigma := G \cup \{\text{yes}\}$ be its alphabet. In the first step all heads move one cell to the right. In the next $2 \cdot k$ steps, only the 0th head is active: it (nondeterministically) writes k elements of G on its tape, say $b_1, \ldots, b_k$ (the elements of the intended dominating set), and goes back to the cell containing b_1. In the next k steps the 0th head again reads these elements; at the same time, in the jth step, the ith head checks whether $a_i = b_j$ or Ea_ib_j; in the affirmative case the ith head prints "yes" and moves to the right, in the negative case it neither prints nor moves; finally, the machine moves all heads one cell to the left and accepts, if the heads on the tapes number 1 to n read "yes". Clearly,

$$(\mathcal{G}, k) \in \text{p-DS} \iff (M, 3 \cdot (k+1)) \in \text{p-HPNMT}.$$ □

5 Complete Halting Problems for the W-Hierarchy

As already mentioned at the beginning of Section 4, the halting problem for nondeterministic Turing machines, parameterized by the number of steps, is a quite generic W[1]-complete problem. The corresponding halting problems for alternating machines yield complete problems for the classes of the A-hierarchy. Indeed, for $t \geq 1$, we have $A[t] = [\text{p-HPA}_t]^{\text{FPT}}$ (cf. [8]), where

p-HPA$_t$	*Input:*	An alternating Turing machine M whose initial state is existential.
	Parameter:	$k \in \mathbb{N}$.
	Problem:	Does M accept the empty word in at most k steps with at most $t-1$ alternations?

One would expect that in order to obtain complete problems for the classes of the W-hierarchy one has to bound the number of steps of all non-alternating blocks but the first one. More precisely, for $t, u \geq 1$, consider the following problem:

p-HPA$_{t,u}$ *Input:*	An alternating Turing machine M whose initial state is existential.
Parameter:	$k \in \mathbb{N}$.
Problem:	Does M accept the empty word in at most k steps with at most $t-1$ alternations, where every block of steps without alternation, besides the first one, has length $\leq u$?

Essentially along the lines of the proof of p-HPA$_t \in$ A$[t]$ in [8], one can show that p-HPA$_{t,u} \in$ W$[t]$. But the corresponding hardness proof does not seem to go through. Recall that in order to establish that p-MC(GRAPH, Σ_t) $\leq^{\text{FPT}}$ p-HPA$_t$, given a graph and a Σ_t-sentence φ, one constructs an alternating Turing machine that first associates values to the quantifiers in φ (the values for existentially quantified variables are chosen in existential states, the values for universally quantified variables in universal states) and that then checks whether the selected variables satisfy the quantifier-free part of φ, the *quantifier-free check*. Of course, a $\Sigma_{t,u}$-prefix would yield an alternation sequence according to p-HPA$_{t,u}$, but the number of steps needed for the quantifier-free check cannot be bounded in terms of t and u as required in p-HPA$_{t,u}$, it depends on φ. Therefore, we add suitable oracles to the Turing machines that carry out the quantifier-free check in a single step. Of course, we have to add them in such a way that the corresponding halting problem still is in W$[t]$.

As in an alternating machine, the set Q of states of an *alternating Turing machine M with oracle* is the disjoint union of the set of universal states Q_u and the set of existential states Q_e. The "oracle states" $q_?$, q_y, and q_n are all contained in Q_u or all in Q_e. Let Σ be the vocabulary of M and let $O \subseteq \Sigma^*$ be a language, the "oracle language". M^O, the machine M with oracle O, in state $q_?$ will check if the word to the left of the cell scanned by its head is in O and will change to the "yes state" q_y or to the "no state" q_n (without printing a letter nor moving its head).

For a graph $\mathcal{G} = (G, E^{\mathcal{G}})$ and a first-order formula $\psi(x_1, \ldots, x_p)$, let $O(\mathcal{G}, \psi)$ be the following set of words over G, $O(\mathcal{G}, \psi) \subseteq G^*$:

$$O(\mathcal{G}, \psi) := \{a_1 \ldots a_p \mid \mathcal{G} \models \psi(a_1, \ldots, a_p)\}.$$

Due to space limitations we omit the proof of the following theorem:

Theorem 2. *For $t \geq 1$,*

$$\text{W}[t] = [\{\text{p-HPAO}_{t,u} \mid u \geq 1\}]^{\text{FPT}},$$

where p-HPAO$_{t,u}$ is the parameterized halting problem for alternating Turing machines with oracle:

p-HPAO$_{t,u}$ *Input:*	An alternating Turing machine M with oracle whose initial state is existential and a graph $\mathcal{G} = (G, E^{\mathcal{G}})$ such that G is a subset of the alphabet of M.
Parameter:	$k \in \mathbb{N}$ and a quantifier-free formula $\psi(x_1, \ldots, x_p)$.
Problem:	Does $M^{O(\mathcal{G},\psi)}$ accept the empty word in at most k steps with at most $t-1$ alternations, where every block of steps without alternation, besides the first one, has length $\leq u$?

6 Conclusions

Most standard complexity classes like LOGSPACE, NLOGSPACE, PTIME, NPTIME, the classes of the polynomial hierarchy, or PSPACE have definitions in terms of machines. By contrast, originally nearly all parameterized complexity classes containing intractable problems were defined via complete problems. In [4], machine characterizations of W[1], W[P], and of A[t] for $t \geq 1$ were presented; in this paper we derive such characterizations for W[t] for $t \geq 2$.

As mentioned at the beginning of Section 3, in [4] AW-programs for alternating RAMs were introduced, which have "unrestricted access" to the guessed numbers. For $t \geq 1$, denote by L[t] the class that satisfies Theorem 1 if we replace W-RAM by alternating RAM, i.e., L[t] is the class of parameterized problems Q such that there is a computable function h and an AW-program $\mathbb{P}$ for an alternating RAM deciding Q such that for every run of $\mathbb{P}$ on an instance (x, y) of Q as input (with $k = |y|$)

- all existential and universal steps are among the last $h(k)$ steps of the computation,
- there are at most $t-1$ alternations between existential and universal states, and the first guess step is existential,
- every block without alternations, besides the first one, contains at most u guess steps.

Clearly, by [4] and Theorem 1, we have

$$\mathrm{W}[t] \subseteq \mathrm{L}[t] \subseteq \mathrm{A}[t].$$

We do not know, if W[t] = L[t] or if L[t] = A[t]. Let an *f-vocabulary* be a finite set of relation symbols, function symbols, and constant symbols. By an appropriate refinement of the proof of Theorem 1 one can show:

Theorem 3. *For $t \geq 1$,*

$$\mathrm{L}[t] = [\{\text{p-MC}(\mathrm{STR}, \Sigma_{t,u}[\tau]) \mid u \geq 1, \tau \textit{ f-vocabulary}\}]^{\mathrm{FPT}}.$$

For every f-vocabulary τ, the classical problem MC(STR, $\Sigma_{t,u}[\tau]$) is in NP. Therefore, FPT $\subseteq$ W[t] $\subseteq$ L[t] $\subseteq$ para-NP (compare [9] for the definition of para-NP). As FPT $\neq$ para-NP is equivalent to P $\neq$ NP (cf. [9]), we obtain from the preceding theorem:

Corollary 2. *If* $\mathrm{W}[t] \neq \mathrm{L}[t]$ *for some* $t \geq 1$, *then* $\mathrm{P} \neq \mathrm{NP}$.

References

1. L. Cai, J. Chen, R.G. Downey, and M.R. Fellows. On the parameterized complexity of short computation and factorization. *Archive for Mathematical Logic*, 36:321–337, 1997.
2. M. Cesati. The Turing way to parameterized intractability, 2001. Submitted for publication.
3. M. Cesati and M. Di Ianni. Computation models for parameterized complexity. *Math. Log. Quart.*, 43:179–202, 1997.
4. Y. Chen, J. Flum, and M. Grohe. Bounded nondeterminism and alternation in parameterized complexity theory. To appear in *Proc. of the Conference on Computational Complexity*, 2003. Available at `http://www.dcs.ed.ac.uk/home/grohe/pub.html`.
5. R.G. Downey and M.R. Fellows. Threshold dominating set and an improved characterization of W[2]. *Theoretical Computer Science*, 209:123–140, 1996.
6. R.G. Downey and M.R. Fellows. *Parameterized Complexity*. Springer-Verlag, 1999.
7. R.G. Downey, M.R. Fellows, and K. Regan. Descriptive complexity and the *W*-hierarchy. In P. Beame and S. Buss, editors, *Proof Complexity and Feasible Arithmetic*, volume 39 of *AMS-DIMACS Volume Series*, pages 119–134. AMS, 1998.
8. J. Flum and M. Grohe. Fixed-parameter tractability, definability, and model checking. *SIAM Journal on Computing*, 31(1):113–145, 2001.
9. J. Flum and M. Grohe. Describing parameterized complexity classes. In H. Alt and A. Ferreira, editors, *Proceedings of the 19th Annual Symposium on Theoretical Aspects of Computer Science*, volume 2285 of *Lecture Notes in Computer Science*, pages 359–371. Springer-Verlag, 2002.
10. C.H. Papadimitriou. *Computational Complexity*. Addison-Wesley, 1994.

Extending the Dolev-Yao Intruder for Analyzing an Unbounded Number of Sessions*

Yannick Chevalier[1], Ralf Küsters[2], Michaël Rusinowitch[1], Mathieu Turuani[1], and Laurent Vigneron[1]

[1] LORIA-INRIA-Universités Henri Poincaré, Nancy 2
54506 Vandoeuvre-les-Nancy cedex, France
{chevalie,rusi,turuani,vigneron}@loria.fr
[2] Department of Computer Science
Stanford University, Stanford CA 94305, USA
kuesters@theory.stanford.edu

Abstract. We propose a protocol model which integrates two different ways of analyzing cryptographic protocols: i) analysis w.r.t. an unbounded number of sessions and bounded message size, and ii) analysis w.r.t. an a priori bounded number of sessions but with messages of unbounded size. We show that in this model secrecy is DEXPTIME-complete. This result is obtained by extending the Dolev-Yao intruder to simulate unbounded number of sessions.

1 Introduction

Formal analysis has been very successful in finding flaws in published cryptographic protocols [7]. Even fully automatic analysis of such protocols is possible, based on models for which security is decidable or based on approximations (see, e.g., [13, 19, 18, 1, 16], and [17, 9] for an overview of the different approaches, decidability, and complexity theoretic results).

The decidability of security, or more precisely secrecy, of protocols heavily depends on whether in the analysis an unbounded number of sessions of a protocol is taken into account or only an a priori bounded number. In the former case, secrecy is in general undecidable [1, 13, 14], with only a few exceptions [13, 11, 2, 8]. One such exception, which is of particular interest in this paper, is that secrecy is DEXPTIME-complete when the message size is bounded and nonces, i.e., newly generated constants, are disallowed [13]. In what follows, let us call this setting the *bounded message model.* In the latter case, in which the number of sessions is bounded, secrecy is known to be NP-complete [19], even when there is no bound on the size of messages, complex keys are allowed, i.e., keys that may be complex messages, and messages can be paired to form larger messages. We will refer to this setting as the *unbounded message model.*

* This work was partially supoorted by PROCOPE and IST AVISPA. The second author was also supported by the DFG.

M. Baaz and J.A. Makowsky (Eds.): CSL 2003, LNCS 2803, pp. 128–141, 2003.

In this paper, we integrate the two models — the bounded and the unbounded message model, and thus, integrate two different approaches for protocol analysis: i) analysis w.r.t. an unbounded number of sessions, which has the advantage that the exact sessions to be analyzed do not need to be provided beforehand, but where a bound on the size of messages is put, and ii) analysis which is rather detailed since the size of messages is not bounded, but where only explicitly given sessions are analyzed. More precisely, we consider a protocol model in which there are two kinds of principals, *bounded message* and *unbounded message* principals, or *bounded* and *unbounded* principals for short, which only accept messages of bounded size from the environment or messages of unbounded size, respectively. Conversely, in a protocol run, bounded principals may be involved in an unbounded number of sessions while unbounded principals run in at most one session. The communication between the principals is controlled by the standard Dolev-Yao intruder, in particular, the size of the messages the intruder may produce is not bounded. Just as in the bounded and unbounded message model, the principals and the intruder are not allowed to generate nonces. Our model, in what follows referred to as *integrated model*, comprises both the bounded and the unbounded message model: If in the integrated model the set of bounded principals is empty, then the model coincides with the unbounded message model, and if the set of unbounded principals is empty, then this gives the bounded message model.

The main result shown in this paper is that secrecy in the integrated model is DEXPTIME-complete. The main difficulty is to establish the complexity upper bound. The key idea is as follows: To deal with the bounded principals in the integrated model, and thus, the unbounded number of sessions, the bounded principals are turned into intruder rules, and thus, they extend the ability of the intruder to derive new messages. These intruder rules can be applied by the intruder an arbitrary number of times and in this way simulate the unbounded number of sessions. More precisely, we will extend the standard Dolev-Yao intruder by oracle rules, i.e., intruder rules that satisfy certain properties, and show that insecurity w.r.t. a set of *unbounded* principals and the extended intruder is in NP (given an oracle for applying the oracle rules). — This result is obtained in a similar way as the one in [5], although the kind of oracle rules considered in [5] is quite different from the rules studied here. — We then turn the bounded principals into oracle rules, show that these rules in fact simulate the bounded principals, and prove that the rules can be applied in exponential time. These steps are non-trivial. From this, we conclude the desired complexity upper bound, i.e., obtain a deterministic exponential time algorithm for deciding secrecy in the integrated model.

As we will see in Section 3, the integrated model is not more powerful than the unbounded message model in the sense that from every protocol in the integrated model one can construct a protocol in the unbounded message model such that one protocol preserves secrecy only if the other one does. Moreover, feeding this constructed protocol into an algorithm for analyzing protocols in the unbounded message model, yields an alternative way of deciding secrecy in the integrated model. However, since the number of (unbounded) principals in the constructed protocol grows exponentially, using the NP-completeness result shown in [19], this reduction only provides an NEXPTIME decision algorithm. (Note that, together with the main result of this paper and the result shown in [19], the existence of a polynomial time reduction from the integrated model to the unbounded message model would imply NP=EXPTIME.) More importantly, the constructed protocol is too big for current analysis tools in the unbounded messages model, e.g., [3, 18], since they can only handle a small number of principals. Conversely, in our decision algorithm, we not only reduce secrecy in the integrated model to secrecy in the unbounded message model but in addition extend the Dolev-Yao intruder to simulate the bounded principals. In this way, we avoid creating new (unbounded) principals. In addition to the improved complexity theoretic result, this approach seems to be much better amenable to practical implementations. In fact, in [6] an implementation is presented for an intruder with capabilities similar to those needed here.

Structure of the paper. In the following section, the protocol and intruder model is presented. Then we state the main result (Section 3). In Section 4, the intruder extended by oracle rules is introduced and it is shown that insecurity is in NP given an oracle for applying oracle rules. We then, Section 5, turn bounded principals into oracle rules, and by applying the result from Section 4 establish the complexity upper bound. We conclude in Section 6.

We refer the reader to our technical report [4] for full proofs and a formal description of the Three-Pass Mutual Authentication ISO Protocol in our model.

2 Problem Definition

We now provide a formal definition of our protocol and intruder model. We first define terms and messages, then protocols, and finally the intruder and attacks.

2.1 Terms and Messages

Terms are defined according to the following grammar:

$$term ::= \mathcal{A} \,|\, \mathcal{V} \,|\, \langle term, term\rangle \,|\, \{term\}^s_{term} \,|\, \{term\}^p_{\mathcal{K}}$$

where $\mathcal{A}$ is a finite set of constants (*atomic messages*), containing principal names, nonces, keys, and the atomic messages secret and I (the intruder's name); $\mathcal{K}$ is a subset of $\mathcal{A}$ denoting the set of public and private keys; and $\mathcal{V}$ is a finite set of variables. We assume that there is a bijection $\cdot^{-1}$ on $\mathcal{K}$ which maps every public (private) key k to its corresponding private (public) key k^{-1}. The binary symbol $\langle \cdot, \cdot \rangle$ is called *pairing*, the binary symbol $\{\cdot\}^s_{\cdot}$ is called *symmetric encryption*, the binary symbol $\{\cdot\}^p_{\cdot}$ is *public key encryption*. Note that a symmetric key can be any term and that for public key encryption only atomic keys (namely, public and private keys from $\mathcal{K}$) can be used.

Variables are denoted by x, y, terms are denoted by s, t, u, v, and finite sets of terms are written $E, F, \ldots$, and decorations thereof, respectively. We abbreviate $E \cup F$ by E, F, the union $E \cup \{t\}$ by E, t, and $E \setminus \{t\}$ by $E \setminus t$. The cardinality of a set S is denoted by $\mathsf{card}(S)$.

For a term t and a set of terms E, $Var(t)$ and $Var(E)$ denote the set of variables occurring in t and E, respectively.

A *ground term* (also called *message*) is a term without variables. A *(ground) substitution* is a mapping from $\mathcal{V}$ to the set of (ground) terms. The application of a substitution σ to a term t (a set of terms E) is written $t\sigma$ ($E\sigma$), and is defined as usual.

The set of *subterms* of a term t, denoted by $Sub(t)$, is defined as follows:

- If $t \in \mathcal{A} \cup \mathcal{V}$, then $Sub(t) = \{t\}$.
- If $t = \langle u, v \rangle$, $\{u\}_v^s$, or $\{u\}_v^p$, then $Sub(t) = \{t\} \cup Sub(u) \cup Sub(v)$.

Let $Sub(E) = \bigcup_{t \in E} Sub(t)$. We define the size of a term and a set of terms basically as the size of the representation as a dag. That is, the *size* $|t|$ ($|E|$) of a term t (a set of terms E) is $\mathsf{card}(Sub(t))$ ($\mathsf{card}(Sub(E))$).

2.2 Protocols

We now define principals and protocols.

Definition 1. *A* principal Π *is a finite linear ordering of rules of the form* $R \to S$ *where* R *and* S *are terms. We assume that every variable in* S *occurs in* R *or on the left-hand side of a rule preceding* $R \to S$*. The rules are called* principal rules.

A protocol P *is a tuple* $(\mathcal{F}_u, \mathcal{F}_b, E_I, \mathcal{D})$ *where* $\mathcal{F}_u$ *and* $\mathcal{F}_b$ *are finite unions of principals, and thus, partially ordered sets,* E_I *is a finite set of messages with* $I \in E_I$*, and* $\mathcal{D}$ *is some representation of a finite set of messages such that the dag size of messages in the set represented by* $\mathcal{D}$ *is lineary bounded in the size of the representation of* $\mathcal{D}$*.*

Given a protocol P, in the following we will assume that $\mathcal{A}$ is the set of constants occurring in P. We define the *size* $|P|$ of P as the number of different subterms in $\mathcal{F}_u$, $\mathcal{F}_b$, and E_I plus the size of the representation of $\mathcal{D}$. For instance, $\mathcal{D}$ may be a non-negative integer n (encoded in unary) representing the set of all messages of dag size $\leq n$. This implies that the dag size of the set of messages represented by $\mathcal{D}$ is exponentially bounded in the size $|P|$ of the protocol. We define $Var(P)$ to be the set of variables occurring in P.

The idea behind the definition of a protocol is as follows. In an attack on P the intruder may use every principal in $\mathcal{F}_u$ at most once but the principals in $\mathcal{F}_b$ maybe used as often as the intruder wishes. In other words, the principals in $\mathcal{F}_b$ may be involved in an unbounded number of sessions in one attack while the principals in $\mathcal{F}_u$ only participate in at most one session. It is well-known that deciding the security of a protocol w.r.t. an intruder who may use an unbounded number of sessions and may produce messages of unbounded size is undecidable

[13, 2]. For this reason, we will restrict the messages that can be substituted for variables of rules in $\mathcal{F}_b$ to belong to the finite domain $\mathcal{D}$. However, we put no restrictions on the variables of rules in $\mathcal{F}_u$, i.e., these variables can be substituted by messages of unbounded size. We therefore refer to principals in $\mathcal{F}_u$ as *unbounded* and to those in $\mathcal{F}_b$ as *bounded.* A rule of an unbounded principal is called *unbounded* and analogously a rule of a bounded principal is *bounded.* In the following section, attacks are defined formally and the relationship to other models is further discussed. As mentioned, our technical report [4] contains a formal description of a protocol in our protocol model.

2.3 The Intruder and Attacks

Our intruder model follows the Dolev-Yao intruder [12]. That is, the intruder has complete control over the network and he can derive new messages from his initial knowledge and the messages received from honest principals during protocol runs. To derive a new message, the intruder can compose and decompose, encrypt and decrypt messages, in case he knows the key. What distinguishes our model from most other models in which security is decidable is that the intruder may use the (bounded) principals as often as he wishes to perform his attack. As mentioned in the introduction, to deal with this, in Section 4 we will extend the intruder by so-called oracle rules.

The intruder derives new messages from a given (finite) set of messages by applying rewrite rules. A *rewrite rule* (or *t-rule*) L is of the form $M \to t$ where M is a finite set of messages and t is a message. Given a finite set E of messages, the rule L *can be applied* to E if $M \subseteq E$. We define the *step relation* $\to_L$ induced by L as a binary relation on finite sets of messages. For every finite set of messages E: $E \to_L E, t$ (recall that E, t stands for $E \cup \{t\}$) if L is a t-rule and L can be applied to E. If $\mathcal{L}$ denotes a (finite or infinite) set of intruder rules, then $\to_{\mathcal{L}}$ denotes the union $\bigcup_{L \in \mathcal{L}} \to_L$ of the step relations $\to_L$ with $L \in \mathcal{L}$. With $\to^*_{\mathcal{L}}$ we denote the reflexive and transitive closure of $\to_{\mathcal{L}}$.

The set of rewrite rules the intruder can use is listed in Table 1. These rules are called *(Dolev-Yao) intruder rules.* In the table, a, b denote (arbitrary) messages, K is an element of $\mathcal{K}$, and E is a finite set of messages.

The intruder rules are denoted as shown in Table 1. We consider $L_{p1}(\langle a, b\rangle)$, $\ldots, L_{sd}(\{a\}^s_b)$ and $L_c(\langle a, b\rangle), \ldots, L_c(\{a\}^s_b)$ as singletons. Note that the number of decomposition and composition rules is always infinite since there are infinitely many messages a, b.

We further group the intruder rules as follows. In the following, t ranges over all messages.

- $L_d(t) := L_{p1}(t) \cup L_{p2}(t) \cup L_{ad}(t) \cup L_{sd}(t)$. In case, for instance, $L_{p1}(t)$ is not defined, i.e., the head symbol of t is not a pair, then $L_{p1}(t) = \emptyset$; analogously for the other rule sets,
- $L_d := \bigcup_t L_d(t)$, $L_c := \bigcup_t L_c(t)$,
- $\mathcal{L}_{DY} := L_d \cup L_c$ (where DY stands for "Dolev and Yao").

Table 1. Intruder Rules

	Decomposition rules	Composition rules
Pair	$L_{p1}(\langle a,b\rangle)$: $\langle a,b\rangle \to a$ $L_{p2}(\langle a,b\rangle)$: $\langle a,b\rangle \to b$	$L_c(\langle a,b\rangle)$: $a,b \to \langle a,b\rangle$
Asymmetric	$L_{ad}(\{a\}^p_K)$: $\{a\}^p_K, K^{-1} \to a$	$L_c(\{a\}^p_K)$: $a, K \to \{a\}^p_K$
Symmetric	$L_{sd}(\{a\}^s_b)$: $\{a\}^s_b, b \to a$	$L_c(\{a\}^s_b)$: $a, b \to \{a\}^s_b$

The set of messages the intruder can derive from a (finite) set E of messages is:

$$d_{DY}(E) := \bigcup\{E' \mid E \to^*_{\mathcal{L}_{DY}} E'\}.$$

Before we can define attacks on a protocol $P = (\mathcal{F}_u, \mathcal{F}_b, E_I, \mathcal{D})$, we need some new notions.

Given a partially ordered set $\mathcal{F}$ of principal rules with associated ordering $<$, an *execution ordering* π for $\mathcal{F}$ is a bijective mapping from some subset $\mathcal{F}'$ of $\mathcal{F}$ into $\{1, \ldots, \mathsf{card}(\mathcal{F}')\}$ such that $L < L'$ implies $\pi(L) < \pi(L')$ for every $L, L' \in \mathcal{F}'$. The *size* of π is $\mathsf{card}(\mathcal{F}')$.

The *partially ordered set of instantiations of the bounded principals in* P is $\mathcal{F}_b^{\mathcal{D}} := \{\Pi\sigma' \mid \Pi \in \mathcal{F}_b \text{ and } \sigma' : Var(\Pi) \to \mathcal{D}\}$. The *partially ordered set induced by* P is $\mathcal{F}_P := \mathcal{F}_u \cup \mathcal{F}_b^{\mathcal{D}}$.

We are now prepared to define attacks. In an attack, a principal Π performs his sequence (linear ordering) of principal rules $R_1 \to S_1, \ldots, R_n \to S_n$ one after the other. Note that the different rules may share variables which are subsituted by the same message and in this way model the (unbounded) memory of a principal. When in step i a message m is received, then m is matched against R_i yielding a matcher σ (if any) with $R_i\sigma = m$ and Π returns $S_i\sigma$ as output. Variables in R_i and S_i which occurred in a previous step, and thus, have been assigned a message already, are substituted by this message. As mentioned in Section 2.2, the intruder may use an unbounded principal, i.e., a principal in $\mathcal{F}_u$, at most once, and he may use every bounded principal, i.e., a principal in $\mathcal{F}_b$, as often has he wishes (any time with a possibly different matching). The difference between unbounded and bounded principals is as follows: While an unbounded principal accepts every message as long as it matches the current input pattern R_i, a bounded principal expects that the variables are filled with elements of the domain represented by $\mathcal{D}$. Thus, $\mathcal{F}_b^{\mathcal{D}}$ is the set of instances of bounded principals the intruder may use to perform an attack. Note that subsequent use of an instance after the first time does not yield new knowledge. Therefore, we assume w.l.o.g. that bounded principal instances in $\mathcal{F}_b^{\mathcal{D}}$ are used only once. Note, however, that $\mathcal{F}_b^{\mathcal{D}}$ contains different (an exponential number of) instances of one bounded principal. Altogether, the intruder may use every principal in $\mathcal{F}_P$ once. For a subset of these principals he (nondeterministically) chooses some execution ordering and then tries to produce input messages for the principal rules. These input messages are derived from the intruder's initial knowledge and the output messages produced by executing the principal rules.

The aim of the intruder is to derive the message secret. Formally, attacks are defined as follows.

Definition 2. *Let $P = (\mathcal{F}_u, \mathcal{F}_b, E_I, \mathcal{D})$ be a protocol and let $\mathcal{F}_P$ be the partially ordered set induced by P. An $\mathcal{L}_{DY}$-*attack *(or simply* attack*) on P is a tuple (π, σ) where π is an execution ordering on $\mathcal{F}_P$, of size k, and σ is a ground substitution of the variables occurring in P such that*

$$R_i\sigma \in d_{DY}(S_0, S_1\sigma, ..., S_{i-1}\sigma)$$

for every $i \in \{1, \ldots, k\}$ where $R_i \rightarrow S_i = \pi^{-1}(i)$, and

$$\mathsf{secret} \in d_{DY}(S_0, S_1\sigma, ..., S_k\sigma).$$

The decision problem we are interested in is the following set of protocols where we assume the terms occurring in a protocol to be given as dags.

$$\textsc{Insecure} := \{P \mid \text{ there exists an } \mathcal{L}_{DY}\text{-attack on } P\}.$$

If we restrict the set $\mathcal{F}_b$ of bounded principals to be the empty set (and in this case we do not need $\mathcal{D}$), then this is the case of protocol analysis w.r.t. a bounded number of sessions and unbounded message size as considered, for instance, in [15, 19, 18], and called unbounded message model in the introduction. On the other hand, if we restrict $\mathcal{F}_u$ to be an empty set, then this is basically the case of protocol analysis w.r.t. an unbounded number of sessions but with bounded message size as studied in [13], and called bounded message model in the introduction. We note, however, that in contrast to [13], here we allow the intruder to derive messages of arbitrary size, only the size of messages accepted by the (bounded) principals is bounded. Also, we allow complex rather than only atomic keys.

Summing up, with the protocol and the intruder model considered here, we integrate the bounded and the unbounded message models.

3 Main Result

The main result of this paper is:

Theorem 1. *The problem* Insecure *is DEXPTIME-complete.*

In [13], it has been shown that deciding secrecy in the bounded model, i.e., an unbounded number of sessions but bounded messages, is DEXPTIME-complete. Since here we extend this setting, it is not surprising that for Insecure we also obtain DEXPTIME-hardness. In fact, one can use the same reduction, namely a reduction from the recognition problem for Datalog programs [10], as in [13].

In [19], it has been shown that deciding Insecure for protocols P without bounded principals (i.e., $\mathcal{F}_b = \emptyset$) is NP-complete. We can use this result to also obtain an upper bound for Insecure in the general case: Let $P = (\mathcal{F}_u, \mathcal{F}_b, E_I, \mathcal{D})$. Observe that $P \in$ Insecure iff $P' = (\mathcal{F}_u \cup \mathcal{F}_b^{\mathcal{D}}, \emptyset, E_I, \emptyset) \in$

INSECURE. The protocol P' can be handled with the algorithm proposed in [19]. However, since P' may be of size exponential in the size of P this only shows that INSECURE is in NEXPTIME. Thus, the main problem in proving Theorem 1 is to establish the tight upper bound.

The main idea of this proof is as follows: We first extend capabilities of the Dolev-Yao intruder by so-called oracle rules, i.e., intruder rules which satisfy certain conditions. For this extended intruder we show that insecurity for protocols without bounded principals is in NP given an oracle for performing oracle rules (Theorem 2). We then turn the set $\mathcal{F}_b^{\mathcal{D}}$ of instantiated bounded principals into intruder rules and show that these rules are in fact oracle rules. This will yield the claimed exponential time upper bound (Section 5).

In the following section oracle rules are introduced and the NP-decision algorithm is presented.

4 A General Framework

We now extend the Dolev-Yao intruder by oracle rules, which are intruder rules satisfying certain conditions, and show that insecurity in presence of such an extended intruder for protocols without bounded principals is in NP given a procedure for applying oracle rules. We first introduce oracle rules and then present the NP algorithm.

4.1 Extending the Dolev-Yao Intruder by Oracle Rules

In the rest of this paper, let L_o denote a (finite or infinite) set of rewrite rules of the form $M \to t$ where M is a finite set of messages and t is a message. In Definition 4, we will impose restrictions on this set and then call it the set of oracle rules. The subset of L_o consisting of t-rules is denoted by $L_o(t)$. The union of the Dolev-Yao intruder rules and the oracle rules is denoted by $\mathcal{L}_{DYO} := \mathcal{L}_{DY} \cup L_o$ and called *oracle intruder rules*. Define $\mathcal{L}_c := L_c \cup L_o$ to be the set of composition rules, $\mathcal{L}_c(t) := L_c(t) \cup L_o(t)$, and $\mathcal{L}_d(t)$ to be the set of all decomposition t-rules in Table 1.

The set $d_{DYO}(E)$ of messages the intruder can derive from E using the rules $\mathcal{L}_{DYO}$ is defined analogously to $d_{DY}(E)$. Also, $\mathcal{L}_{DYO}$*-attacks* are defined analogously to $\mathcal{L}_{DY}$-attacks.

Given finite sets of messages E, E', an *($\mathcal{L}_{DYO}$-)derivation* D of length n, $n \geq 0$ from E to E' is a sequence of steps of the form $E \to_{L_1} E, t_1 \to_{L_2} \cdots \to_{L_n} E, t_1, \ldots, t_n$ with messages $t_1, \ldots, t_n$, $E' = E \cup \{t_1, \ldots, t_n\}$, and $L_i \in \mathcal{L}_{DYO}$ such that $E, t_1, \ldots, t_{i-1} \to_{L_i} E, t_1, \ldots, t_i$ and $t_i \notin E \cup \{t_1, \ldots, t_{i-1}\}$, for every $i \in \{1, \ldots, n\}$. The rule L_i is called the *ith rule* in D and the step $E, t_1, \ldots, t_{i-1} \to_{L_i} E, t_1, \ldots, t_i$ is called the *ith step* in D. We write $L \in D$ to say that $L \in \{L_1, \ldots, L_n\}$. If $\mathcal{L}'$ is a set of rewrite rules, then we write $\mathcal{L}' \notin D$ to say $\mathcal{L}' \cap \{L_1, \ldots, L_n\} = \emptyset$. The message t_n is called the *goal* of D.

We also need *well formed* derivations which are derivations where every message generated by an oracle intruder rule is a subterm of the goal or a subterm of a term in the initial set of messages.

Definition 3. *Let* $D = E \to_{L_1} \ldots \to_{L_n} E'$ *be a derivation with goal* t. *Then,* D *is* well formed *if for every* $L \in D$ *and every* t': $L \in \mathcal{L}_c(t')$ *implies* $t' \in Sub(E,t)$, *and* $L \in \mathcal{L}_d(t')$ *implies* $t' \in Sub(E)$.

We can now define oracle rules. Condition 1. in the following definition requires the existence of well formed derivations. This will allow us to bound the length of derivations and the size of messages needed in derivations. The remaining conditions are later used to bound the size of the substitution σ of an attack.

Definition 4. *Let* L_o *be a (finite or infinite) set of rules and* P *be a protocol. Then,* L_o *is a* set of oracle rules *(w.r.t.* $L_c \cup L_d$ *as defined above) iff there exists a polynomial* $p(\cdot)$ *such that:*

1. *For every message* t, *if* $t \in d_{DYO}(E)$, *then there exists a well formed derivation from* E *with goal* t.
2. *If* $F \to_{L_o(t)} F, t$ *and* $F, t \to_{L_d(t)} F, t, a$, *then there exists a derivation* D *from* F *with goal* a *such that* $L_d(t) \notin D$.
3. *For every rule* $F \to t \in L_o(t)$ *we have* $|t| \leq p(|P|)$ *and for all* $t' \in F$, $|t'| \leq p(|P|)$.

In what follows, we always assume that L_o is a set of oracle rules. We call a protocol P of the form $(\mathcal{F}_u, \emptyset, E_I, \emptyset)$ *restricted.* We want to decide the insecurity of a restricted protocol w.r.t. an intruder using $\mathcal{L}_{DYO}$, i.e., the Dolev-Yao intruder rules plus the oracle rules. Formally, the decision problem we are interested in is the following set of *restricted* protocols P:

INSECURE$\mathcal{O}$:= $\{P \mid$ there exists an $\mathcal{L}_{DYO}$-attack on the restricted protocol $P\}$

4.2 An NP Decision Algorithm

The following theorem is used to prove Theorem 1.

Theorem 2. *Let* L_o *be a set of oracle rules. Given a procedure (an oracle) for deciding* $E \to_{L_o} t$ *for every finite set* E *of messages and message* t *in constant time,* INSECURE$\mathcal{O}$ *can be decided by a nondeterministic polynomial time algorithm.*

The NP decision procedure is given in Figure 1. In (1) and (2) of the procedure, an attack (π, σ) is guessed of size polynomially bounded in n. Then, it is checked whether this is in fact an attack.

Obviously, the procedure is sound. As for completeness, one needs to show that it suffices to only consider substitutions bounded as done in the procedure. This is proved in Section 4.3, Theorem 3.

To show that the procedure is in fact an NP procedure given a procedure for deciding $E \to t \in L_o$, we prove that (3) and (4) can be decided by an NP algorithm. Given that $|R_i\sigma, S_0\sigma, \ldots, S_{i-1}\sigma|$ is polynomially bounded in $|P|$ for every $i \leq k$ (see Corollary 1), it suffices to show that the following problem belongs to NP (given the decision procedure for $E \to t \in^? L_o$):

DERIVE := $\{(E,t) \mid$ there exists an $\mathcal{L}_{DYO}$-derivation from E with goal t $\}$.

Input: restricted protocol $P = (\mathcal{F}_u, \emptyset, S_0, \emptyset)$ with $n = p(|P|)$, where $p(\cdot)$ is the polynomial associated to the oracle rules, and $V = Var(P)$.

1. Guess an execution ordering π for P. Let k be the size of π. Let $R_i \rightarrow S_i = \pi^{-1}(i)$ for $i \in \{1, \ldots, k\}$.
2. Guess a normalized ground substitution σ such that $|\sigma(x)| \leq 3n^2$ for all $x \in V$.
3. Test that $R_i\sigma \in d_{DYO}(\{S_0\sigma, \ldots, S_{i-1}\sigma\})$ for every $i < k$.
4. Test that secret $\in d_{DYO}(\{S_0\sigma, \ldots, S_k\sigma\})$.
5. If each test is successful, then answer "yes", and otherwise, "no".

Fig. 1. NP Decision Procedure for Insecurity

In this problem, E and t are assumed to be represented as dags. The following lemma follows quite easily from the existence of well formed derivations (see [4] for the proof).

Lemma 1. *Given a procedure for deciding $E \rightarrow^? t \in L_o$,* DERIVE *can be decided in nondeterministic polynomial time.*

From Theorem 3 proved in the following section, completeness of the procedure depicted in Figure 1 follows.

4.3 Polynomial Bounds on Attacks

To show completeness of the NP decision algorithm depicted in Figure 1, we need to prove that it suffices to consider substitutions bounded as in the second step of this algorithm. To this end, we consider an attack of minimal size, a so-called normal attack, and show that the size of this attack can be bounded as stated in the algorithm.

Given an attack (π, σ) on a protocol P define $|\sigma| := \Sigma_{x \in Var(P)} |\sigma(x)|$. We say that the attack (π, σ) is *normal* if $|\sigma|$ is minimal, i.e., for every attack (π', σ'), $|\sigma| \leq |\sigma'|$. Clearly, if there is an attack, there is a normal attack. Note also that a normal attack is not necessarily uniquely determined.

The next lemma says that normal attacks can always be constructed by linking subterms that are initially occurring in the problem specification or by terms bounded by $p(|P|)$. This will allow us to bound the size of attacks as desired (Theorem 3 and Corollary 1). To state the lemma, we need some notation.

Let P, R_i, S_i, (π, σ), V, $p(\cdot)$, and k be defined as in Figure 1. Let $\mathcal{SP} = Sub(\{R_j | j \in \{1, \ldots, k\}\} \cup \{S_j | j \in \{0, \ldots, k\}\})$. We recall that $\mathcal{A} \subseteq \mathcal{SP}$.

Definition 5. *Let t and t' be two terms and θ a ground substitution. Then, t is a θ-*match *of t', denoted $t \sqsubseteq_\theta t'$, if t is not a variable, and $t\theta = t'$.*

In [4], we prove:

Lemma 2. *Given a normal attack (π, σ), for all variables x: $|\sigma(x)| \leq p(|P|)$ or there exists $t \in \mathcal{SP}$ such that $t \sqsubseteq_\sigma \sigma(x)$.*

Using this lemma, it is now easy to bound the size of every $\sigma(x)$ (see [4] for the proof):

Theorem 3. *For every protocol P, if (π, σ) is a normal attack on P, then $|\{\sigma(x) \mid x \in Var\}| \leq 3 \cdot p(|P|)^2$, where $|P|$ is the size of P as defined in Section 2.2 and $p(\cdot)$ is the polynomial associated to the set of oracle rules.*

From this, we easily obtain:

Corollary 1. *For every protocol P and normal attack (π, σ) on P: $|R_i\sigma, S_0\sigma, \ldots, S_{i-1}\sigma|$ and $|Secret, S_0\sigma, \ldots, S_k\sigma|$ can be bounded by a polynomial in $|P|$ for every $i \in \{1, \ldots, k\}$.*

5 Proof of the Complexity Upper Bound

We now show the complexity upper bound claimed in Theorem 1. In what follows, let $P = (\mathcal{F}_u, \mathcal{F}_b, E_I, \mathcal{D})$ be a protocol.

The idea of the proof is to turn the partially ordered set $\mathcal{F}_b^{\mathcal{D}}$ of instantiated bounded principals into oracle rules and then use Theorem 2.

The conversion of $\mathcal{F}_b^{\mathcal{D}}$ is carried out in two steps. First, this set is turned into a set of so-called aggregated rules. Then, the rules are turned into oracle rules.

5.1 Aggregated Rules

The set $\mathcal{F}_b^{\mathcal{D}}$ consists of a finite set of (instantiated) principals Π. Assume that the linear ordering associated to Π is $<$, $\Pi = \{R_0 \rightarrow S_0, \ldots, R_{n-1} \rightarrow S_{n-1}\}$, and $R_i \rightarrow S_i < R_j \rightarrow S_j$ for every $i < j$.

Now, replace every $R_i \rightarrow S_i$ in Π by a rewrite rule $\{R_0, \ldots, R_i\} \rightarrow S_i$. We denote the resulting set by Π_{agg} and call this set the *aggregated version of Π*. Let $\mathcal{F}_{agg}$ denote the set obtained from $\mathcal{F}_b^{\mathcal{D}}$ by replacing every principal by its aggregated version. We call this set *the set of aggregated rules induced by P*. Define $\mathcal{L}_{agg} := \mathcal{L}_{DY} \cup \mathcal{F}_{agg}$, the *set of aggregated intruder rules (induced by P)*. Note that $\mathcal{L}_{agg}$ depends on P. However, for simplicity, we omit P in the notation of this set.

The set of terms the intruder can derive from E using $\mathcal{L}_{agg}$ is defined as:

$$d_{agg}(E) := \bigcup \{E' \mid E \rightarrow^*_{\mathcal{L}_{agg}} E'\}.$$

An *$\mathcal{L}_{agg}$-attack* on P is defined analogously to $\mathcal{L}_{DY}$-attacks.

The following lemma states that there is an $\mathcal{L}_{DY}$-attack on P iff there exists an $\mathcal{L}_{agg}$-attack on P when the bounded principals of P are removed. In other words, the bounded principals are moved to the intruder. The proof of this lemma is straightforward.

Lemma 3. *There exists an $\mathcal{L}_{DY}$-attack on $P = (\mathcal{F}_u, \mathcal{F}_b, E_I, \mathcal{D})$ iff there exists an $\mathcal{L}_{agg}$-attack on $(\mathcal{F}_u, \emptyset, E_I, \emptyset)$.*

From this and if $\mathcal{F}_{agg}$ were oracle rules (in the sense of Definition 4), INSECURE $\in$ *DEXPTIME* would immediately follow from Theorem 2. In general the set $\mathcal{F}_{agg}$ does not meet the restrictions on oracle rules. Therefore, we define *principal oracle rules* meeting the restrictions on oracle rules. In what follows, they are formally defined and it is shown that whether $E \rightarrow t$ is such a rule can be decided in exponential time. Then, we show that these rules can replace aggregated rules and that they are oracle rules. Together with Theorem 2, this will yield Theorem 1.

5.2 Principal Oracle Rules

Let $\mathsf{Sub}_r(\mathcal{F}_{agg})$ denote the set of subterms occurring on the right hand-side of rewrite rules in $\mathcal{F}_{agg}$.

Definition 6. *A* principal oracle rule *induced by a protocol P is a rewrite rule of the form $E \rightarrow t$ where E is some finite set of ground terms with $|u| \leq |P|^2$ for every $u \in E$ and $t \in \mathsf{Sub}_r(\mathcal{F}_{agg})$ such that $t \in d_{agg}(E)$. Let $\mathcal{F}_p$ denote the set of principal oracle rules induced by P.*

Note that in the above definition $|t| \leq |P|^2$ and $\mathcal{F}_{agg} \subseteq \mathcal{F}_p$.

We now show that principal oracle rules can be decided in exponential time.

Proposition 1. *For every E and t, it can be decided in exponential time in the dag size of E and P whether $E \rightarrow t \in \mathcal{F}_p$.*

The key to the proof of this proposition is the following lemma, which is proved in [4]. Intuitively, it states that $\mathcal{L}_{agg}$-derivations are well-formed in the sense that the messages produced in each step of the derivation are subterms of a certain set of messages. Note that $\mathcal{L}_{agg}$-derivations are derivations, as defined in Section 4.1, which use only intruder rules from $\mathcal{L}_{agg}$. In what follows, let $\mathcal{H}$ denote the set of subterms of $\mathcal{F}_{agg}$.

Lemma 4. *Assume that $E \rightarrow t \in \mathcal{F}_p$. Let D denote a derivation from E with goal t over $\mathcal{L}_{agg}$ of minimal length. Then, $u \in Sub(E, t, \mathcal{H})$ for every message u such that there exists a u-rule in D.*

Now to test whether $E \rightarrow t \in \mathcal{F}_p$ one can iteratively apply rules in $\mathcal{L}_{agg}$ to E that create subterms of $E, t, \mathcal{H}$. Let E' be the resulting set of terms. Then, Lemma 4 ensures that $t \in E'$ iff $E \rightarrow t \in \mathcal{F}_p$. It is easy to see that E' can be computed in time exponential in the size of P. This completes the proof of Proposition 1.

5.3 Principal Oracle Rules can Replace Aggregated Rules

Let $\mathcal{L}_p := \mathcal{L}_{DY} \cup \mathcal{F}_p$ be the set of *principal intruder rules*. The set of terms the intruder can derive from E using $\mathcal{L}_p$ is defined as $d_p(E) := \bigcup\{E' \mid E \rightarrow^*_{\mathcal{L}_p} E'\}$. An $\mathcal{L}_p$*-attack* on P is defined analogously to $\mathcal{L}_{DY}$-attacks.

Obviously, $d_{agg}(E) = d_p(E)$ for every finite set E of messages. As an immediate consequence, we obtain:

Lemma 5. *Let $P = (\mathcal{F}_u, \mathcal{F}_b, E_I, \mathcal{D})$ be a protocol and let $\mathcal{L}_{agg}$ and $\mathcal{L}_p$ be the aggregated and principal intruder rules induced by P. Then, there exists an $\mathcal{L}_{agg}$-attack on $(\mathcal{F}_u, \emptyset, E_I, \emptyset)$ iff there exists an $\mathcal{L}_p$-attack on this protocol.*

Together with Lemma 3 this yields:

Proposition 2. *Let $P = (\mathcal{F}_u, \mathcal{F}_b, E_I, \mathcal{D})$ be a protocol and let $\mathcal{L}_p$ be the principal intruder rules induced by P. Then, there exists an $\mathcal{L}_{DY}$-attack on P iff there exists an $\mathcal{L}_p$-attack on $(\mathcal{F}_u, \emptyset, E_I, \emptyset)$.*

5.4 Principal Oracle Rules Are Oracle Rules

In what follows, we identify L_o with $\mathcal{F}_p$, and show that L_o is a set of oracle rules. By definition of $\mathcal{F}_p$, the last condition on oracle rules (Definition 4, 3.) is met with $p(n) = n^2$.

The following lemma shows the second condition in the definition of oracle rules.

Lemma 6. *If $F \rightarrow_{L_o(t)} F, t$ and $F, t \rightarrow_{L_d(t)} F, t, a$, then there exists a derivation D from F with goal a such that $L_d(t) \notin D$.*

Proof. The proof is obvious. It suffices to observe that $a \in d_p(F) \cap \mathsf{Sub}_r(\mathcal{F}_{agg})$, and thus, $F \rightarrow a \in \mathcal{F}_p$. □

The next lemma, shown in [4], states that if a derivation exists, then also a well formed derivation.

Lemma 7. *If $t \in d_p(E)$, then there exists a well formed derivation with goal t.*

The two lemmas imply:

Proposition 3. *The set L_o of principal oracle rules is a set of oracle rules.*

Now, together with Theorem 2 and Proposition 1 this shows the complexity upper bound claimed in Theorem 1.

6 Conclusion

We have proposed a protocol model which integrates what we have called the unbounded and the bounded message models, and we have shown that deciding secrecy in our model is EXPTIME-complete. For this purpose we have extended the Dolev-Yao intruder in a general framework by oracle rules and applied this framework to handle an unbounded number of sessions.

In future work, we will investigate in how far this framework can be applied to yield other interesting extensions of the Dolev-Yao intruder. Another question is whether the oracle rules introduced here can be combined with those considered in [5], with the potential of even more powerful intruders, e.g., those combining unbounded number of sessions with the XOR operator.

References

1. R.M. Amadio and W. Charatonik. On Name Generation and Set-Based Analysis in the Dolev-Yao Model. In *CONCUR 2002*, LNCS 2421, pages 499–514. Springer-Verlag, 2002.
2. R.M. Amadio, D. Lugiez, and V. Vanackère. On the symbolic reduction of processes with cryptographic functions. Technical Report RR-4147, INRIA, 2001.
3. A. Armando, D. Basin, M. Bouallagui, Y. Chevalier, L. Compagna, S. Mödersheim, M. Rusinowitch, M. Turuani, L. Viganò, and L. Vigneron. The AVISS Security Protocol Analysis Tool. In *CAV 2002*, LNCS 2404, pages 349–353. Springer, 2002.
4. Y. Chevalier, R. Küsters, M. Rusinowitch, M. Turuani, and L. Vigneron. Extending the Dolev-Yao Intruder for Analyzing an Unbounded Number of Sessions. Technical Report available at `http://www.inria.fr/rrrt/liste-2003.html`. To appear.
5. Y. Chevalier, R. Küsters, M. Rusinowitch, and M. Turuani. An NP decision procedure for protocol insecurity with XOR. In *Proceedings of LICS 2003*, 2003. To appear.
6. Y. Chevalier and L. Vigneron. Automated unbounded verification of security protocols. In *CAV 2002*, Springer, 2002.
7. J. Clark and J. Jacob. *A Survey of Authentication Protocol Literature*, 1997. Web Draft Version 1.0 available from http://citeseer.nj.nec.com/.
8. H. Comon-Lundh and V. Cortier. *New decidability results for fragments of first-order logic and application to cryptographic protocols.* In *Proc. 14th Int. Conf. Rewriting Techniques and Applications (RTA'2003)*, Valencia, Spain, June 2003, volume 2706 of LNCS. To appear.
9. H. Comon and V. Shmatikov. Is it possible to decide whether a cryptographic protocol is secure or not? *Journal of Telecommunications and Information Technology*, 2002. To appear.
10. E. Dantsin, T. Eiter, G. Gottlob, and A. Voronkov. Complexity and expressive power of logic programming. In *CCC'97*, pages 82–101. IEEE Computer Society, 1997.
11. D. Dolev, S. Even, and R.M. Karp. On the Security of Ping-Pong Protocols. *Information and Control*, 55:57–68, 1982.
12. D. Dolev and A.C. Yao. On the Security of Public-Key Protocols. *IEEE Transactions on Information Theory*, 29(2):198–208, 1983.
13. N.A. Durgin, P.D. Lincoln, J.C. Mitchell, and A. Scedrov. Undecidability of bounded security protocols. In *Workshop on Formal Methods and Security Protocols (FMSP'99)*, 1999.
14. S. Even and O. Goldreich. On the Security of Multi-Party Ping-Pong Protocols. In *FOCS'83*, pages 34–39, 1983.
15. A. Huima. Efficient infinite-state analysis of security protocols. In *Workshop on Formal Methods and Security Protocols (FMSP'99)*, 1999.
16. R. Küsters. On the decidability of cryptographic protocols with open-ended data structures. In *CONCUR 2002*, LNCS 2421, pages 515–530. Springer-Verlag, 2002.
17. C. Meadows. Open issues in formal methods for cryptographic protocol analysis. In *DISCEX 2000*, pages 237–250. IEEE Computer Society Press, 2000.
18. J. K. Millen and V. Shmatikov. Constraint solving for bounded-process cryptographic protocol analysis. In *CCS 2001*, pages 166–175. ACM Press, 2001.
19. M. Rusinowitch and M. Turuani. Protocol Insecurity with Finite Number of Sessions is NP-complete. In *CSFW-14*, pages 174–190, 2001.

On Relativisation and Complexity Gap for Resolution-Based Proof Systems

Stefan Dantchev[1] and Søren Riis[2]

[1] Dept. of Mathematics and Computer Science, University of Leicester
University Road, Leicester, LE1 7RH, UK
dantchev@mcs.le.ac.uk

[2] Dept. of Computer Science, Queen Mary, University of London
Mile End Road, London E1 4NS, UK
smriis@dcs.qmul.ac.uk

Abstract. We study the proof complexity of *Taut*, the class of Second-Order Existential (SO$\exists$) logical sentences which fail in all finite models. The Complexity-Gap theorem for Tree-like Resolution says that the shortest Tree-like Resolution refutation of any such sentence Φ is either fully exponential, $2^{\Omega(n)}$, or polynomial, $n^{O(1)}$, where n is the size of the finite model. Moreover, there is a very simple model-theoretics criteria which separates the two cases: the exponential lower bound holds if and only if Φ holds in some infinite model.

In the present paper we prove several generalisations and extensions of the Complexity-Gap theorem.

1. For a natural subclass of *Taut*, *Rel* (*Taut*), there is a gap between polynomial Tree-like Resolution proofs and sub-exponential, $2^{\Omega(n^{\varepsilon})}$, general (DAG-like) Resolution proofs, whilst the separating model-theoretic criteria is the same as before. *Rel* (*Taut*) is the set of all sentences in *Taut*, relativised with respect to a unary predicate.
2. The gap for stronger systems, Res* (k), is between polynomial and $\exp\left(\Omega\left(\frac{\log k}{k} n\right)\right)$ for every k, $1 \leq k \leq n$. Res* (k) is an extension of Tree-like Resolution, in which literals are replaced by terms (i.e. conjunctions of literals) of size at most k. The lower bound is tight.
3. There is (as expected) no gap for any propositional proof system (including Tree-like Resolution) if we enrich the language of SO logic by a built-in order.

1 Introduction

In [1] a new kind of results for propositional logic was introduced. Expressed somewhat informally, it was shown that any sequence ψ_n of tautologies which expresses the validity of a fixed combinatorial principle either is "easy" i.e. has polynomial size tree-resolution proofs or is "difficult" i.e requires full exponential size tree-resolution proofs. It was shown that the class of tautologies which are hard (for tree-resolution) is identical to the class of tautologies which are based on combinatorial principles which are violated for infinite sets.

M. Baaz and J.A. Makowsky (Eds.): CSL 2003, LNCS 2803, pp. 142–154, 2003.

According to this result the proof complexity of a combinatorial principle never have intermediate growth rates like for example $2^{\log^k(n)}, k = 2, 3, \ldots$. In this paper we extend this result to a number of related resolution-based systems.

A central question in the theory of proof complexity concerns to the amount of resources (usually proof length) which is needed to prove a certain sequence of tautologies. Usually, the sequence of tautologies consists of Tautologies which are *similar* except for their size. The paper is organised as follows:

Firstly, we consider the resolution proof system in a setting of DAG-like proofs rather than tree-like structure. In a DAG-like proof a once derived disjunction can be used any number of times later in the proof. This cannot happen in the tree like cases. Thus a given tautology might have a DAG-like proof which is substantially shorter than the shortest tree-like proof [2].

The class of combinatorial problems which are hard for DAG-like resolution differs from the class of combinatorial problems which are hard for tree like resolution. "Minimal element" is a principle separates the two systems [3].

In this paper we show the class of combinatorial problems which are hard for DAG resolution is identical to the class which is hard for tree resolution *provided the combinatorial principles are being relativised.* This answers an open question by Krajicek [4].

Secondly, we consider *Res*(k) which is similar but stronger than resolution. In *Res*(k) clauses i.e. disjunctions, are replaced by disjunctions of conjunctions of $\leq k$ literals. The rules are strengthened so one can resolve not just a single variable (like in resolution), but also conjunctions of up to k variables. An easy extension of [1] gives a complexity jump from polynomial to $\exp(\Omega(\frac{n}{k}))$ for these problems (when proofs are represented as trees). This lower bound is also implicit in [4] by Krajicek. We improve this, and show that the jump is from polynomial to $\exp(\Omega(\frac{n \log k}{k}))$. Moreover, we show that the lower bound is tight.

Finally, we show that there is no complexity gap if we enrich the language of SO logic by a built-in order. Even though this result is expected it has a less obvious consequence. It allows us to answer an open question from [5] by showing that there is no complexity gap for tree resolution above $2^{n \log(n)}$. We expect that there is a complexity gap above 2^n, but are not sure if the gap jumps the whole way up to $2^{n \log(n)}$.

2 Preliminaries

2.1 Resolution with Bounded Conjuction, Res(k)

In this section we recall some of the basic concepts related to resolution proofs.

A *literal* is a propositional variable or the negation of a propositional variable. A k-conjunction $\bigwedge_j l_j$ is a conjunction of at most k literals. A k-DNF is a disjunction of k-conjunctions.

The Res(k) proof system is a refutation system designed to provide certificates (i.e. proofs) that a system of k-DNF's is unsatisfiable. This is done by means of the following four derivation rules. The $\wedge$-*introduction rule* is

$$\frac{C_1 \vee \bigwedge_{j \in J_1} l_j \quad C_2 \vee \bigwedge_{j \in J_2} l_j}{C_1 \vee C_2 \vee \bigwedge_{j \in J_1 \cup J_2} l_j},$$

provided that $|J_1 \cup J_2| \leq k$. The *cut (or resolution) rule* is

$$\frac{C_1 \vee \bigvee_{j \in J} l_j \quad C_2 \vee \bigwedge_{j \in J} \neg l_j}{C_1 \cup C_2}.$$

The two *weakening rules* are

$$\frac{C}{C \vee \bigwedge_{j \in J} l_j},$$

provided that $|J| \leq k$, and

$$\frac{C \vee \bigwedge_{j \in J_1 \cup J_2} l_j}{C \vee \bigwedge_{j \in J_1} l_j}.$$

Here C's are k-DNFs, and l's are literals.

The given DNF's are often referred to as *axioms*, and the task is to derive the empty clause (the contradiction) from the axioms. In *Tree-like* Res(k), denoted Res*(k), the proof is organised as a binary tree with the axioms in the leaves and the empty clause in the root. In *DAG-like* Res(k) denoted just Res(k), the proof is given as a linear sequence $C_1, C_2, \ldots, C_u$ of clauses, where each clause either is an axiom or can be obtained by means of the resolution rule (applied to two already derived clauses). In a Resolution proof, clauses can be reused more the once. A tree-like proof do not allow this.

2.2 Proving Lower Bounds for Resolution and Tree-Like Res(k)

We will first describe the *search problem*, associated to an *inconsistent set of clauses* as defined in [6]: Given a truth assignment, find a clause, falsified under the assignment.

We can use a refutation of the set of clauses to solve the search problem as follows. We first turn around all the edges of the graph of the proof. The contradiction now becomes the only root (source) of the new graph, and the axioms and the initial formulae become the leaves (sinks). We perform a search in the new graph, starting from the root, which is falsified by any assignment, and always going to a vertex which is falsified under the given assignment. Such a vertex always exists as the inference rules are sound. We end up at a leaf, which is one of the initial clauses.

Thus, if we want to prove the *existence of a particular kind of clauses in any proof*, we can use an *adversary argument against the refutation, solving the search problem as described above.* The argument is particularly nice for Resolution (Res(1)) as developed by Pudlak in [7]. There are two players, named *Prover* and *Adversary*. An unsatisfiable set of clauses is given. Adversary claims wrongly that there is a satisfying assignment. *Prover* holds a Resolution refutation, and uses it to solve the search problem. A *position* in the game is a partial assignment of the propositional variables. The positions can be viewed as conjunctions of literals. All the possible positions in the game are exactly negations of all the clauses in the Prover's refutation. The game start from the empty position (which corresponds to $\top$, the negation of the empty clause). Prover has two kind of moves:

1. She queries a variable, whose value is unknown in the current position. Adversary answers, and the position then is extended with the answer.
2. She forgets a value of a variable, which is known. The current position is then reduced, i.e., the variable value becomes unknown.

The game is over, when the current partial assignment falsifies one of the clauses. Prover then wins, having shown a contradiction.

We will be interested in *deterministic Adversary's strategies* which allows to prove the existence of certain kind of clauses in a Resolution refutation.

In order to prove *lower bounds on the size of a Resolution proof*, we will use the known technique, "bottleneck counting". It was introduced by Haken in his seminal paper [8] (for the modern treatment see [9]). We first define the concept of *big clause.* We then design *random restrictions*, so that they "kill" (i.e. evaluate to $\top$) any big clause with high probability (whp). By the union bound principle, if there are few big clauses, there is a restriction which kills them all. We now consider the *restricted set of clauses*, and using *Prover-Adversary game*, show that there has to be *at least one big clause in the restricted proof*, which is a contradiction and completes the argument.

The case of Tree-like proofs, either Resolution or Res(d), is much simpler as a *tree-like proof* of a given set of clause *is equivalent* to a *decision tree, solving the search problem* [6]. We can use pretty straightforward adversary argument against a decision tree, in order to show that it has to have many nodes.

2.3 Relativising Combinatorial Principles

Let Ψ be a SO$\exists$ logical sentence. Informally, the statement Ψ states that some property holds for the whole universe. Informally, the relativised sentence $Rel(\Psi)$ say that the principle Ψ holds for any non-empty subset if the universe. The relativised principle not only state the validity of the principle for models of size n, but also implies that the principles holds for all models of size $\leq n$. Thus the relativised principle is in general harder to prove than its non-relativised counter part.

We will briefly describe the translation of a SO$\exists$ sentence into a set of clauses. Assume first that there is a single relation symbol F which is quantified existentially, and the FO part of the sentence is in a prenex normal form, i.e.

$$\forall x_1 \exists y_1 \ldots \forall x_m \exists y_m \, F(x_1, y_1, \ldots x_m, y_m) \, .$$

We first introduce *Skolem relations* s,

$$\bigvee_{k=1}^{n} s^i_{j_1,\ldots j_i,k} \quad \text{for all } 1 \leq i \leq m \text{ and all } j_1, \ldots j_i \in [n] \, ,$$

and then the sentence translates into

$$\bigwedge_{i=1}^{m} s^{i}_{j_1,j_2,\ldots j_i,k_i} \to f_{j_1,k_1,\ldots j_m,k_m} \quad \text{for all } j_1,k_1,\ldots j_m,k_m \in [n] .$$

The case when the quantifier-free part of the Ψ, F is not an atomic formula is as easy as this one. We just rewrite F in CNF, and then the clause above becomes a set of clauses. The number of these clauses is a constant, i.e. independent from the size of the universe, n.

Let us now consider the relativisation, $Rel(\Psi)$. It is

$$\forall x_1 \in R \exists y_1 \in R \ldots \forall x_m \in R \exists y_m \in R\, F(x_1, y_1, \ldots x_m, y_m),$$

and can be rewritten as

$$\forall x_1 (R(x_1) \to \exists y_1 (R(y_1) \wedge \ldots \forall x_m (R(x_m) \to \exists y_m (R(y_m) \wedge F(x_1, \ldots y_m))))) .$$

It is not hard to see that the latter formula translates into the following sets of clauses:

$$\bigwedge_{i=1}^{l} r_{j_i} \wedge s^{l}_{j_1,j_2,\ldots j_l,k} \to r_k \quad \text{for all } 1 \le l \le m \text{ and all } j_1, \ldots j_l, k \in [n]$$

and

$$\bigwedge_{i=1}^{m} r_{j_i} \wedge \bigwedge_{i=1}^{m} s^{i}_{j_1,j_2,\ldots j_i,k_i} \to f_{j_1,k_1,\ldots j_m,k_m} \quad \text{for all } j_1,k_1,\ldots j_m,k_m \in [n] .$$

Finally we add the clause saying that the unary predicate, we relativise with respect to, should not define the empty set, i.e.

$$\bigvee_{i=1}^{n} r_i .$$

3 Complexity Gap for *Rel* (*Taut*)

Let us first introduce some conventions. Recall that φ_n is built upon two kinds of variables, r's and s's. *r-variables correspond to the relation symbols* in the original sentence Φ. Suppose there are m such variables, $r^1, r^2, \ldots r^m$, having arities $p_1, p_2, \ldots p_m$, and let us denote $p = \max_i p_i$. Given the r-variable $r^{i}_{j_1,j_2,\ldots j_{p_i}}$, we say it *mentions* the elements $j_1, j_2, \ldots j_{p_i}$. *s-variables correspond to the Skolem relations* we use to encode Φ as a set of clauses. Suppose there are l such variables $s^1, s^2, \ldots s^l$. Given the s-variable $s^{i}_{j_1,j_2,\ldots j_{q_i},j_0}$, we say it *mentions only its arguments* $j_1, j_2, \ldots j_{q_i}$, *but not the witness*, j_0. We also define its arity to be q_i, not $q_i + 1$. Let us denote $q = \max_i q_i$ as well as $t = \max\{p, q\}$.

We will first prove that we need big clauses to refute Φ.

Lemma 1. *Any Resolution refutation of φ_n contains a clause which mentions at least $n^{1/q}/\left(2l^{1/q}\right) - t$ elements.*

Proof. We will describe a deterministic Adversary's strategy which enforces a big clause. As usual Adversary holds an infinite model M which satisfies Φ. We say that an element is *busy* in the current position (i.e. clause) iff it is mentioned by the clause. We say an element is *hidden* iff it is not busy, but is the witness of a Skolem relation, having all its arguments busy. An element, which is neither busy nor hidden, is said to be *free*. At each stage in the Prover-Adversary game Adversary maintains two disjoint sets B and H. B is the set of all the busy elements, and H is the set of the hidden elements.

The Adversary's strategy is now clear. At any stage in the game all the elements from the disjoint union $B \uplus H$ have interpretations in M. Initially $B = H = \emptyset$. There are two kinds of Prover's moves:

1. She queries a new propositional variable. The easier case is when the variable can be evaluated under the current interpretation. Adversary replies with the value, and does not change B and H. If the variable cannot be evaluated, Adversary needs to enlarge B with all the new elements mentioned by the variable. Note that there is a constant number of such elements, namely at most t. H then has to be enlarged as well with the witnesses of all the new tuples in B.
2. She forgets a propositional variable. Some elements from B may then become non-busy. Adversary removes these from B. Some of them may become hidden, i.e. they are witnesses of a Skolem relation with all its arguments from B, and go to H. The rest become free. Some elements from H, namely the witnesses of the tuples which contained at least one of the just forgotten elements, become free as well.

In any case no contradiction can be achieved as far as $|B| + |H| \leq n$. On the other hand there are l Skolem relation, each with arity at most q, so at any time $|H| \leq l\,|B|^q$. Thus, without loss of generality, we can assume that at any stage in the game, before a contradiction is achieved, we have

$$|B| + l\,|B|^q \leq n.$$

Since $|B| = n^{1/q}/\left(2l^{1/q}\right)$ satisfies the inequality, and $|B|$ increases by at most t after any stage, there should be a point where $|B| > n^{1/q}/\left(2l^{1/q}\right) - t$ as claimed. □

Let us now consider $Rel\,(\varphi_n)$. We will first describe a distribution of *random restrictions* which kills any big clause in any Resolution refutation of $Rel\,(\varphi_n)$ with high probability (whp). The idea is to randomly divide the universe $U = [n]$ into two approximately equal parts. One of them, R, will represent the predicate, we relativise by, R^0; all the variables within it will remain unset. The rest, C, will be the "chaotic" part; all the variables within C and most of the variables between C and R will be set entirely at random.

More precisely, the random restrictions are as follows.

1. We first set all the variables r^0_j to either $\top$ or $\bot$ independently at random with equal probabilities, 1/2. Let us denote the set of variables with $r^0_j = \top$ by R, and the set of variables with $r^0_j = \bot$ by C, $C = U \setminus R$.
2. We now set all the variables $r^i_{j_1,j_2,\ldots j_{p_i}}$, $i > 0$, which mention at least one element of C, i.e. $\{j_1, j_2, \ldots j_{p_i}\} \cap C \neq \emptyset$, to either $\top$ or $\bot$ independently at random with equal probabilities, 1/2.
3. We set all the variables $s^i_{j_1,j_2,\ldots j_{q_i},j_0}$, which mention at least one element of C, i.e. $\{j_1, j_2, \ldots j_{q_i}\} \cap C \neq \emptyset$, to either $\top$ or $\bot$ independently at random with equal probabilities, 1/2.
4. We finally set to $\bot$ all the variables $s^i_{j_1,j_2,\ldots j_{q_i},j_0}$ which mention only elements from R, but the witness is in C, i.e. $\{j_1, j_2, \ldots j_{q_i}\} \subseteq R$ and $j_0 \in C$.

It is very important to note that the variables and clauses, which survive the random restriction, define exactly Φ on R, i.e. $\varphi_{|R|}$.

There are few minor problems. The third case of the above description may violate an axiom as well as the first case may make R very small. We can however see that these bad events happen with exponentially small probability.

Lemma 2. *The probability that the random restrictions are inconsistent with the axioms or $|R| \leq n/4$ is at most $ln^q/2^n + 1/e^{n/16}$.*

Proof. Indeed, an axiom is violated iff in the third case there is a q_i-tuple $j_1, j_2, \ldots j_{q_i}$ such that for every j_0, $s^i_{j_1,j_2,\ldots j_{q_i},j_0} = \bot$, i.e. there is no witness for the tuple. The probability for this is $1/2^n$ and there are at most ln^q such tuples, so that the union-bound gives $ln^q/2^n$. By the Chernoff bound the probability that $|R| \leq n/4$ is at most $e^{-n/16}$. □

The next step is to show that the random restrictions kill any clause with exponential probability in the number of the elements mentioned.

Lemma 3. *Given a clause, which mentions at least k elements, the probability it does not evaluate to $\top$ under the random restrictions is at most $(3/4)^{k/t}$.*

Proof. Let us denote the clause by A, and perform the following experiment. We pick up a literal l_1 from A. The probability that at least one of the elements, mentioned by l_1, is in the chaotic set C is at least 1/2. Given such an element, the probability that l_1 evaluates to $\top$ under the random restrictions is 1/2. Thus the probability l_1 does not evaluate to $\top$ is at most 3/4. We now take all the elements, mentioned by l_1 and mark them.

We pick another literal l_2 from A which mentions at least one unmarked element, and proceed as we have done with l_1. We then pick yet another literal l_3 and so on.

The clause A mentions at least k elements, whilst after having considered a literal we mark at most t elements, so that we can repeat the above procedure at least k/t times. These trials have been independent by the construction of the random restrictions. Therefore the probability that A does not evaluate to $\top$ is at most $(3/4)^{k/t}$ as claimed. □

We can now prove the main result of the section.

Theorem 1. *A SO$\exists$ sentence Φ is given which fails in all finite models, but holds in some infinite model. Let us denote Rel (Φ) the relativisation of Φ with respect to a unary predicate. Let Rel (φ_n) be the translation of the latter into set of clauses, assuming a finite model of size n. Then there is a constant ε, depending only on Φ, such that any Resolution refutation of Rel (φ_n) is of size* $\exp\left(\Omega\left(n^{\varepsilon}\right)\right)$.

Proof. Let us denote $k = \frac{n^{1/p}}{2(4l)^{1/p}} - t$, and say a clause is *big*, iff it mentions at least k elements. We will prove that any Resolution refutation of *Rel* (φ_n) contains exponentially many in k big clauses, which would give the desired lower bound.

Assume, for the sake of contradiction, there is a refutation Γ which contains at most $(4/3)^{k/2}$ big clauses. We hit Γ by random restrictions. By the lemma 2 and the lemma 3 + union-bound, the probability that the restrictions are "bad" is at most

$$\frac{ln^q}{2^n} + \frac{1}{e^{n/16}} + \left(\frac{3}{4}\right)^{n^{1/p}/\left(2(4l)^{1/p}\right)-t} .$$

As this quantity is smaller than 1 for big enough n (recall that l, p, q and t depend on Φ, but not on n), there is a set of good restrictions, "good" meaning that they kill all the big clauses, and moreover what survives is φ_m for some $m > n/4$. But now by the lemma 1 there has to be at least one big clause in the restricted refutation which is a contradiction. □

Note that we do not claim that the bound proven, $\exp\left(\Omega\left(n^{1/p}\right)\right)$ is tight. As a matter of fact, we believe the right lower bound is $\exp\left(\Omega\left(n\right)\right)$, but we also believe this might be very hard to prove, say as hard as proving the Complexity Gap theorem for (DAG-like) Resolution.

4 Complexity Gap for Res* (k)

Theorem 2. *A SO$\exists$ sentence Φ is given which fails in all finite models, but holds in some infinite model. Let us denote by φ_n the translation of Φ into propositional logic, assuming a finite model of size n. Then for any k, $2 \leq k \leq n$, any $R^*(k)$ refutation of φ_n is of size* $\exp\left(\Omega\left(\frac{\log k}{k} n\right)\right)$.

Proof. We will describe the modifications of the original proof, Section 4 of [1]. Recall that the proof goes as follows. Prover holds the Tree-like Resolution refutation which is equivalent to a decision tree solving the search problem for φ_n. Adversary's task to force a big subtree in the Prover's tree. In doing so, he uses an infinite model M in which Φ holds. At any stage in the game some elements of the universe are interpreted in M, and it is clear that no contradiction can be achieved by Prover, unless she forces Adversary to interpret all the n elements. When a variable is queried by Prover, Adversary answer as follows.

1. If the truth value can be derived from the current interpretation, i.e. the partial assignment, he replies with the value. The current interpretation is not extended as this was a forced question.
2. If the value cannot be derived from the current interpretation, it follows that both $\top$ and $\bot$ answers are consistent with some extension of it. Thus Adversary is free to choose an answer, and in both possible cases there is a consistent extension of the current interpretation by at most r new elements, where r is a constant, the maximal arity of relation symbols in φ_n.

This shows that Prover's decision tree has to contain a complete binary subtree of height n/r which implies a $2^{n/r}$ lower bound.

Let us now consider the case of *Res** (k) instead of Tree-like Resolution. A Prover's query is a k-disjunction instead of a single variable. Adversary first simplifies the query, using the current interpretation. That is, if a literal evaluates to $\bot$ it vanishes; if all the literals vanish, the query itself is forced, and the answer is $\bot$. If a literal evaluates to $\top$ so does the entire disjunction; the query is forced, and the answer is $\top$.

The non-trivial case is when the query is not forced, i.e. after having been simplified, it can still be answered both $\top$ and $\bot$. For the positive answer it is enough to force a single literal to $\top$,and therefore to interpret at most r new elements. For the negative answer Adversary should force all the literals to $\bot$, and therefore he has to interpret all the mentioned elements which are at most kr.

We will show that a subtree rooted at a given node can be lower-bounded by a function S in the number of free elements at the node. If the number of free elements at the current node is u, the $\top$ successor has at least $u - r$ such elements whilst the $\bot$ successor has at most $u - kr$. Thus we have

$$S(u) \geq S(u - r) + S(u - kr) + 1.$$

We will prove that $S(u) \geq x_k^{u/r} - 1$ where x_k is the biggest positive real root of the equation

$$x^k - x^{k-1} - 1 = 0.$$

The induction step is trivial. By the induction hypothesis we get

$$\begin{aligned} S(u) &\geq \left(x_k^{(u/r)-1} - 1\right) + \left(x_k^{(u/r)-k} - 1\right) + 1 \\ &= x_k^{u/r} - 1. \end{aligned}$$

Let us now observe that there are positive constants a and b such that for every $k \geq 1$, x_k is in the interval $\left(1 + a\frac{\ln k}{k}, 1 + b\frac{\ln k}{k}\right)$. Indeed, let us denote $f(x) = x^k - x^{k-1} - 1$. We have $f\left(1 + c\frac{\ln k}{k}\right) = \left(1 + \frac{c \ln k}{k}\right)^{k-1} \frac{c \ln k}{k} - 1$ where c is a constant. Since $\lim_{k \to \infty} \left(1 + \frac{c \ln k}{k}\right)^{k-1} k^{-c} = 1$ we can conclude that for every constant $\varepsilon > 0$ there is k_ε so that for every $k \geq k_\varepsilon$, $f\left(1 + (1 - \varepsilon)\frac{\ln k}{k}\right) < 0$ and $f\left(1 + (1 + \varepsilon)\frac{\ln k}{k}\right) > 0$.

It is now clear how to get the desired lower bound. At the root of every decision tree we have all the n elements free, so that the decision tree has to be of size at least $\left(1+a\frac{\ln k}{k}\right)^{n/r}-1$ which is $\sim e^{\frac{an\ln k}{rk}}$ and therefore $\exp\left(\Omega\left(\frac{\log k}{k}n\right)\right)$. What remains to be checked is the basis case, $n < kr$. In this case the size of the tree is at least n/r (as this is the minimal number of queries required to force interpretation of all the elements), whilst the lower bound expression is $\leq e^{a\ln k} = k^a$. Clearly $n/r > k^a$ for big enough n as $k \leq n$ and r, a, $a < 1$ are constants independent from n. This completes the proof. □

It is important to note that the lower bound we have proven is tight. The SO∃ sentence which shows this is *Minimal Element Principle* (MEP_n), saying that *a finite* (partially) *ordered n*-element *set* has a *minimal element.* Its negation is

$$\begin{aligned}\exists P \quad & ((\forall x\, \neg P(x,x)) \\ & \wedge (\forall x,y,z\ (P(x,y) \wedge P(y,z)) \rightarrow P(x,z)) \\ & \wedge (\forall x \exists y\, P(y,x)))\,.\end{aligned}$$

It is not hard to see that there is a $Res^*(k)$ refutation of MEP_n of size $\exp\left(\Omega\left(\frac{\log k}{k}n\right)\right)$ for any k, $2 \leq k \leq n-1$. Note also that the $Res^*(n-1)$ proof of MEP_n is essentially the same as the (DAG-like) Resolution proof of the principle, so that our result is consistent with the known fact that MEP_n is hard for Tree-like Resolution, but easy for Resolution.

5 Built-in Order "Kills" the Gap

We will first show that there is no Complexity gap for Tree-like Resolution if we enrich the SO∃-language with a built-in order predicate.

Theorem 3. *There is no tree-resolution complexity gap for the logical sentences in the language SO∃ + built-in order.*

Proof. Let us first describe the argument very informally. Assume a finite model of size n. There are know tautologies which requires size $2^{\Omega(n)}$ to refute in Tree-like Resolution. The most natural of them is *Minimal Element Principle (MEP)* which has already been mentioned (note that the partial order, defined by MEP, is entirely independent from the built-in total order predicate).

As we have built-in order, we can interpret the elements of the universe as the first n natural numbers. We can define in the SO∃ language a broad class of functions, such as $\log x$, $x^{p/q}$ (p and q integers) and so on. We will show that it is easy, i.e. polynomially size doable, to verify that a given element k of the universe is $f(n)$, where f is a function from the class. Then we will restrict Minimal Element Principle to the first k elements (in the built-in order) to get a sentence in the SO∃ language + built-in order predicate whose optimal refutation is of size $2^{\Omega(f(n))}$, provided $f(n) = \Omega(\log n)$.

What remains is to show how to define and verify $f(.)$ in polynomial in n size. As an example, we will give the definitions of $\sqrt{\cdot}$ and $\log(\cdot)$.

In what follows, all the relation symbols are quantified existentially and the free variables are quantified universally. Moreover, the definitions are nested in the order they appear, starting with the above definition of a total order, being outermost one. We denote by L the built-in order predicate, i.e. $L(x, y)$ stands for $x < y$.

We first define the successor function as the relation $S(x, y)$ standing for y is the successor of x.

$$S(x,y) \equiv (L(x,y) \wedge (\forall z \neg (L(x,z) \wedge L(z,y))))$$

We can define any constants, by the relations $C_a(x)$, meaning $x = a$.

$$C_0(x) \equiv (\forall y \neg S(y,x))$$

$$C_a(x) \equiv (\exists y \ (C_{a-1}(y) \wedge S(y,x)))$$

We are now ready to define addition and multiplication recursively by the following relations, $A(x, y, z)$ and $M(x, y, z)$ standing for $x + y = z$ and $x \times y = z$, respectively.

$$\begin{aligned} A(x,y,z) \equiv & ((C_0(x) \wedge C_0(y) \wedge C_0(z)) \\ & \vee \exists u, v \ (S(v,z)) \\ & \quad \wedge ((S(u,x) \wedge A(u,y,v)) \\ & \quad \vee (S(u,y) \wedge A(x,u,v)))) \end{aligned}$$

$$\begin{aligned} M(x,y,z) \equiv & (((C_0(x) \vee C_0(y)) \wedge C_0(z)) \\ & \vee \exists u, v \ ((S(u,x) \wedge A(y,v,z) \wedge M(u,y,v)) \\ & \vee (S(u,y) \wedge A(x,v,z) \wedge M(x,u,v)))) \end{aligned}$$

We can now define $y = \lfloor \sqrt{x} \rfloor$ by $y = \lfloor \sqrt{x} \rfloor$ if and only if either $y^2 \le x < (y+1)^2$ or $y^2 \le x$, but $(y+1)^2 > n$.

$$\begin{aligned} R_2(x,y) \equiv & (M(y,y,x) \\ & \vee \exists u \ (M(y,y,u) \wedge L(u,x) \\ & \wedge (\exists z, w \ (S(y,z) \wedge M(z,z,w) \wedge L(x,w)) \\ & \quad \vee \forall z, w \ (S(y,z) \rightarrow \neg M(z,z,w))))) \end{aligned}$$

We also define $y = \lfloor \frac{x}{2} \rfloor$ as

$$\begin{aligned} D_2(x,y) \equiv & \exists u, v, w \ (M(u,x,v) \wedge A(v,w,u) \\ & \wedge C_2(u) \wedge (C_0(v) \vee C_1(v))), \end{aligned}$$

and finally $y = \lfloor \log x \rfloor$, using the recursion

$$\lfloor \log x \rfloor = \begin{cases} 0 & x = 1 \\ 1 + \lfloor \log \lfloor \frac{x}{2} \rfloor \rfloor & x > 1 \end{cases}$$

$$L_2(x,y) \equiv (C_1(x) \wedge C_0(y) \\ \vee (\exists u, v\ (S(v,y) \wedge D_2(x,u) \wedge L_2(u,v))))\,.$$

We shall not formally prove that it is possible (in a straightforward way) to verify all the functions defined above within the decision tree computational model. The sketch of the proof is, however, pretty clear: as all the definitions are inductive and there are no mutual recursive relations, we check the functions in the order they are defined, starting from the basis cases. Thus all the queries are forced in the sense that if the assignment does not satisfy the definition of the given relation then the "wrong" answer leads to an immediate contradiction. Thus the size of the initial part of the tree, verifying the definitions of these functions, is polynomial in n as claimed. □

Of course, the result holds for many other propositional proof systems as Tree-like Resolution is a very weak system, and can be polynomially-simulated by them. As an important consequence we get the following

Theorem 4. *There is no Tree-like Resolution Complexity gap above* $2^{\theta(n \log n)}$.

Proof. Let us first observe that there are $SO\exists$ statements with optimal proofs of size $2^{\theta(n^p)}$ for every $p \geq 1$. It is enough to take MEP_{n^p}, i.e. defined on p-tuples instead of single elements of the universe. To get the intermediate complexities, we could restrict the last element of any such p-tuple to be less than or equal to $f(n)$ where f is some function and n is the size of the universe (model). The optimal tree resolution proof of the obtained in this way statement would be $2^{\theta\left(n^{p-1}f(n)\right)}$. However we have not any predefined functions in our language. The definition of f therefore should be a part of the sentence and the proof should "verify" the definition of the function f. We shall use the same argument as we have done in the previous proof. However we have not total order either, so we have to define it within the SO$\exists$ language, and to verify it by a decision tree. In the rest of the section we will show how to do this, which would complete the proof. □

The total order can be defined as

$$\exists L \quad (\forall x\, \neg L(x,x) \\ \wedge\, \forall x,y\ ((x=y) \vee L(x,y) \vee L(y,x)) \\ \wedge\, \forall x,y,z\ ((L(x,y) \wedge L(y,z)) \rightarrow L(x,z)))\,,$$

and we complete the argument by proving the following lemma.

Lemma 4. *The relation L can be* optimally *verified by a decision tree of size* $2^{\theta(n \log n)}$.

Proof. The sentence translates into the following set of clauses:

$$\neg l_{ii} \quad i$$

$$l_{ij} \vee l_{ji} \quad i,j$$

$$\neg l_{ij} \vee \neg l_{jk} \vee l_{ik} \quad i,j,k.$$

It is clear that every permutation π of $\{1, 2, \ldots n\}$ defines a satisfying assignment by setting l_{ij} to $\top$ if and only if $\pi(i) < \pi(j)$ and vice versa. This observation immediately implies a lower bound of $n! = 2^{\Omega(n \log n)}$.

A decision tree which verifies L can be constructed by incrementally ordering the elements of the universe. Suppose we already have a decision tree which orders the first j elements, i.e. each leaf corresponds either to a contradiction or to a permutation of $\{1, 2, \ldots j\}$ as explained above. We can now expand the latter leaves by finding the place of the $j+1$-th element. In doing so, we use binary search which uses $O(\log(j+1))$ *free-choice* queries. Once the place of the $j+1$-th element has been found, all the queries involving it and some of the previous j elements are *forced*, i.e. one of the answers leads to an immediate contradiction. Thus the forced queries contribute a polynomial factor to the size of the subtree consisting of the free-choice queries only. The depth of this subtree is $\sum_{j=1}^{n-1} O(\log(j+1)) = O(n \log n)$, and therefore its size and the size of the entire decision tree is $2^{O(n \log n)}$. □

References

1. Riis, S.: A complexity gap for tree-resolution. Computational Complexity **10** (2001) 179–209
2. Ben-Sasson, E., Impagliazzo, R., Wigderson, A.: Near-optimal separation of general and tree-like resolution. (Combinatorica) to appear.
3. Bonet, M., Galesi, N.: A study of proof search algorithms for resolution and polynomial calculus. In: Proceedings of the 40th IEEE Symposium on Foundations of Computer Science, IEEE (1999)
4. Krajicek, J.: Combinatorics of first order structures and propositional proof systems. (2001)
5. Dantchev, S., Riis, S.: Tree resolution proofs of the weak pigeon-hole principle. In: Proceedings of the 16th annual IEEE Conference on Comutational Complexity, IEEE (2001)
6. Krajîĉek, J.: Bounded Arithmetic, Propositional Logic, and Complexity Theory. Cambridge University Press (1995)
7. Pudlák, P.: Proofs as games. American Mathematical Monthly (2000) 541–550
8. Haken, A.: The intractability of resolution. Theoretical Computer Science **39** (1985) 297–308
9. Beame, P., Pitassi, T.: Simplified and improved resolution lower bounds. In: Proceedings of the 37th annual IEEE symposium on Foundation Of Computer Science. (1996) 274–282

Strong Normalization of the Typed λ_{ws}-Calculus

René David[1] and Bruno Guillaume[2]

[1] Université de Savoie, Campus Scientifique, F-73376 Le Bourget du Lac
david@univ-savoie.fr
[2] LORIA / INRIA Lorraine, Campus Scientifique, F-54506 Vandœuvre-lès-Nancy
Bruno.Guillaume@loria.fr

Abstract. The λ_{ws}-calculus is a λ-calculus with explicit substitutions introduced in [4]. It satisfies the desired properties of such a calculus: step by step simulation of β, confluence on terms with meta-variables and preservation of the strong normalization. It was conjectured in [4] that simply typed terms of λ_{ws} are strongly normalizable. This was proved in [7] by Di Cosmo & al. by using a translation of λ_{ws} into the proof nets of linear logic. We give here a direct and elementary proof of this result. The strong normalization is also proved for terms typable with second order types (the extension of Girard's system F). This is a new result.

1 Introduction

Explicit substitutions provide an intermediate formalism which, by decomposing the β rule of the λ-calculus into more atomic steps, gives a better understanding of the execution models. The pioneer calculus with explicit substitutions, $\lambda\sigma$, was introduced by Curien & al. in [1] as a bridge between the classical λ-calculus and concrete implementations of functional programming languages. Since Melliès [6] has shown that this calculus does not preserve strong normalization, even for typed terms, finding a system satisfying the following properties became a challenge:

- step by step simulation of β,
- confluence on terms with meta-variables,
- strong normalization of the calculus of substitutions,
- preservation of strong normalization of the β-reduction.

During the last decade, various systems were presented in the literature but none of them satisfied simultaneously the previous properties. λ_{ws}, the calculus we introduced in [4], has been the first satisfying all of them. In addition to explicit substitutions, the terms of λ_{ws} are decorated with "labels". The typed version of the calculus (also introduced in [4]) shows that there is a strong link between the computational and the logical points of view: substitutions correspond to cuts and labels to weakenings. The proof that any pure λ-term which is β-strongly normalizable is still strongly normalizable in the λ_{ws}-calculus was highly technical and uses ad-hoc methods. We conjectured that the typed

M. Baaz and J.A. Makowsky (Eds.): CSL 2003, LNCS 2803, pp. 155–168, 2003.

terms are strongly normalizable (SN). Di Cosmo, Kesner and Polonovsky [7] understood the relation between λ_{ws} and linear logic and, by using a translation of λ_{ws} into proof nets, they proved this conjecture. We give here a direct and arithmetical proof of SN for simply typed terms. This proof is based on the one for the (usual) λ-calculus due to the first author [2, 3]. We also prove, by using the standard notion of reducibility candidates, that terms typable with second order types (the extension of Girard's system F) are strongly normalizable. This result is new.

The general idea of the proofs is the following. We first give a simple characterization of strongly normalizing terms (theorem 3). This result, which is only concerned with the *untyped* calculus, is interesting by itself and may be used to prove other results on λ_{ws}. It can be seen as a kind of standardization result. Theorem 3 mainly consists of commutation results. Note that permutation of rules is also the main ingredient in the proof of [7]. Then, for $\mathcal{S}$, we use this characterization to prove, by a tricky induction, a substitution lemma (theorem 6) from which the result follows immediately. For $\mathcal{F}$, we use this characterization to prove that if a term is typed then it belongs to the interpretation of its type.

The paper is organized as follows. Section 2 gives the main notations. In section 3 we introduce some useful notions and we prove the key technical result. It is used in section 4 to prove SN for simply typed terms and in section 5 for second order types.

2 The λ_{ws}-Calculus

2.1 The Untyped Calculus

We define here a variant of λ_{ws} which is equivalent to the one in [4]: $\langle k\rangle$ is no more primitive but becomes the abbreviation of $\langle\rangle \ldots \langle\rangle$, k many times and n is coded by $\langle n\rangle 0$. Since the strong normalization of both formulations are equivalent (see proposition 1 below) and the proof is a bit simpler for the new one, we introduce here this calculus.

Definition 1. *The set of terms of λ_{ws} is defined by the following grammar:*

$$T = 0 \mid \lambda T \mid (T\ T) \mid \langle\rangle T \mid [i/T, j]T \text{ where } i, j \in \mathbb{N}.$$

and the reduction rules of the λ_{ws}-calculus are given in fig.1.

Remark 1.
- The "logical" meaning of $\langle\rangle$ and $[i/u, j]t$ is given by the typing rules. The "algorithmic" meaning is, intuitively, the following: $\langle k\rangle t$ means that each de Bruijn index in t is increased by k (as a consequence, there is no variable with de Brujin indices less than k in t) and $[i/u, j]t$ represents the term t in which the variable indexed by i is substituted by u with a re-indexing commanded by j.
- It is clear that the version of λ_{ws} presented here is a restriction of the one in [4]. For self completeness the terms and the rules of this calculus are given in the appendix. The translation ϕ from the latter to the present one is given by: $\phi(t)$ is obtained from t by replacing n by $\langle n\rangle 0$ and then $\langle k\rangle$ by $\langle\rangle \ldots \langle\rangle$, k many times. In particular, $\langle 0\rangle$ is empty.

b	$(\langle k\rangle \lambda t\, u) \longrightarrow [0/u,k]t$	
l	$[i/u,j]\lambda t \longrightarrow \lambda[i+1/u,j]t$	
a	$[i/u,j](t\, v) \longrightarrow (([i/u,j]t)\ ([i/u,j]v))$	
e_1	$[0/u,j]\langle\rangle t \longrightarrow \langle j\rangle t$	
e_2	$[i/u,j]\langle\rangle t \longrightarrow \langle\rangle[i-1/u,j]t$	$i>0$
n_1	$[i/u,j]0 \longrightarrow 0$	$i>0$
n_2	$[0/u,j]0 \longrightarrow u$	
c_1	$[i/u,j][k/v,l]t \longrightarrow [k/[i-k/u,j]v, j+l-1]t$	$k \leq i < k+l$
c_2	$[i/u,j][k/v,l]t \longrightarrow [k/[i-k/u,j]v,l][i-l+1/u,j]t$	$k+l \leq i$

Fig. 1. Reduction rules of λ_{ws}

– Note that, in this variant, the reduction rules become a bit simpler and some of them (m and n_3 in the original calculus) even disappear. Also note that rules b_1 and b_2 give a unique rule b which is in fact a *family* of rules since $\langle k\rangle$ represents a family of symbols.

Proposition 1. *If $t \to t'$ then $\phi(t) \to^+ \phi(t')$. In particular, the strong normalization of both versions of λ_{ws} are equivalent.*

Proof. Straightforward. □

2.2 The Typed Calculus

Definition 2. *Let $\mathcal{V}$ be a set of type variables.*

- *The set $\mathcal{S}$ of simple types is defined by: $\mathcal{S} ::= \mathcal{V} \mid \mathcal{S} \to \mathcal{S}$*
- *The set $\mathcal{F}$ of second-order types is defined by: $\mathcal{F} ::= \mathcal{V} \mid \mathcal{F} \to \mathcal{F} \mid \forall\mathcal{V}.\mathcal{F}$*

Definition 3. – *A basis Γ is an (ordered) list of types. The length of Γ is denoted by $\|\Gamma\|$.*

– *The typing rules for $\mathcal{F}$ are the given in fig.2. Note that the first element (on the left) of Γ corresponds to the variable with de Bruijn index 0. For $\mathcal{S}$, just forget $\forall_i$ and $\forall_e$.*

Proposition 2. *Both systems have subject reduction: if $\Gamma \vdash t : A$ and $t \to u$, then $\Gamma \vdash u : A$.*

Proof. We have to check that, for each rule, the typing is preserved after reduction. We give below the example of rule b. The proof is detailed in [5] for the original version of the calculus.

The typing of the b-redex $(\langle k\rangle \lambda t\ u)$ is given on the left and the typing of its reduct $[0/u,k]t$ is given on the right. We assume that $\|\Gamma\| = k$ and the last element of Γ is C.

$$\frac{}{A,\Gamma \vdash 0 : A}(Ax) \qquad \frac{\Gamma \vdash t : A}{B,\Gamma \vdash \langle\rangle t : A}(Weak)$$

$$\frac{A,\Gamma \vdash t : B}{\Gamma \vdash \lambda t : A \to B}(\to_i) \qquad \frac{\Gamma \vdash t : A \to B \quad \Gamma \vdash u : A}{\Gamma \vdash (t\, u) : B}(\to_e)$$

$$\frac{\Gamma, A, \Phi \vdash t : B \quad \Delta, \Phi \vdash u : A}{\Gamma, \Delta, \Phi \vdash [i/u, j]t : B}(Cut) \text{ where } i = \|\Gamma\| \text{ and } j = \|\Delta\|$$

$$\frac{\Gamma \vdash t : A}{\Gamma \vdash t : \forall\alpha.A}(\forall_i) \text{ if } \alpha \notin \Gamma \qquad \frac{\Gamma \vdash t : \forall\alpha.A}{\Gamma \vdash t : A\{\alpha := B\}}(\forall_e)$$

Fig. 2. Typing rules of the λ_{ws}-calculus

$$\cfrac{\cfrac{\cfrac{\cfrac{\cfrac{A,\Delta \vdash t : B}{\Delta \vdash \lambda t : A \to B}(\to_i)}{C,\Delta \vdash \langle\rangle\lambda t : A \to B}(Weak)}{\vdots}(Weak)}{\Gamma,\Delta \vdash \langle k\rangle\lambda t : A \to B}(Weak) \quad \Gamma,\Delta \vdash u : A}{\Gamma,\Delta \vdash (\langle k\rangle\lambda t\, u) : B}(\to_e)$$

$$\frac{A,\Delta \vdash t : B \qquad \Gamma,\Delta \vdash u : A}{\Gamma,\Delta \vdash [0/u,k]t : B}(Cut)$$

□

3 Characterization of Strongly Normalizable Terms

This section gives a characterization (Theorem 3) of strongly normalizable terms. This is the key of the proof of the strong normalization for both systems. We first need some definitions.

3.1 Some Definitions

Definition 4. *The set S of substitutions and the set Σ are defined by the following grammars:*

$$S ::= \emptyset \mid [i/T,j]S \qquad \Sigma ::= \emptyset \mid \langle\rangle\Sigma \mid [i/T,j]\Sigma$$

Definition 5. *Some particular contexts are defined by the following grammars where $*$ denotes a hole and, if H is a context, $H[t]$ denotes the term obtained by replacing $*$ by t in H.*

$$C_i ::= * \mid \Sigma C_i \mid \lambda C_i \qquad C_e ::= * \mid \Sigma C_e \mid (C_e\, T) \qquad C ::= C_i[C_e]$$

Note that these contexts have a unique hole at the leftmost position. The elements of C (resp. C_i, C_e) are called head contexts (resp. i-contexts, e-contexts). Elements of T (resp. S, Σ, C) will be denoted by t, u, v, w (resp. by s, by σ, by H, K).

Notation 1 *1. We denote by $\rightarrow$ the least congruence on $T \cup C$ containing the rules of fig.1. As usual, $t \rightarrow^* t'$ (resp. $t \rightarrow^+ t'$) means that t reduces to t' by some steps (resp. at least one step) of reduction.*
2. The set of strongly normalizable terms (i.e. such that every sequence of $\rightarrow$ reductions is finite) is denoted by SN.

Lemma 1 (and notation). *Every term in T can uniquely be written as $H[0]$ or $H[(\sigma\lambda u\ v)]$ where H is an head context. The* head *of t (denoted by $\mathbf{hd}(t)$) is:*

$$\mathbf{hd}(H[0]) = H \qquad \mathbf{hd}(H[(\sigma\lambda u\ v)]) = H[(\sigma * \ v)]$$

Proof. Straightforward. □

Notation 2 *Say $t \rightarrow_{\mathbf{r}} t'$ if $t \rightarrow t'$ with the following restrictions: use only the rules a, e, c and only in $\mathbf{hd}(t)$ either at the top level or, recursively, for $[i/u, j]$ in $\mathbf{hd}(t)$, only in $\mathbf{hd}(u)$. The rule l is also permitted but only in H_i where $\mathbf{hd}(t) = H_i[H_e]$ with $H_i \in C_i$ and $H_e \in C_e$.*

Example 1.

$$[0/b,0](\lambda[0/c,0]1\ \ [0/d,1]0) \rightarrow^*_{\mathbf{r}} ([0/b,0]\lambda[0/c,0]1\ [0/[0/b,0]d,1]0)$$

$$[0/a,0]\lambda(b\ c) \rightarrow^*_{\mathbf{r}} \lambda([1/a,0]b\ [1/a,0]c)$$

$$[0/[0/a,0]\lambda c,0]b \not\rightarrow^*_{\mathbf{r}} [0/\lambda[1/a,0]c,0]b$$

$$(\lambda[0/a,0]\langle\rangle b\ c) \not\rightarrow^*_{\mathbf{r}} (\lambda b\ c)$$

Lemma 2 (and notation). *The reduction $\rightarrow_{\mathbf{r}}$ is locally confluent and thus is confluent for terms such that $\mathbf{hd}(t) \in SN$. The* $\mathbf{r}$*-normal form of t will be denoted by $\mathbf{r}(t)$.*

Proof. Straightforward. □

Remark 2. The $\mathbf{r}$-reduction is actually strongly normalizing for every term and thus confluent. This follows immediately from the strong normalization of the calculus of substitution (i.e. all the rules except b) which is proved in [4]. We have stated the previous lemma in this way to keep this paper self contained, i.e. our proof does not need this result. Thus, in the rest of the paper, when we use $\mathbf{r}(t)$ or the confluence of $\mathbf{r}$ we have to check that $\mathbf{hd}(t) \in SN$. We will not mention this since this is always straightforward.

Definition 6. *1. Let H be an head context. Let $R(H) \subset T$, $L(H) \in C_i$ and, if $H \in C_i$, $I(H) \in T$ be defined by the following rules:*
- *$R(*) = \emptyset$, $R(\lambda H) = \lambda R(H)$, $R(\sigma H) = \sigma R(H)$ and $R((H\ t)) = R(H) \cup \{t\}$.*
- *$L(*) = *$, $L(\lambda H) = \lambda L(H)$, $L(\sigma H) = \sigma L(H)$ and $L((H\ t)) = L(H)$.*
- *$I(*) = 0$, $I(H[\lambda *]) = I(H[\langle\rangle *]) = I(H)$ and $I(H[[i/u,j]*]) = H[\langle i\rangle u]$.*

2. An head context is pure *if $L(H)$ has no substitutions.*

3. *Let t be a term in T. The set* $\mathbf{arg}(t) \subset T$ *is defined by:*
   - $\mathbf{arg}(H[0]) = R(H) \cup \{I(L(H))\}$
   - $\mathbf{arg}(H[(\sigma\lambda u\ v)]) = R(H[(*\ v)]) \cup L(H)[\sigma\lambda u]$

Remark 3. In the previous definition, the equation $R(\lambda H) = \lambda R(H)$ actually means, since $R(H)$ is a set of terms, $R(\lambda H) = \{\lambda t \mid t \in R(H)\}$ and similarly for $R(\sigma H) = \sigma R(H)$.

Example 2.

$$\mathbf{arg}([4/0, j](\langle 2\rangle\lambda 3\ 0)) = \{[4/0, j]\langle 2\rangle\lambda 3, [4/0, j]0\}$$

$$\mathbf{arg}([2/0, j][0/v, 2]\langle\rangle 0) = \{[2/0, j]v\}$$

Lemma 3. – *Let $t = H[0]$. Then $\mathbf{r}(t)$ can be uniquely written as $K[s0]$ where K is pure.*
- *Let $t = H[(\sigma\lambda u\ v)]$. Then $\mathbf{r}(t)$ can be uniquely written as $K[(\langle k\rangle s\lambda u\ v_1)]$ where K is pure.*

Proof. Straightforward. □

Definition 7. *Let $s \in S$ be a substitution, we define $s^+ \in S$ and $s^\downarrow \in T$ as follows:*

- *s^+ is defined by: $\emptyset^+ = \emptyset$ and $([i/u, j]s)^+ = [i + 1/u, j]s^+$.*
- *$s^\downarrow$ is defined by: $\emptyset^\downarrow = 0$ and $(s[i/u, j])^\downarrow = su$ if $i = 0$ and $s^\downarrow$ otherwise.*

Definition 8. *Let t be a term in T. The head reduct of t (denoted as* $\mathbf{hred}(t)$*) is defined as follows:*

- *If $t = H[0]$ and $\mathbf{r}(t) = K[s0]$ then $\mathbf{hred}(t) = K[s^\downarrow]$.*
- *If $t = H[(\sigma\lambda u\ v)]$ and $\mathbf{r}(t) = K[(\langle k\rangle s\lambda u\ v_1)]$ then $\mathbf{hred}(t) = K[[0/v_1, k]s^+u]$.*

Example 3. With terms as in the previous example, we have:

$$\mathbf{hred}([4/0, j](\langle 2\rangle\lambda 3\ 0)) = [0/[4/0, j]0, 2][3/0, j]3$$

$$\mathbf{hred}([2/0, j][0/v, 2]\langle\rangle 0) = 2$$

Theorem 3. *Let $t \in T$ be such that* $\mathbf{arg}(t) \subset SN$.

1. *Assume $t \to_{\mathbf{r}}^* t'$ and $t' \in SN$. Then $t \in SN$.*
2. *Assume* $\mathbf{hred}(t) \in SN$*. Then $t \in SN$.*

3.2 Proof of Theorem 3

We first need some notations and lemmas.

Notation 4 1. *If $t \in SN$, $\eta(t)$ is the length of the longest reduction starting from t and $\eta_0(t)$ is the maximum number of b or n steps in a reduction starting from t.*

2. *The complexity of a term t (denoted by $cxty(t)$) is defined by:* $cxty(*) = cxty(0) = 0$, $cxty(\lambda t) = cxty(\langle\rangle t) = cxty(t) + 1$, $cxty((t\ t')) = cxty(t) + cxty(t') + 1$ *and finally* $cxty([i/t', j]t) = cxty(t) + cxty(t') + i + 1$.

Note that the unusual definition of $cxty([i/t', j]t)$ is due to the fact that $cxty(\langle k \rangle) = k$. It ensures that $cxty([i/u, j]) > cxty(\langle i \rangle u)$ and thus, except for $t = 0$, $cxty(u) < cxty(t)$ for any $u \in \mathbf{arg}(t)$.

Lemma 4. *Let H be an head context, u be a term and $w \in \mathbf{arg}(H[u])$. Then,*

- *either $w \in R(H)$,*
- *or $w = L(H)[v]$ for some $v \in \mathbf{arg}(u)$,*
- *or H is* not *an i-context, $u = \sigma\lambda u'$ and $w = L(H)[u]$.*

Proof. Straightforward. □

Lemma 5. *Let $H \in C$ be pure.*

1. *If $t = H[u] \in SN$ and $s \in SN$, then $H[[0/u, j]s^+0] \in SN$.*
2. *If $t = H[[0/v, k]s^+u] \in SN$, then $H[(\langle k \rangle s\lambda u\ v)] \in SN$.*

Proof. By induction on $\eta(t) + \eta(s^+0)$ for (1) and $\eta(t) + cxty(s)$ for (2). □

Lemma 6. *Let K be an head context. Assume that*

- *either $k \geq i + j$ and $w = [i/[k - i/v, l]u, j][k - j + 1/v, l]K \to_{\mathrm{r}}^* w_1 = K_1[[0/[k - i/v, l]u, j]s_1^+*]$*
- *or $i \leq k < i + j$ and $w = [i/[k - i/v, l]u, k + j - 1]K \to_{\mathrm{r}}^* w_1 = K_1[[0/[k - i/v, l]u, j]s_1^+*]$.*

Then, there is an head context K_2 such that $[i/u, j]K \to_{\mathrm{r}}^* K_2[[0/u, j_2]s_2^+*]$ and $[k/v, l]K_2[[0/u, j_2]s_2^+*] \to_{\mathrm{r}}^* w_1$.

Proof. By induction on the length of the reduction $w \to_{\mathrm{r}}^* w_1$. □

Lemma 7. *Assume $w = [i/u, j]K_1[[k/v', l]K_2] \to_{\mathrm{r}}^* w_1 = K_3[[0/u, j]s^+*]$ and $v \to v'$. Then, $[i/u, j]K_1[[k/v, l]K_2] \to_{\mathrm{r}}^* K_4[[0/u, j]s_1^+*] \to^* K_3[[0/u, j]s^+*]$ for some K_4, s_1.*

Proof. By induction on the length of the reduction $w \to_{\mathrm{r}}^* w_1$. □

Lemma 8. 1. *Assume $t = H[(\sigma\lambda u\ v)] \to^* t_0 = H_0[(\langle k_0 \rangle\lambda u_0\ v_0)]$. Then, there is a term $t_1 = H_1[(\langle k_1 \rangle s_1\lambda u\ v_1)]$ such that $t \to_{\mathrm{r}}^* t_1 \to^* t_0$.*
2. *Assume $t = H[0] \to^* t_0 = H_0[[0/u_0, j_0]s_0^+0]$. Then, H can be written as $K[[i/u, j]K_0]$ such that $[i/u, j]K_0 \to_{\mathrm{r}}^* K_1' = K_1[[0/u, j]s^+*]$ and $t_1 = K[K_1'][0] \to^* t_0$.*

Proof. First note that we should be a bit more precise in the terms of the lemma: we implicitly assume that the potential b-redex (resp. n-redex) at the end of the left branch of t is not reduced during the reduction $t \to^* t_0$. The lemma is proved by induction on the length of the reduction $t \to^* t_0$. We give some details only for (2). They are similar and simpler for (1).

The result is clear for $t = t_0$. Assume $t \to^+ t_0$. By the induction hypothesis, $H \to H_1 = K[[i/u, j]K_0]$ for some K, u, K_0 such that $[i/u, j]K_0 \to_{\mathrm{r}}^* K_1' = K_1[[0/u, j]s^+*]$ and $t_1 = K[K_1'][0] \to^* t_0$. H can be written as $K_3[[i/u_1, j_1]K_2]$

- if $K_3 \to K$ or $u_1 \to u$ the result is trivial,
- if $K_2 = (* \, v)$ and $K = K_3[(* \, [i/u, j]v)]$ the result is trivial,
- if $[i/u, j]K_2 \to_{\mathbf{r}} [i/u, j]K_0$ the result is trivial,
- if $K_2 \to K_0$ but the reduction is not an **r**-reduction, the result follows from lemma 7,
- if $K_3 = K[[k/v, l]*]$ and, either $[i/u, j] = [i/[k - i/v, l]u_1, j_1]$ and $K_0 = [k - j_1 + 1/v, l]K_2]$, or $[i/u, j] = [i/[k - i/v, l]u_1, l + j_1 - 1]]$ and $K_0 = K_2$, the result follows from lemma 6. □

Lemma 9. *1. Assume* $t_1 = H_1[(\sigma_1 \lambda u_1 \, v_1)] \to^* t_0 = H_0[(\langle k_0 \rangle \lambda u_0 \, v_0)]$. *Then,* $H_1[[0/v_1, k_1]s_1^+ u_1] \to^* H_0[[0/v_0, k_0]u_0]$ *where* $\mathbf{r}(\sigma_1) = \langle k_1 \rangle s_1$.
2. Assume $t_1 = H_1[[0/u_1, j_1]s_1^+ 0] \to^* t_0 = H_0[[0/u_0, j_0]s_0^+ 0]$. *Then,* $H_1[u_1] \to^* H_0[u_0]$.

Proof. By induction on the length of the reduction $t_1 \to^* t_0$. Look at the first reduction. Note that there is no simple relation between the original and the resulting reduction sequence and, in particular, the latter may be longer than the original. □

Lemma 10. *1. Assume* $H[(\sigma \lambda u \, v)] \to_{\mathbf{r}}^* t_0$. *Then* t_0 *has the form* $H_0[(\sigma_0 \lambda u \, v_0)]$ *and* $H[[0/v, k]s^+ u] \to_{\mathbf{r}}^* H_0[[0/v_0, k_0]s_0^+ u]$ *where* $\mathbf{r}(\sigma) = \langle k \rangle s$ *and* $\mathbf{r}(\sigma_0) = \langle k_0 \rangle s_0$.
2. Assume $H_0[0/u, j]s_0^+ 0 \to_{\mathbf{r}}^* t_0$. *Then* t_0 *has the form* $H_1[[0/u_1, j_1]s_1^+ 0]$ *where* $H_0 \to_{\mathbf{r}}^* H_1[s_2 *]$ *for some* s_2 *such that* $s_2[0/u, j]s_0^+ \to_{\mathbf{r}}^* [0/u_1, j_1]s_1^+$.

Proof. Straightforward. □

Lemma 11. *Let* K *be an i-context. Then,* $K \in SN$ *iff* $I(K) \in SN$ *and, in this case,* $\eta_0(I(K)) \leq \eta_0(K)$.

Proof. This follows immediately from the following result. Let K be an i-context, then: $K[[i/u, j]*] \in SN \Leftrightarrow K[\langle i \rangle u] \in SN$ and, in this case, $\eta_0(K[\langle i \rangle u]) \leq \eta_0(K[[i/u, j]*])$.

⇒ Prove, by induction on $(\eta(t), cxty(K))$ that if $t = K[s[i/u, j]*] \in SN$ then $K[d(s, i)u] \in SN$ where $d(s, i)$ is the result of moving down s through $\langle i \rangle$. It is enough to prove that, if $K[d(s, i)u] \to t'$ then $t' \in SN$. This is done by a straightforward case analysis.

⇐ This is proved by showing that to any sequence of reductions of $t' = K[\langle i \rangle u]$ corresponds a sequence of reductions of t with the same b or n steps. Define for $s \in S$, $\delta(s) \in \mathbb{Z}$ by: $\delta(\emptyset) = 0$ and $\delta([k/v, l]s) = \delta(s) + l - 1$.
We show that, to a term of the form $K'[\langle i' \rangle u']$ coming from t' corresponds, for some s such that $\delta(s) < i'$, the term $K'[s[i' - \delta(s)/u', l]*]$ coming from t. This is done by a straightforward case analysis. For example, if $t' \to^* K'[[k/v, l]\langle i' \rangle u'] \to K'[\langle l + i' - 1 \rangle u']$ then $t \to^* K'[[k/v, l]s[i' - \delta(s)/u', l]*] = K'[s'[i' - \delta(s')/u', l]*]$ where $s' = [k/v, l]s$.
It is important to note that the result on η_0 would not be true with η. This is essentially because $[k/v, l]$ can always go through $\langle i \rangle$ whereas $[k/v, l]$ cannot move down in $[i/u, j]$ if $k < i$. □

Lemma 12. *Let $t \in T$ be such that $\mathbf{arg}(t) \subset SN$ and $t \notin SN$. Then,*

1. *If $t = H[(\sigma\lambda u\ v)]$, there is a term $t_1 = H_1[(\langle k_1\rangle s_1\lambda u\ v_1)]$ such that $t \to_{\mathbf{r}}^* t_1$ and $H_1[[0/v_1, k_1]s_1^+ u] \notin SN$.*
2. *If $t = H[0]$ there is a term $t_1 = K[K_1[[0/u, j]s_1^+ 0]]$ such that $t \to_{\mathbf{r}}^* t_1$, t can be written as $K[[i/u, j]K_0][0]$ and $K[K_1][u] \notin SN$.*

Proof. 1. Since $\mathbf{arg}(t) \subset SN$, the potential b-redex must be reduced in an infinite reduction of t and thus such a reduction looks like: $t \to^* H_0[(\langle k_0\rangle\lambda u_0\ v_0)] \to H_0[[0/v_0, k_0]u_0] \to ...$ and the result follows from lemmas 8 and 9.

2. Since $\mathbf{arg}(t) \subset SN$ and thus, by lemma 11, $H \in SN$, an infinite reduction of t looks like: $t \to^* H_0[[0/u_0, j_0]s_0^+ 0] \to H_0[u_0] \to ...$ and the result follows from lemma 8 and 9. □

Proof of theorem 3

1. By induction on $(\eta_0(t'), cxty(t))$. Note that the proof is by contradiction. We tried to find a constructive proof but we have been unable to find a correct one.
   - Assume first $t = H[(\sigma\lambda u\ v)]$ and $t \notin SN$. By lemma 12, let $t \to_{\mathbf{r}}^* t_0 = H_0[(\langle k_0\rangle s_0\lambda u\ v_0)]$ be such that $t_1 = H_0[[0/v_0, k_0]s_0^+ u] \notin SN$. By the confluence of $\to_{\mathbf{r}}^*$, let t_0' be such that $t' \to_{\mathbf{r}}^* t_0'$ and $t_0 \to_{\mathbf{r}}^* t_0'$. By lemma 10 with the reduction $t_0 \to_{\mathbf{r}}^* t_0'$, $t_0' = H'[(\sigma'\lambda u\ v')]$. Let $t_1' = H'[[0/v', k']s'^+ u]$ where $\mathbf{r}(\sigma') = \langle k'\rangle s'$. Then $\eta_0(t_1') < \eta_0(t')$ and, by lemma 10, $t_1 \to_{\mathbf{r}}^* t_1'$. It is thus enough to show that $\mathbf{arg}(t_1) \subset SN$ to get a contradiction from the induction hypothesis.
     Let $w_1 \in \mathbf{arg}(t_1)$. By lemma 4, either $w_1 \in \mathbf{arg}(t_0)$ and the result is trivial or $w_1 = L(H_0)[[0/v_0, k_0]s_0^+ w]$ for some $w \in \mathbf{arg}(u)$ or H is not an i-context and $w_1 = L(H_0)[[0/v_0, k_0]s_0^+ u]$.
     Since the second case is similar, we consider only the first one. Let $a = L(H)[(\sigma\lambda w\ v)]$ and $a' = L(H')[(\sigma'\lambda w\ v')]$. Then, $a \to_{\mathbf{r}}^* a'$ and $\eta_0(a') \le \eta_0(t')$ (use lemma 11 for the difficult case, i.e. when $u = K[0]$ and $w = I(L(K))$). If it is *not* the case that H is an i-context and $u = 0$, then $cxty(a) < cxty(t)$ and, by the induction hypothesis, $a \in SN$ and the result follows since $a \to^* w_1$. Otherwise, the result is trivial since it is easily seen (by induction on $(\eta(H), cxty(H))$) that, if $t = H[(\sigma\lambda 0\ v)]$ (where H is an i-context), $\mathbf{r}(\sigma) = \langle k\rangle s$ and $\mathbf{arg}(t) \subset SN$, then $H[[0/v, k]s^+ 0] \in SN$.
   - Assume $t = H[0]$ and $t \notin SN$. By lemma 12, let $t = K[[i/u, j]K_0][0] \to_{\mathbf{r}}^* t_0 = K[H_0][[0/u, j]s_0^+ 0]$ be such that $t_1 = K[H_0][u] \notin SN$. By the confluence of $\to_{\mathbf{r}}^*$, let t_0' be such that $t' \to_{\mathbf{r}}^* t_0'$ and $t_0 \to_{\mathbf{r}}^* t_0'$. By lemma 10 with the reduction $t_0 \to_{\mathbf{r}}^* t_0'$, $t_0' = H'[[0/u', j']s'^+ 0]$ where $K[H_0] \to_{\mathbf{r}}^* H'[s_1 *]$ for some s_1 such that $s_1[0/u, j]s_0^+ \to_{\mathbf{r}}^* [0/u', j']s'^+$. Let $t_1' = H'[u']$. Then $\eta_0(t_1') < \eta_0(t')$ and, by lemma 10, $t_1 \to_{\mathbf{r}}^* t_1'$. It is thus enough to show that $\mathbf{arg}(t_1) \subset SN$ to get a contradiction from the induction hypothesis.

Let $w_1 \in \mathbf{arg}(t_1)$. By lemma 4 either $w_1 \in \mathbf{arg}(t_0)$ and the result is trivial or $w_1 = L(K[H_0])[w]$ for some $w \in \mathbf{arg}(u)$ or H is not an i-context and $w_1 = L(K[H_0])[u]$.
Since the second case is similar, we consider only the first one. Let $a = L(K[[i/w,j]K_0])[0]$. Since $s_1 u \to_{\mathbf{r}}^* u'$, it is easy to find w' such that $s_1 w \to_{\mathbf{r}}^* w'$ and, letting $a' = L(H')[[0/w',j']s'^{+}0]$, $a \to_{\mathbf{r}}^* a'$ and $\eta_0(a') \leq \eta_0(t')$ (use lemma 11 for the difficult case, i.e. when $u = K[0]$ and $w = I(L(K)))$. Since $cxty(a) < cxty(t)$ (except if H is an i-context and $u = 0$ but in this case again the result is trivial), by the induction hypothesis, $a \in SN$ and the result follows since $a \to^* w_1$.

2. This follows immediately from (1) and lemma 5. □

4 Strong Normalization for $\mathcal{S}$

Theorem 5 below has first been proved in [7] by Di Cosmo & al. It is of course a trivial consequence of theorem 7 of section 5. However, the proof presented below is interesting in itself because it is purely arithmetical whereas the one of section 5 is not.

Theorem 5. *Typed terms of T are strongly normalizing.*

Proof. By induction on $cxty(t)$. The cases $t = 0$, $t = \lambda t'$ and $t = \langle\rangle t'$ are immediate. The case $t = [i/u,j]t'$ follows immediately from theorem 6 below. The remaining case is $t = (u\ v)$. By the induction hypothesis, u and $(0\ \langle 1\rangle v)$ are in SN. Thus, by theorem 6, $[0/u,0](0\ \langle 1\rangle v) \in SN$ and since $[0/u,0](0\ \langle 1\rangle v) \to^* t$ it follows that $t \in SN$. □

Theorem 6. *Assume $u, t \in T \cap SN$. Then $[i/u,j]t \in SN$.*

Proof. We prove the following. Let $u \in T \cap SN$. Then,
(1) If $t' \in T \cap SN$, then $[i/u,j]t' \in SN$.
(2) If $H \in C \cap SN$ is *pure*, then $H[u] \in SN$.
This is done by simultaneous induction on $(type(u), \eta_0(v), cxty(v), \eta_0(u))$ where $type(u)$ is the number of $\to$ in the type of u and $v = t'$ for (1) (resp. $v = H$ for (2)). The induction hypothesis will be denoted by IH.

1. $t = [i/u,j]t'$. The fact that $\mathbf{arg}(t) \subset SN$ follows immediately from the IH. By theorem 3, it is thus enough to show that $\mathbf{hred}(t) \in SN$.
   (a) If $t' = H[(\sigma\lambda v_1\ v_2)]$: since $\eta_0(\mathbf{hred}(t')) < \eta_0(t')$, it follows from the IH that $[i/u,j]\mathbf{hred}(t') \in SN$ and the result follows since $[i/u,j]\mathbf{hred}(t') \to^* \mathbf{hred}(t)$.
   (b) If $t' = H[0]$: let $\mathbf{r}(t') = K[s0]$.
      - If $s^{\downarrow} \neq 0$: since $\eta_0(\mathbf{hred}(t')) < \eta_0(t')$, it follows from the IH that $[i/u,j]\mathbf{hred}(t') \in SN$ and the result follows since $[i/u,j]\mathbf{hred}(t') \to^* \mathbf{hred}(t)$.

- Otherwise, let $\mathbf{r}(t) = K'[s'0]$. If $s'^{\downarrow} = 0$ the result is trivial. Otherwise $s'^{\downarrow} = u'$ for some u' such that $u \to_{\mathrm{r}}^{*} u'$ and thus $t_1 = \mathbf{hred}(t) = K'[u']$. If K' is an i-context the result is trivial. Otherwise $K' = H'[(\langle k \rangle * \ t_0)]$. Then $t_1 = H'[(\langle k \rangle u' \ t_0)]$. It is clear that $\mathbf{arg}(t_1) \subset SN$. It is thus enough to show that $\mathbf{hred}(t_1) \in SN$.
  - If $u' = \langle k' \rangle \lambda u'_0$ and thus $\mathbf{hred}(t_1) = H'[w]$ where $w = [0/t_0, k + k']u'_0$. Since $type(t_0) < type(u)$, by the IH, $w \in SN$. By the IH, $H'[w] \in SN$ since $type(w) < type(u)$. Note that, here, we use (2).
  - Else $\mathbf{hred}(t_1) = H'[(\langle k \rangle \mathbf{hred}(u') \ t_0)] = \mathbf{hred}([i/\mathbf{hred}(u'), j]t')$. If $u' \to^{+} \mathbf{hred}(u')$, the result follows from the IH. Otherwise, the result is trivial.

2. $t = H[u]$. If H is a i-context, the result is immediate. Otherwise, $H = H'[(\langle k \rangle * \ t')]$. It is clear that $\mathbf{arg}(t) \subset SN$. It remains to prove that $\mathbf{hred}(t) \in SN$.
   (a) If $u = \sigma \lambda u'$: then $\mathbf{hred}(t) = \mathbf{r}(H')[[0/t', k + k']s^{+}u']$ where $\mathbf{r}(\sigma) = \langle k' \rangle s$. Since $u \in SN$, $s^{+}u' \in SN$. By the IH since $type(t') < type(u)$, $[0/t', k + k']s^{+}u' \in SN$. Finally $\mathbf{hred}(t) \in SN$ since $type([0/t', k + k']s^{+}u') < type(u)$.
   (b) Otherwise $\mathbf{hred}(t) = H[\mathbf{hred}(u)]$. If $u \to^{+} \mathbf{hred}(u)$ the result follows from the IH and otherwise the result is trivial. □

Remark 4. We need (2) in the proof of (1) for the following reason: we cannot always find H' and i, j such that $[i/v, 0]H'[\langle j \rangle 0] \to^{*} H[v]$. By choosing i large enough and j conveniently it is not difficult to get $[i/v, 0]H[\langle j \rangle 0] \to^{*} H[\langle j \rangle v]$ but we do not know how to get rid of $\langle j \rangle$. This is rather strange since, in the λ-calculus, this corresponds to the trivial fact that $(u \ v)$ can be written as $(x \ v)[x := u]$ where x is a fresh variable.

5 Strong Normalization for $\mathcal{F}$

The proof uses the same lines as the one for the (ordinary) λ-calculus. We first define the candidates of reducibility and show some of their properties. Then, we define the interpretation of a type and we show that if t has type A then t belongs to the interpretation of A.

Definition 9. *1. If X and Y are subsets of T, $X \to Y$ denotes the set of t such that, for all $u \in X$, $(t \ u) \in Y$.*
2. The set $\mathcal{C}$ of candidates of reducibility is the smallest set which contains SN and is closed by $\to$ and intersection.
3. N_0 is the set of terms of the form $(0 \ u_1 ... u_n)$ where $u_i \in SN$ for each i.

Lemma 13. *Assume $C \in \mathcal{C}$. Then, $N_0 \subset C \subset SN$.*

Proof. By induction on C. □

Definition 10. *An* interpretation I *is a function from* $\mathcal{V}$ *to* $\mathcal{C}$. I *is extended to* $\mathcal{F}$ *by:* $|\alpha|_I = I(\alpha)$, $|A \to B|_I = |A|_I \to |B|_I$ *and* $|\forall\alpha.A|_I = \bigcap_{C\in\mathcal{C}} |A|_{I\{\alpha:=C\}}$ *(where* $J = I\{\alpha := C\}$ *is such that* $J(\alpha) = C$ *and* $J(\beta) = I(\beta)$ *for* $\beta \neq \alpha$*).*

Definition 11.
- *Let* $u_0, \ldots, u_{n-1}$ *be a sequence of terms. We denote by* $[i/\boldsymbol{u}]$ *the substitution* $[i/u_0, 0][i+1/u_1, 0]\ldots[i+n-1/u_{n-1}, 0]$.
- *For* $\Gamma = A_0, ..., A_{n-1}$, $\boldsymbol{u} \in |\Gamma|_I$ *means that* $u_i \in |A_i|_I$ *for all* i.
- *A substitution* s *is regular if it is of the form* $[i/\boldsymbol{u}]$ *and* $u_i \in SN$ *for each* i.

Lemma 14. *Let* $\boldsymbol{w}$ *be a sequence of terms in* SN, $s \in S$ *be regular and* $C \in \mathcal{C}$. *Assume either* $t' \to_{\mathbf{r}}^* t$ *or* $t' = [0/t, j]s^+0$ *or* $t' = (s\lambda u\ v)$ *and* $t = [0/v, 0]s^+u$. *If* $(t\ \boldsymbol{w}) \in C$, *then* $(t'\ \boldsymbol{w}) \in C$.

Proof. By induction on C. The case $C = SN$ follows immediately from theorem 3. The other cases are straightforward. □

Lemma 15. $|A\{\alpha := B\}|_I = |A|_{I\{\alpha:=|B|_I\}}$ *and thus* $|A|_{I\{\alpha:=B\}} = |A|_I$ *if* $\alpha \notin A$.

Proof. Straightforward. □

Lemma 16. *Let* I *be an interpretation. Assume* $\Gamma \vdash t : B$ *and* $\boldsymbol{u} \in |\Gamma|_I$ *then* $[0/\boldsymbol{u}]t \in |B|_I$.

Proof. By induction on $\Gamma \vdash t : B$. For simplicity, we write $|A|$ instead of $|A|_I$. Assume $\boldsymbol{u} \in |\Gamma|$ and look at the last rule used in the typing derivation:

- rule Ax:

$$\frac{}{A, \Gamma \vdash 0 : A}$$

  Let $v \in |A|$. By lemma 13, $v, \boldsymbol{u} \in SN$ and the result follows from lemma 14.
- rule $\to_i$:

$$\frac{A, \Gamma \vdash t : B}{\Gamma \vdash \lambda t : A \to B}$$

  Let $v \in |A|$ and $w = ([0/\boldsymbol{u}]\lambda t\ v)$. By the IH, $[0/v, 0][1/\boldsymbol{u}]t \in |B|$ and the result follows from lemma 14.
- rule $\to_e$:

$$\frac{\Gamma \vdash t : A \to B \quad \Gamma \vdash v : A}{\Gamma \vdash (t\, v) : B}$$

  By the IH, $[0/\boldsymbol{u}]t \in |A \to B|$ and $[0/\boldsymbol{u}]v \in |A|$. Thus $([0/\boldsymbol{u}]t\ [0/\boldsymbol{u}]v) \in |B|$ and the result follows from lemma 14.
- rule $Weak$:

$$\frac{\Gamma \vdash t : A}{B, \Gamma \vdash \langle\rangle t : A}$$

  Let $v \in |B|$. By the IH $[0/\boldsymbol{u}]t \in |A|$ and the result follows from lemma 14.

– rule *Cut*:

$$\frac{\Gamma, A, \Phi \vdash t : B \quad \Delta, \Phi \vdash v : A}{\Gamma, \Delta, \Phi \vdash [i/v, j]t : B} \text{ where } i = \|\Gamma\| \text{ and } j = \|\Delta\|$$

Let $\boldsymbol{u}_1 \in |\Delta|$, $\boldsymbol{u}_2 \in |\Phi|$ and $w' = [0, \boldsymbol{u}][i/\boldsymbol{u}_1][i+j/\boldsymbol{u}_2][i/v, j]t$. By the IH (on the second premise), $[0/\boldsymbol{u}_1][j/\boldsymbol{u}_2]v \in |A|$. By the IH (on the first premise), $w = [0/\boldsymbol{u}][i/[0/\boldsymbol{u}_1][j/\boldsymbol{u}_2]v, 0][i+1/\boldsymbol{u}_2]t \in |B|$. Since $w' \to_{\mathbf{r}}^* w$, The result follows from lemma 14.

– rule $\forall_i$:

$$\frac{\Gamma \vdash t : A}{\Gamma \vdash t : \forall \alpha . A} \text{ if } \alpha \notin \Gamma$$

Let $C \in \mathcal{C}$. Since $\alpha \notin \Gamma$, by lemma 15, $\boldsymbol{u} \in |\Gamma|_{I\{\alpha:=C\}}$ and thus, by the IH, $[0/\boldsymbol{u}]t \in |A|_{I\{\alpha:=C\}}$. It follows that $[0/\boldsymbol{u}]t \in |\forall\alpha.A|_I$.

– rule $\forall_e$:

$$\frac{\Gamma \vdash t : \forall \alpha . A}{\Gamma \vdash t : A\{\alpha := B\}}$$

By the IH, $[0/\boldsymbol{u}]t \in |\forall\alpha.A|_I$ and thus $[0/\boldsymbol{u}]t \in |A|_{I\{\alpha:=|B|_I\}} = |A\{\alpha := B\}|_I$ (by lemma 15). □

Theorem 7. *Every typed term is strongly normalizing.*

Proof. Assume $\Gamma \vdash t : B$. By lemma 13, $\mathbf{0} \in |\Gamma|$ and thus, by lemma 16, $[0/\mathbf{0}]t \in |B|$. By lemma 13, $[0/\mathbf{0}]t \in SN$ and thus, since SN is closed by subterms, $t \in SN$. □

References

1. M. Abadi, L. Cardelli, P.-L. Curien, and J.-J. Lévy. Explicit substitutions. *Journal of Functional Programming*, 1(4):375–416, 1991.
2. R. David. A short proof of the strong normalization of the simply typed λ-calculus. available: www.lama.univ-savoie.fr/~david.
3. R. David. Normalization without reducibility. *Annals of Pure and Applied Logic*, 107:121–130, 2001.
4. R. David and B. Guillaume. A λ-calculus with explicit weakening and explicit substitution. *Mathematical Structures for Computer Science*, 11:169–206, 2001.
5. B. Guillaume. *Un calcul de substitutions avec étiquettes.* Phd thesis, Université de Savoie, 1999. available: http://www.loria.fr/~guillaum/publications/Gui99.ps.gz.
6. P.-A. Melliès. Typed λ-calculi with explicit substitutions may not terminate. *Proceedings of Typed Lambda Calculi and Applications 95* in *Lecture Notes in Computer Science*, 902:328–334, 1995.
7. D. Kesner R. Di Cosmo and E. Polonovsky. Proof nets and explicit subsitutions. In *Fossacs'2000 and LNCS, 1784 : 63-81*, 2000.

A Appendix

The set of terms and the reduction rules of the original calculus of [4] are:

Terms

$$T = \underline{n} \mid \lambda T \mid (T\ T) \mid \langle k \rangle T \mid [i/T, j]T \text{ where } n, k, i, j \in \mathbb{N}.$$

Rules

$$
\begin{array}{llll}
b_1 & (\lambda t\ u) & \longrightarrow [0/u, 0]t & \\
b_2 & (\langle k \rangle \lambda t\ u) & \longrightarrow [0/u, k]t & \\
l & [i/u, j]\lambda t & \longrightarrow \lambda [i+1/u, j]t & \\
a & [i/u, j](t\ v) & \longrightarrow (([i/u, j]t)\ ([i/u, j]v)) & \\
e_1 & [i/u, j]\langle k \rangle t & \longrightarrow \langle j+k-1 \rangle t & i < k \\
e_2 & [i/u, j]\langle k \rangle t & \longrightarrow \langle k \rangle [i-1/u, j]t & k \leq i \\
n_1 & [i/u, j]\underline{n} & \longrightarrow \underline{n} & n < i \\
n_2 & [i/u, j]\underline{n} & \longrightarrow \langle i \rangle u & n = i \\
n_3 & [i/u, j]\underline{n} & \longrightarrow \underline{n+j-1} & i < n \\
c_1 & [i/u, j][k/v, l]t & \longrightarrow [k/[i-k/u, j]v, j+l-1]t & k \leq i < k+l \\
c_2 & [i/u, j][k/v, l]t & \longrightarrow [k/[i-k/u, j]v, l][i-l+1/u, j]t & k+l \leq i \\
m & \langle i \rangle \langle j \rangle t & \longrightarrow \langle i+j \rangle t &
\end{array}
$$

A Fixed-Point Logic with Symmetric Choice*

Anuj Dawar and David Richerby

University of Cambridge Computer Laboratory,
William Gates Building,
J.J. Thomson Avenue,
Cambridge, CB3 0FD, UK
{Anuj.Dawar,David.Richerby}@cl.cam.ac.uk

Abstract. Gire and Hoang introduce a fixed-point logic with a 'symmetric' choice operator that makes a nondeterministic choice from a definable set of tuples at each stage in the inductive construction of a relation, as long as the set of tuples is an automorphism class of the structure. We present a clean definition of the syntax and semantics of this logic and investigate its expressive power. We extend the logic of Gire and Hoang with parameterized and nested fixed points and first-order combinations of fixed points. We show that the ability to supply parameters to fixed points strictly increases the power of the logic. Our logic can express the graph isomorphism problem and we show that, on almost all structures, it captures $\mathbf{P}^{\mathrm{GI}}$, the class of problems decidable in polynomial time by a deterministic Turing machine with an oracle for graph isomorphism.

1 Introduction

Descriptive complexity classifies problems according to the richness of the logical language required to describe them, offering a view of complexity that is independent of any machine model. Fagin's result that the problems describable in the existential fragment of second-order logic are exactly those in **NP** [Fag74] invites the natural question of whether there is a logic for **P**. First-order logic can only express queries of low computational complexity. This is because it is unable to express many fundamental algorithmic tools such as iteration, counting, arithmetic and the selection of a single element with some property.

Fixed-point logics such as LFP and IFP address the first of these deficiencies. With the proviso that every structure be equipped with a linear ordering of its vertices, these logics capture the class **P** [Imm86,Var82]. (A logic $\mathcal{L}$ is said to *capture* a complexity class $\mathcal{C}$ if every problem in $\mathcal{C}$ is definable by a formula in $\mathcal{L}$ and the problem of evaluating $\mathcal{L}$-formulae is, itself, in $\mathcal{C}$.) Given a linear order, iteration allows counting, arithmetic and choice to be performed. The successive elements of the ordering can be used to simulate numbers and choice from a set can be achieved by taking the element that is least according to the order.

The imposition of an extrinsic linear ordering is undesirable as it allows the expression of queries such as, 'there are no edges to the last vertex in the graph,'

* Research supported by EPSRC grant GR/S06721.

M. Baaz and J.A. Makowsky (Eds.): CSL 2003, LNCS 2803, pp. 169–182, 2003.

which is as much a property of the ordering as it is of the graph. A compromise would be to allow only those queries which are invariant under changes to the ordering but this attempt fails because the set of formulae giving rise to order-invariant queries is undecidable. In the absence of a linear order, LFP and IFP cannot express simple counting properties like the parity of a set.

The ability to count alone does not suffice as first-order logic with counting quantifiers cannot define the class of connected graphs. However, the combination of inflationary fixed points and counting is a widely-studied and reasonably powerful logic. Immerman had conjectured that IFP + C would capture **P** but this was refuted by Cai, Fürer and Immerman [CFI92]. Nonetheless, the logic does capture **P** on many classes of graphs, such as trees [IL90], planar graphs [Gro98], and graphs of bounded tree width [GM99] or genus [Gro00].

In [DR], we considered the combination of fixed points and nondeterministic choice, extending the work that appears in [AB87,GH98]. Iteration with choice can be used to construct a linear order on the universe of a structure but, in general, formulae are nondeterministic: they define sets of relations rather than single relations. The queries defined by deterministic formulae (those defining a single relation) are exactly **P** but this set of formulae is undecidable so cannot reasonably be called a logic. We also considered several notions of satisfaction for nondeterministic formulae and showed that the resulting logics capture a wide range of complexity classes including **NP** and **co-NP**.

Assuming that $\mathbf{P} \neq \mathbf{NP}$, unrestricted choice is, therefore, too powerful to capture **P**. In [GH98], Gire and Hoang consider a combination of fixed points and symmetric choice, a restriction in which choices may be made only from automorphism classes of the structure on which the formula is being evaluated (that is, sets in which every pair of elements is exchanged by some automorphism). They define a fixed-point operator that takes as arguments formulae φ and ψ. At each stage in evaluating the fixed point, ψ defines a set of tuples, the choice set. If this set is an automorphism class, one tuple is nondeterministically chosen from it; otherwise, no tuple is chosen. The chosen tuple (or lack of such) is then used by φ to add tuples to the relation being constructed.

This fixed-point operator is 'semideterministic': while an induction may still define a set of relations, these relations are pairwise isomorphic. This allows the definition of a logic that we call SC-IFP. This is still not a serious candidate for capturing **P** as, even if it were shown to express all polynomial-time properties, evaluating formulae requires testing that sets are automorphism classes and there is no known polynomial-time algorithm for doing this.

Gire and Hoang deal with this problem by defining a sublogic in which the fixed-point operator takes as an argument an extra formula that is supposed to provide a witness automorphism for each pair of tuples in the choice set; if the witness formula does not show that the choice set is an automorphism class, no element is chosen. This guarantees that formulae can be evaluated in polynomial time while maintaining semideterminism of the fixed-point operator. It is this logic (or, rather, its closure under an operator that performs interpretations between structures) that Gire and Hoang show to be strictly more expressive than IFP + C while still defining only polynomial-time properties. While it seems unlikely that this logic captures **P**, this remains to be proved.

In this paper, we concentrate on SC-IFP. Showing that there are polynomial-time properties not definable in this logic would settle the status of the sublogic and, with this aim in mind, we investigate the power of SC-IFP. The logic we define is broadly similar to that of Gire and Hoang but there are some important differences in the syntax (we allow parameterized and nested fixed points) and semantics (Gire and Hoang require the choice set to be an automorphism class of the structure; we require it only to be an automorphism class up to those relations mentioned in the formula). In particular, we show that the ability to supply first-order parameters to fixed points strictly increases the power of the logic. This is significant as it involves the first inexpressibility result for SC-IFP and the usual techniques for proving such results (embedding into infinitary logics or using games) seem to be unavailable. We proceed by choosing a class of structures on which the parameterless fragment is weak.

The expressive power of SC-IFP is closely related to the graph isomorphism problem, GI, which it can express. GI is clearly in **NP** (guess a mapping between two graphs and check that it is an isomorphism) but not known to be either **NP**-complete, in **P** or even **P**-hard. There are, nonetheless, many results concerning the complexity of this problem. For example, it is known to be **NL**-hard under logarithmic space reductions [Tor00] and it is unlikely to be **NP**-complete as, if it were, the polynomial hierarchy would collapse at its second level [BHZ87,Sch88].

Our main result is that, on almost all structures, the closure of SC-IFP under first-order reductions captures $\mathbf{P}^{\mathrm{GI}}$, the class of problems decidable in polynomial time by a Turing machine with an oracle for GI. In particular, it captures $\mathbf{P}^{\mathrm{GI}}$ on any class of structures on which IFP + C captures **P**. Independent of GI, it is known from [GH98] that the logic is strictly more expressive than IFP + C.

The rest of this section contains background definitions. SC-IFP is introduced in Section 2 and we discuss the effects of allowing parameterized definitions in Section 3. In Section 4, we investigate the expressive power of the logic.

Preliminaries. All structures in this paper are finite and all vocabularies are finite and purely relational, though constant symbols are omitted for notational convenience only. We write $|\mathfrak{A}|$ for the universe of structure $\mathfrak{A}$ and $\|\mathfrak{A}\|$ for the cardinality of $|\mathfrak{A}|$.

Classes of structures are assumed to be isomorphism-closed: if a structure is in a class, all images of that structure under isomorphisms are in the class. If $\mathcal{C}$ is a class of structures, a *k-ary* query Q on $\mathcal{C}$ maps each structure $\mathfrak{A} \in \mathcal{C}$ to a k-ary relation on $|\mathfrak{A}|$ such that if $\rho : \mathfrak{A} \to \mathfrak{B}$ is an isomorphism, $Q(\mathfrak{B}) = \rho(Q(\mathfrak{A}))$. The case $k = 0$ is known as a *boolean query*; $\emptyset$ represents false and $\{\langle\,\rangle\}$, the relation containing the unique empty tuple, represents true. A boolean query Q on a class $\mathcal{C}$ may be associated with the class $\{\mathfrak{A} \in \mathcal{C} : Q(\mathfrak{A}) \text{ is true}\}$.

A query Q on a class $\mathcal{C}$ of structures is *$\mathcal{L}$-definable* for some logic $\mathcal{L}$ if there is a formula $\varphi \in \mathcal{L}$ such that $\varphi^{\mathfrak{A}} = Q(\mathfrak{A})$ for all $\mathfrak{A} \in \mathcal{C}$. We write $\mathcal{L}_1 \leqslant \mathcal{L}_2$ if every $\mathcal{L}_1$-definable query is also $\mathcal{L}_2$-definable and $\mathcal{L}_1 = \mathcal{L}_2$ if $\mathcal{L}_1 \leqslant \mathcal{L}_2$ and $\mathcal{L}_2 \leqslant \mathcal{L}_1$. We denote by $\mathcal{L}^k$ the fragment of first-order logic in which no formula contains more than k distinct variables and we assume familiarity with the conventional fixed-point logics LFP and IFP and with logics with counting. See [EF99] for an introduction to these topics.

2 Symmetric Choice

In this section, we introduce the **sc-ifp** fixed-point operator and establish some of its basic properties. **sc-ifp** takes two operands, respectively the induction formula and the choice formula. We consider the fixed point as being built up in a series of stages. At each stage of evaluation on some structure $\mathfrak{A}$, the choice formula is evaluated to produce a set of tuples, the *choice set.* If the choice set is an automorphism class of $\mathfrak{A}$, one of its tuples is nondeterministically chosen; if it is not, no element is chosen. The induction formula is then evaluated using the chosen tuple, if any, along with the relation built up so far to define the tuples to be added to the relation being constructed.

To avoid circularity of definition, we define the operator to take a general class of maps from relations to relations as its arguments. We also introduce parameters to the definitions from the outset as the definitions are more sensitive to parameters than the definitions of conventional fixed-point logics. This complicates the definitions a little but the reader may treat all tuples of parameters as empty at a first reading. Throughout, we use the notation $\bar{z}$ to denote variables that are treated as parameters.

On a structure $\mathfrak{A}$ and for any interpretation of the r-ary relation X and the n-tuple of parameters $\bar{z}$, a formula $\varphi(X, \bar{x}\bar{z})$, where $|\bar{x}| = r$, defines an r-ary relation. We can, therefore, associate with φ a map $f^{\mathfrak{A}} : |\mathfrak{A}|^n \times \mathcal{P}(|\mathfrak{A}|^r) \to \mathcal{P}(|\mathfrak{A}|^r)$ and we shall define our operator in terms of maps such as this.

Definition 1. *Let $\mathfrak{A}$ be a structure and let X and Y be new relation symbols of arity r and s, respectively. A pair of maps $f^{\mathfrak{A}}$ and $g^{\mathfrak{A}}$ is* appropriate *for X and Y on $\mathfrak{A}$ if:*

- *both have domain $|\mathfrak{A}|^n \times \mathcal{P}(|\mathfrak{A}|^r) \times \mathcal{P}(|\mathfrak{A}|^s)$;*
- *$f^{\mathfrak{A}}$ has range $\mathcal{P}(|\mathfrak{A}|^r)$ and $g^{\mathfrak{A}}$ has range $\mathcal{P}(|\mathfrak{A}|^s)$;*
- *for any fixed $\bar{a} \in |\mathfrak{A}|^n$ and interpretations for X and Y in $\mathfrak{A}$, $f^{\mathfrak{A}}(\bar{a}, X, Y)$ and $g^{\mathfrak{A}}(\bar{a}, X, Y)$ are invariant under any automorphism of $(\mathfrak{A}, X, Y, \bar{a})$.*

To cope with parameterized definitions, we need to restrict attention to automorphisms that fix the interpretations of the parameters and we define automorphism classes appropriately. This is a technical condition which we will explain in Example 7 after we have defined the **sc-ifp** operator.

Definition 2. *Let $\mathfrak{A}$ be a structure and $n, s \in \mathbb{N}$. For a tuple $\bar{a} \in |\mathfrak{A}|^n$, an* $\bar{a}$-respecting automorphism class *of $|\mathfrak{A}|^s$ is a maximal set $C \subseteq |\mathfrak{A}|^s$ such that, for all $\bar{b}, \bar{c} \in C$, there is an automorphism ρ of $\mathfrak{A}$ such that $\rho(\bar{a}) = \bar{a}$ and $\rho(\bar{b}) = \bar{c}$.*

Definition 3. *Let $\mathfrak{A}$ be a structure and let $f^{\mathfrak{A}}$ and $g^{\mathfrak{A}}$ be maps appropriate for new relation symbols X and Y on $\mathfrak{A}$. For any fixed tuple of parameters $\bar{a} \in |\mathfrak{A}|^n$, let $\mathcal{T}^{\mathfrak{A},\bar{a}}_{f,g}$ be the least tree with the following properties:*

- *the root is labelled* $\langle \emptyset, \emptyset \rangle$;
- *if* $g^{\mathfrak{A}}(\bar{a}, R, S)$ *is a non-trivial* $\bar{a}$*-respecting automorphism class of* $(\mathfrak{A}, R, S)$, *a node labelled* $\langle R, S \rangle$ *has a child* $\langle R \cup f^{\mathfrak{A}}(\bar{a}, R, \{\bar{s}\}), \{\bar{s}\} \rangle$ *for each* $\bar{s} \in g^{\mathfrak{A}}(\bar{a}, R, S)$;
- *otherwise, a node labelled* $\langle R, S \rangle$ *has a child labelled* $\langle R \cup f(\bar{a}, R, \emptyset), \emptyset \rangle$.

Call a path in $\mathcal{T}^{\mathfrak{A},\bar{a}}_{f,g}$ R-dense *if infinitely many of its nodes have* R *as the first component of their label and define* **sc-ifp** $(f^{\mathfrak{A}}; g^{\mathfrak{A}})(\bar{a})$ *to be the set of relations* R *for which there is an* R*-dense path in the tree.*

The tree is equivalent to a nondeterministic computation. At each stage, if the map $g^{\mathfrak{A}}$ defines a single automorphism class of $\mathfrak{A}$, an element of that class is nondeterministically chosen and fed into the map $f^{\mathfrak{A}}$ to generate the next stage. Of course, there is no guarantee that the process defines a unique relation.

Example 4. On a pure set $\mathfrak{A}$, consider the maps $f^{\mathfrak{A}}$ and $g^{\mathfrak{A}}$ defined by

$$\varphi(X, Y, x_1x_2) \equiv Y(x_2) \wedge \big(x_1 = x_2 \vee X(x_1x_1)\big)$$
$$\psi(X, Y, y) \equiv \neg X(yy),$$

respectively. **sc-ifp** $(f^{\mathfrak{A}}; g^{\mathfrak{A}})$ defines the set of linear orders on $|\mathfrak{A}|$. Initially, the relations X and Y are empty. At the first iteration, ψ defines the whole of $|\mathfrak{A}|$, which is an automorphism class of the structure $\langle |\mathfrak{A}|, X, Y \rangle$ so an element is nondeterministically selected — call it '1' — and we set $Y = \{1\}$. We then add to the relation X all tuples satisfying φ. In this case, there is just one such tuple, 11. At the second iteration, ψ defines all members of $|\mathfrak{A}|$ other than 1. So long as $\mathfrak{A}$ is not a singleton, this is an automorphism class, so an element that we shall call '2' is nondeterministically selected and Y is set to $\{2\}$. Now φ is satisfied by the tuples 12 and 22 and these are added to X. At each subsequent stage, the choice set contains all elements that have not yet been ordered. If this set is non-empty, it is an automorphism class so one unordered element is selected to be the new maximal element of the ordering. Once all elements have been ordered, the choice set will always be empty so no more tuples will be added.

Although Definition 3 refers to labels occurring infinitely often in infinite trees, on a finite structure $\mathfrak{A}$, we need consider only a finite portion of the tree.

Lemma 5 (Finite Evaluation Lemma). *Let* $\mathfrak{A} \in \mathrm{STRUC}[\sigma]$ *and* $\bar{a} \in |\mathfrak{A}|^n$. *There is an* R*-dense path in* **sc-ifp** $(f^{\mathfrak{A}}; g^{\mathfrak{A}})(\bar{a})$ *if, and only if, there is an* S *such that* $\langle R, S \rangle$ *labels two nodes on a path in the tree between depth* d *and* $2d$, *for some* d *bounded by a polynomial in* $\|\mathfrak{A}\|$. □

This follows from the finiteness of $\mathfrak{A}$ and the fact that if $\langle R_i, S_i \rangle_{i \geqslant 0}$ labels a maximal path in the tree, $R_i \subseteq R_{i+1} \subseteq |\mathfrak{A}|^r$ for some r and $\|S_i\| \leqslant 1$ for all i. We omit the details as the proof is identical to that of Lemma 17 in [DR].

While **sc-ifp** does not, in general, define a unique relation, the relations defined by an application are pairwise isomorphic: the operator is *semideterministic* in the terminology of Gyssens, Van den Bussche and Van Gucht [GVV94].

Proposition 6. *If* $R, R' \in \mathbf{sc\text{-}ifp}\,(f^{\mathfrak{A}}_{;}\, g^{\mathfrak{A}})$ *then* $(\mathfrak{A}, R) \cong (\mathfrak{A}, R')$.

Proof. We show that, whenever $\langle R, S \rangle$ and $\langle R', S' \rangle$ label nodes at depth d in the tree, $(\mathfrak{A}, R, S) \cong (\mathfrak{A}, R', S')$. For $d = 0$, this is trivial.

Suppose $\langle R, S \rangle$ and $\langle R', S' \rangle$ label two (not necessarily distinct) nodes at depth d and that the former has a child labelled $\langle T, U \rangle$ and the latter a child labelled $\langle T', U' \rangle$. By the inductive hypothesis, there is an isomorphism $\rho : (\mathfrak{A}, R, S) \to (\mathfrak{A}, R', S')$. (We assume that all isomorphisms and automorphisms mentioned in this proof fix any first-order parameters to the maps.)

Since $g^{\mathfrak{A}}$ is invariant under automorphisms, it defines an automorphism class on $(\mathfrak{A}, R, S)$ if, and only if, it does on $(\mathfrak{A}, R', S')$. If it does not define an automorphism class, $U = U' = \emptyset$. If it does, we have $U = \{\bar{u}\}$ and $U' = \{\bar{u}'\}$ and there must be an automorphism α of $(\mathfrak{A}, R, S)$ such that $\rho(\alpha(\bar{u})) = \bar{u}'$. In both cases, $(\mathfrak{A}, R, U) \cong (\mathfrak{A}, R', U')$.

Since $f^{\mathfrak{A}}$ is invariant under automorphisms, it follows that $(\mathfrak{A}, T, U) = (\mathfrak{A}, R \cup f^{\mathfrak{A}}(\bar{a}, R, U), U) \cong (\mathfrak{A}, R' \cup f^{\mathfrak{A}}(\bar{a}, R', U'), U') = (\mathfrak{A}, T', U')$. □

The restriction to automorphisms that fix parameters is necessary in order to guarantee semideterminism. This is illustrated by the following example.

Example 7. Consider a pure set $\mathfrak{A}$ with at least two elements and the maps $f^{\mathfrak{A}}$ and $g^{\mathfrak{A}}$ defined, respectively, by the formulae

$$\varphi(X, Y, xz) \equiv Y(x) \vee x = z \qquad \text{and} \qquad \psi(X, Y, x) \equiv \text{true},$$

treating z as a parameter to φ. Fix an interpretation $a \in |\mathfrak{A}|$ for z. Initially, X and Y are both empty. At every iteration of the fixed point, ψ defines the whole of the set. At the first iteration, this is an automorphism class of the structure $\langle |\mathfrak{A}|, X, Y \rangle$ so some element 'b' is nondeterministically chosen and Y is set to $\{b\}$. φ is now satisfied by b and a (which are not necessarily distinct) and these elements are added to X. At subsequent iterations, the structure $\langle |\mathfrak{A}|, X, Y \rangle$ is not a single automorphism class so nothing more will happen: no more elements will be chosen and no more elements will be added to X. Therefore, $\mathbf{sc\text{-}ifp}\,(f^{\mathfrak{A}}_{;}\, g^{\mathfrak{A}})$ contains the relation $\{a\}$ and the relation $\{a, b\}$ for each $b \neq a \in |\mathfrak{A}|$: these relations are not isomorphic to each other.

We may use the **sc-ifp** operator to define a deterministic logic but, in doing so, we need to be careful. Since the semideterminism of the operator guarantees that the relations defined are isomorphic to each other, a sentence such as $\exists \bar{x}\, R(\bar{x})$, where the variables occurring in an atomic formulae involving R are quantified, is either true in $(\mathfrak{A}, R)$ for all $R \in \mathbf{sc\text{-}ifp}\,(f^{\mathfrak{A}}_{;}\, g^{\mathfrak{A}})$ or false for all of them. However, suppose the formula of Example 4 is evaluated in a graph $G = (V, E)$. The relation E does not appear in the formula, but the meaning of the formula would appear to be different in G than in its reduct to the empty vocabulary, as G may have fewer automorphisms than its reduct. Thus, the logic would fail to have Ebbinghaus's 'reduct property' [Ebb85].

To avoid defining something which, by Ebbinghaus's criteria, is not a logic at all, our semantics of the **sc-ifp** operator requires invariance under automorphisms of the relevant reduct of the structure. That is, if the operands only mention vocabulary τ, the expression is evaluated in $\mathfrak{A}\restriction\tau$. This means that the operator is only semideterministic with respect to this reduct but this is still sufficient to guarantee that formulae are deterministic. We denote by $\operatorname{voc}\varphi$ the vocabulary of relations mentioned in φ.

Definition 8. *Let σ be a vocabulary. The formulae of* SC-IFP$[\sigma]$ *are the least set containing all atomic σ-formulae and closed under first-order operations and the following rule.*

- *If $\varphi, \psi \in$ SC-IFP$[\sigma, X, Y]$, $\bar{x}$ and $\bar{y}$ are tuples of variables with $|\bar{x}| = \operatorname{ar}(X) = r$ and $|\bar{y}| = \operatorname{ar}(Y)$, $\bar{u}$ is an r-tuple of variables drawn from $u_1 \dots u_k$ and each $\mathbf{Q}_i$ is a quantifier $\exists$ or $\forall$ then $\mathbf{Q}_1 u_1 \cdots \mathbf{Q}_k u_k\, \mathbf{sc\text{-}ifp}_{X,Y,\bar{x}\bar{y}}(\varphi;\, \psi)(\bar{u})$ is a formula of* SC-IFP$[\sigma]$. *The free variables of this formula are the free variables of φ and ψ except those in $\bar{x}$ or $\bar{y}$.*

The semantics is that of first-order logic with the addition of the following rule.

- *Let $\Phi \equiv \bar{\mathbf{Q}}\bar{u}\, \mathbf{sc\text{-}ifp}_{X,Y,\bar{x}}(\varphi;\, \psi)(\bar{u})$ with free variables $z_1 \dots z_n$ and let $\tau = \operatorname{voc}\Phi$. If $\bar{a} \in |\mathfrak{A}|^n$, we write $(\mathfrak{A}, \bar{a}) \models \Phi$ if, and only if, $(\mathfrak{A}, R) \models \bar{\mathbf{Q}}\bar{u}\, R(\bar{u})$ for all $R \in \mathbf{sc\text{-}ifp}\,(f_\varphi^{\mathfrak{A}\restriction\tau}(\bar{a}, -, -), f_\psi^{\mathfrak{A}\restriction\tau}(\bar{a}, -, -))$.*

Note that, since **sc-ifp** is semideterministic, if $(\mathfrak{A}, R) \models \bar{\mathbf{Q}}\bar{u}\, R(\bar{u})$ for at least one R defined by the fixed point, then it is true for all of them.

The logic we have defined is broadly the same as FO + IFP$_{c,s}$ defined by Gire and Hoang in [GH98], but there are a number of important differences. Firstly, their logic could be more properly denoted IFP$_{c,s}$(IFP) as it has formulae only of the form $\bar{\mathbf{Q}}\bar{u}\, \mathbf{sc\text{-}ifp}_{X,Y,\bar{x}}(\varphi;\, \psi)(\bar{u})$ where φ and ψ are IFP formulae. That is, they do not allow either nesting or first-order combinations of fixed points, whereas our definitions permit both freely. Secondly, FO + IFP$_{c,s}$ does not allow parameters to fixed-point expressions which, as we show in the next section, has a significant effect on the expressive power of the logic. Thirdly, Gire and Hoang consider automorphisms of the whole structure rather than the reduct to the vocabulary of the formula. This has the undesirable effect that the meaning of a formula may depend on relations not mentioned in the formula.

In addition to the formal differences above, it is also convenient to allow simultaneous fixed-points, such as $\bar{\mathbf{Q}}\bar{u}\, \mathbf{sc\text{-}ifp}_{X_1,\dots,X_n,Y,\bar{x}}(\varphi_1, \dots, \varphi_n;\, \psi)(\bar{u})$. As usual, the fixed point defines relation X_1 with $X_2, \dots, X_n$ considered as auxiliary relations. Note that we permit only one choice formula, as making more than one choice per stage would destroy semideterminism. (Consider making two simultaneous choices from a pure set; the effect is essentially the same as in Example 7.) Simultaneous definitions of the form indicated do not affect the expressive power of the language, as can be shown by standard techniques for combining several relations into a single relation of wider arity, though we must use a coding, such as that in [DR], that does not interfere with automorphisms of the structure. We shall freely use simultaneous definitions from this point.

In [DR], we showed that the logics C-IFP, NIO and IFP + δ (from [BG00]) have equal expressive power. It is natural to ask why we consider here the symmetric version only of C-IFP. The answer is that the symmetric versions of the other two do not lead to sensible logics. IFP with symmetric δ is not semideterministic as it is possible to write a formula that makes two simultaneous choices with the effects discussed in the previous paragraph. On the other hand, a symmetric **nio** operator seems to be too restricted. When constructing a relation, at each stage, it would only be able to add a tuple from a definable automorphism class: it is not even obvious that such an operator could simulate IFP.

3 The Rôle of Parameters

In the previous section, we defined the logic SC-IFP, allowing fixed-point expressions to take first-order parameters and observed that Gire and Hoang's logic did not allow such parameters. Here, we show that the parameterless fragment of SC-IFP is much weaker than the whole logic. We denote by pSC-IFP this fragment, which is defined as the set of formulae of SC-IFP in which no occurrence of the fixed-point operator takes first-order parameters.

Consider structures over a vocabulary with a single binary relation $\approx$, interpreted as an equivalence relation. Call an equivalence relation *even* if all of its classes are of even cardinality. The following formula defines the class of even equivalence relations, by saying that no element z is in a class of odd cardinality.

$$\chi \equiv \theta \wedge \neg \exists z \, \mathbf{sc\text{-}ifp}_{P,Q,R,xy}(\varphi_P, \varphi_Q; \psi_R),$$

where θ states that $\approx$ is an equivalence relation, P, Q and R are new relation symbols of arity 0, 1 and 2 respectively and

$$\begin{aligned} \varphi_P &\equiv \forall u \left(u \approx z \to Q(u)\right) \\ \varphi_Q(x) &\equiv x = z \vee \exists u \left(R(ux) \vee R(xu)\right) \\ \psi_R(xy) &\equiv x \neq y \wedge x \neq z \wedge y \neq z \wedge x \approx y \approx z \wedge \neg Q(x) \wedge \neg Q(y). \end{aligned}$$

At each stage of the evaluation, a pair of elements xy is selected such that x, y and z are equivalent but distinct and neither x nor y is in Q. They and z are added to Q. If all elements of z's equivalence class are added to Q, it follows that the number of members of the class distinct from z is even, i.e., that the class is of odd cardinality. In contrast, the main result of this section is:

Theorem 9. *The class of even equivalence relations is not* pSC-IFP*-definable.*

From this and the formula χ above, it is immediate that:

Corollary 10. pSC-IFP $<$ SC-IFP. □

Towards a proof of Theorem 9, call an equivalence relation k*-large* if it has more than k equivalence classes and each class contains more than k elements.

Lemma 11. *On the class of k-large equivalence relations, every $\mathcal{L}^k$ formula is equivalent to one without quantifiers.*

Proof. Let $\mathfrak{A}$ be a k-large equivalence relation and $\varphi(\bar{x}) \in \mathcal{L}^k$. Since $\mathcal{L}^k$ cannot distinguish between $\mathfrak{A}$'s equivalence classes, if $(\mathfrak{A}, \bar{a}) \models \varphi$, the same is true for all $\bar{b}$ of the same atomic type so φ is equivalent to a disjunction of atomic types. □

Lemma 12. *Let $\varphi(\bar{x}) \in \mathcal{L}^k$. Either there is a formula in the language of equality to which φ is equivalent on all k-large equivalence relations or $\approx^{\mathfrak{A}}$ is uniformly $\mathcal{L}^k$-definable in $\langle\, |\mathfrak{A}|, \varphi^{\mathfrak{A}} \,\rangle$ for all k-large equivalence relations $\mathfrak{A}$.*

Proof. We may assume $R = \varphi^{\mathfrak{A}}$ is at least binary: if it is unary, it must be either empty or $|\mathfrak{A}|$ as $\mathcal{L}^k$ cannot distinguish between the elements of $\mathfrak{A}$. By the previous lemma, we may assume that φ is a disjunction of quantifier-free types, which we may write as $\bigvee_i (\eta_i \wedge \bigvee_j \alpha_{ij})$, where the η_i are an enumeration of all possible equality types of the relevant arity and the α_{ij} are $\approx$-types.

If all the $\bigvee_j \alpha_{ij}$ are trivial, R is a union of equality types; otherwise, there must be i and $m \neq n$ such that $\mathfrak{A} \models \forall \bar{x} \left[(\varphi_1(\bar{x}) \rightarrow \varphi(\bar{x})) \wedge (\varphi_2(\bar{x}) \rightarrow \neg\varphi(\bar{x})) \right]$, where $\varphi_1(\bar{x}) \equiv \eta_i \wedge x_m \approx x_n \wedge \beta$ and $\varphi_2(\bar{x}) \equiv \eta_i \wedge x_m \not\approx x_n \wedge \beta$ or vice-versa, for some quantifier-free formula β. We may assume that $m = 1$ and $n = 2$.

Because φ_1 and φ_2 are both satisfiable, it follows from the symmetry of $\mathfrak{A}$ that any pair $a_1 \approx a_2$ can be extended to a tuple $\bar{a}$ satisfying φ_1 (and, hence, satisfying φ) and any pair $a_1 \not\approx a_2$ to a tuple satisfying φ_2 (and, hence, not satisfying φ) or vice-versa. So, on $\langle\, |\mathfrak{A}|, R \rangle$, the following formula, which is equivalent to one in $\mathcal{L}^k$, defines either $\approx$ or $\not\approx$: $\exists \bar{u} \left[\eta_i(xy\bar{u}) \wedge R(xy\bar{u}) \;\wedge\; \exists z \left(\eta_i(xz\bar{u}) \wedge \neg R(xz\bar{u}) \right) \right]$. □

We are now ready to prove Theorem 9. We show that, for any k-variable pSC-IFP formula Φ, there are k-large equivalence relations $\mathfrak{A}$, which is even, and $\mathfrak{A}'$, which is not, between which Φ cannot distinguish. This is done by showing that the relations defined by every stage of the evaluation of every fixed point in Φ are defined by the same $\mathcal{L}^k$ formulae on both equivalence relations. Since no formula of $\mathcal{L}^k$ can distinguish between $\mathfrak{A}$ and $\mathfrak{A}'$, Φ cannot either. We shall denote by pSC-IFP^k_n the fragment of pSC-IFP in which formulae contain at most k distinct variables and at most n nested fixed points.

Proof (Theorem 9). Fix $k \in \mathbb{N}$. Let $\mathfrak{A}$ and $\mathfrak{A}'$ be k-large equivalence relations, with $\mathfrak{A}$ having classes of size $2k$, $2k+2$, …, $4k$ and $\mathfrak{A}'$ having classes of size $2k-1$, $2k+3$, $2k+4$, $2k+6$, …, $4k$. Note that $\mathfrak{A}$ is even and $\mathfrak{A}'$ is not.

Let $\varphi_1, \ldots, \varphi_r \in \mathcal{L}^k$ and $\mathfrak{A}^* = \langle \mathfrak{A}, \varphi_1^{\mathfrak{A}}, \ldots, \varphi_r^{\mathfrak{A}} \rangle$; define $\mathfrak{A}'^*$ similarly. We claim that, for any formula $\Phi \in$ pSC-IFP^k_n and any r, there is an $\mathcal{L}^k$ formula $\widehat{\Phi}$ such that $\mathfrak{A}^* \models \Phi \leftrightarrow \widehat{\Phi}$ and $\mathfrak{A}'^* \models \Phi \leftrightarrow \widehat{\Phi}$. This is proved by induction on the depth n of nesting of fixed points, assuming inductively that any relations defined so far by fixed points within which Φ is nested are defined by the φ_i. The base case $n = 0$ is trivial since pSC-IFP$^k_0 = \mathcal{L}^k$.

Assume the claim is true for pSC-IFP^k_n and consider $\Phi \in$ pSC-IFP$^k_{n+1}$. Let $\Psi \equiv \bar{\mathbf{Q}}\bar{u}\, \textbf{sc-ifp}_{X,Y,\bar{x}}(\varphi;\, \psi)(\bar{u})$ be a subformula of Φ. We may assume that $\approx$ is either mentioned in Ψ or is $\mathcal{L}^k$-definable from the relations that are. If not, by Lemma 12, the relations mentioned in Ψ are unions of equality types so Ψ cannot possibly distinguish between $\mathfrak{A}^*$ and $\mathfrak{A}'^*$ and we are done.

To evaluate Ψ on $\mathfrak{A}^*$, we must consider automorphisms of $\mathfrak{B} = \mathfrak{A}^* \restriction \mathrm{voc}\,\Psi$. The automorphisms of $\mathfrak{B}$ are exactly those of $\mathfrak{A}$ as $\mathfrak{A}$ and $\mathfrak{B}$ are $\mathcal{L}^k$-definable in each other. Since no two equivalence classes of $\mathfrak{A}$ are the same size, two tuples $\bar{a}, \bar{b} \in |\mathfrak{A}|^k$ are in the same automorphism class if, and only if, they have the same equality type and $a_i \approx b_i$ for $1 \leqslant i \leqslant k$.

As the automorphisms of $\mathfrak{B}$ are exactly the automorphisms of $\mathfrak{A}$, it suffices to consider the evaluation of Ψ in $\mathfrak{A}$ (and, of course, $\mathfrak{A}'$) and we shall assume that all references in Ψ to relations defined by the φ_i are replaced with the formulae defining them. After this substitution, Ψ still has at most k distinct variables.

Let $\langle P_i, Q_i \rangle_{i \geqslant 0}$ be the labels on a maximal path in $\mathcal{T}_\Psi^{\mathfrak{A}}$ and $\langle R_i, S_i \rangle_{i \geqslant 0}$ the labels on a maximal path in $\mathcal{T}_\Psi^{\mathfrak{A}'}$. Our second claim is that, for all $i \geqslant 0$, $Q_i = S_i = \emptyset$ and there are formulae $\theta_i \in \mathcal{L}^k$ such that $P_i = \theta_i^{\mathfrak{A}}$ and $R_i = \theta_i^{\mathfrak{A}'}$.

This is trivial for $i = 0$ as $P_0 = Q_0 = R_0 = S_0 = \emptyset$ by definition. Suppose $P_i = \theta_i^{\mathfrak{A}}$, $R_i = \theta_i^{\mathfrak{A}'}$ and $Q_i = S_i = \emptyset$. Q_{i+1} is the result of evaluating ψ in the structure $(\mathfrak{A}, P_i, Q_i)$. Since $\psi \in \mathrm{pSC\text{-}IFP}_n^k$, it follows by the inductive hypothesis of the first claim that it is equivalent to some $\widehat{\psi} \in \mathcal{L}^k$ which, by Lemma 11, we may assume to be quantifier-free. As no quantifier-free formula defines an automorphism class of $\mathfrak{A}$, $Q_{i+1} = \emptyset$. Similarly, $S_{i+1} = \emptyset$.

P_{i+1} is the result of evaluating φ in $(\mathfrak{A}, P_i, Q_{i+1})$. By the inductive hypothesis of the first claim, φ is equivalent on $\mathfrak{A}$ and $\mathfrak{A}'$ to some $\widehat{\varphi} \in \mathcal{L}^k$; by the inductive hypothesis of the second claim, $P_i = \theta_i^{\mathfrak{A}}$. It follows that $P_{i+1} = (\theta_i \vee \widehat{\varphi})^{\mathfrak{A}}$. Similarly, $R_{i+1} = (\theta_i \vee \widehat{\varphi})^{\mathfrak{A}'}$ and the second claim is proven.

By Lemma 5, there is a d such that, for all $i \geqslant d$, $P_i = P_d$ and $R_i = R_d$. Therefore, $\Psi \equiv \bar{\mathsf{Q}}\bar{u}\, \theta_d(\bar{u})$ and the first claim is proven. A simple pebble-game argument shows that no $\mathcal{L}^k$ formula can distinguish between $\mathfrak{A}$ and $\mathfrak{A}'$. □

It can be shown, using similar techniques, that there is no formula of $\mathrm{SC\text{-}IFP}_1$ (even with parameters) that defines the class of graphs all of whose components are of even cardinality. This class can, however, be defined by nesting applications of fixed-point operators. Therefore, $\mathrm{SC\text{-}IFP}_1 < \mathrm{SC\text{-}IFP}_2$: the ability to nest fixed points also increases the expressive power of the logic. We conjecture that the SC-IFP nesting hierarchy is strict, i.e., that $\mathrm{SC\text{-}IFP}_n < \mathrm{SC\text{-}IFP}_{n+1}$ for all n. There is no obvious way to simulate a formula using n nested fixed points with one using fewer as each level of nesting may mention a different vocabulary and hence be sensitive to automorphisms of different reducts of the structure.

The results of this section are in contrast to, for example, LFP, IFP, IFP + C, C-IFP and NIO, where neither the ability to supply first-order parameters to fixed points nor to nest fixed points increases the expressive power of the resulting logic (see, e.g., [DR]; the proofs are essentially the same for all of the logics mentioned). Parameters and nesting alter the expressiveness of SC-IFP because they alter the number of automorphisms available to formulae by fixing certain elements or increasing or reducing the number of relevant relations. Although SC-IFP is somewhat unusual in this respect, it remains a reasonable logic. It is, for example, regular in the sense of Ebbinghaus [Ebb85].

4 Expressive Power

It is easy to see that SC-IFP is strictly more powerful than IFP. Any formula of IFP is equivalent to one of the form $\exists y\,(\mathbf{ifp}_{X,\bar{x}}\,\varphi)(y\ldots y)$ for some first-order φ and this can be translated directly as $\exists y\,\mathbf{sc\text{-}ifp}_{X,Y,\bar{x}}(\varphi;\,\psi)(y\ldots y)$, for any syntactically appropriate ψ, as the choice part of the induction is ignored. We saw in the previous section that the class of even equivalence relations is SC-IFP-definable but it is easy to show that it is not definable in IFP.

In the remainder of this section, we consider the close relationship between SC-IFP and the graph isomorphism problem, GI. This is the following query:

Definition 13. *Let* $\sigma = \langle A^1, B^1, E^2 \rangle$ *and let* $\mathfrak{A}$ *and* $\mathfrak{B}$ *be graphs on disjoint vertex sets. Denote by* $\mathfrak{A}+\mathfrak{B}$ *the* σ*-structure* $\langle |\mathfrak{A}| \cup |\mathfrak{B}|, |\mathfrak{A}|, |\mathfrak{B}|, E^{\mathfrak{A}} \cup E^{\mathfrak{B}} \rangle$*. The* graph isomorphism problem *is the boolean query* $\{\mathfrak{A}+\mathfrak{B} : \mathfrak{A} \cong \mathfrak{B}\} \subset \mathrm{STRUC}[\sigma]$.

For our purposes, nothing is lost in restricting our attention to graph isomorphism rather than considering the isomorphism problem for general structures. The reader who would prefer to think in terms of general structure isomorphism may view graphs as a notational convenience.

Lemma 14. *Let* σ *be a vocabulary. There is a formula* $\rho_\sigma(\bar{u},\bar{v})$ *of* SC-IFP *such that, for any structure* $\mathfrak{A}$ *in a vocabulary including* σ*,* $(\mathfrak{A},\bar{a},\bar{b}) \models \rho_\sigma$ *if, and only if,* $\bar{a}$ *and* $\bar{b}$ *are exchanged by some automorphism of* $\mathfrak{A}\restriction\sigma$*.*

Proof. Let $|\bar{u}| = |\bar{v}| = k$. To determine whether there is an automorphism of $\mathfrak{A}\restriction\sigma$ that exchanges $\bar{u}$ and $\bar{v}$, we define a new k-ary relation P containing just those two tuples and ask if this relation is an automorphism class. Since P is a new relation, it will be an automorphism class of $(\mathfrak{A}\restriction\sigma, P)$ if, and only if, there is an automorphism of $\mathfrak{A}\restriction\sigma$ that exchanges $\bar{u}$ and $\bar{v}$.

Let X, Y, P and Z be new relation symbols of arity 0, 0, k and k, respectively.

$$\begin{aligned}\rho_\sigma(\bar{u},\bar{v}) &\equiv \mathbf{ifp}_{X,P,\bar{x}}(\theta, \bar{x}=\bar{u} \vee \bar{x}=\bar{v})\\ \theta &\equiv \mathbf{sc\text{-}ifp}_{Y,Z,\bar{y}}(\mathrm{true} \wedge \exists\bar{x}\,Z(\bar{x});\ P(\bar{y})),\end{aligned}$$

where 'true' is some tautology mentioning all of the relation symbols in σ. θ attempts to choose a tuple from P: this will succeed only if P is a non-trivial automorphism class of $\mathfrak{A}\restriction\sigma$. (Note that $\bar{u}$ and $\bar{v}$ are not parameters to θ so we are not restricted to automorphisms that fix these tuples.) Before the first iteration of ρ_σ, X is false and P is empty. Since P is not a non-trivial automorphism class, X remains false after the first iteration but P is set to $\{\bar{u},\bar{v}\}$. At the second iteration, X will become true if P is an automorphism class but nothing else will change at any future iteration. Hence, ρ_σ is satisfied if, and only if, the tuples interpreting $\bar{u}$ and $\bar{v}$ are exchanged by some automorphism of $\mathfrak{A}\restriction\sigma$. □

Theorem 15. *The graph isomorphism problem is* SC-IFP*-definable.*

Proof (sketch). Let $\mathfrak{A}$ be a structure in the vocabulary of Definition 13. $\mathfrak{A}$ is an instance of the graph isomorphism problem if $A^{\mathfrak{A}}$ and $B^{\mathfrak{A}}$ partition $|\mathfrak{A}|$ and every edge in $E^{\mathfrak{A}}$ is within one partition: this is first-order definable. To determine whether $\mathfrak{A}$ is a yes instance, we first, construct the relation $C = (A^{\mathfrak{A}} \times A^{\mathfrak{A}}) \cup (B^{\mathfrak{A}} \times B^{\mathfrak{A}})$. $\mathfrak{A}$ is a yes instance if, and only if, $\langle |\mathfrak{A}|, C, E \rangle$ contains two vertices, one in each C-component, that are exchanged by an automorphism. This is SC-IFP-definable by the previous lemma. □

Recall that a query Q is *Cook-reducible* to a query R if Q is decided by a polynomial-time Turing machine with an oracle for R, i.e., $Q \in \mathbf{P}^R$. We denote by $\mathbf{P}^{\mathrm{GI}}$ the class of problems that are Cook-reducible to graph isomorphism. SC-IFP $\leqslant \mathbf{P}^{\mathrm{GI}}$ as any SC-IFP formula requires at most a polynomial number of iterations, each requiring polynomially many calls to the oracle.

Let IFP(GI) be the logic obtained by closing first-order logic under inflationary fixed points and vectorized generalized quantifiers for graph isomorphism (see [Daw95,EF99] for a definition of vectorized quantifiers). By [Daw95, Theorem 5.6], IFP(GI) captures $\mathbf{P}^{\mathrm{GI}}$ on ordered structures. It is immediate from Theorem 15 that IFP(GI) $\leqslant$ SC-IFP*, the closure of SC-IFP under first-order reductions. (We require closure under reductions to deal with vectorizations; it is not clear whether SC-IFP is so closed.) We conjecture that SC-IFP* is strictly more powerful than IFP(GI) as there seems to be no way for the latter to simulate choice on arbitrary structures. However, on ordered structures, choice can be simulated by taking the least element of any set. This gives,

Theorem 16. *On ordered structures,* SC-IFP* = IFP(GI) = $\mathbf{P}^{\mathrm{GI}}$. □

What, then, is the expressive power of SC-IFP* on unordered structures? Suppose we can define an order of the automorphism classes of a structure: that is, a linear pre-order whose equivalence classes are precisely the automorphism classes. The elements of each automorphism class can be linearly ordered using the techniques of Example 4 so the pre-order can be refined to a linear order. This linear order can be used to express any $\mathbf{P}^{\mathrm{GI}}$ property of the structure.

It is easy to see that SC-IFP* captures $\mathbf{P}^{\mathrm{GI}}$ on pure sets and on any class of structures with only unary relations. We have seen in Example 4 how to linearly order a pure set. In the second case, the atomic types of elements are automorphism classes and, for any fixed vocabulary, there are only finitely many atomic types and they can be ordered even by a first-order formula.

Results on IFP + C allow us to show that SC-IFP* captures $\mathbf{P}^{\mathrm{GI}}$ on large classes of graphs. A linear ordering of the $\mathcal{C}^k$ types realized in a structure can be produced by an iterated refinement scheme known as the $(k-1)$-dimensional Weisfeiler–Lehman method, which is expressible in IFP + C [CFI92,Ott97]. It can be shown that this ordering of $\mathcal{C}^k$ types is also SC-IFP-definable.

A class of $\mathcal{G}$ graphs *contains almost all graphs* if, as $n \to \infty$, the proportion of all n-vertex graphs that are in $\mathcal{G}$ tends to 1. A logic captures a complexity class $\mathcal{C}$ on almost all graphs if it captures $\mathcal{C}$ on a class containing almost all graphs.

Theorem 17. SC-IFP* *captures* $\mathbf{P}^{\mathrm{GI}}$ *on almost all graphs.*

Proof (sketch). Immerman and Lander show that the $\mathcal{C}^2$-equivalence classes of almost all graphs are exactly the automorphism classes [IL90]. For any k, the ordering of the C^k types is SC-IFP-definable and the result follows from the observations above. □

Theorem 18. SC-IFP* *captures* $\mathbf{P}^{\mathrm{GI}}$ *on any class of structures which is closed under disjoint unions and on which* IFP + C *captures* **P**.

Proof. It follows from [Ott97, Theorem 4.22] that, if IFP + C captures **P** on a class of structures which is closed under disjoint unions, there is a k such that the C^k-equivalence classes of k-tuples are automorphism classes. □

It follows from Theorems 17 and 18 that SC-IFP* captures $\mathbf{P}^{\mathrm{GI}}$ on many natural classes of graphs, such as trees [IL90], planar graphs [Gro98] and graphs of bounded tree width [GM99] or genus [Gro00].

Any class of structures on which IFP + C captures **P** has a polynomial time isomorphism algorithm (namely, the k-dimensional Weisfeiler–Lehman method, for some k) so one might hope that Theorem 18 could be improved to SC-IFP* capturing **P** on such a class of structures. Although the isomorphism problem for these structures is in **P**, SC-IFP formulae may construct new structures internally and there might not be a polynomial-time isomorphism algorithm for these structures. For example, IFP + C captures **P** on ordered graphs but we have already shown that SC-IFP* captures $\mathbf{P}^{\mathrm{GI}}$ on that class. In particular, an SC-IFP formula can take an ordered instance of GI, 'forget' the ordering and ask if the two graphs encoded in the resulting structure are isomorphic. Hence, we cannot expect SC-IFP* to capture **P** on all classes of structures on which IFP + C does, unless graph isomorphism is itself in **P**.

Conclusion. Gire and Hoang's specified symmetric choice logic is a possible candidate for a logic for isomorphism-invariant polynomial-time properties. That is, it is one of a few logics that have been shown to properly extend IFP + C while being contained in **P**. There is no known example of a polynomial-time property that it cannot express. Towards a better understanding of this logic, in this paper we develop the semantics of symmetric choice and investigate its expressive power. We show that it is closely related to the complexity class $\mathbf{P}^{\mathrm{GI}}$. We also develop some techniques for establishing lower bounds, using them to demonstrate the power of parameters and of nesting.

References

[AB87] V. Arvind and S. Biswas. Expressibility of first order logic with a nondeterministic inductive operator. In *Proceedings of 4th Annual Symposium on Theoretical Aspects of Computer Science*, vol. 247 of *Lecture Notes in Computer Science*, pp. 323–335. Springer-Verlag, 1987.

[BG00] A. Blass and Y. Gurevich. The logic of choice. *Journal of Symbolic Logic*, 65(3):1264–1310, 2000.

[BHZ87] R.B. Boppana, J. Håstad and S. Zachos. Does co-NP have short interactive proofs? *Information Processing Letters*, 25(2):127–132, 1987.

[CFI92] J.-Y. Cai, M. Fürer and N. Immerman. An optimal lower bound on the number of variables for graph identification. *Combinatorica*, 12(4):389–410, 1992.

[Daw95] A. Dawar. Generalized quantifiers and logical reducibilities. *Journal of Logic and Computation*, 5(2):213–226, 1995.

[DR] A. Dawar and D.M. Richerby. Fixed-point logics with nondeterministic choice. *Journal of Logic and Computation.* To appear.

[Ebb85] H.-D. Ebbinghaus. Extended logics: The general framework. In J. Barwise and S. Feferman, editors, *Model-Theoretic Logics*, Perspectives in Mathematical Logic, pp. 25–76. Springer-Verlag, 1985.

[EF99] H.-D. Ebbinghaus and J. Flum. *Finite Model Theory*, 2nd edition. Springer-Verlag, 1999.

[Fag74] R. Fagin. Generalized first-order spectra and polynomial-time recognizable sets. In R. Karp, editor, *Complexity of Computation*, vol. 7 of *SIAM-AMS Proceedings*, pp. 43–73. SIAM-AMS, 1974.

[GH98] F. Gire and H.K. Hoang. An extension of fixpoint logic with a symmetry-based choice construct. *Information and Computation*, 144:40–65, 1998.

[GM99] M. Grohe and J. Mariño. Definability and descriptive complexity on databases of bounded tree-width. In *Proceedings of 7th International Conference on Database Theory*, vol. 1540 of *Lecture Notes in Computer Science*, pp. 70–82. Springer-Verlag, 1999.

[Gro98] M. Grohe. Fixed-point logics on planar graphs. In *Proceedings of 13th IEEE Annual Symposium on Logic in Computer Science*, pp. 6–15. IEEE Computer Society, 1998.

[Gro00] M. Grohe. Isomorphism testing for embeddable graphs through definability. In *Proceedings of 32nd ACM Symposium on Theory of Computing*, pp. 63–72. ACM, 2000.

[GVV94] M. Gyssens, J. Van den Bussche and D. Van Gucht. Expressiveness of efficient semi-deterministic choice constructs. In *Proceedings of 21st International Colloquium on Automata, Languages and Programming*, vol. 820 of *Lecture Notes in Computer Science*, pp. 106–117. Springer-Verlag, 1994.

[IL90] N. Immerman and E.S. Lander. Describing graphs: A first-order approach to graph canonization. In A. Selman, editor, *Complexity Theory Retrospective*, pp. 59–81. Springer-Verlag, 1990.

[Imm86] N. Immerman. Relational queries computable in polynomial time. *Information and Control*, 68(1–3):86–104, 1986.

[Ott97] M. Otto. *Bounded Variable Logics and Counting — A Study in Finite Models*, vol. 9 of *Lecture Notes in Logic*. Springer-Verlag, 1997.

[Sch88] U. Schöning. Graph isomorphism is in the low hierarchy. *Journal of Computer and System Sciences*, 37(3):312–323, 1988.

[Tor00] J. Torán. On the hardness of graph isomorphism. In *Proceedings of 41st Annual Symposium on Foundations of Computer Science*, pp. 180–186. IEEE Computer Society, 2000.

[Var82] M.Y. Vardi. Complexity of relational query languages. In *Proceedings of 14th ACM Symposium on Theory of Computing*, pp. 137–146. ACM, 1982.

Positive Games and Persistent Strategies*

Jacques Duparc

LuFG Mathematische Grundlagen der Informatik,
RWTH-Aachen, D-52056 Aachen
Fax: +49/241/8022-215
duparc@informatik.rwth-aachen.de

Abstract. At CSL 2002, Jerzy Marcinkowsi and Tomasz Truderung presented the notions of positive games and persistent strategies [8]. A strategy is persistent if, given any finite or infinite run played on a game graph, each time the player visits some vertex already encountered, this player repeats the decision made when visiting this vertex for the first time. Such strategies require memory, but once a choice is made, it is made for ever. So, persistent strategies are a weakening of memoryless strategies.
The same authors established a direct relation between positive games and the existence of persistent winning strategies. We give a description of such games by means of their topological complexity. In games played on finite graphs, positive games are unexpectedly simple. On the contrary, infinite game graphs, as well as infinite alphabets, yield positive sets involved in non determined games.
Last, we discuss positive Muller winning conditions. Although they do not help to discriminate between memoryless and LAR winning strategies, they bear a strong topological characterization.

1 Introduction

The theoretical framework of infinite two-player games has found growing interest in theoretical computer science. With applications to verification and synthesis of reactive programs, the usual setting are the finite-state games where two players move a token along the edges of a finite game graph. The player (**0** or **1**) to whom the current vertex belongs, pushes the token to one of the successors of this vertex. Practical applications imposing simple winning conditions, these games are usually determined, and one can decide the winner, and compute a winning strategy. Here, for algorithmic concerns, how much memory is needed to win plays a crucial role. Indeed, memoryless (or positional) strategies for Muller games can be verified in polynomial time, provided winning conditions for the opponent, together with the game graph itself, are given as input.

Recently, a weakening of the notion of memoryless strategy was introduced to obtain results on complexity of deciding graph games with winning conditions defined by formulas from fragments of *LTL* [8].

* The author sincerely thanks Erich Grädel for numerous remarks and corrections.

M. Baaz and J.A. Makowsky (Eds.): CSL 2003, LNCS 2803, pp. 183–196, 2003.

Definition 1. *A strategy σ for player P is* persistent *if for each play $v_1v_2\dots v_k$, played according to σ, $v_i = v_j$ and P is to move at v_i, then $v_{i+1} = v_{j+1}$.*

A persistent strategy behaves like a becoming positional one. Uncertainty is confined to the vertices not encountered yet. But, once every vertex has been visited, the strategy definitively becomes positional.

It is very useful to know of some general property of the winning conditions that gives the existence of persistent strategies among the winning ones. In [8]. such a very effective property is pointed out.

Theorem 1 (Marcinkowski & Truderung). *If a game on a finite graph is positive for player P, and his opponent has a winning strategy, then he has a* persistent *winning strategy.*

Here, a game is *positive*, if for any two infinite plays x, y, where x is a subsequence of y, whenever x is winning, y is also winning.

Intuitively, theorem 1 says that, in a positive game for the opponent, if, in the picture below, the left infinite path is winning for the other player, then the right one also.

In other words, there exists a persistent winning strategy if removing loops that are subpaths still preserves the winning condition.

Not all games that admit persistent winning strategies are positive (for one of the players). For example, most parity games are not positive for either player, however the winner has a memoryless - hence persistent - winning strategy.

This paper gives a description of these positive games, by means of their topological complexity. We show that they are quite simple - $\boldsymbol{\Pi}^0_2$ - when the game graph is finite. On the contrary, when the constraint of finite graphs is abandonned, these games can be extremely complicated, and can even give rise to non determined games. Last, we study the prominent example of Muller winning condition. In this context, we show that positive games and $\boldsymbol{\Pi}^0_2$ games coincide.

2 Preliminaries

Given two - finite or infinite - sequences x, y, the relation $x \sqsubseteq y$ means that x is a subsequence of y. We write $x \sqsubset y$ when $x \sqsubseteq y$ holds, but $y \sqsubseteq x$ fails.

Unless mentionned, throughout this paper we consider infinite two-player games played on a directed graph $\mathcal{G} = (v_{\text{it}}, V_0, V_1, E)$, where $\{V_0, V_1\}$ is a partition (between the two players $\mathbf{0}$ and $\mathbf{1}$) of the set of vertices V. The initial vertex where the play starts is v_{it}. An infinite play x is nothing but an infinite sequence of vertices: $x = v_0v_1v_2\ldots v_iv_{i+1}\ldots$ such that $v_0 = v_{\text{it}}$, and for every integer i, $(v_i, v_{i+1}) \in E$. The choice of v_{i+1} being made by the player to whom v_i belongs to.

We set $\mathcal{G}^*$ as the tree of all possible positions inside $\mathcal{G}$.

$$\mathcal{G}^* = \{v_0v_1v_2\ldots v_n \in V^* : \ v_0 = v_{\text{it}} \ \wedge \ \forall i < n \ \ (v_i, v_{i+1}) \in E\}.$$

We write $\mathcal{G}^\omega$ for the set of infinite plays (the set of infinites branches of $\mathcal{G}^*$). With these notations, the definition of a positive game becomes.

Definition 2 (Marcinkowski & Truderung [8]). *Let $A \subseteq \mathcal{G}^\omega$,*

$$A \textit{ is } \text{positive} \iff_{def} \forall x, y \in \mathcal{G}^\omega \ \ (x \in A \wedge x \sqsubseteq y) \longrightarrow y \in A.$$

We equip $\mathcal{G}^\omega$ with the usual topology - that is the topology induced by the usual topology on V^ω: the product topology of the discrete topology on V. In other words, non empty open sets of $\mathcal{G}^\omega$ take the form $WV^\omega \cap \mathcal{G}^\omega$ for some set $W \subseteq \mathcal{G}^*$ of finite words.

We recall that the *finite Borel Hierarchy* is a sequence $\boldsymbol{\Sigma}^0_1, \boldsymbol{\Pi}^0_1, \boldsymbol{\Sigma}^0_2, \boldsymbol{\Pi}^0_2, \ldots$ of classes of ω-languages over spaces of the form $\mathcal{G}^\omega$ inductively defined by:

- $\boldsymbol{\Sigma}^0_1 = \{\textit{Open sets}\}$
- $\boldsymbol{\Pi}^0_n = \left\{A^{\complement} : \ A \in \boldsymbol{\Sigma}^0_n\right\}$ $(n > 0)$, where $A^{\complement}$ stands for the complement of A.
- $\boldsymbol{\Sigma}^0_{n+1} = \left\{\bigcup_{i \in \mathbb{N}} A_i : \ \forall i \in \mathbb{N} \ \ A_i \in \boldsymbol{\Pi}^0_n\right\}$ - countable unions of sets in $\boldsymbol{\Pi}^0_n$.

This hierarchy classifies sets with respect to their topological complexity, it is ordered by inclusion in the sense that $\boldsymbol{\Sigma}^0_n$ and $\boldsymbol{\Pi}^0_n$ are both strictly included in $\boldsymbol{\Sigma}^0_{n+1}$ and $\boldsymbol{\Pi}^0_{n+1}$, while $\boldsymbol{\Sigma}^0_n$ and $\boldsymbol{\Pi}^0_n$ are incomparable for inclusion. The class $\boldsymbol{\Sigma}^0_n \cap \boldsymbol{\Pi}^0_n$ is denoted $\boldsymbol{\Delta}^0_n$.

3 Finite Game Graphs

For the general study of positive games, we need to consider the following basic sets.

Definition 3. *Let $\alpha = (u, \Lambda) \in \mathcal{G}^* \times \mathcal{P}_f(V)$,*

$$\Omega_\alpha = \bigcup_{u \sqsubseteq v} (v\mathbb{I}_\Lambda \cap \mathcal{G}^\omega) = \left(\bigcup_{u \sqsubseteq v} v\mathbb{I}_\Lambda\right) \cap \mathcal{G}^\omega$$

where $\mathbb{I}_\Lambda = \{y \in V^\omega : \ \forall a \in \Lambda \ \ a \text{ occurs infinitely often in } y\}$.

So, Ω_α is the set of all infinite sequences in $\mathcal{G}^\omega$ that contain u as a subsequence, and infinitely many times all vertices of Λ. It is ω-rational and topologically simple:

Proposition 1. *Let $\alpha = (u, \Lambda) \in \mathcal{G}^* \times \mathcal{P}_f(V)$,*

$$\Omega_\alpha \in \boldsymbol{\Pi}^0_2.$$

Proof of proposition 1: We assume $\Lambda = \{a_0, \dots, a_k\}$. To show that $\Omega_\alpha \in \boldsymbol{\Pi}^0_2$, it is enough to see that

$$\Omega_\alpha = \left(\bigcap_{n\in\omega} \left(\bigcup_{u\sqsubseteq v\in V} v\,(V^* a_0 V^* a_1 \cdots V^* a_k V^*)^n\, V^\omega \right)\right) \cap \mathcal{G}^\omega .$$

$\dashv$ 1

We equip $\mathcal{G}^* \times \mathcal{P}_+(V)$ with the following partial ordering.

Definition 4. *Let* $\alpha = (u, \Lambda)$, $\beta = (v, \Gamma) \in \mathcal{G}^* \times \mathcal{P}_+(V)$,

$$\alpha \leq_s \beta \iff_{def} u \sqsubseteq v \ \wedge\ \Lambda \subseteq \Gamma .$$

Lemma 1.

$$\langle \mathcal{G}^* \times \mathcal{P}_+(V), \leq_s \rangle \textit{ is a well quasi ordering (WQO).}$$

Proof of lemma 1: We first recall Higman's lemma ([5][11]) which states that given $\tilde{\leq}$ any WQO on some alphabet Σ, the partial ordering on Σ^* defined by $u \tilde{\sqsubseteq} v \iff_{\text{def}} x_i \tilde{\leq} y_{\phi(i)}$ *holds for some strictly increasing* $\phi: lh(u) \longmapsto lh(v)$, is also a WQO.

Clearly, $\langle V, = \rangle$ is a WQO (only because V is finite!). Hence, by Higman's lemma, $\sqsubseteq$ is a WQO on V^*, from which we derive that $\sqsubseteq$ is a WQO on $\mathcal{G}^*$.

- Towards a contradiction, we assume that there exists an infinite decreasing sequence $\alpha_0 >_s \alpha_1 >_s \alpha_2 >_s \alpha_3 >_s \dots$, Since $\mathcal{P}_+(V)$ is finite, there is an infinite subsequence $\alpha_{i_0} >_s \alpha_{i_1} >_s \alpha_{i_2} >_s \alpha_{i_3} >_s \dots$ where all elements have same second projection. Let u_j denotes the first projection of α_j, we obtain $u_{i_0} \sqsupset u_{i_1} \sqsupset u_{i_2} \sqsupset u_{i_3} \sqsupset \dots$, which contradicts the fact $\sqsubseteq$ is a WQO on V^*.
- Towards a contradiction, let us assume that there exists $(\alpha_i)_{i\in\mathbb{N}}$ an infinite antichain for $\leq_s$. Since $\mathcal{P}_+(V)$ is finite, there exists an infinite subset $I \subseteq \mathbb{N}$ such that all elements in $(\alpha_i)_{i\in I}$ have same second projection. Let u_i denotes the first projection of α_i, we obtain that $(u_i)_{i\in I}$ is an infinite antichain for $\sqsubseteq$, which also contradicts the fact $\sqsubseteq$ is a WQO on V^*.

So $\leq_s$ is a WQO on $V^* \times \mathcal{P}_+(V)$, therefore $\leq_s$ is also a WQO on $\mathcal{G}^* \times \mathcal{P}_+(V)$.
$\dashv$ 1

Definition 5. *We define the mapping* $x \longmapsto \alpha_x$ *from* $\mathcal{G}^\omega$ *to* $\mathcal{G}^* \times \mathcal{P}_+(V)$ *by:*

$$\alpha_x = (u, \Lambda) \in \mathcal{G}^* \times \mathcal{P}_+(V) \textit{ where:}$$

- Λ *is the non empty set of all letters occuring infinitely often in* x.
- u *is the shortest sequence such that* $u^{-1}x \in \Lambda^\omega$. *(i.e.* $x = uy$ *with* $y \in \Lambda^\omega$*). Or equivalently,* u *is the longest prefix of* u *that ends with a letter not in* Λ.

This mapping, the sets Ω_α, and the WQO $\leq_s$, are all we need to prove:

Theorem 2. *Let $A \subseteq \mathcal{G}^\omega$ be positive, and $x \in \mathcal{G}^\omega$,*

1. $x \in A \Longrightarrow x \in \Omega_{\alpha_x} \subseteq A$,
2. $A = \bigcup_{x \in A} \Omega_{\alpha_x}$,
3. *there exists some finite set* $\{x_0, \ldots, x_k\} \subseteq A$ *such that* $A = \bigcup_{i \leq k} \Omega_{\alpha_{x_i}}$,
4. $A \in \boldsymbol{\Pi}^0_2$, *and A is an ω-rational set.*

Proof of theorem 2: (1) and (2) are immediate from the definition of both α_x and Ω_α. (3) is an immediate consequence of lemma (1): we set

$$B = \{\alpha_x : \ x \in A\}.$$

We let Min_B be the set of all $\leq_s$-minimal elements of B. clearly Min_B forms an antichain for $\leq_s$, hence, by lemma 1, it is finite. Since $\Omega_\beta \subseteq \Omega_\alpha$ clearly follows from $\alpha \leq_s \beta$, we obtain

$$A = \bigcup_{x \in A} \Omega_{\alpha_x} = \bigcup_{\alpha \in B} \Omega_\alpha \subseteq \bigcup_{\alpha \in Min_B} \Omega_\alpha \subseteq A.$$

Finally (4) comes from the fact $A = \bigcup_{\alpha \in Min_B} \Omega_\alpha$, where each Ω_α is both ω-rational and a $\boldsymbol{\Pi}^0_2$-set. Hence A is ω-rational and $\boldsymbol{\Pi}^0_2$ as a finite union of ω-rational $\boldsymbol{\Pi}^0_2$-sets. ⊣ 2

This yields the following characterization of positive games played on finite graphs:

Proposition 2. *Let $A \subseteq \mathcal{G}^\omega$,*

$$A \text{ positive} \iff A = \bigcup_{i \leq k} \Omega_{\alpha_i} \text{ for finitely many } \alpha_0, \ldots, \alpha_k \in \mathcal{G}^* \times \mathcal{P}_+(V).$$

Proof of proposition 2: ($\Longrightarrow$) is case (3) of previous theorem. The other direction relies on:

- the class of *positive* subsets of $\mathcal{G}^\omega$ is closed under union,
- each Ω_α is positive.

⊣ 2

Similarly to computational complexity, there are notions of reduction and completeness for topological complexity. What plays there the role of polynomial time reduction, is continuous reduction: given two ω-languages, A and B, one says A continuously reduces to B (denoted $A \leq_W B$ since introduced by Wadge in 1974 [15]) if there is a continuous mapping φ such that $x \in A \iff \varphi(x) \in B$ holds. Where continuity for such a mapping $\varphi : \ \mathcal{G}^\omega \longmapsto \mathcal{G}'^\omega$, means that every inverse image of an open set is open - for any $W_B \subseteq \mathcal{G}'^*$ there exists some $W_A \subseteq \mathcal{G}^*$ such that $y \in (W_A V^\omega) \cap \mathcal{G}^\omega \iff \varphi(y) \in (W_B V'^\omega) \cap \mathcal{G}'^\omega$ holds for all $y \in \mathcal{G}'^\omega$.

The relation $\leq_W$ is a partial ordering which measures the topological complexity. Intuitively $A \leq_W B$ means that A is less complicated than B with regard to the topological structure. This notion of continuous reduction yields the one of completeness for topological classes. Indeed, a set A is $\boldsymbol{\Sigma}^0_n$-complete (resp. $\boldsymbol{\Pi}^0_n$-complete) if it belongs to the class $\boldsymbol{\Sigma}^0_n$ (resp. $\boldsymbol{\Pi}^0_n$) and reduces all sets in the class. For example, $0^*1\{0,1\}^\omega$ is $\boldsymbol{\Sigma}^0_1$-complete, and the set $\mathbb{Q}$ of all infinite words over the alphabet $\{0,1\}$ that have finitely many 1 is $\boldsymbol{\Sigma}^0_2$-complete. As a matter of fact, if A is $\boldsymbol{\Sigma}^0_n$-complete, then its complement $A^{\complement}$ is $\boldsymbol{\Pi}^0_n$-complete, and both A and $A^{\complement}$ are incomparable for $\leq_W$. We write $A \equiv_W B$ when both $A \leq_W B$ and $B \leq_W A$ hold; and $A <_W B$ when $A \leq_W B$ is verified but not $B \leq_W A$. (For background see e.g. [6]).

The main device in working with this measure of complexity is due to Wadge [15]. It is a game that links the existence of a winning strategy for a player to the existence of a continuous function that witnesses the relation $A \leq_W B$:

Definition 6 (Wadge game). *Given $A \subseteq \Sigma_A^\omega$, $B \subseteq \Sigma_B^\omega$, $\mathbb{W}(A,B)$ is an infinite two player game between players I and II, where players take turns. Player I plays letters in Σ_A, and II plays finite words over the alphabet Σ_B. At the end of an infinite play (in ω moves), I has produced an ω-sequence $x \in \Sigma_A^\omega$ of letters and II has produced an ω-sequence of finite words which concatenated give rise to a finite or ω-word $y \in \Sigma_B^* \cup \Sigma_B^\omega$. The winning condition on the resulting play, denoted here $x\hat{\ }y$, is the following:*

$$\text{II wins the play } x\hat{\ }y \quad \Longleftrightarrow_{def} \ y \text{ is infinite} \wedge (x \in A \longleftrightarrow y \in B).$$

Proposition 3 ([15]). *II has a winning strategy in $\mathbb{W}(A,B) \iff A \leq_W B$.*

We recall that an ordinal is the set of its predecessors. i.e. $\xi+1 = \{0,1,2,\ldots,\xi\}$. In order to get a finer characterization of the positive sets, we introduce the following sets.

Definition 7. *Given ξ any countable ordinal, $i \in \{0,1,(0,1)\}$, we define $\mathcal{D}_i^\xi$ by:*

- *$\mathcal{D}_0^\xi = \{x \in (\xi+1)^\omega : \mathit{parity}(\mathit{min}_x) \text{ is even}\}$, where min_x is the least ordinal in the longest decreasing initial sequence of x, and α is even iff $\alpha = \omega \cdot \beta + n$, for some even $n \in \mathbb{N}$.*
- *$\mathcal{D}_1^\xi = (\xi+1)^\omega \setminus \mathcal{D}_0^\xi$, and*
- *$\mathcal{D}_{(0,1)}^\xi = 0\mathcal{D}_0^\xi \cup 1\mathcal{D}_1^\xi$.*

The space $(\xi+1)^\omega$ is equipped with the usual topology, that is basic open sets are of the form $u(\xi+1)^\omega$ for some finite sequence u of ordinals below $\xi+1$.

We need two technical results. For $u \in \mathcal{G}^\omega$, and $u \in \mathcal{G}^*$, $A_{(u)}$ denotes the set of infinite words x such that $ux \in A$

Lemma 2. *Let $A \subseteq \Sigma^\omega$,*

$$A \in \boldsymbol{\Delta}^0_2 \iff \exists! \xi < \omega_1 \ \exists! i \in \{0,1,(0,1)\} \ \left(A \equiv_W \mathcal{D}_i^\xi\right).$$

Lemma 3. *Let $A \subseteq \mathcal{G}^\omega$, $\theta < \lambda$ be countable ordinals with λ limit, $i \in \{0,1,(0,1)\}$, and $n \in \mathbb{N}$,*

$$\mathcal{D}_0^\theta \leq_W A \equiv_W \mathcal{D}_i^{\lambda+n} \implies \exists u \in \mathcal{G}^* \begin{cases} A_{(u)} \equiv_W \mathcal{D}_0^{\theta+1} \\ \quad or \\ A_{(u)} \equiv_W \mathcal{D}_0^\theta. \end{cases}$$

The previous two lemmas are very similar to results published in [3]. For completeness, we give direct proofs in the appendix. $\boldsymbol{D}_n(\boldsymbol{\Sigma}_1^0)$ is the class of sets $A = (A_{n-1} \setminus A_{n-2}) \cup (A_{n-3} \setminus A_{n-4}) \cup \ldots$ for open sets $A_0 \subseteq A_1 \subseteq \ldots \subseteq A_{n-1}$.

Lemma 4. *Let $A \subseteq \mathcal{G}^\omega$ be positive,*

$$A \in \boldsymbol{\Delta}_2^0 \implies \exists n \in \mathbb{N} \;\; A \in \boldsymbol{D}_n(\boldsymbol{\Sigma}_1^0).$$

(i.e. $A \equiv_W \mathcal{D}_i^n$ for some $i \in \{0,1,(0,1)\}$ and $n \in \mathbb{N}$).

Proof of lemma 4: Towards a contradiction, we assume that $A \in \boldsymbol{D}_n(\boldsymbol{\Sigma}_1^0)$ fails for all integer n. This is equivalent to say $\mathcal{D}_0^n \leq_W A$ holds for any n. From proposition 2, we can assume that there exists an integer k such that:

$$A = \bigcup_{i \leq k} \Omega_{\alpha_i} \text{ with } \alpha_0 = (u_0, \Lambda_0), \ldots, \alpha_k = (u_k, \Lambda_k) \in \mathcal{G}^* \times \mathcal{P}_+(V).$$

Since $A \in \boldsymbol{\Delta}_2^0 \setminus \left(\bigcup_{n \in \mathbb{N}} \boldsymbol{D}_n(\boldsymbol{\Sigma}_1^0) \right)$, there exists some limit cardinal λ, some integer n and some $i \in \{0,1,(0,1)\}$ such that $A \equiv_W \mathcal{D}_i^{\lambda+n}$. By lemma 3, given some non limit ordinal $\theta < \lambda$, there exists a sequence $(w_i)_{i \in \mathbb{N}}$ of finite words of $\mathcal{G}^*$, together with a sequence $(c_i)_{i \in \mathbb{N}} \in 2^\omega$ such that $A_{(w_i)} \equiv_W \mathcal{D}_{c_i}^{\theta+i}$. So, in particular $A_{(w_i)} < A_{(w_j)}$ holds for any $i < j$. It follows that there exists an infinite subset $I \subseteq \mathbb{N}$ such that, given any two $i, j \in I$:

$$w_i(lh(w_i)-1) = w_j(lh(w_j)-1) \text{ and } \forall l \leq k \;\; \forall w \subseteq u_l \;\; (w \sqsubseteq w_i \iff w \sqsubseteq w_j).$$

In other words, w_i and w_j both end with the same vertex (let us call it v_{end}), and they extend exactly the same prefixes of the same words among $\{u_0, \ldots, u_k\}$.

Now consider two different $i, j \in I$, such that $i < j$, one has $A_{(w_i)} <_W A_{(w_j)}$. Get σ any winning strategy for $I\!I$ in the game $\mathbb{W}(\mathcal{D}_{c_j}^{\theta+j}, A_{(w_j)})$. We claim that σ is winning for $I\!I$ in the game $\mathbb{W}(\mathcal{D}_{c_j}^{\theta+j}, A_{(w_i)})$. For this, we let $\mathcal{G}_{v_{\text{end}}}^\omega$ denote the set of infinite paths in the graph $\mathcal{G}$ with initial vertex v_{end}. It is then enough to notice the following:

$$\begin{aligned} \forall x \in \mathcal{G}_{v_{\text{end}}}^\omega \quad x \in A_{(w_j)} &\iff \exists l \leq k \;\; (u_l \sqsubseteq w_j x \;\wedge\; x \in \mathbb{I}_{\Lambda_l}) \\ &\iff \exists l \leq k \;\; (u_l \sqsubseteq w_i x \;\wedge\; x \in \mathbb{I}_{\Lambda_l}). \end{aligned}$$

Which proves $I\!I$ wins $\mathbb{W}(\mathcal{D}_{c_j}^{\theta+j}, A_{(w_i)})$, hence $A_{(w_j)} \leq_W \mathcal{D}_{c_j}^{\theta+j} \leq_W A_{(w_i)}$ holds, which contradicts $A_{(w_i)} <_W A_{(w_j)}$.

⊣ 4

These technical results put together give the following characterization of positive games.

Theorem 3. *Let V be finite, and $A \subseteq \mathcal{G}^\omega$,*

$$A \text{ positive} \implies \begin{cases} A \equiv_W \mathcal{D}_i^n & \text{for some } n \in \mathbb{N}, \text{ and } i \in \{0,1,(0,1)\} \\ or & \\ A \equiv_W \mathbb{Q}^{\mathsf{C}}. & \end{cases}$$

Proof of theorem 3: From theorem 2(4), we know that if A is positive, then A belongs to $\boldsymbol{\Pi}_2^0$, hence $A \leq_W \mathbb{Q}^{\mathsf{C}}$ holds. If $A <_W \mathbb{Q}^{\mathsf{C}}$ is verified, then $A \in \boldsymbol{\Delta}_2^0$, and by lemma 4, this implies that A is a finite difference of open sets. Hence, the only possibilities are $A \equiv_W \mathcal{D}_i^n$ for some integer n and some $i \in \{0,1,(0,1)\}$.
⊣ 3

To say it differently, any infinite play will sooner or later be restricted to some Sticly Connected Component (SCC) of the game graph. Assume ux and uy are two infinite plays with final parts x and y taking place in the same SCC. The first condition $A \equiv_W \mathcal{D}_i^n$ means that $ux \in A$ iff $uy \in A$. The second condition $A \equiv_W \mathbb{Q}^{\mathsf{C}}$ means that for at least one SCC, there exists such particular u, x, y such that $ux \notin A$, $uy \in A$, and $x \sqsubseteq y$.

Remark 1. The other direction of the implication in theorem 3 does not hold: consider the open set 0102^ω inside the following graph. It is not positive, and its complement is not positive either.

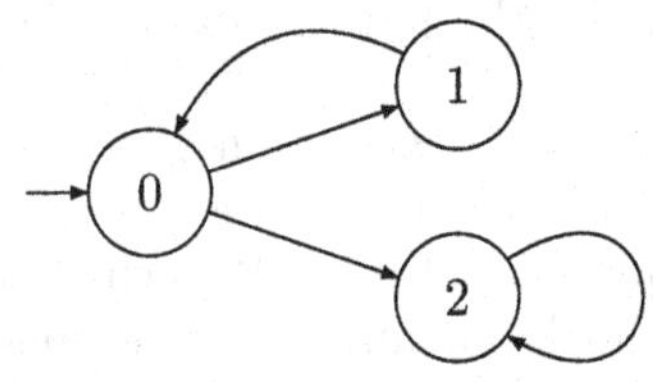

To conclude with finite graphs, we would like to focus on the particular case of positive subsets of Σ^ω, for some finite alphabet Σ. For the rest of this section, positive set means set that satisfy $\forall x, y \in \Sigma^\omega \quad ((x \in A \wedge x \sqsubseteq y) \longrightarrow y \in A)$. Replacing $\mathcal{G}$ by Σ, Theorem 1 still holds for such winning sets. Amazingly, getting rid of the constraint of the graph yields even sharper results. First we need two easy lemmas.

Lemma 5. *Let $A \subseteq \Sigma^\omega$ be positive, $B \subseteq \Gamma^\omega$ any non empty set, and $b \notin \Gamma$,*

$$B \leq_W A \implies (B \cup \Gamma^* b \Gamma^\omega) \leq_W A$$

where B' is the subset of $(\Gamma \cup \{b\})^\omega$, defined by $B' = (B \cup \Gamma^* b \Gamma^\omega)$.

In particular, $A \equiv_W \mathcal{D}_1^\xi$ always fails for $\xi > 0$.

Proof of lemma 5: Since B is non empty, A must be non empty, hence there exists some $\alpha = (u, \Lambda) \in \Sigma^* \times \mathcal{P}_+(\Sigma)$ such that $\Omega_\alpha \subseteq A$. Now consider any winning strategy σ for II in $\mathbb{W}(B, A)$, the following strategy is also winning for II in $\mathbb{W}(B', A)$: play σ as long as II does not play b, play some sequence $y \in u\mathbb{I}_\Lambda$ when I plays b.
⊣ 5

Lemma 6. *Let $A \subseteq \Sigma^\omega$,*

$$A \text{ positive } \Longrightarrow A \not\leq_W A^C.$$

Proof of lemma 6: First of all, in $\boldsymbol{\Sigma}^0_1 \cup \boldsymbol{\Pi}^0_1$ the only sets that verify $A \leq_W A^C$ (called *self-dual*) are the $\boldsymbol{\Delta}^0_1$-complete ones [2][15]. But if A is both positive, and satisfies $u\Sigma^\omega \subseteq A^{\complement}$ (for some finite word u), then $A = \emptyset$. This shows that there exists no positive self-dual set inside $\boldsymbol{\Sigma}^0_1 \cup \boldsymbol{\Pi}^0_1$. Since $\boldsymbol{\Pi}^0_2$-complete sets are non self-dual (see [2][15]), the only remaining problem are sets in $\boldsymbol{\Delta}^0_2 \setminus (\boldsymbol{\Sigma}^0_1 \cup \boldsymbol{\Pi}^0_1)$. By lemma 4, this case is the one of the sets of the form $A \in \boldsymbol{D}_n(\boldsymbol{\Sigma}^0_1)$ for some integer n. So, towards a contradiction, we assume that both $A \leq_W A^{\complement}$, and $A \in \bigcup_{n\in\mathbb{N}} \boldsymbol{D}_n(\boldsymbol{\Sigma}^0_1) \setminus (\boldsymbol{\Sigma}^0_1 \cup \boldsymbol{\Pi}^0_1)$ hold. It follows that there exists some integer $n > 1$ such that $\mathcal{D}^n_{(0,1)} \equiv_W A$. By lemma 5, this leads to $\mathcal{D}^n_{(0,1)}{}' \leq_W A$, where $\mathcal{D}^n_{(0,1)}{}'$ is the subset of $\{0, 1, \ldots, n, b\}^\omega$, defined by $\mathcal{D}^n_{(0,1)}{}' = \mathcal{D}^n_{(0,1)} \cup \{0, 1, \ldots, n, \}^* b\{0, 1, \ldots, n, b\}^\omega$. But clearly, $\mathcal{D}^n_{(0,1)}{}' \equiv_W \mathcal{D}^{n+1}_0$ holds, which contradicts the fact $A \equiv_W \mathcal{D}^n_{(0,1)}$. ⊣ 6

Theorem 4. *Let Σ be finite, and $A \subseteq \Sigma^\omega$,*

$$A \text{ positive } \Longrightarrow \begin{cases} A \equiv_W \emptyset \\ \quad or \\ A \equiv_W \mathcal{D}^n_0 \quad (n \in \mathbb{N}) \\ \quad or \\ A \equiv_W \mathbb{Q}^{\complement}. \end{cases}$$

Proof of theorem 4: From remark 2(4), we know that if A is positive, then A belongs to $\boldsymbol{\Pi}^0_2$, hence $A \leq_W \mathbb{Q}^{\complement}$ holds. If $\mathbb{Q}^{\complement} \leq_W A$ fails, then $A \in \boldsymbol{\Delta}^0_2$. Hence by lemma 4, A is a finite difference of open sets, or in other words: $A \equiv_W \mathcal{D}^n_i$ for some integer n and some $i \in \{0, 1, (0,1)\}$. But lemmas 6 and 5 prohibit respecetively the case $i = (0,1)$, and the case $i = 1$ (except for $n = 0$, wich corresponds to $A = \emptyset$). So the only remaining possibilities are $A \equiv_W \mathcal{D}^n_0$ for all $n \in \mathbb{N}$, with $A = \emptyset \equiv_W \mathcal{D}^0_1$. Examples for the first and last case are $A = \emptyset$, and $A = \mathbb{Q}^{\complement}$. The only one example for the case $A \equiv_W \mathcal{D}^0_0$ is $A = \Sigma^\omega$. Finally, for $A \equiv_W \mathcal{D}^n_0$ with $n > 0$, set $A = \Omega_{\alpha_0} \cup \Omega_{\alpha^n_1}$ where α_0 and α^n_1 are defined by:

$$\alpha_0 = (u_0, \Lambda_0) \text{ with } u_0 = \varepsilon,\ \Lambda_0 = \{0\},$$
$$\alpha^n_1 = (u^n_1, \Lambda_1) \text{ with } u^n_1 = \underbrace{\ldots 01010}_{n \text{ letters}},\ \Lambda_1 = \{1\}.$$

By $u^n_1 = \overbrace{\ldots 01010}^{n \text{ letters}}$, we mean $u^1_1 = 0$, $u^{2k+1}_1 = 0u^{2k}_1$, and $u^{2k+2}_1 = 1u^{2k+1}_1$. The set $A = \Omega_{\alpha_0} \cup \Omega_{\alpha^n_1}$ is both positive, and satisfies $A \equiv_W \mathcal{D}^n_0$. ⊣ 4

4 Infinite Game Graphs

While working with finite graphs, we saw that positive sets are very simple objects. On the contrary, when working with infinite game graphs, a first remark is that any set is homeomorphic to a positive one. Precisely, given any alphabet Σ, one can design a game graph $\mathcal{G} = (v_{\text{it}}, V_0, V_1, E)$ such that $V = \Sigma^*$, $v_{\text{it}} = \varepsilon$ (the empty word) and $E = \{(u, ua) : \text{ any } u \in \Sigma^*, a \in \Sigma\}$ - the partition $V = \{V_0, V_1\}$ doesn't matter. Clearly $\mathcal{G}^\omega$ and Σ^ω are homeomorphic, and for two infinite sequences $x, y \in \mathcal{G}^\omega$, $x \sqsubseteq y$ implies $x = y$; so that any subset of $\mathcal{G}^\omega$ is positive. Therefore working with positive sets in this general setting is totally useless.

The fundamental reason of this vacuity is the capacity of the graph to control positiveness. So we may ask the question in a different setting. We saw that positive subsets of Σ^ω are simpler than subsets of $\mathcal{G}^\omega$ when the alphabet, respectively the graph, is simple. Therefore, the idea of considering infinite alphabets instead of infinite graphs should here also, lead to a completely different answer. It certainly does, however, even if the condition $\forall x, y \in \Sigma^\omega \ \ ((x \in A \wedge x \sqsubseteq y) \longrightarrow y \in A)$ is rarely satisfied, even these positive sets may be extremely complicated. Amazingly, when the alphabet is finite, all such sets are in $\boldsymbol{\Pi}^0_2$, but as soon as Σ becomes infinite, they reach such high levels of complexity that they make determinacy fail.

We recall that, given a tree T on an alphabet Σ, and a winning set $A \subseteq \Sigma^\omega$, the Gale-Stewart game $\mathbb{G}(T, A)$ is an infinite two player game, where players (I and II) take turn playing letters in Σ. Player I begins, and all positions must remain inside T (otherwise the first player to exit this tree loses). Then, if both players only reach positions inside T, the play ($x \in \Sigma^\omega$) is an infinite branch of this tree. Player I wins if $x \in A$, player II wins otherwise. It is well known that this game is determined as soon as A is Borel [9], and for sets of higher level in the projective hierarchy, determinacy was shown to be equivalent to large cardinal hypotheses [10].

Proposition 4. *(in ZF+BPI) there exists a positive set $A \subseteq \mathbb{N}^\omega$, and a tree $T \subseteq \mathbb{N}^*$ such that*

$$\mathbb{G}(T, A) \ \text{ is not determined.}$$

The proof - widely inspired by folklore - takes place in usual Set Theory (*ZF*), with some additional choice, namely *BPI* (Boolean prime ideal theorem). BPI stands for the assumption that *every Boolean algebra has a prime ideal.* It is a common statement in algebra, much weaker than the full Axiom of Choice. However it is elsewhere independant from *DC* (Dependant Choice) another weak version of the axiom of choice, used to prove large amounts of mathematics.

Proof of proposition 4: We describe the tree T, and the winning set A later. First, we let $\mathcal{U}$ be a free ultrafilter over $\mathbb{N}$. We recall that *filters* over $\mathbb{N}$ are subsets $\mathcal{F} \subseteq \mathcal{P}(N)$ that satisfy:

- $U \in \mathcal{F} \ \wedge \ V \in \mathcal{F} \implies U \cap V \in \mathcal{F}$
- $U \in \mathcal{F} \ \wedge \ U \subseteq V \implies V \in \mathcal{F}$
- $\emptyset \notin \mathcal{F}$

An *ultrafilter* is a *filter* with the additional property that for all $U \subseteq \mathbb{N}$, either $U \in \mathcal{F}$ or $(\mathbb{N} \setminus U) \in \mathcal{F}$ holds.

The filter $\{U \subseteq \mathbb{N} : \mathbb{N} \setminus U \text{ is finite}\}$ of all co-finite subsets of $\mathbb{N}$ is called the *Fréchet filter*, and any *ultrafilter* that contains (extends) the *Fréchet filter* is called a *free* ultrafilter. We need the Boolean Prime Ideal theorem to ensure that the *Fréchet filter* can be embedded into an ultrafilter. So, we assume that there exists $\mathcal{U}$, some free ultrafilter over $\mathbb{N}$. We set A as the following set

$$A = \{y \in \mathbb{N}^\omega : \Im(y) \in \mathcal{U}\},$$

where $\Im(y)$ denotes the image of $y \in \mathbb{N}^\omega$ (regarded as a function from $\mathbb{N}$ to $\mathbb{N}$).

It is immediate to see that A is positive, since given any two infinite sequence $x, y \in \mathbb{N}^\omega$, $x \sqsubseteq y$ implies $\Im(x) \subseteq \Im(y)$. Therefore, if x is in A, then $\Im(x)$ is in $\mathcal{U}$, hence $\Im(y)$ is also in $\mathcal{U}$, therefore $y \in A$.

Instead of formally describing the tree T, we explicit the rules of the play that it allows and forbids:

- As first move, player I is allowed to play any integer x_0. But once this is done, the next x_0 moves of both players are restricted. Player I must precisely play $x_0 - 1, x_0 - 2, \dots, 0$, while player II can only play $0, 0, \dots, 0$.
- Then player II can play any $x_1 > x_0$, forcing player I to play $x_2 > x_1$ as next move. From this point, the next $x_1 - x_0$ moves are also restricted to $0, 0, \dots, 0$ for player II, and precisely $x_2 - 1, x_2 - 2, \dots, x_1$, for player I.
- Then player I can play any $x_3 > x_2$, forcing player II to play $x_4 > x_2$. The next $x_4 - x_3$ moves being $0, 0, \dots, 0$ for player II, and $x_4 - 1, x_4 - 2, \dots, x_3$ for player I.

And so on... We remark that

- if one skips all occurences of 0, it appears that II played exactly the infinite strictly increasing subsequence $x_1 < x_3 < x_5 < \dots$
- I played $x_0,\ x_0 - 1,\ \dots,\ 0,\ x_2,\ x_2 - 1,\ \dots,\ x_1,\ x_4,\ x_4 - 1,\ \dots,\ x_3, x_5 \dots$
- The integers played during this run x are precisely $[0, x_0] \cup \bigcup_{i \in \mathbb{N}} [x_{2i+1}, x_{2i+2}]$.

In the rest of the proof we simply forget about the part of the play that is totally imposed. We concentrate on the subsequence $x_0, x_1, x_2, \dots$ - the full play being easily reconstructed from it.

To show that $\mathbb{G}(T, A)$ is not determined, we first assume, towards a contradiction, that I has a winning strategy σ. We let II play as follows: after I plays x_0, II plays any $x_1 > x_0$, then I answers with x_2, but then II considers a second play - let's call it *fake* - of the game $\mathbb{G}(T, A)$ where I still applies σ, but II, in this *fake* game, *copies* I's answers in the *original* game with a shift, skipping x_0. Similarly, player II, in the *original* game, *copies* I's answers in the *fake* game, with here also a shift. See next picture where x'_i stands for $1 + x_i$.

Since σ is a winning strategy for I, it follows: $\Im(x_{original}) = [0, x_0] \cup [x_1, x_2] \cup \bigcup_{i \in \mathbb{N}} [x'_{2i+3}, x_{2i+4}] \in \mathcal{U}$, and $\Im(x_{fake}) = [0, x_0] \cup \bigcup_{i \in \mathbb{N}} [x'_{2i+2}, x_{2i+3}] \in \mathcal{U}$. Since $\mathcal{U}$ is a filter, $\Im(x_{original}) \cap \Im(x_{fake}) = [0, x_0] \in \mathcal{U}$. Elsewhere $\mathbb{N} \setminus [0, x_0]$ belongs to $\mathcal{U}$ by very definition. This yields $[0, x_0] \bigcap (\mathbb{N} \setminus [0, x_0]) = \emptyset \in \mathcal{U}$, a contradiction.

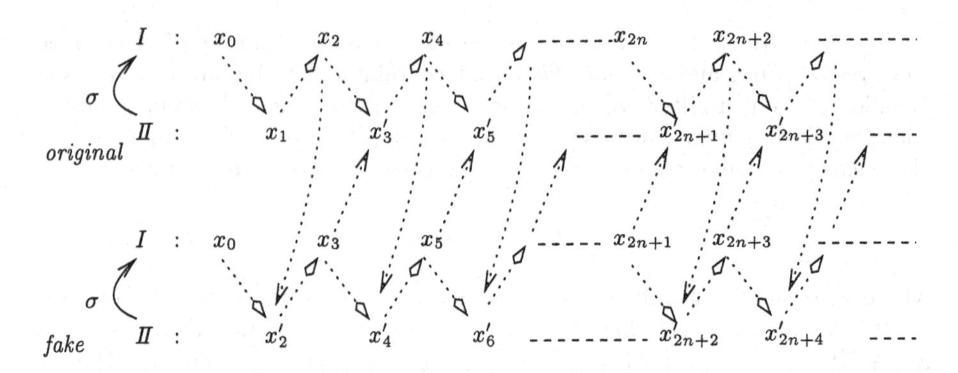

We have proved that player I cannot have a winning strategy, the assumption that II has a w.s. in $\mathbb{G}(T, A)$ leads exactly to the same contradiction. ⊣ 4

Remark 2. In a game on some graph $\mathcal{G}$ with infinitely many vertices, if a player P has a winning strategy, and his opponent's winning set is positive, then P does not necessarily have a *persistent* winning strategy. Consider the following counterexample where there is only one vertex (the initial one) for player **0**, and player **1** wins if and only if the sequence of **1**-vertices (the ones with two circles) is not strictly increasing:

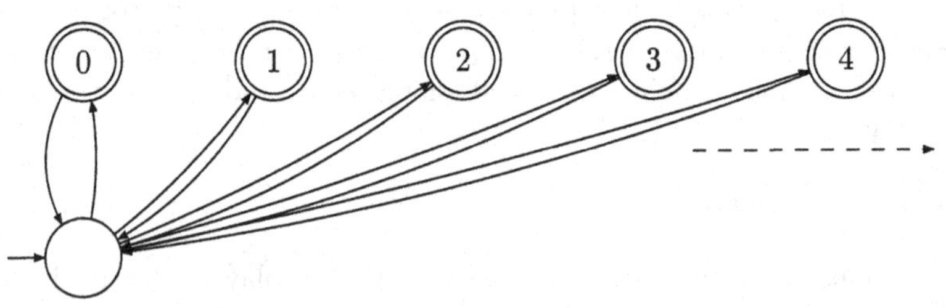

Since any subsequence of a strictly increasing sequence is also strictly increasing, A (the winning set for player **1**) satisfies $\forall x, y \in \mathcal{G}^\omega \;\; (x \sqsubseteq y \wedge x \in A) \Longrightarrow y \in A$, hence this game is positive. Clearly, player **0** has a winning strategy, but certainly do not have any *persistent* winning strategy. In fact, every winning strategy must be *persistently changing.*

5 Muller Acceptance Conditions

For the games where the winning set is defined in terms of Muller acceptance condition - i.e. an infinite play is accepted iff the set of vertices visited infinitely often belongs to some set of the form $\mathcal{F} \subseteq \mathcal{P}(V)$ [12][13] - topological properties do help.

Theorem 5. *Let V be finite, and $A \subseteq \mathcal{G}^\omega$ defined by Muller acceptance conditions $\mathcal{F} \subseteq \mathcal{P}(V)$, the following are equivalent:*

1. *A is positive ,*
2. *$A \equiv_W \mathbb{Q}^{\mathbf{C}}$ or $A \equiv_W \mathcal{D}_i^n$ for some $n \in \mathbb{N}$, and $i \in \{0, 1, (0,1)\}$,*
3. *$A \in \boldsymbol{\Pi}_2^0$.*

In particular, this theorem gives an amazing proof of an already known result that all $\boldsymbol{\Delta}^0_2$-sets defined in terms of Muller acceptance conditions are *finite* differences of open sets (see [16], [3]).
Proof of theorem 5: The direction (1)$\Longrightarrow$(2) is theorem 3. For (2)$\Longrightarrow$(3), it is enough to say that if $A \equiv_W \mathbb{Q}^{\mathsf{C}}$, then A is $\boldsymbol{\Pi}^0_2$-complete, and if $A \equiv_W \mathcal{D}^n_i$, then A is in $\boldsymbol{\Delta}^0_2$. So that in both cases, A belongs to $\boldsymbol{\Pi}^0_2$.

For (3)$\Longrightarrow$(1), towards a contradiction, we assume both $A \in \boldsymbol{\Pi}^0_2$, and A is not positive. This means, in particular, that there exists two infinite sequences $x, y \in \mathcal{G}^\omega$, that verify $x \sqsubseteq y$, $x \in A$, and $y \notin A$. Let Λ_x, and Λ_y be the set of vertices visited infinitely often by respectively x and y. Since $x \sqsubseteq y$ holds, we infer that $\Lambda_x \subsetneq \Lambda_y$ holds too. So there is at least one vertex $v \in \Lambda_x$ that is accessible from v_{it} and from which there is, inside $\mathcal{G}$, both a path w_x going through precisely all vertices in Λ_x and coming back to v, and another path w_y going through precisely all vertices in Λ_y and also coming back to v. This yields the following winning strategy for player $I\!I$ in $\mathbb{W}(\mathbb{Q}, A)$:

first reach vertex v, then

- each time player I plays a 0, play the whole path w_x,
- each time player I plays a 1, play the whole path w_y,

The sequence played by I belongs to $\mathbb{Q}$ iff it contains finitely many 1, in which case the set of vertices visited infinitely often by $I\!I$ is exactly Λ_x, showing that the sequence played by $I\!I$ also belongs to A.

If I's play contains infinitely many 1, then it is not in $\mathbb{Q}$, but in this case $I\!I$ visited infinitely often precisely all vertices in Λ_y, so that $I\!I$ also wins this case.

We have shown that $\mathbb{Q} \leq_W A$ holds. Since $\mathbb{Q}$ is $\boldsymbol{\Sigma}^0_2$-complete, it follows that $A \notin \boldsymbol{\Pi}^0_2$, a contradiction.

$\dashv$ 5

Remark 3. If a player's winning set is a $\boldsymbol{\Sigma}^0_2$-set defined in terms of Muller acceptance conditions, and this player has a winning strategy, then this player also has a persistent winning strategy. However, this fact does not help much, since this player not only has a persistent winning strategy but also a memoryless one.

Lemma 7. *Let V be finite, and $A \subseteq \mathcal{G}^\omega$ defined by Muller acceptance conditions $\mathcal{F} \subseteq \mathcal{P}(V)$,*

1. *A positive $\Longrightarrow$ $\mathcal{F}$ is closed under union.*
2. *A positive winning set for player* **1** *(resp.* **0***), and player* **0** *(resp.* **1***) has a w.s. $\Longrightarrow$ player* **0** *(resp.* **1***) has a memoryless winning strategy.*

Proof of lemma 7:

1. It is immediate to see that given $Q, Q' \in \mathcal{F}$ such that $Q \cup Q' \notin \mathcal{F}$, and given any sequence x that visits infinitely often the set of vertices Q, one can build a sequence y that visits infinitely often all vertices in $Q \cup Q'$ and satisfies $x \sqsubseteq y$. Which leads to both $x \in A$ and $y \notin A$, a contradiction.
2. This is an immediate consequence of theorem 17 in [17].

$\dashv$ 7

References

1. J. Duparc, *La forme normale des Boréliens de rangs finis.* Thèse de Doctorat, Université Denis Diderot Paris VII. Juillet 95.
2. J. Duparc, *Wadge hierarchy and Veblen hierarchy: part I Borel sets of finite rank.* J. Symbolic Logic 66 (2001), no. 1, 56–86.
3. J. Duparc, *A Hierarchy of Deterministic Context-Free ω-languages.* Theoretical Computer Science, 290 (2003), no. 3, 1253-1300.
4. J. Duparc, O. Finkel, J-P. Ressayre, *Computer Science and the fine Structure of Borel Sets.* Theoretical Computer Science, 257 (1-2) (2001) pp. 85-105.
5. G. Higman, *Ordering by divisibility in abstract algebras.* Proc. London Math. Soc. (3) 2, (1952). 326–336.
6. A.S. Kechris, *Classical descriptive set theory.* Graduate texts in mathematics; vol 156. Springer Verlag (1994)
7. A. Louveau, *Some Results in the Wadge Hierarchy of Borel Sets.* Cabal Sem 79-81, Lecture Notes in Mathematics (1019) 28-55.
8. J. Marcinkowski, T. Truderung, *Optimal Complexity Bounds for Positive LTL Games.* Proceedings of the 11th Annual Conference of the European Association for Computer Science Logic, CSL 2002, Lecture Notes in Computer Science 2471. Springer, 2471 (2002), 262-275
9. D. A. Martin, *Borel determinacy.* Ann. of Math. (2) 102 (1975), no. 2, 363–371.
10. D. A. Martin, J. R. Steel, *A proof of projective determinacy.* J. Amer. Math. Soc. 2 (1989), no. 1, 71–125.
11. C.St.J.A. Nash-Williams, *On well-quasi-ordering finite trees.* Proceedings of the Cambridge Philosophical Society 59 (1963) 833–835.
12. W. Thomas, *Automata on Infinite Objects.* Handbook of Theoretical Computer Science, ch. 4; edited by J. van Leeuwen Elsevier Science Publishers B. V., 1990.
13. W. Thomas, *Languages, automata, and logic.* Handbook of formal languages, Vol. 3, 389–455, Springer, Berlin, 1997.
14. W.W. Wadge, *Degrees of complexity of subsets of the Baire space.* Notice A.M.S. (1972), A-714.
15. W.W. Wadge, *Ph.D. Thesis*, Berkeley
16. K. W. Wagner, *On ω-regular sets.* Inform. and Control 43 (1979), no. 2, 123–177.
17. W. Zielonka, *Infinite games on finitely coloured graphs with applications to automata on infinite trees.* Theoret. Comput. Sci. 200 (1998), no. 1-2, 135–183.

Generating All Abductive Explanations for Queries on Propositional Horn Theories*

Thomas Eiter[1] and Kazuhisa Makino[2]

[1] Institut für Informationssysteme, Technische Universität Wien,
Favoritenstraße 9-11, A-1040 Wien, Austria
eiter@kr.tuwien.ac.at

[2] Division of Systems Science, Graduate School of Engineering Science, Osaka University,
Toyonaka, Osaka, 560-8531, Japan
makino@sys.es.osaka-u.ac.jp

Abstract. Abduction is a fundamental mode of reasoning, which has taken on increasing importance in Artificial Intelligence (AI) and related disciplines. Computing abductive explanations is an important problem, and there is a growing literature on this subject. We contribute to this endeavor by presenting new results on computing multiple resp. all of the possibly exponentially many explanations of an abductive query from a propositional Horn theory represented by a Horn CNF. Here the issues are whether a few explanations can be generated efficiently and, in case of all explanations, whether the computation is possible in *polynomial total time* (or *output-polynomial time*), i.e., in time polynomial in the combined size of the input and the output. We explore these issues for queries in CNF and important restrictions thereof. Among the results, we show that computing all explanations for a negative query literal from a Horn CNF is not feasible in polynomial total time unless $P = NP$, which settles an open issue. However, we show how to compute under restriction to acyclic Horn theories polynomially many explanations in input polynomial time and all explanations in polynomial total time, respectively. Complementing and extending previous results, this draws a detailed picture of the computational complexity of computing multiple explanations for queries on Horn theories.

Keywords: Computational logic, abduction, propositional logic, Horn theories, polynomial total time computation, NP-hardness.

1 Introduction

Abduction is a fundamental mode of reasoning, which was extensively studied by C.S. Peirce [19]. It has taken on increasing importance in Artificial Intelligence (AI) and related disciplines, where it has been recognized as an important principle of common-sense reasoning (see e.g. [3]). Abduction has applications in many areas of AI and Computer Science including diagnosis, database updates, planning, natural language understanding, learning etc. (see e.g. references in [10]), where it is primarily used for generating explanations.

* This work was supported in part by the Austrian Science Fund (FWF) Project Z29-N04, by a TU Wien collaboration grant, and by the Scientific Grant in Aid of the Ministry of Education, Science, Sports, Culture and Technology of Japan.

M. Baaz and J.A. Makowsky (Eds.): CSL 2003, LNCS 2803, pp. 197–211, 2003.

In a logic-based setting, abduction can be viewed as the task to find, given a set of formulas Σ (the *background theory*) and a formula χ (the *query*), a minimal set of formulas E (an *explanation*) from a set of hypotheses H such that Σ plus E is satisfiable and logically entails χ. Often considered is a scenario where Σ is a propositional Horn theory, χ is a single literal or a conjunction of literals, and H contains literals (see [24, 10] and references therein). For use in practice, the computation of abductive explanations in this setting is an important problem, for which well-known early systems such as Theorist [20] or ATMS solvers [6, 22] have been devised. Since then, there has been a growing literature on this subject, indicating the need for efficient abductive procedures. We refer to [18], which gives an excellent survey on intimately closely related problems in computational logic. Note that much effort has been spent on studying various input restrictions, cf. [14, 4, 13, 25, 8, 7, 10, 23, 24].

While computing *some* explanation of a query χ has been studied extensively in the literature, the issue of computing multiple or even *all* explanations for χ has received less attention. This problem is important since often one would like to select one out of a set of alternative explanations according to a preference or plausibility relation; this relation may be based on subjective intuition and thus difficult to formalize. As easily seen, exponentially many explanations may exist for a query, and thus computing all explanations inevitably requires exponential time in general, even in propositional logic. However, it is of interest whether the computation is possible in *polynomial total time* (or *output-polynomial time*), i.e., in time polynomial in the combined size of the input and the output. Furthermore, if exponential space is prohibitive, it is of interest to know whether a few explanations (e.g., polynomially many) can be generated in polynomial time, as studied by Selman and Levesque [24].

Computing some explanation for a query χ which is a literal from a Horn theory is a well-known polynomial problem. Selman and Levesque conjectured [24] that generating $O(n)$ many explanations for a positive literal is NP-hard, where n is the number of propositional atoms in the language, even if it is guaranteed that there are only few explanations overall. As shown in [11], this conjecture is not true unless P=NP. This follows from the result of [11] that all explanations for an atom can be generated in polynomial total time.

The status of generating all explanations for a negative literal $\chi = \overline{q}$ from a Horn CNF φ, however, remained open in [11]. Moreover, it was unclear whether a resolution-style procedure similar to the one for query atoms in [11] could solve the problem in polynomial total time. In this paper, we provide a negative answer to this question, by showing that given a collection of explanations for a query $\chi = \overline{q}$ from a Horn CNF φ, deciding whether there is an additional explanation is NP-complete. Consequently, the existence of a polynomial total time algorithm for computing all explanations implies P=NP. However, for the well-known class of acyclic Horn theories (see e.g. [5, 24, 21, 1]) we present an algorithm which enumerates all explanations for $\overline{q}$ with incremental polynomial delay (i.e., in time polynomial in the size of the input and output so far), and thus solves the problem in polynomial total time. Compared to explanations for an atomic query q, intuitively cyclic dependencies between atoms make the problem difficult. For completeness, a resolution-style procedure as in [11] needs to consider besides the input and output clauses also auxiliary clauses (see Example 7), whose derivation may cause a lot of overhead, since it is not a priori clear which such clauses are needed.

We furthermore address computing all explanations for queries χ beyond literals, where we consider CNF and important special cases such as a clause and a term (i.e., a conjunction of literals). Note that the explanations for single clause queries correspond to the minimal support clauses for a clause in Clause Management Systems [22]. In the light of the negative results from above, we aim at elucidating the tractability frontier and present positive as well as negative results for such queries.

Our results shed new light on the computational nature of abduction and Horn theories in particular. They imply that, e.g., generating all minimal support clauses for a given clause (cf. [22]) from an acyclic Horn CNF is feasible in polynomial total time. The intractability result for negative literal queries $\overline{q}$ is somewhat unexpected, and the tractability result for acyclic Horn theories is more difficult to obtain than in case of atomic queries. As a byproduct, we also obtain results for computing all prime implicates of Horn theories containing a certain literal, which complement and refine previous results for computing all prime implicates of a Horn theory [2].

For space reasons, some proofs are omitted; we refer to the extended version [12].

2 Preliminaries and Notation

We assume a standard propositional language with atoms $x_1, x_2, \ldots, x_n$ from a set At, where each x_i takes either value 1 (true) or 0 (false). Negated atoms are denoted by $\overline{x}_i$, and the opposite of a literal ℓ by $\overline{\ell}$. Furthermore, we use $\overline{A} = \{\overline{\ell} \mid \ell \in A\}$ for any set of literals A and set $Lit = At \cup \overline{At}$.

A clause is a disjunction $c = \bigvee_{p \in P(c)} p \vee \bigvee_{p \in N(c)} \overline{p}$ of literals, where $P(c)$ and $N(c)$ are the sets of atoms occurring positively and negated in c and $P(c) \cap N(c) = \emptyset$. Dually, a term is conjunction $t = \bigwedge_{p \in P(t)} p \wedge \bigwedge_{p \in N(t)} \overline{p}$ of literals, where $P(t)$ and $N(t)$ are similarly defined. We also view clauses and terms as sets of literals $P(c) \cup N(c)$ and $P(t) \cup N(t)$, respectively. A clause c is *Horn*, if $|P(c)| \leq 1$; *definite*, if $|P(c)| = 1$; and *negative* (resp., *positive*), if $|P(c)| = 0$ (resp., $|N(c)| = 0$). A conjunctive normal form (CNF) is a conjunction of clauses. It is *Horn* (resp., *definite, negative, positive*), if it contains only Horn clauses (resp., definite, negative, positive clauses). A *theory* Σ is any finite set of formulas; it is *Horn*, if it is a set of Horn clauses. As usual, we identify Σ with $\varphi = \bigwedge_{c \in \Sigma} c$, and write $c \in \varphi$ etc.

A *model* is a vector $v \in \{0,1\}^n$, whose i-th component is denoted by v_i. For $B \subseteq \{1, \ldots, n\}$, we let x^B be the model v such that $v_i = 1$, if $i \in B$ and $v_i = 0$, if $i \notin B$, for $i \in \{1, \ldots, n\}$. Satisfaction $v \models \varphi$ and logical consequence $\varphi \models c$, $\varphi \models \psi$ etc. are defined as usual (i.e., $\varphi(v) = 1$ etc.).

Example 1. The CNF $\varphi = (\overline{x}_1 \vee \overline{x}_4) \wedge (\overline{x}_4 \vee \overline{x}_3) \wedge (\overline{x}_1 \vee x_2) \wedge (\overline{x}_4 \vee \overline{x}_5 \vee x_1) \wedge (\overline{x}_2 \vee \overline{x}_5 \vee x_3)$ over $At = \{x_1, x_2, \ldots, x_5\}$ is Horn. The vector $u = (0, 1, 0, 1, 0)$ is a model of φ. □

The following proposition is well-known.

Proposition 1. *Given a Horn CNF φ and a clause c, deciding whether $\varphi \models c$ is possible in polynomial time (in fact, in linear time, cf.* [9]*).*

Recall that two clauses c and c' resolve on a pair of literals $x, \overline{x}$ if $x, \overline{x} \in c \cup c'$ and $c \cup c' \setminus \{x, \overline{x}\}$ is a legal clause (thus, x must occur in exactly one of c and c', and same for $\overline{x}$); c and c' resolve if their is a pair of literals $x, \overline{x}$ on which they resolve. Note that this pair, if it exists, is unique. In that case, we denote by $c \oplus c'$ the clause $c \cup c' \setminus \{x, \overline{x}\}$, which is their *resolvent* (otherwise, $c \oplus c'$ is undefined). A *resolution proof* of a clause c from a CNF φ is a sequence $c_1, c_2, \ldots, c_l$ of clauses such that $c_l = c$ and, for all $i = 1, \ldots, l$, either $c_i \in \varphi$ or $c_i = c_j \oplus c_k$ for clauses c_j and c_k such that $j, k < i$. It is well-known that resolution proofs are sound and complete with respect to clause inference in the following sense (cf. [18]): For any CNF φ and clause c, $\varphi \models c$ holds iff there is a clause $c' \subseteq c$ which has a resolution proof from φ. For further background on resolution, we refer to [17, 16].

2.1 Abductive Explanations

The notion of an abductive explanation can be formalized as follows (cf. [24, 10]).

Definition 1. Given a (Horn) theory Σ, called the background theory, a CNF χ (called *query*), an *explanation of* χ is a minimal set of literals $E \subseteq Lit$ such that

(i) $\Sigma \cup E \models \chi$, and
(ii) $\Sigma \cup E$ is satisfiable.

Example 2. Reconsider the Horn CNF $\varphi = (\overline{x}_1 \vee \overline{x}_4) \wedge (\overline{x}_4 \vee \overline{x}_3) \wedge (\overline{x}_1 \vee x_2) \wedge (\overline{x}_4 \vee \overline{x}_5 \vee x_1) \wedge (\overline{x}_2 \vee \overline{x}_5 \vee x_3)$ from above. Suppose we want to explain $\chi = x_2$ from $A = \{x_1, x_4\}$. Then, we find that $E = \{x_1\}$ is an explanation. Indeed, $\Sigma \cup \{x_1\} \models x_2$, and $\Sigma \cup \{x_1\}$ is satisfiable; moreover, E is minimal. On the other hand, $E' = \{x_1, \overline{x}_4\}$ satisfies (i) and (ii) for $\chi = x_2$, but is not minimal. □

More restricted forms of explanations require that E must be formed over a given set of abducible letters (cf. [24]); however, in such a setting, generating all explanations is easily seen to be coNP-hard for the cases that we consider from results in the literature.

The following characterization of explanations is immediate by the monotonicity of classical logic.

Proposition 2. *For any theory Σ, any query χ, and any $E \subseteq Lit$, E is an explanation for χ from Σ iff the following conditions hold: (i) $\Sigma \cup E$ is satisfiable, (ii) $\Sigma \cup E \models \chi$, and (iii) $\Sigma \cup (E \setminus \{\ell\}) \not\models \chi$, for every $\ell \in E$.*

From Proposition 2, we thus obtain the following easy lemma.

Lemma 1. *Given a Horn CNF φ, a set $E \subseteq Lit$, and a CNF query χ, deciding whether E is an explanation for χ w.r.t. A is possible in polynomial time.*

3 Intractability of Negative Literal Queries

In this section, we show that computing all explanations of a negative query $\chi = \overline{q}$ is not possible in polynomial total time unless P = NP. This result follows by standard arguments from the following theorem.

Theorem 1. *Given a Horn CNF φ, a query $\chi = \overline{q}$, and explanations $E_1, E_2, \ldots, E_k$ for χ, deciding whether χ has some additional explanation E_{k+1} different from each E_i, $1 \leq i \leq k$, is* NP*-complete. Hardness holds even if φ is definite Horn.*

In the proof of Theorem 1, we use the following well-known lemma, which links prime implicates of a theory to explanations. Recall that a *prime implicate* of a theory Σ is a minimal (w.r.t. inclusion) clause c such that $\Sigma \models c$. Let us call an explanation E for a literal query $\chi = \ell$ *trivial*, if $E = \{\ell\}$.

Lemma 2 (cf. [22,15]). *Given a theory Σ, a set $E \subseteq Lit$ is a nontrivial explanation of a query literal χ iff the clause $c = \bigvee_{\ell \in E} \overline{\ell} \vee \chi$ is a prime implicate of Σ.*

Note $\chi = \ell$ has a trivial explanation iff $\Sigma \not\models \ell$ and $\Sigma \not\models \overline{\ell}$, which can be checked in polynomial time. Hence, as for NP-hardness we can without loss of generality focus on generating the nontrivial explanations of $\overline{q}$, i.e., all prime implicates containing $\overline{q}$.

Proof of Theorem 1. As for membership in NP, an additional explanation E_{k+1} can be guessed and, by Lemma 1, be verified in polynomial time.

We show the NP-hardness by a reduction from 3SAT. Let $\gamma = c_1 \wedge \cdots \wedge c_m$, $m \geq 2$, be a 3CNF over atoms $x_1, \ldots, x_n$, where $c_i = \ell_{i,1} \vee \ell_{i,2} \vee \ell_{i,3}$. We introduce for each clause c_i a new atom y_i, for each x_j a new atom x'_j (which intuitively corresponds to $\overline{x}_i$), and special atoms q and z. The Horn CNF φ contains the following clauses:

1. $c_{i,j} = \overline{q} \vee \overline{\ell^{\star}_{i,j}} \vee y_i$, for all $i = 1, \ldots, m$ and $j = 1, 2, 3$;
2. $d_{i,j} = \overline{\ell^{\star}_{i,j}} \vee y_i \vee \overline{y}_{i+1}$, for all $i = 1, \ldots, m$ and $j = 1, 2, 3$;
3. $\overline{x}_i \vee \overline{x'_i} \vee z$, for all $i = 1, \ldots, n$;
4. $e = \overline{y}_1 \vee \overline{y}_2 \vee \cdots \vee \overline{y}_m \vee z$,

where $\ell^{\star}_{i,j} = x_k$ if $\ell_{i,j} = x_k$ and $\ell^{\star}_{i,j} = x'_k$ if $\ell_{i,j} = \overline{x}_i$, and $y_{m+1} = y_1$.

Note that φ is definite Horn, and thus all prime implicates of φ are definite Horn. Informally, the clauses $c_{i,j}$ and $d_{i,j}$ stand for selection of literal $\ell_{i,j}$ in clause c_i. The clause in 4., which is needed to produce any negative prime implicate c containing $\overline{q}$, and the minimality of a prime implicate will effect that a literal is chosen from each clause c_i, and the clauses in 3. will ensure that the choice is consistent, such that γ is satisfied. Since the positive prime implicates containing $\overline{q}$ are just the clauses $c_{i,j}$, the a further prime implicate of φ containing $\overline{q}$ exists iff γ is satisfiable.

We establish the following properties of φ.

Lemma 3. *Any prime implicate c of φ such that $q \in N(c)$ and $P(c) \neq \{z\}$ is of the form $c_{i,j}$, where $i \in \{1, \ldots, m\}$ and $j \in \{1, 2, 3\}$.*

Lemma 4. *Any prime implicate c of φ such that $P(c) = \{z\}$ and $q \in N(c)$ satisfies (i) $\{x_i, x'_i\} \not\subseteq N(c)$, for all $i = 1, \ldots, n$, and (ii) $y_i \notin N(c)$, for all $i = 1, \ldots, m$.*

From Lemma 3, it is now easy to see that all prime implicates of φ given by the clauses in 1. correspond to nontrivial explanations of $\overline{q}$, and from Lemma 4 that an additional nontrivial explanation for $\overline{q}$ exists if and only if some prime implicate of φ of form $c = \overline{q} \vee \bigvee_{x_i \in X} \overline{x}_i \vee \bigvee_{x'_i \in X'} \overline{x'}_i \vee z$ exists iff the CNF γ is satisfiable. As for the last equivalence, note that for each smallest (w.r.t. $\subseteq$) choice $\ell_i \in c_i$ of a consistent collection

of literals $\ell_1, \ldots, \ell_m$, we have an additional prime implicate of φ of form $\overline{q} \vee \bigvee_{\ell_i} \overline{\ell_i^\star} \vee z$. Conversely, each additional prime implicate c containing $\overline{q}$ gives rise to a consistent set of literals $\{x_j \mid x_{i,j} \in N(c)\} \cup \{\overline{x}_j \mid x'_{i,j} \in N(c)\}$ which satisfies γ.
Clearly, φ is constructible in polynomial time from γ. Since φ is definite, this proves the NP-hardness under the asserted restriction. □

We note that φ in the hardness proof of Theorem 1 remains Horn upon switching the polarity of z. From this easily NP-completeness of deciding the existence of an explanation for $\chi = \overline{q}$ formed of only positive literals follows, even if all other explanations are given. This contrasts with respective tractability results for acyclic theories (implied by the next section) and for atomic queries $\chi = q$ on arbitrary Horn CNFs [11].

4 Negative Literal Queries on Acyclic Horn Theories

Since as shown in the previous section, a polynomial total time procedure for generating all explanations of a negative literal query is infeasible in general unless P=NP, it becomes an issue to find restricted input classes for which this is feasible. In this section, we show a positive result for the important class of acyclic Horn theories, which has been studied extensively in the literature (see, e.g., [5, 24, 21, 1]).

We first recall the concept of acyclic Horn theories (see e.g. [5, 24]).

Definition 2. *For any Horn CNF φ over atom set At, its dependency graph is the directed graph $G(\varphi) = (V, E)$, where $V = At$ and $E = \{x_i \to x_j \mid c \in \varphi, x_i \in N(c), x_j \in P(c)\}$, i.e., E contains an arc from each atom in a negative literal to the positive literal in a clause (if such a literal exists). A Horn CNF φ is* acyclic *if $G(\varphi)$ has no directed cycle.*

Example 3. As easily seen, the edges of $G(\varphi)$ for the CNF φ in Examples 1 and 2 are $x_1 \to x_2$, $x_4 \to x_1$, $x_5 \to x_1$, $x_2 \to x_3$, and $x_5 \to x_3$. Hence, φ is acyclic. □

Since the trivial explanation $E = \{\overline{q}\}$ can be easily generated (if it applies), we focus on generating all nontrivial explanations. For a negative query on an acyclic Horn theory, this is accomplished by Algorithm N-EXPLANATIONS in Figure 1. It first converts the input into an equivalent prime Horn CNF $\varphi^\star$, and then applies a restricted resolution procedure, in which pairs (c, c') of clauses are considered of which at least one is a prime implicate containing $\overline{q}$ and the other is either a clause of this form or a clause from the converted input $\varphi^\star$. In case their resolvent $d := c \oplus c'$ exists and, as implied by condition (ii) in Definition 1, includes only prime implicates containing $\overline{q}$, any such prime implicate d' is computed. If d' is recognized as a new prime implicate which has not been generated so far, a corresponding explanation is output and the set of candidate pairs is enlarged.

Example 4. Reconsider $\varphi = (\overline{x}_1 \vee \overline{x}_4) \wedge (\overline{x}_4 \vee \overline{x}_3) \wedge (\overline{x}_1 \vee x_2) \wedge (\overline{x}_4 \vee \overline{x}_5 \vee x_1) \wedge (\overline{x}_2 \vee \overline{x}_5 \vee x_3)$, and apply N-EXPLANATIONS for $\chi = \overline{x}_1$. All clauses of φ are prime except $\overline{x}_4 \vee \overline{x}_5 \vee x_1$, which contains the prime implicate $\overline{x}_4 \vee \overline{x}_5$. Thus, $\varphi^\star = (\overline{x}_1 \vee \overline{x}_4) \wedge (\overline{x}_4 \vee \overline{x}_3) \wedge (\overline{x}_1 \vee x_2) \wedge (\overline{x}_4 \vee \overline{x}_5) \wedge (\overline{x}_2 \vee \overline{x}_5 \vee x_3)$, and S contains the clauses $\overline{x}_1 \vee \overline{x}_4$ and $\overline{x}_1 \vee x_2$; the corresponding explanations $E_1 = \{x_4\}$ and $E_2 = \{\overline{x}_2\}$ are output. In

Algorithm N-EXPLANATIONS
Input: An acyclic Horn CNF φ and an atom q.
Output: All nontrivial explanations of the query $\chi = \overline{q}$ from φ.

Step 1. $\varphi^\star := \emptyset$, $S := \emptyset$, and $O := \emptyset$;
Step 2. **for each** $c \in \varphi$ **do**
add any prime implicate $c' \subseteq c$ of φ to $\varphi^\star$;
for each $c' \in \varphi^\star$ with $q \in N(c')$ and $c' \notin S$ **do**
begin output $(\{\overline{\ell} \mid \ell \in c' \setminus \{\overline{q}\}\}$;
$S := S \cup \{c'\}; O := O \cup \{(c, c') \mid c \in \varphi^\star, q \notin P(c)\}$
end;
Step 3. **while** some $(c, c') \in O$ exists **do**
begin $O := O \setminus \{(c, c')\}$;
if (1) c and c' resolve and (2) $\varphi^\star \not\models (c \oplus c' \setminus \{\overline{q}\})$
then begin $d := c \oplus c'$;
compute any prime implicate $d' \subseteq d$ of φ;
if $d' \notin S$ **then**
begin output $(\{\overline{\ell} \mid \ell \in d' \setminus \{\overline{q}\}\}$; $S := S \cup \{d'\}$;
$O := O \cup \{(d'', d') \mid d'' \in \varphi^\star, q \notin P(d'')\} \cup \{(d'', d') \mid d'' \in S\}$
end
end
end. □

Fig. 1. Algorithm computing all nontrivial explanations of a query $\chi=\overline{q}$ on an acyclic Horn theory

Step 2, the pair $(\overline{x}_2 \vee \overline{x}_5 \vee x_3, \overline{x}_1 \vee x_2)$ is found in O which satisfies condition (i) in Def. 1. Moreover, (ii) is satisfied, since $\varphi^\star \not\models \overline{x}_5 \vee x_3$. Thus, a prime implicate within $d = \overline{x}_1 \vee \overline{x}_5 \vee x_3$ is computed; in fact, d is prime. Therefore, $E_3 = \{x_5, \overline{x}_3\}$ is output and d is added to S, and then O is enlarged. Eventually, the pair $(\overline{x}_3 \vee \overline{x}_4, \overline{x}_1 \vee \overline{x}_5 \vee x_3)$ from O, which satisfies condition (i), will be considered. However, $\varphi^\star \models \overline{x}_4 \vee \overline{x}_5$, and thus S remains unchanged. Hence, the output of N-EXPLANATIONS is E_1, E_2, and E_3. As can be seen, these are all nontrivial explanations for $\overline{x}_1$ from φ. □

We remark that our algorithm is similar in spirit to an algorithm for computing all prime implicates of a Horn CNF in polynomial total time [2]. Our algorithm solves a more constrained problem, though.

In the rest of this section, we show that Algorithm N-EXPLANATIONS generates all explanations in polynomial total time. For that, we first show its correctness, which splits into a soundness and completeness part, and then analyze its time complexity.

As for soundness, it is easily seen that Algorithm N-EXPLANATIONS produces output only if d' is some prime implicate of $\varphi^\star$ (and thus of φ) such that $q \in N(d')$. Thus, from Lemma 2, we immediately obtain

Lemma 5 (Soundness of N-EXPLANATIONS). *Algorithm* N-EXPLANATIONS *outputs only nontrivial explanations for* $\overline{q}$ *from* φ.

It is much more difficult to show the completeness, i.e., that Algorithm N-EXPLANATIONS actually generates all nontrivial explanations. Intuitively, the difficulty stems from the fact that the restricted resolution procedure retains only prime clauses containing $\overline{q}$, and, moreover, may skip relevant prime implicates $d' \subseteq c \oplus c'$ in Step 3 if condition (ii) fails, i.e., $c \oplus c'$ is an implicate of φ (which is tantamount to the condition that $c \oplus c'$ contains some prime implicate of φ that does not contain $\overline{q}$). To see that no explanation is missed requires a careful analysis of how the desired explanations are generated, and leads to a nontrivial argument which takes the complex interaction between clauses into account.

We need a number of preliminary technical lemmas on which our proof builds, which are interesting in their own right. In what follows, we call a Horn clause c *definite*, if $P(c) \neq \emptyset$. Furthermore, for any literal ℓ, a clause c is a *ℓ-clause* if c contains ℓ.

The following propositions are well-known.

Proposition 3. *Let c_1, c_2 be Horn implicates of a Horn CNF φ that resolve. Then, $c = c_1 \oplus c_2$ is Horn, and if c_1 contains a negative implicate of φ, then also $c_1 \oplus c_2$ contains a negative implicate of φ.*

Proposition 4 (cf. [2]). *Every prime implicate c of a Horn CNF φ has an input resolution proof from it, i.e., a resolution proof $c_1, c_2, \ldots, c_l (= c)$ such that either $c_i \in \varphi$ or $c_i = c_j \oplus c_k$ where $j, k < l$ and either $c_j \in \varphi$ or $c_k \in \varphi$, for all $i \in \{1, \ldots, l\}$.*

We start with the following lemma.

Lemma 6. *Let φ be a prime Horn CNF, and let c be any prime implicate of φ such that $c \notin \varphi$. Then, $c = c_1 \oplus c_2$, where c_1 is a prime implicate contained in φ, and either (i) c_2 is a prime implicate of φ, or (ii) $c_2 = c \cup \{\ell\}$ where $c_1 \setminus \{\overline{\ell}\} \subset c$ and c is the unique prime implicate of φ contained in c_2.*

Note that item (ii) is needed in this lemma, as shown by the following example.

Example 5. Consider the Horn CNF $\varphi = (\overline{x}_0 \vee \overline{x}_1 \vee x_2)(\overline{x}_2 \vee \overline{x}_3)(\overline{x}_3 \vee x_0)$. As easily checked, φ is prime and has a further prime implicate $\overline{x}_1 \vee \overline{x}_3$, which can not be derived as the resolvent of any two prime implicates of φ. Note that φ is acyclic. □

Next we state some important properties of acyclic Horn CNFs under resolution.

Proposition 5. *Let φ be an acyclic Horn CNF, and let $c = c_1 \oplus c_2$ where $c_1, c_2 \in \varphi$. Then, $\varphi' = \varphi \wedge c$ is acyclic Horn, and the dependency graphs $G(\varphi)$ and $G(\varphi')$ have the same transitive closure. Furthermore, any subformula $\varphi'' \subseteq \varphi$ is acyclic Horn.*

Thus, adding repeatedly clauses derived by resolution preserves the acyclicity of a CNF, and, moreover, the possible topological sortings of the dependency graph.

The following proposition captures that for an acyclic Horn CNF, resolution cannot be blocked because of multiple resolving pairs of x_i and $\overline{x}_i$ of literals.

Proposition 6. *Let φ be an acyclic Horn CNF, and let c_1 and c_2 be any implicates of φ derived from φ by resolution. Then, c_1 and c_2 do not resolve iff $P(c_1) \cap N(c_2) = \emptyset$ and $P(c_2) \cap N(c_1) = \emptyset$.*

We define an ordering on Horn clauses as follows. Suppose that $\leq$ imposes a total ordering on the atoms ($x_{i_1} \leq x_{i_2} \leq \cdots \leq x_{i_n}$). Then, for any Horn clauses c_1 and c_2, define $c_1 \leq c_2$ iff $c_1 = c_2$ or one of the following conditions holds:

(i) $P(c_1) \neq \emptyset$ and $P(c_2) = \emptyset$;
(ii) $P(c_1) = \{x_i\}$ and $P(c_2) = \{x_j\}$ and $x_i < x_j$;
(iii) $P(c_1) = P(c_2)$ and $\max N(c_1) \triangle N(c_2) \in c_1$, where "$\triangle$" denotes standard symmetric difference (i.e., $S_1 \triangle S_2 = (S_1 \cup S_2) \setminus (S_1 \cap S_2)$).

As usual, we write $c_1 < c_2$ if $c_1 \leq c_2$ and $c_1 \neq c_2$, $c_1 > c_2$ for $c_2 < c_1$ etc. Note that $\leq$ orders first all definite Horn clauses along their positive literals, followed by the negative clauses. Notice that $c_1 \subset c_2$ implies $c_2 < c_1$, for any Horn clauses c_1 and c_2.

The following proposition is not difficult to establish:

Proposition 7. *Every total ordering $\leq$ of the atoms At induces a total ordering $\leq$ of all Horn clauses over At as described.*

With respect to acyclic Horn CNFs φ, in the rest of this paper we assume an arbitrary but fixed total ordering $\leq$ of the atoms which is compatible with some topological sorting of the dependency graph $G(\varphi)$.

Proposition 8. *Let c_1 and c_2 be Horn clauses such that $c = c_1 \oplus c_2$ exists. Then, $c_1 < c$ and $c_2 < c$ hold.*

Corollary 1. *Let φ be an acyclic Horn CNF, and let c, c_1 and c_2 be any implicates of φ derived from φ such that $c \subseteq c_1 \oplus c_2$. Then, $c > c_1$ and $c > c_2$ holds.*

Consequently, in any input resolution proof of a clause from an acyclic Horn CNF the derived clauses increase monotonically. As for the derivation of prime implicates, we find for such CNFs a more general form than in Lemma 6:

Lemma 7. *Let φ be an acyclic prime Horn CNF, and let c be any prime implicate of φ such that $c \notin \varphi$. Then, there are prime implicates c_1 and c_2 of φ and, for some $k \geq 0$, prime implicates $d_1, d_2, \ldots, d_k$ and literals $\ell_1, \ell_2, \ldots, \ell_k$, respectively, such that: (i) $c_1, d_1, \ldots, d_k \in \varphi$, and (ii) $c = c_1 \oplus e_1$, where $e_i = c \cup \{\ell_i\} = d_i \oplus e_{i+1}$, for $i \in \{1, \ldots, k\}$, and $e_{k+1} = c_2$, such that e_i contains the single prime implicate c.*

An immediate consequence of this result is that prime implicates of an acyclic Horn CNF can be generated from two prime implicates as follows.

Corollary 2. *Let φ be an acyclic prime Horn CNF, and let c be any prime implicate of φ such that $c \notin \varphi$. Then, there exist prime implicates c_1 and c_2 of φ which resolve such that either (i) $c = c_1 \oplus c_2$ or (ii) $c_1 \oplus c_2 = c \cup \{\ell\}$, where $\ell \notin c$ and c is the unique prime implicate of φ contained in $c_1 \oplus c_2$.*

In Example 5, the further prime implicate $\overline{x}_1 \vee \overline{x}_3$ can be derived as in case (ii) of Corollary 2: For $c_1 = \overline{x}_0 \vee \overline{x}_1 \vee x_2$ and $c_2 = \overline{x}_2 \vee \overline{x}_3$, we have $c_1 \oplus c_2 = \overline{x}_1 \vee \overline{x}_3 \vee \overline{x}_0$, and $c = \overline{x}_1 \vee \overline{x}_3$ is the unique prime implicate of φ contained in $c_1 \oplus c_2$.

After the preparatory results, we now show that Algorithm N-EXPLANATIONS is complete. Using an inductive argument on clause orderings, we show that all explanations are generated by taking into account possible derivations of prime implicates as established in Lemma 7 and Corollary 2. However, an inductive proof along $\leq$ encounters two major difficulties: First, the resolvent $c = c_1 \oplus c_2$ of two clauses is *larger* than c_1 and c_2, thus we cannot simply rearrange resolution steps and appeal to smaller clauses.

Second, Algorithm N-EXPLANATIONS does not generate prime implicates d' by a resolution step alone, but using *minimization* in Step 3; that is, a prime implicate *included* in the resolvent $d = c \oplus c'$. A respective statement is much more difficult to prove than the one if d were prime.

In order to overcome these difficulties, we use a more sophisticated ordering of clause pairs (c, c') and establish as a stepping stone the following key lemma. For ease of reference, let us say that resolvable implicates c_1 and c_2 of a Horn CNF φ satisfy the (technical) property $(*)$, if the following conditions hold:

1. At least one of c_1 and c_2 is prime.
2. If c_i is not prime, then it is of form $c_i = c' \cup \{\ell\}$, where c' is the unique prime implicate of φ contained in c_i ($i \in \{1, 2\}$), and c_i occurs in some derivation of c' as in Lemma 7.
3. There is no implicate $c_1' \subset c_1$ (resp., $c_2' \subset c_2$) of φ such that $c = c_1' \oplus c_2$ (resp., $c = c_1 \oplus c_2'$).

Lemma 8 (Key Lemma). *Let φ be a prime acyclic Horn CNF, and let c_1 and c_2 be resolvable clauses satisfying $(*)$ such that $\overline{q} \in c := c_1 \oplus c_2$. Suppose that $c_i \in S$ if c_i is prime and $q \in N(c_i)$ (resp., $c_i' \in S$ if $c_i = c_i' \cup \{\ell_i\}$ where c_i' is prime and $q \in N(c_i')$) for $i \in \{1, 2\}$. Then at least one of the following conditions hold:* (i) *$c \setminus \{\overline{q}\}$ is an implicate of φ, or* (ii) *c contains a $\overline{q}$-clause from S.*

Proof. (Outline) We prove the statement using an inductive argument which involves clause orderings and takes into account how the clauses c_1 and c_2 are recursively generated. Depending on the shape of c_1 and c_2, we consider different cases.

Consider first the case in which both c_1 and c_2 contain $\overline{q}$. Then, w.l.o.g. $c = \overline{q} \vee \overline{a} \vee \overline{b}\,(\vee x_i)$, $c_1 = \overline{q} \vee \overline{a} \vee \overline{x}\,(\vee x_i)$, and $c_2 = \overline{b} \vee x \vee \overline{q}$. Here, $\overline{a}$ and $\overline{b}$ are disjunctions of negative literals, while x is a single atom; "$(\vee x_i)$" means the optional presence of x_i.

Both c_1 and c_2 contain a unique prime implicate c_1' resp. c_2' of φ (where possibly $c_i' = c_i$). If $\overline{q} \in c_i'$, then by assertion we have $c_i' \in S$. Thus, if both c_1' and c_2' contain $\overline{q}$, Algorithm N-EXPLANATIONS considers $c_1' \oplus c_2'$, which implies the statement. No other cases are possible, since either c_1' or c_2' must contain $\overline{q}$ (since c_1 or c_2 is prime) and condition 3 of $(*)$ excludes that exactly one of c_1' and c_2' contains $\overline{q}$. This proves the statement if both c_1 and c_2 contain $\overline{q}$.

For the other cases, assume that $\overline{q} \in c_1$ and $\overline{q} \notin c_2$ and prove the statement by induction along the lexicographic ordering of the pairs (c_1, c_2), where the clauses c_1 are in *reverse* ordering $\geq$ and the clauses c_2 in regular ordering $\leq$. We distinguish the following cases:

Definite/Negative Case 1 (DN1): $c = \overline{q} \vee \overline{a} \vee \overline{b}\,(\vee x_i)$, $c_1 = \overline{q} \vee \overline{a} \vee \overline{x}\,(\vee x_i)$, and $c_2 = \overline{b} \vee x$. That is, the $\overline{q}$-clause c is generated by resolving a $\overline{q}$-clause c_1 with a non-$\overline{q}$-clause c_2, where the positive resolution literal x is in c_2.

Definite/Negative Case 2 (DN2): $c = \overline{q} \vee \overline{a} \vee \overline{b}\,(\vee x_i)$, $c_1 = \overline{q} \vee \overline{a} \vee x$, and $c_2 = \overline{b} \vee \overline{x}\,(\vee x_i)$. That is, the $\overline{q}$-clause c is generated by resolving a $\overline{q}$-clause c_1 with a non-$\overline{q}$-clause c_2, where the positive resolution literal x is in c_1.

The statement is shown by a careful analysis of parent clauses of c_1 and c_2, and by reordering and adapting resolution steps. DN1 recursively only involves cases of the same kind (in fact, for negative c_1 we need to appeal only to smaller instances (c_1', c_2') where c_1' is negative), while DN2 recursively involves itself as well as DN1. □

By combining Lemma 8 with Proposition 8 and Corollary 2, we obtain by an inductive argument on the clause ordering $\leq$ the desired completeness result.

Lemma 9 (Completeness of N-EXPLANATIONS). *Algorithm* N-EXPLANATIONS *outputs all nontrivial explanations for a query $\chi = \overline{q}$ from an acyclic Horn CNF φ.*

Proof. We prove by induction on $\leq$ that S contains each $\overline{q}$-prime implicate c of φ.
(Basis) Let c be the least prime implicate of φ which contains $\overline{q}$. From Proposition 8 and Corollary 2, we conclude that $c \in \varphi$ must hold. Hence, $c \in S$.
(Induction) Suppose the claim holds for all $\overline{q}$-prime implicates c' of φ such that $c' < c$, and consider c. By Corollary 2, there exist prime implicates c_1 and c_2 such that either (i) $c = c_1 \oplus c_2$ or (ii) c is the unique prime implicate contained in $c_1 \oplus c_2 = c \cup \{\ell\}$ where $\ell \notin c$. By Proposition 8 and the induction hypothesis, we have $c_i \in S$ if $q \in N(c_i)$ holds for $i \in \{1, 2\}$. Consequently, c_1 and c_2 satisfy the conditions of Lemma 8. Hence, either (a) $c_1 \oplus c_2 \setminus \{\overline{q}\}$ is an implicate of φ, or (b) $c_1 \oplus c_2$ contains a $\overline{q}$-clause c' from S. Since $q \in N(c)$ and c is the unique prime implicate contained in $c_1 \oplus c_2$, we have (b). It follows from the uniqueness of c that $c' = c$, which proves the statement. □

We are now in a position to establish the main result of this section. Let $||\varphi||$ denote the size (number of symbols) of any CNF φ.

Theorem 2. *Algorithm* N-EXPLANATIONS *incrementally outputs, without duplicates, all nontrivial explanations of $\chi = \overline{q}$ from φ. Moreover, the next output (respectively termination) occurs within $O(s \cdot (s + m) \cdot n \cdot ||\varphi||)$ time, where m is the number of clauses in φ, n the number of atoms, and s the number of explanations output so far.*

Proof. By Lemmas 5 and 9, it remains to verify the time bound. Computing a prime implicate $c' \subseteq c$ and $d' \subseteq d$ of φ in Steps 2 and 3, respectively, is feasible in time $O(n \cdot ||\varphi||)$ (cf. Proposition 1), and thus the outputs in Step 2 occur with $O(m \cdot n \cdot ||\varphi||)$ delay. As for Step 3, note that O contains only pairs (c, c') where $c \in \varphi^\star \cup S$ and $c' \in S$ such that the explanation corresponding to c' was generated, and each such pair is added to O only once. Thus, the next output or termination follows within $s \cdot (s + m)$ runs of the while-loop, where s is the number of solutions output so far. The body of the loop can be done, using proper data structures, in $O(n \cdot ||\varphi||)$ time (for checking $d' \notin S$ efficiently, we may store S in a prefix tree). Thus, the time until the next output resp. termination is bounded by $O(s \cdot (s + m) \cdot n \cdot ||\varphi||)$. □

Corollary 3. *Computing polynomially many explanations for a negative query $\chi = \overline{q}$ from an acyclic Horn CNF φ is feasible in polynomial time (in the size of the input).*

We conclude this section with some remarks on Algorithm N-EXPLANATIONS.

(1) As for implementation, standard data structures and marking methods can be used to realize efficient update of the sets O and S, to determine resolvable clauses, and to eliminate symmetric pairs (c, c') an (c', c) in O.

(2) Algorithm N-EXPLANATIONS is incomplete for cyclic Horn theories, as shown by the following example.

Example 6. Consider the Horn CNF $\varphi = (\overline{x}_0 \vee \overline{x}_1 \vee x_2)(\overline{x}_0 \vee \overline{x}_1 \vee x_3)(\overline{x}_1 \vee \overline{x}_2 \vee x_3)(\overline{x}_1 \vee x_2 \vee \overline{x}_3)(\overline{x}_2 \vee \overline{x}_3 \vee x_4)$ over $x_0, \ldots, x_4$. Note that all clauses in φ are prime, and that x_2 and x_3 are symmetric. There are three further prime implicates, viz. $c_1 = \overline{x}_1 \vee \overline{x}_2 \vee x_4$, $c_2 = \overline{x}_1 \vee \overline{x}_3 \vee x_4$, and $c_3 = \overline{x}_0 \vee \overline{x}_1 \vee x_4$. Thus, $\overline{q} = \overline{x}_0$ has the nontrivial explanations $E_1 = \{x_1, \overline{x}_2\}$, $E_2 = \{x_1, \overline{x}_3\}$, and $E_3 = \{x_1, \overline{x}_4\}$. Apply then algorithm N-EXPLANATIONS on input φ and $q = x_0$. While it outputs E_1 and E_2, it misses explanation E_3. □

Algorithm N-EXPLANATIONS may be extended to handle this example and others correctly by adding in Step 2 prime implicates to φ^* which are generated in polynomial time (e.g., by minimizing clauses derived by resolution proofs from φ^* whose number of steps is bounded by a constant).

(3) Algorithm N-EXPLANATIONS is no longer complete if we constrain the resolution process to input resolution, i.e., consider only pairs (c, c') in Step 3 where at least one of c and c' is from φ (which means that in the update of O in Step 3, the part "$\{(d'', d') \mid d'' \in S\}$" is omitted). This is shown by the following example.

Example 7. Consider the Horn CNF $\varphi = (\overline{x}_0 \vee x_1)(\overline{x}_1 \vee \overline{x}_2 \vee x_3)(\overline{x}_1 \vee \overline{x}_3 \vee x_4)$ over $x_0, \ldots, x_4$. As easily seen, φ is acyclic. Moreover, φ is prime. There are three further prime implicates containing $\overline{x}_0$, viz. $c_1 = \overline{x}_0 \vee \overline{x}_2 \vee x_3$, $c_2 = \overline{x}_0 \vee \overline{x}_3 \vee x_4$, and $c_3 = \overline{x}_0 \vee \overline{x}_2 \vee x_4$. Hence, $\overline{q} = \overline{x}_0$ has the nontrivial explanations $E_1 = \{\overline{x}_1\}$, $E_2 = \{x_2, \overline{x}_3\}$, $E_3 = \{x_3, \overline{x}_4\}$, and $E_4 = \{x_2, \overline{x}_4\}$. If at least one of the clauses (c, c') in Step 3 must be from φ, then E_2 and E_3 are generated from $(\overline{x}_1 \vee \overline{x}_2 \vee x_3, \overline{x}_0 \vee x_1)$ and $(\overline{x}_1 \vee \overline{x}_3 \vee x_4, \overline{x}_0 \vee x_1)$, respectively, while E_4 is missed: The pairs $(\overline{x}_1 \vee \overline{x}_3 \vee x_4, \overline{x}_0 \vee \overline{x}_2 \vee x_3)$ and $(\overline{x}_1 \vee \overline{x}_2 \vee x_3, \overline{x}_0 \vee \overline{x}_3 \vee x_4)$ yield the same resolvent $\overline{x}_0 \vee \overline{x}_1 \vee \overline{x}_2 \vee x_4$, for which $\varphi^* \not\models (c \oplus c' \setminus \{\overline{q}\})$ fails since $\overline{x}_1 \vee \overline{x}_2 \vee x_4$, which is the resolvent of the last two clauses in φ, is an implicate. Note that E_4 is generated from each of the excluded symmetric pairs $(\overline{x}_0 \vee \overline{x}_2 \vee x_3, \overline{x}_0 \vee \overline{x}_3 \vee x_4)$ and $(\overline{x}_0 \vee \overline{x}_3 \vee x_4, \overline{x}_0 \vee \overline{x}_2 \vee x_3)$. □

In terms of generating prime implicates, this contrasts with the cases of computing all prime implicates of a Horn CNF and all prime implicates that contain a positive literal q, for which input-resolution style procedures are complete, cf. [2, 11].

5 Compound Queries

In this section, we consider generating all explanations for queries beyond literals. Theorem 1 implies that this problem is intractable for any common class of CNF queries which admits a negative literal. However, also for positive CNFs, it is intractable.

Theorem 3. *Deciding whether a given CNF χ has an explanation from a Horn CNF φ is* NP*-complete. Hardness holds even if χ is positive and φ is negative (thus acyclic).*

Proof. Membership in NP easily follows from Lemma 1. Hardness is shown via a reduction from the classical EXACT HITTING SET problem. Let $\mathcal{S} = \{S_1, \ldots, S_m\}$ be a collection of subsets $S_i \subseteq U$ of a finite set U. Construct $\chi = \bigwedge_i \left(\bigvee_{u \in S_i}\right)$ and $\varphi = \bigwedge_i \bigwedge_{x \neq y \in S_i} (\overline{x} \vee \overline{y})$. Then χ has an explanation from φ iff there exists an exact hitting set for $\mathcal{S}$, i.e., a set $H \subseteq U$ such that $|H \cap S_i| = 1$ for all $i \in \{1, \ldots, m\}$. □

For important special cases of positive CNFs, we obtain positive results. In particular, this holds if the query χ is restricted to be a clause or a term.

Theorem 4. *Computing polynomially many (resp., all) explanations for a query χ which is either a positive clause or a positive term from a Horn CNF φ is feasible in polynomial time (resp., polynomial total time).*

Proof. Let us first consider the case in which χ is a positive clause $c = \bigvee_{x \in P(c)} x$. Then let $\varphi^* = \varphi \wedge \bigwedge_{x \in P(c)} (\overline{x} \vee x^*)$, where x^* is a new letter. As easily seen, φ^* is a Horn CNF and there is a one-to-one correspondence between explanations for a query χ from φ and the ones for x^* form φ^* (except for a trivial explanation x^*). This, together with the result in [11] that all explanations for a query $\chi = q$ where q is an atom from a Horn CNF can be generated with incremental polynomial delay, proves the theorem.

Similarly, if χ is a positive term $t = \bigwedge_{x \in P(t)} x$, one can consider explanations for x^* from the Horn CNF $\varphi^* = \varphi \wedge (\bigvee_{x \in P(t)} \overline{x} \vee x^*)$, where x^* is a new letter. □

In case of acyclic Horn theories, the positive result holds also in the case where negative literals are present in a clause query.

Theorem 5. *Computing polynomially many (resp., all) explanations for a query $\chi = c$ where c is a clause from an acyclic Horn CNF φ is feasible in polynomial time (resp., polynomial total time).*

Proof. Let $\chi = \bigvee_{x \in P(c)} x \vee \bigvee_{x \in N(c)} \overline{x}$. Then let $\varphi^* = \varphi \wedge \bigwedge_{x \in P(c)} (\overline{x} \vee \overline{x^*}) \wedge \bigwedge_{x \in N(c)} (x \vee \overline{x^*})$, where x^* is a new letter. It is not difficult to see that φ^* is an acyclic Horn CNF, and there is a ono-to-one correspondence between explanations for a query χ from φ and the ones for $\overline{x^*}$ from φ^* (except for a trivial explanation $\overline{x^*}$). This together with Theorem 2 proves the theorem. □

Note that explanations for a single clause query $\chi = c$ correspond to the minimal support clauses for c as used in Clause Management Systems [22]. Thus, from Theorems 1 and 5 we obtain that while in general, generating all minimal support clauses for a given clause c is not possible in polynomial total time unless P = NP, it is feasible with incremental polynomial delay for acyclic Horn theories.

The presence of negative literals in a query $\chi = t$ for a term t from an acyclic Horn theory is more involved; a similar reduction technique as for a clause to a single literal seems not to work. We can show that generating all nontrivial explanations E (i.e., $E \cap \chi = \emptyset$) for a term is intractable; the case of all explanations is currently open.

6 Conclusion

We considered computing all abductive explanations for a query χ from a propositional Horn CNF φ, which is an important problem that has many applications in AI and Computer Science. We presented a number of new complexity results, which complement and extend previous results in the literature; they are compactly summarized in Table 1.

We showed the intractability of computing all abductive explanations for a negative literal query χ from a general Horn CNF φ (thus closing an open issue), while we

Table 1. Complexity of computing all abductive explanations for a query χ from a Horn theory (PTT = polynomial total time, NPTT = not polynomial total time unless P=NP)

Query χ	CNF		single literal		single clause		single term	
Horn theory Σ, by	general	positive	atom q	$\overline{q}$	positive	general	positive	general
Horn CNF φ	NPTT	NPTT	PTT[a]	NPTT	PTT	NPTT	PTT	NPTT
Acyclic Horn CNF φ	NPTT	NPTT	PTT[a]	PTT	PTT	PTT	PTT	–

[a] By the results of [11].

presented a polynomial total time algorithm for acyclic Horn CNFs. Since this amounts to computing all prime implicates of φ which contain $\overline{q}$, we have obtained as a byproduct also new results on computing all such prime implicates from a Horn CNF. Note that our intractability result contrasts with the result in [2] that all prime implicates of a Horn CNF are computable in polynomial total time. Furthermore, our results on clause queries imply analogous results for generating all minimal support clauses for a clause in a Clause Management System [22].

It remains for further work to complete the picture and to find further meaningful input classes of cyclic Horn theories which permit generating a few resp. all explanations in polynomial total time. For example, this holds for clause queries from quadratic Horn CNFs (i.e., each clause is Horn and has at most 2 literals) and for literal queries from Horn CNFs in which each clause contains the query literal. Another issue is a similar study for the case of predicate logic.

Acknowledgments

We thank the reviewers for helpful suggestions.

References

1. K. Apt and M. Bezem. Acyclic programs. *New Generation Comp.*, 9(3-4):335–364, 1991.
2. E. Boros, Y. Crama, and P. L. Hammer. Polynomial-time inference of all valid implications for Horn and related formulae. *Ann. Math. and Artif. Int.*, 1:21–32, 1990.
3. G. Brewka, J. Dix, and K. Konolige. *Nonmonotonic Reasoning – An Overview*. Number 73 in CSLI Lecture Notes. CSLI Publications, Stanford University, 1997.
4. T. Bylander. The monotonic abduction problem: A functional characterization on the edge of tractability. In *Proc. KR-91*, 70-77. Morgan Kaufmann, 1991.
5. L. Console, D. Theseider Dupré, and P. Torasso. On the relationship between abduction and deduction. *J. Logic and Computation*, 1(5):661–690, 1991.
6. J. de Kleer. An assumption-based truth maintenance system. *Artif. Int.*, 28:127–162, 1986.
7. A. del Val. On some tractable classes in deduction and abduction. *Artif. Int.*, 116(1-2):297–313, 2000.
8. A. del Val. The complexity of restricted consequence finding and abduction. In *Proc. National Conference on Artificial Intelligence (AAAI-2000)*,pp. 337–342, 2000.
9. W. Dowling and J. H. Gallier. Linear-time algorithms for testing the satisfiability of propositional Horn theories. *J. Logic Programming*, 3:267–284, 1984.

10. T. Eiter and G. Gottlob. The complexity of logic-based abduction. *JACM*, 42(1):3–42, 1995.
11. T. Eiter and K. Makino. On computing all abductive explanations. In *Proc. AAAI-02*, pp. 62–67. AAAI Press, July 2002.
12. T. Eiter and K. Makino. Generating all abductive explanations for queries on propositional Horn theories. Tech. Rep. INFSYS RR-1843-03-09, Institute of Inf.Sys., TU Vienna, 2003.
13. K. Eshghi. A tractable class of abduction problems. In *Proc. IJCAI-93*, pp. 3–8, 1993.
14. G. Friedrich, G. Gottlob, and W. Nejdl. Hypothesis classification, abductive diagnosis, and therapy. In *Proc. Int'l Workshop on Expert Sys. in Engineering*, LNCS 462, pp. 69–78, 1990.
15. K. Inoue. Linear resolution for consequence finding. *Artif. Int.*, 56(2-3):301–354, 1992.
16. H. Kleine Büning and T. Lettmann. *Aussagenlogik - Deduktion und Algorithmen*. B.G. Teubner, Stuttgart, 1994.
17. A. Leitsch. *The Resolution Calculus*. Springer, 1997.
18. P. Marquis. Consequence finding algorithms. In D. Gabbay, Ph. Smets (eds), *Handbook of Defeasible Reasoning and Uncertainty Management Systems* (V), pp. 41-145. Kluwer, 2000.
19. C. S. Peirce. Abduction and induction. In J. Buchler, editor, *Philosophical Writings of Peirce*, chapter 11. Dover, New York, 1955.
20. D. Poole. Explanation and prediction: An architecture for default and abductive reasoning. *Computational Intelligence*, 5(1):97–110, 1989.
21. D. Poole. Probabilistic Horn abduction and Bayesian networks. *Artif. Int.*, 64:81-130, 1993.
22. R. Reiter and J. de Kleer. Foundations of assumption-based truth maintenance systems: Preliminary report. In *Proc. AAAI-87*, pp. 183–188, 1982.
23. B. Selman and H. J. Levesque. Abductive and default reasoning: A computational core. In *Proc. AAAI-90*, pp. 343–348, 1990.
24. B. Selman and H. J. Levesque. Support Set Selection for Abductive and Default Reasoning. *Artif. Int.*, 82:259–272, 1996.
25. B. Zanuttini. New Polynomial Classes for Logic-Based Abduction. *J. Artificial Intelligence Research*, 2003. To appear.

Refined Complexity Analysis of Cut Elimination

Philipp Gerhardy

BRICS*
Department of Computer Science
University of Aarhus
Ny Munkegade
DK-8000 Aarhus C, Denmark
peegee@daimi.au.dk

Abstract. In [1, 2] Zhang shows how the complexity of cut elimination depends on the nesting of quantifiers in cut formulas. By studying the role of contractions we can refine that analysis and show how the complexity depends on a *combination* of contractions and quantifier nesting. With the refined analysis the upper bound on cut elimination coincides with Statman's lower bound. Every non-elementary growth example must display a combination of nesting of quantifiers and contractions similar to Statman's lower bound example. The upper and lower bounds on cut elimination immediately translate into bounds on Herbrand's theorem. Finally we discuss the role of quantifier alternations and show an elementary upper bound for the $\forall - \wedge$-case (resp. $\exists - \vee$-case).

1 Introduction

The most commonly used proofs of cut elimination by Schwichtenberg[3] and Buss[4] give an estimate of the depth, resp. size, of a cut free proof in terms of the (logical) depth of the largest cut formula and the depth, resp. size, of the original proof. As shown by Zhang[1, 2] this bound can be refined by distinguishing between propositional connectives and quantifiers in the complexity analysis of cut elimination, showing that the elimination of quantifiers causes the non-elementary complexity. We derive a proof of Zhang's improved result from Buss's (clear and understandable) proof of cut elimination in [4]. In [5, 6] Luckhardt discusses simplifications of Zhang's result, which originally motivated the proof given below.

Analysing contractions we can show that all non-elementary growth examples for cut elimination must display both nested quantifiers *and also* contractions on ancestors of cut formulas. If the proof is contraction free or if one restricts contraction w.r.t. the nesting of quantifiers in contracted formulas, cut elimination can be shown to be elementary. We discuss how Statman's lower bound example displays an optimal combination of nested quantifiers and contractions, and how Statman's lower and our refined upper bound coincide. These bounds translate directly into bounds on the size of a Herbrand disjunction, i.e. the number of

* Basic Research in Computer Science (www.brics.dk), funded by the Danish National Research Foundation.

M. Baaz and J.A. Makowsky (Eds.): CSL 2003, LNCS 2803, pp. 212–225, 2003.

different formula instances in the disjunction. Finally we discuss a further refinement of cut elimination based upon eliminating arbitrary $\forall - \wedge$-cuts (resp. $\exists - \vee$-cuts) with exponential complexity.

2 Notation and Definitions

2.1 The Calculus LK

We use a version of the sequent calculus **LK**. We write x, y, z for bound and α, β, γ for free variables. Terms and formulas are defined in the usual way. An **LK**-proof is a rooted tree in which the nodes are sequents. The root of the tree is called the *end-sequent*, the leaves $A \vdash A$ for atomic formulas A are called *initial sequents*. All nodes but the initial sequents must be inferred by one of the rules below. Proofs are denoted by ϕ, χ and ψ. Γ, Δ, Π and Λ serve as metavariables for multisets of formulas. The multiset of formulas to the left of the separation symbol $\vdash$ is called *antecedent*, the multiset to the right is called *succedent*.

The logical rules are:

$$\frac{A, B, \Gamma \vdash \Delta}{A \wedge B, \Gamma \vdash \Delta} \wedge : l \qquad \frac{\Gamma \vdash \Delta, A \quad \Pi \vdash \Lambda, B}{\Gamma, \Pi \vdash \Delta, \Lambda, A \wedge B} \wedge : r$$

$$\frac{A, \Gamma \vdash \Delta \quad B, \Pi \vdash \Lambda}{A \vee B, \Gamma, \Pi \vdash \Delta, \Lambda} \vee : l \qquad \frac{\Gamma \vdash \Delta, A, B}{\Gamma \vdash \Delta, A \vee B} \vee : r$$

$$\frac{\Gamma \vdash \Delta, A \quad B, \Pi \vdash \Lambda}{A \rightarrow B, \Gamma, \Pi \vdash \Delta, \Lambda} \rightarrow : l \qquad \frac{A, \Gamma \vdash \Delta, B}{\Gamma \vdash \Delta, A \rightarrow B} \rightarrow : r$$

$$\frac{\Gamma \vdash \Delta, A}{\neg A, \Gamma \vdash \Delta} \neg : l \qquad \frac{A, \Gamma \vdash \Delta}{\Gamma \vdash \Delta, \neg A} \neg : r$$

$$\frac{\Gamma, A\{x \leftarrow t\} \vdash \Delta}{\Gamma, (\forall x)A \vdash \Delta} \forall : l \qquad \frac{\Gamma \vdash \Delta, A\{x \leftarrow \alpha\}}{\Gamma \vdash \Delta, (\forall x)A} \forall : r,\ \alpha \notin FV(\forall x A(x))$$

$$\frac{\Gamma \vdash \Delta, A\{x \leftarrow t\}}{\Gamma \vdash \Delta, (\exists x)A} \exists : r \qquad \frac{\Gamma, A\{x \leftarrow \alpha\} \vdash \Delta}{\Gamma, (\exists x)A \vdash \Delta} \exists : l,\ \alpha \notin FV(\exists x A(x))$$

$\forall : r$ (resp. $\exists : l$) must fulfill the eigenvariable condition, i.e. the free variable α does not occur in $\Gamma \vdash \Delta$. In $\forall : l$ (resp. $\exists : r$) t may be an arbitrary term, but admitting only free variables.

The structural rules are:

$$\frac{\Gamma \vdash \Delta}{A, \Gamma \vdash \Delta} w : l \qquad \frac{\Gamma \vdash \Delta}{\Gamma \vdash \Delta, A} w : r$$

$$\frac{A, A, \Gamma \vdash \Delta}{A, \Gamma \vdash \Delta} c : l \qquad \frac{\Gamma \vdash \Delta, A, A}{\Gamma \vdash \Delta, A} c : r$$

$$\frac{\Gamma \vdash \Delta, A \quad A, \Pi \vdash \Lambda}{\Gamma, \Pi \vdash \Delta, \Lambda} cut$$

Weakening and contraction inferences are called *weak* inferences. All other inferences are called *strong*. In the contraction free case the above multiplicative version of **LK** corresponds to the multiplicative fragment of affine linear logic[7, 8], which also is discussed as direct predicate calculus in [9, 10].

2.2 Definitions: Properties of Terms, Formulas and Proofs

Definition 1. *(free variable normal form). A proof ϕ is in* free variable normal form *if no variable that is free in the end-sequent is used as an eigenvariable somewhere in the proof and every other free variable appearing in ϕ is used exactly once as an eigenvariable and appears in ϕ only in sequents above the inference in which it is used as an eigenvariable.*

Definition 2. *mid-sequent A proof ϕ is in* mid-sequent form *if it can be divided in an upper, quantifier-free part and a lower part, consisting only of quantifier inferences and contractions. The last sequent before the first quantifier inference is called the* mid-sequent.

Definition 3. *(depth and size of a proof). The depth $|\phi|$ of a proof is the depth of the proof tree, counting only strong inferences and axioms. The size $||\phi||$ of a proof ϕ is the number of strong inferences and axioms in the proof.*

Definition 4. *(depth of a formula). The depth $|A|$ of an atomic formula A is defined as $|A| = 0$. For formulas A and B we define $|A \vee B| = |A \wedge B| = |A \rightarrow B| = max(|A|, |B|) + 1$, and $|\neg A| = |\forall x A(x)| = |\exists x A(x)| = |A| + 1$.*

Definition 5. *(cut-rank ρ of a proof). The cut-rank $\rho(\phi)$ of a proof ϕ is the supremum of the depths of the cut formulas in ϕ.*

Definition 6. *(nested quantifier depth of formulas and proofs). The nested quantifier depth $nqf(A)$ of an atomic formula A is defined as $nqf(A) = 0$. For formulas A and B we define $nqf(A \vee B) = nqf(A \wedge B) = nqf(A \rightarrow B) = max(nqf(A), nqf(B))$, $nqf(\neg A) = nqf(A)$, and $nqf(\forall x A(x)) = nqf(\exists x A(x)) = nqf(A) + 1$. The nested quantifier depth $nqf(\phi)$ of a proof ϕ is the supremum of the nested quantifier depths of the cut formulas in ϕ.*

Definition 7. *(propositional depth of formulas). Let B be a formula occurrence in A and let A be constructed from B and other formulas by propositional connectives only. Then the propositional depth of B in A is the number of propositional connectives which have to be removed from A to obtain B.*

Definition 8. *(deepest quantified formulas and proofs). For a formula A we define $dqf(A)$ as the supremum over the propositional depths of subformulas B of A which have $nqf(B) = nqf(A)$. Let ϕ be a proof of $\Gamma \vdash \Delta$. Then $dqf(\phi)$ is the supremum over $dqf(A_i)$ for cut formulas A_i in ϕ which have $nqf(A_i) = nqf(\phi)$.*

Definition 9. *(the hyper-exponential function). The hyper-exponential function 2^x_n is defined as $2^x_0 \equiv x$ and $2^x_{n+1} \equiv 2^{2^x_n}$. We write 2_n for 2^0_n.*

3 Cut Elimination

3.1 Zhang's Refined Bound on Cut Elimination

To prove Zhang's refined upper bound on cut elimination we state the following Refined Reduction Lemma:

Refined Reduction Lemma. *Let ϕ be an* **LK**-*proof of a sequent $\Gamma \vdash \Delta$ with the final inference a cut with cut formula A. Then if for all other cut formulas B*

(i) $nqf(A) \geq nqf(B)$ and $dqf(\phi) = dqf(A) > dqf(B)$, then there exists a proof ϕ' of the same sequent with $dqf(\phi') \leq dqf(\phi) - 1$ and $|\phi'| \leq |\phi| + 1$.

(ii) $nqf(\phi) = nqf(A) > nqf(B)$ and $dqf(A) = 0$, then there exists a proof ϕ' of the same sequent with $nqf(\phi') \leq nqf(\phi) - 1$ and $|\phi'| < 2 \cdot |\phi|$.

If the cut formula A is atomic and both subproofs are cut free, then there is a cut free proof ϕ' with $|\phi'| < 2 \cdot |\phi|$.

Proof: The proof is by cases on the structure of the cut formula A. We assume w.l.o.g. that ϕ is is free variable normal form and that for no cut formula all direct ancestors in either subproof are introduced by weakenings only, as the reduction is trivial in that case. We implicitly skip weakening inferences that introduce a direct ancestor to a cut formula. This can be done without harming the validity of the proof transformations given below.

Case 1: $A \equiv \neg B$. We then have the final inference of ϕ:

$$\frac{\overset{\chi}{\Gamma \vdash \Delta, \neg B} \quad \overset{\psi}{\neg B, \Pi \vdash \Lambda}}{\Gamma, \Pi \vdash \Delta, \Lambda}\, cut$$

We transform the proofs χ and ψ into proofs χ' and ψ' of the sequents $B, \Gamma \vdash \Delta$ and $\Pi \vdash \Lambda, B$ respectively. In χ we find all $\neg : r$ inferences which introduce a direct ancestor of $\neg B$. By skipping these inferences, and replacing contractions to the right on $\neg B$ with contractions to the left on B, we obtain a proof of $B, \Gamma \vdash \Delta$. The transformation of ψ into ψ' is similar. The final inference of the modified proof ϕ' is then:

$$\frac{\overset{\psi'}{\Pi \vdash \Lambda, B} \quad \overset{\chi'}{B, \Gamma \vdash \Delta}}{\Gamma, \Pi \vdash \Delta, \Lambda}\, cut$$

Trivially $dqf(\phi') \leq dqf(\phi) - 1$ and $|\phi'| \leq |\phi|$.

Case 2a: $A \equiv B \vee C$. We then have the final inference of ϕ:

$$\frac{\overset{\chi}{\Gamma \vdash \Delta, B \vee C} \quad \overset{\psi}{B \vee C, \Pi \vdash \Lambda}}{\Gamma, \Pi \vdash \Delta, \Lambda}\, cut$$

We transform the proofs χ and ψ into proofs χ' of the sequent $\Gamma \vdash \Delta, B, C$, and into proofs ψ_B and ψ_C of the sequents $B, \Pi \vdash \Lambda$ and $C, \Pi \vdash \Lambda$ respectively. In

χ we skip the $\vee : r$ inferences where direct ancestors of the cut formula $B \vee C$ are introduced and replace contractions on $B \vee C$ by contractions on B and C. For the transformed proof $|\chi'| \leq |\chi|$. To transform ψ into ψ_B we replace all inferences where a direct ancestor of the cut formula $B \vee C$ has been introduced

$$\dfrac{\overset{\pi_0}{B, \Pi_0 \vdash \Lambda_0} \quad \overset{\pi_1}{C, \Pi_1 \vdash \Lambda_1}}{B \vee C, \Pi' \vdash \Lambda'} \vee : l$$

(with $\Pi' = \Pi_0 \cup \Pi_1$ and $\Lambda' = \Lambda_0 \cup \Lambda_1$) with inferences

$$\dfrac{\overset{\pi_0}{B, \Pi_0 \vdash \Lambda_0}}{B, \Pi' \vdash \Lambda'} w : l, r$$

We replace contractions on $B \vee C$ with contractions on B. The proof for ψ_C is obtained in a similar way. In both instances no strong inferences have been added to the proof, so $|\psi_B| \leq |\psi|$ and $|\psi_C| \leq |\psi|$. The transformed proofs can then be combined to a new proof ϕ' in the following way:

$$\dfrac{\dfrac{\overset{\chi'}{\Gamma \vdash \Delta, B, C} \quad \overset{\psi_B}{B, \Pi \vdash \Lambda}}{\Gamma, \Pi \vdash \Delta, \Lambda, C} cut \quad \overset{\psi_C}{C, \Pi \vdash \Lambda}}{\Gamma, \Pi \vdash \Delta, \Lambda} cut, c : r, l$$

Then $dqf(\phi') \leq dqf(\phi) - 1$ and $|\phi'| = sup(|\chi'|, |\psi_B|, |\psi_C|) + 2 \leq |\phi| + 1$.

Case 2b+c: $A \equiv B \wedge C$, $A \equiv B \rightarrow C$. These cases are symmetrical to 2a.

Case 3a: $A \equiv \exists x B(x)$. We then have the final inference of the proof:

$$\dfrac{\overset{\chi}{\Gamma \vdash \Delta, \exists x B(x)} \quad \overset{\psi}{\exists x B(x), \Pi \vdash \Lambda}}{\Gamma, \Pi \vdash \Delta, \Lambda} cut$$

In the subproof χ we find the k $\exists : r$ inferences where direct ancestors of the formula $\exists x B(x)$ have been introduced and enumerate these (for $1 \leq i \leq k$)

$$\dfrac{\Gamma_i \vdash \Delta_i, B(t_i)}{\Gamma_i \vdash \Delta_i, \exists x B(x)} \exists : r$$

Similarly in ψ we enumerate all l $\exists : l$ inferences which introduce a direct ancestor of the cut formula as (for $1 \leq j \leq l$)

$$\dfrac{B(\alpha_j), \Pi_j \vdash \Lambda_j}{\exists x B(x), \Pi_j \vdash \Lambda_j} \exists : r$$

We obtain proofs ψ_i of the sequents $B(t_i), \Pi \vdash \Lambda$ by replacing in ψ all l variables α_j with the term t_i. This is unproblematic since the proof is in free variable normal form. We replace the contractions on $\exists x B(x)$ by contractions on $B(t_i)$.

We construct a proof ϕ' by replacing the k enumerated quantifier introductions in χ with cuts, so that the inferences become

$$\frac{\overset{\chi_i}{\Gamma_i \vdash \Delta_i, B(t_i)} \quad \overset{\psi_i}{B(t_i), \Pi \vdash \Lambda}}{\Gamma_i, \Pi \vdash \Delta_i, \Lambda} \; \exists : r$$

We contract multiple copies of Π and Λ in the end sequent to make ϕ' a proof of $\Gamma, \Pi \vdash \Delta, \Lambda$. Trivially $nqf(\phi') \leq nqf(\phi) - 1$ and since $|\chi_i| \leq |\chi| < |\phi|$ and $|\psi_i| \leq |\psi| < |\phi|$ and $|\phi'| \leq |\chi| + sup\{|\psi_i|\}$, also $|\phi'| < 2 \cdot |\phi|$.

Case 3b: $A \equiv \forall x B(x)$**.** This case is symmetrical to 3a.

Case 4: A is atomic. We then have the final inference:

$$\frac{\overset{\chi}{\Gamma \vdash \Delta, A} \quad \overset{\psi}{A, \Pi \vdash \Lambda}}{\Gamma, \Pi \vdash \Delta, \Lambda} \; cut$$

By assumption χ and ψ are cut free. We obtain the desired proof by modifying the proof ψ of $A, \Pi \vdash \Lambda$. We remove from ψ all direct ancestors of the cut formula A and add Γ and Δ to the antecedent, resp. succedent, of each sequent, where an ancestor of the cut formula A has been removed. We then ensure all initial sequents in ψ have valid proofs. The initial sequents $B \vdash B$ for $B \neq A$ do not need new proofs. For $\Gamma \vdash \Delta, A$ we use the (cut free) subproof χ, thus obtaining a proof of $\Gamma^k, \Pi \vdash \Delta^k, \Lambda$, where k is the number of initial sequents we had to "repair". By contractions we arrive at a proof ϕ' of $\Gamma, \Delta \vdash \Pi, \Lambda$, where ϕ' is cut free and $|\phi'| < 2 \cdot |\phi|$. □

We use the following lemmas to prove Zhang's refined cut elimination theorem:

Lemma 10. *Let ϕ be an **LK**-proof of a sequent $\Gamma \vdash \Delta$. If $dqf(\phi) = d > 0$, then there is a proof ϕ' of the same sequent with $dqf(\phi') = 0$ and $|\phi'| \leq 2^d \cdot |\phi|$.*

Proof: By the Refined Reduction Lemma and induction on d. □

Lemma 11. *Let ϕ be an **LK**-proof of a sequent $\Gamma \vdash \Delta$. If $dqf(\phi) = 0$ and $nqf(\phi) = d > 0$, then there is a proof ϕ' of the same sequent with $nqf(\phi') \leq d-1$ and $|\phi'| < 2^{|\phi|}$.*

Proof: By the Refined Reduction Lemma and induction on $|\phi|$. □

First Refined Cut Elimination Theorem. *Let ϕ be an **LK**-proof of a sequent $\Gamma \vdash \Delta$. If $nqf(\phi) = d > 0$, then there is a proof ϕ' of the same sequent and a constant c, depending only on the propositional nesting of the cut formulas, so that $nqf(\phi') \leq d - 1$ and $|\phi'| \leq 2^{c \cdot |\phi|}$.*

Proof: Assume we have a proof with $dqf(\phi) = k$ and $nqf(\phi) = d$. Then by Lemma 10 we get a proof ϕ'' with $dqf(\phi'') = 0$ and $|\phi''| \leq 2^k \cdot |\phi|$. Let $c \geq 2^k$, then by Lemma 11 we get a proof ϕ' with $nqf(\phi') \leq d - 1$ and $|\phi'| \leq 2^{c \cdot |\phi|}$. □

Corollary 12. *Let ϕ be an **LK**-proof of a sequent $\Gamma \vdash \Delta$ and let $nqf(\phi) = d$. Then there is a constant c, depending only on the propositional nesting of the cut formulas, and a proof ϕ' of the same sequent where ϕ' is cut free and $|\phi'| \leq 2_{d+1}^{c \cdot |\phi|}$.*

Proof: By the First Refined Cut Elimination Theorem and induction on d. □

3.2 The Role of Contractions in Cut Elimination

We show that if no direct ancestor of a cut formula has been contracted, the cut can be reduced by a mere rearrangement of the proof. We define *cnqf*, the contracted nested quantifier depth, of formulas and proofs.

Definition 13. *Let A be a cut formula in a proof and let $B_1, B_2, \ldots, B_k$ be ancestors of A, s.t. B_i is a principal formula in a contraction inference. Then the contracted nested quantifier depth $cnqf(A)$, is the supremum over $nqf(B_i)$. For a proof ϕ $cnqf(\phi)$ is the supremum over $cnqf(A_i)$ for cut formulas A_i.*

We state the following variant of the reduction lemma:

Lemma 14. *Let ϕ be a proof with the final inferencea cut with cut formula A, where there have been no contractions on direct ancestors of the cut formula. Then if for all other cut formulas B*

(i) $nqf(A) > nqf(B)$ and $dqf(A) = 0$, then we can find a proof ϕ' of the same sequent with $nqf(\phi') \leq nqf(\phi) - 1$

(ii) $nqf(A) \geq nqf(B)$ and $dqf(A) > dqf(B)$, then we can find a proof ϕ' of the same sequent with $nqf(\phi') = nqf(\phi)$ and $dqf(\phi') \leq dqf(\phi) - 1$

(iii) A is atomic and the subproofs are cut free, then we can find a cut free proof ϕ' of the same sequent
where in all cases no new contractions have been added to ϕ' and $||\phi'|| \leq ||\phi||$.

Proof: The proof is by cases on the structure of A.

Case 1: $A \equiv \neg B$. We use the same proof transformation as before. No new contractions have been added to ϕ', so $dqf(\phi') \leq dqf(\phi)$ and $||\phi'|| \leq ||\phi||$.

Case 2a: $A \equiv B \vee C$. Since no contractions have been made on direct ancestors of the cut formula, we can find the unique inferences where the $\vee$-connective has been introduced. Those inferences are (in χ, resp.ψ):

$$\frac{\overset{\chi_0}{\Gamma' \vdash \Delta', B, C}}{\Gamma' \vdash \Delta', B \vee C}\ \vee : r \qquad \frac{\overset{\psi_0}{B, \Pi_0 \vdash \Lambda_0} \quad \overset{\psi_1}{C, \Pi_1 \vdash \Lambda_1}}{B \vee C, \Pi' \vdash \Lambda'}\ \vee : r$$

where $\Pi' \equiv \Pi_0 \cup \Pi_1$ and $\Lambda' \equiv \Lambda_0 \cup \Lambda_1$. We construct the following new proof:

$$\frac{\dfrac{\overset{\chi_0}{\Gamma' \vdash \Delta', B, C} \quad \overset{\psi_0}{B, \Pi_0 \vdash \Lambda_0}}{\Gamma', \Pi_0 \vdash \Delta', \Lambda_0, C}\ cut \quad \overset{\psi_1}{C, \Pi_1 \vdash \Lambda_1}}{\Gamma', \Pi' \vdash \Delta', \Lambda'}\ cut$$

The remaining steps to get a proof of $\Gamma, \Pi \vdash \Delta, \Lambda$ are as in the subproofs χ and ψ respectively. In the new proof ϕ' we replaced two $\vee$-introductions and a cut by two new cuts. No new contractions have been added to ϕ', so $dqf(\phi') \leq dqf(\phi)-1$ and $||\phi'|| \leq ||\phi||$.

Case 2b+c: $A \equiv B \wedge C$, $A \equiv B \rightarrow C$. These cases are symmetrical to 2a.

Case 3a: $A \equiv \exists x B(x)$. We again know that in χ there is a unique inference where the quantified variable x has been introduced for some term t:

$$\frac{\Gamma' \vdash \Delta', B(t)}{\Gamma' \vdash \Delta', \exists x B(x)} \exists : r$$

The subproof ψ can be transformed to a proof ψ' of $B(t), \Pi \vdash \Lambda$ as described earlier. These proofs can now be combined to a new proof ϕ' by replacing the inference in χ with a cut

$$\frac{\overset{\chi'}{\Gamma' \vdash \Delta', B(t)} \quad \overset{\psi'}{B(t), \Pi \vdash \Lambda}}{\Gamma', \Pi \vdash \Delta', \Lambda} \, cut$$

and then continuing with χ as before. We skip some inferences and then rearrange the proof, adding no new contractions, so $nqf(\phi') \leq nqf(\phi)-1$ and $||\phi'|| \leq ||\phi||$.

Case 3b: $A \equiv \forall x B(x)$. This case is symmetrical to 3a.

Case 4: A is atomic. Again we know there is exactly one ancestor to the cut formula in each subproof. We showed earlier how to obtain a cut free proof of $\Gamma^k, \Pi \vdash \Delta^k, \Lambda$, where k was the number of direct ancestors of the cut formula A in the left subproof ψ. Since now $k = 1$ we automatically have a proof ϕ' of $\Gamma, \Pi \vdash \Delta, \Lambda$, and $||\phi'|| \leq ||\phi||$. □

Contraction Lemma. *Let ϕ be an **LK**-proof of a sequent $\Gamma \vdash \Delta$, with $nqf(\phi) > cnqf(\phi)$ then there is proof ϕ' of the same sequent with $nqf(\phi') = cnqf(\phi')$ and $||\phi'|| \leq ||\phi||$. As a consequence also $|\phi'| \leq 2^{|\phi|}$*

Proof: By induction on the number of uncontracted cuts and Lemma 14 we get a proof ϕ' s.t. $nqf(\phi') = cnqf(\phi')$ and $||\phi'|| \leq ||\phi||$. By $||\phi|| \leq 2^{|\phi|}$ and $|\phi'| \leq ||\phi'||$ we get $|\phi'| \leq 2^{|\phi|}$. □

Second Refined Cut Elimination Theorem. *Let ϕ be an **LK**-proof of a sequent $\Gamma \vdash \Delta$. Then there is a constant c depending only on the propositional nesting of the cut formulas and a cut free proof ϕ' of the same sequent where $|\phi'| \leq 2^{c \cdot |\phi|}_{cnqf(\phi)+2}$.*

Proof: By the Contraction Lemma and Corollary 12. □

Remark. *In the contraction free case, i.e. for the multiplicative fragment of affine linear logic, we get an exponential upper bound on cut elimination. A comparable result was shown by Bellin and Ketonen in [10].*

If we compare the results of this section with previous results by Schwichtenberg[3], Buss[4] and Zhang[1, 2], we get the following table:

bound	*complexity* $\lvert\cdot\rvert$	*complexity* $\lVert\cdot\rVert$
Schwichtenberg	$2^{\lvert\phi\rvert}_{\rho(\phi)+1}$	$2^{\lVert\phi\rVert}_{\rho(\phi)+2}$
Buss	–	$2^{\lVert\phi\rVert}_{2\cdot\rho(\phi)+2}$
Zhang	$2^{c\cdot\lvert\phi\rvert}_{nqf(\phi)+1}$	$2^{c\cdot\lVert\phi\rVert}_{nqf(\phi)+2}$
Gerhardy	$2^{c\cdot\lvert\phi\rvert}_{cnqf(\phi)+2}(*)$	$2^{c\cdot\lVert\phi\rVert}_{cnqf(\phi)+3}(*)$

(*) In the case that $cnqf(\phi) = nqf(\phi)$ we can, as discussed, use the estimates based on $nqf()$ to match Zhang's bound.

3.3 Statman's Theorem

A Kalmar non-elementary lower bound on Herbrand's theorem, and hence also cut elimination, was first proved by Statman[11]. Statman's theorem states that there exist simple sequents S_n, for which we have short proofs (of depth linear in n), but for which every cut free proof must have depth at least 2_n. The original example used by Statman is formulated in relational logic. Later presentations are due to Orevkov[12] and Pudlak[13]. We will present Pudlak's version of the theorem, which uses a simple fragment T of arithmetic

Let the theory T be the language with function symbols $+, \cdot, 2^{(\cdot)}$, relation symbols $=, I(.)$ and constants $0, 1$. The theory T has the following non-logical axioms: $(x+y)+z = x+(y+z)$, $y+0=y$, $2^0 = 1$, $2^x + 2^x = 2^{1+x}$, $I(0)$ and $I(x) \rightarrow I(1+x)$.

We want to prove the sequents $S_n \equiv A_{T+eq} \vdash I(2_n)$, expressing that T with some equality proves 2_n is an integer. Here $\bigwedge A_{T+eq}$ is the universal closure of a finite conjunction of axioms from T and equality axioms over T. For the short proofs of S_n we make use of recursively defined relations R_i : $R_0 :\equiv I$, $R_{i+1}(x) :\equiv \forall y(R_i(y) \rightarrow R_i(2^x + y))$. The proofs will employ subproofs ϕ_i and ψ_i as presented in [14]. Both ϕ_i and ψ_i have short, almost cut-free proofs, as only some cuts on instances of equality axioms are needed.

We give a sketch of short proofs π_n of the sequents S_n. For convenience and readability A_{T+eq} is only written in the end-sequent, though it is used throughout the proof. The cut and contraction inferences that are critical for the non-elementary complexity are marked with a *.

$$
\dfrac{\dfrac{\psi_0}{\vdash R_0(0)} \qquad \dfrac{\dfrac{\dfrac{\psi_1}{\vdash R_1(0)} \qquad \dfrac{\dfrac{\cdots}{\vdash R_2(2_{n-2})} \qquad \dfrac{\phi_1}{R_2(2_{n-2}), R_1(0) \vdash R_1(2_{n-1})}}{R_1(0) \vdash R_1(2_{n-1})}\,cut}{\vdash R_1(2_{n-1})}\,\mathbf{cut(*)} \qquad \dfrac{\phi_0}{R_1(2_{n-1}), R_0(0) \vdash R_0(2_n)}}{R_0(0) \vdash R_0(2_n)}\,cut}{A_{T+eq} \vdash R_0(2_n)}\,\mathbf{cut(*)}
$$

The proof continues in a symmetric fashion, ending with the top line:

$$\dfrac{\overset{\psi_n}{\vdash R_n(2_{n-n=0})} \quad \overset{\phi_{n-1}}{R_n(2_0), R_{n-1}(0) \vdash R_{n-1}(2_1)}}{R_{n-1}(0) \vdash R_{n-1}(2_1)}\ cut$$

which can be simplified as

$$\overset{\psi_n^*}{R_{n-1}(0) \vdash R_{n-1}(2_1)}$$

The subproofs for ϕ_i and ψ_i are simple proofs of depth linear in n. We show the proof of ψ_i, as it is in these proofs that ancestors of cut formulas $R_i(0)$ are contracted. The proofs ϕ_i are contraction free. The proof ψ_i is:

$$\dfrac{\dfrac{\dfrac{\dfrac{\dfrac{\dfrac{\dfrac{R_{i-2}(2^\beta+\alpha) \vdash R_{i-2}(2^\beta+\alpha) \quad R_{i-2}(2^{1+\beta}+\alpha) \vdash R_{i-2}(2^{1+\beta}+\alpha)}{R_{i-2}(2^\beta+\alpha) \to R_{i-2}(2^{1+\beta}+\alpha), R_{i-2}(2^\beta+\alpha) \vdash R_{i-2}(2^{1+\beta}+\alpha)}\ \to: l}{R_{i-1}(\beta), R_{i-2}(2^\beta+\alpha) \vdash R_{i-2}(2^{1+\beta}+\alpha)}\ \forall: l \quad R_{i-2}(\alpha) \vdash R_{i-2}(\alpha)}{R_{i-1}(\beta), R_{i-2}(\alpha) \to R_{i-2}(2^\beta+\alpha) \vdash R_{i-2}(\alpha) \to R_{i-2}(2^{1+\beta}+\alpha)}\ \to: l, r}{R_{i-1}(\beta), R_{i-1}(\beta) \vdash R_{i-1}(1+\beta)}\ \forall: l, r}{R_{i-1}(\beta) \vdash R_{i-1}(1+\beta)}\ \text{c:l(*)}}{\vdash R_{i-1}(\beta) \to R_{i-1}(2^0+\beta)}\ \to: r}{A_{T+eq} \vdash R_i(0)}\ \forall: r$$

The proof ψ_n^* necessary for the simplification of the top line of the proof π_n is obtained from ψ_n by skipping the last two inferences and substituting 0 for β throughout the proof. For more details on the proofs ϕ_i and ψ_i see [14] and [15].

Proposition 15. *There exist short proofs π_n of sequents S_n s.t. $|\pi_n| \leq \Theta(n)$.*

Proof: See Pudlak[13], Baaz/Leitsch [14] or Gerhardy[15]. □

To estimate the lower bound for the complexity of cut elimination we will use a property of mid-sequents. It is common to extract mid-sequents from cut free proofs, but it can be shown that proofs with all cuts quantifier free suffice.

Proposition 16. *Let ϕ be a proof of a prenex formula A with at most quantifier free cuts, then we can extract a mid-sequent and bound the number of non-identical formula instances of A in the mid-sequent by $2^{|\phi|}$.*

Proof: In Appendix. □

Since we can derive a Herbrand disjunction from a mid-sequent we immediately have the following theorem:

Theorem 17. *Let ϕ be a proof of a prenex formula A, then for some constant c the number of formula instances of A in a Herbrand disjunction can be bounded by $2^{c \cdot |\phi|}_{cnqf(\phi)+2}$, resp. $2^{c \cdot |\phi|}_{nqf(\phi)+1}$.*

Proof: By the Second Refined Cut Elim. Theorem and Proposition 16. □

Lemma 18. *Let $S_n \equiv \vdash \exists \ldots \exists (\bigwedge A_{T+eq} \to I(2_n))$ be the sequents as described above. Let χ_n be a proof in mid-sequent form of the sequent S_n. Then the mid-sequent must contain at least 2_n different instances of the axiom $I(x) \to I(1+x)$.*

Proof: See Pudlak[13]. □

Theorem 19. *Let the sequents S_n be as described above. Then for every proof χ_n of S_n with quantifier free cuts $|\chi_n| \geq 2_{n-1}$.*

Proof: By Proposition 16 we obtain a mid-sequent from a proof ϕ with quantifier free cuts. By Lemma 18 a mid-sequent for S_n must have at least 2_n different instances of the axiom $I(x) \to I(1+x)$, and hence the proof must have depth at least 2_{n-1}. □

To compare the obtained lower bound for cut elimination with our upper bound for cut elimination, we state the following theorem:

Theorem 20. *There are proofs χ_n of the sequents S_n with at most quantifier free cuts, so that the depth of χ_n can be bounded by $2_{n-1}^{\Theta(n)}$, i.e. $|\chi_n| \leq 2_{n-1}^{\Theta(n)}$.*

Proof: By Proposition 15 there exist short proofs π_n, of depth linear in n, of the sequents S_n. The cut formulas involving R_i, $i \leq n-1$ in these short proofs have $nqf(R_i) \leq n-1$. There are no cut formulas with deeper nesting of quantifiers in the proof. From the proof of Corollary 12 it follows that we can bound the depth of a proof with quantifier free cuts by $|\chi_n| \leq 2_{n-1}^{\Theta(n)}$. □

The Second Improved Cut Elimination Theorem givesno further improvement of the upper bound, since sufficiently complex subformulas of cut formulas $R_i(0)$ are contracted in the subproofs ψ_i, i.e. $cnqf(\pi_n) = nqf(\pi_n)$. This demonstrates, how efficient these short proofs are in their use of cuts and contractions. Every non-elementary growth example must employ a similar combination of nesting of quantifiers and contractions, as we have shown that cut elimination in the absence of either becomes elementary.

Remark. *In [16] Carbone defines the so-called bridge-groups of proofs which capture the structure of a proof w.r.t. cuts and contractions. It is shown that short proofs of Statman's example correspond to the so-called Gersten-group for which distortion into cyclic subgroups is hyper-exponential. Carbone conjectures that for bridge-groups the complexity of distortion into cyclic subgroups characterizes the complexity of cut elimination for the corresponding proofs. Since the Gersten group is the only known 1-relator group with hyper-exponential distortion this further suggests that Statman's lower bound example is a characteristic example of the non-elementary complexity of cut elimination.*

In comparison to the refined upper bound, the bounds obtained from Schwichtenberg's and Buss's cut elimination theorem both estimate the complexity in the logical depth of the cut formulas R_i, where $|R_i| = 2i$. Comparing Statman's lower bound to Schwichtenberg's and Buss's upper bounds, we notice a substantial gap between the upper and the lower bound. With the improved analysis of cut elimination the gap between our upper and Statman's lower bound is now closed. In tabular form the bounds on proofs with quantifier free cuts compare as follows. Note that the bounds on the size of a proof equivalently give a bound on the size of a Herbrand disjunction:

	$\lvert\cdot\rvert$	$\lVert\cdot\rVert$(*Herbrand*)
Statman/Pudlak (lower bound)	2_{n-1}	2_n
Schwichtenberg (upper bound)	$2_{2n-2}^{\Theta(n)}$	$2_{2n-1}^{\Theta(n)}$
Buss (upper bound)	–	$2_{4n-3}^{\Theta(n)}$
Zhang/Gerhardy (upper bound)	$2_{n-1}^{\Theta(n)}$	$2_n^{\Theta(n)}$

Remark. *In [1] Zhang also states that the upper and lower bound coincide. His statement relies on two propositions: that there are, as shown above, proofs of the sequents S_n of depth linear in n with $nqf() \leq n-1$, and that every cut free proof of the sequents S_n must have hyper-exponentially in n many quantifier inferences* in serial. *For the first proposition no proof is given in Zhang's paper, for the second proposition a counterexample is presented in [15].*

3.4 Further Improvement of the Cut Elimination Theorem

We can show that if all cut formulas are composed exclusively of atomic formulas and $\forall$ and $\wedge$ connectives (resp. $\exists$ and $\vee$), cut elimination is exponential, regardless of the nesting of $\forall$ (resp. $\exists$) quantifiers. For arbitrary cut formulas the complexity depends on the number of alternating $\forall - \wedge$ and $\exists - \vee$ blocks, as was shown in [2]. We discuss two propositions that naturally lead to the result.

Proposition 21. *Assume we have in a proof ϕ a cut inference with the cut formula $\forall\bar{x}B(\bar{x})$. Then we can eliminate that block of quantifiers* simultaneously, *replacing the cut with a number of cuts with $B(\bar{t})$, for appropriate terms $\bar{t}$. For the modified proof ϕ' we have that $|\phi'| < 2\cdot|\phi|$.*

Proof: Let the cut inference be:

$$\frac{\overset{\psi}{\Gamma\vdash\Delta,\forall\ldots\forall\bar{x}B(\bar{x})}\quad\overset{\chi}{\forall\ldots\forall\bar{x}B(\bar{x}),\Pi\vdash\Lambda}}{\Gamma,\Pi\vdash\Delta,\Lambda}\,cut$$

In χ we enumerate the l inferences where an innermost $\forall$ quantifier of an ancestor of the cut formula is introduced as (for $1\leq i\leq l$)

$$\frac{B(t_{i,1},\ldots,t_{i,n}),\Pi_i\vdash\Lambda_i}{\forall x_nB(t_{i,1},\ldots,t_{i,n-1},x_n),\Pi_i\vdash\Lambda_i}\,\forall:l$$

Similar to the earlier proof transformation in the $\forall$-case, we can obtain proofs ψ_i of the sequents $\Gamma\vdash\Delta,B(t_{i,1},\ldots,t_{i,n})$. Replacing the enumerated inferences in χ by cuts with $B(t_{i,1},\ldots,t_{i,n})$ we obtain the desired result. □

Proposition 22. *Assume we have in a proof ϕ a cut inference with the cut formula $\bigwedge_{i=1}^{n}B_i$. Then we can eliminate all those conjunctions* simultaneously, *replacing that cut with a number of cuts with B_i. For the modified proof ϕ' we have that $|\phi'| < 2\cdot|\phi|$.*

Proof: The proof is almost identical to the proof of Proposition 21, but instead eliminating cuts where the "innermost" conjunct is introduced. □

The crucial insight is that both for the $\forall$-quantifier and the $\wedge$-connective, the projection to a subformula of the cut formula occurs in the left subproof, leading to new cut inferences in the right subproof χ. Therefore, if all cut formulas are composed exclusively of some formulas B_i and the connectives $\forall$ and $\wedge$, all connectives can be eliminated in one step.

Lemma 23. *Let ϕ be a proof of a sequent $\Gamma \vdash \Delta$ with the last inference a cut. Let the cut formula be constructed from formulas $B_1, \dots B_n$ by the connectives $\forall$ and $\wedge$ only (resp. $\exists$ and $\vee$). Then we can replace that cut by a number of smaller cuts with cut formulas B_i. For the resulting proof ϕ' we have $|\phi'| < 2 \cdot |\phi|$.*

Proof: Combining the reductions for $\forall$ and $\wedge$ described in Lemma 21 and Lemma 22 we replace the $\wedge$ or $\forall$-introductions by cuts. For the resulting proof ϕ' we get $|\phi'| < 2 \cdot |\phi|$. The $\exists - \vee$-case is handled symmetrically, with roles of the left and right subproofs reversed. □

This immediately gives us a double-exponential bound for cut elimination if all cut formulas are composed of $\forall$, $\wedge$ and atomic formulas. However, we can eliminate the $\forall$ and $\wedge$ connectives and the remaining atomic cuts in one go, by combining the above lemmas with the technique to eliminate atomic cuts. We state the following theorem:

Theorem 24. *Let ϕ be a proof of a sequent $\Gamma \vdash \Delta$ with all cut formulas $\forall - \wedge$ (resp. $\exists - \vee$). Then there is a cut-free proof ϕ' of the same sequent with $|\phi'| \leq 2^{|\phi|}$.*

Proof: In Appendix. □

Remark. *The result still holds, if we allow negation of atomic formulas in the cut formulas, as elimination of negation can easily be combined with the above cut elimination technique.*

The above technique can easily be generalized to arbitrary cut formulas, if we restrict the logical connectives in the cut formulas to $\forall, \wedge, \exists, \vee$ and $\neg$, with negation only appearing in front of atomic formulas. If we alternatingly eliminate $\forall - \wedge$-connectives and $\exists - \vee$-connectives we get a non-elementary upper bound for cut elimination, where the height of the tower of exponentials depends on the number of alternations of $\forall - \wedge$ and $\exists - \vee$ blocks in the cut formulas. Again the short proofs π_n of sequents S_n perfectly exploit this, as the forall-implication structure of the cut formulas R_i corresponds to such alternations, i.e. the number of alternations in R_i is i.

Already the elimination of $\forall - \wedge$ cuts with elementary complexity has applications. In [17] Baaz and Leitsch prove an exponential upper bound for cut elimination in a fragment of **LK** called $\mathcal{QMON}$. At the heart of the proof is a technique for eliminating $\forall - \wedge$ cuts similar to the one presented above. The rest of their proof relies on special properties of the fragment $\mathcal{QMON}$.

Acknowledgements

This paper describes results from my master's thesis[15], written under the supervision of Ulrich Kohlenbach. I am greatly indebted to Ulrich Kohlenbach for many discussions on the subject and for useful comments on drafts of this paper. Furthermore I am grateful to the referees for many helpful suggestions which led to an improved presentation of the results.

References

1. Zhang, W.: Cut elimination and automatic proof procedures. Theoretical Computer Science **91** (1991) 265–284
2. Zhang, W.: Depth of proofs, depth of cut-formulas and complexity of cut formulas. Theoretical Computer Science **129** (1994) 193–206
3. Schwichtenberg, H.: Proof Theory: Some Applications of Cut-Elimination. In Barwise, J., ed.: Handbook of Mathematical Logic. North-Holland Publishing Company (1977) 868–894
4. Buss, S.R.: An Introduction to Proof Theory. In Buss, S.R., ed.: Handbook of Proof Theory. Elsevier Science B.V. (1998) 2–78
5. Luckhardt, H.: Komplexität der Beweistransformation bei Schnittelimination. Handwritten notes (1991)
6. Kohlenbach, U.: Notes to [5]. Handwritten notes (1995)
7. Girard, J.Y.: Linear logic. Theoretical Computer Science **50** (1987) 1–102
8. Blass, A.: A game semantics for linear logic. Annals of Pure and Applied Logic **56** (1992) 183–220
9. Ketonen, J., Weyrauch, R.: A Decidable Fragment of Predicate Calculus. Theoretical Computer Science **32** (1984) 297–307
10. Bellin, G., Ketonen, J.: A Decision Procedure Revisited: Notes on Direct Logic, Linear Logic and its Implementation. Theoretical Computer Science **95** (1992) 115–142
11. Statman, R.: Lower Bounds on Herbrand's Theorem. Proc. of the Amer. Math. Soc. **75** (1979) 104–107
12. Orevkov, V.P.: Complexity of Proofs and Their Transformations in Axiomatic Theories. Volume 128 of Translations of Mathematical Monographs. American Mathematical Society, Providence, Rhode Island (1993)
13. Pudlak, P.: The Length of Proofs. In Buss, S.R., ed.: Handbook of Proof Theory. Elsevier Science B.V. (1998) 548–637
14. Baaz, M., Leitsch, A.: On Skolemizations and Proof Complexity. Fundamenta Informatica **20/4** (1994) 353–379
15. Gerhardy, P.: Improved Complexity Analysis of Cut Elimination and Herbrand's Theorem. Master's thesis (2003)
16. Carbone, A.: Asymptotic Cyclic Expansion and Bridge Groups of Formal Proofs. Journal of Algebra **242** (2001) 109–145
17. Baaz, M., Leitsch, A.: Cut Normal Forms and Proof Complexity. Annals of Pure and Applied Logic **97** (1999) 127–177

Comparing the Succinctness of Monadic Query Languages over Finite Trees

Martin Grohe and Nicole Schweikardt*

Laboratory for Foundations of Computer Science
University of Edinburgh, Scotland, UK
{grohe,v1nschwe}@inf.ed.ac.uk

Abstract. We study the *succinctness* of monadic second-order logic and a variety of monadic fixed point logics on trees. All these languages are known to have the same expressive power on trees, but some can express the same queries much more succinctly than others. For example, we show that, under some complexity theoretic assumption, monadic second-order logic is non-elementarily more succinct than monadic least fixed point logic, which in turn is non-elementarily more succinct than monadic datalog.
Succinctness of the languages is closely related to the combined and parameterized complexity of query evaluation for these languages.

Keywords: Finite Model Theory, Monadic Second-Order Logic, Fixed Point Logics, μ-Calculus, Monadic Datalog, Tree-like structures, Succinctness

1. Introduction

A central topic in finite model theory has always been a comparison of the expressive power of different logics on finite relational structures. In particular, the expressive power of fragments of monadic second-order logic and various fixed-point logics has already been investigated in some of the earliest papers in finite model theory [Fag75,CH82]. One of the main motivations for such studies was an interest in the expressive power of query languages for relational databases.

In recent years, the focus in database theory has shifted from relational to semi-structured data and in particular data stored as XML-documents. A lot of current research in the database community is concerned with the design and implementation of XML query languages (see, for example, [FSW00,HP00,GK02] or the monograph [ABS99] for a general introduction into semi-structured data and XML). The languages studied in the present paper may be viewed as node-selecting query languages for XML. They all contain the core of the language XPath, which is an important building block of several major XML-related technologies. Recently, monadic datalog has been proposed as a node-selecting query language with a nice balance between expressive power and very good algorithmic properties [GK02,Koc03].

XML-documents are best modelled by trees, or more precisely, finite labelled ordered unranked trees. It turns out that when studying node-selecting query languages for

* Supported by a fellowship within the Postdoc-Programme of the German Academic Exchange Service (DAAD)

M. Baaz and J.A. Makowsky (Eds.): CSL 2003, LNCS 2803, pp. 226–240, 2003.

XML-documents, expressive power is not the central issue. Quite to the contrary: Neven and Schwentick [NS02] proposed to take the expressive power of monadic second-order logic (MSO) as a benchmark for node-selecting XML-query languages and, in some sense, suggested that such languages should at least have the expressive power of MSO. However, even languages with the same expressive power may have vastly different complexities. For example, monadic datalog and MSO have the same expressive power over trees [GK02]. However, monadic datalog queries can be evaluated in time linear both in the size of the datalog program and the size of the input tree [GK02], and thus the combined complexity of monadic datalog is in polynomial time, whereas the evaluation of MSO queries is PSPACE complete. The difference becomes even more obvious if we look at parameterized complexity: Unless PTIME $\neq$ NP, there is no algorithm evaluating a monadic second-order query in time $f(\text{size of query})p(\text{size of tree})$ for any elementary function f and polynomial p [FG03]. Similar statements hold for the complexity of the satisfiability problem for monadic datalog and MSO over trees. The reason for this different behaviour is that even though the languages have the same expressive power on trees, in MSO we can express queries much more *succinctly*. Indeed, there is no elementary translation from a given MSO-formula into an equivalent monadic datalog program. We also say that MSO is *non-elementarily more succinct* than monadic datalog. Just to illustrate the connection between succinctness and complexity, let us point out that if there was an elementary translation from MSO to monadic datalog, then there would be an algorithm evaluating a monadic second-order query in time $f(\text{size of query})p(\text{size of tree})$ for an elementary function f and a polynomial p.

In this paper, we study the succinctness (in the sense just described) of a variety of fixed point logics on finite trees. Our main results are the following:

1. MSO *is non-elementarily more succinct than monadic least fixed point logic* MLFP *(see Theorem 2). Unfortunately, we are only able to prove this result under the odd, but plausible complexity theoretic assumption that for some $i \geqslant 1$,* NP *is not contained in* $\text{DTIME}(n^{\log^{(i)}(n)})$*, where $\log^{(i)}$ denotes the i times iterated logarithm.*
2. MLFP *is non-elementarily more succinct than its* 2*-variable fragment* MLFP^2 *(see Corollary 3).*
3. MLFP^2 *is exponentially more succinct than the full modal μ-calculus, that is, the modal μ-calculus with future and past modalities (see Theorem 3, Example 2, and Theorem 4).*
4. *The full modal μ-calculus is at most exponentially more succinct than stratified monadic datalog, and conversely, stratified monadic datalog is at most exponentially more succinct than the full modal μ-calculus (see Theorem 7 and 8). Furthermore, stratified monadic datalog is at most exponentially more succinct than monadic datalog (see Theorem 6).*
   The exact relationship between these three languages remains open.

Of course we are a not the first to study the succinctness of logics with the same expressive power. Most known results are about modal and temporal logics. The motivation for these results has not come from database theory, but from automated verification and model-checking. The setting, however, is very similar. For example, Kamp's well know theorem states that first-order logic and linear time temporal logic have the same expressive power on strings [Kam68], but there is no elementary translation from first-order logic to linear time temporal logic on strings. Even closer to our results, monadic second-order logic and the modal μ-calculus have the same expressive power on (ordered) trees, but again is well-known that there is no elementary translation from the former to the latter. Both of these results can be proved by simple automata theoretic arguments. More refined results are known for various temporal logics [Wil99,AI00,AI01,EVW02]. By and large, however, succinctness has received surprisingly little attention in the finite model theory community. Apart from automata theoretic arguments, almost no good techniques for proving lower bounds on formula sizes are known. A notable exception are Adler and Immerman's [AI01] nice games for proving such lower bounds. Unfortunately, we found that these games (adapted to fixed point logic) were of little use in our context. So we mainly rely on automata theoretic arguments. An exception is the, complexity theoretically conditioned, result that MSO is non-elementarily more succinct than MLFP. To prove this result, we are building on a technique introduced in [FG03].

The paper is organised as follows: In Section 2 we fix the basic notations used throughout the paper. Section 3 concentrates on the translation from MSO to MLFP. In Section 4 we present our results concerning the two-variable fragment of MLFP and the full modal μ-calculus. In Section 5 we concentrate on monadic datalog, stratified monadic datalog, and their relations to finite automata and to MLFP. Finally, Section 6 concludes the paper by pointing out several open questions.

Due to space limitations, we had to defer detailed proofs of our results to the full version of this paper [GS03].

2. Preliminaries

2.1. Basic Notations. Given a set Σ we write Σ^* to denote the set of all finite strings over Σ, and we use ε to denote the empty string. We use $\mathbb{N}$ to denote the set $\{0, 1, 2, ..\}$ of natural numbers. We use lg to denote the logarithm with respect to base 2. With a function f that maps natural numbers to real numbers we associate the corresponding function from $\mathbb{N}$ to $\mathbb{N}$ defined by $n \mapsto \lceil f(n) \rceil$. For simplicity we often simply write $f(n)$ instead of $\lceil f(n) \rceil$.

The function *Tower* : $\mathbb{N} \to \mathbb{N}$ is inductively defined via $\mathit{Tower}(0) := 1$ and $\mathit{Tower}(h{+}1) = 2^{\mathit{Tower}(h)}$, for all $h \in \mathbb{N}$. I.e., $\mathit{Tower}(h)$ is a tower of 2s of height h.

We say that a function $f : \mathbb{N} \to \mathbb{N}$ has bound $f(m) \leqslant \mathit{Tower}(o(h(m)))$, for some function $h : \mathbb{N} \to \mathbb{N}$, if there is a function $g \in o(h)$ and a $m_0 \in \mathbb{N}$ such that for all $m \geqslant m_0$ we have $f(m) \leqslant \mathit{Tower}(g(m))$. Note that, in particular, every *elementary* function f has bound $f(m) \leqslant \mathit{Tower}(o(m))$. Indeed, for every elementary function f there is a $h \in \mathbb{N}$ such that, for all $n \in \mathbb{N}$, $f(n)$ is less than or equal to the tower of 2s of height h with an n on top.

2.2. Structures. A *signature* τ is a finite set of relation symbols and constant symbols. Each relation symbol $R \in \tau$ has a fixed arity $ar(R)$. A τ-structure $\mathcal{A}$ consists of a set $\mathcal{U}^{\mathcal{A}}$ called the *universe* of $\mathcal{A}$, an interpretation $c^{\mathcal{A}} \in \mathcal{U}^{\mathcal{A}}$ of each constant symbol $c \in \tau$, and an interpretation $R^{\mathcal{A}} \subseteq (\mathcal{U}^{\mathcal{A}})^{ar(R)}$ of each relation symbol $R \in \tau$. *All structures considered in this paper are assumed to have a finite universe.*

The main focus of this paper lies on the class *Trees* of finite binary trees. Precisely, finite binary trees are particular structures over the signature

$$\tau_{\textit{Trees}} := \{\textit{Root},\ 1^{st}\textit{Child},\ 2^{nd}\textit{Child},\ \textit{Has-No-}1^{st}\textit{Child},\ \textit{Has-No-}2^{nd}\textit{Child}\},$$

where *Root*, *Has-No-*1st*Child*, *Has-No-*2nd*Child* are unary relation symbols and 1st*Child*, 2nd*Child* are binary relation symbols. We define *Trees* to be the set of all $\tau_{\textit{Trees}}$-structures T that satisfy the following conditions:

1. $\mathcal{U}^T \subset \{1,2\}^*$ and for every string $si \in \mathcal{U}^T$ with $i \in \{1,2\}$ we also have $s \in \mathcal{U}^T$.
2. $\textit{Root}^T$ consists of the empty string ε.
3. $1^{st}\textit{Child}^T$ consists of the pairs $(s, s1)$, for all $s1 \in \mathcal{U}^T$.
4. $2^{nd}\textit{Child}^T$ consists of the pairs $(s, s2)$, for all $s2 \in \mathcal{U}^T$.
5. $\textit{Has-No-}1^{st}\textit{Child}^T$ consists of all strings $s \in \mathcal{U}^T$ with $s1 \notin \mathcal{U}^T$.
6. $\textit{Has-No-}2^{nd}\textit{Child}^T$ consists of all strings $s \in \mathcal{U}^T$ with $s2 \notin \mathcal{U}^T$.

For $T \in \textit{Trees}$ and $t \in \mathcal{U}^T$ we write T_t to denote the subtree of T with root t.

A *schema* σ is a set of unary relation symbols each of which is distinct from *Has-No-*1st*Child*, *Has-No-*2nd*Child*, *Root*. A σ-labelled tree is a $(\tau_{\textit{Trees}} \cup \sigma)$-structure consisting of some $T \in \textit{Trees}$ and additional interpretations $P^T \subseteq \mathcal{U}^T$ for all symbols $P \in \sigma$. We sometimes write $\textit{label}(t)$ to denote the set $\{P \in \sigma : t \in P^T\}$ of labels at vertex t in T.

We identify a string $w = w_0 \cdots w_{n-1}$ of length $|w| = n \geqslant 1$ over an alphabet Σ with a σ-labelled tree T^w in the following way: We choose σ to consist of a unary relation symbol P_a for each letter $a \in \Sigma$, we choose T^w to be the (unique) element in *Trees* with universe $\mathcal{U}^{T^w} = \{\varepsilon, 1, 11, .., 1^{n-1}\}$, and we choose $P_a^{T^w} := \{1^i : w_i = a\}$, for each $a \in \Sigma$. This corresponds to the conventional representation of strings by structures in the sense that $\langle \mathcal{U}^{T^w}, 1^{st}\textit{Child}, (P_a^{T^w})_{a\in\Sigma}\rangle$ is isomorphic to the structure $\langle \{0, .., n-1\}, \textit{Succ}, (P_a^w)_{a\in\Sigma}\rangle$ where *Succ* denotes the binary successor relation on $\{0, .., n-1\}$ and P_a^w consists of all positions of w that carry the letter a. When reasoning about strings in the context of first-order logic, we sometimes also need the linear ordering $<$ on $\{0, .., n-1\}$ (respectively, the transitive closure of the relation 1st*Child*). In these cases we explicitly write FO($<$) rather than FO to indicate that the linear ordering is necessary.

XML-documents are usually modelled as ordered unranked trees and not as binary trees. Here *ordered* refers to the fact that the order of the children of a vertex is given. However, a standard representation of ordered unranked trees as relational structures uses binary relations 1st*Child*, *Next-Sibling* and unary relations *Root*, *Leaf*, *Last-Sibling* (for details, see [GK02]) and thus essentially represents ordered unranked trees as binary trees. Therefore, all our results also apply to ordered unranked trees.

2.3. Logics and Queries. We assume that the reader is familiar with *first-order logic*, for short: FO, and with *monadic second-order logic*, for short: MSO (cf., e.g., the textbooks [EF99,Imm99]). We use FO(τ) and MSO(τ), respectively, to denote the class of all first-order formulas and monadic second-order formulas, respectively, of signature

τ. We write $\varphi(x_1,..,x_k,X_1,..,X_\ell)$ to indicate that the free first-order variables of the formula φ are $x_1,..,x_k$ and the free set variables are $X_1,..,X_\ell$. Sometimes we use $\overline{x}$ and $\overline{X}$ as abbreviations for sequences $x_1,..,x_k$ and $X_1,..,X_\ell$ of variables.

A formula $\varphi(x)$ of signature τ defines the *unary query* which associates with every τ-structure $\mathcal{A}$ the set of elements $a \in \mathcal{U}^{\mathcal{A}}$ such that $\mathcal{A} \models \varphi(a)$, i.e., $\mathcal{A}$ satisfies φ when interpreting the free occurrences of the variable x by the element a. A *sentence* φ of signature τ (i.e., a formula that has no free variables) defines the *Boolean query* that associates the answer "yes" with all τ-structures that satisfy φ and the answer "no" with all other τ-structures.

Apart from FO and MSO we will also consider *monadic least fixed point logic* MLFP which is the extension of first-order logic by unary least fixed point operators. We refer the reader to [EF99] for the definition of MLFP (denoted by FO(M-LFP) there).

2.4. Formula Size and Succinctness. In a natural way, we view formulas as finite trees, where leaves correspond to the atoms of the formulas and inner vertices correspond to Boolean connectives, quantifiers, and fixed-point operators. We define the *size* $||\varphi||$ of a formula φ to be the number of vertices of the tree that corresponds to φ.

Note that this measure of formula size is a *uniform cost measure* in the sense that it accounts just 1 cost unit for each variable and relation symbol appearing in a formula, no matter what its index is. An alternative is to define the size of a formula as the length of a binary encoding of the formula. Such a *logarithmic cost measure* is, for example, used in [FG03]. Switching between a uniform and a logarithmic measure usually involves a logarithmic factor.

Definition 1 (Succinctness). *Let L_1 and L_2 be logics, let F be a class of functions from $\mathbb{N}$ to $\mathbb{N}$, and let C be a class of structures.*

We say that L_1 is F-succinct in L_2 on C *if there is a function $f \in F$ such that for every formula $\varphi_1 \in L_1$ there is a formula $\varphi_2 \in L_2$ of size $||\varphi_2|| \leqslant f(||\varphi_1||)$ which is equivalent to φ_1 on all structures in C.* □

Intuitively, a logic L_1 being F-succinct in a logic L_2 means that F gives an upper bound for the size of L_1-formulas needed to express *all* of L_2. This definition may seem slightly at odds with the common use of the term "succinctness" in statements such as "L_2 is exponentially *more succinct* than L_1" meaning that there is *some* L_2-formula that is not equivalent to any L_1-formula of subexponential size. In our terminology, we would rephrase this last statement as "L_1 is *not* $2^{o(n)}$-succinct in L_2" (here we interpret subexponential as $2^{o(n)}$, but of course this is not the issue). The reason for defining F-succinctness the way we did is that it makes the formal statements of our results much more convenient. We will continue to use statements such as "L_2 is exponentially more succinct than L_1" in informal discussions.

Example 1. MLFP is $\mathcal{O}(m)$-succinct in MSO on the class of all finite structures, because every formula $[\mathrm{LFP}_{x,X}\varphi(x,X,\overline{y},\overline{Y})](z)$ is equivalent to $\forall X\,\big(Xz \vee \exists x\,\neg Xx \wedge \varphi(x,X,\overline{y},\overline{Y})\big)$. □

3. From MSO to MLFP

By the standard translation from MSO-logic to tree automata (cf., e.g., [Tho96]) one knows that every MSO-sentence Φ can be translated into a nondeterministic tree automaton with $Tower(\mathcal{O}(||\Phi||))$ states that accepts exactly those labelled trees that satisfy Φ. This leads to

Theorem 1 (Folklore). *MSO-sentences are $Tower(\mathcal{O}(m))$-succinct in* MLFP *on the class of all labelled trees.* □

To show that we cannot do essentially better, i.e., that there is no translation from MSO to MLFP of size $Tower(o(m))$ we need a complexity theoretic assumption that, however, does not seem to be too far-fetched. Let SAT denote the NP-complete satisfiability problem for propositional formulas in conjunctive normal form. Until now, all known deterministic algorithms that solve SAT have worst-case complexity $2^{\Omega(n)}$ (cf., [DGH+02]). Although not answering the P vs. NP question, the exposition of a deterministic algorithm for SAT with worst-case complexity $\leqslant n^{\lg n}$ would be a surprising and unexpected breakthrough in the SAT-solving community.

In the following, we write $\lg^{(i)}$ to denote the i times iterated logarithm, inductively defined by $\lg^{(1)}(n) := \lg(n)$ and $\lg^{(i+1)}(n) := \lg(\lg^{(i)}(n))$. Moreover, we we write $\lg^*$ to denote the "inverse" of the *Tower* function, that is, the (unique) integer valued function with $Tower(\lg^*(n){-}1) < n \leqslant Tower(\lg^*(n))$.

Theorem 2. *Unless* SAT *is solvable by a deterministic algorithm that has, for every $i \in \mathbb{N}$, time bound $||\gamma||^{\lg^{(i)}(n)}$ (where γ is the input formula and n the number of propositional variables occurring in γ),* MSO *is not $Tower(o(m))$-succinct in* MLFP *on the class of all finite strings.* □

The overall proof idea is to assume that the function f specifies the size of the translation from MSO to MLFP and to exhibit a SAT-solving algorithm which

- constructs a string w that represents the SAT-instance γ,
- constructs an MSO-formula $\Phi(z)$ of extremely small size that, when evaluated in w, specifies a canonical satisfying assignment for γ (if γ is satisfiable at all),
- tests, for all MLFP-formulas $\Psi(z)$ of size $\leqslant f(||\Phi||)$, whether Ψ specifies a satisfying assignment for γ.

Before presenting the proof in detail we provide the necessary notations and lemmas:

It is straightforward to see

Lemma 1. *There is an algorithm that, given an* MLFP*-formula $\Psi(z)$, a string w, and a position p in w, decides in time $|w|^{\mathcal{O}(||\Psi||)}$ whether $w \models \Psi(p)$.* □

Let us now concentrate on the construction of a string w that represents a SAT-instance γ and of an MSO-formula $\Phi(z)$ that specifies a canonical satisfying assignment of γ (provided that γ is satisfiable at all). Since we want Φ to be extremely short, we cannot choose w to be the straightforward string-representation of γ. Instead, we use the following, more complicated, representation of [FG03]:

For all $h \geqslant 1$ let $\Sigma_h := \{0, 1, \texttt{<1>}, \texttt{</1>}, .., \texttt{<h>}, \texttt{</h>}\}$. The "tags" <i> and </i> represent single letters of the alphabet and are just chosen to improve readability.

For every $n \geqslant 1$ let $L(n)$ be the length of the binary representation of the number $n-1$, i.e., $L(0) = 0$, $L(1) = 1$, and $L(n) = \lfloor \lg(n-1) \rfloor + 1$, for all $n \geqslant 2$. By $\mathrm{bit}(i, n)$ we denote the i-th bit of the binary representation of n, i.e., $\mathrm{bit}(i, n)$ is 1 if $\lfloor \frac{n}{2^i} \rfloor$ is odd, and $\mathrm{bit}(i, n)$ is 0 otherwise.

We encode every number $n \in \mathbb{N}$ by a string $\mu_h(n)$ over the alphabet Σ_h, where $\mu_h(n)$ is inductively defined as follows: $\mu_1(0) :=$ `<1></1>`, and

$\mu_1(n) :=$ `<1>` $\mathrm{bit}(0, n-1)\, \mathrm{bit}(1, n-1) \cdots \mathrm{bit}(L(n)-1, n-1)$ `</1>`,

for $n \geqslant 1$. For $h \geqslant 2$ we let $\mu_h(0) :=$ `<h></h>` and

$\mu_h(n) :=$ `<h></h>`$\mu_{h-1}(0)\, \mathrm{bit}(0, n-1) \cdots$ `</h>`$\mu_{h-1}(L(n)-1)\, \mathrm{bit}(L(n)-1, n-1)$`</h>`,

for $n \geqslant 1$. Here empty spaces and line breaks are just used to improve readability.

To encode a CNF-formula γ by a string we use an alphabet Σ'_h that extends Σ_h by the symbols $+, -, \star$ and a number of additional tags. Let $i \in \mathbb{N}$ and let X_i be a propositional variable. The literal X_i is encoded by the string

$\mu_h(X_i) :=$ `<lit>` $\mu_h(i) +$ `</lit>`,

and the literal $\neg X_i$ is encoded by $\mu_h(\neg X_i) :=$ `<lit>` $\mu_h(i) -$ `</lit>`.

A clause $\delta := \lambda_1 \vee \cdots \vee \lambda_r$ of literals is encoded by

$\mu_h(\delta) :=$ `<clause>` $\mu_h(\lambda_1) \cdots \mu_h(\lambda_r)$ `</clause>`.

A CNF-formula $\gamma := \delta_1 \wedge \cdots \wedge \delta_m$ is encoded by the string

$\mu_h(\gamma) :=$ `<cnf>` $\mu_h(\delta_1) \cdots \mu_h(\delta_m)$ `</cnf>`.

We write $\mathrm{CNF}(n)$ to denote the class of all CNF-formulas the propositional variables of which are among $X_0, .., X_{n-1}$. To provide the "infrastructure" for specifying a truth assignment, we use the string

$\mu_h(X_0, .., X_{n-1}) :=$ `<ass><val>` $\mu_h(0) \star$ `</val>` $\cdots$
`<val>` $\mu_h(n-1) \star$ `</val></ass>`.

Remark 1. There is a 1–1-correspondence between assignments $\alpha : \{X_0, .., X_{n-1}\} \to \{\mathit{true}, \mathit{false}\}$, on the one hand, and sets P of positions of $\mu_h(X_0, .., X_{n-1})$ that carry the letter $\star$, on the other hand: Such a set P specifies the assignment α^P that, for each $i < n$, maps the variable X_i to the value *true* iff the $\star$-position directly after the substring $\mu_h(i)$ in $\mu_h(X_0, .., X_{n-1})$ belongs to P. Conversely, a given assignment α specifies the set P^α consisting of exactly those $\star$-positions of $\mu_h(X_0, .., X_{n-1})$ that occur directly after a substring $\mu_h(i)$ where $\alpha(X_i) = \mathit{true}$. □

Finally, we encode a formula $\gamma \in \mathrm{CNF}(n)$ by the string

$$\mu_h(\gamma, \star) := \mu_h(\gamma)\, \mu_h(X_0, .., X_{n-1}).$$

$\mu_h(\gamma, \star)$ is the string w that we will furtheron use as the representative of a SAT-instance γ. We use the following result of [FG03]:

Lemma 2.

(a) *There is an algorithm that, given $h \in \mathbb{N}$ and $\gamma \in CNF(n)$, computes (a binary representation of) the string $\mu_h(\gamma,\star)$ in time $\mathcal{O}(h \cdot (\lg h) \cdot (\lg n)^2 \cdot (||\gamma|| + n))$ (cf., [FG03, Lemma 9]).*
The string $\mu_h(\gamma,\star)$ has length $|\mu_h(\gamma,\star)| = \mathcal{O}(h \cdot (\lg n)^2 \cdot (||\gamma|| + n))$.

(b) *There is an algorithm that, given $h \in \mathbb{N}$, computes (the binary representation of) a* FO(<)*-formula $\varphi_h(Z)$ in time $\mathcal{O}(h \cdot \lg h)$, such that for all $n \leqslant Tower(h)$, for all $\gamma \in CNF(n)$, and for all sets P of $\star$-positions in the string $\mu_h(\gamma,\star)$ we have*

$$\mu_h(\gamma,\star) \models \varphi_h(P) \quad \textit{iff} \quad \alpha^P \textit{ is a satisfying assignment for } \gamma$$

(cf., [FG03, Lemma 10]). The formula $\varphi_h(Z)$ has size[1] $||\varphi_h(Z)|| = \mathcal{O}(h)$. □

Given a CNF(n)-formula γ and its representative $\mu_h(\gamma,\star)$, we now specify a *canonical* satisfying assignment of γ, provided that γ is satisfiable at all. As observed in Remark 1, every assignment $\alpha : \{X_0, .., X_{n-1}\} \to \{\textit{true}, \textit{false}\}$ corresponds to a set P^α of positions in $\mu_h(\gamma,\star)$ that carry the letter $\star$. P^α, again, can be identified with the 0-1-string of length $|\mu_h(\gamma,\star)|$ that carries the letter 1 exactly at those positions that belong to P^α. Now, the lexicographic ordering of these strings gives us a linear ordering on the set of all assignments $\alpha : \{X_0, .., X_{n-1}\} \to \{\textit{true}, \textit{false}\}$. As the canonical satisfying assignment of γ we choose the lexicographically smallest satisfying assignment.

Lemma 3. *There is an algorithm that, given $h \in \mathbb{N}$, computes (the binary representation of) an* MSO*-formula $\Phi_h(z)$ in time $\mathcal{O}(h \cdot \lg h)$, such that for all $n \leqslant Tower(h)$, for all $\gamma \in CNF(n)$, and for all positions p of $\mu_h(\gamma,\star)$ that carry the letter $\star$, we have*

$\mu_h(\gamma,\star) \models \Phi_h(p)$ *iff in the lexicographically smallest satisfying assignment for γ, the propositional variable corresponding to position p is assigned the value true.*

The formula $\Phi_h(z)$ has size $||\Phi_h|| = \mathcal{O}(h)$. □

Finally, we are ready for the Proof of Theorem 2:

Proof of Theorem 2.
Let $f : \mathbb{N} \to \mathbb{N}$ be a function such that there is, for every MSO-formula $\Phi(z)$, a MLFP-formula $\Psi(z)$ of size $||\Psi|| \leqslant f(||\Phi||)$ which defines the same query as Φ on the class of all finite strings (recall that such an f does indeed exist, because MSO and MLFP have the same expressive power over the class of finite strings).

Consider the algorithm displayed in Figure 1, which decides if the input formula γ is satisfiable.

The correctness of this algorithm directly follows from Lemma 3 and from the fact that at least one of the formulas $\Psi(z)$ of size $\leqslant f(||\Phi_h||)$ defines the same query as $\Phi_h(z)$.

It remains to determine the worst-case running time of the algorithm. Let γ be an input CNF-formula for the algorithm, let n be the number of propositional variables of γ, and let $h := \lg^*(n)$.

[1] In [FG03], an additional factor $\lg h$ occurs because there a logarithmic cost measure is used for the formula size, whereas here we use a uniform measure (cf., Section 2.4).

Input: a SAT-instance γ in CNF

1. Count the number n of propositional variables occurring in γ, and modify γ in such a way that only the propositional variables $X_0, .., X_{n-1}$ occur in it.
2. Compute $h := \lg^*(n)$, i.e., choose $h \in \mathbb{N}$ such that $\mathit{Tower}(h-1) < n \leqslant \mathit{Tower}(h)$.
3. Construct the string $\mu_h(\gamma, \star)$ that represents γ (see Lemma 2 (a)).
4. Construct an MSO-formula $\Phi_h(z)$ that has the following property:
   Whenever p is a position in $\mu_h(\gamma, \star)$ that carries the letter $\star$, we have

   $\mu_h(\gamma, \star) \models \Phi_h(p)$ iff in the lexicographically smallest satisfying assignment for γ, the propositional variable corresponding to position p is assigned the value *true*

   (cf., Lemma 3).
5. For all MLFP-formulas $\Psi(z)$ of size $||\Psi|| \leqslant f(||\Phi_h||)$ do:
   (a) Initialise the assignment $\alpha := \emptyset$.
   (b) For all positions p in $\mu_h(\gamma, \star)$ that carry the letter $\star$ do
       check whether $\mu_h(\gamma, \star) \models \Psi(p)$;
       if so, then insert the propositional variable corresponding to p into α.
   (c) Check whether α is a *satisfying* assignment for γ;
       if so, then STOP with output "γ is satisfiable via assignment α".
6. STOP with output "γ is not satisfiable".

Fig. 1. A SAT-solving algorithm.

The steps 1–4 of the algorithm will be performed within a number of steps polynomial in $||\gamma||$, and the MSO-formula $\Phi_h(z)$ produced in step 4 will have size $||\Phi_h|| \leqslant c \cdot h$, for a suitable constant $c \in \mathbb{N}$ (cf., Lemma 2 (a) and Lemma 3).

The loop in step 5 will be performed for $2^{c_1 \cdot f(||\Phi_h||) \cdot \lg(f(||\Phi_h||))}$ times, for a suitable constant $c_1 \in \mathbb{N}$. To see this, note that formulas of length $\leqslant f(||\Phi_h||)$ use at most $f(||\Phi_h||)$ different first-order variables and at most $f(||\Phi_h||)$ different set variables. I.e., these formulas can be viewed as strings of length $f(||\Phi_h||)$ over an alphabet of size $c_2 + 2 \cdot f(||\Phi_h||)$, for a suitable constant $c_2 \in \mathbb{N}$. Therefore, the number of such formulas is $\leqslant (c_2 + 2 \cdot f(||\Phi_h||))^{f(||\Phi_h||)} \leqslant 2^{c_1 \cdot f(||\Phi_h||) \cdot \lg(f(||\Phi_h||))}$.

Each performance of the loop in step 5 will take a number of steps polynomial in

$$|\mu_h(\gamma, \star)|^{O(f(||\Phi_h||))} \leqslant (c_3 \cdot h \cdot (\lg n)^2 \cdot ||\gamma||)^{c_4 \cdot f(c \cdot h)},$$

for suitable constants $c_3, c_4 \in \mathbb{N}$ (cf., Lemma 1 and Lemma 2 (a)). Altogether, for suitable constants $c, d \in \mathbb{N}$, the algorithm will perform the steps 1–6 within

$$||\gamma||^{d \cdot f(c \cdot h) \cdot \lg(f(c \cdot h))}$$

steps.

Now let us suppose that f has bound $f(m) \leqslant \mathit{Tower}\big(o(m)\big)$. From Lemma 4 below we then obtain that our SAT-solving algorithm has, for every $i \in \mathbb{N}$, time bound $||\gamma||^{\lg^{(i)}(n)}$. This finally completes the proof of Theorem 2. ∎

Lemma 4. *Let $f : \mathbb{N} \to \mathbb{N}$ be a function with bound $f(m) \leqslant Tower(o(m))$, and let $c, d \in \mathbb{N}$. For every $i \in \mathbb{N}$ there is an $n_0 \in \mathbb{N}$ such that for all $n \geqslant n_0$ we have*

$$d \cdot f(c \cdot \lg^*(n)) \cdot \lg\left(f(c \cdot \lg^*(n))\right) \leqslant \lg^{(i)}(n).$$

□

4. The Two-Variable Fragment of MLFP and the Full Modal μ-Calculus

Defining the 2-variable fixed-point logics requires some care: MLFP^2 is the fragment of MLFP consisting of all formulas with just 2 individual variables and no parameters in fixed point operators, i.e., for all subformulas of the form $[\text{LFP}_{x,X}\varphi](y)$, x is the only free first-order variable of φ. This is the monadic fragment of the standard 2-variable least fixed-point logic logic (cf. [GO99]). Without the restriction on free variables in fixed-point operators, we obtain full MLFP even with just two individual variables (we prove this in the full version of this paper [GS03]).

We first note that MLFP^2, and actually FO^2, the two variable fragment of first-order logic, is doubly exponentially more succinct than nondeterministic automata on the class of all finite strings:

Example 2. Let $\sigma := \{L, R, P_1, .., P_n\}$ and

$$\varphi_n := \forall x\Big(Lx \to \exists y\big(Ry \wedge \bigwedge_{i=1}^{n}(P_i x \leftrightarrow P_i y)\big)\Big).$$

We claim that every nondeterministic finite automaton accepting precisely those strings over alphabet 2^σ that satisfy φ has at least 2^{2^n} states. To see this, for every $S \subseteq 2^{\{1,..,n\}}$, we define strings $X_n(S)$ and $Y_n(S)$ such that

- $L^{X_n(S)} = \mathcal{U}^{X_n(S)}$ and $R^{Y_n(S)} = \mathcal{U}^{Y_n(S)}$
- For all $s \in S$ there exists an $x \in \mathcal{U}^{X_n(S)}$ and an $y \in \mathcal{U}^{Y_n(S)}$ such that $s = \{i \mid x \in P_i^{X_n(S)}\} = \{i \mid y \in P_i^{Y_n(S)}\}$.

Let $W_n(S,T) := X_n(S)Y_n(T)$ be the concatenation of $X_n(S)$ and $Y_n(T)$. Then $W_n(S,T) \models \varphi \iff S \subseteq T$. Clearly, a nondeterministic finite automaton accepting precisely those strings $W_n(S,T)$ with $S \subseteq T$ needs at least 2^{2^n} states. □

Let us return to binary trees now. Following Vardi [Var98], we define the *full modal μ-calculus* FL_μ on binary trees as follows: For each schema σ, an FL_μ-formula of schema σ is either:

- *true*, *false*, P, or $\neg P$, where $P \in \sigma \cup \{Root,\ Has\text{-}No\text{-}1^{st}Child,\ Has\text{-}No\text{-}2^{nd}Child\}$;
- $\Phi_1 \wedge \Phi_2$ or $\Phi_1 \vee \Phi_2$, where Φ_1 and Φ_2 are FL_μ-formulas of schema σ;
- X, where X is a propositional variable;
- $\langle R\rangle\,\Phi$ or $[R]\,\Phi$, where $R \in \{1^{st}Child,\ 2^{nd}Child,\ 1^{st}Child^{-1},\ 2^{nd}Child^{-1}\}$ and Φ is an FL_μ-formula of schema σ;
- $\mu X.\Phi$ or $\nu X.\Phi$, where X is a propositional variable and Φ is an FL_μ-formula of schema σ.

The semantics of FL_μ is defined in the usual way interpreting the binary relations over trees. The following is starightforward:

Proposition 1. FL_μ *is* $\mathcal{O}(m)$*-succinct in* MLFP^2.

Our next result is that there also is a reverse translation from MLFP^2 to FL_μ which only incurs an exponential blow-up in size:

Theorem 3. MLFP^2 *is* $2^{poly(m)}$*-succinct in* FL_μ *on the class of labelled trees.* □

Theorem 4 (Vardi [Var98]). *For every formula* Φ *of the full modal* μ*-calculus* FL_μ *there is a nondeterministic tree automaton of size* $2^{poly(||\Phi||)}$ *that accepts exactly those labelled trees in which* Φ *holds at the root.* □

As a matter of fact, Vardi [Var98] proved a stronger version of this theorem for infinite trees and parity tree automata. But on finite trees, a parity acceptance condition can always be replaced by a normal acceptance for finite tree automata.

The Theorems 3 and 4 directly imply

Corollary 1. *For every* MLFP^2*-formula* $\varphi(x)$ *there is a nondeterministic tree automaton of size* $2^{2^{poly(||\varphi||)}}$ *that accepts exactly those labelled trees in which* φ *holds at the root.* □

5. Monadic Datalog and Stratified Monadic Datalog

We assume that the reader is familiar with *datalog*, which may be viewed as logic programming without function symbols (cf., e.g., the textbook [AHV95]). A datalog program is *monadic* if all its IDB-predicates (i.e., its intensional predicates that appear in the head of some rule of the program) are unary. In this paper we restrict attention to monadic datalog programs that are interpreted over labelled trees. A monadic datalog program of schema σ may use as EDB-predicates (i.e., extensional predicates which are determined by the structure the program is interpreted over) the predicates in τ_{Trees}, the predicates in σ, and a predicate $\neg P$ for every $P \in \sigma$ which is interpreted as the complement of P. We use $\mathrm{IDB}(\mathcal{P})$ to denote the set of IDB-predicates of $\mathcal{P}$, and we write MonDatalog to denote the class of all monadic datalog programs.

More formally, a monadic datalog program $\mathcal{P}$ of schema σ is a finite set of rules of the form $X(x) \leftarrow \gamma(x, \overline{y})$, where γ is a conjunction of atomic formulas over the signature $\tau_{Trees} \cup \sigma \cup \{\neg P : P \in \sigma\} \cup \mathrm{IDB}(\mathcal{P})$. Every program has a distinguished *goal* IDB-predicate that determines the query defined by the program.

We define the *size* $||\mathcal{P}||$ of $\mathcal{P}$ in the same way as we defined the size of formulas.

In [GK02] it was shown that MonDatalog can define the same unary queries on the class of labelled trees as monadic second-order logic. In the remainder of this section we will compare the succinctness of MonDatalog, S-MonDatalog, FL_μ, MLFP, and a particular kind of tree automaton.

5.1. From MonDatalog to Finite Automata. Several mechanisms have been proposed in the literature for specifying *unary* queries by finite automata operating on labelled trees (cf., [NS02]). One such mechanism, introduced in [Nev99] and further investigated in [FGK03,Koc03], is the *selecting tree automaton*:

Definition 2 (STA). *Let* σ *be a schema. A selecting* σ*-tree automaton (*σ*-STA, for short) is a tuple* $\mathfrak{A} = (Q, 2^\sigma, F, \delta, S)$*, where* $S \subseteq Q$ *is the set of* selecting states *and*

$(Q, 2^\sigma, F, \delta)$ is a conventional nondeterministic bottom-up tree automaton (cf., e.g., [Tho96]) with finite state space Q, input alphabet 2^σ, accepting states $F \subseteq Q$, and transition function

$$\delta \;:\; 2^\sigma \,\cup\, \big(\{1\} \times Q \times 2^\sigma\big) \,\cup\, \big(\{2\} \times Q \times 2^\sigma\big) \,\cup\, \big(Q \times Q \times 2^\sigma\big) \;\to\; 2^Q \,.$$

A run *of $\mathfrak{A}$ on a σ-labelled tree T is a mapping $\rho : \mathcal{U}^T \to Q$ that has the following property, for all vertices $t, t_1, t_2 \in \mathcal{U}^T$: if t has no children then $\rho(t) = \delta(label(t))$; if $1^{st}Child(t, t_1) \wedge Has\text{-}No\text{-}2^{nd}Child(t)$ then $\rho(t) \in \delta\big(1, \rho(t_1), label(t)\big)$; if $2^{nd}Child(t, t_2) \wedge Has\text{-}No\text{-}1^{st}Child(t)$ then $\rho(t) \in \delta\big(2, \rho(t_2), label(t)\big)$; if $1^{st}Child(t, t_1) \wedge 2^{nd}Child(t, t_2)$ then $\rho(t) \in \delta\big(\rho(t_1), \rho(t_2), label(t)\big)$.*

A run ρ of $\mathfrak{A}$ on T is said to be accepting *if it maps the root of T to a state in F. The* unary query *defined by $\mathfrak{A}$ is the query which maps every σ-labelled tree T to the set of those vertices $t \in \mathcal{U}^T$ that satisfy the following condition: $\rho(t) \in S$ for every* accepting *run ρ of $\mathfrak{A}$ on T.* □

It was shown in [FGK03,Nev99] that STAs can define exactly those unary queries on the class of labelled trees that are definable in monadic-second order logic.

Theorem 5 ([FGK03,GK02]). MonDatalog *is $2^{\mathcal{O}(m)}$-succinct in STAs on the class of labelled trees.* □

It is not hard to show that this result is asymptotically optimal, that is, that MonDatalog is not $2^{o(m)}$-succinct in STAs on the class of labelled trees (see [GS03] for details).

5.2. From S-MonDatalog to MonDatalog. In this section we show that S-MonDatalog-programs can be translated into MonDatalog-programs of at most exponential size. It remains open if the exponential size is indeed necessary or if, on the contrary, for every S-MonDatalog-program $\mathcal{P}$ there exists an equivalent MonDatalog-program $\mathcal{P}'$ of size polynomial in $||\mathcal{P}||$.

Lemma 5. *For every σ-STA $\mathfrak{A} = (Q, 2^\sigma, F, \delta, S)$ there is a* MonDatalog-*program $\mathcal{P}$ of size $\mathcal{O}\big(|Q|^3 \cdot |2^\sigma| + |\sigma| \cdot |2^\sigma|\big)$ that defines the complement of the query defined by $\mathfrak{A}$ on the class of all σ-labelled trees.* □

Using Theorem 5 and Lemma 5 one easily obtains

Proposition 2. *For every* MonDatalog-*program $\mathcal{P}$ there is a* MonDatalog-*program $\mathcal{P}'$ of size $2^{\mathcal{O}(||\mathcal{P}||)}$ that defines the complement of the query defined by $\mathcal{P}$ on the class of labelled trees.* □

Using the above proposition, it is not difficult to prove

Theorem 6. S-MonDatalog *is $2^{\mathcal{O}(m)}$-succinct in* MonDatalog *on the class of labelled trees.* □

5.3. S-MonDatalog vs FL$_\mu$. From Theorem 4 and Lemma 5 one directly obtains

Theorem 7. FL$_\mu$ *is $2^{poly(m)}$-succinct in* S-MonDatalog *on the class of labelled trees.* □

Conversely, it is not hard to show the following

Theorem 8. S-MonDatalog *is* $2^{O(m \cdot \lg m)}$*-succinct in* FL_μ *on the class of labelled trees.* □

It remains open whether the above bounds are optimal.

5.4. From MLFP to S-MonDatalog. Similarly to Theorem 1 one easily obtains

Theorem 9 (Folklore). MLFP*-sentences are* $\mathit{Tower}(\mathcal{O}(m))$*-succinct in* S-MonDatalog *on the class of labelled trees.* □

The aim of this section is to show that there are no essentially smaller translations from MLFP to S-MonDatalog. We will use the following well-known observation:

Proposition 3 (Folklore). *There is no function* $f : \mathbb{N} \to \mathbb{N}$ *with bound* $f(m) \leqslant \mathit{Tower}(o(m))$ *such that for every* FO($<$)*-sentence* φ *there is a nondeterministic finite automaton* $\mathfrak{A}$ *with at most* $f(||\varphi||)$ *states that accepts exactly those strings that satisfy* φ. □

Using Proposition 3 and the results of the Sections 5.1 and 5.2, one obtains the following:

Theorem 10. *There is no function* $f : \mathbb{N} \to \mathbb{N}$ *with bound* $f(m) \leqslant \mathit{Tower}(o(m))$ *such that for every* FO($<$)*-sentence* φ *there is a* S-MonDatalog*-program* $\mathcal{P}$ *of size* $||\mathcal{P}|| \leqslant f(||\varphi||)$ *and a designated goal predicate* $X \in \mathit{IDB}(\mathcal{P})$ *such that* $(\mathcal{P}, X)$ *defines the same Boolean query as* φ *on the class of all finite strings.* □

Since FO($<$) is included in MLFP, the above theorem directly implies the following:

Corollary 2. MLFP *is not* $\mathit{Tower}(o(m))$*-succinct in* S-MonDatalog *on the class of all finite strings.* □

It remains open if this result remains valid when replacing MLFP with MLFP^2. Note, however, that for the proof of Proposition 3 a small number k of first-order variables suffices. I.e., Proposition 3 remains valid when replacing FO($<$) with $\mathrm{FO}^k(<)$, and Corollary 2 remains valid when replacing MLFP with MLFP^k.

Together with Corollary 1 and Lemma 5, the above Corollary 2 implies

Corollary 3. MLFP *is not* $\mathit{Tower}(o(m))$*-succinct in* MLFP^2 *on the class of all finite strings.* □

6. Conclusion

We studied the succinctness of a number of fixed point logics on trees. We believe that the analysis of succinctness, which may be viewed as a refined, "quantitative" analysis of expressive power, is a very interesting topic that deserves much more attention.

Even though we were able to get a good overall picture of the succinctness of monadic fixed point logics on trees, a number of questions remain open. Let us just mention a few of them:

- The exact relationship between monadic datalog, stratified monadic datalog, and the full modal μ-calculus remains unclear. In particular: Is the class of all queries whose complements can be defined by monadic datalog programs polynomially succinct in monadic datalog, or is there an exponential lower bound? (Recall that in Proposition 2 we prove an exponential upper bound.)

- Our proof that MSO is not *Tower*$(o(m))$-succinct in MLFP relies on a complexity theoretic assumption. Is it possible to prove this result without such an assumption?
- We have only considered the 2-variable fragment of MLFP here. What about the k-variable fragments, for $k \geqslant 3$? Do they form a strict hierarchy with respect to succinctness?

References

[ABS99] S. Abiteboul, P. Buneman, and D. Suciu. *Data on the Web: From Relations to Semistructured Data and XML.* Morgan Kaufmann, 1999.

[AHV95] Serge Abiteboul, Richard Hull, and Victor Vianu. *Foundations of databases.* Addison-Wesley, 1995.

[AI00] N. Alechina and N. Immerman. Reachability logic: An efficient fragment of transitive closure logic. *Logic Journal of the IGPL*, 8(3):325–338, 2000.

[AI01] M. Adler and N. Immerman. An $n!$ lower bound on formula size. In *Proceedings of the 16th IEEE Symposium on Logic in Computer Science*, pages 197–206, 2001.

[CH82] A. Chandra and D. Harel. Structure and complexity of relational queries. *Journal of Computer and Systems Sciences*, 25:99–128, 1982.

[DGH+02] E. Dantsin, A. Goerdt, E.A. Hirsch, R. Kannan, J. Kleinberg, C. Papadimitriou, P. Raghavan, and U. Schöning. A deterministic $(2-2/(k+1))^n$ algorithm for k-SAT based on local search. *Theoretical Computer Science*, 289(1):69–83, 2002. Revised version of: Deterministic algorithms for k-SAT based on covering codes and local search, ICALP'00, LNCS volume 1853.

[EF99] Heinz-Dieter Ebbinghaus and Jörg Flum. *Finite model theory.* Springer, New York, second edition, 1999.

[EVW02] K. Etessami, M. Y. Vardi, and Th. Wilke. First-order logic with two variables and unary temporal logic. *Information and Computation*, 179(2):279–295, 2002.

[Fag75] R. Fagin. Monadic generalized spectra. *Zeitschrift für mathematische Logik und Grundlagen der mathematik*, 21:89–96, 1975.

[FG03] Markus Frick and Martin Grohe. The complexity of first-order and monadic second-order logic revisited. Journal version of LICS'02 paper. Available at http://www.dcs.ed.ac.uk/home/grohe/pub.html, 2003.

[FGK03] Markus Frick, Martin Grohe, and Christoph Koch. Query evaluation on compressed trees. In *18th IEEE Symposium on Logic in Computer Science (LICS'03)*, Ottawa, Canada, June 2003.

[FSW00] M. F. Fernandez, J. Siméon, and Ph. Wadler. An algebra for XML query. In S. Kapoor and S. Prasad, editors, *Proceedings of the 20th Conference on Foundations of Software Technology and Theoretical Computer Science (FSTTCS'00)*, volume 1974 of *Lecture Notes in Computer Science*, pages 11–45. Springer-Verlag, 2000.

[GK02] Georg Gottlob and Christoph Koch. Monadic datalog and the expressive power of web information extraction languages. *Submitted*, November 2002. Journal version of PODS'02 paper. Available as CoRR report arXiv:cs.DB/0211020.

[GO99] E. Grädel and M. Otto. On Logics with Two Variables. *Theoretical Computer Science*, 224:73–113, 1999.

[GS03] M. Grohe and N. Schweikardt. Comparing the succinctness of monadic query languages over finite trees. Technical Report EDI-INF-RR-0168, School of Informatics, University of Edinburgh, 2003.

[HP00] H. Hosoya and B. C. Pierce. XDuce: A typed XML processing language (preliminary report). In D. Suciu and G. Vossen, editors, *International Workshop on the Web and Databases*, 2000. Reprinted in *The Web and Databases, Selected Papers*, Springer LNCS volume 1997, 2001.

[Imm99] Neil Immerman. *Descriptive complexity*. Springer, New York, 1999.

[Kam68] H. Kamp. *Tense Logic and the theory of linear order*. PhD thesis, University of California, Los Angeles, 1968.

[Koc03] Christoph Koch. Efficient processing of expressive node-selecting queries on XML data in secondary storage: A tree-automata based approach. In *Proceedings of the 29th Conference on Very Large Data Bases*, 2003. To appear.

[Nev99] Frank Neven. *Design and Analysis of Query Languages for Structured Documents – A Formal and Logical Approach.* PhD thesis, Limburgs Universitair Centrum, 1999.

[NS02] Frank Neven and Thomas Schwentick. Query automata over finite trees. *Theoretical Computer Science*, 275(1-2):633–674, 2002. Journal version of PODS'00 paper.

[Tho96] Wolfgang Thomas. Languages, automata, and logic. In G. Rozenberg and A. Salomaa, editors, *Handbook of formal languages*, volume 3. Springer, New York, 1996.

[Var98] Moshe Y. Vardi. Reasoning about the past with two-way automata. In K.G. Larsen, S. Skyum, and G. Winskel, editors, *25th International Colloquium on Automata, Languages and Programming (ICALP'98)*, volume 1443 of *Lecture Notes in Computer Science*, pages 628–641. Springer-Verlag, 1998.

[Wil99] T. Wilke. CTL+ is exponentially more succinct than CTL. In C. P. Rangan, V. Raman, and R. Ramanujam, editors, *Proceedings of the 19th Conference on Foundations of Software Technology and Theoretical Computer Science*, volume 1738 of *Lecture Notes in Computer Science*, pages 110–121. Springer-Verlag, 1999.

The Arithmetical Complexity of Dimension and Randomness

John M. Hitchcock[1,*], Jack H. Lutz[1,*], and Sebastiaan A. Terwijn[2,**]

[1] Department of Computer Science, Iowa State University, Ames, IA 50011, USA
{jhitchco,lutz}@cs.iastate.edu
[2] Technische Universität Wien, Wiedner Hauptstrasse 8-10, A-1040 Vienna, Austria.
terwijn@logic.at.

Abstract. Constructive dimension and constructive strong dimension are effectivizations of the Hausdorff and packing dimensions, respectively. Each infinite binary sequence A is assigned a dimension $\dim(A) \in [0,1]$ and a strong dimension $\mathrm{Dim}(A) \in [0,1]$.
Let DIM^α and $\mathrm{DIM}^\alpha_{\mathrm{str}}$ be the classes of all sequences of dimension α and of strong dimension α, respectively. We show that DIM^0 is properly Π^0_2, and that for all Δ^0_2-computable $\alpha \in (0,1]$, DIM^α is properly Π^0_3.
To classify the strong dimension classes, we use a more powerful effective Borel hierarchy where a co-enumerable predicate is used rather than a enumerable predicate in the definition of the Σ^0_1 level. For all Δ^0_2-computable $\alpha \in [0,1)$, we show that $\mathrm{DIM}^\alpha_{\mathrm{str}}$ is properly in the Π^0_3 level of this hierarchy. We show that $\mathrm{DIM}^1_{\mathrm{str}}$ is properly in the Π^0_2 level of this hierarchy.
We also prove that the class of Schnorr random sequences and the class of computably random sequences are properly Π^0_3.

1 Introduction

Hausdorff dimension – the most extensively studied fractal dimension – has recently been effectivized at several levels of complexity, yielding applications to a variety of topics in theoretical computer science, including data compression, polynomial-time degrees, approximate optimization, feasible prediction, circuit-size complexity, Kolmogorov complexity, and randomness [14, 15, 3, 1, 8, 5, 7, 17]. The most fundamental of these effectivizations is *constructive dimension*, which is closely related to Kolmogorov complexity and algorithmic randomness. For every subset $\mathcal{X}$ of $\mathbf{C}$, the *Cantor space* of all infinite binary sequences, a constructive dimension $\mathrm{cdim}(\mathcal{X}) \in [0,1]$ is assigned. Informally, this dimension is determined by the maximum rate of growth that a lower semicomputable martingale can achieve on all sequences in $\mathcal{X}$.

* This research was supported in part by National Science Foundation Grant 9988483.
** Supported by the Austrian Research Fund (Lise Meitner grant M699-N05). Part of this author's research was done while visiting the second author at Caltech in January 2002.

M. Baaz and J.A. Makowsky (Eds.): CSL 2003, LNCS 2803, pp. 241–254, 2003.

Just as Martin-Löf [16] used constructive measure to define the randomness of individual sequences, Lutz [15] used constructive dimension to define the dimensions of individual sequences. Each sequence $A \in \mathbf{C}$ is assigned a *dimension* $\dim(A) \in [0,1]$ by $\dim(A) = \mathrm{cdim}(\{A\})$. Every Martin-Löf random sequence has dimension 1, but there are nonrandom sequences with dimension 1. For every real number $\alpha \in [0,1]$, there is a sequence with dimension α.

It is useful to understand the arithmetical complexity of a class of sequences. For example, knowing that RAND, the class of Martin-Löf random sequences, is a Σ^0_2-class allows the application of Kreisel's Basis Lemma [12, 18] to give a short proof [25] that

$$\mathrm{RAND} \cap \Delta^0_2 \neq \emptyset. \tag{1.1}$$

For any $\alpha \in [0,1]$, let

$$\mathrm{DIM}^\alpha = \{A \in \mathbf{C} \mid \dim(A) = \alpha\}.$$

Lutz [15] showed that

$$\mathrm{DIM}^\alpha \cap \Delta^0_2 \neq \emptyset \tag{1.2}$$

for any Δ^0_2-computable $\alpha \in [0,1]$. As these dimension classes do not appear to be Σ^0_2, Lutz was unable to apply the Basis Lemma to them, so he used different techniques to prove (1.2).

We investigate the complexities of these dimension classes in terms of the arithmetical hierarchy of subsets of $\mathbf{C}$. We show that DIM^0 is properly Π^0_2, and for all Δ^0_2-computable $\alpha \in (0,1]$ we show that DIM^α is properly Π^0_3. Therefore, the proof for (1.1) using Kreisel's Basis Lemma cannot directly be used to establish (1.2). (See however the comments made after Corollary 4.10.)

More recently, packing dimension, another important fractal dimension, has also been effectivized by Athreya, Hitchcock, Lutz, and Mayordomo [2]. At the constructive level, this is used in an analogous way to define the *strong dimension* $\mathrm{Dim}(A) \in [0,1]$ for every sequence A. For any $\alpha \in [0,1]$, let

$$\mathrm{DIM}^\alpha_{\mathrm{str}} = \{A \in \mathbf{C} \mid \mathrm{Dim}(A) = \alpha\}.$$

To classify these strong dimension classes, we use introduce a more powerful effective Borel hierarchy where a co-enumerable predicate is used rather than a enumerable predicate in the definition of the Σ^0_1 level. We show that $\mathrm{DIM}^1_{\mathrm{str}}$ is properly in the Π^0_2 level of this stronger hierarchy. For all Δ^0_2-computable $\alpha \in [0,1)$, we show that $\mathrm{DIM}^\alpha_{\mathrm{str}}$ is properly in the Π^0_3 level of this hierarchy.

Our techniques for classifying the dimension and strong dimension classes include Baire category, Wadge reductions, and Kolmogorov complexity. We also classify some effective randomness classes.

Section 2 gives an overview of the randomness and dimension notions used in this paper. In Section 3 we introduce the stronger effective Borel hierarchy that we use for the strong dimension classes. Section 4 presents the classification of DIM^α and $\mathrm{DIM}^\alpha_{\mathrm{str}}$.

2 Background on Randomness and Dimension

This section provides an overview of the notions of randomness and dimension used in this paper. We write $\{0,1\}^*$ for the set of all finite binary *strings* and $\boldsymbol{C}$ for the *Cantor space* of all infinite binary *sequences*. In the standard way, a sequence $A \in \boldsymbol{C}$ can be identified with the subset of $\{0,1\}^*$ or $\mathbb{N}$ for which it is the characteristic sequence, or with a real number in the unit interval. The length of a string $w \in \{0,1\}^*$ is $|w|$. The string consisting of the first n bits of $x \in \{0,1\}^* \cup \boldsymbol{C}$ is denoted by $x \upharpoonright n$. We write $w \sqsubseteq x$ if w is a prefix of A.

2.1 Martin-Löf Randomness

Martin-Löf [16] introduced the notion of a *constructive null set*. A set is constructively null if it can be covered by a uniform sequence of c.e. open sets that are shrinking in size. That is, $\mathcal{A} \subseteq \boldsymbol{C}$ is constructive null if $\mathcal{A} \subseteq \bigcap_i \mathcal{U}_i$, where $\{\mathcal{U}_i\}_{i\in\mathbb{N}}$ is uniformly c.e. such that $\mu(\mathcal{U}_i) \leq 2^{-i}$. The sequence $\{\mathcal{U}_i\}_{i\in\mathbb{N}}$ is called a *Martin-Löf test*. An individual sequence $A \in \boldsymbol{C}$ is *Martin-Löf random* if $\{A\}$ is not constructively null. The Martin-Löf random sequences play an important role in algorithmic information theory, see e.g. Li and Vitányi [13].

Schnorr [20], following Ville [26], characterized constructive null sets in terms of martingales. A function $d : \{0,1\}^* \to [0,\infty)$ is a *martingale* if for every $w \in \{0,1\}^*$, d satisfies the averaging condition

$$2d(w) = d(w0) + d(w1),$$

and d is a *supermartingale* if it satisfies

$$2d(w) \geq d(w0) + d(w1).$$

The *success set of* d is

$$S^\infty[d] = \left\{ A \in \boldsymbol{C} \,\middle|\, \limsup_{n\to\infty} d(A \upharpoonright n) = \infty \right\},$$

i.e., it is the set of all sequences on which d has unbounded value. We say that d succeeds on a class $\mathcal{A} \subseteq \boldsymbol{C}$ if $\mathcal{A} \subseteq S^\infty[d]$.

Ville [26] proved that a set $\mathcal{A} \subseteq \boldsymbol{C}$ has Lebesgue measure 0 if and only if there is a martingale d that succeeds on $\mathcal{A}$. Schnorr [20] showed that $\mathcal{A}$ is constructively null if and only if d can be chosen to be lower semicomputable, that is, if d can be computably approximated from below. We call such a d *constructive.*

Martin-Löf [16] proved that there is a universal constructive null set. That is, he proved that there is a Martin-Löf test $\{\mathcal{U}_i\}_i$ such that for every other test $\{\mathcal{V}_i\}$ it holds that $\bigcap_i \mathcal{V}_i \subseteq \bigcap_i \mathcal{U}_i$. By Schnorr's analysis this implies that there is also a universal constructive supermartingale $\boldsymbol{d}$. That is, for any constructive supermartingale d' there is a $c > 0$ such that $\boldsymbol{d}(w) \geq cd'(w)$ for all $w \in \{0,1\}^*$. We will use this universal supermartingale in section 4. We denote the complement of $S^\infty[\boldsymbol{d}]$ by RAND, so that RAND consists of all the Martin-Löf random sequences.

2.2 Schnorr Randomness

Schnorr [20] criticized the notion of constructive null for an actual lack of constructiveness, and introduced the more constructive notion of a *Schnorr null set*, which is defined by requiring that the measure of the levels $\mathcal{U}_i$ in a Martin-Löf test be computably approximable to within any given precision. It is easy to see that this is equivalent to the following: $\mathcal{A}$ is Schnorr null if $\mathcal{A} \subseteq \bigcap_i \mathcal{U}_i$, where $\{\mathcal{U}_i\}_{i\in\mathbb{N}}$ is uniformly c.e. such that $\mu(\mathcal{U}_i) = 2^{-i}$. The sequence $\{\mathcal{U}_i\}_{i\in\mathbb{N}}$ is called a *Schnorr test.*

Following Schnorr [20], we call an unbounded nondecreasing function $h : \{0,1\}^* \to \{0,1\}^*$ an *order.* (N.B. An "Ordnungsfunktion" in Schnorr's terminology is always computable, whereas we prefer to leave the complexity of orders unspecified in general.) For any order h and martingale d, we define the *order h success set of d* as

$$S^h[d] = \left\{ A \in \mathbf{C} \,\middle|\, \limsup_{n\to\infty} \frac{d(A \upharpoonright n)}{h(n)} \geq 1 \right\}.$$

Schnorr pointed out that the rate of success of a constructive martingale d can be so slow that it cannot be computably detected. Thus rather than working with constructive null sets of the form $S^\infty[d]$ with d constructive, he worked with null sets of the form $S^h[d]$, where both d and h are computable. He proved that a set $\mathcal{A}$ is Schnorr null if and only if it is included in a null set of the form $S^h[d]$, with d and h computable.

A sequence $A \in \mathbf{C}$ is *Schnorr random* if $\{A\}$ is not Schnorr null. This is related the notion of computable randomness. A sequence A is *computably random* if for any computable martingale d, $A \notin S^\infty[d]$.

We write $\mathrm{RAND}_{\mathrm{Schnorr}}$ for the class of all Schnorr random sequences and $\mathrm{RAND}_{\mathrm{comp}}$ for the class of all computably random sequences. By definition we have that

$$\mathrm{RAND} \subseteq \mathrm{RAND}_{\mathrm{comp}} \subseteq \mathrm{RAND}_{\mathrm{Schnorr}}.$$

The first inclusion was proved strict by Schnorr [20], and the second inclusion was proved strict by Wang [27].

2.3 Constructive Dimension

Hausdorff [6] introduced the concept of null covers that "succeed exponentially fast" to define what is now commonly called Hausdorff dimension, the most widely used dimension in fractal geometry. Basically, this notion allows one to discern structure in classes of measure zero, and to calibrate them. As for constructive measure, already Schnorr drew special attention to null sets of "exponential order", although he did not make an explicit connection to Hausdorff dimension.

Lutz [14, 15] gave a characterization of Hausdorff dimension in terms of *gales*, which are a generalization of martingales. Let $s \in [0,\infty)$. An *s-gale* is a function $d : \{0,1\}^* \to [0,\infty)$ that satisfies the averaging condition

$$2^s d(w) = d(w0) + d(w1) \tag{2.1}$$

for every $w \in \{0,1\}^*$. Similarly, d is an *s-supergale* if (2.1) holds with $\geq$ instead of equality. The success set $S^\infty[d]$ is defined exactly as was done for martingales above. Lutz showed that for any class $\mathcal{A} \subseteq \boldsymbol{C}$, the Hausdorff dimension of $\mathcal{A}$ is

$$\dim_{\mathrm{H}}(\mathcal{A}) = \inf\left\{ s \,\middle|\, \begin{array}{l}\text{there exists an } s\text{-gale} \\ d \text{ for which } \mathcal{A} \subseteq S^\infty[d]\end{array} \right\}. \tag{2.2}$$

Lutz [15] effectivized this characterization to define the constructive dimensions of sets and sequences. An s-(super)gale is called *constructive* if it is lower semicomputable. The *constructive dimension* of a class $\mathcal{A} \subseteq \boldsymbol{C}$ is

$$\mathrm{cdim}(\mathcal{A}) = \inf\left\{ s \,\middle|\, \begin{array}{l}\text{there exists a constructive } s\text{-gale} \\ d \text{ for which } \mathcal{A} \subseteq S^\infty[d]\end{array} \right\} \tag{2.3}$$

and the constructive dimension of an individual sequence $A \in \boldsymbol{C}$ is

$$\dim(A) = \mathrm{cdim}(\{A\}).$$

(Supergales can be equivalently used in place of gales in both (2.2) and (2.3) [14, 9, 4].)

Constructive dimension has some remarkable properties. For example, Lutz [15] showed that for any class $\mathcal{A}$,

$$\mathrm{cdim}(\mathcal{A}) = \sup_{A \in \mathcal{A}} \dim(A). \tag{2.4}$$

Also, Mayordomo [17] established a strong connection with *Kolmogorov complexity*: for any $A \in \boldsymbol{C}$,

$$\dim(A) = \liminf_{n \to \infty} \frac{K(A \upharpoonright n)}{n}, \tag{2.5}$$

where $K(A \upharpoonright n)$ is the size of the smallest program that causes a fixed universal self-delimiting Turing machine to output the first n bits of A. (For comments on the relation of this result to earlier results, see the report [23] by Staiger and section 6 of [15]. For more details on Kolmogorov complexity, we refer to [13].)

One can also characterize constructive dimension using the Schnorr null sets (see Section 2.2) of exponential order. The following proposition was observed by several authors, including those of [1, 24].

Proposition 2.1. *Let $\boldsymbol{d}$ be the universal constructive supermartingale. For any $\mathcal{A} \subseteq \boldsymbol{C}$,*

$$\mathrm{cdim}(\mathcal{A}) = \inf\{s \in \mathbb{Q} : \mathcal{A} \subseteq S_{2^{(1-s)n}}[\boldsymbol{d}]\}.$$

2.4 Constructive Strong Dimension

More recently, Athreya, Hitchcock, Lutz, and Mayordomo [2] also characterized *packing dimension*, another important fractal dimension, in terms of gales. For this, the notion of *strong success* of an s-gale d was introduced. The *strong success set of* d is

$$S_{\mathrm{str}}^{\infty}[d] = \left\{ A \in \mathbf{C} \,\middle|\, \liminf_{n\to\infty} d(A \upharpoonright n) = \infty \right\}.$$

Analogously to what was done for Hausdorff dimension, packing dimension can be characterized using strong success sets of gales. Effectivizing this in the same way leads to the definition of the *constructive strong dimension* of a class $\mathcal{A} \subseteq \mathbf{C}$ as

$$\mathrm{cDim}(\mathcal{A}) = \inf \left\{ s \,\middle|\, \begin{array}{l} \text{there exists a constructive } s\text{-gale} \\ d \text{ for which } \mathcal{A} \subseteq S_{\mathrm{str}}^{\infty}[d] \end{array} \right\}.$$

The constructive strong dimension of a sequence $A \in \mathbf{C}$ is

$$\mathrm{Dim}(A) = \mathrm{cDim}(\{A\}).$$

A pointwise stability property analogous to (2.4) also holds for strong dimension, as well as a Kolmogorov complexity characterization [2]:

$$\mathrm{Dim}(A) = \limsup_{n\to\infty} \frac{K(A \upharpoonright n)}{n} \tag{2.6}$$

for any $A \in \mathbf{C}$.

3 Borel Hierarchies

We use $\boldsymbol{\Sigma}_n^0$ and $\boldsymbol{\Pi}_n^0$ to denote the levels of the Borel hierarchy for subsets of Cantor space. The levels of the arithmetical hierarchy (the corresponding effective hierarchy) are denoted by Σ_n^0 and Π_n^0.

We will also make use of the following more general hierarchy definition.

Definition. Let $\mathcal{P}$ be a class of predicates, let $n \geq 1$, and let $\mathcal{X} \subseteq \mathbf{C}$.

- $\mathcal{X} \in \Sigma_n^0[\mathcal{P}]$ if for some predicate $P \in \mathcal{P}$,

$$A \in \mathcal{X} \iff (\exists k_n)(\forall k_{n-1}) \cdots (Qk_1) P(k_n, \ldots, k_2, A \upharpoonright k_1),$$

  where $Q = \exists$ if n is odd and $Q = \forall$ if n is even.
- $\mathcal{X} \in \Pi_n^0[\mathcal{P}]$ if for some predicate $P \in \mathcal{P}$,

$$A \in \mathcal{X} \iff (\forall k_n)(\exists k_{n-1}) \cdots (Qk_1) P(k_n, \ldots, k_2, A \upharpoonright k_1),$$

  where $Q = \forall$ if n is odd and $Q = \exists$ if n is even.

If we take $\mathcal{P}$ to be Δ_1^0 (decidable predicates), then the above definition is equivalent to the standard arithmetical hierarchy; that is,

$$\Sigma_n^0 = \Sigma_n^0[\Delta_1^0]$$

and

$$\Pi_n^0 = \Pi_n^0[\Delta_1^0]$$

hold for all n. Also, if ALL is the class of all predicates, then we obtain the classical Borel hierarchy:

$$\boldsymbol{\Sigma}_n^0 = \Sigma_n^0[\text{ALL}]$$

and

$$\boldsymbol{\Pi}_n^0 = \Pi_n^0[\text{ALL}].$$

In this paper, we will also be interested in the cases where $\mathcal{P}$ is Σ_1^0 (enumerable predicates) or Π_1^0 (co-enumerable predicates). In some cases, the classes in the generalized hierarchy using these sets of predicates are no different that the standard arithmetical hierarchy classes. If n is odd, then $\Sigma_n^0 = \Sigma_n^0[\Sigma_1^0]$ as the existential quantifier in the Σ_1^0 predicate can be absorbed into the last quantifier in the definition of $\Sigma_n^0[\Delta_1^0] = \Sigma_n^0$. Analogously, $\Pi_n^0 = \Pi_n^0[\Pi_1^0]$ for odd n, and for even n we have $\Sigma_n^0 = \Sigma_n^0[\Pi_1^0]$ and $\Pi_n^0 = \Pi_n^0[\Sigma_1^0]$. On the other hand, using the complementary set of predicates defines an effective hierarchy that is distinct from and interleaved with the arithmetical hierarchy.

Proposition 3.1. *1. If n is odd, then*

$$\Sigma_n^0 \subsetneq \Sigma_n^0[\Pi_1^0] \subsetneq \Sigma_{n+1}^0$$

and

$$\Pi_n^0 \subsetneq \Pi_n^0[\Sigma_1^0] \subsetneq \Pi_{n+1}^0.$$

2. If n is even, then

$$\Sigma_n^0 \subsetneq \Sigma_n^0[\Sigma_1^0] \subsetneq \Sigma_{n+1}^0$$

and

$$\Pi_n^0 \subsetneq \Pi_n^0[\Pi_1^0] \subsetneq \Pi_{n+1}^0.$$

Intuitively, the classes $\Sigma_1^0[\Pi_1^0]$, $\Pi_1^0[\Sigma_1^0]$, $\Sigma_2^0[\Sigma_1^0]$, $\Pi_2^0[\Pi_1^0], \ldots$ are slightly more powerful than their respective counterparts in the arithmetical hierarchy because they use one additional quantifier that is limited to the predicate. We now give a simple example of a class that is best classified in this hierarchy: the class of all 1-generic sequences is $\Pi_2^0[\Pi_1^0]$ but not Π_2^0.

Example 3.2. Recall that a sequence $X \in \boldsymbol{C}$ is 1-generic (see e.g. Jockusch [10]) if

$$(\forall e)(\exists \sigma \sqsubset X)\big[\,\{e\}^\sigma(e)\downarrow \;\vee\; (\forall \tau \sqsupset \sigma)[\{e\}^\tau(e)\uparrow]\big]$$

From this definition it is immediate that the class $\mathcal{G} = \{X \mid X \text{ is 1-generic}\}$ is in $\Pi_2^0[\Pi_1^0]$. To show that $\mathcal{G}$ is not Π_2^0, suppose that it is. Then there is a computable predicate R such that

$$X \in \mathcal{G} \iff (\forall n)(\exists m)\big[R(n, X \upharpoonright m)\big].$$

As $\mathcal{G}$ is dense, we can now easily construct a computable element of it by a computable finite extension argument, which gives a contradiction. (Given σ at stage n, search for extension $\sigma' \sqsupseteq \sigma$ such that $R(n, \sigma')$. Such extension will be found by density. Take this extension and proceed to stage $n+1$.)

Staiger has pointed out to us that the class $\Pi_1^0[\Sigma_1^0]$ already occured under a different guise in his paper [22], where it was called $\mathfrak{P}$, and several presentations were proven to be equivalent to it. The following definitions are from [21]. For any set W of initial segments define

$$\lim W = \{A \in \boldsymbol{C} \mid (\forall \sigma \sqsubseteq A)\sigma \in W\}$$
$$W^\sigma = \{A \in \boldsymbol{C} \mid (\forall^\infty \sigma \sqsubseteq A)\sigma \in W\}.$$

Staiger proved that the classes in $\Pi_1^0[\Sigma_1^0]$ are those of the form $\lim W$, for $W \in \Sigma_1^0$, and the classes in $\Sigma_2^0[\Sigma_1^0]$ are those of the form W^σ, for $W \in \Sigma_1^0$.

4 Classification of DIM^α and $\mathrm{DIM}^\alpha_{\mathrm{str}}$

In this section we investigate the arithmetical complexity of the following dimension and strong dimension classes.

$$\begin{aligned}
\mathrm{DIM}^\alpha &= \{A \in \boldsymbol{C} \mid \dim(A) = \alpha\} \\
\mathrm{DIM}^{\leq\alpha} &= \{A \in \boldsymbol{C} \mid \dim(A) \leq \alpha\} \\
\mathrm{DIM}^{\geq\alpha} &= \{A \in \boldsymbol{C} \mid \dim(A) \geq \alpha\} \\
\mathrm{DIM}^\alpha_{\mathrm{str}} &= \{A \in \boldsymbol{C} \mid \mathrm{Dim}(A) = \alpha\} \\
\mathrm{DIM}^{\leq\alpha}_{\mathrm{str}} &= \{A \in \boldsymbol{C} \mid \mathrm{Dim}(A) \leq \alpha\} \\
\mathrm{DIM}^{\geq\alpha}_{\mathrm{str}} &= \{A \in \boldsymbol{C} \mid \mathrm{Dim}(A) \geq \alpha\}
\end{aligned}$$

Let $\alpha \in [0,1]$ be Δ_2^0-computable. For any such α, it is well known that there is a computable function $\hat{\alpha} : \mathrm{N} \to \mathbb{Q}$ such that $\lim_{n\to\infty} \hat{\alpha}(n) = \alpha$. Using (2.5), we have

$$\begin{aligned}
\dim(X) \leq \alpha &\iff \liminf_{n\to\infty} \frac{K(X \upharpoonright n)}{n} \leq \alpha \\
&\iff (\forall k)(\forall N)(\exists n \geq N)K(X \upharpoonright n) < (\hat{\alpha}(n) + 1/k)n,
\end{aligned}$$

so $\mathrm{DIM}^{\leq\alpha}$ is a Π_2^0-class. Also,

$$\begin{aligned}
\dim(X) \geq \alpha &\iff \liminf_{n\to\infty} \frac{K(X \upharpoonright n)}{n} \geq \alpha \\
&\iff (\forall k)(\exists N)(\forall n \geq N)K(X \upharpoonright n) > (\hat{\alpha}(N) - 1/k)n,
\end{aligned}$$

so $\mathrm{DIM}^{\geq\alpha}$ is a Π_3^0-class. Therefore we have the following.

Proposition 4.1. *1. The class* DIM^0 *is* Π_2^0.
2. For all Δ_2^0*-computable* $\alpha \in (0,1]$, DIM^α *is a* Π_3^0*-class.*
3. For arbitrary $\alpha \in (0,1]$, DIM^α *is a* $\boldsymbol{\Pi}_3^0$*-class.*

The situation is slightly more complicated for strong dimension. By (2.6), we have

$$\begin{aligned}\mathrm{Dim}(X) \le \alpha &\iff \limsup_{n\to\infty} \frac{K(X \upharpoonright n)}{n} \le \alpha \\ &\iff (\forall k)(\exists N)(\forall n \ge N) K(X \upharpoonright n) < (\hat{\alpha}(N) + 1/k)n \\ &\iff (\forall k)(\exists N)(\forall n \ge N)(\exists \langle \pi, t\rangle)|\pi| < (\hat{\alpha}(N) + 1/k)n \\ &\qquad \text{and } U(\pi) = X \upharpoonright n \text{ in } \le t \text{ computation steps,}\end{aligned}$$

where U is the fixed universal self-delimiting Turing machine used to define K. From this it is clear that $\mathrm{DIM}_{\mathrm{str}}^{\le\alpha} \in \Pi_4^0$. However, the "$(\exists\langle\pi, t\rangle)$" quantifier is local to the defining predicate, so we have $\mathrm{DIM}_{\mathrm{str}}^{\le\alpha} \in \boldsymbol{\Pi}_3^0$, and in fact, it is a $\Pi_3^0[\Sigma_1^0]$-class. Also,

$$\begin{aligned}\mathrm{Dim}(X) \ge \alpha &\iff \limsup_{n\to\infty} \frac{K(X \upharpoonright n)}{n} \ge \alpha \\ &\iff (\forall k)(\forall N)(\exists n \ge N) K(X \upharpoonright n) > (\hat{\alpha}(n) - 1/k)n,\end{aligned}$$

so $\mathrm{DIM}_{\mathrm{str}}^{\ge\alpha}$ is a $\Pi_2^0[\Pi_1^0]$-class. This establishes the following analogue of Proposition 4.1.

Proposition 4.2. *1. The class* $\mathrm{DIM}_{\mathrm{str}}^1$ *is* $\Pi_2^0[\Pi_1^0]$.
2. For all Δ_2^0*-computable* $\alpha \in [0, 1)$, $\mathrm{DIM}_{\mathrm{str}}^\alpha$ *is a* $\Pi_3^0[\Sigma_1^0]$*-class.*
3. For arbitrary $\alpha \in [0, 1)$, $\mathrm{DIM}_{\mathrm{str}}^\alpha$ *is a* $\boldsymbol{\Pi}_3^0$*-class.*

In the remainder of this section we prove that the classifications in Propositions 4.1 and 4.2 cannot be improved in their respective hierarchies.

4.1 Category Methods

Recall that a class $\mathcal{X}$ is *meager* if it is included in a countable union of nowhere dense subsets of $\boldsymbol{C}$, and *comeager* if its complement $\overline{\mathcal{X}}$ is meager. The following lemma (implicit in Rogers [19, p341]) will be useful.

Lemma 4.3. *If* $\mathcal{X} \in \boldsymbol{\Sigma}_2^0$ *and* $\overline{\mathcal{X}}$ *is dense then* $\mathcal{X}$ *is meager.*

The class RAND of Martin-Löf random sets can easily be classified with category methods.

Theorem 4.4. (folk) RAND *is a* Σ_2^0*-class, but not a* $\boldsymbol{\Pi}_2^0$*-class.*

As DIM^0 and $\mathrm{DIM}_{\mathrm{str}}^1$ are dense $\boldsymbol{\Pi}_2^0$-classes that have dense complements, an argument similar to the one used for Theorem 4.4 shows that they are not $\boldsymbol{\Sigma}_2^0$-classes.

Theorem 4.5. *The classes* DIM^0 *and* $\mathrm{DIM}_{\mathrm{str}}^1$ *are not* $\boldsymbol{\Sigma}_2^0$*-classes.*

We now develop category methods for the other DIM^α classes. For every rational s, define the computable order $h_s(n) = 2^{(1-s)n}$. Let $\boldsymbol{d}$ be the optimal constructive supermartingale.

Lemma 4.6. *For every rational $s \in (0,1)$, $S^{h_s}[\boldsymbol{d}]$ is a comeager Π_2^0-class.*

Proof. Notice that $\overline{S^{h_s}[\boldsymbol{d}]} \in \Sigma_2^0$ and $S^{h_s}[\boldsymbol{d}]$ is dense. Now apply Lemma 4.3. □

Lemma 4.7. *For all $\alpha \in (0,1]$, DIM^α is meager.*

Proof. Let $s < \alpha$ be rational. Lutz [15] showed that $\boldsymbol{d}^{(s)}(w) = 2^{(s-1)|w|}\boldsymbol{d}(w)$ is an optimal constructive s-supergale. It follows that for any $A \in \boldsymbol{C}$, $A \in S^{h_s}[\boldsymbol{d}] \Rightarrow \dim(S) < \alpha$. Therefore $\mathrm{DIM}^\alpha \subseteq \overline{S^{h_s}}$, so DIM^α is meager by Lemma 4.6. □

Proposition 4.8. *For all $\alpha \in (0,1]$, DIM^α is not a Π_2^0-class.*

Proof. If $\mathrm{DIM}^\alpha \in \Pi_2^0$, then Lemma 4.3 implies that DIM^α is comeager, contradicting Lemma 4.7. □

To strengthen Proposition 4.8 to show that DIM^α is not Σ_3^0, we now turn to Wadge reductions.

4.2 Wadge Reductions

Let $\mathcal{A}, \mathcal{B} \subseteq \boldsymbol{C}$. A *Wadge reduction* of $\mathcal{A}$ to $\mathcal{B}$ is a function $f : \boldsymbol{C} \to \boldsymbol{C}$ that is continuous and satisfies $\mathcal{A} = f^{-1}(\mathcal{B})$, i.e., $X \in \mathcal{A} \iff f(X) \in \mathcal{B}$. We say that $\mathcal{B}$ is *Wadge complete* for a class $\boldsymbol{\Gamma}$ of subsets of $\boldsymbol{C}$ if $\mathcal{B} \in \boldsymbol{\Gamma}$ and every $\mathcal{A} \in \boldsymbol{\Gamma}$ Wadge reduces to $\mathcal{B}$. As the classes of the Borel hierarchy are closed under Wadge reductions, Wadge completeness can be used to properly identify the location of a subset of $\boldsymbol{C}$ in the hierarchy.

We now prove that DIM^1 is Wadge complete for $\boldsymbol{\Pi}_3^0$. We will then give Wadge reductions from it to DIM^α for the other values of α.

Theorem 4.9. DIM^1 *is Wadge complete for $\boldsymbol{\Pi}_3^0$. Therefore* DIM^1 *is not a $\boldsymbol{\Sigma}_3^0$-class, and in particular is not a Σ_3^0 -class.*

Proof. One could prove this by reducing a known $\boldsymbol{\Pi}_3^0$-complete class to DIM^1, e.g. the class of sets that have a limiting frequency of 1's that is 0 (this class was proved to be $\boldsymbol{\Pi}_3^0$-complete by Ki and Linton [11]), but it is just as easy to build a direct reduction from an arbitrary $\boldsymbol{\Pi}_3^0$-class.

Let $\boldsymbol{d}$ be the universal constructive supermartingale. Note that we have (cf. Proposition 2.1)

$$S^{2^n}[\boldsymbol{d}] \subsetneq \ldots \subsetneq S^{2^{\frac{1}{k}n}}[\boldsymbol{d}] \subsetneq S^{2^{\frac{1}{k+1}n}}[\boldsymbol{d}] \subsetneq \ldots \subsetneq \mathrm{DIM}^1.$$

Let $\bigcup_k \bigcap_s \mathcal{O}_{k,s}$ be a $\boldsymbol{\Sigma}_3^0$-class. Without loss of generality $\mathcal{O}_{k,s} \supseteq \mathcal{O}_{k,s+1}$ for all k,s. We define a continuous function $f : \boldsymbol{C} \to \boldsymbol{C}$ such that

$$\forall k \left(X \in \bigcap_s \mathcal{O}_{k,s} \iff f(X) \in S^{2^{\frac{1}{k}n}}[\boldsymbol{d}] \right) \tag{4.1}$$

so that we have

$$X \notin \bigcup_k \bigcap_s \mathcal{O}_{k,s} \iff \forall k \left(f(X) \notin S^{2^{\frac{1}{k}n}}[d] \right)$$
$$\iff f(X) \in \mathrm{DIM}^1.$$

The image $Y = f(X)$ is defined in stages, $Y = \bigcup_s Y_s$, such that every initial segment of X defines an initial segment of Y.

At stage 0 we define Y_0 to be the empty sequence.

At stage $s > 0$ we consider $X \upharpoonright s$, and for each k we define $t_{k,s}$ to be the largest stage $t \leq s$ such that $X \upharpoonright s \in \mathcal{O}_{k,t}$. (Let $t_{k,s} = 0$ if such a t does not exist.) Define k to be *expansionary* at stage s if $t_{k,s-1} < t_{k,s}$. Now we let $k(s) = \min\{k : k \text{ is expansionary at } s\}$. There are two substages.

Substage (a). First consider all strings σ extending Y_{s-1} of minimal length with $d(\sigma) \geq 2^{\frac{1}{k(s)}|\sigma|}$, and take the leftmost one of these σ's. Such σ's exist because $S^{2^{\frac{1}{k(s)}n}}[d]$ is dense. If $k(s)$ does not exist, let $\sigma = Y_{s-1}$.

Substage (b). Next consider all extensions $\tau \sqsupseteq \sigma$ of minimal length such that $d(\tau \upharpoonright i) \leq d(\tau \upharpoonright (i-1))$ for every $|\sigma| < i < |\tau|$, and $d(\tau) \leq |\tau|$. Clearly such τ exist, by direct diagonalization against d. Define Y_s to be the leftmost of these τ. This concludes the construction.

So Y_s is defined by first building a piece of evidence σ that d achieves growth rate $2^{\frac{1}{k(s)}n}$ on Y and then slowing down the growth rate of d to the order n. Note that f is continuous. If $X \in \bigcup_k \bigcap_s \mathcal{O}_{k,s}$, then for the minimal k such that $X \in \bigcap_s \mathcal{O}_{k,s}$, infinitely many pieces of evidence σ witness that d achieves growth rate $2^{\frac{1}{k}n}$ on Y, so $Y \notin \mathrm{DIM}^1$. On the other hand, if $X \notin \bigcup_k \bigcap_s \mathcal{O}_{k,s}$ then for every k only finitely often $d(Y_s) \geq 2^{\frac{1}{k}|Y_s|}$ because in substage (a) the extension σ is chosen to be of minimal length, so $Y \notin S_{h_k}[d]$. Hence $Y \in \mathrm{DIM}^1$. □

As RAND is a Σ^0_2-class, we have the following corollary (which can also be proved by a direct construction).

Corollary 4.10. (Lutz [15]) RAND *is a proper subset of* DIM^1.

In order to establish the existence of Δ^0_2-computable sequences of any Δ^0_2-computable dimension $\alpha \in [0,1)$, Lutz [15] defined a *dilution function* $g_\alpha : \boldsymbol{C} \to \boldsymbol{C}$ that is computable and satisfies $\dim(g_\alpha(X)) = \alpha \cdot \dim(X)$ for all $X \in \boldsymbol{C}$. Applying this to any Δ^0_2-computable Martin-Löf random sequence (which must have dimension 1) establishes the existence theorem. (We note that $g_\alpha(X)$ has the same Turing degree as X. Since by the Low Basis Theorem of Jockusch and Soare [18, Theorem V.5.32] there are Martin-Löf random sets of low degree, we immediately obtain that there are low sets of any Δ^0_2-computable dimension α.) As g_α is continuous, it is a Wadge reduction from DIM^1 to DIM^α if $\alpha > 0$. Combining this with the previous theorem, we have that DIM^α is Wadge complete for $\boldsymbol{\Pi}^0_3$ for all Δ^0_2-computable $\alpha \in (0,1)$. We now give a similar dilution construction that will allow us to prove this for *arbitrary* $\alpha \in (0,1)$.

Let $X \in \boldsymbol{C}$ and let $\alpha \in (0,1)$. Write $X = x_1 x_2 x_3 \ldots$ where $|x_n| = 2n-1$ for all n, noting that $|x_1 \cdots x_n| = n^2$. For each n, let $k_n = \lceil n\frac{1-\alpha}{\alpha} \rceil$ and $y_n = 0^{k_n}$. We then define $f_\alpha(X) = x_1 y_1 x_2 y_2 \cdots x_n y_n \cdots$. Observe that f_α is a continuous function mapping $\boldsymbol{C}$ to $\boldsymbol{C}$. We now show that it modifies the dimension of X in a controlled manner.

Lemma 4.11. *For any $X \in \boldsymbol{C}$ and $\alpha \in (0,1)$, $\dim(f_\alpha(X)) = \alpha \cdot \dim(X)$ and $\mathrm{Dim}(f_\alpha(X)) = \alpha \cdot \mathrm{Dim}(X)$.*

The function f_α establishes the completeness of DIM^α.

Theorem 4.12. *For all $\alpha \in (0,1)$, DIM^α is Wadge complete for $\boldsymbol{\Pi}^0_3$. Therefore it is not a $\boldsymbol{\Sigma}^0_3$-class, and in particular not a Σ^0_3-class.*

Proof. By Lemma 4.11, f_α is a Wadge reduction from DIM^1 to DIM^α. Therefore DIM^α is Wadge complete for $\boldsymbol{\Pi}^0_3$ by composing f_α with the reduction from Theorem 4.9. □

For lack of space, we state the following theorems without proof.

Theorem 4.13. *For all $\alpha \in [0,1)$, $\mathrm{DIM}^\alpha_{\mathrm{str}}$ is Wadge complete for $\boldsymbol{\Pi}^0_3$. Therefore $\mathrm{DIM}^\alpha_{\mathrm{str}}$ is not a $\boldsymbol{\Sigma}^0_3$-class, and in particular is not a $\Sigma^0_3[\Pi^0_1]$-class.*

Theorem 4.14. $\mathrm{RAND}_{\mathrm{Schnorr}}$ *is a Π^0_3-class, but not a $\boldsymbol{\Sigma}^0_3$-class.*

Theorem 4.15. $\mathrm{RAND}_{\mathrm{comp}}$ *is a Π^0_3-class, but not a $\boldsymbol{\Sigma}^0_3$-class.*

4.3 Ad Hoc Methods

When the level of the class in the effective hierarchy is not the same as the level in the classical hierarchy one often needs to resort to ad hoc arguments. One might think that the notion of effective Wadge reduction, or recursive functional, would be the proper notion to use in classifying classes of reals in the effective hierarchy. However, this notion is rarely useful for the following reason. Let $\mathcal{X}$ be a class without computable elements, such as the class of Martin-Löf random sets or the class of 1-generic sets. Then $\mathcal{X}$ cannot be proven to be complete for any level of the effective hierarchy by a recursive Wadge reduction f. For if X is recursive, then so is $f(X)$, so we can never have $X \in \boldsymbol{C} \iff f(X) \in \mathcal{X}$. So we see that "easy" classes like $\boldsymbol{C}$ that contain recursive elements cannot be reduced in such a way to many "difficult" classes, which renders the notion rather useless.

We have left open the question whether $\mathrm{DIM}^1_{\mathrm{str}}$ is not in Π^0_2, and whether $\mathrm{DIM}^\alpha_{\mathrm{str}}$ is not in Π^0_3 for any Δ^0_2-computable $\alpha \in [0,1)$. We have no answer to the second question, but we provide an answer to the first in the next theorem. We make use of the following lemma:

Lemma 4.16. *If $\mathcal{X} \in \Pi^0_2$ is dense then there is a computable $X \in \mathcal{X}$.*

Proof. This is an easy finite extension argument. Suppose that $\mathcal{X} = \{X : (\forall m)(\exists k)R^X(m,k)\downarrow = 1\} \in \Pi^0_2$ is dense. (Here R is a computable predicate. Note that R does not have to be defined with oracles X that are not in $\mathcal{X}$.) Given any initial segment τ such that

$$(\forall n < m)(\exists k)R^\tau(m,k)\downarrow = 1,$$

we show how to compute an extension $\sigma \sqsupset \tau$ such that

$$(\exists k)R^\sigma(m,k)\downarrow = 1. \tag{4.2}$$

Because $\mathcal{X}$ is dense, there are $X \sqsupset \tau$ and k such that $R^X(m,k)\downarrow = 1$. Let u be the use of this computation, i.e. the part of the oracle X used in it. Now define $\sigma = \max\{X \restriction u, \tau\}$. Then $\sigma \sqsupseteq \tau$ satisfies (4.2).

Now it is clear that for every m we can compute appropriate extensions σ_m such that $X = \bigcup_m \sigma_m$ is computable and $(\forall m)(\exists k)R^{\sigma_m}(m,k)\downarrow = 1$, so that $X \in \mathcal{X}$. □

Theorem 4.17. $\mathrm{DIM}^1_{\mathrm{str}}$ *is not a* Π^0_2*-class. Hence it is properly* $\Pi^0_2[\Pi^0_1]$.

Proof. Suppose that $\mathrm{DIM}^1_{\mathrm{str}}$ is Π^0_2. Then, since clearly $\mathrm{DIM}^1_{\mathrm{str}}$ is dense, by Lemma 4.16 it contains a computable real, contradicting that every computable real has strong dimension 0. □

We conclude this section by summarizing its main results in the following table.

	DIM^α	$\mathrm{DIM}^\alpha_{\mathrm{str}}$
$\alpha = 0$	$\Pi^0_2 - \Sigma^0_2$	$\Pi^0_3[\Sigma^0_1] - \Sigma^0_3$
$\alpha \in (0,1) \cap \Delta^0_2$	$\Pi^0_3 - \Sigma^0_3$	$\Pi^0_3[\Sigma^0_1] - \Sigma^0_3$
$\alpha = 1$	$\Pi^0_3 - \Sigma^0_3$	$\Pi^0_2[\Pi^0_1] - (\Sigma^0_2 \cup \Pi^0_2)$
arbitrary $\alpha \in (0,1)$	$\boldsymbol{\Pi}^0_3 - \boldsymbol{\Sigma}^0_3$	$\boldsymbol{\Pi}^0_3 - \boldsymbol{\Sigma}^0_3$

Question 4.18. Is it the case that $\mathrm{DIM}^\alpha_{\mathrm{str}}$ is not in Π^0_3 for any Δ^0_2-computable $\alpha \in [0,1)$?

References

1. K. Ambos-Spies, W. Merkle, J. Reimann, and F. Stephan. Hausdorff dimension in exponential time. In *Proceedings of the 16th IEEE Conference on Computational Complexity*, pages 210–217, 2001.
2. K. B. Athreya, J. M. Hitchcock, J. H. Lutz, and E. Mayordomo. Effective strong dimension, algorithmic information, and computational complexity. Technical Report cs.CC/0211025, Computing Research Repository, 2002.
3. J. J. Dai, J. I. Lathrop, J. H. Lutz, and E. Mayordomo. Finite-state dimension. *Theoretical Computer Science*. To appear.

4. S. A. Fenner. Gales and supergales are equivalent for defining constructive Hausdorff dimension. Technical Report cs.CC/0208044, Computing Research Repository, 2002.
5. L. Fortnow and J. H. Lutz. Prediction and dimension. In *Proceedings of the 15th Annual Conference on Computational Learning Theory*, pages 380–395, 2002.
6. F. Hausdorff. Dimension und äusseres Mass. *Mathematische Annalen*, 79:157–179, 1919.
7. J. M. Hitchcock. Fractal dimension and logarithmic loss unpredictability. *Theoretical Computer Science*. To appear.
8. J. M. Hitchcock. MAX3SAT is exponentially hard to approximate if NP has positive dimension. *Theoretical Computer Science*, 289(1):861–869, 2002.
9. J. M. Hitchcock. Gales suffice for constructive dimension. *Information Processing Letters*, 86(1):9–12, 2003.
10. C. G. Jockusch. Degrees of generic sets. In *Recursion Theory: its Generalizations and Applications*, volume 45 of *London Mathematical Society Lecture Notes Series*, pages 110–139. Cambridge University Press, 1980.
11. H. Ki and T. Linton. Normal numbers and subsets of $\mathbb{N}$ with given densities. *Fundamenta Mathematicae*, 144:163–179, 1994.
12. G. Kreisel. Note on arithmetical models for consistent formulae of the predicate calculus. *Fundamenta Mathematicae*, 37:265–285, 1950.
13. M. Li and P. M. B. Vitányi. *An Introduction to Kolmogorov Complexity and its Applications*. Springer-Verlag, Berlin, 1997. Second Edition.
14. J. H. Lutz. Dimension in complexity classes. *SIAM Journal on Computing*. To appear.
15. J. H. Lutz. The dimensions of individual strings and sequences. *Information and Computation*. To appear.
16. P. Martin-Löf. The definition of random sequences. *Information and Control*, 9:602–619, 1966.
17. E. Mayordomo. A Kolmogorov complexity characterization of constructive Hausdorff dimension. *Information Processing Letters*, 84(1):1–3, 2002.
18. P. Odifreddi. *Classical recursion theory*, volume 125 of *Studies in Logic and the Foundations of Mathematics*. North-Holland, 1989.
19. H. Rogers, Jr. *Theory of Recursive Functions and Effective Computability*. McGraw - Hill, New York, N.Y., 1967.
20. C. P. Schnorr. Zufälligkeit und Wahrscheinlichkeit. *Lecture Notes in Mathematics*, 218, 1971.
21. L. Staiger. Recursive automata on infinite words. In *Proceedings of the 10th Annual Symposium on Theoretical Aspects of Computer Science*, pages 629–639, 1993.
22. L. Staiger. On the power of reading the whole infinite input tape. In C. S. Calude and Gh. Paun, editors, *Finite Versus Infinite: Contributions to an Eternal Dilemma*, pages 335–348. Springer-Verlag, 2000.
23. L. Staiger. Constructive dimension equals Kolmogorov complexity. Technical Report CDMTCS-210, University of Auckland, January 2003.
24. S. A. Terwijn. Complexity and randomness. Technical Report CDMTCS-212, University of Auckland, March 2003. Notes for a course given at the University of Auckland.
25. M. van Lambalgen. *Random Sequences*. PhD thesis, Department of Mathematics, University of Amsterdam, 1987.
26. J. Ville. *Étude Critique de la Notion de Collectif*. Gauthier–Villars, Paris, 1939.
27. Y. Wang. *Randomness and Complexity*. PhD thesis, Department of Mathematics, University of Heidelberg, 1996.

Towards a Proof System for Admissibility

Rosalie Iemhoff*

Technical University Vienna, Wiedner Hauptstrasse 8–10, A-1040 Vienna, Austria
iemhoff@logic.at
http://www.logic.at/people/iemhoff

Abstract. In [4] a basis for the admissible rules of intuitionistic propositional logic IPC was given. Here we strengthen this result by presenting a system ADM that has the following two properties. $A \vdash_{\mathsf{ADM}} B$ implies that A admissibly derives B. ADM is complete in the sense that for every formula A there exists a formula $A \vdash_{\mathsf{ADM}} \Lambda_A$ such that the admissibly derivable consequences of A are the (normal) consequences of Λ_A. This work is related to and partly relies upon research by Ghilardi on projective formulas [2] [3]....

1 Introduction

An interesting meta-mathematical property of intuitionistic propositional logic IPC is the existence of non-derivable admissible rules for this logic. That is, there exist rules that, when added to IPC, do not lead to new theorems, but that are not derivable in IPC either. We write $A \mathrel{|\!\sim} B$ if A/B is an admissible rule. Some results during the last decade have shed some light on the structure of the admissible derivability relation $\mathrel{|\!\sim}$. Rybakov [6] showed that $\mathrel{|\!\sim}$ is decidable and Ghilardi [3] presented a transparent algorithm. In [4] a simple syntactical characterization for $\mathrel{|\!\sim}$ was given. This result implied that Visser's rules $V = \{Vn \mid n = 1, 2, \ldots\}$, where

$$Vn \quad (\bigwedge_{i=1}^{n}(A_i \to B_i) \to A_{n+1} \vee A_{n+2}) \mathrel{|\!\sim} \bigvee_{j=1}^{n+2}(\bigwedge_{i=1}^{n}(A_i \to B_i) \to A_j),$$

form a basis for the admissible rules of IPC. Intuitively, this means that all admissible rules of IPC can be obtained from Visser's rules via derivability in IPC. Here we strengthen this result in the following way. Our aim is to present a decent proof system for $\mathrel{|\!\sim}$, i.e. a system that given $A \mathrel{|\!\sim} B$, tells what are the syntactical manipulations one has to apply to A to obtain B. With decent we mean something like "as cut-free as possible". The mentioned characterization does not fulfill this aim since it contains a full cut rule: if $A \mathrel{|\!\sim} B$ and $B \mathrel{|\!\sim} C$, then $A \mathrel{|\!\sim} C$. The form of V does not seem to suggest the existence of a neat cut-free Gentzen calculus for $\mathrel{|\!\sim}$, as the rules Vn violate the subformula property.

* Research supported by a Marie Curie fellowship of the European Union under grant HPMC-CT-2001-01383.

M. Baaz and J.A. Makowsky (Eds.): CSL 2003, LNCS 2803, pp. 255–270, 2003.

Although cut elimination might not be possible in full extent, it might however still be possible to reach the above mentioned goal in a satisfying way. Here we present a result in this direction. We present a proof system ADM such that $A \vdash_{\mathsf{ADM}} B$ implies $A \mathrel{|\!\sim} B$. The converse does not hold. However, the system is complete in the sense that for every formula A there exists a unique formula $A \vdash_{\mathsf{ADM}} \Lambda_A$ such that

$$\forall B: \ A \mathrel{|\!\sim} B \text{ iff } \Lambda_A \vdash B.$$

Thus, given A, once one has obtained Λ_A via ADM, $A \mathrel{|\!\sim} B$ is reduced to $\Lambda_A \vdash B$. The system ADM is "cut-free" in the sense that it consists solely of "rewriting" rules and applications of V. With "rewriting" rules we mean in this context simple rules that e.g. infer $(A \to B) \wedge (A \to C)$ from $(A \to B \wedge C)$. In contrast to V, all these rules of ADM are derivable admissible rules. Note that this result implies that V is a basis for the admissible rules of IPC. However, it is stronger in the sense that it provides more information about $\mathrel{|\!\sim}$.

These results are intimately linked with and inspired by results from Ghilardi's papers [2] and [3]. In fact, our main theorem heavily relies on results in [3]. His results stem from research on unification in intermediate and modal logics. To this end he defines in [2] the notion of a projective formula. A formula A is called *projective* if there exists a substitution σ such that

$$\vdash \sigma A \text{ and } \forall B\ (A \vdash \sigma B \leftrightarrow B).$$

Projective formulas are very useful in the context of admissible rules because of the following property: for every projective formula A, for all formulas B: $A \mathrel{|\!\sim} B$ iff $A \vdash B$. Ghilardi shows in [2] that every formula A has a *projective approximation* $\Pi(A)$. $\Pi(A)$ is a finite set of projective formulas that derive A and such that for every projective C that derives A there exists $D \in \Pi(A)$ such that $C \vdash D$. It follows from [2] that, like for Λ_A, it holds that $A \mathrel{|\!\sim} B$ iff $\bigvee \Pi(A) \vdash B$. However, for our aim of giving a proof system that derives Λ_A from A, it seems that the notion of admissible projective sets instead of projective approximations is more useful. An *admissible projective set* for a formula A is a set of formulas X such that all formulas in X are either inconsistent or projective and

$$A \mathrel{|\!\sim} \bigvee X \vdash A.$$

A formula A is inconsistent if $A \vdash \bot$. The special behavior of projective formulas w.r.t. admissibility clearly holds for inconsistent formulas too:

For every projective or inconsistent formula A, for all formulas B: $A \mathrel{|\!\sim} B$ iff $A \vdash B$. If X is an admissible projective set for a formula A, then for all formulas B: $A \mathrel{|\!\sim} B$ iff $\bigvee X \vdash B$.

Now the main theorems of the paper, Theorem 1 and 2, state the following:
Theorem 1: $A \vdash_{\mathsf{ADM}} B$ then $A \mathrel{|\!\sim} B$.
Theorem 2: For every formula A there exists a unique formula $\Lambda_A = \bigvee dj(\Lambda_A)$, such that $dj(\Lambda_A)$ is an admissible projective set for A and $A \vdash_{\mathsf{ADM}} \Lambda_A$.

These results have various consequences for other intermediate logics. The most interesting one being that V is a basis for the admissible rules for any intermediate logic for which it is admissible. Due to lack of space we are not able to include these results here. They are discussed in [5]. The system ADM is not yet as elegant and neat, and whence as useful, as one would hope and expect. In particular, instead of replacing previously derived expressions, it just adds expressions to the initial formula. We hope to be able to present a better system in the near future.

2 Preliminaries

In this paper we will only be concerned with intuitionistic propositional logic IPC. We write $\vdash$ for derivability in IPC. The letters A, B, C, D, E, F, H range of formulas, the letters p, q, r, s, t, range over propositional variables. We assume $\top$ and $\bot$ to be present in the language. $\neg A$ is defined as $(A \to \bot)$. We omit parentheses when possible; $\wedge$ binds stronger than $\vee$, which in turn binds stronger than $\to$. A *substitution* σ will in this paper always be a map from propositional formulas to propositional formulas that commutes with the connectives. A *(propositional) admissible rule* of IPC is a rule A/B such that adding the rule to the logic does not change the theorems of IPC, i.e. $\forall\sigma : \vdash \sigma A$ implies $\vdash \sigma B$. We write $A \mathrel{|\!\sim} B$ if A/B is admissible. The rule is called *derivable* if $A \vdash B$ and *proper* if $A \nvdash B$. We say that a collection $\mathcal{R}$ of rules, e.g. V, is admissible if all rules in $\mathcal{R}$ are admissible. We write $A \vdash^{\mathcal{R}} B$ if B is derivable from A in the logic consisting of IPC extended with the rules $\mathcal{R}$, i.e. there are $A = A_1, \dots, A_n = B$ such that for all $i < n$, $A_i \vdash A_{i+1}$ or there exists a σ such that $\sigma B_i/\sigma B_{i+1} = A_i/A_{i+1}$ and $B_i/B_{i+1} \in \mathcal{R}$. A set $\mathcal{R}$ of admissible rules is a *basis for the admissible rules* if for every admissible rule $A \mathrel{|\!\sim} B$ we have $A \vdash^{\mathcal{R}} B$.

3 The System ADM

The aim of the system ADM is to derive a formula Λ_A, which is the disjunction of an admissible projective set for A, by applying some syntactical rules to A. Intuitively it works as follows. First A is rewritten into a disjunction of conjunctions. Under all other rewriting rules that follow it will remain in this form, i.e. in each step the algorithm will only change one of the conjuncts of some disjunct of the formula. We write these conjuncts as sequents, which are interpreted in the usual way as implications. For the outer conjunctions and disjunctions we use different symbols, to distinguish them from the symbols in the sequent. The symbol "$\cdot$" is interpreted as conjunction, the symbol "$|$" as disjunction, where $|$ binds stronger than $\cdot$. Thus e.g. $S_1 \mid S_2 \cdot S_3$ stands for $S_1 \vee (S_2 \wedge S_3)$. The interpretation of $|$ as disjunction is in accordance with its interpretation in hypersequent calculi, see e.g. [1]. Our sequents will be of the form $\Gamma, \Pi \Rightarrow \Delta$, where Γ, Π and Δ are sets of formulas. We denote these sets without the {'s and the comma's. Thus e.g. $A\,(B \wedge C), D \Rightarrow E\ F$ is a sequent in the sense defined above, and means $\{A, B \wedge C\}, \{D\} \Rightarrow \{E, F\}$. Also, if we write $\Gamma\ \{A_i \mid i = 1, 2\}, \Pi \Rightarrow \Delta$ we mean $\Gamma\ A_1\ A_2, \Pi \Rightarrow \Delta$. We let S range over sequents, Θ over expressions of the form $S_1 \cdot S_2 \cdot \ \dots\ \cdot S_n$, and Λ over expressions of the form $\Theta_1 \mid \Theta_2 \mid \ldots \mid \Theta_n$. The Θ are called *conjunction expressions* and the Λ are called *disjunction expressions.* Both Θ and Λ are considered as sets which components are respectively sequents and conjunction expressions. Thus e.g. $S_1 \cdot S_2 \cdot S_1 = S_2 \cdot S_1$. If $\Theta = S_1 \cdot S_2 \cdot \ \ldots \ \cdot S_n$, the S_i are called the *conjuncts* of Θ, and Θ is interpreted as the formula $S_1' \wedge \ldots \wedge S_n'$, where S_i' is the formula corresponding to the sequent S_i. We write $S \in \Theta$ if Θ is of the form $\Theta' \cdot S$, i.e. if S is a conjunct of Θ. For $\Lambda = \Theta_1 \mid \Theta_2 \mid \ldots \mid \Theta_n$, the Θ_i are called the *disjuncts* of Λ, and Λ is interpreted as the formula $(\bigwedge \Theta_1') \vee \ldots \vee (\bigwedge \Theta_n')$, where Θ_i' is the formula corresponding to Θ_i. Sometimes we use sequents and conjunction and disjunction expressions also for their corresponding propositional formula. If we write a sequent but mean a formula, we often leave out the comma between Γ and Π since it plays no role in the context of formulas: $\Gamma\Pi \Rightarrow \Delta$ has the same interpretation as $\Gamma, \Pi \Rightarrow \Delta$. We write $\Gamma\Pi$ for $\Gamma \cup \Pi$. The empty conjunction is taken to be $\top$ and the empty disjunction is $\bot$. For a formula A we define its corresponding disjunction expression (the s for succedent):

$$\Lambda_{sA} \equiv_{def} (\ , \ \Rightarrow A).$$

Note that $\vdash A \leftrightarrow \Lambda_{sA}$. In fact, following our convention, Λ_{sA} is interpreted as the formula $(\top \to A)$ in $\vdash A \leftrightarrow \Lambda_{sA}$.

Definition 1. *We associate the following sets with a given set Γ:*

$$\begin{array}{ll}
\Gamma^a & = \{A \mid \exists B\ (A \to B \in \Gamma)\} \\
\Gamma^p & = \{A \in \Gamma \mid \exists B\ (A = p \to B)\} \\
C^{A\to B} & = \begin{cases} B \to D & \textit{if } C = (A \to B) \to D \\ C & \textit{otherwise} \end{cases} \\
C^{A\wedge} & = \begin{cases} A \to (A' \to B) & \textit{if } C = (A \wedge A' \to B) \\ C & \textit{otherwise} \end{cases} \\
C^{\wedge A} & = \begin{cases} A \to (A' \to B) & \textit{if } C = (A' \wedge A \to B) \\ C & \textit{otherwise} \end{cases} \\
C^{A\vee A'} & = \begin{cases} (A \to B) \wedge (A' \to B) & \textit{if } C = (A \vee A' \to B) \\ C & \textit{otherwise} \end{cases}
\end{array}$$

For a set Γ we define $\Gamma^{A\to B} = \{C^{A\to B} \mid C \in \Gamma\}$. Similarly for the other superscripts.

ADM

$$L\bot: \frac{\Lambda \mid \Theta \cdot \Gamma\bot, \Pi \Rightarrow \Delta}{\Lambda \mid \Theta} \qquad R\bot: \frac{\Lambda \mid \Theta \quad (\Gamma, \Pi \Rightarrow \bot\Delta \in \Theta)}{\Lambda \mid \Theta \cdot \Gamma, \Pi \Rightarrow \Delta}$$

$$L\top: \frac{\Lambda \mid \Theta \quad (\Gamma\top, \Pi \Rightarrow \Delta \in \Theta)}{\Lambda \mid \Theta \cdot \Gamma, \Pi \Rightarrow \Delta} \qquad R\top: \frac{\Lambda \mid \Theta \cdot \Gamma, \Pi \Rightarrow \top\Delta}{\Lambda \mid \Theta}$$

$$L\wedge: \frac{\Lambda \mid \Theta \quad (\Gamma A \wedge B, \Pi \Rightarrow \Delta \in \Theta)}{\Lambda \mid \Theta \cdot \Gamma A B, \Pi \Rightarrow \Delta} \qquad L\vee: \frac{\Lambda \mid \Theta \quad (\Gamma A \vee B, \Pi \Rightarrow \Delta \in \Theta)}{\Lambda \mid \Theta \cdot \Gamma A, \Pi \Rightarrow \Delta \cdot \Gamma B, \Pi \Rightarrow \Delta}$$

$$L\Gamma\rightarrow:\ \frac{\Lambda \mid \Theta \quad (\Gamma A \rightarrow B, \Pi \Rightarrow \Delta \in \Theta)}{\Lambda \mid \Theta \cdot \Gamma A \rightarrow B, \Pi \Rightarrow A\Delta \cdot \Gamma B, \Pi \Rightarrow \Delta}$$

$$L\Pi\rightarrow:\ \frac{\Lambda \mid \Theta \quad (\Gamma, A \rightarrow B \Pi \Rightarrow \Delta \in \Theta)}{\Lambda \mid \Theta \cdot \Gamma, A \rightarrow B \Pi \Rightarrow A\Delta \cdot \Gamma B, \Pi \Rightarrow \Delta}$$

$$R\wedge:\ \frac{\Lambda \mid \Theta \quad (\Gamma, \Pi \Rightarrow A \wedge B \Delta \in \Theta)}{\Lambda \mid \Theta \cdot \Gamma^{A\wedge}, \Pi \Rightarrow A\Delta \cdot \Gamma^{\wedge B}, \Pi \Rightarrow B\Delta}$$

$$R\vee:\ \frac{\Lambda \mid \Theta \quad (\Gamma, \Pi \Rightarrow A \vee B \Delta \in \Theta)}{\Lambda \mid \Theta \cdot \Gamma^{A\vee B}, \Pi \Rightarrow AB\Delta}$$

$$R\rightarrow:\ \frac{\Lambda \mid \Theta \quad (\Gamma, \Pi \Rightarrow A \rightarrow B \Delta \in \Theta)}{\Lambda \mid \Theta \cdot A\Gamma^{A\rightarrow B}, \Pi \Rightarrow B\Delta}$$

$$Res:\ \frac{\Lambda \mid \Theta \quad (\Gamma, \Pi \Rightarrow p\Delta \in \Theta,\ p\Gamma', \Pi' \Rightarrow \Delta' \in \Theta)}{\Lambda \mid \Theta \cdot (\Gamma \backslash \Gamma^p)\Gamma', \Gamma^p \Pi\Pi' \Rightarrow \Delta\Delta'}$$

$$Vg:\ \frac{\Lambda \mid \Theta \quad (\Gamma, \Pi \Rightarrow \Delta \in \Theta,\ n > 0,\ \Gamma \text{ only implications},\ \Gamma^a \subseteq \Delta = A_0..A_n)}{\Lambda \mid \Theta \cdot \Gamma\Pi, \Rightarrow A_0 \mid \ldots \mid \Theta \cdot \Gamma\Pi, \Rightarrow A_n}$$

Res stands for resolution. The rule is called after an analogous rule in Ghilardi's algorithm CHECK-PROJECTIVITY. The expressions Λ are called *side expressions.* The conjunction expressions in the inferences that do not occur in Λ are the *principal conjunction expressions* of the inference. The sequents that are explicitly indicated in the rules above, are called *principal sequents* of the inference. Note that rules have in general more than one principal formula, sequent and conjunction expression. If Λ/Λ' is an instance of a rule R, then we write $\Lambda \vdash^R_{\mathsf{ADM}} \Lambda'$ or $\Lambda \vdash^1_{\mathsf{ADM}} \Lambda'$. If there are $\Lambda_1 \ldots \Lambda_n$ such that $\Lambda = \Lambda_1 \vdash^1_{\mathsf{ADM}} \cdots \vdash^1_{\mathsf{ADM}} \Lambda_n = \Lambda'$, we write $\Lambda \vdash_{\mathsf{ADM}} \Lambda'$. The depth of a disjunction expression Λ in a derivation $\mathcal{D} = \Lambda_1, \ldots, \Lambda_n$ is the minimal i such that $\Lambda = \Lambda_i$. The depth of a conjunction expression Θ in $\mathcal{D}$ is the minimal i such that Θ is a conjunct of Λ_i.

We will be only interested in derivations $\Lambda_{sA} \vdash_{\mathsf{ADM}} \Lambda$, where the first expression is of the special form Λ_{sA} for some formula A. The Λ that occur in these derivations have some special properties, as we will see in Proposition 1. We leave it to the reader to check that the following holds. If $\Lambda_{sA} \vdash_{\mathsf{ADM}} \Lambda$, then for all sequents $\Gamma, \Pi \Rightarrow \Delta$ that occur in Λ, the set Π consists only of implications.

Definition 2. *A set Γ is called* simple *if all its elements are implications or propositional variables. A sequent $\Gamma, \Pi \Rightarrow \Delta$ is called simple if $\Gamma\Pi\Delta$ is simple. A derivation in* ADM *is in* normal form *if the principal sequents in any Res or Vg inference are simple. If $\mathcal{D}$ is the smallest normal form derivation $\Lambda_{sA} \vdash_{\mathsf{ADM}} \Lambda$ such that for all $\Lambda \vdash^1_{\mathsf{ADM}} \Lambda'$ we have $\Lambda' = \Lambda$, then we denote Λ by Λ_A. Observe that Λ_A is unique. We denote the set of all disjuncts of a disjunction expression Λ by $dj(\Lambda)$. Thus $\vdash \Lambda \leftrightarrow \bigvee dj(\Lambda)$. We define*

$$p(\Lambda) = \{\Theta \mid \Theta \in dj(\Lambda),\ \Theta \textit{ has no conjunct } (\Gamma, \Pi \Rightarrow \bot) \textit{ for which } \Gamma^a = \{\bot\}\}.$$

3.1 Properties of ADM

All rules of ADM except $R\wedge$, $R\vee$ and $R\rightarrow$ have a subformula property: the formulas that occur in the sequents of the lower expression are subformulas of the formulas in the sequents in the upper expression. The reason that e.g. $R\vee$ is not formulated in the naive way $\Gamma, \Pi \Rightarrow A \vee B\Delta / \Gamma, \Pi \Rightarrow A\,B\,\Delta$ (we have left out the side expressions), in which case it would have the subformula property, is best explained by an example. Consider a formula $A = (B \rightarrow p \vee q)$ where $B = (p \vee q \rightarrow r)$. Formulating $R\vee$ in the naive way would give the following derivation.

$$
Vg\,\dfrac{R\vee\,\dfrac{R\rightarrow\,\dfrac{, \Rightarrow A}{B, \Rightarrow p \vee q}}{B, \Rightarrow pq}}{B, \Rightarrow p \vee q \mid B, \Rightarrow p \mid B, \Rightarrow q}
$$

It is not difficult to see that this implies that Λ_A will contain a disjunct that is equivalent to $A \wedge (r \rightarrow p \vee q)$. The semantical characterization of projective formulas given by Ghilardi [2], and recalled in the appendix, shows that neither is this formula projective, nor is it inconsistent. Therefore, the naive formulation of $R\vee$ does not suffice for our purposes. In our case the above derivation will become:

$$
L\Gamma\rightarrow\,\dfrac{Vg\,\dfrac{R\vee\,\dfrac{R\rightarrow\,\dfrac{, \Rightarrow A}{B, \Rightarrow p \vee q}}{(p \rightarrow r)(q \rightarrow r), \Rightarrow pq}}{B, \Rightarrow p \mid B, \Rightarrow q}}{B, \Rightarrow p \;\cdot\; r \Rightarrow p \mid B, \Rightarrow q \cdot r \Rightarrow q}
$$

This shows that the non-projective disjunct will not appear in this way. Similar reasons explain the form of $R\wedge$ and $R\rightarrow$.

Although there is no subformula property, the situation is not that bad. All formulas in the sequents that appear in a derivation from Λ_{sA} belong to the smallest set $\mathcal{A}$ that contains the subformulas of A and that is closed under the operations $()^{B\rightarrow C}$, $()^{B\vee C}$, $()^{B\wedge}$, $()^{\wedge B}$. Clearly, this set is finite. Note that the finiteness of $\mathcal{A}$ implies that Λ_A exists.

The rule Vg is different from the other rules in that it adds a sequent S to a conjunct Θ that is in general not derivable from Θ. However, as we will see in Theorem 1, the upper expression of the Vg rule admissibly derives the lower expression, at least in derivations that start with a sequent of the form Λ_{sA}. If in the application of the Vg rule to a sequent $\Gamma, \Pi \Rightarrow \Delta$, the set Π is empty, this follows already from the admissibility of the rule V mentioned in the introduction. If Π is non-empty, Vg is a generalization of the rule V. In the remainder of this section we show that the admissibility of V implies the admissibility of Vg. This is the main ingredient of Theorem 1 in which we show that $A \mathrel{|\!\sim} \Lambda_A \vdash A$. We need this generalization of V to make the proof of Theorem 2 in which we show that $dj(\Lambda_A)$ is an admissible projective set, work.

Remark 1. We will need the following consequences of the admissibility of V, the verification of which we leave to the reader. If Γ and Σ are two sets of implications, it holds, if at least one of the conditions stated below is satisfied, that

$$\Gamma\Sigma \Rightarrow \Delta \mathrel{|\!\sim} \bigvee_{A\in\Delta\cup\Sigma^a} (\Gamma\Sigma \Rightarrow A).$$

The conditions are as follows. (1) If $\Gamma^a \subseteq \Delta$. (2) If for all $A \in \Gamma^a$ there exists a $B \in \Delta$ such that $A \vdash B$. (3) If $\Gamma^a = \bigvee \Delta$. For in this case $\bigwedge \Gamma$ is equivalent (over IPC, and whence over L) to $\bigwedge\{A \to B \mid \bigvee \Delta \to B \in \Gamma, A \in \Delta\}$.

Definition 3. *For a set of formulas I we denote $\{A \to B \in \Pi \mid A \in I\}$ by Π^I. For a sequent $(\Gamma, \Pi \Rightarrow C)$ we define*

$$(\Gamma, \Pi \Rightarrow \Delta)^I = \Gamma\{A \in \Pi^a \mid A \in I\}, \Pi\backslash\Pi^I \Rightarrow \Delta.$$

*For a sequent S we say that another sequent $(\Gamma, \Pi \Rightarrow \Delta)$ is S-*correct *if $\forall I \subseteq \Pi^a : \; S \vdash (\Gamma, \Pi \Rightarrow \Delta)^I$. For a set of implications Π we define*

$$\Pi_\Delta = \{\bigvee \Delta \to D \mid \exists C(C \to D \in \Pi)\},$$

i.e. Π_Δ is the result of replacing all antecedents of the implications in Π by $\bigvee \Delta$. $\Pi^I_\Delta = (\Pi^I)_\Delta$.

Remark 2. If $\Gamma, \Pi \Rightarrow \Delta$ is S-correct then $S \vdash \Gamma, \Pi \Rightarrow \Delta$. If Π is empty, then $\Gamma, \Pi \Rightarrow \Delta$ is S-correct iff S derives $\Gamma, \Pi \Rightarrow \Delta$.

The reader is advised to first read the proof of Proposition 1 before embarking on the following two lemma's, that may otherwise seem somewhat mysterious.

Lemma 1. *If the sequent $\Gamma, \Pi \Rightarrow \Delta$ is S-correct then for all $I \subseteq \Pi^a$ the sequent $\Gamma\Pi^I_\Delta, \Pi\backslash\Pi^I \Rightarrow \Delta$ is S-correct too.*

Proof. We use induction on $|I|$. The case $|I| = 0$ follows from the S-correctness of $\Gamma, \Pi \Rightarrow \Delta$. Suppose $|I| > 0$. Let $\Sigma = \Pi\backslash\Pi^I$. We have to show that for all $J \subseteq \Sigma^a$

$$S \vdash \Gamma\Pi^I_\Delta\{A \in \Sigma^a \mid A \in J\}, \Sigma\backslash\Sigma^J \Rightarrow \Delta. \tag{1}$$

Pick a $C \in I$ and let $I' = I\backslash\{C\}$ and $\Sigma' = \Pi\backslash\Pi^{I'}$. Note that $C \notin \Sigma^a$, whence $C \notin J$ and $\Sigma^a \cup \{C\} = (\Sigma')^a$. The induction hypothesis says that $\Gamma\Pi^{I'}_\Delta, \Pi\backslash\Pi^{I'} \Rightarrow \Delta$ is S-correct. This implies that we have for all $H \subseteq (\Sigma')^a$

$$S \vdash \Gamma\Pi^{I'}_\Delta\{A \in (\Sigma')^a \mid A \in H\}, \Sigma'\backslash(\Sigma')^H \Rightarrow \Delta. \tag{2}$$

Thus we have $S \vdash \Gamma\Pi^{I'}_\Delta\{A \in (\Sigma')^a \mid A \in J\}, \Sigma'\backslash(\Sigma')^J \Rightarrow \Delta$ by taking $H = J$. We show that

$$S \vdash \Gamma\Pi^I_\Delta\{A \in \Sigma^a \mid A \in J\}, \Sigma\backslash\Sigma^J \Rightarrow \bigwedge \Gamma\Pi^{I'}_\Delta\{A \in (\Sigma')^a \mid A \in J\}, \Sigma'\backslash(\Sigma')^J.$$

Then (1) will follow from (2). Clearly, $S \vdash \Gamma\Pi^I_\Delta \Rightarrow \bigwedge \Pi^{I'}_\Delta$. Since $\Sigma^a \cap J = (\Sigma')^a \cap J$ also $S \vdash \{A \in \Sigma^a \mid A \in J\} \Rightarrow \bigwedge\{A \in (\Sigma')^a \mid A \in J\}$. It remains to show that $S \vdash \Gamma\Pi^I_\Delta\{A \in \Sigma^a \mid A \in J\}, \Sigma\backslash\Sigma^J \Rightarrow \bigwedge \Sigma'\backslash(\Sigma')^J$. Observe that

$\Sigma'\backslash(\Sigma')^J = \Sigma\backslash\Sigma^J \cup \{C \to D \mid C \to D \in \Pi\}$ since $C \notin J$. Therefore, it suffices to show that for all $(C \to D) \in \Pi$ we have

$$\Gamma\Pi^I_\Delta\{A \in \Sigma^a \mid A \in J\}, \Sigma\backslash\Sigma^J \Rightarrow C \to D.$$

The following derivation shows that this holds.

$$\begin{aligned} S \vdash\ & \Gamma\Pi^{I'}_\Delta C\{A \in (\Sigma')^a \mid A \in J\}, \Sigma'\backslash(\Sigma')^{J\cup\{C\}} \Rightarrow \Delta \\ & \Gamma\Pi^{I'}_\Delta\{A \in (\Sigma')^a \mid A \in J\}, \Sigma'\backslash(\Sigma')^{J\cup\{C\}} \Rightarrow C \to \bigvee\Delta \\ & \Gamma\Pi^{I'}_\Delta(\bigvee\Delta \to D)\{A \in (\Sigma')^a \mid A \in J\}, \Sigma'\backslash(\Sigma')^{J\cup\{C\}} \Rightarrow C \to D \\ & \Gamma\Pi^{I}_\Delta\{A \in \Sigma^a \mid A \in J\}, \Sigma'\backslash(\Sigma')^{J\cup\{C\}} \Rightarrow C \to D \\ & \Gamma\Pi^{I}_\Delta\{A \in \Sigma^a \mid A \in J\}, \Sigma\backslash\Sigma^J \Rightarrow C \to D \end{aligned}$$

The first step is (2) with $H = J \cup \{C\}$. The step from the 3th to 4th line holds, as $\Sigma^a \cap J = (\Sigma')^a \cap J$. The step from the 4th to 5th line holds, as $\Sigma\backslash\Sigma^J = \Sigma'\backslash(\Sigma')^{J\cup\{C\}}$. This proves the lemma.

Lemma 2. *If Π consist of implications only and the sequent $\Gamma, \Pi \Rightarrow \Delta$ is S-correct, then for all $I \subseteq \Pi^a$:*

$$S \wedge \bigwedge\{\Gamma\Pi \Rightarrow A \mid A \in I\} \mathrel{|\!\sim} \bigvee\{\Gamma\Pi \Rightarrow A \mid A \in \Delta \cup \Gamma^a \cup (\Pi^a\backslash I)\}.$$

Proof. Consider $I \subseteq \Pi^a$. Let S' denote $S \wedge \bigwedge\{\Gamma\Pi \Rightarrow A \mid A \in I\}$. Let Σ denote $\Gamma\Pi^I_\Delta, \Pi\backslash\Pi^I$ and consider the sequent $T = (\Sigma \Rightarrow \Delta)$. By Lemma 1, T is S-correct, whence it is S'-correct. In particular, $S' \vdash T$. By Remark 1 $S' \mathrel{|\!\sim} \bigvee\{\Sigma \Rightarrow A \mid A \in \Delta \cup \Gamma^a \cup \Pi^a\backslash I\}$. Whence we are done once we can show that $S' \vdash \Gamma\Pi \Rightarrow A$ for all $A \in \Sigma$. Clearly, $S' \vdash \Gamma\Pi \Rightarrow A$ for all $A \in I$. Therefore, for all $(A \to B) \in \Pi$ for which $A \in I$, $S' \vdash \Gamma\Pi \Rightarrow B$, and whence $S' \vdash \Gamma\Pi \Rightarrow \bigvee\Delta \to B$. Therefore, $S' \vdash \Gamma\Pi \Rightarrow A$ for all $A \in \Sigma$. This proves the lemma.

Proposition 1. *If Γ and Π are sets of implications such that the sequent Γ, $\Pi \Rightarrow \Delta$ is S-correct and $\Gamma^a \subseteq \Delta$, then $S \mathrel{|\!\sim} \bigvee\{\Gamma\Pi \Rightarrow D \mid D \in \Delta\}$.*

Proof. The lemma follows from the following derivation.

$$\begin{aligned} S \mathrel{|\!\sim}\ & \textstyle\bigvee_{D\in\Delta}(\Gamma\Pi \Rightarrow D) \vee \bigvee_{A\in\Pi^a}(\Gamma\Pi \Rightarrow A) && (\mathit{Remark}\ 1\ \mathit{and}\ 2) \\ & \textstyle\bigvee_{D\in\Delta}(\Gamma\Pi \Rightarrow D) \vee \bigvee_{A\in\Pi^a}(\Gamma\Pi \Rightarrow A) \wedge \\ & \quad \textstyle\wedge(\bigvee_{D\in\Delta}\Gamma\Pi \Rightarrow D \vee \bigvee_{B\in\Pi^a, B\neq A}(\Gamma\Pi \Rightarrow B)) && (\mathit{Lemma}\ 2) \\ & \textstyle\bigvee_{D\in\Delta}(\Gamma\Pi \Rightarrow D) \vee \bigvee_{A,B\in\Pi^a, A\neq B}(\Gamma\Pi \Rightarrow A) \wedge (\Gamma\Pi \Rightarrow B) \\ & \vdots \\ & \textstyle\bigvee_{D\in\Delta}(\Gamma\Pi \Rightarrow D) \vee \bigwedge_{A\in\Pi^a}(\Gamma\Pi \Rightarrow A) \\ & \textstyle\bigvee_{D\in\Delta}(\Gamma\Pi \Rightarrow D) && (\mathit{Lemma}\ 2) \end{aligned}$$

Lemma 3. *If Π consist of implications only, $S' = (\Gamma, \Pi \Rightarrow \bot)$ is S-correct, and $\Gamma^a = \{\bot\}$, then $S \vdash \bot$.*

Proof. Let $n = |\Pi^a|$. With induction to $n - |I|$ we show that for all $I \subseteq \Pi^a$

$$S \vdash \bigwedge\{A \in \Pi^a \mid A \in I\} \to \bot. \tag{3}$$

Let us first see that this implies the lemma. For all $B \in \Pi^a$, (3) gives $S \vdash \neg B$, by taking $I = \{B\}$. This implies $S \vdash \bigwedge \Pi$. Also, $S \vdash \bigwedge \Gamma$ because $\Gamma^a = \{\bot\}$. The S-correctness of S' gives $S \vdash (\bigwedge \Gamma \Pi \Rightarrow \bot)$, by taking $I = \emptyset$. Whence $S \vdash \bot$.

It remains to prove (3). The case $|I| = n$ follows from the S-correctness of $(\Gamma, \Pi \Rightarrow \bot)$ and the fact that $S \vdash \bigwedge \Gamma$. The case that $|I| < n$. By the induction hypothesis we have for all $B \in \Pi^a \backslash I$ that $S \vdash B \wedge \bigwedge\{A \in \Pi^a \mid A \in I\} \to \bot$. Whence

$$\forall B \in \Pi^a \backslash I : S \vdash \bigwedge\{A \in \Pi^a \mid A \in I\} \to \neg B. \tag{4}$$

This implies that $S \vdash \bigwedge\{A \in \Pi^a \mid A \in I\} \to \bigwedge(\Pi \backslash \Pi^I)$. By S-correctness we have

$$S \vdash \bigwedge \Gamma \wedge \bigwedge\{A \in \Pi^a \mid A \in I\}, \bigwedge(\Pi \backslash \Pi^I) \to \bot.$$

Therefore, $S \vdash \bigwedge\{A \in \Pi^a \mid A \in I\} \to \bot$.

Lemma 4. *If $\Lambda_{sA} \vdash_{\mathsf{ADM}} \Lambda$ and $\Lambda = \Theta_1 \mid \ldots \mid \Theta_n$, then for all $i \leq n$, all $S \in \Theta_i$ are Θ_i-correct.*

Proof. We use induction to the depth d of Λ in the derivation $\mathcal{D}$. The case $d = 0$ follows from Remark 2. Assume $d > 0$. We distinguish by cases according to the last rule of $\mathcal{D}$. Observe that if $\Gamma, \Pi \Rightarrow \Delta$ is S-correct and if $S \vdash \Gamma' \Rightarrow \Gamma$ and $S \vdash \Gamma\Gamma'\Delta \Rightarrow \Delta'$, then $\Gamma', \Pi \Rightarrow \Delta'$ is S-correct. This observation suffices for all cases except $L\Pi \to$, *Vg* and *Res*. We leave $L\Pi \to$ to the reader. For the case *Vg*, apply Remark 2. We treat the case *Res*. In this case the one but last disjunction expression in $\mathcal{D}$ is $\Lambda' = \Lambda'' \mid \Theta'' \cdot \Gamma, \Pi \Rightarrow p\Delta \cdot p\Gamma', \Pi' \Rightarrow \Delta'$ and

$$\Lambda = \Lambda'' \mid \Theta'' \cdot \Gamma, \Pi \Rightarrow p\Delta \cdot p\Gamma', \Pi' \Rightarrow \Delta' \cdot (\Gamma \backslash \Gamma^p)\Gamma', \Gamma^p \Pi \Pi' \Rightarrow \Delta\Delta'.$$

Let $\Theta' = \Theta'' \cdot \Gamma, \Pi \Rightarrow p\Delta \cdot p\Gamma', \Pi' \Rightarrow \Delta'$ and

$$\Theta = \Theta' \cdot (\Gamma \backslash \Gamma^p)\Gamma', \Gamma^p \Pi \Pi' \Rightarrow \Delta\Delta'.$$

For the sequents in Λ'', Θ'' and the sequents $S_1 = (\Gamma, \Pi \Rightarrow p\Delta)$ and $S_2 = (p\Gamma', \Pi' \Rightarrow \Delta')$ the induction hypothesis applies. Whence they are Θ'-correct, and thus also Θ-correct. We consider the remaining sequent of Λ. Let $\Sigma = \Gamma^p \Pi \Pi'$. We have to show that for all $I \subseteq \Sigma^a$,

$$\Theta \vdash (\Gamma \backslash \Gamma^p)\Gamma'\{A \in \Sigma^a \mid A \in I\}, \Sigma \backslash \Sigma^I \Rightarrow \Delta\Delta'. \tag{5}$$

Let $I_1 = I \cap \Pi^a$ and $I_2 = I \cap (\Pi')^a$. Observe that $\Pi \backslash \Pi^{I_1}$ and $\Pi' \backslash (\Pi')^{I_2}$ are subsets of $\Sigma \backslash \Sigma^I$. The fact that S_2 is Θ-correct implies that we have $\Theta \vdash p\Gamma'\{A \in (\Pi')^a \mid A \in I_2\}, \Pi' \backslash (\Pi')^{I_2} \Rightarrow \Delta'$. Whence

$$\Theta \vdash p(\Gamma \backslash \Gamma^p)\Gamma'\{A \in \Sigma^a \mid A \in I\}, \Sigma \backslash \Sigma^I \Rightarrow \Delta\Delta'. \tag{6}$$

If $p \in I$, then $p \in \{A \in \Sigma^a \mid A \in I\}$. Hence (5) follows from (6). If $p \notin I$ we use the fact that S_1 is Θ-correct, i.e. $\Theta \vdash \Gamma\{A \in \Pi^a \mid A \in I_1\}, \Pi\backslash\Pi^{I_1} \Rightarrow p\,\Delta$. Since $p \notin I$, $\Gamma^p \subseteq \Sigma\backslash\Sigma^I$. Therefore it follows that

$$\Theta \vdash (\Gamma\backslash\Gamma^p)\Gamma'\{A \in \Sigma^a \mid A \in I\}, \Sigma\backslash\Sigma^I \Rightarrow p\,\Delta. \tag{7}$$

Combining (7) with (6) gives (5).

Note that the previous lemma does not hold without the assumption that $\Lambda_{sA} \vdash_{\mathsf{ADM}} \Lambda$. The special form of the first expression plays a role in the induction.

Theorem 1. *If $\Lambda_{sA} \vdash_{\mathsf{ADM}} \Lambda \vdash_{\mathsf{ADM}} \Lambda'$ then $\Lambda \mathrel{|\!\sim} \Lambda' \vdash \Lambda$. In particular, $A \mathrel{|\!\sim} \Lambda_A \vdash A$. In fact, for all rules R except Vg, even $\Lambda \vdash^R_{\mathsf{ADM}} \Lambda'$ implies $\Lambda \vdash \Lambda' \vdash \Lambda$.*

Proof. We use induction to the depth d of the derivation. The proof for $\Lambda' \vdash \Lambda$ is trivial. For $\Lambda \mathrel{|\!\sim} \Lambda'$, in the case $d = 0$ we have $\Lambda = \Lambda'$. For $d > 0$ we distinguish by cases according to the last rule that is applied. We only treat the case Vg. Suppose Λ derived from the application of the rule Vg to Λ. Suppose $\Lambda = \Lambda'' \mid \Theta \cdot S$, where $S = (\Gamma, \Pi \Rightarrow \Delta)$ with $\Delta = A_0 \ldots A_n$ and Γ a set of implications such that $\Gamma^a \subseteq \Delta$. As observed before, the fact that $\Lambda_{sA} \vdash_{\mathsf{ADM}} \Lambda$ implies that Π consists of implications only. For $1 \leq i \leq n$, let $S_i = (\Gamma\Pi, \Rightarrow A_i)$. Whence

$$\Lambda' = \Lambda'' \mid \Theta \cdot S_0 \mid \ldots \mid \Theta \cdot S_n.$$

Therefore, to prove the lemma it suffices to show that $\Theta \wedge S \mathrel{|\!\sim} S_0 \vee \ldots \vee S_n$. This follows from Lemma 4 in combination with Proposition 1.

Note that in the previous theorem, the fact that Λ and Λ' appear in derivations from Λ_{sA} is crucial: in general, $\Lambda \vdash^{Vg}_{\mathsf{ADM}} \Lambda'$ does not imply $\Lambda \mathrel{|\!\sim} \Lambda'$. The previous theorem together with the second main theorem, Theorem 2 proved in the next section, show that ADM has the properties we want it to have, as discussed in the introduction.

3.2 Properties of Λ_A

In this section we discuss only IPC. Whence $\vdash$ and $\mathrel{|\!\sim}$ stand again for $\vdash_{\mathsf{IPC}}$ and $\mathrel{|\!\sim}_{\mathsf{IPC}}$. Recall from Theorem 1 that $A \mathrel{|\!\sim} \Lambda_A \vdash A$. The last fact implies that for all disjuncts Θ of Λ_A we have $\Theta \vdash A$.

Lemma 5. *For every formula A, for every $\Theta \in dj(\Lambda_A)\backslash p(\Lambda_A)$, it holds that $\vdash \neg\Theta$.*

Proof. Let Θ be a disjunct of Λ_A that is not in $p(\Lambda_A)$. Since $\Theta \notin p(\Lambda_A)$ it follows that Θ contains a conjunct $(\Gamma, \Pi \Rightarrow \bot)$ for which $\Gamma^a = \{\bot\}$. By Lemma 4 $(\Gamma, \Pi \Rightarrow \bot)$ is Θ-correct. Thus by Lemma 3 we have $\Theta \vdash \bot$.

Lemma 6. *For any formula A, every formula in $p(\Lambda_A)$ is projective.*

Proof. Let Θ be one of the disjuncts of Λ_A. We show that if $\Theta \in p(\Lambda_A)$, then Θ is projective. We will use Ghilardi's algorithm, described in the appendix, to show this. As Ghilardi remarks in [3], the order in which the steps of the algorithm are applied is irrelevant. We will apply them in the following order.

$S1$ Apply $T\top$, $T\bot$, Tp, $T\wedge$, $T\vee$, $T\!\!\to$, $T\neg$ and the two Simplification Rules.
$S2$ Apply $F\top$, $F\bot$, Fp, $F\wedge$, $F\vee$, $F\!\!\to$ and $F\neg$.
$S3$ Apply the Resolution Rule.

We apply these steps $S1-S3$ as long as possible, and we will not apply $S3$ ($S2$) as long as we can apply $S2$ ($S1$). When computing the algorithm on a formula A, we will initialize with $\{FA\}$. The input formulas of the algorithm are conjunctions of implications. Thus in this case it would e.g. be $(\top \to A)$. According to the Initialization we should then start with $\{T\top, FA\}$. After one application of $T\top$ we have $\{FA\}$ instead. Thus we see that starting with $\{FA\}$ gives the same result as starting with the Initialization according to the algorithm. We do this only to obtain a smooth induction. Namely, we will show in Claim 1 that when applying the algorithm in this order to Θ, for every set O that is created there exists a conjunct $S_O = (\Gamma, \Pi \Rightarrow \Delta)$ of Θ with certain properties. Then we show, Claim 2 and Claim 3, that this shows that either all output sets of Θ contain atomic modalities or $\Theta \notin p(\Lambda_A)$. By Theorem 4.2 of [3] (cited here as Theorem 4) all formulas for which all output sets contain atomic modalities are projective. This then proves the theorem. We denote $\{A \in \Gamma \mid TA \notin O\}$ by $\Gamma\backslash TO$. For $X, Y, Z \in \{T, T^+, T_c\}$ or $X, Y, Z \in \{F, F^+, F_c\}$, XYO denotes $XO \cup YO$ and $XYZO$ denotes $XO \cup YO \cup ZO$.

Claim. For every O that is created and not removed there exists a conjunct $S_O = (\Gamma, \Pi \Rightarrow \Delta)$ of Θ that satisfies the following properties:

(*a*) $TT^+O \subseteq \Gamma\Pi$
(*b*) $\Delta = FF^-F_cO$
(*c*) $\forall B \in (\Gamma\Pi\backslash TO)$ (B is an implication and $\Theta \vdash \bigwedge T_cO \to B$)
(*d*) $\forall (B \to C) \in (\Gamma\backslash TO)(B \in \Delta \cup \{\bot\})$.

Proof. With induction to the number of sets that were created before O.

The Initialization is clear. We consider the induction step. We distinguish by cases according to the step Si in which O is created. Assume O is the result of an application of a rule to O' or, in the case of *Res* to (O', O''). The induction hypothesis for O' implies that there is a conjunct $S_{O'}$ of Θ that satisfies $(a)-(d)$ w.r.t. O'. Clearly, the Simplification Rules do not have to be considered. Observe that we can also omit the rules $T\neg$ and $F\neg$. As they can be seen as consecutive applications of $T\to$ and $T\bot$ or $F\to$ and $F\bot$ respectively.

$S1$. O is the result of an application of one of the rules $T\top$, $T\bot$, Tp, $T\wedge$, $T\vee$ and $T\!\!\to$ to some formula in O'. In the cases $T\top$ and Tp we take $S_O = S_{O'}$. In the case $T\bot$, O' is removed, so O does not exist. We treat the cases $T\wedge$ and $T\!\!\to$ and leave $T\vee$ to the reader. In the first case $T\wedge$ is applied to some formula $T(B \wedge C) \in O'$. By (a), $S_{O'} = (\Gamma B \wedge C, \Pi \Rightarrow \Delta)$ for some Γ, Π, Δ. We let $S_O = (\Gamma BC, \Pi \Rightarrow \Delta)$. This sequent is a conjunct of Θ by the rule $L\wedge$. We leave it to the reader to check that S_O satisfies $(a)-(d)$. For the case $T\!\rightarrow$, O is the result of an application of $T\!\rightarrow$ to some formula $T(B\rightarrow C)$ in O'. By (a), $S_{O'}=(\Gamma\, B\rightarrow C,\Pi\Rightarrow\Delta)$ or $S_{O'}=(\Gamma, B\rightarrow C\,\Pi\Rightarrow\Delta)$. Assume we are in the first situation. The second one is analogous. We have

$$O=O'\backslash\{T(B\rightarrow C)\}\cup\{T_c(B\rightarrow C),FB\} \text{ or } O=O'\backslash\{T(B\rightarrow C)\}\cup\{TC\}.$$

In the first case, take $S_O=(\Gamma\, B\rightarrow C,\Pi\Rightarrow B\,\Delta)$. In the second case take $S_O=(\Gamma\, C,\Pi\Rightarrow\Delta)$. The rule $L\Gamma\rightarrow$ in ADM guarantees that S_O is a conjunct of Θ. We leave it to the reader to check that in both cases S_O satisfies $(a)-(d)$.

$S2$. O is the result of an application of $F\top$, $F\bot$, Fp, $F\wedge$, $F\vee$ or $F\rightarrow$ to some formula in O'. We treat the cases $F\wedge$, $F\rightarrow$. The case $F\vee$ is similar to the case $F\wedge$. First, the case $F\wedge$. Then $F\wedge$ is applied to some $F(B\wedge C)\in O'$. By (b) $S_{O'}=(\Gamma,\Pi\Rightarrow B\wedge C\,\Delta)$. Thus

$$O=O'\backslash\{FB\wedge C\}\cup\{FB\} \text{ or } O=O'\backslash\{FB\wedge C\}\cup\{FC\}.$$

For reasons of symmetry we only have to consider one case, suppose the first one. The rule $R\wedge$ in ADM implies that Θ has a conjunct $\Gamma^{B\wedge},\Pi\Rightarrow B\,\Delta$. Let S_O be this conjunct. To see that (a) holds, note that since $S1$ cannot be applied, there are no formulas TC in O' and whence in O. It is easy to see that (b) holds. For (c), observe that formulas in Γ, if replaced in going from $S_{O'}$ to S_O, are only replaced by equivalent formulas. Whence the induction hypothesis applies. For (d), note that formulas $(B\wedge C\rightarrow D)\in\Gamma$ are replaced by $(B\rightarrow(C\rightarrow D))\in\Gamma^{B\wedge}$.

Second, the case $F\rightarrow$. There is some $F(B\rightarrow C)\in O$ such that

$$O=O'\backslash\{FB\rightarrow C\}\cup\{TB,FC\} \text{ or } O=O'\backslash\{FB\rightarrow C\}\cup\{F_c(B\rightarrow C)\}.$$

By (b), $S_{O'}=(\Gamma,\Pi\Rightarrow B\rightarrow C\,\Delta)$. In the first case take $S_O=(B\,\Gamma^{B\rightarrow C},\Pi\Rightarrow C\,\Delta)$. This is a conjunct of Θ by $R\rightarrow$. To see that (a) holds: since $S1$ cannot be applied to O', there are no formulas of the form TD in O except TB. (b) is easy. The only problematic case for (c) might be formulas $(B\rightarrow C)\rightarrow D\in\Gamma$. However, note that the fact that $\Theta\vdash T_cO'\rightarrow((B\rightarrow C)\rightarrow D)$ implies $\Theta\vdash T_cO\rightarrow(C\rightarrow D)$. For (d), observe that formulas $(B\rightarrow C)\rightarrow D\in\Gamma$ are replaced by $(C\rightarrow D)$ in $\Gamma^{B\rightarrow C}$. In the second case take $S_O=S_{O'}$. Here we leave the verification of $(a)-(d)$ to the reader.

$S3$. O is the result of applying the Resolution Rule to (O',O''). By (a) and (b) we may assume that $S_{O'}$ and $S_{O''}$ have the following form.

$$S_{O'}=(\Gamma,\Pi\Rightarrow p\,\Delta) \text{ and } S_{O''}=(p\,\Gamma',\Pi'\Rightarrow\Delta').$$

Note that since $S1$ and $S2$ cannot be applied, both sequents are simple. We take $S_O=(\Gamma\backslash\Gamma^p)\Gamma',\Gamma^p\Pi\Pi'\Rightarrow\Delta\Delta'$. Note that this sequent is a conjunct of Θ by the rule *Res* and the fact that $S_{O'}$ and $S_{O''}$ are simple (recall that we assume the proof to be in normal form). We leave the verification of (a), (b) and (c) to the reader. For (d), observe that the only possibly problematic case would be formulas $(p\rightarrow A)\in\Gamma$. However, for these formulas the property (d) does not have to hold anymore, because of their place in S_O. This finishes the proof of the claim.

Claim. If an algorithm set O contains only context modalities, and F_cO is non-empty, then O is removed by the Simplification Rule.

Proof. Let $S_O = (\Gamma, \Pi \Rightarrow \Delta)$. By (*c*) this implies that $\Gamma\Pi$ consists of implications only. By (*b*) and the fact that O contains only context modalities, Δ consists of implications and propositional variables only. In particular, S_O is simple. Note that (*b*) implies that Δ is non-empty. By (*d*), $\Gamma^a \subseteq \Delta$. Whence the rule Vg implies that there exists an $A \in \Delta$ such that $(\Gamma\Pi, \Rightarrow A)$ is a conjunct of Θ. By (*c*), $\Theta \vdash T_cO \to \bigwedge \Gamma\Pi$. Whence $\Theta \vdash T_cO \to A$. Since $A \in F_cO$, this shows that O will be removed. This proves the claim.

Claim. If an algorithm set O contains only context modalities, and F_cO is empty, then Θ contains a conjunct $(\Gamma, \Pi \Rightarrow\ )$ with $\Gamma^a = \{\bot\}$.

Proof. Let $S_O = (\Gamma, \Pi \Rightarrow \Delta)$. By (*d*), $\Gamma^a = \{\bot\}$. By (*b*), Δ is empty. This proves the claim.

Observe that if Θ contains a conjunct $(\Gamma, \Pi \Rightarrow\ )$, then it also contains a conjunct $(\Gamma, \Pi \Rightarrow \bot)$. Since $\Theta \in p(\Lambda_A)$, it does not contain such conjuncts. Therefore, Claim 3 implies that if an algorithm set of Θ contains only context modalities, then F_cO is non-empty. Claim 2 implies that this set will be removed. This proves that all output sets of Θ contain atomic modalities.

Theorem 2. *For every formula A, the set $dj(\Lambda_A)$ is an admissible projective set for A.*

Proof. By the previous lemma and theorem, all formulas in $dj(\Lambda_A)$ are either projective or inconsistent. By Theorem 1 we have $A \mathrel{|\!\sim} \bigvee dj(\Lambda_A) \vdash A$.

References

1. Avron, A.: The method of hypersequents in the proof theory of propositional non-classical logics. Logic: from Foundations to Applications, European Logic Colloquium (1996) 1–32
2. Ghilardi, S.: Unification in intuitionistic logic. J. of Symb. Logic **64** (1999) 859–880
3. Ghilardi, S.: A resolution/tableaux algorithm for projective approximations in *IPC*. Logic Journal of the IGPL **10** (2002) 229–243
4. Iemhoff, R.: On the admissible rules of Intuitionistic Propositional Logic. J. of Symb. Logic **66** (2001) 281–294
5. Iemhoff, R.: Intermediate logics and Visser's rules. Manuscript (2003)
6. Rybakov, V.: Admissibility of Logical Inference Rules. (1997)

A Appendix

In [2] introduced the notion of a projective formula. The paper contains the following useful semantical characterization of projective formulas. For Kripke models $K_1, \ldots, K_n$ we denote by $(\sum_i K_i)'$ the Kripke model which is the result of attaching one new node below all nodes in $K_1, \ldots, K_n$, at which no propositional variable is valid. We say that two rooted Kripke models are *variants* of each other when they have the same domain and partial order, and their forcing relations only possibly differ at the roots. A class of rooted Kripke models has the *extension property* when for every finite set of Kripke models $K_1, \ldots, K_n$ in this class, there is a variant of $(\sum_i K_i)'$ which is in this class as well.

Theorem 3. *(Ghilardi [2]) A formula is projective if and only if its class of models has the extension property.*

In [3] Ghilardi presents an algorithm called CHECK-PROJECTIVITY that checks if a formula is projective. We will use this algorithm in the next section to show that the formulas in $p(\Lambda_A)$ are projective. In this section we will state the algorithm and cite the results in [3] that we will need in the next section. Thus this section is a recapturing of results in [3]. Only at one point, which we will indicate, our notation differs slightly from the one used by Ghilardi. In this section $\vdash$ stands again for $\vdash_{\mathsf{IPC}}$.

An input formula A of the CHECK-PROJECTIVITY algorithm is assumed to have the form $\bigwedge(A_i \to A_i')$ (w.l.o.g. one may assume that all formulas have this form). The algorithm manipulates sets O, O' of so-called *signed subformulas* of A. A signed formula is an expression of the following six kinds: TB, FB, T_cB, F_cB, p^+, p^-, where B is a subformula of A and p is an atomic formula. (Ghilardi uses x for atomic formulas, and Δ and Γ for the sets we denote by O). Signed formulas of the form TB, FB are called *truth modalities*, signed formulas of the form T_cB, F_cB are called *context modalities* and signed formulas of the form p^+ or p^- are called *atomic modalities.* Signed formulas express conditions on models K: TB (FB) means that B is true (false) at the root of K, T_cB (F_cB) means that B is true in the context of K (the *context* of K are all nodes different from the root), p^+ is a synonymous of Tp and p^- means that p is false at the root and true in the context of K. If O is a set of signed subformulas then $K \models O$ means that K matches all the requirements expressed by O. A set O is inconsistent iff it contains either $F_c\top$ or TB and FB, or TB and F_cB, or p^+ and p^-, or Tp and p^- or p^+ and Fp, for some B or some atomic formula p. For sets O, TO is the set of truth modalities of O, T^+O is the set of atomic modalities of O and T_cO is the set of context modalities of O. For $X, Y, Z \in \{T, T^+, T_c\}$, XYO denotes $XO \cup YO$ and $XYZO$ denotes $XO \cup YO \cup ZO$. The sets FO, F^-O and F_cO are defined in a similar way.

The algorithm analyzes reasons for which we may have that A is false in the root and true in the context of a given model. From the rules of the algorithm given below it is easy to deduce that the following Soundness lemma holds. We will not need this lemma, but thought it instructive to state since it captures the idea behind the algorithm.

Lemma 7. *Let A be the input formula and $O_1, \ldots, O_n$ a stage of the algorithm. If K is a model such that $k \models A$ for all $k \in K$ different from the root, then $K \not\models A$ iff ($K \models O_1$ or ... or $K \models O_n$).*

The CHECK-PROJECTIVITY Algorithm. (The names that are given to the rules do not come from Ghilardi, but for our purposes it is convenient to have them available.)

Initialization

$$\{TA_1, FA'_1\}, \ldots, \{TA_n, FA'_n\}$$

Tableaux Rules

$$T\wedge\ \frac{O_1, \ldots, O_j \cup \{TB_1 \wedge B_2\}, \ldots, O_n}{O_1, \ldots, O_j \cup \{TB_1, TB_2\}, \ldots, O_n}$$

$$T\top\ \frac{O_1, \ldots, O_j \cup \{T\top\}, \ldots, O_n}{O_1, \ldots, O_j, \ldots, O_n}$$

$$T\vee\ \frac{O_1, \ldots, O_j \cup \{TB_1 \vee TB_2\}, \ldots, O_n}{O_1, \ldots, O_j \cup \{TB_1\}, O_j \cup \{TB_2\}, \ldots, O_n}$$

$$T\bot\ \frac{O_1, \ldots, O_j \cup \{T\bot\}, \ldots, O_n}{O_1, \ldots, O_{j-1}, O_{j+1}, \ldots, O_n}$$

$$F\wedge\ \frac{O_1, \ldots, O_j \cup \{FB_1 \wedge FB_2\}, \ldots, O_n}{O_1, \ldots, O_j \cup \{FB_1\}, O_j \cup \{FB_2\}, \ldots, O_n}$$

$$F\top\ \frac{O_1, \ldots, O_j \cup \{F\top\}, \ldots, O_n}{O_1, \ldots, O_{j-1}, O_{j+1}, \ldots, O_n}$$

$$F\vee\ \frac{O_1, \ldots, O_j \cup \{FB_1 \vee B_2\}, \ldots, O_n}{O_1, \ldots, O_j \cup \{FB_1, FB_2\}, \ldots, O_n}$$

$$F\bot\ \frac{O_1, \ldots, O_j \cup \{F\bot\}, \ldots, O_n}{O_1, \ldots, O_j, \ldots, O_n}$$

$$T\rightarrow\ \frac{O_1, \ldots, O_j \cup \{TB_1 \rightarrow B_2\}, \ldots, O_n}{O_1, \ldots, O_j \cup \{FB_1, T_cB_1 \rightarrow B_2\}, O_j \cup \{TB_2\} \ldots, O_n}$$

$$F\rightarrow\ \frac{O_1, \ldots, O_j \cup \{FB_1 \rightarrow B_2\}, \ldots, O_n}{O_1, \ldots, O_j \cup \{F_cB_1 \rightarrow B_2\}, O_j \cup \{TB_1, FB_2\}, \ldots, O_n}$$

$$T\neg\ \frac{O_1, \ldots, O_j \cup \{T\neg B\}, \ldots, O_n}{O_1, \ldots, O_j \cup \{FB, T_c\neg B\}, \ldots, O_n}$$

$$F\neg\ \frac{O_1, \ldots, O_j \cup \{F\neg B\}, \ldots, O_n}{O_1, \ldots, O_j \cup \{F_c\neg B\}, O_j \cup \{TB\}, \ldots, O_n}$$

$$Tp \frac{O_1, \ldots, O_j \cup \{Tp\}, \ldots, O_n}{O_1, \ldots, O_j \cup \{p^+\}, \ldots, O_n}$$

$$Fp \frac{O_1, \ldots, O_j \cup \{Fp\}, \ldots, O_n}{O_1, \ldots, O_j \cup \{p^-\}, O_j\{F_c p\}, \ldots, O_n}$$

Resolution Rule

$$\frac{O_1, \ldots, O_j \cup \{p^+\}, \ldots, O_i \cup \{p^-\}, \ldots, O_n}{O_1, \ldots, O_j \cup \{p^+\}, \ldots, O_i \cup \{p^-\}, \ldots, O_n, O_j \cup O_i \cup \{T_c p\}}$$

Simplification Rule

$$\frac{O_1, \ldots, O_j, \ldots, O_n}{O_1, \ldots, O_{j-1}, O_{j+1}, \ldots, O_n}$$

provided $A \wedge \bigwedge T_c O_j \vdash C$, for some $C \in F_c(O_j)$.

Auxiliary Simplification Rule

$$\frac{O_1, \ldots, O_j, \ldots, O_n}{O_1, \ldots, O_{j-1}, O_{j+1}, \ldots, O_n}$$

provided $O_j \supseteq O_i$ for some $i \neq j$ or provided O_j is inconsistent.

Ghilardi shows that the algorithm terminates after a finite number of steps. The following theorem is the one we will need. It is Theorem 4.2 of [3].

Theorem 4. *(Ghilardi [3]) A is projective iff each of its output sets contains at least one atomic modality.*

Program Complexity of Dynamic LTL Model Checking*

Detlef Kähler and Thomas Wilke

Christian-Albrechts University, 24098 Kiel, Germany
{kaehler,wilke}@ti.informatik.uni-kiel.de

Abstract. Using a recent result by Hesse we show that for any fixed linear-time temporal formula the dynamic model checking problem is in Dyn-TC^0, a complexity class introduced by Hesse, Immerman, Patnaik, containing all dynamic problems where the update after an operation has been performed can be computed by a DLOGTIME-uniform constant-depth threshold circuit. The operations permitted to modify the transition system to be verified include insertion and deletion of transitions and relabeling of states.

1 Introduction

Usually, a model checking problem is defined as follows. Given a transition system and a system specification, also called property, check whether the system has the property. This is motivated by the fact that in order to ensure that a piece of hardware or software is correct one checks that it has certain desirable properties. Looking more closely one observes that as the system evolves during the design process the desirable properties are checked over and over again, because properties once true can be made false by a modification and properties once false hopefully become true. This gives rise to a *dynamic* view of model checking and motivates a dynamic version of the model checking problem: recompute the truth value of the property in question after a modification of the system. Clearly, the recomputation may be facilitated by keeping around auxiliary data, but then this must also be adapted as the system is modified. From a theoretical point of view it is now interesting to determine how difficult the recomputation is (both the mere recomputation of the truth value of the property and the update of the auxiliary data), that is, how efficiently the recomputation can be carried out. In the present paper, this is studied in a very specific setting, namely when the properties are specified in linear-time temporal logic.

We work in the logical framework of dynamic complexity provided by Hesse, Immerman, and Patnaik [11, 8]. In this framework, hereafter called the HIP approach, the problem instances as well as the auxiliary data are represented by relational structures, and the changes to the auxiliary data are described in an appropriate logic (of low complexity). For instance, Hesse and Immerman

* Work supported by German-Israeli Foundation for Scientific Research and Development, Project No. I-638-95.6/1999

M. Baaz and J.A. Makowsky (Eds.): CSL 2003, LNCS 2803, pp. 271–284, 2003.

show in [8] that a dynamic version of the circuit value problem is complete for Dyn-FO, the class of dynamic problems where the update can be described by a first-order formula. The most sophisticated result, which is also the basis for our classification of the model checking problem, is Hesse's Theorem [7], which says that dynamic graph reachability is in Dyn-TC^0 or, equivalently, Dyn-FO(Count) [9]. This means that the update can be carried out by threshold circuits of constant depth or first-order logic with counting, respectively.

Graph reachability is almost like LTL model checking; recall that in the automata-theoretic approach to LTL model checking [17] a given formula (its negation) is converted into a Büchi automaton, then a product of this automaton and the transition system is built, and finally an emptiness test for the resulting Büchi automaton, which amounts to a number of reachability tests, is carried out. This explains briefly how we obtain our main theorem that says that the formula complexity of dynamic LTL model checking is in Dyn-TC^0.

The situation with model checking is somewhat more complicated than with graph reachability, because we also consider the possibility of relabeling states (which is similar to adding and deleting nodes) whereas Hesse is mainly interested in adding and removing edges.

To the best of our knowledge, there are as yet no non-trivial results on the dynamic complexity of model checking. (This is somewhat different for the related area of dynamic graph problems and the dynamic evaluation of database queries with transitive closure, see, e.g., [3, 5, 10, 12].) Concrete dynamic model checking algorithms (as opposed to complexity results) have been dealt with in various papers, see, e.g., [14–16]. The most recent results on dynamic graph reachability are by Roditty and Zwick [13].

This paper is organized as follows. In Sect. 2, we review the HIP approach, in Sect. 3 we review LTL model checking, Sect. 4 describes how dynamic model-checking can be reduced to dynamic selective transitive closure, a problem specifically tailored for our purposes, and, finally, Sect. 5 explains why dynamic selective transitive closure is in Dyn-TC^0, which also proves that dynamic LTL model checking is in Dyn-TC^0.

For background on circuit complexity, descriptive complexity, and model checking, see [18], [9], and [2], respectively.

2 Dynamic Complexity Framework

In this section, we provide the necessary background on dynamic complexity by looking at examples; for detailed information, see [8]. As stated in the introduction, we use the HIP approach and modify it slightly in order to be able to treat computation problems—in [8], only decision problems are considered.

2.1 Background on Descriptive Complexity

Our entire approach is based on descriptive complexity, and we start off with some background on this.

We are only interested in relational structures $\mathcal{A}$ where the universe, denoted $|\mathcal{A}|$, is an initial segment of the natural numbers, that is, $|\mathcal{A}| = \{0, 1, \ldots, n-1\}$. In addition, we assume that every structure is provided with the built-in predicate $\leq$ (with the natural interpretation) and the built-in predicate $\mathrm{BIT}^{(2)}$, which may be used to query the binary representation of the numbers building the universe: $\mathrm{BIT}^{\mathcal{A}}(i, j)$ holds if the jth bit of i is 1. Moreover, we assume the structures have at least 2 elements, and we identify 0 with false and 1 with true. Structures of this kind will be referred to as *arithmetic structures*. Given a relational vocabulary τ, we write $\mathrm{Struc}[\tau]$ for the set of all arithmetic structures with vocabulary τ (more precisely, with vocabulary $\tau \cup \{\leq, \mathrm{BIT}\}$) and $\mathrm{Struc}_n[\tau]$ for all such structures with n elements.

For specifying properties of arithmetic structures, we only consider first-order logic, denoted FO, and extensions thereof. Our main theorem involves FO[Count], the logic obtained from FO by adding so-called *counting quantifiers* $\exists ix\phi$ with the meaning "there exists i many x such that ϕ holds" and where the occurrence of i is free.

Observe that already in FO, with the built-in predicates $\leq$ and BIT, it is easy to express addition and multiplication on the universe, see [9], as well as the boolean operations on the subuniverse $\{0, 1\}$. This is the reason why in the following we will use $+$ and $\cdot$ freely in our first-order formulas as well as boolean operations on terms meant to denote boolean values.

An important fact, which ties the logical approach to the computational side, is the following.

Theorem 1 ([1, 4]) *In* FO[Count], *one can express exactly the properties that can be computed by DLOGTIME-uniform polynomial-size constant-depth threshold circuits.*

2.2 Dynamic Problems

The basic idea is that a dynamic problem is specified by (1) a set of *operations* that can be used to build instances of the problem and (2) for every sequence of operations a *solution* to the instance represented by this sequence. A particular aspect of the HIP approach is that the problems are parametrized according to the size of the input; we will use n to denote this size.

As an example, we consider the problem dynamic transitive closure, DynTransClos, where we are interested in computing (maintaining) the transitive closure of a binary relation, which we view as the edge set of a directed graph. The graph can (dynamically) be constructed by deleting and inserting edges. Therefore, we use the vocabulary $\Sigma^{tc} = \{\mathsf{Insert}^{(2)}, \mathsf{Delete}^{(2)}\}$ of two binary operators. For the parameter n, the set of operations for constructing the instances of DynTransClos is then given by

$$\Sigma_n^{tc} = \{\mathsf{Insert}(i, j) \mid i, j < n\} \cup \{\mathsf{Delete}(i, j) \mid i, j < n\} \ . \tag{1}$$

The transitive closure of a graph is simply a binary relation on the vertex set of the given graph, which is why for representing the solutions to the instances we will use structures over the vocabulary $\tau^{tc} = \{C^{(2)}\}$. For any n, the solutions are described by the mapping

$$s_n^{tc}\colon u \mapsto \mathcal{A}_u^{tc}, \quad u \in (\Sigma_n^{tc})^*, \quad \mathcal{A}_u^{tc} \in \mathrm{Struc}_n[\tau^{tc}] \tag{2}$$

where A_u^{tc} represents the transitive closure of the graph obtained from the empty graph with n vertices by inserting and deleting edges as described by u.

Formally, our *dynamic problem* is now given by

$$\textsc{DynTransClos} = (\Sigma^{tc}, \tau^{tc}, \{s_n^{tc}\}_n) \ . \tag{3}$$

In this paper, we will be interested in two other dynamic problems as well, namely selective transitive closure, DynSelTransClos, which we introduce below, and a family of model checking problems, for each LTL formula ϕ a problem DynLTLModCheck$_\phi$, which we introduce in the subsequent section.

DynSelTransClos is a modification of DynTransClos where we try to model that vertices can be added and removed. To this end, we imagine that the vertices of the graph in question can be in two different states, selected or unselected. In determining the transitive closure only edges incident with selected vertices can be used. In other words, we only consider paths with selected nodes on them, also called *active paths*. Formally, we set

$$\Sigma^{stc} = \Sigma^{tc} \cup \{\mathsf{Select}^{(1)}, \mathsf{Deselect}^{(1)}\} \ , \tag{4}$$

$$\tau^{stc} = \tau^{tc} \ , \tag{5}$$

and have

$$s_n^{stc}\colon u \mapsto \mathcal{A}_u^{stc}, \quad u \in (\Sigma_n^{stc})^*, \quad \mathcal{A}_u^{stc} \in \mathrm{Struc}_n[\tau^{stc}] \tag{6}$$

where A_u^{stc} represents the transitive closure of the graph obtained from the empty graph with n vertices by performing the operations listed in u.

In fact, we are interested in a variant of DynSelTransClos, denoted DynSelTransClos$_\psi$, in which the set of selected vertices of the initial graph is described by a fixed first-order formula ψ. That is, the operation sequence $u \in (\Sigma_n^{stc})^*$ is applied to the graph on n vertices with no edges and with vertex i selected iff $\psi(i)$ holds.

2.3 Dynamic Programs

An algorithm for solving a dynamic problem will usually maintain some auxiliary data structure where crucial information is stored. For instance, we will see that for computing the transitive closure Hesse maintains counts of paths of certain lengths. In the HIP approach, the auxiliary data structure is modeled by additional relations. The computation itself—how the data structure is updated when an operation is performed—is then described by appropriate formulas.

Consider a very simple example, DynDivBy6, where an element of a structure is either selected or not and suppose we want to determine if the number of selected elements is divisible by 6. In analogy to before, we choose

$$\Sigma^{div6} = \{\mathsf{Select}^{(1)}, \mathsf{Deselect}^{(1)}\} \ , \tag{7}$$

$$\tau^{div6} = \{d_6^{(0)}\} \ , \tag{8}$$

and we have a description of the solutions,

$$s_n^{div6} \colon u \mapsto A_u^{div6}, \quad u \in (\Sigma_n^{div6})^*, \quad A_u^{div6} \in \mathrm{Struc}_n[\tau^{div6}] \ . \tag{9}$$

In order to be able to determine (efficiently) whether or not the number of selected vertices is divisible by six we (have to) remember which vertices are selected and which aren't and their count modulo 6. For this purpose, our auxiliary data structure will have a unary relation, denoted S, and six boolean constants $m_0, \ldots, m_5$. The updates of the auxiliary data and the updates of the solution relation d_6 are described by simple formulas. For $\mathsf{Select}(i)$, we have

$$S'(x) :\equiv i = x \vee S(x) \ , \tag{10}$$

$$m_0' :\equiv (S(i) \wedge m_0) \vee (\neg S(i) \wedge m_5) \ , \tag{11}$$

$$m_1' :\equiv (S(i) \wedge m_1) \vee (\neg S(i) \wedge m_0) \ , \tag{12}$$

$$m_2' :\equiv (S(i) \wedge m_2) \vee (\neg S(i) \wedge m_1) \ , \tag{13}$$

$$m_3' :\equiv (S(i) \wedge m_3) \vee (\neg S(i) \wedge m_2) \ , \tag{14}$$

$$m_4' :\equiv (S(i) \wedge m_4) \vee (\neg S(i) \wedge m_3) \ , \tag{15}$$

$$m_5' :\equiv (S(i) \wedge m_5) \vee (\neg S(i) \wedge m_4) \ , \tag{16}$$

$$d_6' :\equiv (S(i) \wedge m_0) \vee (\neg S(i) \wedge m_5) \ . \tag{17}$$

Note that $:\equiv$ is to be read as an assignment where the unprimed variables refer to the relations and constants before the operation $\mathsf{Select}(i)$ has been carried out, while the primed variables refer to the relations and constants after the operation has been carried out. Note also that (17) simply says that d_6 and m_0 are the same. The entire block of assignments, (10)–(17), is called the *dynamic procedure* for Select. For Deselect, we could set up a very similar dynamic procedure. In addition to this, we have to specify the initial interpretation of the auxiliary data structure and the result we expect before any operation has been carried out:

$$S(x) :\equiv \textsc{False} \ , \tag{18}$$

$$m_0 :\equiv \textsc{True} \ , \qquad m_1 :\equiv \textsc{False} \ , \qquad m_2 :\equiv \textsc{False} \ , \tag{19}$$

$$m_3 :\equiv \textsc{False} \ , \qquad m_4 :\equiv \textsc{False} \ , \qquad m_5 :\equiv \textsc{False} \ , \tag{20}$$

$$d_6 :\equiv \textsc{True} \ . \tag{21}$$

This block of assignments, (18)–(21), is called the *initialization procedure*. The assignments all-together constitute our *dynamic program* for DynDivBy6.

2.4 Dynamic Complexity

In the HIP approach, dynamic complexity classes are made up from logics. Given a logic $\mathcal{L}$, the complexity class Dyn-$\mathcal{L}$ consists of all dynamic problems that can be solved via a dynamic program (as above) where the initial settings and the updates are described by formulas from $\mathcal{L}$.

Hesse's result, which we rely on here heavily, gives an upper bound for the complexity of dynamic transitive closure:

Theorem 2 **([7])** DynTransClos, *is in Dyn*-FO[Count] *(= Dyn*-TC^0*).*

The equality Dyn-FO[Count] = Dyn-TC^0 is justified by Theorem 1.

2.5 Reductions

Just as in ordinary complexity theory the notion of a reduction is quite useful, both for defining completeness and for ease in proving membership in a certain complexity class.

As a simple example, consider the problem DynDivBy3, the obvious modification of DynDivBy6. Clearly, divisibility by three can be deduced from divisibility by six if one can double: 3 divides x iff 6 divides $2x$. Therefore, in our reduction from DynDivBy3 to DynDivBy6, we will double the universe.— One might be tempted to use the simple formula $d_0 \vee m_3$ as a reduction; this does not work because m_3 is not part of the specification of DynDivBy6.)

In general, a *reduction* from a dynamic problem $P = (\Sigma, \tau, \{s_n\}_n)$ to a dynamic problem $Q = (\Gamma, \sigma, \{t_n\}_n)$ is composed of

— a *universe expansion function,* $e\colon \mathbf{N} \to \mathbf{N}$, a polynomial with $e(n) \geq n$ for all n,

— a family $\{h_n\}_n$ of *reduction homomorphisms* $h_n\colon \Sigma_n^* \to \Gamma_{e(n)}^*$, and

— for each symbol $R^{(k)} \in \tau$ a first-order *result definition* $R(x_0, \ldots, x_{k-1}) :\equiv \phi_R(x_0, \ldots, x_{k-1})$ where ϕ is in the vocabulary σ.

It is required that the family $\{h_n\}_n$ is bounded, that is, there must exist some k such that for all n and for all $a \in \Sigma_n$, $|h_n(a)| \leq k$. Moreover, all images $h_n(a)$ must be specified in a uniform way, say, by a first-order formula.

In ordinary complexity theory, for a reduction f from L to L' we need that $x \in L$ iff $f(x) \in L$ for all strings x. This translates to our dynamic setting as follows. For every n, every sequence $u \in \Sigma_n$, every k-ary relation symbol $R \in \tau$, and $a_0, \ldots, a_{k-1} < n$, we have $s_n(u) \models R(a_0, \ldots, a_{k-1})$ iff $t_{e(n)}(h_n(u)) \models \phi_R(a_0, \ldots, a_{k-1})$.

For the concrete reduction from DynDivBy3 to DynDivBy6 we choose the above components as follows. The expansion function is given by $e(n) = 2n$, the reduction homomorphisms are determined by $h_n(\mathsf{Select}(i)) = \mathsf{Select}(i)\mathsf{Select}(n+i)$ and $h_n(\mathsf{Deselect}(i)) = \mathsf{Deselect}(i)\mathsf{Deselect}(n+i)$, and the result definition is simply $d_3 :\equiv d_6$.

The important lemma we will need is the following.

Lemma 1 *If P and P' are dynamic problems such that $P' \in$ Dyn*-FO[Count] *and P can be reduced to P', then $P \in$ Dyn*-FO[Count].

The proof is similar to the corresponding lemma from [8]; the fact that we deal with computation problems is only a technical point.

3 LTL Model Checking and Main Result

We first recall ordinary LTL model checking, then define dynamic LTL model checking, and conclude with the description of the automata-theoretic method for solving the ordinary LTL model checking problem.

3.1 Ordinary LTL Model Checking

In ordinary LTL model checking we are given (1) a transition system

$$T = (S, P, R, L, s_I) \tag{22}$$

with state set S, set of propositional variables P, edge relation $R \subseteq S \times S$, labeling function $L\colon S \to 2^P$, and initial state s_I, and (2) a temporal formula ϕ, built from the propositional variables in P, boolean connectives and temporal operators such as "next", "eventually", and "until", and we ask whether $T \models \phi$.

The relation $T \models \phi$ holds if each infinite computation $s_0 s_1 s_2 \ldots$ through T (meaning $s_0 = s_I$, $(s_i, s_{i+1}) \in R$ for each i) satisfies ϕ, where, in turn, this means that $L(s_0)L(s_1)L(s_2)\cdots \models \phi$ holds in the usual sense.

3.2 Dynamic LTL Model Checking and Main Result

In our dynamic version of the problem defined above, we allow the transition system to be changed, but leave the LTL formula fixed. This means we are interested in the program complexity of LTL model checking [17].

The operations of the dynamic version of the problem defined above with fixed formula ϕ, DynLTLModCheck$_\phi$, are insertion and deletion of edges and relabeling of the states, more precisely,

$$\Sigma^{mc_\phi} = \{\mathsf{Insert}^{(2)}, \mathsf{Delete}^{(2)}\} \cup \{\mathsf{SetVar}_U^{(1)} \mid U \subseteq P\} \tag{23}$$

where the intended meaning of $\mathsf{SetVar}_U(i)$ is that the ith state of the transition system gets label U. The solution to the model checking problem is represented by a boolean constant, v (for verified), that is

$$\tau^{mc_\phi} = \{v^0\} \ , \tag{24}$$

and finally,

$$s_n^{mc_\phi}\colon u \mapsto \mathcal{A}_u^{mc_\phi}, \quad u \in (\Sigma_n^{mc_\phi})^*, \quad \mathcal{A}_u^{mc_\phi} \in \mathrm{Struc}_n[\tau^{mc_\phi}] \tag{25}$$

where $s_n^{mc_\phi}(u) \models v$ iff ϕ holds in the transition system with state set $\{0, \ldots, n-1\}$, initial state 0 and edge relation and labeling according to u.

The main result of this paper is:

Theorem 3 *For every LTL formula ϕ, dynamic model checking of the property ϕ,* DynLTLModCheck$_\phi$, *is in Dyn-*FO[Count] *(= Dyn-*TC^0*).*

The remainder of the paper explains our proof of this theorem. We start with a brief review of the automata-theoretic approach to model checking.

3.3 The Automata-Theoretic Method

In the automata-theoretic approach to LTL model checking [17], given an LTL formula ϕ, one first constructs a Büchi automaton

$$A_\phi = (Q, 2^P, q_I, \delta, F) \tag{26}$$

with finite state set Q, alphabet 2^P, initial state q_I, nondeterministic transition function $\delta\colon Q \times 2^P \to 2^Q$, and accepting state set F, which recognizes exactly all infinite strings over 2^P which do not satisfy ϕ.

Then, in the second step, one constructs the product of the given transition system T (as above):

$$A_{T\times\phi} = (S \times Q, 2^P, s_I \times q_I, \delta', S \times F), \tag{27}$$

where for $(s,q) \in S \times Q$ and $U \subseteq P$,

$$(s',q') \in \delta'((s,q),U) \quad \text{iff} \quad L(s) = U \wedge (s,s') \in R \wedge q' \in \delta(q,U). \tag{28}$$

This construction guarantees that if the language of $A_{T\times\phi}$ is empty, one can conclude no computation of T violates ϕ, which means $T \models \phi$, and if the language is not empty, one can conclude there exists a computation of T violating ϕ, which means $T \not\models \phi$. So, when we denote the language recognized by a Büchi automaton A by $L(A)$, we can state:

Lemma 2 *$T \models \phi$ iff $L(A_{T\times\phi}) = \emptyset$.*

As a consequence, in the third and last step of the automata-theoretic approach, one checks $L(A_{T\times\phi})$ for emptiness.

Observe that in general, the language of a Büchi automaton $(Q, \Sigma, q_I, \delta, F)$ is non-empty iff there exists $q \in F$ such that q is reachable from q_I and q is reachable from itself (via a non-trivial path). This can easily be expressed using the transitive closure of the underlying transition graph.

3.4 Outline of Proof

Our proof of Theorem 3 now goes as follows. The first step is to show that DynLTLModCheck$_\phi$ can be reduced to DynSelTransClos$_\psi$, for a certain formula ψ, using the automata-theoretic approach just described. We then show that DynSelTransClos$_\psi$ is in Dyn-FO[Count], which, together with Lemma 1, will complete the proof of Theorem 3.

4 Reducing Dynamic LTL Model Checking to Selective Transitive Closure

We first show how to model the product Büchi automaton $A_{T\times\phi}$ as a selective graph G in such a way that we will be able to reduce DynLTLModCheck$_\phi$ to

DynSelTransClos$_\psi$. On the one hand we want to decide whether $L(A_{T\times\phi}) = \emptyset$ by considering the transitive closure of G, on the other hand the selective graph G should be easily updatable according to the possible changes the transition system T may undergo. The main problem here is to handle the relabeling of states in T. To cope with this we had to introduce the notion of selective graphs.

4.1 The Product Automaton as a Selective Graph

For determining if a transition system T is a model of an LTL-formula ϕ it suffices to know the transitive closure of the transition graph of the product Büchi automaton $A_{T\times\phi}$, see Section 3. The edge relation of this transition graph is described by (28). In order to reduce DynLTLModCheck$_\phi$ to DynSelTransClos we have to simulate each operation of Σ^{mc_ϕ} by a bounded sequence of operations from Σ^{stc}. An insertion or deletion of a transition in T can only cause at most $|Q|^2$ changes in the transition graph of $A_{T\times\phi}$ and $|Q|^2$ is independent of the size n of the transition system T. But if we change the label of state i of T we may have to change up to n^2 edges in the transition graph of $A_{T\times\phi}$. As a consequence, the operations SetVar_U cannot be simulated by a uniformly bounded sequence of insertions and deletions in the transition graph associated with $A_{T\times\phi}$.

With selective graphs, we can handle relabeling as well. We split up (28) into two parts: (i) $L(i) = U$ and (ii) $(i, i') \in R \wedge q' \in \delta(q, U)$, and model the second part by transitions and the first part by selecting vertices.

The selective graph G representing the product Büchi automaton

$$A_{T\times\phi} = (S \times Q, 2^P, s_I \times q_I, \delta', S \times F), \tag{29}$$

has vertex set $V = S \times 2^P \times Q$. There is an edge in G from vertex (i, U, q) to vertex (i', U', q') if (ii) holds. So the edges of G represent all possible labelings of T at once. Which labeling is actually present is described by the selected vertices in G. At any point during the computation, a vertex $(i, U, q) \in V$ will be selected iff the label of i in T (at that point) is U, reflecting condition (i).

Then the following lemma holds.

Lemma 3 $L(A_{T\times\phi}) \neq \emptyset$ *iff there exists* $q \in F$ *and* $s \in S$ *such that there is an active path from* $(0, L(0), q_I)$ *to* $(s, L(s), q)$ *and an active path from* $(s, L(s), q)$ *to* $(s, L(s), q)$.

4.2 Reducing DynLTLModCheck to DynSelTransClos

We may assume that $Q = \{q_0, \ldots, q_{2^p-1}\}$, that is, the size of $S\times 2^P\times Q$ is $n\cdot 2^{m+p}$. We want to encode the vertices of G as numbers in order to encode it as an instance of DynSelTransClos$_\psi$. For this purpose, we identify each element U of 2^P with a unique number from $\{0, \ldots, 2^m-1\}$, where, by convention, 0 represents the empty labeling. A vertex (i, U, q_j) is encoded as the number $i\cdot 2^{m+p} + U\cdot 2^p + j$. We represent the vertex set V of G by the set $\{0, \ldots, n \cdot 2^{m+p} - 1\}$. Initially, the vertices of the selective graph G correspond to the initial labeling of transition system T, i. e., all vertices i with $L(i) = \emptyset$ are selected. For the reduction from $\text{DynLTLModCheck}_\phi$ to $\text{DynSelTransClos}_\psi$ we have to provide (i) the formula ψ, (ii) the reduction homomorphisms that map operation sequences of $\text{DynLTLModCheck}_\phi$ to operation sequences of $\text{DynSelTransClos}_\psi$, and (iii) the formula that expresses the solution of $\text{DynLTLModCheck}_\phi$ in terms of the solution of the corresponding problem in $\text{DynSelTransClos}_\psi$. This is what we will do in the remainder of this section.

(i) According to what was said above, we set

$$\psi(i) \equiv \forall j(p \leq j < p+m \rightarrow \neg \text{BIT}(i,j)) \ . \tag{30}$$

(ii) If we insert the edge (i,j) into T we have to insert the uniform set $\{((i,U,q),(j,U',q')) \mid q' \in \delta(q,U)\}$ into the corresponding selective graph G. If we delete edge (i,j) the above set of edges has to be removed. So we can simulate insertion and deletion of an edge in the transition system T by a bounded sequence of insertions and deletions in the corresponding selective graph:

$$h(\mathsf{Insert}(i,j)) = \mathsf{Insert}(i \cdot 2^{p+m} + u_1, j \cdot 2^{p+m} + v_1) \ldots \\ \ldots \mathsf{Insert}(i \cdot 2^{p+m} + u_t, j \cdot 2^{p+m} + v_t) \ , \tag{31}$$

$$h(\mathsf{Delete}(i,j)) = \mathsf{Delete}(i \cdot 2^{p+m} + u_1, j \cdot 2^{p+m} + v_1) \ldots \\ \ldots \mathsf{Delete}(i \cdot 2^{p+m} + u_t, j \cdot 2^{p+m} + v_t) \ , \tag{32}$$

where $\{(u_1,v_1),\ldots,(u_t,v_t)\} = \{(2^p \cdot U + i, 2^p \cdot U' + j) \mid q_j \in \delta(q_i,U)\}$. Note that the operation sequences are indeed bounded because t is a fixed number not depending on n.

If we change the label of state i of T to U, we first deselect all vertices of the form $(i,\cdot,\cdot)$ and then select all vertices of the form $(i,U,\cdot)$. This leads to

$$h(\mathsf{SetVar}_U(i)) = \alpha\beta \tag{33}$$

where

$$\alpha = \mathsf{Deselect}(i \cdot 2^{p+m}) \ldots \mathsf{Deselect}(i \cdot 2^{p+m+1} - 1) \ , \tag{34}$$

$$\beta = \mathsf{Select}(i \cdot 2^{p+m} + 2^p \cdot U) \ldots \mathsf{Select}(i \cdot 2^{p+m} + 2^p \cdot U + 2^m - 1) \ . \tag{35}$$

(iii) By Lemma 3 we can express $v \in \tau^{mc}$ in terms of $C \in \tau^{stc}$:

$$v :\equiv \neg \exists x \exists y(\rho_{\mathsf{initial}}(x) \wedge \rho_{\mathsf{final}}(y) \wedge (C(x,y) \wedge C(y,y)), \tag{36}$$

where

$$\rho_{\mathsf{initial}}(x) \equiv \forall i(i < p \rightarrow (\neg \text{BIT}(x,i)) \wedge \forall i(i > m+p \rightarrow (\neg \text{BIT}(x,i)) \ , \tag{37}$$

$$\rho_{\mathsf{final}}(x) \equiv \bigvee_{q_j \in F} \forall i(i < p \rightarrow (\text{BIT}(j,i) \leftrightarrow \text{BIT}(x,i)) \ . \tag{38}$$

Taking everything into account, we have shown the following lemma.

Lemma 4 *For any fixed LTL-formula ϕ the problem* $\text{DynLTLModCheck}_\phi$ *is reducible to* $\text{DynSelTransClos}_\psi$.

5 DynSelTransClos Is in Dyn-TC^0

The last step of our proof is to show that DynSelTransClos$_\psi$ is in Dyn-TC^0. We do this by extending Hesse's proof which shows that DynTransClos is in Dyn-TC^0. Without loss of generality, we only consider the non-parametrized version DynSelTransClos in the following.

The main idea of Hesse's proof is to count the number of paths up to a certain length between any two vertices of the considered graph. These "path numbers" are stored as polynomials whose coefficients stand for the number of paths. To use these "path counting polynomials" in the dynamic setting one needs update formulas according to the possible graph operations. Hesse, in his paper, provides update formulas for inserting and deleting edges and deleting vertices. In order to use the idea of the path counting polynomials for DynSelTransClos we give an update formula for selecting (and deselecting) vertices. To place DynTransClos in Dyn-TC^0 Hesse represents the counting polynomials as large integers and uses the results from [6] to show that one can perform the updates on this number representation of the path counting polynomials by a Dyn-TC^0 (or Dyn-FO[Count]) procedure.

In the following, we explain Hesse's ideas in more detail and present our lemma for modeling the selection of vertices in selective graphs.

5.1 Path Counting Polynomials

For vertices s, t of a directed graph G we represent the number of paths between s and t by the path counting polynomial

$$f_{s,t}(x) = \sum_{k=0}^{\infty} p_{s,t}(k)x^k \tag{39}$$

where $p_{s,t}(k)$ is the number of directed paths from s to t of length k.

If we insert or delete edges in G the coefficients $p_{s,t}(k)$ of the path counting polynomials have to be recomputed. Hesse gives update formulas for the path counting polynomials for three operations: insertion and deletion of edges and for deleting all edges incident with a single vertex at once. As an example we give the update formula for inserting a new edge (i, j) from vertex i to j. By $f'_{s,t}(x)$ we denote the path counting polynomial after insertion of edge (i, j). With this notation, we can state one of Hesse's results:

$$f'_{s,t}(x) = f_{s,t}(x) + f_{s,i}(x) \left(\sum_{k=0}^{\infty} (f_{j,i}(x)x)^k \right) x f_{j,t}(x) \ . \tag{40}$$

In DynSelTransClos we consider selective graphs, which means we count only active paths between vertices: in our setting, the coefficient $p_{s,t}(k)$ is the number of active paths between s and t of length k. In DynSelTransClos there are four operations to be considered: insertion and deletion of edges and selection and deselection of vertices. For the insertion and deletion of edges and deselection of vertices we can apply the update formulas given by Hesse mentioned above. For solving DYNSELTRANSCLOS we prove the following lemma, where, for a vertex i we denote by $\langle i]$ and $[i\rangle$ the set of all predecessors and successors of i, respectively.

Lemma 5 *Let i be a non-active vertex of graph G and G' the graph G with activated vertex i and edge (i,i) not contained in G. For all vertices s,t of G, let $f_{s,t}(x)$ and $f'_{s,t}(x)$ be the path counting power series of G and G', respectively:*

$$f_{s,t}(x) = \sum_{k=0}^{\infty} p_{s,t}(k)x^k , \qquad f'_{s,t}(x) = \sum_{k=0}^{\infty} p'_{s,t}(k)x^k. \tag{41}$$

Then, for $i \notin \{s,t\}$,

$$f'_{s,t}(x) = f_{s,t}(x) + f_{s,\langle i]}(x)x^2 \left(\sum_{k=0}^{\infty} (f_{[i\rangle,\langle i]}(x)x^2)^k \right) f_{[i\rangle,t}(x) \tag{42}$$

where

$$f_{s,\langle i]}(x) = \sum_{u \in \langle i]} f_{s,u}(x) , \tag{43}$$

$$f_{[i\rangle,t}(x) = \sum_{v \in [i\rangle} f_{v,t}(x) , \tag{44}$$

$$f_{[i\rangle,\langle i]}(x) = \sum_{(v,u) \in [i\rangle \times \langle i]} f_{v,u}(x) . \tag{45}$$

Proof (following the proof of [7, Lemma 2]). The main idea is to write $p'_{s,t}(l)$ as a sum where the j-th addend is the number of active paths of length l from s to t containing i exactly j times.

Each active path from s to t in G' of length l containing i exactly once ($j = 1$) can be decomposed into three parts: an active path from s to an active predecessor u of i in G of length m, edges (u,i) and (i,v) for an active successor of i in G, and an active path from v to t of length $l-2-m$. Thus, the number of active paths from s to t in G' containing vertex i exactly once is:

$$\sum_{m=0}^{l-2} p_{s,\langle i]}(m)p_{[i\rangle,t}(l-2-m). \tag{46}$$

This is the coefficient of x^l in $f_{s,\langle i]}(x)x^2 f_{[i\rangle,t}(x)$.

An active path from s to t using i exactly twice ($j = 2$) can be decomposed into an active path from s to an active predecessor u of i, edges (u,i) and (i,v) for an active successor v of i, an active path from v to an active predecessor u' of i, edges (u',i) and (i,v') for an active successor v' of i, and an active path from v' to t in G. The number of such paths of length l is:

$$\sum_{m=0}^{l-4} \sum_{o=0}^{l-4-m} p_{s,\langle i]}(m)p_{[i\rangle,\langle i]}(o)p_{[i\rangle,t}(l-4-m-o) . \tag{47}$$

This is the coefficient of x^l in $f_{s,\langle i]}(x)x^2 f_{[i\rangle,\langle i]}(x)x^2 f_{[i\rangle,t}(x)$. Generalizing this to arbitrary j and summing the respective expressions will yield an expression identical with the right-hand side of (42). □

5.2 Integer Representation of the Polynomials

In order to decide whether vertex t is reachable from vertex s by an active path we only need to know if there is an active path of length less than n from s to t. Thus, we only have to maintain the first n coefficients of the polynomials $f_{s,t}(x)$, and these can be extracted from the single number

$$a_{s,t} = \sum_{k=0}^{n-1} p_{s,t}(k) r^k \tag{48}$$

where r is large enough. In fact, as Hesse points out, $r = 2^{n^2}$, because then the binary representation of $a_{s,t}$ is merely the concatenation of the binary representations of the $p_{s,t}(k)$'s with appropriate padding.

The update formulas for the path counting polynomials of Section 5.1 can be applied directly to compute the updates of their integer representations by computing mod r^n. Now (40) turns into

$$a'_{s,t} = a_{s,t} + a_{s,i} \left(\sum_{k=0}^{n-2} (a_{j,i} r)^k \right) r a_{j,t} \mod r^n \ . \tag{49}$$

Using the result of [6] that the product of $n^{O(1)}$ numbers in binary of $n^{O(1)}$ bits can be computed by a TC^0 (or FO[Count]) procedure we show just as Hesse [7] that the update formulas for the integers $a_{s,t}$ can be computed by a TC^0 program. Together with Lemmas 4 and 1 this completes our proof of Theorem 3.

6 Conclusion

In this paper, we have shown that dynamic LTL model checking is in Dyn-TC^0. This is a first step towards a dynamic treatment of model checking problems from a complexity-theoretic point of view. Other model checking problems should be considered, and, of course, related problems. For instance, we have not dealt with counter-examples (error traces) in this paper, and it is not clear to us how to maintain counter-examples explicitly, say, by a function which on input i returns the ith state on an error trace.

References

1. David A. Mix Barrington, Neil Immerman, and Howard Straubing. On uniformity within NC^1. *Journal of Computer and System Sciences*, 41(3):274–306, 1990.
2. Edmund M. Clarke, Orna Grumberg, and Doron A. Peled. *Model Checking.* MIT Press, 1999.

3. Guozhu Dong and Jianwen Su. Incremental and decremental evaluation of transitive closure by first-order queries. *Information and Computation*, 120(1):101–106, 1995.
4. Kousha Etessami. Counting quantifiers, successor relations, and logarithmic space. In *IEEE Structure in Complexity Theory Conference*, pages 2–11, Minneapolis, Minnesota, 1995.
5. Kousha Etessami. Dynamic tree isomorphism via first-order updates to a relational database. In *ACM Symposium on Principles of Database Systems (PODS)*, pages 235–243, Seattle, Washington, 1998.
6. William Hesse. Division is in uniform TC^0. In *International Colloquium on Automata, Languages and Programming (ICALP)*, pages 104–114, Crete, Greece, 2001.
7. William Hesse. The dynamic complexity of transitive closure is in $DynTC^0$. *Theoretical Computer Science*, 296:473–485, March 2003.
8. William Hesse and Neil Immerman. Complete problems for dynamic complexity classes. In *IEEE Symposium on Logic in Computer Science (LICS)*, pages 313–324, Copenhagen, Denmark, 2002.
9. Neil Immerman. *Descriptive Complexity.* Graduate Texts in Computer Science. Springer, 1999.
10. Leonid Libkin and Limsoon Wong. Incremental recomputation of recursive queries with nested sets and aggregate functions. In *Database Programming Languages (DBPL)*, pages 222–238, Estes Park, Colorado, 1997.
11. Sushant Patnaik and Neil Immerman. Dyn-FO: A parallel, dynamic complexity class. In *ACM Symposium on Principles of Database Systems (PODS)*, pages 210–221, Minneapolis, Minnesota, 1994.
12. G. Ramalingam and Thomas Reps. On the computational complexity of dynamic graphs problems. *Theoretical Computer Science*, 158:233–277, 1996.
13. Liam Roditty and Uri Zwick. Improved dynamic reachability algorithms for directed graphs. In *IEEE Symposium on Foundations of Computer Science (FOCS)*, pages 679–690, Vancouver, Canada, 2002.
14. Oleg V. Sokolsky and Scott A. Smolka. Incremental model checking in the modal mu-calculus. In *International Computer Aided Verification Conference*, pages 351–363, Stanford, California, 1994.
15. Gitanjali Swamy. Incremental methods for FSM traversal. In *International Conference on Computer Design (ICCD)*, pages 590–595, Austin, Texas, 1995.
16. Gitanjali Swamy. *Incremental Methods for Formal Verification and Logic Synthesis.* PhD thesis, University of California, Berkeley, 1996.
17. Moshe Y. Vardi and Pierre Wolper. An automata-theoretic approach to automatic program verification. In *IEEE Symposium on Logic in Computer Science (LICS)*, pages 322–331, Cambridge, Massachusetts, 1986.
18. Heribert Vollmer. *Introduction to Circuit Complexity : a Uniform Approach.* Springer, 1999.

Coping Polynomially with Numerous but Identical Elements within Planning Problems

Max Kanovich[1] and Jacqueline Vauzeilles[2]

[1] University of Pennsylvania, 200 S. 33rd St, Philadelphia, PA 19104
kanovich@saul.cis.upenn.edu
[2] LIPN, UMR CNRS 7030, Paris 13, 99 Av.J.-B.Clément, 93430 Villetaneuse, France
jv@lipn.univ-paris13.fr

Abstract. Since the typical AI problem of making a plan of the actions to be performed by a robot so that it could get into a set of final situations, if it started with a certain initial situation, is generally exponential (it is even EXPTIME-complete in the case of games 'Robot against Nature'), the planners are very sensitive to the number of variables, the inherent symmetry of the problem, and the nature of the logic formalisms being used. The paper shows that linear logic provides a convenient tool for representing planning problems. In particular, the paper focuses on planning problems with an unbounded number of functionally identical objects. We show that for such problems linear logic is especially effective and leads to dramatic contraction of the search space (polynomial instead of exponential). The paper addresses the key issue: "How to automatically recognize functions similarity among objects and break the extreme combinatorial explosion caused by this symmetry," by means of replacing the unbounded number of specific names of objects with one generic name and contracting thereby the exponential search space over 'real' objects to a small polynomial search space but over the 'generic' one, with providing a more abstract formulation whose solutions are proved to be directly translatable into (optimal) polytime solutions to the original planning problem.

Keywords: linear logic.

1 Motivating Examples

The aim of this paper is to show that the linear logic formalism can automatically exploit peculiarities of the AI systems under consideration, and achieve a significant speedup over the traditional ones by decreasing the combinatorial costs associated with searching large spaces.

There are a number of logical formalisms for handling the typical AI problem of *making a plan* of the actions to be performed by a robot so that it could get into a set of *final situations*, if it started with a certain *initial situation* (e.g., see [17, 4, 16, 9, 14, 13]). Since the planning problem is generally PSPACE-hard (it is even EXPTIME-complete in the case of games 'Robot against Nature' [13]), the planners are very sensitive to the number of variables, functions similarity among objects, and the nature of the logic formalism being used. E.g., the

M. Baaz and J.A. Makowsky (Eds.): CSL 2003, LNCS 2803, pp. 285–298, 2003.

computer-aided planners run into difficulties caused by the combinatorial costs associated with searching large spaces for the following planning problems in AIPS Planning Competitions [1, 2] (their *common sense* solutions are obvious !).

Example 1. "Gripper" [1]: There is a robot with two grippers. It can carry a ball in each. The goal is to take N balls from one room to another[1].

Example 2. "Elevator" [2]: There is an elevator such that only 6 people can be board on it at a time. The goal is to move N passengers to their destination.

The situation becomes much more complicated in the case of the *knowledge acquisition* games where we have to look for *winning strategies* against Nature, as in the following folklore example.

Example 3. "Shoes": You have N pairs of shoes in your closet, all identical! And your hands are numb so you can't tell if the shoes that you grab are right or left handed (footed). The challenge is to pick out enough shoes to make sure you have at least one shoe for each foot.

Our paper focuses on planning problems with an unbounded number of functionally identical objects. To address the key issue: "How to automatically recognize functions similarity among objects and break the combinatorial explosion caused by this symmetry," we show that a linear logic approach can automatically aid in detecting symmetries in planning domains, with providing a radical reduction of the number of variables, and thereby *polynomial* solutions to such planning problems.

The main idea is as follows. First, having detected symmetry within a given system, we break it by replacing the *unbounded number* of specific names of objects with *one* 'generic' name, so that one can solve a *mock* 'generic' problem with a drastically smaller state space (polynomial instead of exponential) but over the 'generic' object. Secondly, each of the *mock* solutions dealing with one 'generic' object is proved to be directly translatable into an (optimal) polytime solution to the original planning problem dealing with the unbounded number of 'real' objects. In other words, for a problem in which N objects, say b_1,b_2,..,b_N, *cannot be particularized*, first, we treat them as *identical copies* of a single 'generic' object b, with the *set* $\{b_1, b_2, .., b_N\}$ being replaced by the *multiset* $\{b, b, .., b\}$ (which provides a *polynomial* search space but over one 'generic' b). Secondly, having found a solution to the 'generic' planning problem, we have to convert it into a solution to the original problem.

[1] An excerpt from [1]: "STRIPS representation leads to a combinatorial explosion in the search space. All planners obviously suffer from this explosion, with the exception of HSP that does quite well. Interestingly, HSP plans only transport one ball at a time leading to lots of unnecessary move actions ... . IPP has been run with RIFO switched on, which excludes one gripper as irrelevant, ie. non-optimal plans are found using only the left gripper... . IPP and BLACKBOX don't even provide data on the harder problem instances."

Contrary to traditional, a.k.a. set-theoretical, logics, linear logic is capable of direct handling *multiset* constructs. The fact that *"two copies of b have property Q"* can be directly expressed as the formula $(Q(b) \otimes Q(b))$ within the LL framework, with $\otimes$ representing the *coexistent* "and". But the situation is generally more subtle and messy. E.g., a formula of the form $(P(b) \otimes Q(b))$ can be interpreted (*and used!*) in two different ways: *"one and the same copy of b has both properties P and Q"*, or *"one copy of b has property P, and another has property Q"*. In the case of a single *action* of the form $(P(z) \otimes Q(z)) \vdash R(z)$ (which presupposes the *former interpretation*) the above *erasing individuality* trick fails: a 'real' planning problem of the form $(P(b_1) \otimes Q(b_2)) \Rightarrow (R(b_1) \oplus R(b_2))$ is unsolvable, notwithstanding that the 'generic' problem $(P(b) \otimes Q(b)) \Rightarrow R(b)$ has a one-step solution. On the other hand, the *latter interpretation* is more preferable in the case where $(P(b) \otimes Q(b))$ describes a 'generic' *configuration* of the system. To resolve this conflict, we invoke axioms that are *monadic* with respect to such a z. (See Definition 2)

We illustrate our approach with Example 4, which combines basic features of two combinatorially exploded examples "Gripper" and "Elevator" [1, 2].

Example 4. *"Briareus"*[2]: A robot has k grips. It can carry a ball in each. The goal is to take N balls from one room to another.

The number of total cases to be investigated in the planning process seems to be at least *exponential* $\Omega(2^N)$, since each of the balls, say $b_1, b_2, .., b_N$, has at least two states: "in room 1", or "in room 2". But this combinatorial explosion stems primarily from the fact that we are dealing with the *set* $\{b_1, b_2, .., b_N\}$ of N *distinct* names. Whereas the balls are supposed to be *identical* and the initial and final configurations are *symmetric* w.r.t. the ball's individual names. Therefore, we will rather deal with the *multiset* $\{b, b, .., b\}$ consisting of N copies of one 'generic' ball b, bearing in mind that, because of commutativity of multisets, the number of the corresponding 'generic' cases will be *polynomial*. Another source of a combinatorial explosion $\Omega(2^k)$ here is the unbounded number of grips, say $h_1, h_2, .., h_k$. Since the grips are also *indistinguishable* and *interchangeable* within configurations of the system, the collection of all grips could be thought of as the *multiset* $\{h, h, .., h\}$ consisting of k copies of one 'generic' grip h.

Let $\mathtt{R}(x)$ mean *"the robot is in room x"*, $\mathtt{H}(y, z)$ mean *"grip y holds ball z"*, $\mathtt{E}(y)$ mean *"grip y of the robot is empty"*, and $\mathtt{F}(x, z)$ mean *"ball z is on the floor of room x"*. Here x is of *sort* 'room', y is of *sort* 'grip', and z is of *sort* 'ball'.

The 'pick up' action: *"Being in room x and having the empty grip y, grasp and hold ball z with the y"*, is formalized by the following 'Horn clause'

$$\mathtt{pick}(x, y, z) := (\mathtt{R}(x) \otimes \mathtt{E}(y) \otimes \mathtt{F}(x, z)) \vdash (\mathtt{R}(x) \otimes \mathtt{H}(y, z)) \tag{1}$$

The 'put down' action: *"Being in room x and holding ball z in grip y, put the z down on the floor, and leave the y empty"*, is specified as

$$\mathtt{put}(x, y, z) := (\mathtt{R}(x) \otimes \mathtt{H}(y, z)) \vdash (\mathtt{R}(x) \otimes \mathtt{E}(y) \otimes \mathtt{F}(x, z)) \tag{2}$$

[2] During the battle against the Titans, *Briareus*, a hundred-handed giant, took advantage of his one hundred hands by throwing rocks at the Titans. Indeed, Briareus would have failed if he had wasted his time to *particularize* the rocks and hands !

The 'move' action: *"Move from room x_1 to room x_2"*, is axiomatized as

$$\texttt{move}(x_1, x_2) := \texttt{R}(x_1) \vdash \texttt{R}(x_2). \tag{3}$$

Within the original planning problem in Example 4

$$\texttt{In}_{N,k}(h_1, h_2, .., h_k, b_1, b_2, .., b_N) \Rightarrow \texttt{Goal}_N(b_1, b_2, .., b_N) \tag{4}$$

we look for a *plan* of actions (1)-(3) that leads from the initial situation *"The robot is in room 1 with N balls $b_1, b_2, .., b_N$, and its grips $h_1, h_2, .., h_k$ are empty"*:

$$\texttt{In}_{N,k}(h_1, .., h_k, b_1, .., b_N) = (\texttt{R}(1) \otimes \bigotimes_{j=1}^{k} \texttt{E}(h_j) \otimes \bigotimes_{i=1}^{N} \texttt{F}(1, b_i)), \tag{5}$$

into a situation where *"All N balls are in room 2"*:

$$\texttt{Goal}_N(b_1, .., b_N) := \bigotimes_{i=1}^{N} \texttt{F}(2, b_i). \tag{6}$$

Now we *mock* this original (4) with the following 'generic' planning problem

$$\texttt{In}_{N,k}(h, h.., h, b, b, .., b) \Rightarrow \texttt{Goal}_N(b, b, .., b) \tag{7}$$

The advantage of our *erasing individuality* trick is that the number of all 'generic' configurations that can be generated from 'generic' $\texttt{In}_{N,k}(h, h.., h, b, b, .., b)$ by means of actions (1)-(3) turns out to be $O(kN)$, which allows us to find (in `polytime`) the *shortest* plans **but** for the 'generic' problem (7) (See Fig. 1).

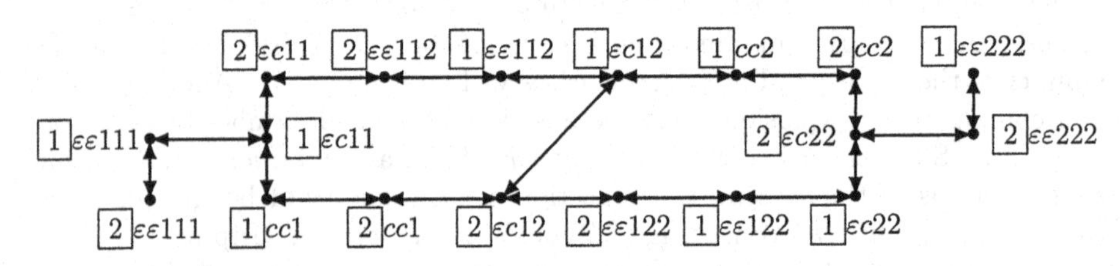

Fig. 1. The search space for $N = 3$, $k = 2$. Here "R(1)" is abbreviated as "1", "R(2)" as "2", "E(h)" as "ε", "H(h, b)" as "c", "F($1, b$)" as "1", and "F($2, b$)" as "2".

To complete (and *justify*) our approach, it still remains to prove that the *mock* 'generic' plans found can be easily converted into plans of the actions within the real world.
Suppose we have to translate the following 'generic' plan:

(1) apply $\texttt{pick}(1, h, b)$ to $(\texttt{R}(1) \otimes \underline{\texttt{E}(h)} \otimes \texttt{E}(h) \otimes \underline{\texttt{F}(1, b)} \otimes \texttt{F}(1, b) \otimes \texttt{F}(1, b))$,
resulting in $(\texttt{R}(1) \otimes \texttt{H}(h, b) \otimes \overline{\texttt{E}(h) \otimes} \texttt{F}(1, b) \otimes \overline{\texttt{F}(1, b)})$;
(2) apply $\texttt{pick}(1, h, b)$ to $(\texttt{R}(1) \otimes \texttt{H}(h, b) \otimes \underline{\texttt{E}(h)} \otimes \underline{\texttt{F}(1, b)} \otimes \texttt{F}(1, b))$,
resulting in $(\texttt{R}(1) \otimes \texttt{H}(h, b) \otimes \texttt{H}(h, b) \otimes \texttt{F}(\overline{1, b}))$;

(3) apply $\texttt{move}(1,2)$; etc., etc., eventually reaching the 'generic' goal
$(\mathtt{R}(2) \otimes \mathtt{E}(h) \otimes \mathtt{E}(h) \otimes \mathtt{F}(2,b) \otimes \mathtt{F}(2,b) \otimes \mathtt{F}(2,b))$.

$$\boxed{1}\,\varepsilon\varepsilon 111 \to \boxed{1}\,\varepsilon c11 \to \boxed{1}\,cc1 \to \cdots \to \boxed{2}\,\varepsilon\varepsilon 222$$

We individualize the 'generic' names, with providing a path through configurations in the real world:

(1) apply $\texttt{pick}(1,h_1,b_1)$ to $(\mathtt{R}(1) \otimes \underline{\mathtt{E}(h_1)} \otimes \mathtt{E}(h_2) \otimes \underline{\mathtt{F}(1,b_1)} \otimes \mathtt{F}(1,b_2) \otimes \mathtt{F}(1,b_3))$, resulting in $(\mathtt{R}(1) \otimes \mathtt{H}(h_1,b_1) \otimes \overline{\mathtt{E}(h_2)} \otimes \mathtt{F}(1,b_2) \overline{\otimes \mathtt{F}(1,b_3)})$;
(2) apply $\texttt{pick}(1,h_2,b_2)$ to $(\mathtt{R}(1) \otimes \mathtt{H}(h_1,b_1) \otimes \underline{\mathtt{E}(h_2)} \otimes \underline{\mathtt{F}(1,b_2)} \otimes \mathtt{F}(1,b_3))$, resulting in $(\mathtt{R}(1) \otimes \mathtt{H}(h_1,b_1) \otimes \mathtt{H}(h_2,b_2) \otimes \mathtt{F}(\overline{1,b_3}))$;
(3) apply $\texttt{move}(1,2)$; etc., etc., eventually reaching the real goal
$(\mathtt{R}(2) \otimes \mathtt{E}(h_1) \otimes \mathtt{E}(h_2) \otimes \mathtt{F}(2,b_1) \otimes \mathtt{F}(2,b_2) \otimes \mathtt{F}(2,b_3))$.

Theorems 1 and 2 provide an easy-to-check syntactical criterion for our *erasing individuality* technique to be *automatically correct* within Example 4. ■

Example 5 illustrates confusing subtleties of *asymmetric* input-outputs even within the same 'symmetric' domain of Example 4.

Example 5. There is a ball in each of two rooms, say ball b_1 in room 1, and ball b_2 in room 2. A one-handed robot is to exchange these two balls.
The 'real' planning problem in Example 5 is as follows:

$$(\mathtt{R}(1) \otimes \mathtt{E}(h_1) \otimes \mathtt{F}(1,b_1) \otimes \mathtt{F}(2,b_2)) \Longrightarrow (\mathtt{F}(2,b_1) \otimes \mathtt{F}(1,b_2)) \tag{8}$$

Erasing individuality yields the 'generic' planning problem

$$(\mathtt{R}(1) \otimes \mathtt{E}(h) \otimes \mathtt{F}(1,b) \otimes \mathtt{F}(2,b)) \Longrightarrow (\mathtt{F}(2,b) \otimes \mathtt{F}(1,b)) \tag{9}$$

which has a trivial solution "Do nothing". But such a 'generic' solution cannot give any clue to the real planning problem (8). ■

Lastly, a more complicated case of a *knowledge acquisition* game 'Robot against Nature' is illustrated with the following example.

Example 6. "Rouge et Noir": A robot deals with N balls wrapped with paper. Each ball is either red or black. The goal is to unwrap enough balls to be sure to obtain exactly n one-colored balls.

Let $\mathtt{W}(z)$ mean *"ball z is wrapped with paper"*, $\mathtt{R}(z)$ mean *"an unwrapped ball z has turned out to be red"*, and $\mathtt{B}(z)$ mean *"an unwrapped ball z is black"*. The 'learn' action: *"Unwrap ball z, with its color being revealed"*, is formalized as

$$\texttt{learn}(z) := \mathtt{W}(z) \vdash (\mathtt{R}(z) \oplus \mathtt{B}(z)); \tag{10}$$

the *disjunctive* $\oplus$-form of which emphasizes that the effect of this unwrapping is *non-deterministic*: the robot does not know, *in advance*, which colour the ball chosen will be of.

For N balls, say b_1,b_2,..,b_N, let $\mathtt{In}_N(b_1,..,b_N) = \bigotimes_{i=1}^{N} \mathtt{W}(b_i)$,
and $\mathtt{Goal}_{N,n}(b_1,..,b_N) = \bigoplus_{\{i_1,..,i_n\}} \bigotimes_{j=1}^{n} \mathtt{R}(b_{i_j}) \oplus \bigoplus_{\{i_1,..,i_n\}} \bigotimes_{j=1}^{n} \mathtt{B}(b_{i_j})$

(here $\{i_1,..,i_n\}$ ranges over all n-element subsets of $\{1,..,N\}$), then the 'real' planning problem in Example 6 is $\mathtt{In}_N(b_1,b_2,..,b_N) \Rightarrow \mathtt{Goal}_{N,n}(b_1,b_2,..,b_N)$.

Any solution to the problem seems to be at least *exponential*, since a plan we are looking for here is a *winning strategy* that should envisage *all* possible reactions of Nature on the road from the *initial position* to a *final position*, and the number of the corresponding 'red-black' distributions is $\Omega(2^N)$.
Our *erasing individuality* trick yields the *mock* 'generic' planning problem

$$\mathtt{In}_N(b,b,..,b) \Rightarrow \mathtt{Goal}_{N,n}(b,b,..,b) \tag{11}$$

with $\mathtt{Goal}_{N,n}(b,b,..,b) = \underbrace{(\mathtt{R}(b) \otimes \cdots \otimes \mathtt{R}(b))}_{n\ times} \oplus \underbrace{(\mathtt{B}(b) \otimes \cdots \otimes \mathtt{B}(b))}_{n\ times}$.

The number of all 'generic' configurations in question is $O(N^3)$. Hence, a *mock* winning strategy to (11), if any, can be assembled in *polytime* in a bottom-up manner (see Theorem 4). For $N{=}3$, $n{=}2$, the result is shown in Fig. 2. Each vertex v prescribes that the robot performs a certain action: its outgoing edges show all effects of the action. E.g., at the *initial position* WWW, $\mathtt{learn}(b)$ is applied, with two outgoing edges showing two possible Nature's reactions: RWW and BWW.

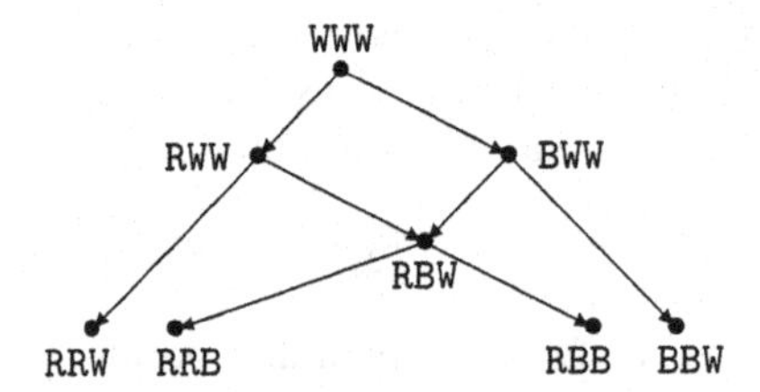

Fig. 2. The 'generic' winning strategy. Here W stands for $\mathtt{W}(b)$, R for $\mathtt{R}(b)$, and B for $\mathtt{B}(b)$.

By Theorem 2, we can convert this 'generic' winning strategy into a winning strategy over the real world (following this strategy, the robot can never ever be punished by Nature's choice of the color of the three balls):

(a) At the *initial position* $(\mathtt{W}(b_1) \otimes \mathtt{W}(b_2) \otimes \mathtt{W}(b_3))$, apply $\mathtt{learn}(b_1)$, resulting either in $(\mathtt{R}(b_1) \otimes \mathtt{W}(b_2) \otimes \mathtt{W}(b_3))$, or in $(\mathtt{B}(b_1) \otimes \mathtt{W}(b_2) \otimes \mathtt{W}(b_3))$.
(b) At a position of the form $(\mathtt{R}(b_1) \otimes \mathtt{W}(b_2) \otimes \mathtt{W}(b_3))$, apply $\mathtt{learn}(b_2)$.
(c) At a position of the form $(\mathtt{B}(b_1) \otimes \mathtt{W}(b_2) \otimes \mathtt{W}(b_3))$, apply $\mathtt{learn}(b_2)$.
(d) At a position of the form $(\mathtt{R}(b_{\pi(1)}) \otimes \mathtt{B}(b_{\pi(2)}) \otimes \mathtt{W}(b_{\pi(3)})$ (for some permutation π), apply $\mathtt{learn}(b_{\pi(3)})$.

2 The Disjunctive Horn Linear Logic

We will use the following syntactical conventions. A number of *sorts* $\tau_1,..,\tau_s$ is assumed. A formula of the form $P_1(t_{1,1},..,t_{1,k_1}) \otimes \cdots \otimes P_m(t_{m,1},..,t_{m,k_m})$, where $P_1,..,P_m$ are predicate symbols, $t_{1,1},..,t_{m,k_m}$ are variables or constants of the corresponding sorts $\tau_1,..,\tau_s$, is called an *LL-monomial* of *degree* m. The *trivial* LL-monomial V of *degree* 0 is allowed, for which $X \otimes V = X$. Formulas of the form $(Z_1 \oplus \cdots \oplus Z_k)$, with $Z_1,..,Z_k$ being LL-monomials, are called *LL-polynomials*. LL-monomials and LL-polynomials are taken modulo commutativity and associativity of $\otimes$ and $\oplus$. A *disjunctive Horn clause* is a sequent of the form $X \vdash (Y_1 \oplus Y_2 \oplus \cdots \oplus Y_\ell)$ where X, $Y_1,Y_2,..,Y_\ell$ are LL-monomials $(\ell \geq 1)$. Intuitively, this Horn clause represents a *move* of a robot based on *precondition* X accomplished by one of the possible *responds* $Y_1,Y_2,..,Y_\ell$ of Nature. A Horn clause of the form $X \vdash Y_1$ will be also called a *pure Horn clause*. Let $\#_z(X)$ denote the number of all occurrences of the z in an LL-monomial X. For LL-monomials $Y_1,Y_2,..,Y_\ell$, $\#_z(Y_1 \oplus Y_2 \oplus \cdots \oplus Y_\ell) := \max\{\#_z(Y_j) \mid j = 1, 2, .., \ell\}$

Given a sort τ, we will write $A(z_1, z_2, .., z_n)$ to indicate that the list $z_1,z_2,..,z_n$ contains all variables of the sort τ that occur in the A. By *symmetric closure* $A^{Sym}(z_1, z_2, .., z_n)$ of a given $A(z_1, z_2, .., z_n)$, we mean the following formula $\bigoplus_\pi A(z_{\pi(1)}, z_{\pi(2)}, .., z_{\pi(n)})$, where π ranges over all permutations of $\{1, .., n\}$. A formula $A(z_1, z_2, .., z_n)$ is *symmetric* if, for any permutation π, a sequent of the form $A(z_1, z_2, .., z_n) \vdash A(z_{\pi(1)}, z_{\pi(2)}, .., z_{\pi(n)})$ is derivable in linear logic.

The desired property "generic plans $\Longrightarrow$ real plans" is represented as follows.

Definition 1. *Let* $\mathtt{Ax}_T$ *be a set of 'proper axioms' specifying the theory* T. *A given* sort τ *is said to be* **generic** *within AL-theory* T, *if the following holds: whatever variables* $z,z_1,z_2,..,z_n$ *of sort* τ, *LL-monomial* $W(z_1, z_2, .., z_n)$, *in which each of the* z_i *has exactly one occurrence, and LL-polynomial* $\widetilde{Z}(z_1, z_2, .., z_n)$ *such that* $\#_{z_i}(\widetilde{Z}) \leq 1$ *for each* z_i, *we take, if a sequent of the form*

$$W(z, z, .., z) \vdash \widetilde{Z}(z, z, .., z) \tag{12}$$

is derivable from $\mathtt{Ax}_T$ *by the rules of affine logic*[3] *then a sequent of the form*

$$W(z_1, z_2, .., z_n) \vdash \widetilde{Z}^{Sym}(z_1, z_2, .., z_n) \tag{13}$$

is also derivable from $\mathtt{Ax}_T$ *by the rules of affine logic.*

(a) The condition on the number of occurrences of z_i makes sense. Let, for instance, $W(z_1, z_2) := (P(z_1) \otimes Q(z_2))$, and $Z(z_1, z_2) := (P(z_1) \otimes Q(z_1))$. Then $W(z, z) \vdash Z(z, z)$ is derivable, but $W(z_1, z_2) \vdash (Z(z_1, z_2) \oplus Z(z_2, z_1))$ is not.

(b) We have need of the permutations in Definition 1. Let, for instance, $W(z_1, z_2) := (P(z_1) \otimes Q(z_2))$, and $Z(z_1, z_2) := (Q(z_1) \otimes P(z_2))$. Then $W(z, z) \vdash Z(z, z)$ is derivable, but $W(z_1, z_2) \vdash Z(z_1, z_2)$ is not.

Proposition 1. *Let* τ *be a generic sort within theory* T. *Let* $z_1,z_2,..,z_n$ *be variables of sort* τ, *and* $W(z_1, z_2, .., z_n)$ *be an LL-monomial in which each of the* z_i *has exactly one occurrence, and* $\widetilde{Z}(z_1, z_2, .., z_n)$ *be a* <u>*symmetric*</u> *LL-polynomial such that* $\#_{z_i}(\widetilde{Z}) \leq 1$ *for each* z_i. *Then for any constants* b, $b_1,b_2,..,b_n$ *of sort* τ *that have no occurrence in* $\mathtt{Ax}_T$, W, *and* $\widetilde{Z}$, *the sequent*

$$W(b_1, b_2, .., b_n) \vdash \widetilde{Z}(b_1, b_2, .., b_n) \tag{14}$$

[3] Affine logic = linear logic + Weakening rule.

is derivable from $\mathtt{Ax}_T$ *in affine logic if and only if a sequent of the 'generic' form*

$$W(b,b,..,b) \vdash \widetilde{Z}(b,b,..,b) \tag{15}$$

is derivable from $\mathtt{Ax}_T$ *by the rules of affine logic.*

Thus, the original *planning problem (14) dealing with a variety of* n *'real objects' can be fully sorted out in terms of the 'generic' planning problem (15) dealing with only* one *'generic object'.*

3 'Generic' Plans $\Longrightarrow$ 'Real' Plans

Definition 2. *Let* $\#_\tau(X)$ *denote the number of all occurrences of variables of sort* τ *in an LL-monomial* X.
$\#_\tau(Y_1 \oplus \cdots \oplus Y_\ell) := \max\{\#_\tau(Y_j) \mid j = 1,2,..,\ell\}$, *for LL-monomials* $Y_1,..,Y_\ell$.
A disjunctive Horn clause $X \vdash (Y_1 \oplus \cdots \oplus Y_\ell)$ *is said to be* τ-monadic *if* $\#_\tau(Y_1 \oplus \cdots \oplus Y_\ell) \le \#_\tau(X) \le 1$.

We will use the following strong version of the *normalization lemma* for Horn linear logic (cf. [12]).

Lemma 1. *Let* $\mathtt{Ax}_T$ *consist of disjunctive Horn clauses.*
Given LL-monomials W, $Z_1,Z_2,..,Z_k$, *let a sequent of the form* $W \vdash \widetilde{Z}$ *where* $\widetilde{Z} = (Z_1 \oplus \cdots \oplus Z_k)$, *be derivable from* $\mathtt{Ax}_T$ *in affine logic. Then* $W \vdash \widetilde{Z}$ *is derivable from* $\mathtt{Ax}_T$ *by means of the following rules*

$$\frac{(V \otimes Y_1) \vdash \widetilde{Z} \quad (V \otimes Y_2) \vdash \widetilde{Z} \quad \ldots \quad (V \otimes Y_\ell) \vdash \widetilde{Z}}{(V \otimes X) \vdash \widetilde{Z}} \tag{16}$$

(for $X \vdash (Y_1 \oplus Y_2 \oplus \cdots \oplus Y_\ell)$ *is an instance of a sequent from* $\mathtt{Ax}_T$, *and* V *is any LL-monomial) and*

$$\frac{}{(Z_i \otimes Z') \vdash \widetilde{Z}} \tag{17}$$

(Z' *is any LL-monomial).*

Theorem 1. *For a given sort* τ, *let* $\mathtt{Ax}_T$ *consist only of* τ*-monadic disjunctive Horn clauses. Then sort* τ *is generic within AL-theory* T, *and, in addition to that, whatever* $z_1,z_2,..,z_n$, *variables of sort* τ, $W(z_1,z_2,..,z_n)$, *an LL-monomial, in which each of the* z_i *has exactly one occurrence, and* $\widetilde{Z}(z_1,z_2,..,z_n)$, *an LL-polynomial such that* $\#_{z_i}(\widetilde{Z}) \le 1$ *for each* z_i, *we take, every AL-proof within theory* T *for (12) can be translated into an AL-proof within theory* T *for (13).*

Proof. Given an AL-proof for (12), first, by Lemma 1 we translate it into a derivation tree with rules (16) and (17). Then we assemble the desired proof for (13) by induction on this derivation. Let us sketch the basic case of rule (16).

Suppose that $X(v) \vdash (Y_1(v) \oplus Y_2(v))$ is a τ-monadic Horn clause from $\mathtt{Ax}_T$, and v is a variable of sort τ, and for some LL-monomial $V(z_2,..,z_n)$, the sequent $V(z,..,z) \otimes X(z) \vdash \widetilde{Z}(z,z,..,z)$ is produced by rule (16) from the sequents

$V(z,..,z) \otimes Y_1(z) \vdash \widetilde{Z}(z,z,..,z)$ and $V(z,..,z) \otimes Y_2(z) \vdash \widetilde{Z}(z,z,..,z)$. By the hypothesis, two sequents of the form $V(u_2,..,u_n) \otimes Y_1(u_1) \vdash \widetilde{Z}^{Sym}(u_1,u_2,..,u_n)$ and $V(v_2,..,v_n) \otimes Y_2(v_1) \vdash \widetilde{Z}^{Sym}(v_1,v_2,..,v_n)$ are derivable by rules (16) and (17), where $u_1,..,u_n$, and $v_1,..,v_n$ are permutations of $z_1,..,z_n$. Since $\widetilde{Z}^{Sym}(z_1, z_2,..,z_n)$ is symmetric, $V(z_2,..,z_n) \otimes Y_1(z_1) \vdash \widetilde{Z}^{Sym}(z_1,z_2,..,z_n)$ and $V(z_2,.., z_n) \otimes Y_2(z_1) \vdash \widetilde{Z}^{Sym}(z_1,z_2,..,z_n)$ are also derivable, and the same rule (16) provides derivability of $V(z_2,..,z_n) \otimes X(z_1) \vdash \widetilde{Z}^{Sym}(z_1,z_2,..,z_n)$. ■

3.1 Plans, a.k.a. Winning Strategies

For the systems with pure deterministic actions, a plan $\mathcal{P}$ is defined as a chain of the actions [17, 4, 16]. In order to cover the general case where the actions with *non-deterministic effects* are allowed, we extend this definition to graph plans $\mathcal{P}$, in which each vertex v prescribes the performance of a certain action, with its outgoing edges showing all possible reactions of Nature.

Definition 3. *Let* $\mathtt{Ax}_T$ *be a set of disjunctive Horn clauses specifying actions in a given robot system.*

A plan $\mathcal{P}$ *for a given problem* $W \Rightarrow (Z_1 \oplus \cdots \oplus Z_k)$ *is a finite directed graph having no directed cycles such that*

(a) each node v *with exactly* $\ell \geq 1$ *sons* $w_1,..,w_\ell$, *is labelled by an instance of a Horn clause from* $\mathtt{Ax}_T$ *(representing the* action *performed at* v*) of the form*

$$X \vdash (Y_1 \oplus \cdots \oplus Y_\ell) \tag{18}$$

and by an LL-monomial (representing the position *at* v*) of the form* $(X \otimes V)$*;*

(b) the outgoing edges (v,w_1), .., (v,w_ℓ) *are labeled by pure Horn clauses* $X \vdash Y_1$, .., $X \vdash Y_\ell$, *respectively;*

(c) the position each w_j *is labeled by must be equal to* $(Y_j \otimes V)$ *(showing all possible effects of action (18));*

(d) there is exactly one node having no incoming edges (the root*); the position at the root is* W*;*

(e) each node v *having no outgoing edges (a* terminal *node) is labeled only by LL-monomial of the form* $(Z_i \otimes Z')$ *(representing a* final position*).*

In many cases Definition 3 yields the exponential size of winning strategies by pure technical reasons, even if they have a uniform structure. Therefore, we make Definition 3 more liberal in the following respect. We will allow to label any node v by a *position pattern* of the form $(X(z_1) \otimes V(z_2,..,z_n))$, meaning that being at a position of the form $(X(c_1) \otimes V(c_2,..,c_n))$, where $c_1,c_2,.., c_n$ is a permutation of $b_1,b_2,.., b_n$, the robot should apply the corresponding version of the action this node v is labeled by. (See Example 6)

Theorem 2. *For a given sort* τ, *let* $\mathtt{Ax}_T$ *consist only of* τ*-monadic disjunctive Horn clauses. Let* $z_1,z_2,..,z_n$ *be variables of sort* τ, *and* $W(z_1,z_2,..,z_n)$ *be an LL-monomial in which each of the* z_i *has exactly one occurrence, and*

$\widetilde{Z}(z_1, z_2, .., z_n)$ be a <u>*symmetric*</u> *LL-polynomial such that $\#_{z_i}(\widetilde{Z}) \leq 1$ for each z_i. Let b, b_1,b_2,..,b_n be constants of sort τ having no occurrence in $\mathtt{Ax}_T$, W, and $\widetilde{Z}$. Then every cut-free AL-proof within theory T for (15) (dealing with* one *'generic object' b) can be converted (in* `polytime`*) into a solution, a winning strategy, to the planning problem (14) dealing with n 'real objects'.*

Proof Sketch. A given AL-proof is transformed into a derivation D for (15) with rules from Lemma 1. Because of τ-monadicness, each of the sequents in this D is of the form $U(b,..,b) \vdash \widetilde{Z}(b,..,b)$, for some LL-monomial $U(z_1,..,z_n)$.

The desired winning strategy $\mathcal{S}$ is constructed as follows.

The left-hand sides $U(b,..,b)$ of *different* sequents from D are taken as the nodes of $\mathcal{S}$. The *position pattern* the node $U(b,..,b)$ must be labeled by is defined as $U(z_1,..,z_n)$. By D_U we denote one of sub-derivations whose root is of the form $U(b,..,b) \vdash \widetilde{Z}(b,..,b)$ and its height is minimal. Suppose that in D_U its root is produced by rule (16) from the sequents just above

$$U_1(b,..,b) \vdash \widetilde{Z}(b,..,b), \quad \dots \quad U_\ell(b,..,b) \vdash \widetilde{Z}(b,..,b),$$

invoking an instance of a sequent from $\mathtt{Ax}_T$ of the form

$$X \vdash (Y_1 \oplus \cdots \oplus Y_\ell) \tag{19}$$

Then we label this node $U(b,..,b)$ by (19), make arrows from $U(b,..,b)$ to each of the nodes $U_1(b,..,b)$, $\dots$, $U_\ell(b,..,b)$, and label these edges by $X \vdash Y_1$, $\dots$, $X \vdash Y_\ell$, respectively. (See Figure 2)

Notice that the size of the plan $\mathcal{S}$ constructed is bounded by the number of *different* sequents in derivation D. ∎

Corollary 1. *Within Example 4,*
the 'real' planning problem $\mathtt{In}_{N,k}(h_1, h_2, .., h_k,\ b_1, b_2, .., b_n) \vdash \mathtt{Goal}_N(b_1, b_2, .., b_n)$
can be fully sorted out in terms of
the 'generic' planning problem $\mathtt{In}_{N,k}(h, h, .., h,\ b, b, .., b) \vdash \mathtt{Goal}_N(b, b, .., b)$.

4 Complexity

4.1 Weighted Balance $\longrightarrow$ EXPTIME Decidability

The (propositional) disjunctive Horn linear logic is undecidable [12]. By means of the "mixed balance" we show here that the planning problem is decidable for a reasonably wide class of natural robot systems (cf. [13]).

Suppose that sorts τ_1,...,τ_m are fixed. Given an LL-monomial X, we define its mixed weight $\omega_{\tau_1..\tau_m}(X)$ as follows:

> $\omega_{\tau_1..\tau_m}(X) = \#_{\tau_1}(X) + \cdots + \#_{\tau_m}(X) +$ the number of occurrences of atomic formulas in X that do not contain an occurrence of a variable of either of sorts τ_1, ..., τ_m.

E.g., for $Y = (\mathtt{R}(x) \otimes \mathtt{H}(y,z))$, $\omega_{ball,grip}(Y) = 3$;
for $X = (\mathtt{R}(x) \otimes \mathtt{E}(y) \otimes \mathtt{F}(x,z))$, $\omega_{ball,grip}(X) = 3$.

Definition 4. *Let* $\mathtt{Ax}_T$ *consist of disjunctive Horn clauses. We will say that* $\mathtt{Ax}_T$ *is* well-balanced *if one can find sorts* $\tau_1, \ldots, \tau_m$ $(m \geq 0)$ *such that for any sequent* $X \vdash (Y_1 \oplus \cdots \oplus Y_\ell)$ *taken from* $\mathtt{Ax}_T$ *the following holds:*
$\omega_{\tau_1..\tau_m}(Y_1) \leq \omega_{\tau_1..\tau_m}(X) \quad \ldots, \ \omega_{\tau_1..\tau_m}(Y_\ell) \leq \omega_{\tau_1..\tau_m}(X)$.

E.g., the system of axioms in Example 4 turns out to be *well-balanced*, notwithstanding that this system is not well-balanced in the strict sense of [13].

Theorem 3. *Let* $\mathtt{Ax}_T$ *be well-balanced, and* D *be a derivation for a sequent of the form* $W \vdash \widetilde{Z}$ *based on rules (16) and (17) from Lemma 1. Then one can construct (at least in EXPTIME) a solution, a winning strategy, to* $W \Rightarrow \widetilde{Z}$.

Proof. Cf. [13]. ∎

4.2 Monadic & Balanced ⟶ Polytime Planning

Theorem 4. *Let* $\mathtt{Ax}_T$ *be well-balanced, and for a given sort* τ, *let* $\mathtt{Ax}_T$ *consist only of* τ*-monadic disjunctive Horn clauses.*
Let $z_1, z_2, .., z_n$ *be variables of sort* τ, *and* $W(z_1, z_2, .., z_n)$ *be an LL-monomial in which each of the* z_i *has exactly one occurrence, and* $\widetilde{Z}(z_1, z_2, .., z_n)$ *be a* <u>*symmetric*</u> *LL-polynomial such that* $\#_{z_i}(\widetilde{Z}) \leq 1$ *for each* z_i. *Let* $b_1, b_2, .., b_n$ *be constants of sort* τ *having no occurrence in* $\mathtt{Ax}_T$, W, *and* $\widetilde{Z}$. *Then we can determine in polynomial time whether there is a plan for the problem*

$$W(b_1, b_2, .., b_n) \vdash \widetilde{Z}(b_1, b_2, .., b_n) \tag{20}$$

and, if the answer is positive, make such a plan in polynomial time.

Proof Sketch. According to Theorem 2, it suffices to take a derivation D for a "mock" sequent of the form

$$W(b, b, .., b) \vdash \widetilde{Z}(b, b, .., b) \tag{21}$$

Let m be the string length of the formula $W(b, .., b)$. By Lemma 1, we can confine ourselves to derivations D in which all sequents are of the form $U(b, .., b) \vdash \widetilde{Z}(b, .., b)$ and the degree of $U(b, .., b)$ does not exceed m.

The number of all $U(b, .., b)$ whose degree does not exceed m is polynomial. The degree of the polynomial is determined by the number of predicate symbols, their arity, and the number of constants in $\mathtt{Ax}_T$, $W(b, .., b)$, and $\widetilde{Z}(b, b, .., b)$.

Applying the bottom-up technique, we can construct in `polytime` a list of all $U(b, .., b)$ used in the proof of Theorem 2, with providing the desired winning strategy. ∎

5 Yet Another Formalism ...

What kind of arguments could we offer about linear logic superiority as a modeling formalism for the planning domain considered here?
Schematically, our approach involves the following steps:

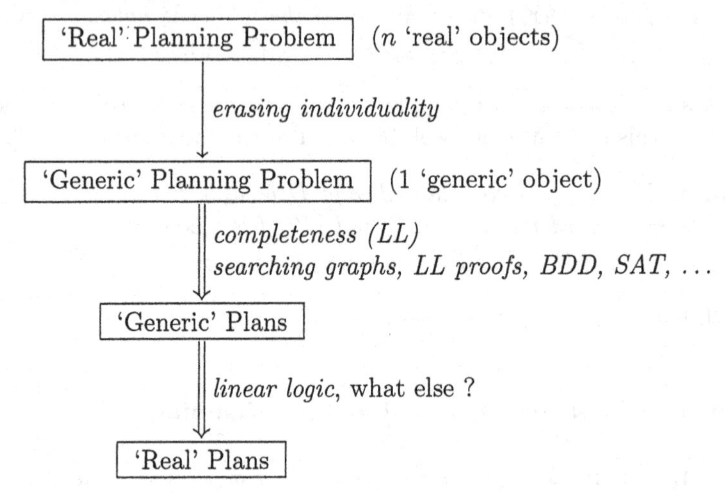

(a) 'Generic' Planning Problem $\Longrightarrow$ 'Generic' Plans
Since the "mock" search space is guaranteed to be polynomial, on the road from a specification of the "mock" generic problem to its solution one can choose among a wide spectrum of techniques: direct searching for the shortest paths, theorem proving, BDDs, SAT, etc. It should be pointed out that we are looking for plans, so that the pure decision procedure is not satisfactory (as a rule, the pure existence of a plan is almost evident). As compared to many existing logic formalisms for planning, the advantage of linear logic here is that there is a clear direct correspondence between proofs and plans.

(b) 'Generic' Plans $\Longrightarrow$ 'Real' Plans
The idea that any "mock" generic problem can be transformed to a plan over the real world seems 'orthogonal' to traditional set-theoretical logical systems but it can be easily stated, justified and explained in terms of linear logic.
Furthermore, the linear logic approach proposed here is *complete* in the sense that any planning problems with symmetries can be specified (or reformulated) in terms of the fragment of linear logic consisting of disjunctive Horn clauses.

5.1 Genericness vs. Bisimulation

Theorem 5. *For a given sort τ, let $\mathtt{Ax}_T$ consist only of τ-monadic pure Horn clauses.*
Then for any variables z, z_1,z_2,..,z_n of sort τ the following holds: whatever LL-monomial $W(z_1, z_2, .., z_n)$ in which each of the z_i has exactly one occurrence,

and LL-polynomial $\widetilde{Z}(z_1, z_2, .., z_n)$ *such that* $\#_{z_i}(\widetilde{Z}) \leq 1$ *for each* z_i, *we take, if a sequent of the form*

$$W(z, z, .., z) \vdash \widetilde{Z}(z, z, .., z) \tag{22}$$

is derivable from $\mathtt{Ax}_T$ *by the rules of affine logic, then for some permutation* π, *a sequent of the form*

$$W(z_1, z_2, .., z_n) \vdash \widetilde{Z}(z_{\pi(1)}, z_{\pi(2)}, .., z_{\pi(n)}) \tag{23}$$

is also derivable from $\mathtt{Ax}_T$ *by the rules of affine logic.*

Theorem 5 allows us to automatically detect a certain bisimulation symmetry, which can be used to construct a quotient bisimulation partition. There have been similar attempts to automatically detect symmetries in other areas such as SAT checking [3]. In brief, their approach is to take an *exponential* state space and partition it with respect to the symmetry detected. But the symmetry by itself does not help to reduce the number of variables (any representative still deals with z_1,z_2,..,z_n). We believe that our approach is more universal, uniform and efficient, since from the very beginning we start from a small "mock" space, and do not invoke the original *exponential* space at all.

Furthermore, Theorem 5 and the bisimulation caused by it fail in a general case:

Example 7. Let $\mathtt{Ax}_T$ consist of the following axioms:

$$\begin{cases} A \otimes P(z) \vdash B \otimes L(z) \oplus C \otimes R(z) \\ B \otimes P(z) \vdash D \otimes R(z) \\ C \otimes P(z) \vdash D \otimes L(z) \end{cases} \tag{24}$$

Then a "mock" sequent of the form $A \otimes P(b) \otimes P(b) \vdash D \otimes L(b) \otimes R(b)$ is derivable from (24), but neither $A \otimes P(b_1) \otimes P(b_2) \vdash D \otimes L(b_1) \otimes R(b_2)$ nor $A \otimes P(b_1) \otimes P(b_2) \vdash D \otimes L(b_2) \otimes R(b_1)$ is derivable from (24) by the linear logic rules.

Indeed, Theorem 1 provides derivability from (24) for a sequent of the 'symmetric' form

$$A \otimes P(b_1) \otimes P(b_2) \vdash D \otimes L(b_1) \otimes R(b_2) \oplus D \otimes L(b_2) \otimes R(b_1).$$

6 Concluding Remarks

We have shown that the linear logic formalism can automatically exploit peculiarities of the AI systems under consideration, and achieve a significant speedup over the traditional ones by decreasing the combinatorial costs associated with searching large spaces.

We have established a clear and easy-to-check syntactical criterion for detecting symmetry in planning domains, and developed techniques to break it by construction of a more abstract formulation whose solution can automatically aid in solving the original problem, with providing, in particular, a radical reduction of the number of variables, and thereby *polynomial* solutions to the original planning problems.

These results in some sense are "orthogonal" to traditional logical systems but are easily specified and handled in terms of linear logic, making bridge from human common-sense reasoning and problem solving to *computer-aided planning* and to the ability of the automated systems to reason effectively in complex but natural domains.

References

1. AIPS'98 Planning Systems Competition. `http://www.informatik.uni-freiburg.de/~koehler/aips.html`
2. AIPS-00 Planning Competition. `http://www.cs.toronto.edu/aips2000/`
3. Proceedings of SAT-2002, Fifth International Symposium on the Theory and Applications of Satisfiability Testing, May 6-9, 2002, Cincinnati, Ohio, USA.
4. W.Bibel. A deductive Solution for Plan Generation, *New Generation Computing* 4 (1986) pp. 115-132.
5. Tom Bylander. The Computational Complexity of Propositional STRIPS Planning, Artificial Intelligence, Volume 69, 1994, pp. 165-204.
6. Tom Bylander. A Linear Programming Heuristic for Optimal Planning, Proceedings of the Fourteenth National Conference on Artificial Intelligence, 1997, pp. 694–699.
7. K. Erol, D. Nau, J. Hendler. "A Critical Look at Critics in HTN Planning." In IJCAI-95, Montreal, August, 1995.
8. K. Erol, D. Nau, V. S. Subrahmanian."Complexity, Decidability and Undecidability Results for Domain-Independent Planning." In Artificial Intelligence, Special Issue on Planning, 76:1-2, July 1995.
9. D.McDermott and J.Hendler. Planning: What it is, What it could be, An introduction to the Special Issue on Planning and Scheduling. *Artificial Intelligence* 76 (1995) pp. 1-16.
10. J.-Y.Girard. Linear logic. *Theoretical Computer Science*, 50:1 (1987) pp.1-102.
11. J.-Y.Girard. Linear logic: its syntax and semantics, In J.-Y. Girard, Y. Lafont, and L. Regnier, editors, *Advances in Linear Logic*, London Mathematical Society Lecture Notes, Vol. 222, pp.1-42. Cambridge University Press, 1995.
12. Max Kanovich. Linear logic as a logic of computations, *Annals of Pure and Applied Logic*, 67 (1994) pp.183-212
13. Max Kanovich and Jacqueline Vauzeilles. The Classical AI Planning Problems in the Mirror of Horn Linear Logic: Semantics, Expressibility, Complexity. *Journal of Mathematical Structures in Computer Science.* (2001), vol. 11, pp. 689-716.
14. H.Kautz and B.Selman, Pushing the envelope: planning, propositional logic, and stochastic search. *Proc. AAAI-1996,* Portland, OR.
15. J.McCarthy. Applications of circumscription to formalizing common-sense knowledge. *Artificial Intelligence* 28(1986) pp. 89-116.
16. M.Masseron, C.Tollu, J.Vauzeilles. Generating Plans in Linear Logic: I. Actions as proofs, *Theoretical Computer Science,* 113 (1993) pp.349-370.
17. N.J.Nilsson, Artificial Intelligence: A New Synthesis, Morgan Kaufmann, 1998.
18. A.Scedrov. Linear logic and computation: a survey. In H.Schwichtenberg, editor, *Proof and Computation, Proc. Marktoberdorf Summer School 1993*, pp.281-298. NATO Advanced Science Institutes, Berlin, 1994.

On Algebraic Specifications of Abstract Data Types

Bakhadyr Khoussainov

Computer Science Department, The University of Auckland, New Zealand
bmk@cs.auckland.ac.nz

Abstract. In this paper we address long standing open problems of Bergstra and Tucker about specifications of abstract data types by means of equations and conditional equations. By an abstract data type we mean the isomorphism type of an algebra. An algebra is algebraically specified if the algebra can be defined uniquely, in a certain precise sense, in terms of a finite number of conditional equations by allowing functions that are not in the original language of the algebra. We provide full solutions to Bergtsra and Tucker problems, explain basic ideas, methods, and the logical dependencies between blocks of proofs used in our solutions.

1 Background

In this paper we outline the basic ideas that lead to complete solutions of long standing open problems of Bergstra and Tucker about specifiability of abstract data types by means of finite number of equations and quasiequations. The problems were formulated in the early 80s (e.g. see [1]). The first problem is concerned with finding equational specifications for abstract data types. In 1987 Kasymov solved the problem in [7] by constructing a specific computably enumerable (c.e.) algebra. In [9], the author revisited Kassymov's proof by providing a natural example of a c.e. algebra, whose construction is based on Kolmogorov complexity, that also solves the problem for equational specifications. However, the second more general problem about specifications of abstract data types by conditional equations (we call them here quasiequations) has since been open. In this paper, we give precise formulations of the problems, a necessary background, and provide basic methods and ideas used in the solutions of the problems.

The modern theory of algebraic specifications has advanced significantly since the early 80s and now possesses its own problems and directions for research. Originally born as a theory for reasoning about abstract data types by means of formal methods and algebraic techniques, the area now covers new methods for specifications and programming concepts (e.g. object-oriented, aspect-oriented, agent-oriented, higher-order logic and functional programming) with the goal of providing foundations for correctness of programs. See for example, recent series of proceedings of the *Workshop on Algebraic Development Techniques* for the current state of the area. However, in the area there are some open problems of foundational character, and the the problems addressed in this paper are among

M. Baaz and J.A. Makowsky (Eds.): CSL 2003, LNCS 2803, pp. 299–313, 2003.

such, that are still needed to be answered. The foundational nature of such problems naturally requires a deep analysis of basic underlying concepts (for this paper these concepts are the precise definitions of absratct data type, equational and quasiequational specification, computation, and deduction) and the use of advanced modern techniques and tools of mathematics for the solutions of the problems (for this paper techniques of modern computability theory and algebra are used). So, we assume that the reader is familair with standard notions from universal algebra (such as congruence relation, finitely generated and finitely presented algebra, homomorphisms, variety and quasiveriety, free algebra), logic (equations, quasiequations, proofs), computability (c.e. sets, simple sets, basics of priority argument constructions), and computable model theory (computable and c.e. algebra, c.e. presentable algebra). All these are in the basic textbooks and introductory chapters from Grätzer [5], Goncharov and Ershov [6], and Soare [11]. Of course, we provide necessary definitions. A good survey on abstract data types and algebraic specifications is Wirsing [12]. A basic paper that underlines the logical and algebraic foundations of the problems is Bergstra and Tucker [2].

Now we outline this paper. Further in this section, we provide the formulation of the problems and some background. For completeness of our exposition Section 2 briefly outlines the proof for the case of equational specifications (which as mentioned above solved one of the questions of Bersgtra and Tucker). Finally, Section 3 is devoted to describing the proof line for the case of quasiequational specifications. Some of the proofs are technical and lengthy, especially the proofs of results in the last section. Therefore we outline basic ideas of the proofs and give appropriate references in which full proofs are presented. We also concentrate on the relationship between the basic blocks of the proofs, and explain methods of algebra and computability that are used in solving the problems.

An *algebra* is a structure of a finite purely functional language (signature). Thus, any algebra $\mathcal{A}$ is of the form $(A; f_0, \dots, f_n)$, where A is a nonempty set called the domain of the algebra, and each f_i is a functional symbol that names a total operation on the domain A. Often the operation named by f_i is also denoted by the same symbol f_i. We refer to the symbols $f_0, \dots, f_n$ as the signature of the algebra. The Presburger arithmetic $(\omega; 0, S, +)$ is an algebra, so are groups, rings, lattices and Boolean algebras. Fundamental structures that arise in computer science such as lists, stacks, queues, trees, and vectors can all be viewed and studied as algebras. A foundational thesis that connects abstract data types (ADTs) with algebras and forms the basis of the work in algebraic specifications is the following stated as a definition.

Definition 1. *An* **abstract data type (ADT)** *is the isomorphism type of an algebra.*

Often abstract data types are defined to be the isomorphism types of many sorted algebras. We note, however, that the specification problems have fully been solved for many sorted algebras (see [1]), and hence for the purpose of this paper we omit the case for algebras with many sorts.

Let σ be a functional finite language (signature) with at least one constant symbol. An **equation** is an expression of the form $p = q$, and a **quasiequation** is an expression of the form $(\&_{i<\omega} p_i = q_i) \rightarrow s = t$, where p, q, s, t are terms of the signature. These terms may contain variables. An **equational specification** is a *finite* set E of equations. A **quasiequational specification** is a *finite* set Q of quasiequations. In this paper the word specification refers to either equational or quasiequational specification unless otherwise stated.

Let $\mathcal{F}(\sigma)$ be the absolutely free algebra generated by the σ–constants. Elements of this algebra are called **ground terms.** Let S be a specification. Ground terms t_1 and t_2 are $\equiv_S$–equivalent if $t_1 = t_2$ can be proved from S (within the first order logic). The relation $\equiv_S$ is a congruence relation on $\mathcal{F}(\sigma)$, and hence one can factorize $\mathcal{F}(\sigma)$ by $\equiv_S$. We denote this factor algebra by $\mathcal{F}(S)$. In universal algebra $\mathcal{F}(S)$ is called the zero generated free algebra of the quasivariety determined by S.

From a programming point of view S is thought to be a specification of an abstract data type that is being (or need to be) implemented. One thinks of S as a set of axioms needed to be satisfied in all implementations of S. Indeed, often when a programmer implements an abstract data type, e.g. stacks or queues or lists, the specification requirements put on the abstract data type are most likely equational or quasiequational in nature. For example, a specification requirement that relates **push** and **pop** operations in a stack abstract data type states that for any stack s and any item i the equaltity $\mathbf{pop}(\mathbf{push}(s,i),i) = s$ is true.

Definition 2. *An algebra $\mathcal{A}$ is* **specified by a specification** *S if $\mathcal{A}$ is isomorphic to the algebra $\mathcal{F}(S)$. If the specification S is an equational specification then the algebra $\mathcal{A}$ is* **equationally specified**. *If S is a quasiequational specification then the algebra $\mathcal{A}$ is* **quasiequationally specified**.

The arithmetic $(\omega, 0, S, +, \times)$ is an equationally specified algebra, and the equations that specify this algebra are the known recursive definitions of $+$ and $\times$ together with commutativity, associativity, and distributivity axioms. Similarly, the term model for combinatory logic is an equationally specified algebra; this is an algebra of the signature $(K, S, I, \cdot)$, where K, S, I are constant symbols and $\cdot$ is a binary operation all satisfying the following equations: $(K \cdot x) \cdot y = x$, $((S \cdot x) \cdot y) \cdot z = (x \cdot z) \cdot (y \cdot z)$, $I \cdot x = x$. Finally, all finitely presented algebras (such groups or rings) are examples of specified algebras.

In order to study specified algebras we need to employ some notions from computable algebra and model theory. Let $\mathcal{A} = (A, f_0, \ldots, f_n)$ be an algebra. For each element $a \in A$ introduce a new constant symbol c_a that names the element a itself. The **atomic diagram of** $\mathcal{A}$ is the set of all expressions of the type $f_i(\bar{c}) = c'$ or $f_i(\bar{c}) \neq c'$ which are true in the algebra $\mathcal{A}$. The **positive atomic diagram of** $\mathcal{A}$ is the set of all expressions of the type $f_i(\bar{c}) = c'$ which are true in the algebra $\mathcal{A}$. The algebra $\mathcal{A} = (A; f_0, \ldots, f_n)$ is **computable** if its atomic diagram is a computable set. The algebra $\mathcal{A} = (A; f_0, \ldots, f_n)$ is **computably enumerable (c.e.)** if its positive atomic diagram is a computably enumerable set. It is not hard to see that computable algebras are the ones isomorphic to algebras of the type $(\omega, f_0, \ldots, f_n)$, where each f_i is a computable function on ω. Clearly, every computable algebra is c.e., and the converse does not generally hold true. Here are some examples of c.e. algebras:

1. Any specified algebra, such as the arithmetic or the term model for combinatory logic.
2. The Lindenbaum Boolean algebra of any c.e. first order theory, such as the Peano arithmetic.
3. Any finitely presented group, and in fact any finitely presented algebra.

A computably enumerable (c.e.) algebra $\mathcal{A}$ can be explained as follows. As the positive diagram of $\mathcal{A}$ can be computably enumerated, the set $E = \{(c_a, c_b) \mid a = b$ is true in algebra $\mathcal{A}\}$ is computably enumerable. Hence the equality relation in $\mathcal{A}$ is c.e. Let f be a basic n-ary operation on $\mathcal{A}$. From the definition of c.e. algebra, the operation f can be thought as a function induced by a computable function, often also denoted by f, that is well-behaved on the E-equivalence classes in the following sense: for all $x_1, \ldots, x_n, y_1, \ldots, y_n$ if $(x_i, y_i) \in E$, then $(f(x_1, \ldots, x_n), f(y_1, \ldots, y_n)) \in E$. Therefore, a natural way to think about $\mathcal{A}$ is that the elements of $\mathcal{A}$ are E-equivalence classes, operations of $\mathcal{A}$ are induced by computable operations. This reasoning suggests another equivalent approach to the definition of c.e. algebra explained in the next paragraph.

Let E be a c.e. equivalence relation on ω. A computable n-ary function f **respects** E if for all natural numbers $x_1, \ldots, x_n$ and $y_1, \ldots, y_n$ so that $(x_i, y_i) \in E$, for $i = 1, \ldots, n$, we have $(f(x_1, \ldots, x_n), f(y_1, \ldots, y_n)) \in E$. Let $\omega(E)$ be the factor set obtained by factorizing ω by E, and let $f_0, \ldots, f_n$ be computable operations on ω that respect the equivalence relation E. An **E-algebra** is then the algebra $(\omega(E), F_0, \ldots, F_n, c_0, \ldots, c_m)$, where each F_i is naturally induced by f_i and each c_j is a constant symbol. It is now not hard to show that an algebra $\mathcal{A}$ is c.e. if and only if $\mathcal{A}$ is an E-algebra for some c.e. equivalence relation E.

An algebra is **computably presentable** if it is isomorphic to a computable algebra. Similarly, an algebra is **c.e. presentable** if it is isomorphic to a c.e. algebra. In the literature c.e. algebras are sometimes called semicomputable data structures [1] or positive algebras [6]. Thus, if one thinks of an ADT as an algebra then a computable or c.e. presentation of it can be identified with a machine-theoretic (or program-theoretic) implementation of the ADT. The theorem below, a classic result of universal and computable algebra, outlines the basic properties of specified algebras:

Theorem A *If the algebra $\mathcal{A}$ is specified by specification S then $\mathcal{A}$ satisfies the following properties:*

1. *For ground terms t_1 and t_2, we have $\mathcal{A} \models t_1 = t_2$ iff S proves $t_1 = t_2$. Hence, $\mathcal{A}$ is a c.e. algebra.*
2. *$\mathcal{A}$ satisfies S.*
3. *For any algebra $\mathcal{B}$ that is generated by σ-constants and satisfies S there exists a unique homomorphism from $\mathcal{A}$ onto $\mathcal{B}$.*
4. *If $\mathcal{B}$ is specified by S then $\mathcal{A}$ and $\mathcal{B}$ are isomorphic.*

The first part of the theorem tell us that all the positive facts about the ADT $\mathcal{A}$ specified by S can be computably enumerated. The second part tells us that $\mathcal{A}$ is a correct implementation of S. The third part tells us that $\mathcal{A}$

is, in a natural sense, the universal implementation among all data structures that satisfy S. Finally, the last part is the uniqueness property of the specified algebra. Therefore, often $\mathcal{A}$ is called the **initial algebra** of S and our approach to specification of abstract data types is called the **initial algebra** semantics method.

There is one fundamental difference between equationally specified algebras and quasiequationally specified algebras. For equationally specified algebras the third part of **Theorem A** can be conversed: if $\mathcal{A}$ is equationally specified by E and $\mathcal{B}$ is a homomorphic image of $\mathcal{A}$ then $\mathcal{B}$ is generated by σ-constants and satisfies E. This difference is the main reason why the proof for equational specification problem, outlined in the next section, fails to solve the problem for quasiequational specifications.

Let us consider the following algebra $(\omega, S, 2^x, 0)$. This algebra is computable and finitely generated but does not have an equational specification (see [2]). An important fact is that one can enrich this algebra by expanding its signature and consider the following expanded algebra $(\omega, S, 2^x, +, \times, 0)$. Now this algebra is equationally specified: the natural recursive definition of $+$, the definition of $\times$ via $+$, and the definition of 2^x via $\times$ do the job.

The observation above suggests the idea of adding more functions to the original signature. This allows one to possess more flexibility in finding a specification of the algebra in an appropriate expansion. We would like to stress that the use of expansions of the original signature is a common and powerful tool that has been employed in pure model theory, algebra, and real programming practice. For example, in model theory expansions are used in constructing models with different properties, e.g. constructing saturated models, finding expansions that have elimination of quantifiers (e.g. expansions by Skolem functions), etc. In programming practice the use of expansions is a usual and natural routine. For instance, after an ADT is implemented by a code it is often the case that one would like to add more functionality to the code. This essentially amounts to adding extra methods to the program which, by its nature, is an expansion of the ADT implemented.

Let $\mathcal{A}$ be a c.e. algebra, and let f_1, ..., f_n be computable functions (well behaved with respect to the equality relation on $\mathcal{A}$). Then the algebra $\mathcal{B} = (\mathcal{A}, f_1, \ldots, f_n)$ obtained by adding the operations f_1, ..., f_n to $\mathcal{A}$ is called an **expansion** of $\mathcal{A}$. The signature $\sigma \cup < f_1, \ldots, f_n >$ is an expansion of the original signature. The algebra $\mathcal{A}$ is called a **reduct** of $\mathcal{B}$. An important result is the following theorem proved by Bergstra and Tucker in [1]:

Theorem B *Any computable algebra $\mathcal{A}$ possesses an expansion that can be equationally specified.*

The proof consists of two steps. The first step assumes that all original functions f of $\mathcal{A}$ are primitive recursive. Hence each f is defined in terms of a sequence of functions (called a **definition** of f) using primitive recursive schemata and the operation of composition successively applied to the basic functions: the successor, projections, and constants. The expansion is then obtained by adding the names of all the functions which participate in the definition of f. This expansion of $\mathcal{A}$ is then specified because the definition of f can be transformed into a specification of f. In the second step it is assumed that there is a non primitive recursive original operation. This is then reduced to the first step by using a primitive recursive enumeration of the graph of the operation. Bergstra and Tucker have a significant improvement of the result: any computable algebra has an equational specification with 6 new functions and 4 equations only [3].

The theorem above tells us that the initial algebra semantics method is sound for the class of all computable algebras. Of course, not every finitely generated c.e. algebra is computable, and by this reason **Theorem B** does not cover the general case. This leads us to the formulation of the following three problems whose solutions are discussed in the next sections. The first question (called subproblem) is a test case as any answer to the question should exclude computable algebras by the theorem above. The last two questions form the problem of Bergstra and Tucker.

Subproblem: *Is it true that any c.e. algebra has an equationally specified expansion?*

Note that the subproblem is not restricted to finitely generated c.e. algebras.

Problem 1 [2]: *Does any finitely generated c.e. algebra have an equationally specified expansion?*

Problem 2 [2]: *Does any finitely generated c.e. algebra possess a quasiequationally specified expansion?*

To the reader the questions may remind the known Higman's embedding theorem stating that any recursively presented group (that is finitely generated c.e. group) can be embedded into a finitely presented (that is equationally specified) one. However, in Higman's theorem the underlying domain of the group is allowed to be extended to a bigger domain; this is prohibited in the questions of Bergstra and Tucker above. In addition, in Higman's theorem the embedding preserves the group structure; this is relaxed in the questions of Bergstra and Tucker as expansions are allowed to be arbitrary computable functions.

Finally, we would like to stress that the specification problems discussed in this paper arose from investigations related to the study of semantics of programs, which is a part of a bigger theme in computer science: correctness of programs (e.g. see [10]). Our methods which solve the specification problems use basics from universal algebra and standard techniques from modern computability theory, thus providing an example of interactions between applied and theoretical computer science, algebra and logic.

2 Solution of Problem 1

We first give our solution to the subproblem proved in [8]. The proof gives a negative answer to the subproblem. We need two notions. An algebra is **locally finite** if any finite subset in it generates a finite subalgebra. A set X of natural numbers is **hypersimple** if it is computably enumerated, co-infinite, and there does not exist a computable function $f : \omega \to \omega$ such that $f(i) > a_i$ for all i, where $a_0.a_1, \ldots$ is the enumeration of the complement of X in strictly increasing order.

Theorem 0 *There exists an infinite algebra $\mathcal{A}$ any c.e. presentable expansion of which is locally finite.*

Proof (*outline*). Let $X \subset \omega$ be a hypersimple set and $1 \notin X$. Consider now the unary algebra (A, f) that satisfies the folloing properties: 1. $f : A \to A$ is a bijection; 2. For every $x \in X$ there exists an n such that $f^n(x) = x$ and all the elements x, $f(x)$, ..., $f^{n-1}(x)$ are pairwise distinct. Call the sequence x, $f(x)$, ..., $f^{n-1}(x)$ an **orbit** of size n. 3. For every $n \in X$ the algebra does not contain an orbit of size n. 4. For every $n \notin X$ the algebra contains exactly one orbit of size n. Clearly, these properties define $\mathcal{A}$ uniquely.

The algebra $\mathcal{A}$ has a c.e. presentation. Indeed, take a computable algebra (ω, f) such that for every $n > 1$ the algebra has exactly one orbit of size n. Now consider the following equivalence relation η on this algebra: $(x, y) \in \eta$ iff either $x = y$ or both x and y belong to orbits O_1 and O_2, respectively such that the of O_1 and the size of O_2 are both in X. The equivalence relation is a c.e. congruence of the algebra. Hence the factor algebra is a c.e. presentation of $\mathcal{A}$. Now, assume that $\mathcal{B}$ is an expansion of $\mathcal{A}$ such that $\mathcal{B}$ is a c.e. algebra. Then if there was a finite subset in $\mathcal{B}$ that generated an infinite subalgebra then the set X would not be hypersimple. □

Note that the theorem above gives us a negative answer to the subproblem in a strongest possible way. Indeed, it shows that *all* c.e. implementations (that is, presentations) of $\mathcal{A}$, and not just one, fail to be equationally specified. The failure is not because c.e. implementations of $\mathcal{A}$ are bad but rather the intrinsic property of the algebra itself.

The next theorem solves Problem 1 in negative. This shows that the initial algebra semantics method with equational specifications is not sound for the class of all finitely generated c.e. algebras. For a detailed proof see [7] or [9]. Recall that a set X of natural numbers is **simple** if it is co-infinite, computably enumerable, whose complement does not contain an infinite c.e. subset.

Theorem 1. *There exists a finitely generated c.e. algebra such that every expansion of the algebra is not equationally specified.*

Proof (*outline*). Let $X \subset \omega$ be a set. Consider the factor set $\omega(X)$ whose elements are the equivalence classes of the equivalence relation $\eta(X)$ defined as follows: $\eta(X) = X^2 \cup \{(x, x) \mid x \in \omega\}$. Thus, each element of $\omega(X)$ either is a singleton or is X^2. Here is the first lemma which we provide without a proof:

Lemma 1. *There exists a finitely generated c.e. algebra $\mathcal{A}(X)$ such that the domain of $\mathcal{A}(X)$ is $\omega(X)$ with X being a simple set.*

The proof can be found either in [7], where X is constructed directly, or in [9] where the algebra $\mathcal{A}(X)$ is constructed by using Kolmogorov complexity. A point of note is that X can not be hypersimple which is consistent with the previous theorem.

The next lemma is due to Malcev. We say that an algebra $\mathcal{A}$ is **residually finite** if for any $a, b \in A$ with $a \neq b$ there is an onto homomorphism $h : \mathcal{A} \to \mathcal{B}$ such that $h(a) \neq h(b)$ and $\mathcal{B}$ is finite. Here is now the next lemma:

Lemma 2. *If an algebra $\mathcal{A}$ is equationally specified and residually finite then the equality relation in the algebra is decidable.*

Note that the equality problem for $\mathcal{A}(X)$ is not decidable because X is not a computable set. The next lemma, given without proof, essentially uses the fact that $\omega \setminus X$ contains no infinite c.e. subset (such sets are called immune sets). The lemma is a good example of interactions between two seemingly not related notions: immune set from computability theory on the one hand, and residually finite algebra from universal algebra on the other. The full proof is in [7] or [9].

Lemma 3. *Any expansion $\mathcal{B}$ of $\mathcal{A}(X)$ is residually finite.*

Thus, the lemmas above show that the algebra $\mathcal{A}(X)$ is a desired one thus proving the theorem. □

An important comment is that for quasiequational specifications the proof of the theorem above fails because, as has been mentioned, the second lemma stated above fails for algebras specified by quasiequations. Therefore, a new point of view is needed for the study of Problem 2 to which we now turn our attention.

3 Solution of Problem 2

Most of the algebras constructed in this section have signature $< c, f, g >$, where c is a constant and f, g are unary function symbols. Therefore the readers can restrict themselves to this signature while reading this section. However, we formulate the results (where possible) for any finite functional signature σ with at least one constant symbol.

Let (a, b) be a pair of an algebra $\mathcal{A}$. Consider the minimal congruence relation, denoted by $\eta(a, b)$ and called a **principal congruence**, containing the pair. We now give an inductive definition of almost free algebras.

Definition 3. *The absolutely free algebra $\mathcal{F}(\sigma)$ is* **almost free**. *Assume that $\mathcal{A}$ is almost free. Then the factor algebra obtained by factorizing $\mathcal{A}$ by $\eta(a, b)$, where (a, b) is any pair in $\mathcal{A}$, is* **almost free**.

Thus $\mathcal{A}$ is almost free if and only if there exists a sequence $\mathcal{A}_1, \ldots, \mathcal{A}_n$ of algebras and the sequence $(a_1, b_1), \ldots, (a_n, b_n)$ of pairs of elements such that $\mathcal{A}_1$ is $\mathcal{F}(\sigma)$, each $\mathcal{A}_{i+1}$ is obtained by factorizing $\mathcal{A}_i$ by $\eta(a_i, b_i)$, and $\mathcal{A}$ is obtained from factorizing $\mathcal{A}_n$ by $\eta(a_n, b_n)$. We call the sequence $(a_1, b_1,), \ldots, (a_n, b_n)$ a **witness sequence** for $\mathcal{A}$ to be almost free. The reason in introducing almost free algebras is the following. The desired counterexample that solves Problem 2 is constructed by stages so that at each stage the construction deals with an almost free algebra, and the construction, if needed, acts by factorizing the algebra with respect to a principal congruence thus again producing an almost free algebra.

In case when $\sigma =< c, f, g >$, almost free algebras can be characterized as explained below. Let $\mathcal{A}$ be an algebra of signature σ and be generated by c. We say that there is an edge from a to b if $f(a) = b$ or $g(a) = b$. Thus, $\mathcal{A}$ is turned into a directed graph. Hence one can employ graph theoretic notions to this algebra viewed as a graph. For example, a distance from a to b is the length of a minimal directed path starting at a and ending at b. A level in $\mathcal{A}$ is the set of all elements that are on the same *distance* from the generator c. Thus, if $\mathcal{A}$ is the absolutely free algebra of the signature σ then there are 2^n number of elements at level n. The following is not hard to see. The algebra $\mathcal{A}$ is almost free if and only if either $\mathcal{A}$ is finite or there is a level L such that for all $x \in L$ the subalgebra $\mathcal{A}_x$ generated by x is ismorphic to $\mathcal{F}(\sigma)$, and for all distinct $x, y \in L$ the domains of $\mathcal{A}_x$ and $\mathcal{A}_y$ have no elemenets in common. Here is a lemma whose proof we omit. The reader can verify the correctness of the lemma in the case (sufficient for our counterexample) when σ is $< c, f, g >$ using the characterization just explained.

Lemma 4. *There exists an algorithm that for any almost free algebra $\mathcal{A}$, a witness sequence $(a_1, b_1), \ldots, (a_n, b_n)$ for $\mathcal{A}$, a first order formula $\Phi(x_1, \ldots, x_n)$, and an m-tuple $(c_1, \ldots, c_m)$ decides whether or not $\mathcal{A} \models \Phi(c_1, \ldots, c_m)$. In particular, the first order theory of any almost free algebra is decidable.*

We use a weaker version of this lemma. First of all, we will deal with the mentioned signature $< c, f, g >$. Secondly, we restrict ourselves to formulas Φ which are quasiequational specifications.

The proof of the next lemma uses a diagonalization argument. The result of this lemma can be obtained without the diagonalization. However, the methods and ideas of this and the next lemmas give a hint towards a solution of the problem.

Lemma 5. *There exists a finitely generated c.e. algebra that has no quasiequational specificationcan in its own signature.*

Proof ***(outline).*** We construct the desired algebra by stages, where at each stage we deal with the almost free algebra obtained at the previous stage. The signature of the algebra $\mathcal{A}$ that we construct is $< c, f, g >$. In order to make $\mathcal{A}$ not quasiequationally specified, our construction needs to make sure that $\mathcal{A}$ is not isomorphic to $\mathcal{F}(C)$ for any given quasiequational specification C. Thus, in order to construct $\mathcal{A}$ the construction must guarantee the satisfaction of the following requirements:

$$R_e : \quad \mathcal{A} \text{ is not isomorphic to } \mathcal{F}(C_e),$$

where $\{C_e\}_{e \in \omega}$ is an effective list of all quasiequational specifications (note that each quasiequational specification is finite). The previous lemma is an important tool guaranteeing that the construction can be carried out effectively. At the initial stage, Stage 0, the construction begins with the algebra is $\mathcal{F}(\sigma)$. Denote this algebra by $\mathcal{A}_0$. At Stage $n+1$, the goal of the construction is to guarantee that the requirement R_n is satisfied. Here is a description of Stage $n+1$.

Stage $n+1$. Assume that $\mathcal{A}_n$ has been constructed, and is almost free. In addition we have its witness for being almost free. Take the quasiequation C_n. Check whether or not $\mathcal{A}_n \models C_n$. This can be checked due to the previous lemma. If $\mathcal{A}_n \models C_n$ then find two distinct elements a, b in $\mathcal{A}_n$ such that the factor algebra $\mathcal{B}$ obtained by factorization of $\mathcal{A}_n$ with respect to the principal congruence $\eta(a,b)$ satisfies $R_0, \ldots, R_n$ and is infinite. Set $\mathcal{A}_{n+1} = \mathcal{B}$. Clearly, $\mathcal{A}_{n+1}$ is almost free. If $\mathcal{A}_n \models \neg C_n$ then find a tuple $\bar{a}$ that witnesses the fact that $\mathcal{A}_n \models \neg C_n$. More presicely, we know that the quasiequational specification C_n is of the form

$$\forall \bar{x}(\&_i(\Phi_i(\bar{x}) \to \Psi_i(\bar{x}))),$$

where each of Φ_i and Ψ_i is an equation between terms. Since $\mathcal{A}_n$ does not satisfy C_n there exists a tuple $\bar{a}$ and i such that $\mathcal{A}_n \models \Phi_i(\bar{a})$ and $\mathcal{A} \models \neg\Psi_i(\bar{a})$. The construction then guarantees that $\neg\Psi_i(\bar{a})$ is always true, by making sure that the requirements C_j of lower priority ($j > n+1$) do not violite the truth value of $\Psi_i(\bar{a})$. In this case $\mathcal{A}_{n+1} = \mathcal{A}_n$.

The desired algebra is now the following. Let η be the congruence relation such that $(a,b) \in \eta$ iff $a = b$ is true in some $\mathcal{A}_n$. Clearly η is a c.e. relation. Set $\mathcal{A}$ to be the factor algebra obtained by factorizing $\mathcal{F}(\sigma)$ with respect to η. It is not hard to show that $\mathcal{A}$ is a desired algebra. □

The lemma suggests the idea of constructing the algebra that gives a negative solution to Problem 2 by trying to diagonalize against all possible quasiequational specifications in *all* possible finite expansions of the signature. Doing this directly seems to be a difficult task because of the following two reasons. Say, the eth quasiequational specification contains a new function symbol ψ. This means that the requirement R_e is equivalent to an infinite list of requirements as now ψ is a new parameter. Hence the list of requirements that correspond to C_e is now this:

$$(\mathcal{A}, \phi_j) \text{ is not isomorphic to } \mathcal{F}(C_e),$$

where $\{\phi_j\}_{j\in\omega}$ is an enumeration of all partial computable functions. Secondly, the behaviour of each ϕ_j is not under our control as we can only control the algebra $\mathcal{A}$ which is being constructed. Therefore it is not clear how to directly construct the desired algebra.

We construct the desired algebra $\mathcal{A}$ indirectly. The basic idea of using an indirect way is implicitly suggested by the two lemmas above. If we can construct $\mathcal{A}$ in such a way that every possible expansion of $\mathcal{A}$ does not produce new functions, in the sense expressed in the definition below, then a modification of the proof of the previous lemma does the job. In order to formally explain this idea we give the following definition. Note that *almost all* means for all but finite.

Definition 4. *A new operation* $\psi : A^n \to A$ *on a c.e. algebra* $\mathcal{A}$ *is* **termal** *if there is a term* $t(x)$ *in the signature of algebra* $\mathcal{A}$ *such that for almost all* $(a_1, \ldots, a_n) \in A^n$ *the equality* $\psi(a_1, \ldots, a_i, \ldots, a_n) = t(a_i)$ *is true. An expansion* $\mathcal{B} = (\mathcal{A}, \psi_1, \ldots, \psi_m)$ *of* $\mathcal{A}$ *is* **termal** *if each new operation* ψ_i *is termal.*

Note that in case of the signature $< c, f, g >$, every term is of the form $f^{n_0} g^{m_0} \ldots^{n_k} g^{m_k}(x)$, where n_i, m_i are natural numbers, and hence contains one variable only.

Thus, the idea is that we want to construct a finitely generated c.e. algebra $\mathcal{A}$ that incorporates the construction of the previous lemma and, in addition, makes sure that any expansion of $\mathcal{A}$ is termal. We single out the following class of c.e. algebras which, we think, is of independent interest.

Definition 5. *We say that an infinite finitely generated computably enumerable algebra* $\mathcal{A}$ *is* **computationally complete** *if any expansion of* $\mathcal{A}$ *is termal.*

The following note is simple. If a finitely generated c.e. algebra is computationally complete then any other c.e. algebra isomorphic to it is also computationally complete. The reason for this is that there is a computable isomorphism between any two finitely generated c.e. algebras. So computational completeness is an isomorphism invariant property for finitely generated c.e. algebras. Thus, for a computational complete algebra every expansion of this algebra, say by a function g, does not give us anything new. The function g already exists in the algebra as g can be expressed as a term apart from finitely many values.

We also note the following. If we omit the requirement that the algebra is not finitely generated, then computational completeness becomes non isomorphism invariant. In particular, one can construct a c.e. presentation of the following trivial algebra (ω, id), where $id(x) = x$ for all $x \in \omega$, such that in that presentation every possible expansion function g becomes identity or a projection function with respect to the equality relation of the presentation.

Finally, in universal algebra there is a notion of primal algebra defined for finite algebras. A finite algebra is primal if any new function $f : A \to A$ can be expressed as a term. For example, the two valued Boolean algebra is primal. Computational complete algebras are infinite analogues of primal algebras.

For the next lemma that shows usefulness of computationally complete algebras we need the following definition.

Definition 6. *A finitely generated algebra* $\mathcal{A}$ *is* **term algebraic** *if for any element* $b \in A$ *the number of ground terms* t *in the language of the algebra such that the values of* t *in* $\mathcal{A}$ *equal* b *is finite.*

Here is now the next lemma:

Lemma 6. *If* $\mathcal{A}$ *is a term algebraic computationally complete algebra then no expansion of* $\mathcal{A}$ *can be quasiequationally specified.*

Proof *(outline).* The idea is the following. Assume that S is a quasiequational specification of an expansion $\mathcal{B} = (\mathcal{A}, \psi_1, \ldots, \psi_n)$ of the algebra $\mathcal{A}$. Then the specification S can be replaced with a new specification S' in the original signature so that S' specifies $\mathcal{A}$. This can be done because, roughly, any new operation ψ_i that is in S can be replaced with a term t_i in the original signature that equals ψ_i almost everywhere. In the transformation of S into S' one needs to use the fact that the algebra is term algebraic. □

Thus, we need to show that computationally complete and term algebraic algebras exist. The next lemma gives a brief outline of the construction. The proof uses a technique borrowed from modern computability theory, a priority construction carried out with a method known as a tree argument construction. For a current full version of the proof see [4].

Lemma 7. *Computationally complete and term algebraic algebras exist.*

Proof *(outline).* We present basic ideas of our construction. The signature of the algebra we construct is our signature $< c, f, g >$. The construction of the algebra is a stagewise construction so that at each stage we deal with an almost free algebra of the given signature. Let $\phi_0, \phi_1, \ldots$ be an effective list of all computable partial functions. The algebra $\mathcal{A}$ that we want to construct must satisfy the following condition. For every $e \in \omega$, if ϕ_e is a total function and expands $\mathcal{A}$ then ϕ_e must be equal to a term t almost everywhere. Thus, to build the desired algebra $\mathcal{A}$, the following list $\{T_e\}_{e \in \omega}$ of requirements must be satisfied:

T_e: If ϕ_e is total and expands the algebra $\mathcal{A}$ then ϕ_e is termal.

We now describe the **basic strategy** that satisfies one requirement T_e. Start constructing the algebra $\mathcal{A}$ by making it isomorphic to the free algebra $\mathcal{F}(\sigma)$. While constructing, wait until for some tuples $\bar{a} = (a_1, \ldots, a_n)$, $\bar{b} = (b_1, \ldots, b_n)$ in $\mathcal{A}$ each of the following occurs:

1. The values $\phi_e(\bar{a})$ and $\phi_e(\bar{b})$ are defined,
2. There is no term t and no i for which $\phi_e(\bar{a}) = t(a_i)$ and $\phi_e(\bar{b}) = t(b_i)$.
3. Viewing $\mathcal{A}$ as a directed graph (described in page 6) there is no directed path neither from a_1 to b_1 nor from b_1 to a_1, where a_1 and b_1 are the first components of the tuples $\bar{a}$ and $\bar{b}$.

If such a pair $(\bar{a}, \bar{b})$ occurs, then the desired algebra is obtained by factorizing $\mathcal{F}(`\sigma)$ by the principal congruence relation $\eta(a_1, b_1)$. Otherwise, the desired algebra remains isomorphic to $\mathcal{F}(\sigma)$.

This strategy satisfies the requirement T_e. Indeed, if $\mathcal{A}$ is isomorphic to $\mathcal{F}(\sigma)$ then either ϕ is undefined somewhere or ϕ equals to some term t almost everywhere. If $\mathcal{A}$ is obtained by factorizing $\mathcal{F}$ by $\eta(a, b)$ then ϕ_e now does not define an operation on $\mathcal{A}$ because there is a tuple at which the value of ϕ is not well-defined. A point here is that strategy makes it sure that the function ϕ_e that acts improperly (that is ϕ_e does not look like a term now), after the action of the construction, does not respect the equality relation of the algebra $\mathcal{A}$ being constructed. Clearly, the algebra is term algebraic

Of course the strategy to satify just one requirment is not a difficult one. Problems arise when one tries to satisfy all of the requirement T_e. The reason is that the requirements T_e interact with each other. Moreover, while satisfying the requirements T_e the construction needs to guarantee that the algebra being constructed does not collapse into a finite algebra. Now we would like to make a few notes of a general character about the construction and present some technical details of the construction to the reader. For a current version of the full construction see [4].

Natirally, first of all we list all the requirments T_e, and say that T_i has a higher priority than T_j if $i < j$. Assume that the requirement T_e is associated with computable partial function ϕ. While constructing the desired algebra $\mathcal{A}$, the construction acts depending on the behavior of ϕ. At any stage there are elements x in $\mathcal{A}$ marked with $\Box_w$ and associated with ϕ. The index w indicates that ϕ is waiting to be defined on these marked x. If ϕ is not defined on some of these elements then the outcome of ϕ at a given stage is that of a waiting state. However, ϕ may be defined on the marked elements. Then ϕ may exhibit two different behaviors. One is that there is a term p and i such that $\phi(x_1, \ldots, x_i, \ldots, x_n) = p(x_i)$ on all marked elements x. The other is that such a term p does not exist. In the former case, the outcome of ϕ is that of "ϕ looks like a term", denoted by t. In the latter case, the construction should act, by using a version of the basic strategy, in such away that ϕ does not respect the equality relation of $\mathcal{A}$. If there is a right environment to do this now then T_e is satisfied (Informally, by a right environment we mean that the action of the construction to satisfy T_e does not destroy the work of those requirements of higher priority). In this case the outcome is that of "ϕ is destroyed now", we denote this outcome by d. If not, then the construction creates an environment so that T_e will have a chance to destroy ϕ at a later stage. In this case the outcome is w_d. A point is that no actions performed due to the requirements of lower priority effect the environment created (However, the environment can be destroyed by satisfying requirements of higher priority in which case T_e will have its chance to be satisfied). Therefore when ϕ is again defined on all elements of the created environment then the construction is able to act and distroy ϕ so that ϕ does respect the equality relation of the algebra being constructed.

The desired algebra $\mathcal{A}$ is built by using a construction on a priority tree. Constructions on priority trees are often used in pure computability theory (see for example [11]). The alphabet of the outcomes of the priority tree is $O = \{t, w_t, w_d, d\}$. The symbol t corresponds to the outcome saying that the function under consideration is termal, w_t and w_d correspond to waiting states. The outcome w_t expresses the fact that the function under consideration seems to be a term, and the construction is waiting for the function to be defined on some elements of the algebra. The outcome w_d expresses the fact that the function seems not to be a termal function; moreover, the construction has created an environment to make it sure that the function does not respect the equality relation of the algebra and is waiting for the function to be defined on certain elements of the algebra. The outcome d corresponds to the fact that the function under consideration does not respect the equality relation on $\mathcal{A}$. The order on these outcomes is $t < w_t < w_d < d$. Consider the tree $T = \{t, w_t, w_d, , d\}^{<\omega}$. To each node β of the tree of length e there corresponds a β-strategy that is devoted to satisfy requirment T_e. The β-strategy is an adaptation of the basic strategy but takes into account the outcomes $\beta(k)$, where $k < e$. In other words, the β-strategy acts by believing that the outcome of the λ-strategy (λ is the root of the priority tree) to satisfy T_0 is $\beta(0)$, the outcome of the $\beta(0)$-strategy to satisfy T_1 is $\beta(1)$, the outcome of the $\beta(0)\beta(1)$-strategy to satisfy R_2 is $\beta(2)$, etc.

At the end of the construction one defines the true path on the tree by induction. Basically, it is the leftmost path f on T so that every node on the right of f acts finitely often and each node in f acts infinitely often. Then one proves, by induction on the lengths of the nodes on the true path, that the construction succeeds along the true path. The fact that the algebra constructed is term algebraic is guaranteed because of the following. Each requirement T_e determines a level (that is, fixes a distance from the generator) that basically says that all the requirement of lower priority are not allowed to work below that level. In other words, once a distance for T_e is fixed, say is d, then all requirements of lower priority than T_e do not affect elements x such that the distance from x to the generator is not greater than d. These levels determined by strategies corresponding to T_e may change finitely many times only. Doing this will guarantee that for each element a of the algebra there will be finitely many ground terms whose values equal to a in the algebra. □

Now we combine constructions in Lemma 5 and Lemma 7 into one to build an algebra that has no quasiequational specification in any expansion. This is guaranteed by Lemma 6. Here is the result:

Theorem 2. *There exists a finitely generated c.e. algebra such that every expansion of the algebra is not quasiequationally specified.*

Proof *(outline)*. Again, the desired algebra is built by using a diagonalization argument. On the one hand, we try to make the algebra computationally complete. On the other, we try ensure that the algebra constructed can not be quasiequationally specified in its own signature. For this we need to satisfy the following lists of requirements:

$$R_e: \quad \mathcal{A} \text{ is not isomorphic to } \mathcal{F}(C_e),$$

where $\{C_e\}_{e\in\omega}$ is an effective list of all quasiequational specifications in the signature $< c, f, g >$, and

$$T_e: \quad \text{If } \phi_e \text{ is total and expands the algebra } \mathcal{A} \text{ then } \phi_e \text{ is termal.}$$

where $\{\phi_e\}_{e\in\omega}$ is an effective list of all computable partial functions. Our construction will be a stagewise construction so that even stages are devoted to satisfy the requirments R_e, and odd stages T_e. Thus, the priority list is the following:

$$R_0, T_0, R_1, T_1, R_2, T_2, \ldots .$$

A point to note here is that the requirements R_e are somewhat independent from the requirements T_e. This is because once we acted to satisfy R_e it can be guaranteed that no requirment of lower priority can injure R_e. The requirements T_e are satisfied in the manner similar to what is described Lemma 7. The requirements R_e are satisfied in the manner as explained in Lemma 5. Naturally, the whole construction is put on a priority tree. At nodes of even length, say $2e$, the requirement R_e is met; and at nodes of odd length length the requirement T_e is met. Each node α of even length has exactly one immediate successor αs, where s stands for "R_e is satisfied", while each node α of odd length has four immediate successors αt, αw_t, αw_d, and αd meaning of which are described in the lemma above. □

References

1. J. Bergstra, J. Tucker. Initial and final algebra semantics for data type specifications: two characterisation theorems. *SIAM Journal on Computing*, V 12, No 2, p. 366-387, 1983.
2. J. Bergstra, J. Tucker. Algebraic Specifications of Computable and Semicomputable Data Types. *Theoretical Computer Science* 50, p.137-181, 1987.
3. J. Bergstra, J. Tucker. The Completeness of the Algebraic Specification Methods for Computable Data Types. *Information and Control*, Vol 54, No 3, p. 186-200, 1982.
4. S. Goncharov and B. Khoussainov. Computationally Complete Algebras, *CDMTCS Research report*, No 206, see Research Report Series in http://www.cs.auckland.ac.nz/CDMTCS/
5. G. Grätzer. Universal Algebra. Second Edition. Springer-Verlag, 1979.
6. Yu. Ershov, S. Goncharov. Constructive Models. Siberian School of Algebra and Logic Series. Consultants Bureau, Kluwer Academic, Plenum, 2000.
7. N. Kasymov. Algebras with Finitely Approximable Positively Representable Enrichments. *Algebra and Logic*, Vol 26, No 6, pp 715-730, 1987.
8. N. Kasymov, B. Khoussainov. Finitely Generated Enumerable and Absolutely Locally Finite Algebras. *Vychisl. Systemy*, No 116, p. 3-15, 1986.
9. B. Khoussainov. Randomness, Computability, and Algebraic Specifications, *Annals of Pure and Applied Logic*, 91, no1, p.1-15, 1998.
10. D. Sannella, A.Tarlecki. Essential Concepts of Algebraic Specification and Program Development. *Formal Aspects of Computing*, 9, p.229-269, 1997.
11. R. Soare. Recursively Enumerable Sets and Degrees. Springer–Verlag, New York, 1987.
12. M. Wirsing. Algebraic Specifications. In J. Van Leeuwen, editor, *Handbook of Theoretical Computer Science*, Vol.B: Formal Model and Semantics, p. 675-788. Elsevier, Amsterdam, 1990.

On the Complexity of Existential Pebble Games

Phokion G. Kolaitis and Jonathan Panttaja*

Computer Science Department
University of California, Santa Cruz
{kolaitis,jpanttaj}@cs.ucsc.edu

Abstract. Existential k-pebble games, $k \geq 2$, are combinatorial games played between two players, called the Spoiler and the Duplicator, on two structures. These games were originally introduced in order to analyze the expressive power of Datalog and related infinitary logics with finitely many variables. More recently, however, it was realized that existential k-pebble games have tight connections with certain consistency properties which play an important role in identifying tractable classes of constraint satisfaction problems and in designing heuristic algorithms for solving such problems. Specifically, it has been shown that strong k-consistency can be established for an instance of constraint satisfaction if and only if the Duplicator has a winnning strategy for the existential k-pebble game between two finite structures associated with the given instance of constraint satisfaction. In this paper, we pinpoint the computational complexity of determining the winner of the existential k-pebble game. The main result is that the following decision problem is EXPTIME-complete: given a positive integer k and two finite structures $\mathbf{A}$ and $\mathbf{B}$, does the Duplicator win the existential k-pebble game on $\mathbf{A}$ and $\mathbf{B}$? Thus, all algorithms for determining whether strong k-consistency can be established (when k is part of the input) are inherently exponential.

1 Introduction and Summary of Results

Combinatorial games are a basic tool for analyzing logical definability and delineating the expressive power of various logics. Typically, a logic L is decomposed into a union $L = \bigcup_{k \geq 1} L(k)$ of fragments according to some syntactic parameter, such as quantifier rank, pattern of quantification, or number of variables. With each fragment $L(k)$, one then seeks to associate a natural combinatorial game $\mathcal{G}(k)$ that captures $L(k)$-equivalence. Specifically, the desired game $\mathcal{G}(k)$ is played between two players, called the *Spoiler* and the *Duplicator*, on two structures $\mathbf{A}$ and $\mathbf{B}$, and has the following property: the Duplicator has a winning strategy for $\mathcal{G}(k)$ on $\mathbf{A}$ and $\mathbf{B}$ if and only if $\mathbf{A}$ and $\mathbf{B}$ satisfy the same $L(k)$-sentences. In the case of first-order logic FO, each such fragment is the set FO(k) of all first-order sentences of quantifier rank at most k, and the game $\mathcal{G}(k)$ is the k-move Ehrenfeucht-Fraïssé-game. Moreover, in the case of the infinitary logic $\mathrm{L}^{\omega}_{\infty\omega}$ with finitely many variables, each such fragment is the infinitary logic $\mathrm{L}^{k}_{\infty\omega}$ with k variables, $k \geq 1$, and the corresponding game is the k-pebble game. As is well known, k-pebble games have turned out to be an indispensable tool in the study of logics with fixed-point operators in finite model theory (see [7] for a survey).

* Research of both authors was partially supported by NSF grant IIS-9907419.

M. Baaz and J.A. Makowsky (Eds.): CSL 2003, LNCS 2803, pp. 314–329, 2003.

Each game $\mathcal{G}(k)$ as above gives rise to the decision problem of determining the winner of this game: given two finite structures $\mathbf{A}$ and $\mathbf{B}$, does the Duplicator have a winning strategy for $\mathcal{G}(k)$ on $\mathbf{A}$ and $\mathbf{B}$? It is easy to show that, for every $k \geq 1$, determining the winner of the k-move Ehrenfeucht-Fraïssé-game is in LOGSPACE (this is a consequence of the fact that each equivalence class of $\mathrm{FO}(k)$-equivalence is first-order definable). The state of affairs, however, is quite different for the k-pebble games. Indeed, Grohe [7] established that, for each $k \geq 2$, determining the winner of the k-pebble game is a P-complete problem, that is, complete for polynomial-time under logarithmic-space reductions. It is also natural to consider the decision problem that arises by taking the parameter k as part of the input (in addition to the structures $\mathbf{A}$ and $\mathbf{B}$). Pezzoli [13] investigated the computational complexity of this problem for the Ehrenfeucht-Fraïssé-game and showed that it is PSPACE-complete. In other words, Pezzoli showed that the following problem is PSPACE-complete: given a positive integer k and two finite structures $\mathbf{A}$ and $\mathbf{B}$, does the Duplicator have a winning strategy for the k-move Ehrenfeucht-Fraïssé-game on $\mathbf{A}$ and $\mathbf{B}$? Thus, when the number of moves is part of the input, an exponential jump occurs in determining the winner of the Ehrenfeucht-Fraïssé-game. It is conjectured that a similar exponential jump in complexity holds for the k-pebble game, when k is part of the input. Specifically, the conjecture is that the following problem is EXPTIME-complete: given a positive integer k and two finite structures $\mathbf{A}$ and $\mathbf{B}$, does the Duplicator have a winning strategy for the k-pebble on $\mathbf{A}$ and $\mathbf{B}$? To date, this conjecture remains unsettled.

In this paper we investigate the computational complexity of the decision problems associated with the class of existential k-pebble games (or, in short, $(\exists, k)$-pebble games), which are an asymmetric variant of the k-pebble games. These games were introduced in [11] as a tool for studying the expressive power of Datalog and of the existential positive infinitary logic $\exists\mathrm{L}^{\omega}_{\infty\omega}$ with finitely many variables. More precisely, $\exists\mathrm{L}^{\omega}_{\infty\omega}$ is the collection of all $\mathrm{L}^{\omega}_{\infty\omega}$-formulas containing all atomic formulas and closed under existential quantification, infinitary conjunction $\bigwedge$, and infinitary disjunction $\bigwedge$. Clearly, $\exists\mathrm{L}^{\omega}_{\infty\omega} = \bigcup_{k\geq 1} \exists\mathrm{L}^{k}_{\infty\omega}$, where $\exists\mathrm{L}^{k}_{\infty\omega}$ is the collection of all $\exists\mathrm{L}^{\omega}_{\infty\omega}$-formulas with at most k distinct variables. The differences between the $(\exists, k)$-pebble game and the k-pebble game played on two structures $\mathbf{A}$ and $\mathbf{B}$ are that in the $(\exists, k)$-pebble game: (1) the Spoiler always plays on $\mathbf{A}$; and (2) the Duplicator strives to maintain a partial homomorphism, instead of a partial isomorphism. The main result of this paper is that determining the winner of the $(\exists, k)$-pebble game, when k is part of the input, is an EXPTIME-complete problem. In contrast, for each fixed $k \geq 2$, determining the winner of $(\exists, k)$-pebble game turns out to be a P-complete problem. Before commenting on the technique used to establish the main result, we discuss the motivation for investigating this problem and the implications of our main result.

Although $(\exists, k)$-pebble games were originally used in database theory and finite model theory, in recent years they turned out to have applications to the study of constraint satisfaction. Numerous problems in several different areas of artificial intelligence and computer science can be modeled as constraint satisfaction problems [4]. In full generality, an instance of the CONSTRAINT SATISFACTION PROBLEM consists of a set of variables, a set of possible values, and a set of constraints on tuples of variables; the question is to determine whether there is an assignment of values to the variables that satisfies the given constraints. Alternatively, as first pointed out by Feder and Vardi [6], the CONSTRAINT SATISFACTION PROBLEM can be identified with the HOMOMORPHISM PROBLEM: given two relational structures $\mathbf{A}$ and $\mathbf{B}$, is there a homomorphism h from $\mathbf{A}$ to $\mathbf{B}$? Intuitively, the structure $\mathbf{A}$ represents the variables and the tuples of variables that participate in constraints, the structure $\mathbf{B}$ represents the domain of values and the tuples of values that the constrained tuples of variables are allowed to take, and the homomorphisms from $\mathbf{A}$ to $\mathbf{B}$ are precisely the assignments of values to variables that satisfy the constraints. The CONSTRAINT SATISFACTION PROBLEM is NP-complete, since it contains BOOLEAN SATISFIABILITY, COLORABILITY, CLIQUE, and many other prominent NP-complete problems as special cases. For this reason, there has been an extensive pursuit of both tractable cases of the CONSTRAINT SATISFACTION PROBLEM and heuristic algorithms for this problem. In this pursuit, a particularly productive approach has been the introduction and systematic use of various *consistency* concepts that make explicit additional constraints implied by the original constraints. The *strong* k*-consistency* property is the most important one among them; intuitively, this property holds when every partial solution on fewer than k variables can be extended to a solution on k variables [5]. Closely related to this is the process of "establishing strong k-consistency", which is the question of whether additional constraints can be added to a given instance of the CONSTRAINT SATISFACTION PROBLEM in such a way that the resulting instance is strongly k-consistent and has the same space of solutions as the original one. Algorithms for establishing strong k-consistency play a key role both in identifying tractable cases of constraint satisfaction and in designing heuristics for this class of problems [2, 5].

In [12], a tight connection was shown to exist between strong k-consistency properties and $(\exists, k)$-pebble games. Specifically, it turns out that strong k-consistency can be established for a given instance of the CONSTRAINT SATISFACTION PROBLEM if and only if the Duplicator has a winning strategy for the $(\exists, k)$-pebble game on the structures $\mathbf{A}$ and $\mathbf{B}$ forming the instance of the HOMOMORPHISM PROBLEM that is equivalent to the given instance of the CONSTRAINT SATISFACTION PROBLEM. This connection was fruitfully exploited in [3], where it was shown that the tractability of certain important cases of constraint satisfaction follows from the fact that the existence of a solution is equivalent to whether the Duplicator can win the $(\exists, k)$-pebble game for some fixed k. Note that, for every fixed k, there is a polynomial-time algorithm to determine whether, given two finite structures $\mathbf{A}$ and $\mathbf{B}$, the Duplicator has a winning strategy for the $(\exists, k)$-pebble game on $\mathbf{A}$ and $\mathbf{B}$ (this had been already observed in [11]). Nonetheless, since many heuristics for constraint satisfaction require testing whether strong k-consistency can be established for arbitrarily large k's, it is important to identify the inherent computational complexity of determining the winner in the $(\exists, k)$-pebble game, when k is part of the input. It is not hard to verify that this problem is solvable in time $O(n^{2k})$, that is, in time exponential in k. Moreover, it was conjectured in [12] that a matching lower bound exists, which means that the following problem is EXPTIME-complete: given a positive integer k and two finite structures $\mathbf{A}$ and $\mathbf{B}$, does the Duplicator have a winning strategy for the $(\exists, k)$-pebble on $\mathbf{A}$ and $\mathbf{B}$?

In this paper, we prove this conjecture by showing that another pebble game, which was known to be EXPTIME-complete, has a polynomial-time reduction to the $(\exists, k)$-pebble game. Specifically, Kasai, Adachi, and Iwata [8] introduced a pebble game, which we will call the KAI game, and showed that it is EXPTIME-complete via a direct reduction from polynomial-space alternating Turing machines (recall that APSPACE = EXPTIME [1]). Our reduction of the KAI game to the $(\exists, k)$-pebble game is quite involved and requires the construction of elaborate combinatorial gadgets. In describing this reduction and establishing its correctness, we will adopt the setup and terminology used by Grohe [7] in showing that, for every $k \geq 2$, the k-pebble game is P-complete. Some of the basic gadgets in our reduction already occurred in Grohe's reduction. However, we will also need to explicitly construct other much more sophisticated gadgets that will serve as "switches" with special properties in the reduction. We note that Grohe also used highly sophisticated gadgets that were graphs with certain homogeneity properties. Grohe's gadgets, however, have size exponential in k and, hence, they cannot be used in a polynomial-time reduction when k is part of the input (this is also the reason why Grohe's reduction does not show that the k-pebble game is EXPTIME-complete, when k is part of the input). An immediate consequence of our main result is that determining whether strong k-consistency can be established, when k is part of the input, is an EXPTIME-complete problem and, thus, inherently exponential. Moreover, this explains why all known algorithms for establishing strong k-consistency are exponential in k (even ones considered to be "optimal", see [2]).

We also address the computational complexity of determining who wins the $(\exists, k)$-pebble game, when k is a fixed positive integer. Kasif [10] showed that determining whether strong 2-consistency can be established is a P-complete problem. From this and the aforementioned connection between strong k-consistency and the $(\exists, k)$-pebble game [12], it follows that determining who wins the $(\exists, 2)$-pebble game is a P-complete problem. Here we give a direct proof to the effect that, for every fixed $k \geq 2$, determining who wins the $(\exists, k)$-pebble game is a P-complete problem. This is done via a reduction from the MONOTONE CIRCUIT VALUE PROBLEM, which we present first as a warm-up to the reduction of the KAI game to the $(\exists, k)$-game, when k is part of the input. Due to space limitations, here we present only outlines of these reductions; complete proofs can be found in the full version of the paper, which is available at http://www.cs.ucsc.edu/~kolaitis/papers/.

2 The Existential k-Pebble Game

Let $\mathbf{A}$ and $\mathbf{B}$ be two relational structures over the same vocabulary. A *homomorphism h from $\mathbf{A}$ to $\mathbf{B}$* is a mapping $h : A \rightarrow B$ from the universe A of $\mathbf{A}$ to the universe B of $\mathbf{B}$ such that, for every relation $R^{\mathbf{A}}$ of $\mathbf{A}$ and every tuple $(a_1, \ldots, a_m) \in R^{\mathbf{A}}$, we have that $(h(a_1), \ldots, h(a_m)) \in R^{\mathbf{B}}$. A *partial homomorphism from $\mathbf{A}$ to $\mathbf{B}$* is a homomorphism from a substructure of $\mathbf{A}$ to a substructure of $\mathbf{B}$.

Let $k \geq 2$ be a positive integer. The *existential k-pebble game* (or, in short, the *$(\exists, k)$-pebble game*) is played between two players, the Spoiler and the Duplicator, on two relational structures $\mathbf{A}$ and $\mathbf{B}$ according to the following rules: each player has k pebbles labeled $1, \ldots, k$; on the i-th move of a round of the game, $1 \leq i \leq k$, the Spoiler places a pebble on an element a_i of A, and the Duplicator responds by placing the pebble with the same label on an element b_i of B. The Spoiler wins the game at the end of that round, if the correspondence $a_i \mapsto b_i$, $1 \leq i \leq k$, is not a homomorphim between the substructures of $\mathbf{A}$ and $\mathbf{B}$ with universes $\{a_1, \ldots, a_k\}$ and $\{b_1, \ldots, b_k\}$, respectively.

Otherwise, the Spoiler removes one or more pebbles, and a new round of the game begins. The Duplicator wins the $(\exists, k)$-pebble game if he has a *winning strategy*, that is to say, a systematic way that allows him to sustain playing "forever", so that the Spoiler can never win a round of the game.

To illustrate this game (and its asymmetric character), let $\mathbf{K}_m$ be the m-clique, that is, the complete undirected graph with m nodes. For every $k \geq 2$, the Duplicator wins the $(\exists, k)$-pebble game on $\mathbf{K}_k$ and $\mathbf{K}_{k+1}$, but the Spoiler wins the $(\exists, k+1)$-pebble game on $\mathbf{K}_{k+1}$ and $\mathbf{K}_k$. As another example, let $\mathbf{L}_s$ be the s-element linear order, $s \geq 2$. If $m < n$, then the Duplicator wins the $(\exists, 2)$-pebble game on $\mathbf{L}_m$ and $\mathbf{L}_n$, but the Spoiler wins the $(\exists, 2)$-pebble game on $\mathbf{L}_n$ and $\mathbf{L}_m$.

Note that the above description of a winning strategy for the Duplicator in the $(\exists, k)$-pebble game is rather informal. The concept of a winning strategy can be made precise, however, in terms of families of partial homomorphisms with appropriate properties. Specifically, a *winning strategy for the Duplicator in the existential k-pebble game on* $\mathbf{A}$ *and* $\mathbf{B}$ is a nonempty family $\mathcal{F}$ of partial homomorphisms from $\mathbf{A}$ to $\mathbf{B}$ such that:

1. For every $f \in \mathcal{F}$, the domain $\mathrm{dom}(f)$ of f has at most k elements.
2. $\mathcal{F}$ is *closed under subfunctions*, which means that if $g \in \mathcal{F}$ and $f \subseteq g$, then $f \in \mathcal{F}$.
3. $\mathcal{F}$ has the *k-forth property*, which means that for every $f \in \mathcal{F}$ with $|\mathrm{dom}(f)| < k$ and every $a \in A$ on which f is undefined, there is a $g \in \mathcal{F}$ that extends f and is defined on a.

Intuitively, the second condition provides the Duplicator with a "good" move when the Spoiler removes a pebble from an element of $\mathbf{A}$, while the third condition provides the Duplicator with a "good" move when the Spoiler places a pebble on an element of $\mathbf{A}$.

3 The $(\exists, k)$-Pebble Game Is P-Complete

In this section, we show that, for every $k \geq 2$, determining the winner of the $(\exists, k)$-pebble game is a P-complete problem. We do this by constructing a reduction from the MONOTONE CIRCUIT VALUE PROBLEM (MCV) in the style of Grohe [7], but with different gadgets. In this reduction, the structures will be undirected graphs with ten unary predicates, called *colors*. So, we actually prove that, for every $k \geq 2$, the $(\exists, k)$-pebble game restricted to such structures is P-complete.

The following concepts and terminology come from Grohe [7].

1. In an undirected graph with colors, a *distinguished pair* of vertices is a pair of vertices that are of the same color, and that color is not used for any other vertex in the graph.
2. A *position* of the $(\exists, k)$-pebble game on A and B, is a set P of ordered pairs such that $P \subseteq A \times B$ and $|P| \leq k$. Often, we will omit the ordered pair notation and use the shorthand $ab \in P$ to mean that $(a, b) \in P$.
3. A *strategy* for the Spoiler is simply a mapping from positions to moves which tells the Spoiler how to play given the current position.
4. We say that the Spoiler *can reach a position* P' *from another position* P *of the* $(\exists, k)$*-pebble game on* A *and* B if the Spoiler has a strategy for the $(\exists, k)$-pebble

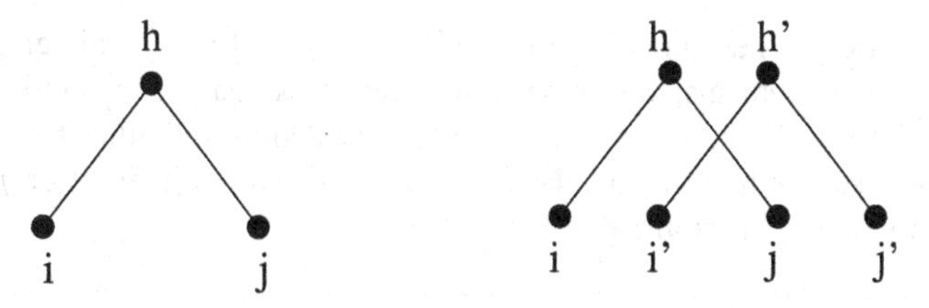

Fig. 1. H Gadget based on the one from [7]. H_S is on the left and H_D is on the right.

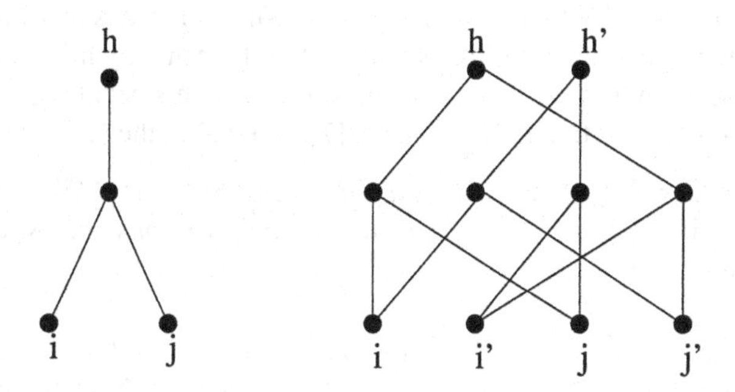

Fig. 2. I Gadget based on the one from [7]. I_S is on the left and I_D is on the right.

game on $\mathbf{A}$ and $\mathbf{B}$ such that, starting from position P, either he wins the game or after a number of moves the game is in a position P'' such that $P' \subseteq P''$.
This concept will be used to combine strategies of the Spoiler on different gadgets in order to construct strategies for the combined game.

5. We say that the Duplicator *can avoid a position P' from another position P of the $(\exists, k)$-pebble game on A and B* if the Duplicator has a winning strategy for the $(\exists, k)$-pebble game on $\mathbf{A}$ and $\mathbf{B}$ such that starting from position P, position P' never occurs.

For each gadget used in the reduction there will be two pieces, one for the Spoiler's structure and one for the Duplicator's structure. For gadget X, we call the Spoiler's side X_S, the Duplicator's side X_D, and the pair (X_S, X_D) simply X.

3.1 The Gadgets H and I

The graphs H_D and I_D, which are both based on gadgets from [7], are going to be used for **and** nodes and **or** nodes respectively. H_D, as seen in Figure 1, consists of six vertices h, h', i, i', j, j'. These six vertices form three distinguished pairs, (h, h'), (i, i'), and (j, j'). There are edges from h to i, and h to j, and edges from h' to i' and h' to j'. This graph has only one non-identity automorphism, which we will call swi, that maps any vertex a to a' and any vertex a' to a. H_S is simply the subgraph of H_D determined by h, i, j. Starting from position hh', the Spoiler can reach both ii' and jj' in the $(\exists, k)$-pebble game on (H_S, H_D).

I_D, seen in figure 2, has ten vertices. It contains the three distinguished pairs (h, h'), (i, i'), and (j, j'), plus four additional nodes which we will name by their connections to the other vertices. These nodes are hij, $h'ij'$, $h'i'j$, and $hi'j'$. This graph has three non-identity automorphisms, which we will refer to as fix_i, fix_j, and fix_h. These automorphism fix i, j, and h respectively, while switching the other two. By playing according to these automorphisms, the Duplicator can avoid either ii' or jj' from hh' but not both in the $(\exists, k)$-pebble game.

3.2 Single Input One-Way Switches

The Single Input One-Way Switches are used to restrict the ways in which the Spoiler can win the game. The basic intuition is that the Spoiler can only make progress in one particular direction; moreover, to do so he must use all of his pebbles.

This lemma is similar to Lemma 14 from [7], adapted to the $(\exists, k)$-pebble game.

Lemma 1. *For every $k \geq 2$ there exists a pair of graphs O^k_S and O^k_D with $O^k_S \subset O^k_D$, $\{x, x', y, y'\} \subset V(O^k_D)$, xx' and yy' distinguished pairs of vertices, and $\{x, y\} \subset V(O^k_S)$, such that:*

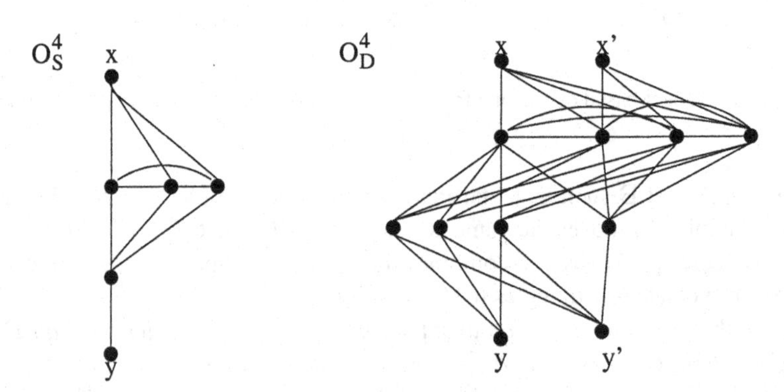

Fig. 3. Single Input One-Way Switch O^4.

1. *The Spoiler can reach yy' from xx' in the $(\exists, k)$-pebble game on (O^k_S, O^k_D).*
2. *There exist two disjoint sets of positions of the $(\exists, k)$-pebble game on (O^k_S, O^k_D), called Pretrapped and Trapped positions such that:*
   (a) Pretrapped and Trapped positions are partial homomorphisms
   (b) The Duplicator can avoid positions that are not Trapped and not Pretrapped from Pretrapped positions
   (c) The Duplicator can avoid positions that are not Trapped from Trapped positions
   (d) The position $\{xx'\}$ is Pretrapped
   (e) If P is Pretrapped and $|P| < k$, then $P \cup \{yy\}$ is Pretrapped
   (f) The positions $\{yy\}$ and $\{yy'\}$ are Trapped
   (g) If P is Trapped and $|P| < k$, then $P \cup \{xx\}$ is Trapped

We call (O^k_S, O^k_D) the Single Input One-Way Switch.

Corollary 1. *The Spoiler can reach $\{(y, y')\}$ from $\{(x, x')\}$ in the $(\exists, k)$-pebble game on the Single Input One-Way Switch, but not in the $(\exists, k-1)$-pebble game on the Single Input One-Way Switch.*

Proof. The first part of the Corollary is simply a restatement of condition 1 from Lemma 1. Assume for the sake of contradiction that the Spoiler can reach $\{(x, x')\}$ from $\{(y, y')\}$ in the $(\exists, k-1)$-pebble game on the Single Input One-Way Switch. Then, from the position $\{(x, x'), (y, y)\}$ of the $(\exists, k)$-pebble game, the Spoiler could reach the position $\{(y, y'), (y, y)\}$ by ignoring the pebbles on (y, y) and playing the $(\exists, k-1)$-pebble game. The position $\{(x, x'), (y, y)\}$ is Pretrapped, but $\{(y, y'), (y, y)\}$ is neither Pretrapped nor Trapped. This is a contradiction because the Duplicator can avoid such positions from Pretrapped positions.

Because of Corollary 1, the Spoiler has to use all of his pebbles in order to make progress. The "One-Way" aspect of the Switch lies in the fact that $\{(y, y')\}$ is Trapped, and $\{(x, x')\}$ is not. This means that the Duplicator can avoid $\{(x, x')\}$ from $\{(y, y')\}$.

3.3 Twisted Switches

The Twisted Switch consists of an H gadget, an I gadget, and two Single Input One-Way Switches in parallel. We use a Twisted Switch in the reduction to initialize the game. The construction of the Twisted Switch is the same as that in Grohe [7], except that we substitute a One-Way Switch in place of what Grohe calls a Threshold Switch.

The following Lemma introduces a set of positions of the Single Input One-Way Switch, called Switched positions. Within the Twisted Switch, the Duplicator uses Switched positions instead of Trapped and Pretrapped positions on the Single Input One-Way Switch.

Lemma 2. *There is a set of positions of the $(\exists, k)$-pebble game on O^k, called Switched positions, such that*

1. *$\{xx', yy\}$ and $\{xx, yy'\}$ are Switched.*
2. *If P is Switched, then either $P \cup \{xx', yy\}$ or $P \cup \{xx, yy'\}$ is Switched.*
3. *The Duplicator can avoid positions that are not Switched from Switched positions in the $(\exists, k)$-pebble game.*

Lemma 3. *On the Twisted Switch, the Spoiler can reach $\{yy'\}$ from $\{xx'\}$ in the $(\exists, k)$-pebble game, and there exists a set of positions of the $(\exists, k)$-pebble game called Twisted positions such that*

1. *The Duplicator can avoid non-Twisted positions from Twisted positions*
2. *$\{yy\}$, and $\{yy'\}$, are Twisted positions.*
3. *If P is a Twisted position, then $P \cup \{xx'\}$ is a Twisted position.*

The Twisted Switch will be used to initialize the game. In order to do this, $x \in T_S$ and $x' \in T_D$ are colored the same color, while $x \in T_D$ is colored a different color. Thus, if the Spoiler plays on x, then the Duplicator must play on x'. From here the Spoiler can play to $\{(y, y')\}$. The Twisted positions allow the Duplicator to avoid $\{(x, x)\}$, which is a losing position.

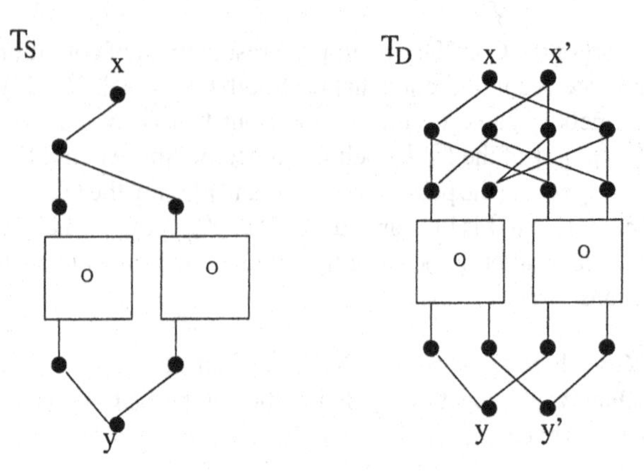

Fig. 4. Twisted Switch [7].

3.4 Reduction from MCV to $(\exists, k)$-Pebble Game

Theorem 1. *For every fixed $k \geq 2$, determining the winner of the $(\exists, k)$-pebble game is P-Complete.*

Proof. (Outline) Given a monotone circuit C, an assignment of values to the input nodes, and a node v, we construct a graph C_D. For each node a in the circuit, there are three choices.

1. If a is an input node, then C_D contains vertices a and a'. If the value of a in C is false, then we color a' black.
2. If a is an **and** node with parents b and c, then C_D contains nodes a and a', a copy of H_D, called H_a with h and h' identified with a and a', as well as two copies of O^k_D, one of which has x and x' connected to i and i', and y and y' connected to b and b' called O_{ab}. The other copy connects j and j' to c and c' in a similar manner and is called O_{ac}.
3. If a is an **or** node with parents b and c, then C_D contains nodes a and a', a copy of I_D called I_a with h and h' identified with a and a', as well as two copies of O^k_D, one of which has x and x' connected to i and i', and y and y' connected to b and b' called O_{ab}. The other copy connects j and j' to c and c' in a similar manner and is called O_{ac}.

In any case, a is colored white.

The construction of C_S is similar, except that we do not add a' and we use the Spoiler version of each gadget and switch instead of the Duplicator version.

Also, there is a Twisted Switch T, such that x is colored a fresh color in C_S, and x' is colored the same color in C_D. Also, in C_S, y in T is connected to v, while in C_D, y is connected to v and y' is connected to v'.

The Spoiler plays from the top of the Twisted Switch through the graph attempting to reach a false input. Because of the properties of the H and I gadgets and the Single Input One-Way Switch , the Spoiler can do this if the value of the goal node is false. If the value of the goal node is true, then the Duplicator can play along indefinitely. He does this by choosing a path down the simulated circuit which leads to a true input. If the Spoiler attempts to depart from the intended mode of play, the Duplicator can use the properties of Trapped and Pretrapped strategies to arrive at a partial homomorphism that is a subset of the identity. ∎

4 The $(\exists, k)$-Pebble Game Is EXPTIME-Complete

Kasai, Adachi and Iwata [8] showed that the following pebble game, which (to avoid confusion in the terminology) we will call the KAI game, is EXPTIME-complete. The game is played between two players, called Player I and Player II. An instance of the KAI game is a quadruple (X, S, R, t), where X is a set of nodes, $R \subseteq X^3$ is a set of *rules*, $S \subseteq X$ is the initial position of the game, and $t \in X$ is the *goal*. There are as many pebbles as nodes in the initial position S; at the start of the game, each node in S has a pebble on it. The two players take turns and in each turn they slide a pebble as follows: if $(x, y, z) \in R$ is a rule, then the current player may slide a pebble from x to z, if there are pebbles on x and y, but not z. The first player to place a pebble on the goal t wins. The problem is, given (X, R, S, t), does Player I have a winning strategy?

We will show that the $(\exists, k)$-pebble game is EXPTIME complete by exhibiting a reduction from the KAI game. The reduction procedes by constructing an instance of the $(\exists, k)$-pebble game such that the Spoiler and the Duplicator simulate the instance of the KAI game. In particular, the Spoiler can only win by simulating a winning strategy for Player I in the KAI game. If there is no winning strategy, then the Spoiler does not gain any advantage by not simulating the KAI game.

In order to perform this simulation, we use a Twisted Switch (Section 3.3) to initialize the game, and new switches to allow Player I and Player II to choose rules, to apply the rules, and to force the Spoiler to simulate the game.

4.1 The Gadgets H^m and I^m

These gadgets allow the Spoiler and Duplicator to choose rules of the KAI game to use. In both the H^m and I^m gadgets, the nodes y^i, y^i_0, y^i_1 are colored the same color, but the color of y^i is different that the color of y^j for $i \neq j$.

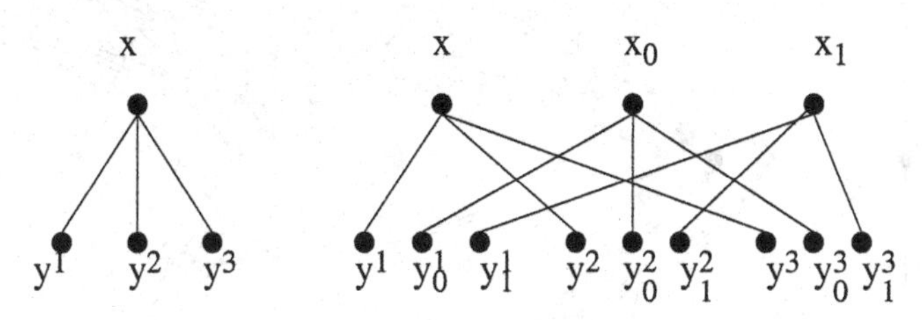

Fig. 5. H^3 Gadget.

Lemma 4. *For every $k \geq 2$, in the $(\exists, k)$-pebble game on (H^m_S, H^m_D), from a position $\{xx_j\}$, $j \in \{0, 1\}$, the Spoiler can reach $\{y^i y^i_j\}$ for any i.*

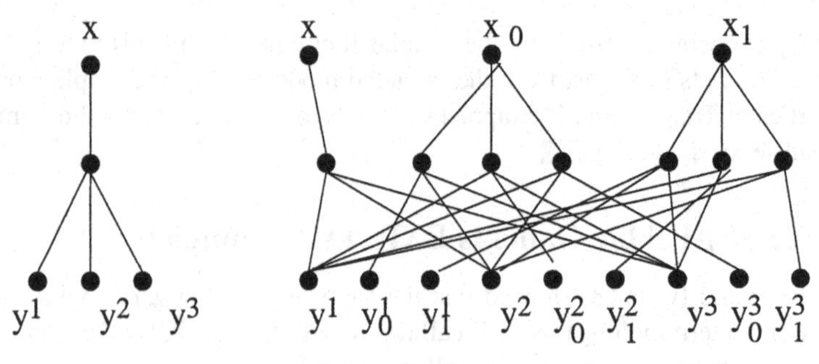

Fig. 6. I^3 Gadget.

Lemma 5. *For every $k \geq 2$, in the $(\exists, k)$-pebble game on I_S^m, I_D^m, from a position $\{xx_j\}$, $j \in \{0,1\}$, the Duplicator can choose any $1 \leq i \leq m$, and avoid $\{y^l y_j^l\}$ for $l \neq i$.*

4.2 Multiple Input One-Way Switches for the $(\exists, k)$-Pebble Game

The idea of the Multiple Input One-Way Switch is to restrict the Spoiler's potential winning strategies. We simulate each node x^i in the KAI game by using three nodes in the Duplicator's graph, x_0^i, x_1^i, x^i. These correspond to not having pebble on x^i in the simulated game, having a pebble on x^i in the simulated game, and no information about x^i, respectively. In the Multiple Input One-Way Switch, the Spoiler can only make progress if he has information about each node in the simulated game. Also, if the Spoiler attempts to play backwards through the Switch, he will end up with no information about any nodes in the simulated game.

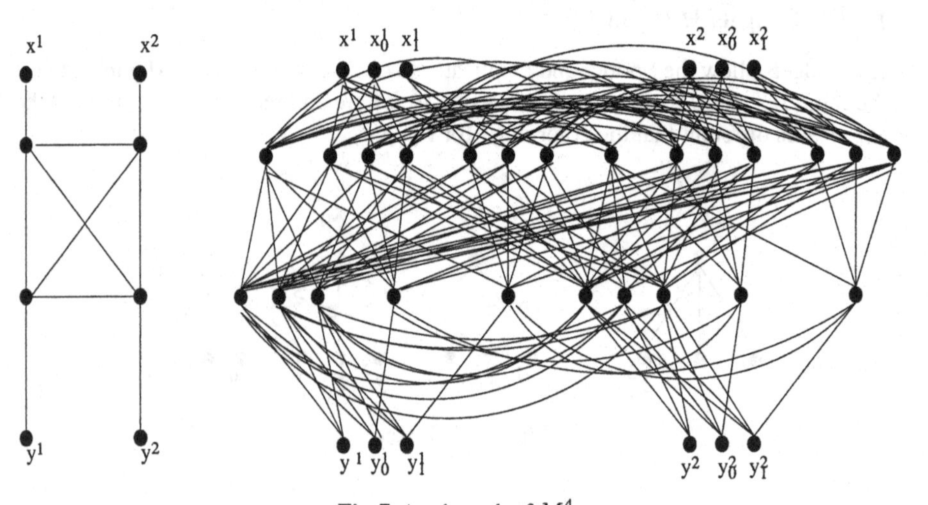

Fig. 7. A subgraph of M^4.

Lemma 6. *For every $k \geq 2$, there exists a pair of graphs M_S^k, and M_D^k such that*

$$\{x^1, x_0^1, x_1^1, \ldots, x^{k-1}, x_0^{k-1}, x_1^{k-1}, y^1, y_0^1, y_1^1, \ldots, y^{k-1}, y_0^{k-1}, y_1^{k-1}\} \subset V(M_D^k),$$

$\{x^1, \ldots, x^{k-1}, y^1, \ldots, y^{k-1}\} \subset V(M_S^k)$ *and the following properties hold:*

1. *From a position $\{x^i x^i_{j_i} | 1 \leq i \leq k-1, j_i \in \{0,1\}\}$, the Spoiler can reach the position $\{y^i y^i_{j_i} | 1 \leq i \leq k-1, j_i \in \{0,1\}\}$ in the $(\exists, k)$-pebble game on M_S^k and M_D^k.*
2. *There exist two disjoint sets of positions of the $(\exists, k)$-pebble game on (M_S^k, M_D^k), called Pretrapped and Trapped positions such that:*
   (a) Pretrapped and Trapped positions are partial homomorphisms
   (b) The Duplicator can avoid positions that are not Pretrapped and not Trapped from Pretrapped positions
   (c) The Duplicator can avoid positions that are not Trapped from Trapped positions
   (d) From any position $P = \{x^i a | 1 \leq i \leq k-1\}$ where $|\{x^i x^i_j \in P | j \in \{0,1\}\}| < k-1$, the Duplicator can avoid $y^i y^i_j$ for all $1 \leq i \leq k-1, j \in \{0,1\}$.
   (e) All positions that are subsets of positions of the form $\{x^i x^i_{j_i} | 1 \leq i \leq k-1, j_i \in \{0,1\}\}$, are PreTrapped.
   (f) If P is Pretrapped and $|P| < k$, then $P \cup \{y^i y^i\}$ is Pretrapped for all i
   (g) Any position in which all of the Spoiler's pebbles are on nodes y^i, is Trapped.
   (h) If P is Trapped and $|P| < k$, then $P \cup \{x^i x^i\}$ is Trapped for all i

Moreover, $|V(M_S^k)|$ is $O(k^2)$ and $|V(M_D^k)|$ is $O(k^2)$.

4.3 The Rule Gadget

The Rule gadgets are used to simulate a move of the KAI game. One rule gadget causes the Spoiler to lose if the rule gadget corresponds to a rule that cannot be applied, and another causes the Duplicator to lose if the rule cannot be applied.

Lemma 7. *If the rule gadget RS^n does not correspond to a legal rule, that is, if one of xx_1, yy_1, zz_0 is not in P, then the Duplicator can avoid $z'z'_1$ in the $(\exists, k)$-pebble game on RS^n.*

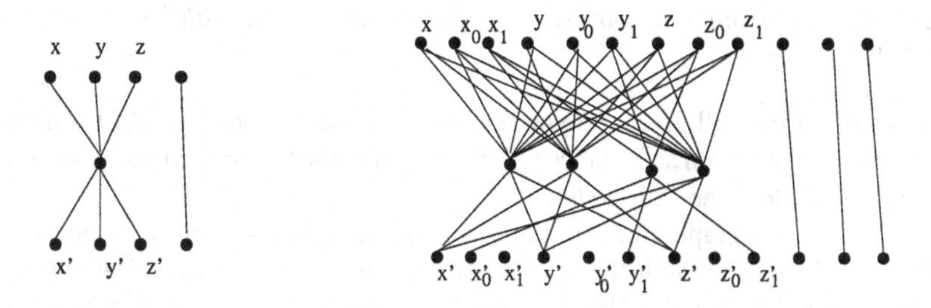

Fig. 8. The Rule Gadget RS^n that penalizes the Spoiler for choosing a bad rule.

Lemma 8. *If the rule gadget RD^n does not correspond to a legal rule, that is, if one of xx_1, yy_1, zz_0 is not in P, then the Spoiler can play to $z'z'_0$, which causes the Duplicator to lose.*

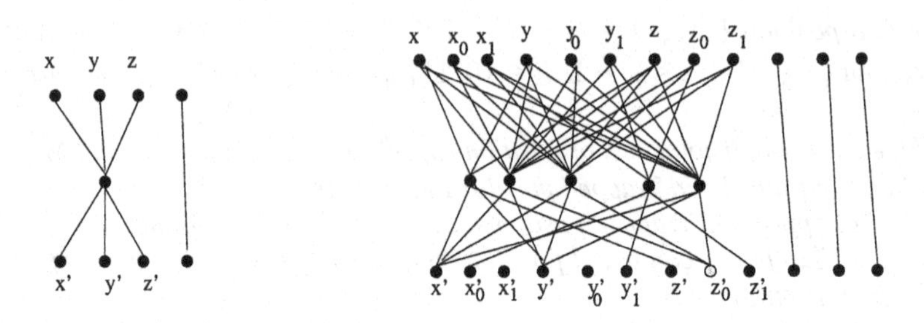

Fig. 9. The Rule Gadget RD^n that penalizes the Duplicator for choosing a bad rule.

4.4 Winning the Game

For each gadget, we define its *boundary* to be the set of nodes which are used to connect that gadget to other gadgets. For M^k, H^m, and I^m, the x and y nodes form the boundary. For RS^n and RD^n, the boundary consists of all the nodes except for the middle nodes. For the Twisted Switch, the boundary consists of y in T_S and y, y' in T_D. In the Twisted Switch, x and x' are never connected to any other gadgets.

Lemma 9. *For each gadget other than the Twisted Switch, starting from a position that is a subset of the identity on the boundary, the Duplicator can avoid any position that is not a subset of the identity on the boundary.*

By combining this lemma with the properties of the Multiple Input One-Way Switch, we obtain a sufficient condition for the Duplicator to win the $(\exists, k)$-pebble game.

4.5 Reduction from KAI Game to $(\exists, k)$-Pebble Game

Theorem 2. *Determining the winner of the $(\exists, k)$-pebble game with k part of the input is EXPTIME-Complete.*

Proof. (Outline) We will give a polynomial-time reduction from the KAI Game to the $(\exists, k)$-pebble game. Given an instance (X, S, R, t) of the KAI game, we form an instance of the $(\exists, k)$-pebble game as follows.

The Duplicator's graph and the Spoiler's graph each have two sides. One side represents Player I's turn in the KAI game, while the other side represents Player II's turn.

First, we build Player I's side of the graph. For each $x^i \in X$, we form three nodes in D, called xs^i, xs^i_0, xs^i_1. These three nodes correspond to specific information about the simulated KAI game. If there is a pebble on xs^i_1, then there is a pebble on x in the KAI game, and xs^i_0 corresponds to no pebble on x. A pebble on xs^i in the Duplicator's

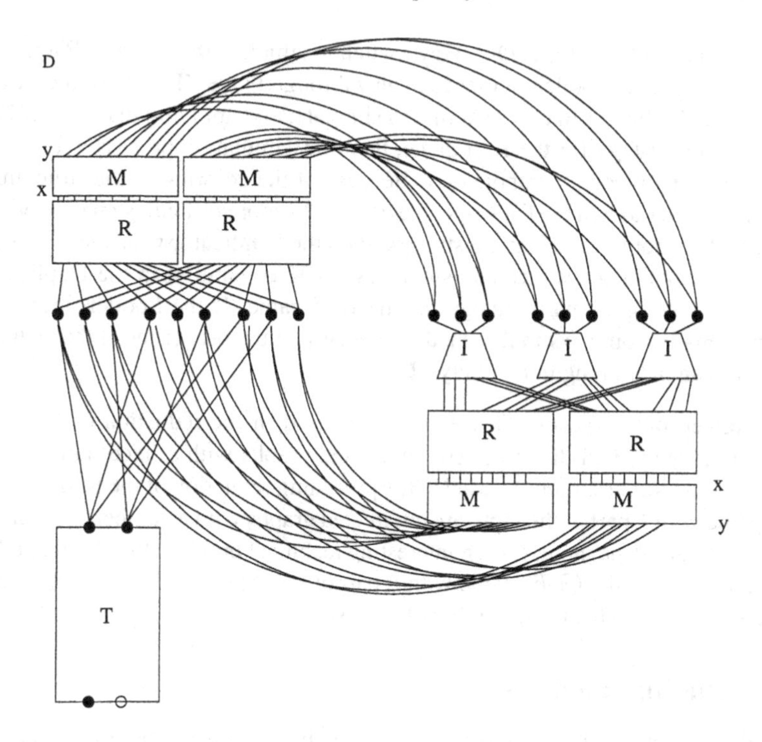

Fig. 10. This is component decomposition of the Duplicator's graph for the reduction.

graph means that the Spoiler has made a mistake. For each $(x, y, z) \in R$, construct a rule gadget to penalize the Spoiler, connected to each xs^i, xs^i_0, xs^i_1 by H^m gadgets. The other end is then connected to a Multiple Input One-Way Switch .

On the other half of the graph, there are another set of nodes xd^i, xd^i_0, xd^i_1. Connecting the xd nodes to Duplicator punishing rule gadgets are a set of I^m gadgets where $m = |R|$. Given any two of these I^m gadgets A and B, there is an edge between $A.y^i_j$ and $B.y^i_k$ for all i, j, k, and an edge between $A.y^i$ and $B.y^i$ for all i. We then connect a Multiple Input One-Way Switch to each of the rule gadgets in the obvious way. The outputs of these are connected back around to the first set of state nodes.

We use a Twisted Switch to set up the initial positions. If $x^i \in S$, then there is an edge from y' of the Twisted Switch to xs^i_1 otherwise there is an edge from y' to xs^i_0. There is an edge from y to xs^i for every i. We then color x' a unique color. In addition we give t_1 a unique color so that if the Spoiler can reach tt_1, then the Duplicator loses.

The Spoiler's graph is constructed in a similar way. The Spoiler gets xs^i and xd^i for every $x^i \in X$. Also, for each rule and each side, there is a corresponding Multiple Input One-Way Switch followed by a rule gadget. For the Twisted Switch, we color x the same color that we colored x' in the Duplicator's graph.

The two players now simulate the KAI game by playing the $(\exists, k)$-pebble game on these structures. The Spoiler initializes the game by playing on the Twisted Switch. From here, the Spoiler uses the H^m gadgets to choose a rule, then plays through the RS^n

gadget and the Multiple Input One-Way Switch to simulate the move of Player I in the KAI game. Then, the Spoiler continues to play through Player II's side of the structures, allowing the Duplicator to choose a rule to simulate, and applying that rule. If Player I has a winning strategy for the KAI game, then this simulation process will eventually simulate Player I placing a pebble on t. Because of the coloring of the structures, this causes the Spoiler to win the $(\exists, k)$-pebble game. The more difficult step is showing that if Player I does not have a winning strategy, then the Duplicator wins the $(\exists, k)$-pebble game. If the Spoiler plays nice and simulates the KAI game, then the Duplicator can avoid t, by playing a smart strategy for the KAI game. If the Spoiler does not play nice, and departs from the simulation, then, because of the properties of the gadgets, the Duplicator can play along indefinitely. ∎

As pointed out in Section 3, the structures used in the reduction of MCV to the $(\exists, k)$-pebble game with fixed k were undirected graphs with a fixed number (ten) of colors. In contrast, the structures used in the preceding reduction of the KAI game to the $(\exists, k)$-game with k part of the input are undirected graphs with a number of colors that is linear in the size of the input. It is an interesting technical problem to exhibit a reduction of the KAI game to the $(\exists, k)$-game with k part of the input in which the structures are undirected graphs with a fixed number of colors.

5 Concluding Remarks

Although in this paper we focused on the $(\exists, k)$-pebble game because of its connections to constraint satisfaction, in database theory there is also interest in the *one-to-one* $(\exists, k)$-*pebble game*, which is the variant of the $(\exists, k)$-pebble game in which the Duplicator strives to maintain one-to-one homomorphisms (see [11]). A perusal of the two reductions presented here reveals that in both these reductions the structures constructed have the property that the Duplicator wins the $(\exists, k)$-pebble game if and only if the Duplicator wins the one-to-one $(\exists, k)$-pebble game. Consequently, determining the winner in the one-to-one $(\exists, k)$-pebble is P-complete for every fixed $k \geq 2$, and is EXPTIME-complete when k is part of the input.

Several problems remain open in this area. Kasai and Iwata [9] proved that the number of pebbles used in the KAI game gives rise to a strict hierarchy on the time complexity of that game. Thus, it is natural to ask whether a similar strict hierarchy result can be proved for the $(\exists, k)$-pebble, for fixed k. This amounts to showing that, for each fixed $k \geq 2$, determining the winner of the $(\exists, k)$-pebble game is not solvable in time $O(n^s)$ for any $s < 2k$. Finally, it remains an open problem to establish that the (two-sided) k-pebble game with k part of the input is an EXPTIME-complete problem.

References

1. A. K. Chandra, D. C. Kozen, and L. J. Stockmeyer. Alternation. *Journal of the ACM*, 28:114–233, January 1981.
2. M. C. Cooper. An optimal k-consistency algorithm. *Artificial Intelligence*, 41:89–95, 1989.
3. V. Dalmau, Ph. G. Kolaitis, and M. Y. Vardi. Constraint satisfaction, bounded treewidth, and finite-variable logics. In *Proc. of Eighth International Conference on Principles and Practice of Constraint Programming*, pages 310–326, 2002.

4. R. Dechter. Constraint networks. In S.C. Shapiro, editor, *Encyclopedia of Artificial Intelligence*, pages 276–285. Wiley, New York, 1992.
5. R. Dechter. From local to global consistency. *Artificial Intelligence*, 55:87–107, 1992.
6. T. Feder and M. Y. Vardi. The computational structure of monotone monadic SNP and constraint satisfaction: a study through Datalog and group theory. *SIAM Journal on Computing*, 28:57–104, 1998.
7. M. Grohe. Equivalence in finite-variable logics is complete for polynomial time. *Combinatorica*, 19(4):507–523, 1999.
8. T. Kasai, A. Adachi, and S. Iwata. Classes of pebble games and complete problems. *SIAM Journal of Computing*, 8(4):574–586, 1979.
9. T. Kasai and S. Iwata. Gradually intractable problems and nondeterminitstic log-space lower bounds. *Mathematical Systems Theory*, 18:153–170, 1985.
10. S. Kasif. On the parallel complexity of some constraint satisfaction problems. In *Proc. of Fifth National Conference on Artificial Intelligencee*, volume 1, pages 349–353, 1986.
11. Ph. G. Kolaitis and M. Y. Vardi. On the expressive power of Datalog: Tools and a case study. *Journal of Computer and System Sciences*, 51:110–134, 1995.
12. Ph. G. Kolaitis and M. Y. Vardi. A game-theoretic approach to constraint satisfaction. In *Proc. of the Seventeenth National Conference on Artificial Intelligence*, pages 175–181, 2000.
13. E. Pezzoli. Computational complexity of Ehrenfeucht-Fraïssé games on finite structures. In *Computer Science Logic. 12th International Workshop, CSL'98.*, pages 159–170, 1999.

Computational Aspects of Σ-Definability over the Real Numbers without the Equality Test*

Margarita Korovina

BRICS**, Department of Computer Science, University of Aarhus,
Ny Munkegade, DK-8000 Aarhus C, Denmark,
A. P. Ershov Institute of Informatics Systems,
Lavrent'ev ave., 6, 630090, Novosibirsk, Russia
korovina@brics.dk
http://www.brics.dk/~korovina

Abstract. In this paper we study the expressive power and algorithmic properties of the language of Σ-formulas intended to represent computability over the real numbers. In order to adequately represent computability, we extend the reals by the structure of hereditarily finite sets. In this setting it is crucial to consider the real numbers without equality since the equality test is undecidable over the reals. We prove Engeler's Lemma for Σ-definability over the reals without the equality test which relates Σ-definability with definability in the constructive infinitary language $L_{\omega_1\omega}$. Thus, a relation over the real numbers is Σ-definable if and only if it is definable by a disjunction of a recursively enumerable set of quantifier free formulas. This result reveals computational aspects of Σ-definability and also gives topological characterisation of Σ-definable relations over the reals without the equality test.

1 Introduction

We start with an informal introduction to problems and solutions considered in this paper. It is well-known that the classical theory of computation, which works with discrete structures, is not suitable for formalisation of computations that operate on real-valued data. Most computational problems in physics and engineering are of this type, e.g. problems relevant to foundation of dynamical and hybrid systems. Since computational processes are discrete in their nature and objects we consider are continuous, formalisation of computability of such objects is already a challenging research problem. Consequently, computability over the reals has become a subject of the great interest and study in Computer Science [4, 11, 15, 18, 21, 24].

* This research was partially supported by the Danish Natural Science Research Council, Grant no. 21-02-0474, RFFI-DFG Grant no. 01-01-04003 and Grant Scientific School-2112.2003.1.

** Basic Research in Computer Science (www.brics.dk), funded by the Danish National Research Foundation.

M. Baaz and J.A. Makowsky (Eds.): CSL 2003, LNCS 2803, pp. 330–344, 2003.

One of the most promising approaches is based on the notion of definability [18], where continuous objects and computational processes involving these objects can be defined using finite formulas in a suitable structure. Definability has been a very successful framework for generalised computability theory [20], descriptive complexity [1,13] and databases [6,22]. One of the most interesting and practically important types of definability is Σ-definability, which generalises recursive enumerability over the natural numbers [2,10]. However, the most developed part of definability theory deals with abstract structures with equality (i.e., the natural numbers, trees, automata, etc.). In the case of continuous data types, such as real numbers, real-valued functions and functionals, it is reasonable to consider the corresponding structures without the equality test. This is motivated by the following natural reason. In all effective approaches to exact real number computation via concrete representations [11,21], the equality test is undecidable. In order to do any kind of computation or to develop a computability theory, one has to work within a structure rich enough for information to be coded and stored. For this purpose we extend the structure $\mathbb{R}$ by the set of hereditarily finite sets $\mathrm{HF}(\mathbb{R})$. The idea that the hereditarily finite sets over a structure form a natural domain for computation is discussed in [2,10,22]. Note that such or very similar extensions of structures are used in the theory of abstract state machines [3], in query languages for hierarchic databases [6], and in Web-like databases [22]. Since Σ-definability in $\mathbf{HF}(\mathbb{R})$ (i.e., definability by Σ-formulas) is a generalisation of recursive enumerability, we are interested in the language of Σ-formulas as in a suitable language intended to express computability over the reals.

In this paper we address the algorithmic properties of the language of Σ-formulas. For this purpose we study the expressive power of Σ-formulas using a suitable fragment of the constructive infinitary language $L_{\omega_1\omega}$. Certain fragments of constructive $L_{\omega_1\omega}$ have been used to study the expressive power of formal approaches to computability such as search computability [19], 'While'-computability [23], dynamic logics [12] and fixed-point logics [8]. One of the most important results in the area is Engeler's Lemma which relates abstract computability with definability in constructive $L_{\omega_1\omega}$. Engeler's Lemma was first proven for formal computability on abstract structures with the equality test (see [9,10,23]). The natural question to ask is: "Does Engeler's Lemma hold for Σ-definability over the reals without the equality test?" In this paper we answer this question positively. Engeler's Lemma, together with Gandy's theorem for $\mathbf{HF}(\mathbb{R})$ (recently proved in [16]) gives an algorithmic characterisation of Σ-definability. We prove that a relation over the real numbers is Σ-definable if and only if it is definable by a disjunction of a recursively enumerable set of quantifier free formulas. Let us note that the proof of the 'if' direction follows from Engeler's Lemma for Σ-definability and the proof of the 'only if' direction uses methods based on Gandy's Theorem. It is worth noting that both of the directions of this characterisation are important. Engeler's Lemma gives us an effective procedure which generates quantifier free formulas approximating Σ-relations. This reveals algorithmic aspects of Σ-definability. The converse direction provides tools for descriptions of the results of effective infinite approximation processes by finite formulas.

The structure of this paper is as follows. In Section 2 we recall the notion of Σ-definability and provide a background necessary to understand the main results. Section 3 introduces a certain fragment of the constructive infinite logic and presents a characterisation of the expressive power of Σ-definability which reveals computational aspects of Σ-definability over the reals without the equality test. We conclude with a discussion of future work.

2 Σ-Definability over the Real Numbers

2.1 Basic Definitions and Notions

In this paper we consider the standard model of the real numbers

$$\langle \mathbb{R}, 0, 1, +, \cdot, < \rangle = \langle \mathbb{R}, \sigma_0 \rangle ,$$

denoted also by $\mathbb{R}$, where $+$ and $\cdot$ are regarded as the usual arithmetic operations on the real numbers. We use the language of strictly ordered rings, so we assume that the predicate $<$ occurs positively in all formulas. In order to develop a computability theory over the reals we extend the structure $\mathbb{R}$ by the set of hereditarily finite sets $\mathrm{HF}(\mathbb{R})$ which is rich enough for information to be coded and stored. We construct the set of hereditarily finite sets, $\mathrm{HF}(A)$ over a structure A, as follows:

1. $\mathrm{HF}_0(A) \rightleftharpoons A$,
2. $\mathrm{HF}_{n+1}(A) \rightleftharpoons \mathcal{P}_\omega(\mathrm{HF}_n(A)) \cup \mathrm{HF}_n(A)$, where $n \in \omega$ and for every set B, $\mathcal{P}_\omega(B)$ is the set of all finite subsets of B.
3. $\mathrm{HF}(A) = \bigcup_{n\in\omega} \mathrm{HF}_n(A)$.

We define $\mathbf{HF}(\mathbb{R})$ as the following model: $\mathbf{HF}(\mathbb{R}) \rightleftharpoons \langle \mathrm{HF}(\mathbb{R}), U, S, \sigma_0, \emptyset, \in \rangle \rightleftharpoons \langle \mathrm{HF}(\mathbb{R}), \sigma \rangle$, where the constant $\emptyset$ stands for the empty set and the binary predicate symbol $\in$ has the set-theoretic interpretation. We also add predicate symbols U for urelements (elements from $\mathbb{R}$) and S for sets. Let $\mathbb{R}^*$ denote the set of all nonempty finite subsets of the reals.
For our convenience, we use variables subject to the following conventions:
$r, x, y, z, \ldots$ range over $\mathbb{R}$ (urelements),
$i, j, k, l, m, \ldots$ range over ω (natural numbers),
$\alpha, \beta, \kappa, \ldots$ range over $\mathrm{HF}(\emptyset)$,
$A, B, C, R, X, Y, Z, \ldots$ range over $\mathbb{R}^*$ (nonempty finite sets over the reals),
$U, V, W, \ldots$ range over $\mathrm{HF}(\mathbb{R}^*)$ (finite sets which do not contain real numbers as members) and
$K, L, M, N, S, \ldots$ range over $\mathrm{HF}(\mathbb{R})$.
We use the same letters as for variables to denote elements from the corresponding structures.

A formula in the form $p_1(\bar{x}) < p_2(\bar{x})$, where p_1 and p_2 are polynomials with coefficients in $\mathbb{N}$, is called an *atomic strict semi-algebraic (s.s.-algebraic) formula.*

The set of atomic formulas is the union of the set of atomic s.s.-algebraic formulas and the formulas of the type $N \in M$, where M ranges over sets.

The closure of the atomic s.s.-algebraic formulas under finite conjunctions and disjunctions forms the set of *s.s.-algebraic formulas.*

The set of *Δ_0-formulas* is the closure of the set of atomic formulas under $\wedge, \vee, \neg$, bounded quantifiers $(\exists M \in S)$ and $(\forall M \in S)$, where $(\exists M \in S)\ \Psi$ denotes $\exists M(M \in S \wedge \Psi)$, $(\forall M \in S)\ \Psi$ denotes $\forall M(M \in S \to \Psi)$ and S ranges over sets.
We assume that predicates from the language σ_0 can occur only positively in our formulas.

With every atomic s.s.-algebraic formula $p_1(x_1, \dots, x_n) < p_2(x_1, \dots, x_n)$ we associate the following formula, called a *lifted atomic s.s.-algebraic formula*:

$$p_1(X_1, \dots, X_n) < p_2(X_1, \dots, X_n) \rightleftharpoons$$
$$(\forall x_1 \in X_1) \dots (\forall x_n \in X_n)\, p_1(x_1, \dots, x_n) < p_2(x_1, \dots, x_n).$$

The terms $p_1(\bar{X})$ and $p_2(\bar{X})$ are called *lifted polynomials.* In a similar way, we can also associate with each s.s.-algebraic formula a *lifted s.s.-algebraic formula*, i.e., the lifted s.s.-algebraic formulas are the closure of lifted atomic s.s.-algebraic formulas under finite conjunctions and disjunctions.

The set of *Σ-formulas* is the closure of the set of Δ_0-formulas under $\wedge$,$\vee$, $(\exists M \in S)$, $(\forall M \in S)$ and $\exists$, where S ranges over sets.

We are interested in Σ-definability of subsets on $\mathbb{R}^n$ which can be considered as a generalisation of recursive enumerability. The analogy of Σ-definable and recursive enumerable sets is based upon the following fact. If we consider the structure $\mathbf{HF} = \langle \mathrm{HF}(\emptyset), \in \rangle$ with the hereditarily finite sets over $\emptyset$ as its universe and membership as its only relation, then the Σ-definable sets are exactly the recursively enumerable sets. The notion of Σ-definability has a natural meaning also in the structure $\mathbf{HF}(\mathbb{R})$.

Definition 1. *1. A relation $B \subseteq \mathrm{HF}(\mathbb{R})^n$ is Δ_0 (Σ)-definable, if there exists a Δ_0 (Σ)-formula Φ such that $\bar{M} \in B \leftrightarrow \mathbf{HF}(\mathbb{R}) \models \Phi(\bar{M})$.*
2. A function is Δ_0 (Σ)-definable if its graph is Δ_0 (Σ)-definable.

Note that the sets $\mathbb{R}$ and ω are Δ_0-definable. This fact makes $\mathbf{HF}(\mathbb{R})$ a suitable domain for studying subsets of $\mathbb{R}^n$.
In the following lemma we introduce some Δ_0-definable and Σ-definable predicates that will be used later.

Lemma 1. *1. The sets $\mathbb{R}$, $\mathbb{R}^*$, $\mathrm{HF}(\mathbb{R}^*)$ and $\mathrm{HF}(n)$ for all $n \in \omega$ are Δ_0-definable.*
2. The following predicates are Δ_0-definable: $M = N$, $M = N \cap S$, $M = N \cup S$, $M = \langle N, S \rangle$, $M = N \setminus S$, where all variables M, N and S range over sets.
3. Every recursive function $f : \omega^n \to \omega^m$ is Σ-definable.
4. Every $B \subset \mathbb{N}^n$ is recursively enumerable if and only if it is Σ-definable.
5. All arithmetic operations on ordinals are Σ-definable.

Proposition 1 (**Σ-reflection principle**). *Every Σ-formula $\Phi(\bar{M})$ is equivalent to a formula of the type $\exists N \Psi(N, \bar{M})$, where Ψ is a Δ_0-formula.*

In the following lemmas we obtain some properties of s.s.-algebraic and lifted s.s.-algebraic formulas that will be used later.

Lemma 2. *If $B \subset \mathbb{R}^n$ is definable by an s.s.-algebraic formula then it is open.*

Lemma 3. *Let $\Phi(\bar{x})$ be an s.s.-algebraic formula and $\Phi_{lifted}(\bar{X})$ be the corresponding lifted s.s.-algebraic formula. If $R_i = \{r_i\}$ for $i = 1 \ldots n$, then*

$$\mathbf{HF}(\mathbb{R}) \models \Phi(\bar{r}) \leftrightarrow \Phi_{lifted}(\bar{R}).$$

Lemma 4. *Let $\Phi(\bar{x})$ be an s.s.-algebraic formula and $\Phi_{lifted}(\bar{X})$ be the corresponding lifted s.s.-algebraic formula. Then for all $\bar{x} \in \mathbb{R}^n$ we have*

$$\mathbf{HF}(\mathbb{R}) \models \Phi(x_1, \ldots, x_n) \leftrightarrow \exists X_1 \ldots \exists X_n \Phi_{lifted}(X_1, \ldots, X_n) \wedge \bigwedge_{i \leq n} (x_i \in X_i).$$

Lemma 5. *Let $\Phi_{lifted}(\bar{X})$ be a lifted s.s.-algebraic formula and $Y_1 \subseteq Z_1$, ..., $Y_n \subseteq Z_n$. If $\mathbf{HF}(\mathbb{R}) \models \Phi_{lifted}(Z_1, \ldots, Z_n)$, then $\mathbf{HF}(\mathbb{R}) \models \Phi_{lifted}(Y_1, \ldots, Y_n)$.*

Lemma 6. *Let $\Phi_{lifted}(\bar{Y})$ be a lifted s.s.-algebraic formula. If for some X_1, ..., X_n we have $\mathbf{HF}(\mathbb{R}) \models \Phi_{lifted}(X_1, \ldots, X_n)$, then for all $m > 0$ there exist $Y_1, \ldots, Y_n$ of cardinality m such that for all $i, j \leq n$ $Y_i = Y_j \leftrightarrow X_i = X_j$ and $\mathbf{HF}(\mathbb{R}) \models \Phi_{lifted}(Y_1, \ldots, Y_n)$.*

Lemma 7. *Let Φ be an existentially quantified s.s.-algebraic formula. Then there exists an s.s.-algebraic formula Ψ such that $\mathbf{HF}(\mathbb{R}) \models \Phi(\bar{x}) \leftrightarrow \Psi(\bar{x})$. Moreover Ψ can be constructed effectively from Φ.*

Let us note that proofs of all lemmas are straightforward (see [17]) except Lemma 7 which follows from the finiteness theorem [5, 7].

2.2 Gandy's Theorem and Inductive Definitions

Let us recall Gandy's Theorem for $\mathbf{HF}(\mathbb{R})$ which will be essentially used in all proofs of the main results. Let $\Phi(a_1, \ldots, a_n, P)$ be a Σ-formula, where P occurs positively in Φ and the arity of Φ is equal to n. We think of Φ as defining an *effective operator* $\Gamma : \mathcal{P}(\mathrm{HF}(\mathbb{R})^n) \to \mathcal{P}(\mathrm{HF}(\mathbb{R})^n)$ given by

$$\Gamma(Q) = \{\bar{a} | \, (\mathbf{HF}(\mathbb{R}), Q) \models \Phi(\bar{a}, P)\}.$$

Since the predicate symbol P occurs only positively, we have that the corresponding operator Γ is monotone, i.e., for any sets B and C, from $B \subseteq C$ follows $\Gamma(B) \subseteq \Gamma(C)$. By monotonicity, the operator Γ has a least (w.r.t. inclusion) fixed point which can be described as follows. We start from the empty set and apply operator Γ until we reach the fixed point: $\Gamma^0 = \emptyset$, $\Gamma^{n+1} = \Gamma(\Gamma^n)$, $\Gamma^\gamma = \cup_{n<\gamma} \Gamma^n$, where γ is a limit ordinal.

One can easily check that the sets Γ^n form an increasing chain of sets: $\Gamma^0 \subseteq \Gamma^1 \subseteq \ldots$. By set-theoretical reasons, there exists the least ordinal γ such that $\Gamma(\Gamma^\gamma) = \Gamma^\gamma$. This Γ^γ is the least fixed point of the given operator Γ.

Theorem 1 (Gandy's Theorem for HF(IR)).
Let $\Gamma : \mathcal{P}(\mathrm{HF}(\mathbb{R})^n) \to \mathcal{P}(\mathrm{HF}(\mathbb{R})^n)$ be an effective operator. Then the least fixed-point of Γ is Σ-definable and the least ordinal such that $\Gamma(\Gamma^\gamma) = \Gamma^\gamma$ is less or equal to ω.

Proof. See [16].

2.3 Universal Σ-Predicate for s.s.-Algebraic Formulas

In order to obtain a result on the existence of a universal Σ-predicate for the s.s.-algebraic formulas, we first construct a universal Σ-predicate for the lifted s.s.-algebraic formulas. For this purpose we prove Σ-definability of the truth of lifted s.s.-algebraic formulas.

In this section we fix a standard effective Gödel numbering of the terms and formulas of the language σ by finite ordinals which are elements of $\mathrm{HF}(\emptyset)$. Let $\lceil\Phi\rceil$, $\lceil p\rceil$ denote the codes of a formula Φ and a term p respectively. It is worth noting that the type of an expression is effectively recognisable by its code. We also can obtain effectively from the codes of expressions the codes of their subexpressions and vice versa. Since equality is Δ_0-definable in $\mathrm{HF}(\emptyset)$, we can use the well-known characterisation which states that all effective procedures over ordinals are Σ-definable. Thus, for example, the following predicates

$$\begin{aligned} Code_{elem}(n,i,j) &\rightleftharpoons n = \lceil X_i < X_j \rceil, \\ Code_{sum1}(n,i,j,k) &\rightleftharpoons n = \lceil p+q < f\rceil \wedge i = \lceil p\rceil \wedge j = \lceil q\rceil \wedge k = \lceil f\rceil, \\ Code_{\wedge}(n,i,j) &\rightleftharpoons n = \lceil \Phi \wedge \Psi\rceil \wedge j = \lceil \Phi\rceil \wedge j = \lceil \Psi\rceil \end{aligned}$$

are Σ-definable. Hence, in Σ-formulas we can use such predicates.

With every element $A \in \mathrm{HF}(\mathbb{R})$ we associate an interpretation γ_A of variables $X_1, X_2, \ldots$ such that

$$\gamma_A(X) = \begin{cases} C & \text{if } \langle \lceil X\rceil, C\rangle \in A \text{ and} \\ & \text{for any } \langle \lceil X\rceil, B\rangle \in A, \text{ we have } B = C \\ \emptyset & \text{otherwise.} \end{cases}$$

Let V be a set of variables. An interpretation γ_A is called *correct for* V if for all $X \in V$ we have $\gamma_A(X) \neq \emptyset$. Let Int denote the set of elements $A \in \mathrm{HF}(\mathbb{R})$ with the following property: if $\langle i, C\rangle$ and $\langle i, B\rangle$ belong to A, then we have $C = B$. It is easy to see that this set is Δ_0-definable by the following formula:

$$\begin{aligned} Int(A) \rightleftharpoons\ & (\forall U \in A)\,(\forall W \in A)\,(\forall V_1 \in U)\,(\forall V_2 \in U)\,(\forall i \in V_1)\,(\forall X \in V_2) \\ & (\forall V_3 \in W)\,(\forall Y \in V_3)\,((U = \langle i, X\rangle \wedge W = \langle i, Y\rangle) \to X = Y)\,. \end{aligned}$$

Theorem 2. *There exists a binary Σ-definable predicate Tr such that for any $n \in \omega$ and $A \in \mathrm{HF}(\mathbb{R})$ we have that $(n, A) \in Tr$ if and only if n is the Gödel number of a lifted s.s.-algebraic formula Φ, γ_A is a correct interpretation for free variables of Φ and* $\mathbf{HF}(\mathbb{R}) \models \Phi[\gamma_A]$.

Proof. The predicate Tr is the least fixed point of the operator defined by the following formula:

$$\Phi(n,U,P) \rightleftharpoons \Phi_{proper}(n) \vee \Phi_{elem}(n,U) \vee \Phi_{sum}(n,U,P) \vee \\ \Phi_{mult}(n,U,P) \vee \Phi_{\wedge}(n,U,P) \vee \Phi_{\vee}(n,U,P),$$

where n, U are free variables and P is a new predicate symbol. The formula $\Phi(n,U,P)$ represents the inductive definition of the truth of the lifted s.s.-algebraic formulas, where the immediate subformulas have the following meaning. The formulas $\Phi_{proper}(n)$ and $\Phi_{elem}(n,U)$ define the basis of the inductive definition. In other words, the formula $\Phi_{proper}(n)$ represents the truth of the proper formulas, i.e., $0<1$ and $1<0$; the formula $\Phi_{elem}(n,U)$ represents the truth of the elementary formulas, i.e., the formulas of the type, $X_i<0$, $0<X_i$, $1<X_i$ and $X_i<1$. The formulas $\Phi_{sum}(n,U,P)$ and $\Phi_{mult}(n,U,P)$ represent the inductive steps for sum and multiplication. Finally, the formulas $\Phi_{\wedge}(n,U,P)$ and $\Phi_{\vee}(n,U,P)$ represent the inductive steps for conjunctions and disjunctions. Let us show how to construct these formulas. The Δ_0-formula $\Phi_{proper}(n)$ is obvious. The Σ-formula $\Phi_{elem}(n,U)$ can be given as follows.

$$\begin{aligned}\Phi_{elem}(n,U) \rightleftharpoons \exists i \exists j \exists A \exists B(\ & (n=\ulcorner X<Y \urcorner \wedge i=\ulcorner X \urcorner \wedge j=\ulcorner Y \urcorner \wedge \\ & U=\{\langle i,A\rangle,\langle j,B\rangle\} \wedge A<B\ ) \vee \\ & (n=\ulcorner X<0 \urcorner \wedge i=\ulcorner X \urcorner \wedge U=\{\langle i,A\rangle\} \wedge A<0) \vee \\ & (n=\ulcorner 0<X \urcorner \wedge i=\ulcorner X \urcorner \wedge U=\{\langle i,A\rangle\} \wedge 0<A) \vee \\ & (n=\ulcorner X<1 \urcorner \wedge i=\ulcorner X \urcorner \wedge U=\{\langle i,A\rangle\} \wedge A<1) \vee \\ & (n=\ulcorner 1<X \urcorner \wedge i=\ulcorner X \urcorner \wedge U=\{\langle i,A\rangle\} \wedge 1<A)\ ).\end{aligned}$$

Now we construct a Σ-formula $\Phi_{sum1}(n,U,P)$ which represents the case when n is the code of a formula of the type: $p+q<f$, where p, q and f are lifted polynomials. Let $nextvar(\ulcorner \Psi \urcorner, l)$ denote the Σ-definable predicate which means that if m is the maximal index of variables which occur in Φ, then $l=m+1$. The formula Φ_{sum} can be given as follows.

$$\begin{aligned}\Phi_{sum1}(n,U,P) \rightleftharpoons \exists i \exists j \exists k \exists l \exists m \exists s \exists V \exists W \exists Y \exists A \exists B \exists C(\ & n=\ulcorner p+q<f \urcorner \wedge \\ & i=\ulcorner p \urcorner \wedge j=\ulcorner q \urcorner \wedge k=\ulcorner f \urcorner \wedge \\ & nextvar(n,l) \wedge m=l+1 \wedge s=m+1 \wedge \\ & P(\ulcorner p<X_l \urcorner, V) \wedge P(\ulcorner q<X_m \urcorner, W) \wedge P(\ulcorner X_s<f \urcorner, Y) \wedge \\ & \langle i,A\rangle \in V \wedge \langle j,B\rangle \in W \wedge \langle k,C\rangle \in Y \wedge A+B<C \wedge \\ & U=(V \cup W \cup Y) \wedge Int(U)\ ).\end{aligned}$$

In a similar way, we can produce a Σ-formula $\Phi_{sum2}(n,U,P)$ which represents the case when n is the code of a formula of the type: $p<q+f$, where p, q and f are lifted polynomials. Put $\Phi_{sum}(n,U,P) \rightleftharpoons \Phi_{sum1}(n,U,P) \wedge \Phi_{sum2}(n,U,P)$. In the same way, we can produce the Σ-formula $\Phi_{mul}(n,U,P)$ which represents the inductive steps for multiplication.

The Σ-formula $\Phi_\wedge$ can be constructed as follows:

$$\Phi_\wedge(n,U,P) \rightleftharpoons \exists i \exists j\, n = \ulcorner \varphi \wedge \psi \urcorner \wedge i = \ulcorner \varphi \urcorner \wedge j = \ulcorner \psi \urcorner \wedge P(i,U) \wedge P(j,U).$$

In a similar way, we can produce the Σ-formula $\Phi_\vee(n,U,P)$ which represents the inductive steps for disjunctions.

From Gandy's theorem (c.f. Section 2.2) it follows that the least fixed point Tr of the effective operator defined by Φ is Σ-definable. □

Theorem 3. *For every $n \in \omega$ there exists a Σ-formula* $\mathrm{Univ}_n^*(m, X_1, \ldots, X_n)$ *such that for any lifted s.s.-algebraic formula $\Phi(X_1,\ldots,X_n)$*

$$\mathbf{HF}(\mathbb{R}) \models \Phi(R_1,\ldots,R_n) \leftrightarrow \mathrm{Univ}_n^*(\ulcorner \Phi \urcorner, R_1,\ldots,R_n).$$

Proof. It is easy to see that the following formula defines a universal Σ-predicate for the lifted s.s.-algebraic formulas of arity n

$$\mathrm{Univ}_n^*(m, X_1, \ldots X_n) \rightleftharpoons \exists U \left(U = \{\langle 1, X_1\rangle, \ldots, \langle n, X_n\rangle\} \wedge Tr(m,U)\right).$$

□

Theorem 4. *For every $n \in \omega$ there exists a Σ-formula $Univ_n(m, x_1, \ldots, x_n)$ such that for any s.s.-algebraic formula $\Phi(x_1,\ldots,x_n)$*

$$\mathbf{HF}(\mathbb{R}) \models \Phi(r_1,\ldots,r_n) \leftrightarrow Univ_n(\ulcorner \Phi \urcorner, r_1,\ldots,r_n).$$

Proof. From the properties of the standard Gödel numbering it follows that the code of an s.s.-algebraic formula can be effectively constructed from the code of the corresponding lifted formula and vice versa. Let $f : \omega \to \omega$ be a recursive function which maps the code of an s.s.-algebraic formula to the code of the corresponding lifted formula. Then the following formula defines a universal Σ-predicate for the s.s.-algebraic formulas of arity n.

$$Univ_n(m,\bar{x}) \rightleftharpoons \exists X_1 \ldots \exists X_n \exists k\, f(m) = k \wedge \mathrm{Univ}_n^*(k, \bar{X}) \wedge \bigwedge_{i \le n} (x_i \in X_i).$$

□

3 Expressive Power of Σ-Definability over the Reals

3.1 Constructive Infinitary Language $L_{\omega_1\omega}$

In order to study the expressive power of Σ-formulas, we will consider a suitable fragment $L^{al}_{\omega_1\omega}$ of the constructive infinitary language $L_{\omega_1\omega}$ (cf. [14]) described below. Informally, the language $L^{al}_{\omega_1\omega}$ is obtained by extending the s.s.-algebraic formulas to allow formulas with effective infinite disjunctions but only finitely many variables; that is, formulas of the form $\bigvee_{i\in A} \Phi_i$, where $\{\Phi_i | i \in A\}$ is an effectively indexed family of s.s.-algebraic formulas, possibly infinite. The meaning of these formulas is as follows: $\mathbf{HF}(\mathbb{R}) \models \bigvee_{i\in A} \Phi_i$ if and only if for at least one $i \in A$ we have $\mathbf{HF}(\mathbb{R}) \models \Phi_i$.

Formally, we propose the inductive definition of formulas as follows. Let Var be a fixed finite set of variables. The set L_{Var} of *formulas over* Var includes the set of s.s.-algebraic formulas all of the variables of which belong to Var. In addition, if $\{\Phi_i | i \in A\}$ is an indexed family of formulas of L_{Var} and A is recursively enumerable, then $\bigvee_{i\in A} \Phi_i$ is a formula of L_{Var}. The language $L^{al}_{\omega_1\omega}$ is the union of all L_{Var} for all finite sets Var of variables which range over $\mathbb{R}$.

3.2 Engeler's Lemma for Σ-Definability

In this section we prove Engeler's lemma for Σ-definability which states that if a relation $B \subset \mathbb{R}^n$ is Σ-definable, then it is definable by a formula of $L^{al}_{\omega_1\omega}$ which can be constructed effectively from the corresponding Σ-formula.

In order to work effectively with elements of the structures HF($\mathbb{R}$) and HF($\mathbb{R}^*$), we represent every element in a regular way. For this, we enrich the language σ to σ' by the additional functions: singleton $\{U\}$ and binary union $U_1 \cup U_2$. Note that these functions will be eliminated in the resulting formulas of Engeler's Lemma. Below a term in the language $\{\{,\},\cup\}$ is called a *structural term.* With every $\alpha \in \mathrm{HF}(n)$ we associate a structured term defined as follows:

$$t_\alpha(N_1,\dots,N_n) = \begin{cases} N_i & \text{if } \alpha = i \\ \{t_{\alpha_1}(\bar{N})\} \cup \dots \cup \{t_{\alpha_k}(\bar{N})\} & \text{if } \alpha = \{\alpha_1,\dots,\alpha_k\}. \end{cases}$$

Lemma 8. *1. For every element $M \in \mathrm{HF}(\mathbb{R})$ there exist a structural term $t_\alpha(N_1,\dots,N_k)$ and a substitution $\tau : \{N_1,\dots,N_k\} \to \mathbb{R}$ such that M is represented by $t_\alpha(\bar{N})\tau$.*

2. For every element $U \in \mathrm{HF}(\mathbb{R}^*)$ there exist a structural term $t_\beta(N_1,\dots,N_l)$ and a substitution $\nu : \{N_1,\dots,N_l\} \to \mathbb{R}^*$ such that U is represented by $t_\beta(\bar{N})\nu$.

If $\tau(N_i) = r_i$ for $i \le k$ and $\nu(N_j) = R_j$ for $j \le l$, then we write $t(r_1,\dots,r_k)$ and $t(R_1,\dots,R_l)$ instead of $t_\alpha(\bar{N})\tau$ and $t_\beta(\bar{N})\nu$ respectively.

Lemma 9. *Let $t_\beta(\bar{X})$ and $t_\gamma(\bar{X})$ be structural terms and $\bar{R} \in \mathrm{HF}(\mathbb{R}^*)^n$. If $t_\beta(\bar{R})$ represents U_1, $t_\gamma(\bar{R})$ represents U_2 and $\mathbf{HF}(\mathbb{R}) \models U_1 \in U_2$, then we have $\mathbf{HF}(\mathbb{R}) \models t_\beta(\bar{N}) \in t_\gamma(\bar{N})$ for every $\bar{N} \in \mathrm{HF}(\mathbb{R})^n$ such that for all $i,j \le n$ $N_i = N_j \leftrightarrow R_i = R_j$.*

With every Σ-formula $\Phi(\bar{x})$ we associate a *lifted Σ-formula* $\Phi_{lifted}(\bar{X})$ obtained from Φ by replacing every variable ranging over HF($\mathbb{R}$) by a variable ranging over HF($\mathbb{R}^*$); every variable ranging over $\mathbb{R}$ by a variable ranging over $\mathbb{R}^*$. For example, $\exists U\, (\forall Y \in U)\, p_1(Y,\bar{X}) < p_2(Y,\bar{X})$ is the lifted Σ-formula corresponding to $\exists M\, (\forall y \in M)\, p_1(y,\bar{x}) < p_2(y,\bar{x})$.

Proposition 2. *Let $\Phi_{lifted}(\bar{X})$ be the lifted Σ-formula corresponding to a Σ-formula $\Phi(\bar{x})$. Then we have*

$$\mathbf{HF}(\mathbb{R}) \models \Phi(x_1 \dots, x_n) \leftrightarrow \exists X_1 \dots \exists X_n \, \Phi_{lifted}(\bar{X}) \wedge \bigwedge_{i \leq n} (x_i \in X_j).$$

Proof. Let $\Phi(\bar{x})$ be a Σ-formula. By the Σ-reflection principle (c.f. Section 2.2.), the formula $\Phi(\bar{x})$ can be represented in the form:

$$\Phi(\bar{x}) \rightleftharpoons \exists M \, \Phi'(M, \bar{x}) \wedge \bigwedge_{i \leq n} (x_i \in M) \rightleftharpoons \exists M \, \Phi''(M, \bar{x}), \tag{1}$$

where Φ' and Φ'' are Δ_0-formulas. The corresponding lifted formula $\Phi_{lifted}(\bar{X})$ is obtained from Φ by replacing every variable ranging over HF($\mathbb{R}$) by a variable ranging over HF($\mathbb{R}^*$); every variable ranging over $\mathbb{R}$ by a variable ranging over $\mathbb{R}^*$, i.e.,

$$\Phi_{lifted}(\bar{X}) \rightleftharpoons \exists U \, \Phi'_{lifted}(U, \bar{X}) \wedge \bigwedge_{i \leq n} (X_i \in U) \rightleftharpoons \exists U \, \Phi''_{lifted}(U, \bar{X}). \tag{2}$$

$\rightarrow$) Suppose (1) is valid in $\mathbf{HF}(\mathbb{R})$ and N, $\bar{r}$ satisfy the formula $\Phi''(M, \bar{x})$. In order to construct some V and $\bar{R}$ which satisfy the corresponding lifted Σ-formula, we use the operation *up* defined by induction:

$$up(K) = \begin{cases} \{K\} & \text{if } K \in \mathbb{R} \\ \{up(K_1), \dots up(K_l)\} & \text{if } K = \{K_1, \dots, K_l\}. \end{cases}$$

We put $V = up(N)$, $R_1 = up(r_1), \dots, R_n = up(r_n)$. By Lemma 3 and Lemma 9, V and $R_1, \dots, R_n$ satisfy the formula $\Phi''_{lifted}(U, \bar{X})$.
$\leftarrow$) Suppose (2) is valid in $\mathbf{HF}(\mathbb{R})$ and V, $R_1, \dots, R_n$ satisfy the given formula $\Phi''_{lifted}(U, \bar{X})$. By Lemma 8, the set V can be represented by $t_\beta(R_1, \dots, R_m)$, where $\{R_1, \dots, R_n\} \subseteq \{R_1, \dots, R_m\}$. From Lemma 9 it follows that if $V = t_\beta(R_1, \dots, R_m)$ satisfies the following requirement:

$$\begin{gathered} \text{there exist } r_1 \in R_1, \dots r_m \in R_m \text{ such that} \\ r_i = r_j \leftrightarrow R_i = R_j \text{ for all } i, j \leq m, \end{gathered} \tag{3}$$

then $t_\beta(\bar{r})$ and $r_1, \dots, r_n$ satisfy $\Phi''(M, \bar{x})$. Let us note that V may not satisfy (3), for example, if $V = \{\{r_1, r_2\}, \{r_1\}, \{r_2\}\}$, where $r_1 \neq r_2 \neq r_3$. The problem here is that the number of elements in the sets is too small to pick different representatives. In this case we construct $R'_1, \dots R'_n$ and V' from $R_1, \dots, R_n$ and V which satisfy the formula $\Phi''_{lifted}(U, \bar{X})$ and the requirement (3). It can be done using Lemma 6. Indeed, there exist $R'_1 \dots, R'_m$ such that

1. for every $i \leq m$ we have $|R'_i| \geq m$;
2. for every $i, j \leq m$ $R'_i = R'_j$ if and only if $R_i = R_j$;

3. for every s.s.-algebraic subformula $\phi_{alg}(\bar{Y},\bar{X})$ of the formula Φ'' and every substitution $\tau : \{\bar{Y},\bar{X}\} \to \{R_1,\dots,R_m\}$ we have

$$\mathbf{HF}(\mathbb{R}) \models \phi_{alg}(\bar{Y},\bar{X})\tau \leftrightarrow \mathbf{HF}(\mathbb{R}) \models \phi_{alg}(\bar{Y},\bar{X})\tau',$$

where $\tau' : \{\bar{Y},\bar{X}\} \to \{R'_1,\dots,R'_m\}$ is a substitution such that $\tau'(Y_k) = R'_j \leftrightarrow \tau(Y_k) = R_j$ and $\tau'(X_l) = R'_i \leftrightarrow \tau(X_l) = R_i$ for all $i,j,k,l \le m$.

By construction, $t_\beta(R'_1,\dots,R'_m)$ and $R'_1,\dots,R'_n$ satisfy $\Phi''_{lifted}(U,\bar{X})$ and (3).

Now every R'_i contains enough elements to choose r_i from R'_i under the condition $r_i = r_j \leftrightarrow R_i = R_j$. It is easy to see that $t_\beta(\bar{r})$ and $r_1,\dots,r_n$ satisfy $\Phi''(M,\bar{x})$. □

In the following proposition the lifted s.s.-algebraic atomic formulas and the Δ_0-formulas of the type $X = Y \rightleftharpoons (\forall x \in X)\, x \in Y \wedge (\forall y \in Y)\, y \in X$ and $\neg X = Y \rightleftharpoons (\exists x \in X)\, x \notin Y \vee (\exists y \in Y)\, y \notin X$ are considered as *basic formulas.*

Proposition 3. *Let $\Phi(X_1,\dots,X_n)$ be a lifted Σ-formula. There exists a constructive infinite formula $\Psi \rightleftharpoons \bigvee_{i\in\omega} \Psi_i$ such that:*

- $\mathbf{HF}(\mathbb{R}) \models \Phi(\bar{X}) \leftrightarrow \Psi(\bar{X})$.
- *Every $\Psi_i(\bar{X})$ is a formula of the form $\exists Y_{m_1} \dots \exists Y_{m_i} \Psi'_i(\bar{Y},\bar{X})$, where Ψ' is a finite conjunction of basic formulas whose quantifiers range over $\mathbb{R}^*$.*

Proof. Let $\Phi(X) \rightleftharpoons \exists U \Phi'(U,\bar{X})$, where Φ' is a Δ_0-formula. In order to obtain the required formula, we first construct an equivalent infinite formula in the language σ' (e.g. $\sigma' = \sigma \cup \{\{,\},\cup\}$) without unbounded quantifiers. Then we prove by induction the existence of an equivalent infinite formula without bounded quantifiers. After that we eliminate $\{,\},\cup,\emptyset,\in$ from the obtained formula. By Lemma 3, every $U \in \mathrm{HF}(\mathbb{R}^*)$ can be represented by $t_\beta(Y_1,\dots Y_n)$ for some $Y_1 \in \mathbb{R}^*,\dots Y_n \in \mathbb{R}^*$. Put

$$\begin{aligned}\Phi^*(X) &\rightleftharpoons \bigvee_{n\in\omega} \bigvee_{\beta\in\mathrm{HF}(n)} \exists Y_1 \dots \exists Y_n \Phi'(t_\beta(Y_1,\dots,Y_n),\bar{X}) \\ &\rightleftharpoons \bigvee_{n\in\omega} \bigvee_{\beta\in\mathrm{HF}(n)} \exists Y_1 \dots \exists Y_n \Phi'_\beta(\bar{Y},\bar{X}),\end{aligned}$$

where every Φ'_β is a Δ_0-formula with quantifiers bounded by subterms of t_β. Since for every U there exists a term t_β which codes the structure of U, the formula Φ^* is equivalent to the given one.

Without loss of generality every formula Φ'_β can be represented as follows.

$$\begin{aligned}\Phi'_\beta(\bar{Y},\bar{X}) &\rightleftharpoons (Q\,U_1 \in t_{\gamma_1}(\bar{Y})) \dots (Q\,U_m \in t_{\gamma_m}(\bar{Y}))\, \phi_\beta(\bar{U},\bar{Y},\bar{X}) \\ &\rightleftharpoons (Q\,U \in t_{\gamma_1}(\bar{Y}))\, \phi'_\beta(\bar{U},\bar{Y},\bar{X}),\end{aligned}$$

where Q is the quantifier $\exists$ or $\forall$ and t_{γ_i} is a subterm of t_β for all $i \le k$. Using induction on the length of the quantifier prefix and the depth of the term which bounds the first quantifier in the quantifier prefix, we show how to obtain an equivalent quantifier free formula. We proceed by induction on the pairs $\langle m, n\rangle$ with the lexicographic order, where m is the length of the quantifier prefix of Φ'_β and n is the depth of t_{γ_1}. Let $\top$ denote a logical truth which can be represented by the formula $0 < 1$ and $\bot$ denote a logical false which can be represented by the formula $1 < 0$.

The cases $\langle 0, 0\rangle$, $\langle 0, n\rangle$ are obvious. In the case $\langle m, 0\rangle$ the formula Φ'_β can be represented in the form $\Phi'_\beta \rightleftharpoons (Q\, U_1 \in \emptyset)\ \phi'_\beta$ or $\Phi'_\beta \rightleftharpoons (Q\, U_1 \in X)\ \phi'_\beta$. If Q is the existential quantifier, then put $\Psi_\beta \rightleftharpoons \bot$. If Q is the universal quantifier, then put $\Psi_\beta \rightleftharpoons \top$.

Consider the inductive step $\langle m, n\rangle \to \langle m, n+1\rangle$. The first possibility is that $\Phi'_\beta \rightleftharpoons \left(\exists U_1 \in t_{\gamma_1}(\bar{Y})\right) \phi'_\beta(U_1, \bar{Y}, \bar{X})$ and $t_{\gamma_1}(\bar{Y}) = t'_{\gamma_1}(\bar{Y}) \cup t''_{\gamma_1}(\bar{Y})$. Let us consider the formula $\left(\exists U_1 \in t'_{\gamma_1}(\bar{Y})\right) \phi'_\beta(U_1, \bar{Y}, \bar{X}) \vee \left(\exists U_1 \in t''_{\gamma_1}(\bar{Y})\right) \phi'_\beta(U_1, \bar{Y}, \bar{X})$ which is equivalent to Φ'_β. The complexity of t'_{γ_1} and t''_{γ_1} is less than $n+1$. By inductive hypothesis, there exists a formula Ψ_β without quantifiers which is equivalent to Φ'_β. The second possibility with a bounded universal quantifier can be considered in a similar way.

Consider the inductive step $\langle m, n\rangle \to \langle m+1, k\rangle$.
We have $\Phi'_\beta \rightleftharpoons \left(\exists U_1 \in \{t(\bar{Y})\}\right) \phi'_\beta(U_1, \bar{Y}, \bar{X})$. Let us consider $\phi'_\beta(t(\bar{Y}), \bar{Y}, \bar{X})$ which is equivalent to Φ'_β. The complexity of the quantifier prefix is less then $m+1$. By inductive hypothesis, there exists a formula Ψ_β without quantifiers which is equivalent to Φ'_β. The case, when $\Phi'_\beta \rightleftharpoons \left(\forall U_1 \in \{t(\bar{Y})\}\right) \phi'_\beta(U_1, \bar{Y}, \bar{X})$, is similar.

Now we eliminate $\{,\}$,$\cup$, $\emptyset$, $\in$ from every formula Ψ_β. It is easy to see that for any terms $t_\gamma(\bar{Y}, \bar{X})$ and $t_\kappa(\bar{Y}, \bar{X})$ of the language σ' it is possible to write effectively $\chi_\in(\bar{Y}, \bar{X})$ and $\chi_=(\bar{Y}, \bar{X})$ such that $\chi_\in(\bar{Y}, \bar{X})$ and $\chi_=(\bar{Y}, \bar{X})$ are finite disjunctions of finite conjunctions of basic formulas and the formula $t_\gamma \in t_\kappa$ is equivalent to $\chi_\in(\bar{Y}, \bar{X})$ and the formula $t_\gamma = t_\kappa$ is equivalent to $\chi_=(\bar{Y}, \bar{X})$. For example, it is easy to see that the formula $\{Y_1\} \cup \{Y_2\} \in \{\{\{Y_3\} \cup \{Y_4\}\}\} \cup \{\{Y_5\}\}$ is equivalent to $(Y_1 = Y_3 \wedge Y_2 = Y_4) \vee (Y_1 = Y_4 \wedge Y_2 = Y_3)$.

Using these formulas we transform every formula Ψ_β into Ψ'_β without occurrences of $\{,\}$,$\cup$, $\emptyset$, $\in$.

Put

$$\Psi(X) \rightleftharpoons \bigvee_{n \in \omega} \bigvee_{\beta \in \mathrm{HF}(n)} \exists Y_1 \ldots \exists Y_n\, \Psi'_\beta(\bar{Y}, X).$$

By construction, Ψ has the required form. □

Now we are ready to prove Engeler's Lemma for Σ-definability over the reals.

Theorem 5 (Engeler's Lemma for Σ-definability). *If a relation $B \subset \mathbb{R}^n$ is Σ-definable, then it is definable by a formula of $L^{al}_{\omega_1\omega}$. Moreover this formula can be constructed effectively from the corresponding Σ-formula.*

Proof. Suppose B is defined by a Σ-formula $\Phi(\bar{x})$. By Proposition 2 and Proposition 3, there exists an effective sequence $\{\Psi_i\}_{i \in \omega}$ such that

$$\mathbf{HF}(\mathbb{R}) \models \Phi(\bar{x}) \leftrightarrow \bigvee_{i \in \omega} \left(\exists X_1 \ldots \exists X_n\, \Psi_i(\bar{X}) \wedge \bigwedge_{j \leq n} x_j \in X_j \right),$$

where every $\Psi_i(\bar{X})$ is a formula of the form $\exists Y_{m_1} \dots \exists Y_{m_i} \Psi'_i(\bar{Y}, \bar{X})$, and Ψ' is a finite conjunction of lifted s.s.-algebraic formulas and formulas of the type $X_j = Y_i$, $Y_i = Y_j$, $\neg X_j = Y_i$ and $\neg Y_i = Y_j$.

Now we show that for every $i \in \omega$ there exists an existential quantified s.s.-algebraic formula φ_i in the language $\sigma_0 = \{0, 1, +, \cdot, <\}$ such that

$$\mathbf{HF}(\mathbb{R}) \models \varphi_i(\bar{x}) \leftrightarrow \mathbf{HF}(\mathbb{R}) \models \left(\exists X_1 \dots \exists X_n \Psi_i(\bar{X}) \wedge \bigwedge_{j \le n} x_j \in X_j \right). \quad (4)$$

For this purpose, in every Ψ_i we first eliminate subformulas of the type $X_j = Y_k$ and $Y_j = Y_k$ in the following way. For all j and k, such that there exists a subformula $X_j = Y_k$, we replace all occurrences of Y_k by X_j, the subformula $X_j = Y_k$ by $\top$ and eliminate the quantifier over Y_i. For all i and k such that, there exists $Y_j = Y_k$ for some j and $k > j$, we replace all occurrences of Y_k by Y_j, the subformula $Y_j = Y_k$ by $\top$ and eliminate the quantifier over Y_k. It is easy to see that the resulting formula $\Psi'_i(\bar{X})$ is equivalent to $\Psi_i(\bar{X})$.

Now if $X_j = X_k$ occurs in $\Psi'_i(\dots, X_j, \dots, X_k, \dots)$ then we replace Ψ'_i by $\Psi'_i(\dots, X_j, \dots, X_j, \dots) \wedge \Psi'_i(\dots, X_k, \dots, X_k, \dots)$ and remove all $X_j = X_k$ from this formula. Let us argue that the obtained formula Ψ''_i is equivalent to Ψ'_i in the following sense

$$\mathbf{HF}(\mathbb{R}) \models \forall x_1 \dots \forall x_n \left[\left(\exists X_1 \dots \exists X_n \Psi'_i(\bar{X}) \wedge \bigwedge_{j \le n} x_j \in X_j \right) \leftrightarrow \right.$$
$$\left. \left(\exists X_1 \dots \exists X_n \Psi''_i(\bar{X}) \wedge \bigwedge_{j \le n} x_j \in X_j \right) \right].$$

Implication from left to right is obvious. In order to prove implication from right to left, we need to show that if for $x_1, \dots, x_n$ there exist $X_1, \dots, X_n$ such that $\Psi''_i(\bar{X}) \wedge \bigwedge_{j \le n} x_j \in X_j$, then there exist $X'_1, \dots, X'_n$ such that $X'_j = X'_k$ and $\Psi'_i(\bar{X}') \wedge \bigwedge_{j \le n} x_j \in X'_j$. For this we can take $X'_j = X'_k = X_j \cup X_k$. In the same way we can eliminate all subformulas of the form $X_j = X_k$ obtaining a formula Ψ'''. We then construct formulas φ_i from Ψ'''_i by replacing X by x, Y_i by y_i and subformulas of the type $\neg x = y$ by $x < y \vee x > y$. From the definition of a lifted s.s.-algebraic formula and Proposition 2 it follows that the formula φ_i satisfies (4). In order to complete the proof, using Lemma 7, we construct an effective sequence $\{\varphi'_i\}_{i \in \omega}$ of s.s.-algebraic formulas such that

$$\mathbf{HF}(\mathbb{R}) \models \Phi(x) \leftrightarrow \bigvee_{i \in \omega} \varphi'_i(x).$$

□

3.3 Characterisation Theorem for Σ-Definability

Let us prove the converse statement of Engeler's Lemma for Σ-definability.

Theorem 6. *If a relation $B \subset \mathbb{R}^n$ is definable by a formula of $L^{al}_{\omega_1\omega}$, then it is Σ-definable. Moreover Σ-formula can be constructed effectively from the corresponding formula of $L^{al}_{\omega_1\omega}$.*

Proof. Let $B \subset \mathbb{R}^n$ be definable by $\bigvee_{\ulcorner\Psi\urcorner\in A}\Psi(\bar{x})$, where A is recursively enumerable. By Theorem 4, there exists a universal Σ-predicate $Univ_n(m,\bar{x})$ for s.s.-algebraic formulas with variables from $\{x_1,\dots,x_n\}$.

Put

$$\Phi(\bar{x}) \rightleftharpoons \exists i\,(i \in A) \wedge Univ^n(i,\bar{x}).$$

It can be shown that Φ is a required Σ-formula. □

Theorem 7 (Characterisation of Σ-definability). *A relation $B \subset \mathbb{R}^n$ is Σ-definable if and only if it is definable by a formula of $L^{al}_{\omega_1\omega}$.*

These results reveal algorithmic aspects of Σ-definability. Indeed, suppose $\Phi(\bar{x})$ is a Σ-formula which defines a relation over the reals and we have $\mathbf{HF}(\mathbb{R}) \models \Phi(\bar{x}) \leftrightarrow \bigvee_{\ulcorner\Psi\urcorner\in A}\Psi(\bar{x})$. Then each s.s.-algebraic formula $\Psi(\bar{x})$, such that $\ulcorner\Psi\urcorner \in A$, represents a simple approximation of the relation definable by $\Phi(\bar{x})$ and there exists a Turing machine that computes these approximations (i.e., enumerates $\Psi(\bar{x})$). A universal Turing machine and a universal Σ-predicate for s.s.-algebraic formulas can then be used to enumerate and check validity of each approximation Ψ.

We also obtain the following topological characterisation of Σ-definability over the reals.

Theorem 8. *1. A set $B \subset \mathbb{R}^n$ is Σ-definable if and only if it is an effective union of open semi-algebraic sets.*
2. A relation $B \subset \mathbb{R}^n$ is Σ-definable if and only if there exists an effective sequence $\{A_i\}_{i\in\omega}$ of open semi-algebraic sets such that
   (a) It monotonically increases: $A_i \subseteq A_{i+1}$, for $i \in \omega$;
   (b) $B = \bigcup_{i\in\omega} A_i$.

Let $\Sigma_{\mathbb{R}}$ denote the set of all Σ-definable subsets of $\mathbb{R}^n$, where $n \in \omega$.

Corollary 1. *1. The set $\Sigma_{\mathbb{R}}$ is closed under finite intersections and effective infinite unions.*
2. The set $\Sigma_{\mathbb{R}}$ is closed under Σ-inductive definitions.
3. The set $\Sigma_{\mathbb{R}}$ is closed under projections.

4 Future Work

In this paper we characterised the expressive power of the language of Σ-formulas over the reals without the equality test. One of the main directions of future work is to develop a logical approach to computability on continuous data types based on this language. So far we have investigated the expressive power of the language of Σ-formulas over the reals. In this respect the following direction of research is of special interest: to propose and study reasonable requirements on the universe and the language of an abstract structure without the equality test under which a similar characterisation can be obtained.

References

1. M. Ajtai. First-order definability on finite structures. *Annals of Pure and Applied Logic*, 45:211–225, 1989.
2. J. Barwise. *Admissible sets and Structures.* Springer Verlag, Berlin, 1975.
3. A. Blass and Y. Gurevich. Background, reserve and Gandy machines. In *Proc. of CSL'2000*, volume 1862 of *Lecture Notes in Computer Science*, pages 1–17, 2000.
4. L. Blum, F. Cucker, M. Shub, and S. Smale. *Complexity and Real Computation.* Springer Verlag, Berlin, 1996.
5. J. Bochnak, M. Coste, and M.-F. Roy. *Real Algebraic Geometry.* Springer Verlag, Berlin, 1999.
6. E. Dahlhaus and J. A. Makowsky. Query languages for hierarchic databases. *Information and Computation*, 101:1–32, 1992.
7. C. N. Delzell. A finiteness theorem for open semi-algebraic sets, with application to Hilbert's 17th problem. In *Ordered Fields and Real Algebraic Geometry*, volume 8 of *Contemp. Math*, pages 79–97. AMS, 1982.
8. H. Ebbinghaus and J. Flum. *Finite Model Theory.* Springer Verlag, Berlin, 1999.
9. E. Engeler. *Formal Languages: Automata and Structures.* Markham Publishing Co, 1968.
10. Yu. L. Ershov. *Definability and computability.* Plenum, New-York, 1996.
11. H. Friedman and K. Ko. Computational complexity of real functions. *Theoretical Computer Science*, 20:323–352, 1992.
12. D. Harel, D. Kozen, and J. Tiuryn. *Dynamic Logic.* The MIT press, Cambridge, MA, 2002.
13. N. Immerman. *Descriptive Complexity.* Springer Verlag, New-York, 1999.
14. C. Karp. *Languages with Expressions of Infinite Length.* Noth-Holland, Amsterdam, 1964.
15. Ulrich Kohlenbach. Proof theory and computational analysis. *Electronic Notes in Theoretical Computer Science*, 13, 1998.
16. M. Korovina. Fixed points on the reals numbers without the equality test. *Electronic Notes in Theoretical Computer Science*, 66(1), 2002.
17. M. Korovina. Recent advances in Σ-definability over continuous data types. BRICS Report Series RS-03-23, http://www.brics.dk/RS/03/23/index.html, 2003.
18. M. Korovina and O. Kudinov. Characteristic properties of majorant-computability over the reals. In *Proc. of CSL'98*, volume 1584 of *Lecture Notes in Computer Science*, pages 188–204, 1999.
19. Y. N. Moschovakis. Abstract first order computability I, II. *Transactions of the American Mathematical Society*, 138:427–504, 1969.
20. Y. N. Moschovakis. *Elementary Induction on Abstract Structures*, volume 77 of *Studies in Logic and the Foundations of Mathematics.* North-Holland, 1974.
21. M. B. Pour-El and J. I. Richards. *Computability in Analysis and Physics.* Springer Verlag, Berlin, 1988.
22. V. Sazonov. Using agents for concurrent querying of web-like databases via hyper-set-theoretic approach. In *Proc. of PSI'01*, volume 2244 of *Lecture Notes in Computer Science*, pages 378–394, 2001.
23. J. V. Tucker and J. I. Zucker. Projections of semicomputable relations on astract data types. *International Journal of the Foundations of Computer Science*, 2:267–296, 1991.
24. Klaus Weihrauch. *Computable Analysis.* Springer Verlag, Berlin, 2000.

The Surprising Power of Restricted Programs and Gödel's Functionals

Lars Kristiansen[1] and Paul J. Voda[2]

[1] Oslo University College, Faculty of Engineering
Cort Adelers gate 30, N-0254 Oslo, Norway
larskri@iu.hio.no
http://www.iu.hio.no/~larskri

[2] Institute of Informatics, Comenius University Bratislava
Mlynská dolina, 842 15 Bratislava, Slovakia
voda@fmph.uniba.sk
http://www.fmph.uniba.sk/~voda

1 Introduction

Consider the following imperative programming language. The programs operate on registers storing natural numbers, the input $\vec{x}$ is stored in certain registers, and a number b, called the *base*, is fixed to $\max(\vec{x}, 1) + 1$ before the execution starts. The single primitive instruction `X+` increases the number stored in the register `X` by 1 modulo b. There are two control structures: the loop `while X {P}` executing the program `P` repeatedly as long as the content of the register `X` is different from 0; the composition `P;Q` executing first the program `P`, then the program `Q`. This is the whole language. The language is natural, extremely simple, yet powerful. We will prove that it captures LINSPACE, i.e. the numerical relations decidable by such programs are exactly those decidable by Turing machines working in linear space. Variations of the language capturing other important deterministic complexity classes, like e.g. LOGSPACE, P and PSPACE are possible (see Kristiansen and Voda [5]).

We will extend the language without destroying its simplicity with variables of finite types holding (finite) array-like data structures which are naturally coded into numbers. The types are defined in the standard way, i.e. $\mathbf{0}$ is a type, and $\sigma \to \tau$ is a type if σ and τ are types. We will prove that the programs compute level by level the functions of the Ritchie hierarchy[1]. A function is Kalmár elementary iff it is computable by a program[2].

We then study a restricted version of Gödel's functionals primitive recursive in finite types which are obtained by removing the successor function and by replacing the constant 0 by 1. Somewhat surprisingly, the restricted functionals have the same power as our programming language.

Numerous schemes have been introduced to restrict recursion in higher types. So-called ramification techniques (e.g. Simmons [11], Leivant [6], Beckmann and

[1] The definition is later in the text.

[2] The classes in the Ritchie hierarchy add up to the class of Kalmár elementary functions.

M. Baaz and J.A. Makowsky (Eds.): CSL 2003, LNCS 2803, pp. 345–358, 2003.

Weiermann [2]) restrict higher type recursion to the Kalmár elementary level. By using so-called linearity constraints in conjunction with ramification techniques, higher type recursion can be restricted further down to the "polytime" level (e.g. Bellantoni and Schwichtenberg [9]). This paper suggests a qualitatively different scheme for restricting the recursion in higher types: Modify the set of initial functions. That the power of higher type recursion is severely weakened when the initial functions are sufficiently weak, is hardly surprising, but it is quite surprising that the functions definable by higher type recursion do not degenerate into an unnatural and uninteresting class, even when the only available initial function is the constant 1.

We have recently found out that Tiuryn [12] has investigated the (finite) functions over finite types obtained when the type **0** is finite. He called them generalized arrays and obtained the same space bounds as we by using families of programming languages having both syntax and semantics somewhat more complicated than ours. He did not investigate the connection to Gödel's functionals and neither did Jones [4] who obtained similar results with an imperative language of functionals over the finite data structure of S-expressions without *cons* (corresponding to the successor function). The finite version of Gödel's functionals was investigated by Goerdt [3]. We stress here that, in contrast to the approach of Jones and Goerdt who thought it necessary to use also the Cartesian types, we work with the original Gödel's functionals (which are infinite and without Cartesian types) where we just remove the successor function.

2 Numbers and Types

Definition. **0** is a *type*; $\sigma \to \tau$ is a type if σ and τ are types. We say a type σ has *degree* n when $\mathrm{dg}(\sigma) = n$ where $\mathrm{dg}(\mathbf{0}) = 0$ and $\mathrm{dg}(\sigma \to \tau) = \max(\mathrm{dg}(\sigma)+1, \mathrm{dg}(\tau))$. We define the *cardinality of type* σ *at base* b, in symbols $|\sigma|_b$, by recursion on the build-up of σ: $|\mathbf{0}|_b = b$ and $|\rho \to \tau|_b = |\tau|_b^{|\rho|_b}$. (We use standard conventions and interpret $\sigma \to \sigma' \to \sigma''$ by associating parentheses to the right, i.e. as $\sigma \to (\sigma' \to \sigma'')$; further, $\sigma, \sigma' \to \sigma$ is alternative notation for $\sigma \to \sigma' \to \sigma''$.) Let $2_0^x \overset{\mathrm{def}}{=} x$ and $2_{n+1}^x \overset{\mathrm{def}}{=} 2^{2_n^x}$. □

Theorem 1. *(i) For any polynomial p and $n > 0$ there exists a type σ of degree n such that $|\sigma|_{\max(x,1)+1} > 2_n^{p(x)}$. (ii) For every type σ of degree n there exists a polynomial p such that $2_n^{p(x)} > |\sigma|_x$.*

Proof. (i) We will prove (i) by induction on n. Assume $n = 1$. Let the types $\xi_0, \xi_1, \ldots$ be defined by $\xi_0 = \mathbf{0}$ and $\xi_{j+1} = \mathbf{0} \to \xi_j$. We have $\mathrm{dg}(\xi_j) = 1$ and $|\xi_j| = x^{(x^j)}$ for every $j > 0$ (*). (This is easily proved by induction on j.) For an arbitrary polynomial p we have $|\xi_k|_{\max(x,2)} > 2^{p(x)}$ for a sufficiently large k. Thus, (i) holds when $n = 1$. We turn to the induction step. Assume that $|\sigma|_{\max(x,1)+1} > 2_n^{p(x)}$ where $\mathrm{dg}(\sigma) = n$. Then, we have

$$2_{n+1}^{p(x)} = 2^{(2_n^{p(x)})} < 2^{|\sigma|_{\max(x,1)+1}} \leq (|\mathbf{0}|_{\max(x,1)+1})^{|\sigma|_{\max(x,1)+1}} = |\sigma \to \mathbf{0}|_{\max(x,1)+1}$$

and $\mathrm{dg}(\sigma \to \mathbf{0}) = \mathrm{dg}(\sigma) + 1$. *(ii)* It is obvious if σ has degree 0. Assume $\sigma = \rho \to \tau$ has degree $n+1$. By the induction hypothesis we have $2_n^{q(x)} > |\rho|_x$ and $2_n^{q(x)} > |\tau|_x$ for some polynomial q. Thus, we have

$$|\sigma|_x = |\tau|_x^{|\rho|_x} < (2_n^{q(x)})^{(2_n^{q(x)})} < (2^{(2_n^{q(x)})})^{(2_n^{q(x)})} = 2^{(2_n^{q(x)})^2} < 2^{(2_n^{p(x)})} = 2_{n+1}^{p(x)}$$

for some suitable polynomial p. □

Definition. The natural number a is a *number of type σ at base b*, in symbols $a{:}\sigma_b$, iff $a < |\sigma|_b$. Let $a{:}(\sigma \to \tau)_b$. Then a can be uniquely written in the form

$$v_0 \; + \; v_1|\tau|_b^1 \; + \cdots + \; v_k|\tau|_b^k$$

where $k = |\sigma|_b - 1$ and $v_j : \tau_b$ for $j \in \{0, \dots, k\}$. We call $v_0, \dots, v_k$ the *digits* in a, and for any $i : \sigma_b$, we denote the i'th digit in a by $a[i]_b$, i.e. $a[i]_b = v_i$. Furthermore, for any $i : \sigma_b$ and $w : \tau_b$, let $a[i := w]_b$ denote the number which is the result of setting the i'th digit in a to w. (Note that $a[i := w]_b$ is a number of type σ at base b.) The notation $a[i_1, \dots, i_n]_b$, where $n \geq 1$, abbreviates $((a[i_1]_b)[i_2]_b)\dots)[i_n]_b$, and we will call $a[i_1, \dots, i_n]_b$ a *sub-digit* of a. (Thus, every digit is a sub-digit, every digit of a digit is sub-digit, every digit of a digit of digit is a sub-digit, and so on.) Further, let

$$a[i_1, \dots, i_{n+1} := w]_b \;\stackrel{\text{def}}{=}\; a[i_1, \dots, i_n := a[i_1, \dots, i_n]_b[i_{n+1} := w]_b]_b$$

for $n \geq 1$. Thus, $a[i_1, \dots, i_n := w]_b$ is the number which is the results of setting the sub-digit $a[i_1, \dots, i_n]_b$ in a to w. □

3 An Imperative Programming Language

We will now define a programming language. An informal explanation of the language follows the definition. It might be a good idea to read the definition and the explanation in parallel.

Definition. First we define the syntax of the programming language. The language has an infinite supply of *program variables* $\mathtt{x}_0^\sigma, \mathtt{x}_1^\sigma, \mathtt{x}_2^\sigma, \dots$ for any type σ. We will use verbatim Latin letters, uppercase or lowercase, with or without subscripts and superscripts, to denote program variables. Any variable of type σ is a *term* of type σ; t`[X]` is an is an term of type τ if `X` is a variable of type σ and t is a term of type $\sigma \to \tau$. We use $\mathtt{a[i_1, \dots, i_n]}$ to abbreviate $\mathtt{a[i_1][i_2]\dots[i_n]}$. (Note that any term is a variable or has the form $\mathtt{a[i_1, \dots, i_n]}$ where $\mathtt{a}, \mathtt{i_1}, \dots, \mathtt{i_n}$ are variables.) The primitive *instruction* t`+` is a *program in* $\vec{\mathtt{X}}$ if t is a term of type $\mathbf{0}$ such that every variable in t occurs in the variable list $\vec{\mathtt{X}}$; the *loop* `while`$\,t\,$`{P}` is a program in $\vec{\mathtt{X}}$ if `P` is a program in $\vec{\mathtt{X}}$, and t is a term of type $\mathbf{0}$ such that every variable in t occurs in the variable list $\vec{\mathtt{X}}$; the *sequence* `P;Q` is a program in $\vec{\mathtt{X}}$ if `P` and `Q` are programs in $\vec{\mathtt{X}}$.

We will now define the semantics of the programming language. Let `P` be a program in $\vec{\mathtt{X}} = \mathtt{X}_1, \dots, \mathtt{X}_n$. The meaning of `P` is a $(2n+1)$-ary *input-output relation* $x_1, \ldots x_n$ $\{\mathtt{P}\}_b$ $y_1, \ldots, y_n$ over the natural numbers. We say that the arguments x_i and y_i (for $i = 1, \ldots, n$) are *assigned to* the program variable $\mathtt{X}_i$ in the list $\vec{\mathtt{X}} = \mathtt{X}_1, \ldots, \mathtt{X}_n$. We define the relation by recursion over the syntactical build-up of P:

- The relation $\vec{x}\{\mathtt{Q;R}\}_b\vec{y}$ holds iff there exists $\vec{z}$ such that $\vec{x}\{\mathtt{Q}\}_b\vec{z}$ and $\vec{z}\{\mathtt{R}\}_b\vec{y}$ holds.
- The relation $\vec{x}\{\mathtt{while}\, t\, \{\mathtt{Q}\}\}_b\vec{y}$ holds iff there exists a sequence $\vec{z}^1, \ldots, \vec{z}^k$ such that
  - $\vec{x} = \vec{z}^1$ and $\vec{y} = \vec{z}^k$
  - $\vec{z}^i\{\mathtt{Q}\}_b\vec{z}^{i+1}$ holds for $i \in \{1, \ldots, k-1\}$.
  - $t^i \neq 0$ for $i \in \{1, \ldots, k-1\}$ and $t^k = 0$, where t^i is the the interpretation of the term t under the assignment $\vec{z}^i$ to the program variables.
- The relation $\vec{x}, z, \vec{y}\{\mathtt{Z+}\}_b\vec{x}, z', \vec{y}$ holds iff z is assigned to the type $\mathbf{0}$ variable Z and $z' = z \oplus_b 1)$.
- The relation $\vec{x}, a, \vec{y}\{\mathtt{a[i_1, \ldots, i_m]+}\}_b\vec{x}, a', \vec{y}$ holds iff a is assigned to a and

$$a' \;=\; a[i_1, \ldots, i_m := (a[i_1, \ldots, i_m]_b \oplus_b 1)]_b \,.$$

This completes the definition of the programming language.

A program P in the variables $\mathtt{X}_1, \ldots, \mathtt{X}_k$ where $\mathtt{X}_1, \ldots, \mathtt{X}_n$ are of type $\mathbf{0}$, computes the number-theoretic function $f(x_1, \ldots, x_n)$ when

$$\forall x_{n+1}, \ldots, x_k\, \exists y_1, \ldots, y_{k-1}\, [\, x_1, \ldots, x_n, x_{n+1}, \ldots, x_k\{\mathtt{P}\}_b\, y_1, \ldots, y_{k-1}, z\,] \quad \text{iff } f(x_1, \ldots, x_n) = z \quad \text{where } b = \max(x_1, \ldots, x_n, 1) + 1$$

(Note that the output variable Z can be of any type.) A program decides a number-theoretic relation if it computes the characteristic function of the relation. A program *variable* X of type σ *has degree* n when σ has degree n. A *program* in the variables $\vec{\mathtt{X}}$ *has degree* n when every variable in $\vec{\mathtt{X}}$ has degree $\leq n$. Let $\mathcal{P}^n$ denote the set of number-theoretic functions computed by the programs of degree n. The *program hierarchy* $\mathcal{P}$ is defined by $\mathcal{P} = \bigcup_{n \in \mathbb{N}} \mathcal{P}^n$.

Finally, we need some notation to develop, specify and reason about programs. Let $\vec{\mathtt{Z}}$ be a list of variables, and let P be a program in $\mathtt{X}_1, \ldots, \mathtt{X}_n, \vec{\mathtt{Z}}$. We will use $\vec{x}\{P\}_b\vec{y}$ to abbreviate $\forall\vec{z}\exists\vec{u}[\, \vec{x}, \vec{z}\{\mathtt{P}\}_b\vec{y}, \vec{u}\,]$. We will develop programs in an informal mnemonic macro notation. The notation should need no further explication for readers familiar with standard programming languages. We will always assume that the local variables in the macros are suitably renamed such that we avoid name conflicts. In the macros we will use uppercase Latin letters X, Y, Z, etc. to denote variables of type $\mathbf{0}$, and lowercase Latin letters u, i, j, etc. to denote variables of higher types. □

In the following we give an explication of the programming language. The execution of program will take place in a certain base b given by $b = \max(\vec{x}, 1) + 1$ where $\vec{x} \in \mathbb{N}$ are the inputs. Program variables of type σ store natural numbers in the set $\{0, 1, \ldots, |\sigma|_b - 1\}$. The only primitive instruction in the language has either the form (i) a[$X_1, \dots, X_n$]+ where a, $X_1, \dots, X_n$ are variables such that a[$X_1, \dots, X_n$] is a type **0** term, or the form (ii) Y+ where Y is a type **0** variable. In case (i) the sub-digit a[$X_1, \dots, X_n$] in a is increased by 1 modulo b, that is, the sub-digit will be increased by 1, except in the case when the sub-digit equals $b-1$, then it is set to 0. In case (ii) Y is increased by 1 modulo b. The language has one control structure, the loop `while t {...}` where t is a term of the same form and type as to those allowed in the primitive instruction. This is a standard while-loop executing its body repeteatedly as long as the value (sub-digit) t refers to is different from 0. The semicolon composing two programs has the standard meaning.

Definition. Let $x \oplus_n y \stackrel{\text{def}}{=} (x+y) \pmod{n}$ and $x \ominus_n y \stackrel{\text{def}}{=} (x-y) \pmod{n}$. Let $sg(x) = 0$ if $x = 0$; otherwise $sg(x) = 1$.

Lemma 1. *Let X, Y and Z denote type* **0** *variables, and assume that x, y and z are assigned to respectively X, Y and Z. Let $b \geq 2$ and assume $x, y, z < b$. The following macros can be implemented as type* **0** *programs.*

- x {X:=0}$_b$ 0 *(assignment of the constant 0)*
- x {X:=1}$_b$ 1 *(assignment of the constant 1)*
- x, y {Y:=X}$_b$ x, x *(assignment)*
- z, x, y {Z:=X+Y}$_b$ $x \oplus_b y, x, y$ *(addition modulo b)*
- z, x, y {Z:=X-Y}$_b$ $x \ominus_b y, x, y$ *(subtraction modulo b)*
- x, y {X:=sg *Y*}$_b$ $sg(y), y$ *(converting numbers to boolean values)*
- x, y {X:=not *Y*}$_b$ x', y *where* $x' = 1$ *if* $y = 0$*; otherwise* $x' = 0$. *(logical* **not***)*
- z, x, y {Z:=*X* or *Y*}$_b$ z', x, y *where* $z' = 0$ *if* $x = y = 0$*; otherwise* $z' = 1$ *(logical* **or***)*
- z, x, y {Z:=*X* and *Y*}$_b$ z', x, y *where* $z' = 1$ *if* $x = y = 1$*; otherwise* $z' = 0$. *(logical* **and***)*

Proof. Let X:=0 $\equiv$ `while X {X+}` and let X:=1 $\equiv$ `X:=0; X+`. Let $\overline{x}$ denote the complement of x modulo b, i.e. $\overline{x}$ is the unique number such that $x \oplus_b \overline{x} = 0$. Let

```
cc(X,Y,Z)  ≡  Y:=0; Z:=0; while X { X+; Y+; Z+ } .
```

Then we have x, y, z cc(X,Y,Z) $0, \overline{x}, \overline{x}$. Note that $\overline{\overline{x}} = x$. Thus, let

```
Y:=X  ≡  cc(X,U,V); cc(U,X,Y) .
```

Further, let Z:=X+Y $\equiv$ `Z:=X; cc(Y,U,V); while U { Z+; Y+; U+ }`. By using the macros we have defined so far we can also easily define the macro Z:=X-Y since $x \ominus_b y = \overline{\overline{x} \oplus_b y}$. Let X:=not Y $\equiv$ `Z:=Y; X:=1; while Z {X:=0; Z+}` and let X:=sg Y $\equiv$ U:=not Y;X:=not U. We see that also Z:=X or Y can be implemented since $sg(sg(x) \oplus_b sg(y)) = 0$ if $x = y = 0$ and 1 otherwise. Use X:=not Y and Z:=X or Y to implement Z:=X and Y. □

Lemma 1 tells that the type **0** fragment of the programming language is powerful. Indeed the fragment yields full Turing computability if the programs always are executed at sufficiently high base, i.e. the base b is set to a sufficiently large number before the execution starts. We will prove that if the base b is set to $\max(\vec{x},1)+1$ (where $\vec{x}$ are the inputs) the type **0** fragment capture the complexity class LINSPACE. Programs containing higher types can compute functions beyond LINSPACE. When we set the base b to $\max(\vec{x},1)+1$, type **0** programs can do arithmetic modulo b, in general, programs of type σ can do do arithmetic modulo $|\sigma|_b$.

Lemma 2. *Let $b \geq 2$ and let $c = |\sigma|_b$. Further, Let* u, v *and* w *be program variables of type σ, and let* X *be a program variable of type* **0**. *Assumed that the numbers u, v, w and x (where $u, v, w < c$ and $x < b$) are assigned to respectively* u,v, w *and* X. *The following macros can be implemented as type σ programs.*

- u {u:=$_\sigma$0}$_b$ 0 *(assignment of 0 to a variable of type σ)*
- u {u+$_\sigma$}$_b$ $u \oplus_c 1$ *(successor modulo $|\sigma|_b$)*
- u {u-$_\sigma$}$_b$ $u \ominus_c 1$ *(predecessor modulo $|\sigma|_b$)*
- u, x {X:=sg(u)$_\sigma$}$_b$ $u, sg(x)$
- u, v, w {u:=$_\sigma$v+w}$_b$ $v \oplus_c w, v, w$ *(addition modulo $|\sigma|_b$)*
- u, v, w {u:=$_\sigma$v-w}$_b$ $v \ominus_c, v, w$ *(subtraction modulo $|\sigma|_b$)*

Proof. We define the three macros u+$_\sigma$, u:=$_\sigma$0 and X:=sg(u)$_\sigma$ simultaneously by recursion over the build up of the type σ. By Lemma 1, we can define the macros when $\sigma \equiv \mathbf{0}$. Now, assume $\sigma \equiv \pi \to \tau$ and that the macros are already defined for the types π and τ. Then we define u:=$_\sigma$0 by

```
i:=π0; X:=1; while X { u[i]:=τ0; i+π; X:=sg(i)π }
```

(Explanation: u is a $|\pi|_b$-digit number in base $|\tau|_b$. The macro sets each digit to 0.) We define X:=sg(u)$_\sigma$ by

```
i:=π0; U:=1;
while U { X:=sg(u[i])τ; i+π; Y:=sg(i)π; Z:=not X; U:=Y and Z }
```

(Explanation: u is a $|\pi|_b$-digit number in base $|\tau|_b$. The macro sets the type **0** variable X to 0 if each digit in u is 0.) We define u+$_\sigma$ by

```
i:=π0; j:=π0; j+π; u[i]+τ; U:=sg(u[i])τ;
while U { j+π; i+π; u[i]+τ;
          X:=sg(u[i])τ; Y:=sg(j)π; Z:=not X; U:=Y and Z }
```

(Explanation: u is a $|\pi|_b$-digit number in base $|\tau|_b$. The macro increases the digit u[0] by 1 modulo $|\tau|_b$. If the digit turns into 0, the macro increase the digit u[1] by 1 modulo $|\tau|_b$. If the digit turns into 0, the the digit u[2] is increased, and so on. Note, when the execution of the loop's body starts, we have i $\oplus_d 1 =$ j where $d = |\tau|_b$.) Given the macros u+$_\sigma$, u:=$_\sigma$0 and X:=sg(u)$_\sigma$, it is easy to define u:=$_\sigma$v+w and u:=$_\sigma$v-w. Implement u:=$_\sigma$v-w and u-$_\sigma$ by computing complements, i.e. take advantage of the equation $x \ominus_c y = \overline{\overline{x} \oplus_c y}$ where $\overline{z}$ denotes the c-complement of z, i.e. the unique number such that $z + \overline{z} = c$. □

Definition. We use standard Turing Machines with a single one-way tape. When computing a number-theoretic function, a Turing machine is assumed to take its input and give its output in binary notation. Let M be a Turing machine computing a number-theoretic function $f(\vec{x})$. We say that M *works in space* $g(\vec{x})$ if the number of cells visited on M's tape during the computation of $f(\vec{x})$ is bounded by $g(\vec{x})$. A Turing machine works in *linear space* if there exists a fixed number k such that the number of tape cells visited under a computation on the input w is bounded by $k|w|$ where w denotes the length of the input.

We define the class $\mathcal{R}^n$ of number-theoretic functions by

- $f \in \mathcal{R}^0$ iff f is computable by a Turing machine working in linear space.
- $f \in \mathcal{R}^{n+1}$ iff $f(\vec{x})$ is computable by a Turing machine working in space $g(\vec{x})$ for some $g \in \mathcal{R}^n$.

We define the Ritchie hierarchy $\mathcal{R}$ by $\mathcal{R} = \bigcup_{n\in\mathbb{N}} \mathcal{R}^n$. □

Lemma 3. *(i) For every function $f \in \mathcal{R}^i$ there exists a polynomial p such that $2_i^{p(\vec{x})} > f(\vec{x})$. (In particular, any function in $\mathcal{R}^0$ is bounded by a polynomial.) (ii) For every polynomial p there exists a function $f \in \mathcal{R}_i$ such that $f(\vec{x}) > 2_i^{p(\vec{x})}$. (In particular, any polynomial is bounded by a function in $\mathcal{R}^0$.)*

Proof. *(i)* We prove (i) by induction on i. It is easy to verify that for any $f \in \mathcal{R}^0$ there exists a polynomial p such that $f(\vec{x}) < p(\vec{x}) = 2_0^{p(\vec{x})}$. We skip the details. Assume $f \in \mathcal{R}^{i+1}$. Then there exist a Turing machine M and $g \in \mathcal{R}^i$ such that M computes f in space $g(\vec{x})$. The induction hypothesis yields a polynomial $q(\vec{x})$ such that M computes f in space $2_i^{q(\vec{x})}$, and thus, the number of bits required to represent the value $f(\vec{x})$ is bounded by $2_i^{q(\vec{x})}$. It follows that $2_{i+1}^{p(\vec{x})} > f(\vec{x})$ for some polynomial p. *(ii)* We prove (ii) by induction on i. We leave the case $i = 0$ to the reader, and focus on the induction step: Let p be an arbitrary polynomial. The induction hypothesis yields a function $h \in \mathcal{R}^i$ such that $h(\vec{x}) > 2_i^{p(\vec{x})}$. Let $f(\vec{x}) = 2^{h(\vec{x})}$. Obviously we have $f(\vec{x}) > 2_{i+1}^{p(\vec{x})}$, and we conclude the proof by argue that $f \in \mathcal{R}^{i+1}$. A Turing machine M can compute f by the following procedure: First M computes the number $h(\vec{x})$ (this can obviously be done in space $g_0(\vec{x})$ for some $g_0 \in \mathcal{R}^i$); then M computes $f(\vec{x})$ by writing down the digit 1 followed by $h(\vec{x})$ copies of the digit 0. Let $g(\vec{x}) = g_0(\vec{x}) + h(\vec{x}) + k$ where k is a sufficiently large fixed number. Then $g \in \mathcal{R}^i$ and M works in space g. Hence, f can be computed by a Turing machine working in space g for some $g \in \mathcal{R}^i$, that is, $f \in \mathcal{R}^{i+1}$. □

The hierarchy $\mathcal{R}$ was introduced in Ritchie [7]. Note that $\mathcal{R}^0 = \mathcal{E}^2$ where $\mathcal{E}^2$ denotes the second Grzegorczyk class; $\mathcal{R} = \mathcal{E}^3$ where $\mathcal{E}^3$ denotes the third Grzegorczyk class also known as the class of Kalmár elementary functions.

Theorem 2. *The program hierarchy and the Ritchie hierarchy match from level 1 and upwards, i.e. $\mathcal{P}^{i+1} = \mathcal{R}^{i+1}$ for any $i \in \mathbb{N}$.*

Proof. We first prove $\mathcal{P}^{i+1} \subseteq \mathcal{R}^{i+1}$. Assume $f \in \mathcal{P}^{i+1}$. Then there exists a program P of degree $i+1$ computing $f(\vec{x})$. By Theorem 1, there exists a polynomial p

such that no variable exceeds $2_{i+1}^{p(\vec{x})}$ during the execution. Thus a Turing machine M representing numbers in binary notation needs $O(2_i^{p(\vec{x})})$ tape cells to simulate the program P. Let c be a constant such that M runs in space $c2_i^{p(\vec{x})}$. Then there exists a polynomial q such that M runs in space $2_i^{q(\vec{x})}$, and by Lemma 3 (ii) there exists $g \in \mathcal{R}^i$ such that M runs in space $g(\vec{x})$. This proves that $f \in \mathcal{R}^{i+1}$.

We will now prove $\mathcal{R}^n \subseteq \mathcal{P}^n$. The proofs splits into two cases (1) $n = i+2$, (2) $n = 1$. *Case (1).* Let $f \in \mathcal{R}^{i+2}$. We can without loss of generality assume that f is a unary function. By Lemma 3, there exist a polynomial p and a Turing machine M computing f such that the number of cells visited on M's tape in the computation of $f(x)$ is bounded by $2_{i+1}^{p(x)}$. We will, uniformly in the Turing machine M, construct a program computing f. Let $|\Sigma|$ denote the cardinality of M's alphabet (including the blank symbol). If $\max(x,1) < |\Sigma|$, the program computes $f(x)$ by a hand tailored algorithm, i.e. the program simply consults a finite table. If $\max(x,1) \geq |\Sigma|$, the program computes $f(x)$ by simulating M. We map each symbol in M's alphabet to a unique value in the set $|\mathbf{0}|_{\max(x,1)+1}$. (This is possible since $\max(x,1) \geq |\Sigma|$) By Lemma 1 there exists a type τ of degree $i+1$ such that $|\tau|_{\max(x,1)+1} > 2_{i+1}^{p(x)}$. Thus, the program can represent M's tape by a variable a of type $\tau \to \mathbf{0}$, and the scanning head by variable X of type τ. The scanning head can be moved one position to the right (left) by executing the program $\texttt{X+}_\tau$ ($\texttt{X-}_\tau$), and the type $\mathbf{0}$-term a[X] yields the scanned symbol. The degree of the program will equal the degree of a, i.e. $i+2$. Thus, f can be computed by a program degree $i+2$. Hence, $f \in \mathcal{P}^{i+2}$. *Case (1).* We skip this case. □

Definition. A function f is *non increasing* when $f(\vec{x}) \leq \max(\vec{x}, 1)$ □

The reason that the Ritchie hierarchy and the program hierarchy do not match at level 0, is simply that a type $\mathbf{0}$ programs compute only non increasing functions. The next theorem says that with respect to non increasing functions the hierarchies also match at level 0. The proof of a similar theorem can be found in [5].

Theorem 3. *We have $f \in \mathcal{P}^0$ iff $f \in \mathcal{R}^0$ for every non increasing function f.*

The following lemma is needed in the succeeding section. We leave the proof to the reader.

Lemma 4. *Let $k > 0$, and let P be a program of degree k computing the total function $f(\vec{x})$. There exists a type σ of degree k such that the number of steps P uses to compute $f(\vec{x})$ is bounded by $|\sigma|_{\max(\vec{x},1)+1}$. (One step corresponds to the execution of one primitive instruction, i.e. an instruction on the form* t+.*)*

4 Restricted Functionals in Finite Types

We will introduce functionals in the style of Shoenfield [10]. This style is more convenient for our purposes then the combinator style used e.g. in [1].

Definition. We will now define the *constants*, *variables*, *terms* and *defining equations* of the system D.

- 1 is a *constant* of type $\mathbf{0}$
- For each type σ we have an infinite supply of variables $x_0^\sigma, x_1^\sigma, x_2^\sigma, \ldots$.
- A variable of type σ is a term of type σ. A constant of type σ is a term of type σ. If t is a term of type $\sigma \to \tau$ and u is a term of type σ, the $t(u)$ is a term of type τ.
- Suppose that $\vec{x}$ are distinct variables of type $\vec{\sigma}$ respectively, and that t is a term of type τ in which no variable other that $\vec{x}$ occurs. We introduce a new constant F of type $\vec{\sigma} \to \tau$ with the defining equation $F(\vec{x}) = t$. We will say that F is defined *explicitly* (over the constants in t).
- Suppose that x is a variable of type $\mathbf{0}$, G is a constant of type σ, and H is a constant of type $\sigma \to (\mathbf{0} \to \sigma)$. We introduce a new constant F of type $\mathbf{0} \to \sigma$ with the defining equations $F(0) = G$ and $F(x+1) = H(F(x))(x)$. We will say that F is defined by *primitive recursion* (from G and H).

The system D provides a syntax for equations defining functionals in finite types. Each constant in D corresponds to a functional. Thus, we will also call the constants in D *functionals* and we will call constants of type $\vec{\mathbf{0}} \to \mathbf{0}$ *functions*. Note that there are no initial constants representing the zero and the successor function in D, and that the left-hand sides in the primitive recursive equations are not terms. We will use the formal system Y of Shoenfield [10] to discuss and compute the functionals defined in D. The language of Y has a constant $\mathbf{k}_n$ for each $n \in \mathbb{N}$ (occasionally we will simply write n instead of $\mathbf{k}_n$). Thus, if $F(0) = G$ and $F(x+1) = H(F(x))(x)$ are the defining equations for F, then $F(\mathbf{k}_0) = G$ and all equations $F(\mathbf{k}_{n+1}) = H(F(\mathbf{k}_n))(\mathbf{k}_n)$ are axioms in Y. Further, $1 = \mathbf{k}_1$ is an axiom. (The axioms induced by the explicit definitions in D are the obvious ones.)

A number-theoretic function f is definable in D when there exists a constant F of type $\vec{\mathbf{0}} \to \mathbf{0}$ such that $\vdash_Y F(\mathbf{k}_{x_1}, \ldots, \mathbf{k}_{x_n}) = \mathbf{k}_{f(x_1,\ldots,x_n)}$ (for all $x_1, \ldots, x_n \in \mathbb{N}$). A number-theoretic relation R is definable in D when there exists a constant F of type $\vec{\mathbf{0}} \to \mathbf{0}$ such that $\vdash_Y F(\mathbf{k}_{x_1}, \ldots, \mathbf{k}_{x_n}) = \mathbf{k}_0$ iff $R(x_1, \ldots x_n)$ holds. □

Lemma 5 (Basic functions). *The following number-theoretic functions are definable in D. (i) 0,1 (constant functions); (ii) for each fixed $k \in \mathbb{N}$ the the function $C_k(x)$ where $C_k(x) = k$ if $x \geq k$, and $C_k(x) = x$ otherwise (so $C_k(x)$ is an almost everywhere constant function yielding k for all but finitely many values of x); (iii) $P(x)$ (predecessor) (iv); $x \dot{-} y$ (modified subtraction); (v) $f(x,y,z)$ such that $f(x,y) = x \oplus_{y+1} 1$ when $x \leq y$; (vi) $c(x,y,z)$ where $c(x,y,z) = x$ if $z = 0$ and $c(x,y,z) = y$ if $z \neq 0$; (vii) $\max(x,y)$.*

Proof. It is easy to see that the constant function 1 is definable in D since it is represented by an initial constant in D. To give equations defining the constant function 0 is slightly nontrivial. Define g by primitive recursion such that $g(x,0) = x$ and $g(x,y+1) = y$. It is easy to verify that g is definable in D. Then we can explicitly define the predecessor P from g by $P(x) = g(x,x)$. Further, we can define the constant function 0 by $0 = P(1)$. This proves that (i) and (iii) holds. (iv) holds since we have $x \dot{-} 0 = x$ and $x \dot{-} (y+1) = P(x \dot{-} y)$. (vi) holds since $c(x, y, 0) = x$ and $c(x, y, z+1) = y$. (vii) holds since $\max(x, y) = c(x, y, 1 \dot{-} (x \dot{-} y))$. (v) holds since $x \oplus_{m+1} 1 = c(0, m \dot{-} ((m \dot{-} x) \dot{-} 1), m \dot{-} x)$. It remains to prove that (ii) holds. Let $M^0(z) = 0$ and $M^{n+1}(z) = M^n(z) \oplus_{z+1} 1$. Now, M^n is definable in D for any $n \in \mathbb{N}$. Further, $M^n(z) = n \pmod{z+1}$. Thus, we have $C_k(x) = c(x, M^k(x), P^k(x))$ where P^k is the predecessor function repeated k times. Hence, (ii) holds. □

Lemma 6 (Iteration functionals). *For all types σ and τ there exists a functional $\mathrm{It}_\sigma^\tau : \mathbf{0}, \tau \to \tau, \tau \to \tau$ such that $\mathrm{It}_\sigma^\tau(\ell, F, X) = F^{|\sigma|_{\ell+1}}(X)$.*

Proof. We prove the lemma by induction on the build-up of σ. Assume $\sigma = \mathbf{0}$. It is easy to see that for any τ we can define $\mathrm{It}_\mathbf{0}^\tau$ such that $\mathrm{It}_\mathbf{0}^\tau(\ell, F, X) = FF^\ell(X)$ Thus, (i) holds since $|\mathbf{0}|_{\ell+1} = \ell + 1$. Assume $\sigma = \sigma_1 \to \sigma_2$. By the induction hypothesis we have $\mathrm{It}_{\sigma_2}^\tau$ and $\mathrm{It}_{\sigma_1}^{\tau\to\tau}$ satisfying the lemma. Define It_σ^τ explicitly from $\mathrm{It}_{\sigma_2}^\tau$ and $\mathrm{It}_{\sigma_1}^{\tau\to\tau}$ such that

$$\mathrm{It}_\sigma^\tau(\ell, F, X) = \mathrm{It}_{\sigma_1}^{\tau\to\tau}(\ell, \mathrm{It}_{\sigma_2}^\tau(\ell), F)(X)\ .$$

We will prove that we indeed have

$$\mathrm{It}_\sigma^\tau(\ell, F, X) = F^{|\sigma|_{\ell+1}}(X) \qquad \text{(Goal)}$$

Let A abbreviate $\mathrm{It}_{\sigma_2}^\tau(\ell)$. Hence, we have

$$A(F, X) = F^{|\sigma_2|_{\ell+1}}(x) \qquad (\dagger)$$

by the induction hypothesis. We need

$$A^k(F, Y) = F^{|\sigma_2|_{\ell+1}^k}(Y) \qquad \text{(Claim)}$$

for any Y of type τ. (Goal) follows from (Claim) since

$$\begin{aligned}
\mathrm{It}_\sigma^\tau(\ell, F, X) &= \mathrm{It}_{\sigma_1}^{\tau\to\tau}(\ell, A, F)(X) && \text{def. of } \mathrm{It}_\sigma^\tau \\
&= A^{|\sigma_1|_{\ell+1}}(F)(X) && \text{ind. hyp. on } \sigma_1 \\
&= F^{|\sigma_2|_{\ell+1}^{|\sigma_1|_{\ell+1}}}(X) && \text{(Claim)} \\
&= F^{|\sigma|_{\ell+1}}(X)\ . && \text{def. of } |\sigma|_{\ell+1}
\end{aligned}$$

(Claim) is proved by induction on k. In the inductive case we need

$$(A^k(F))^n(Y) = F^{|\sigma_2|_{\ell+1}^k \times n}(Y)$$

which is proved by another induction on n. We skip the details. □

Lemma 7 (Transformation functionals). *For any finite list of types $\vec{\rho}$ there exist a type π and transformation functionals $\mathrm{Tr}_{\rho_i}^\pi$ of type $\rho_i \to \pi$, $\mathrm{Tr}_\pi^{\rho_i}$ of type $\pi \to \rho_i$ such that $\mathrm{Tr}_\pi^{\rho_i}(\mathrm{Tr}_{\rho_i}^\pi(u)) = u$ for any u of type ρ_i.*

Proof. The transformation functionals defined in Schwichtenberg [8] satisfy the lemma. We skip the details. □

Lemma 8 (List functionals). *For any finite list of types $\vec{\sigma}$ there exists a type τ and a list functional $\langle \ldots \rangle : \vec{\sigma} \to (\mathbf{0} \to \tau)$ such that $\mathrm{Tr}_{\tau}^{\sigma_i}(\langle X_1, \ldots, X_n \rangle(i)) = X_i$.*

Proof. The functional $\langle \ldots \rangle : \vec{\sigma} \to (\mathbf{0} \to \tau)$ is easily defined from the basic functionals introduced in Lemma 5, and the transformation functionals $\mathrm{Tr}_{\sigma_i}^{\tau}$ introduced in Lemma 7. □

Definition. For a term $t : \sigma$ we define the *value of t at the base b under valuation* $v_1, \ldots, v_k$, in symbols $\mathbf{val}_b^{v_1,\ldots,v_k}(t)$. Here the valuation assigns values to the free variables $x_1, \ldots, x_k$ of t such that to $x_i : \sigma$ we assign $v_i < |\sigma|_b$ and we will have $\mathbf{val}_b^{v_1,\ldots,v_k}(t) < |\sigma|_b$. Recall that for numbers $a : (\rho \to \tau)_b$ and $i : \rho_b$ we denote the i'th digit in a by $a[i]_b$ (see the definition in Section 2).

- Let $\mathbf{val}_b^{v_1,\ldots,v_k}(x_i) = v_i$ for the variable x_i; $\mathbf{val}_b^{v_1,\ldots,v_k}(\mathbf{k}_n) = n$ for $n \in \mathbb{N}$; $\mathbf{val}_b^{v_1,\ldots,v_k}(1) = 1$ for the initial function 1.
- Suppose $t \equiv s(u)$. Let $\mathbf{val}_b^{v_1,\ldots,v_k}(t) = \mathbf{val}_b^{v_1,\ldots,v_k}(s)[\mathbf{val}_b^{v_1,\ldots,v_k}(u)]_b$.
- Suppose $t \equiv F$ where F is a constant explicitly defined by $F(\vec{x}) = a$. Then F is of type $\sigma_1 \to \cdots \sigma_k \to \tau$ for some k, and the variables $x_1, \ldots, x_k$ have types $\sigma_1, \ldots, \sigma_k$ respectively. We define $\mathbf{val}_b^{v_1,\ldots,v_k}(F)$ by recursion on k. If $k = 0$, let $\mathbf{val}_b^{v_1,\ldots,v_k}(F) = \mathbf{val}_b^{v_1,\ldots,v_k}(a)$, else let

$$\mathbf{val}_b^{v_1,\ldots,v_k}(F) \quad = \quad \sum_{i<|\sigma_1|_b} \mathbf{val}_b^{i,v_2,\ldots,v_k}(a) \times |\sigma_2 \to \cdots \to \sigma_k \to \tau|_b^i \,.$$

- Suppose $t \equiv F$ where F is a constant with defining equations $F(0) = G$ and $F(x+1) = H(F(x))(x)$. Then let

$$\mathbf{val}_b^{v_1,\ldots,v_k}(F) \quad = \quad \sum_{i<b} v_i \times |\mathbf{0} \to \sigma|_m^i$$

where $v_0 = \mathbf{val}_b^{v_1,\ldots,v_k}(G)$ and $v_{i+1} = \mathbf{val}_b^{v_1,\ldots,v_k}(H)[v_i]_b$.

□

Lemma 9 (Equality functionals). *For any type σ there exists an equality functional $\mathrm{Eq}_\sigma : \mathbf{0}, \sigma, \sigma \to \mathbf{0}$ such that $\mathrm{Eq}_\sigma(\ell, t, s) = 0$ iff $\mathbf{val}_{\ell+1}(t) = \mathbf{val}_{\ell+1}(s)$ for any variable-free terms $s : \sigma$ and $t : \sigma$.*

Proof. Define $0_0 = 0$ and $0_{\rho \to \tau}(y) = 0_\tau$ for any types ρ and τ. Then we have $\mathbf{val}_b(0_\sigma) = 0$ for any σ and $b > 1$. Further, we need $\mathrm{Le}_\sigma : \mathbf{0}, \sigma, \sigma \to \mathbf{0}$ and $\mathrm{Suc}_\sigma : \mathbf{0}, \sigma \to \sigma$ such that

$$\begin{array}{ll} \mathrm{Le}_\sigma(\ell, t, s) = 0 & \text{if and only if } \mathbf{val}_{\ell+1}(t) \le \mathbf{val}_{\ell+1}(s) \\ \mathrm{Suc}_\sigma(\ell, t) = s & \text{if and only if } \mathbf{val}_{\ell+1}(t) + 1 \ = \ \mathbf{val}_{\ell+1}(s) \pmod{|\sigma|_{\ell+1}} \end{array}$$

for any variable-free terms $t : \sigma$ and $s : \sigma$. Define the functionals Eq_σ, Le_σ and Suc_σ simultaneously by recursion over the build-up of σ. It follows from Lemma

5 that the three functionals are definable when $\sigma = \mathbf{0}$. Assume $\sigma = \rho \to \tau$. Define the *carry predicate* $C{:}\mathbf{0}, \rho \to \tau, \rho \to \mathbf{0}$ such that

$$C(\ell, x, y) = 0 \;\text{ iff }\; \forall z{:}\rho[\,\mathrm{Le}_\rho(\ell, z, y) = 0 \;\to\; \mathrm{Eq}_\tau(\ell, \mathrm{Suc}_\tau(\ell, x(z)), 0_\tau) = 0\,]\,.$$

(Note that although the quantifier ranges over infinitely many functionals $z:\rho$ only their finitely many values at the base $\ell + 1$ matters. The carry predicate is defined from Le_ρ, Eq_τ, Suc_τ, 0_τ using iteration, transformation and list functionals given in Lemma 6, 7 and 8.) Then use the carry predicate to define $\mathrm{Suc}_{\rho\to\tau}$ such that

$$\mathrm{Suc}_{\rho\to\tau}(\ell, x, y) = \begin{cases} \mathrm{Suc}_\tau(\ell, x(y)) & \text{if } C(m, x, y) = 0 \\ x(y) & \text{otherwise.} \end{cases}$$

For the definition of $\mathrm{Le}_{\rho\to\tau}$ define, analogous as C, the characteristic function $A{:}\mathbf{0}, \rho \to \tau, \rho \to \tau, \rho \to \mathbf{0}$ such that

$$A(\ell, x, y, z) = 0 \;\text{ iff }\; \forall w{:}\rho[\,\mathrm{Le}_\rho(\ell, z, w) = 0 \;\to\; \mathrm{Eq}_\tau(\ell, x(w), y(w)) = 0\,]\,.$$

Then define $\mathrm{Le}_{\rho\to\tau}$ to satisfy

$$\mathrm{Le}_{\rho\to\tau}(\ell, x, y) \;\text{ iff }\; \exists z{:}\rho[\,\mathrm{Le}_\tau(\ell, x(z), y(z)) = 0 \;\wedge\; A(\ell, x, y, z)\,]\,.$$

Finally, define $\mathrm{Eq}_{\rho\to\tau}$ to satisfy

$$\mathrm{Eq}_{\rho\to\tau}(\ell, x, y) = 0 \;\text{ iff }\; \mathrm{Le}_{\rho\to\tau}(\ell, x, y) = 0 \wedge \mathrm{Le}_{\rho\to\tau}(\ell, y, x) = 0\,.$$

□

Lemma 10 (Modification functionals). *For any list of types $\vec{\sigma} = \sigma_1, \ldots, \sigma_k$ and type τ there exists a modification functional $\mathrm{Md}_{\vec{\sigma}\to\tau}{:}\mathbf{0}, \vec{\sigma} \to \tau, \vec{\sigma}, \tau \to \vec{\sigma} \to \tau$ such that*

$$\mathbf{val}_{\ell+1}(\mathrm{Md}_{\vec{\sigma}\to\tau}(\ell, F, \vec{a}, v)(\vec{b})) = \begin{cases} \mathbf{val}_{\ell+1}(v) & \textit{if } \mathbf{val}_{\ell+1}(a_i) = \mathbf{val}_{\ell+1}(b_i) \textit{ for } i = 1, \ldots, k \\ \mathbf{val}_{\ell+1}(F(\vec{b})) & \textit{otherwise.} \end{cases}$$

for all variable-free terms $F, \vec{a}, v, \vec{b}$ of proper types.

Proof. We prove the lemma by induction on the length of $\vec{\sigma}$. If $k = 1$ use the equality functional Eq_{σ_1} to define $\mathrm{Md}_{\vec{\sigma}\to\tau}$ as required. If $k > 1$, define

$$\mathrm{Md}_{\vec{\sigma}\to\tau}(\ell, F, \vec{y}, v) = \mathrm{Md}_{\sigma_1\to\tau}(\ell, F, y_1, \mathrm{Md}_{\sigma_2,\ldots,\sigma_k\to\tau}(\ell, F(y_1), y_2, \ldots, y_k, v))\,.$$

□

Theorem 4. *Any relation decidable by a program is also definable in D.*

Proof. Assume $\mathtt{P}$ is a program in $\mathtt{X}_1,\ldots,\mathtt{X}_k$ such that $\mathtt{X}_1,\ldots,\mathtt{X}_n$ and $\mathtt{X}_k$ are type $\mathbf{0}$ variables. We will prove that there exists an n-ary functional $f:\vec{\mathbf{0}}\to\mathbf{0}$ such that

$$\forall x_{n+1},\ldots,x_k\,\exists y_1,\ldots,y_{k-1}\,[\,x_1,\ldots,x_n,x_{n+1},\ldots,x_k\,\{\mathtt{P}\}_b\,y_1,\ldots,y_{k-1},z\,]$$
$$\text{iff}\ \vdash_Y f(\mathbf{k}_{x_1},\ldots,\mathbf{k}_{x_n})=\mathbf{k}_z$$
$$\text{where } b=\max(x_1,\ldots,x_n,1)+1 \qquad (*)$$

for all but finitely many values of $x_1,\ldots,x_n$. (The equivalence (*) will holds when $\max(x_1,\ldots,x_n)\geq k$.) The theorem follows from (*). Let respectively $\mathtt{X}_1,\ldots,\mathtt{X}_k$ be of the types $\sigma_1,\ldots,\sigma_k$. We will assign a functional $F_i^{\mathtt{P}}:\mathbf{0},\vec{\sigma}\to\sigma_i$ to the variable $\mathtt{X}_i$. Then f is given by

$$f(x_1,\ldots,x_n) \;=\; F_k^{\mathtt{P}}(\max(x_1,\ldots,x_n,1),x_1,\ldots,x_n,G_{n+1},\ldots,G_k)$$

where $G_{n+1},\ldots,G_k$ are arbitrary constants of appropriate types. By Lemma 5 (vii), f is definable in D when $F_k^{\mathtt{P}}$ is.

We define $F_1^{\mathtt{P}},\ldots F_k^{\mathtt{P}}$ recursively over the build-up of $\mathtt{P}$. Assume $\mathtt{P}\equiv\mathtt{X}_m\mathtt{+}$. Then $\mathtt{X}_m$ has type $\mathbf{0}$. Let $F_m^{\mathtt{P}}(\ell,\vec{X}) = X_m \oplus_{\ell+1} +1$ and let $F_j^{\mathtt{P}}(\ell,\vec{X}) = X_j$ for $j\neq m$. The functional $F_m^{\mathtt{P}}$ is definable in D by Lemma 5. Assume $\mathtt{P}\equiv \mathtt{X}_m\,\mathtt{[X}_{i_1},\ldots,\mathtt{X}_{i_r}\mathtt{]+}$ where $\mathtt{X}_m$ has type $\sigma_1,\ldots\sigma_r\to\mathbf{0}$. Let

$$F_m^{\mathtt{P}}(\ell,\vec{X}) \;=\; \mathrm{Md}_{\sigma_1,\ldots\sigma_r\to\mathbf{0}}(\ell,X_m,X_{i_1},\ldots,X_{i_r},X_m(X_{i_1},\ldots,X_{i_r})\oplus_{\ell+1}+1)$$

and let $F_j^{\mathtt{P}}(\ell,\vec{X})=X_j$ for $j\neq m$. The functional $F_m^{\mathtt{P}}$ is definable in D by Lemma 5 and Lemma 10. Assume $\mathtt{P}\equiv\mathtt{Q;R}$. Let

$$F_i^{\mathtt{P}}(\ell,\vec{X})=F_i^{\mathtt{R}}(\ell,F_1^{\mathtt{Q}}(\ell,\vec{X}),\ldots,F_k^{\mathtt{Q}}(\ell,\vec{X}))$$

for $i=1,\ldots,k$. Assume $\mathtt{P}\equiv\mathtt{while}\,t\,\mathtt{\{Q\}}$. Let X be a variable of type $\mathbf{0}\to\tau$, and let F of type $\mathbf{0},(\mathbf{0}\to\tau)\to(\mathbf{0}\to\tau)$ be defined by

$$\begin{aligned}F(\ell,X) \;=\; \langle\, & F_1^{\mathtt{Q}}(\ell,\mathrm{Tr}_\tau^{\sigma_1}(X(C_1(\ell))),\ldots,\mathrm{Tr}_\tau^{\sigma_k}(X(C_k(\ell)))),\\ & \ldots,F_k^{\mathtt{Q}}(\ell,\mathrm{Tr}_\tau^{\sigma_1}(X(C_1(\ell))),\ldots,\mathrm{Tr}_\tau^{\sigma_k}(X(C_k(\ell))))\,\rangle\end{aligned}$$

where $\langle\ldots\rangle$ is the list functional given by Lemma 8; $\mathrm{Tr}_\tau^{\sigma_i}$ are the transition functionals given by Lemma 7; C_i are the almost everywhere constant function given by Lemma 5. Further, define $F':\mathbf{0},(\mathbf{0}\to\tau)\to(\mathbf{0}\to\tau)$ such that

$$F'(\ell,X)=\begin{cases}X & \text{if } T(\ell,X)=0\\ F(\ell,X) & \text{otherwise}\end{cases}$$

where $T(\ell,X)=\mathrm{Tr}_\tau^{\mathbf{0}}(X(C_m(\ell)))$ if the loop test t has the form $t\equiv\mathtt{X}_m$, and

$$T(\ell,X)=\mathrm{Tr}_\tau^{\sigma_m}(X(C_m(\ell)))(\mathrm{Tr}_\tau^{\sigma_{i_1}}(X(C_{i_1}(\ell))),\ldots,\mathrm{Tr}_\tau^{\sigma_{i_r}}(X(C_{i_r}(\ell))))$$

if the loop test t has the form $t\equiv\mathtt{X}_m\,\mathtt{[X}_{i_1},\ldots,\mathtt{X}_{i_r}\mathtt{]}$. By Lemma 4 we can pick a type π such that the loop $\mathtt{P}\equiv\mathtt{while}\,t\,\mathtt{\{Q\}}$ terminates within $|\pi|_{\max(\vec{x},1)+1}$ iterations (where $\vec{x}$ is the input to the program). Define $F'':\mathbf{0},(\mathbf{0}\to\tau)\to(\mathbf{0}\to\tau)$

such that $F''(\ell, X) = \mathrm{It}_{\pi}^{\mathbf{0}\to\tau}(\ell, F'(\ell), X)$ where the iteration functional $\mathrm{It}_{\pi}^{\mathbf{0}\to\tau}$ is given by Lemma 6. Finally, let

$$F_i^{\mathrm{p}}(\ell, X_1, \ldots, X_k) = \mathrm{Tr}_{\tau}^{\sigma_i}(F''(\ell, \langle X_1, \ldots, X_k \rangle)(C_i(\ell)))$$

for $i = 1, \ldots, k$. □

Theorem 5. *A relation is definable in D if and only if it is decidable by a program.*

Proof. This follows from the previous theorem and the following argument: Suppose R is definable in D. Then there exists a functional F such that $F(\vec{v}) = 0$ iff $R(\vec{v})$ holds. A Turing machine can decide whether $R(\vec{v})$ holds by computing $\mathbf{val}^{\vec{v}}_{\max(\vec{v},1)+1}(F(\vec{x}))$. Obviously the Turing machine will work in space $|\sigma|_{\max(\vec{v},1)+1}$ for some type σ. Thus, R belongs to the Ritchie hierarchy $\mathcal{R}$. Thus, R is decidable by a program. □

If we adjust the notion of definability in D, and say that a *function* f *is definable* in D when for some type σ there exists a functional $F : \vec{\mathbf{0}} \to \sigma$ such that $f(\vec{v}) = \mathbf{val}^{\vec{v}}_{\max(\vec{v},1)+1}(F(\vec{x}))$, then we have the following result: A number-theoretic function is definable in D if and only if it is computable by a program. (The proof is similar to the proof of Theorem 5.)

References

1. J. Avigad and S. Fererman. *Gödel's functional interpretation.* In "Handbook of Proof Theory" Buss (Ed), Elsevier, 1998.
2. A. Beckmann and A. Weiermann. *Characterizing the elementary recursive functions by a fragment of Gödel's T.* Arch. Math. Logic 39, No.7, 475-491 (2000).
3. A. Goerdt. *Characterizing complexity classes by higher type primitive recursive definitions.* Theor. Comput. Sci. 100, No.1, 45-66 (1992).
4. N.D. Jones. *The expressive power of higher-order types or, life without CONS.* J. Functional Programming 11 (2001) 55-94.
5. L. Kristiansen and P.J. Voda. *Complexity classes and fragments of C.* Submitted. Preprint available from `http://www.iu.hio.no/~larskri/`
6. D. Leivant. *Intrinsic theories and cpmputational complexity.* In "Logical and Computational Complexity. LCC '94", Leivant (ed.), 177-194, Springer, 1995.
7. R.W. Ritchie. *Classes of predictably computable functions.* Trans. Am. Math. Soc. 106, 139-173 (1963).
8. H. Schwichtenberg. *Finite notations for infinite terms* Ann. Pure Appl. Logic 94, No. 1-3, 201-222 (1998).
9. H. Schwichtenberg and S.J. Bellantoni. *Feasible computation with higher types.* Available from Schwichtenberg's home page.
10. J.R. Shoenfield. *Mathematical logic.* Addison-Wesley Publishing Company. VII, 344 p. (1967).
11. H. Simmons. *Derivation and computation. Taking the Curry-Howard correspondence seriously.* Cambridge Tracts in Theoretical Computer Science. 51. Cambridge: Cambridge University (2000).
12. J. Tiuryn. *Higher-order arrays and stacks in programming. An application of complexity theory to logics of programs.* Math. found. of comp. sci. Proc. 12th Symp. Bratislava/Czech. 1986, Lect. Notes Comput. Sci. 233, 177-198 (1986).

Pebble Games on Trees

Łukasz Krzeszczakowski*

Institute of Informatics, Warsaw University, Poland
lkr@mimuw.edu.pl

Abstract. It is well known that two structures $\mathcal{A}$ and $\mathcal{B}$ are indistinguishable by sentences of the infinitary logic with k variables $L^k_{\infty\omega}$ iff Duplicator wins the Barwise game on $\mathcal{A}$ and $\mathcal{B}$ with k pebbles. The complexity of the problem who wins the game is in general unknown if k is a part of the input. We prove that the problem is in $PTIME$ for some special classes of structures such as finite directed trees and infinite regular trees. More specifically, we show an algorithm running in time $log(k)(|A| + |B|)^{O(1)}$.
The algorithm for regular trees is based on a characterization of the winning pairs $(\mathcal{A}, \mathcal{B})$ which is valid also for a more general case of (potentially infinite) rooted trees.

1 Introduction

We say that two structures over the same signature are FO_k-equivalent iff they satisfy the same formulas of first order logic of quantifier depth at most k. Similarly two structures are $L^k_{\infty\omega}$-equivalent iff they satisfy the same $L_{\infty\omega}$ (for finite structures just first order logic) formulas with at most k variables.

It turns out that FO_k-equivalence and $L^k_{\infty\omega}$-equivalence can by expressed in terms of model-theoretic games. Indeed two structures $\mathcal{A}$ and $\mathcal{B}$ are FO_k-equivalent iff Duplicator wins k-moves Ehrenfeucht-Fraïssé game on $\mathcal{A}$ and $\mathcal{B}$; we denote such a game by $EF_k(\mathcal{A}, \mathcal{B})$. Similarly two structures $\mathcal{A}$ and $\mathcal{B}$ are $L^k_{\infty\omega}$-equivalent iff Duplicator wins k-pebbles Barwise game on $\mathcal{A}$ and $\mathcal{B}$; we denote such a game by $Pebble_k(\mathcal{A}, \mathcal{B})$ (J. Barwise introduced these games in [1]). The proofs of these results can be found in [3].

We call the problem of deciding who wins the game $EF_k(\mathcal{A}, \mathcal{B})$ for two finite structures $\mathcal{A}$ and $\mathcal{B}$ and the number k, the *EF problem* and the problem of deciding who wins the game $Pebble_k(\mathcal{A}, \mathcal{B})$ – the *pebble problem*. We are interested in determining the complexity of above problems. Martin Grohe in [2] has shown that if the number of pebbles is fixed, the pebble problem is complete for PTIME. If the number of moves is fixed, then it is easy to see that EF problem is in LOGSPACE. But Elena Pezzoli has shown in [4] and [5] that if the number of moves is a part of the input then the EF problem is PSPACE complete. If the number of pebbles is a part of the input then it can be easily seen that the problem is at least as hard as the Graph Isomorphism problem. Indeed, two

* Supported by Polish KBN grant No. 7 T11C 027 20

M. Baaz and J.A. Makowsky (Eds.): CSL 2003, LNCS 2803, pp. 359–371, 2003.

graphs with k vertices are isomorphic iff Duplicator wins a k-pebble Barwise game on these graphs[1]. On the other hand it is easy to see that the problem is in EXPTIME, but the exact complexity is to our knowledge open.

We are mostly interested in finding classes of structures in which the pebble problem is efficiently computable. For a class $CLASS$ of structures we denote by $Pebble[CLASS]$ the following problem: given two structures $\mathcal{A}, \mathcal{B} \in CLASS$ and $k \in \mathcal{N}$, decide who wins $Pebble_k(\mathcal{A}, \mathcal{B})$. In this paper we show that the problems $Pebble[DTREE]$ and $Pebble[REGTREE]$ are in $PTIME$, where $DTREE$ is the class of finite directed trees and $REGTREE$ is the class of regular trees. Consequently, we can decide in polynomial time if two directed trees, respectively two regular trees, are indistinguishable by $L^k_{\infty\omega}$ formulas.

2 Preliminaries

2.1 Graphs and Trees

Let E be a binary relation symbol. An **(undirected) graph** is an $\{E\}$-structure $\mathcal{G} = \langle G, E^{\mathcal{G}} \rangle$, where $E^{\mathcal{G}}$ is irreflexive and symmetric relation. A **directed graph** is an $\{E\}$-structure $\mathcal{G} = \langle G, E^{\mathcal{G}} \rangle$, where $E^{\mathcal{G}}$ is arbitrary. Then, its **underlying graph** $\mathcal{G}^u$ is given by $\mathcal{G}^u = \langle G, \{(a,b) : a \neq b, E^{\mathcal{G}}(a,b) \text{ or } E^{\mathcal{G}}(b,a)\} \rangle$. A **path** in a graph $\mathcal{G}$ is a finite or infinite sequence $v_1, v_2, \ldots$ of different vertices in G such that $E^{\mathcal{G}}(v_i, v_{i+1})$ for $i = 1, 2, \ldots$. A path in a directed graph is a path in its underlying graph (that is, it need not respect direction of edges).

An **(undirected) tree** is a connected and acyclic graph. A **directed tree** is a directed graph whose underlying graph is a tree.

For a directed tree $\mathcal{T}$ and a vertex $t \in T$ we have a partial order $\leq^{\mathcal{T}}_t$ on T defined by:

$$v \leq^{\mathcal{T}}_t u \text{ iff } v \text{ appears on the (unique) path from } t \text{ to } u$$

A **rooted tree** is a directed tree $\mathcal{T}$ such that $E^{\mathcal{T}}$ is irreflexive and there is an (unique) vertex $t \in T$ such that the transitive and reflexive closure of $E^{\mathcal{T}}$ coincides with $\leq^{\mathcal{T}}_t$. We denote this vertex t by $r^{\mathcal{T}}$ and call it a root of $\mathcal{T}$. We say that v is an (unique) **parent** of w and w is a **child** of v if $E^{\mathcal{T}}(v, w)$. The set of children of v is denoted by $children^{\mathcal{T}}(v)$. We denote by $subtree^{\mathcal{T}}(v)$ the substructure of $\mathcal{T}$ induced by $\{w : v \leq^{\mathcal{T}}_{r^{\mathcal{T}}} w\}$. We denote by $height^{\mathcal{T}}(v)$ the height of the subtree of $\mathcal{T}$ rooted at v, i.e. the length of the longest finite path in $subtree^{\mathcal{T}}(v)$ started with v (if it exists).

For a rooted tree $\mathcal{T}$ we denote by $Dom(\mathcal{T})$ the set $\{subtree(t) : t \in T\}$. It is important to note, that we do not distinguish between the isomorphic structures. Then a **regular tree** is a rooted tree $\mathcal{T}$ such that the set $Dom(\mathcal{T})$ is finite and additionally for every $v \in \mathcal{T}$ the set $\{w : E^{\mathcal{T}}(v, w)\}$ is finite.

A **pointed directed tree** (in short pointed tree) is a structure $\mathcal{T} = \langle T, E^{\mathcal{T}}, c^{\mathcal{T}} \rangle$, where c is a constant symbol, such that $\mathcal{S} = \langle T, E^{\mathcal{T}} \rangle$ is a directed tree. We denote such a structure by $\mathcal{S}(s)$, where $s = c^{\mathcal{T}}$.

[1] The author thanks an anonymous referee for pointing this observation.

Notice that, for a pointed tree $\mathcal{T}(t)$ there exists exactly one rooted tree $\mathcal{S}$ such that $\mathcal{T}^u = \mathcal{S}^u$ and $r^{\mathcal{S}} = t$. Then we write $children^{\mathcal{T}(t)}(v)$ for the set of children of v in $\mathcal{S}$ and $height^{\mathcal{T}(t)}(v)$ for the height of the subtree of $\mathcal{S}$ rooted at v. Furthermore, we denote by $subtree^{\mathcal{T}(t)}(v)$ a pointed tree $\mathcal{T}'$, where $T' = \{w \in T : v \text{ is in the path from } c^{\mathcal{T}} \text{ to } w\}$, $c^{\mathcal{T}'} = v$ and $E^{\mathcal{T}'} = E^{\mathcal{T}} \cap T' \times T'$.

We denote the class of regular trees by $REGTREE$, the class of finite directed trees by $DTREE$ and the class of finite pointed directed trees by $PDTREE$ (recall that we don't distinguish between isomorphic structures).

When a rooted tree (or a pointed tree) is clear from the context, we omit the superscripts in the denotations $children(t)$, $subtree(t)$ and $height(t)$.

2.2 Pebble Games

Let $k \in \mathcal{N}$ and $\mathcal{A}, \mathcal{B}$ be structures over the same signature. A **configuration** of the game $Pebble_k(\mathcal{A}, \mathcal{B})$ is a function $c : \{1, .., k\} \to A \times B \cup \{*\}$, where $*$ is a fixed element not in $A \times B$. If $c(i) = (v, w)$ for some $i \in \{1, .., k\}$, we say that the vertices $v \in A$ and $w \in B$ are **marked**, or that they are **marked by the pebble** p_i. Additionally we denote v and w by $p_i^{\mathcal{A}}$ and $p_i^{\mathcal{B}}$ respectively. For a fixed configuration c, we denote by **Peb** the set $\{i : 1 \leq i \leq k, c(i) \neq *\}$. The **initial configuration** of $Pebble_k(\mathcal{A}, \mathcal{B})$ is c such that $c(i) = *$ for $i \in \{1, .., k\}$.

The game is played by two players: Spoiler and Duplicator. The play consists of (potentially) infinite sequence of rounds. In each round Spoiler selects a structure, $\mathcal{A}$ or $\mathcal{B}$, an element $v \in A$ (or $w \in B$) and $i \in \{1, .., k\}$. In such a case we say that Spoiler places a pebble p_i on v (on w). Then Duplicator selects an element $w \in B$ (or $v \in A$) and we say that he answers (p_i, w) (or (p_i, v)), or that the answer of Duplicator is w (or v). After the round the configuration is changed by $c(i) := (v, w)$.

Duplicator wins the game if after each round the configuration induces a partial isomorphism between $\mathcal{A}$ and $\mathcal{B}$, i.e. the relation $\bigcup_{i \in Peb} c(i) \subseteq A \times B$ is a partial isomorphism.

For a fixed configuration of a play of $Pebble_k(\mathcal{T}_1, \mathcal{T}_2)$, where $\mathcal{T}_1, \mathcal{T}_2 \in REGTREE$ we denote by $\mathbf{spread}^{\mathcal{T}_i}$ for $i \in \{1, 2\}$ a minimal substructure $\mathcal{S}$ of $\mathcal{T}_i$ such that S includes root of $\mathcal{T}_i$ and all marked vertices of $\mathcal{T}_i$.

For a fixed configuration of a play of $Pebble_k(\mathcal{T}_1, \mathcal{T}_2)$, where $\mathcal{T}_1, \mathcal{T}_2 \in DTREE$ and $t_1 \in T_1$, $t_2 \in T_2$, we denote by $\mathbf{spread}^{\mathcal{T}_i(t_i)}$ for $i \in \{1, 2\}$ a pointed tree $\mathcal{S}$ such that $S = \{v \in T_i : v \text{ is in the path from } t_i \text{ to } p_j^{\mathcal{T}_i} \text{ for some pebble } p_j\}$, $c^{\mathcal{S}} = t_i$ and $E^{\mathcal{S}} = E^{\mathcal{T}_i} \cap S \times S$.

3 Pebble Games on Directed Trees

3.1 Characterisation of Pebble Problem on Directed Trees

Let $k \in \mathcal{N}$. We define the relation $=_k \in \mathcal{N} \times \mathcal{N}$ by:

$$m =_k n \text{ iff } m, n \geq k \text{ or } m = n$$

Definition 1 *Let $\mathcal{S}(s), \mathcal{T}(t) \in PDTREE$, let R be an equivalence relation over $PDTREE$ and $k \in \mathcal{N}$. We denote by $\Delta_k(R, \mathcal{S}(s), \mathcal{T}(t))$ the following condition:*

- $E^{\mathcal{S}}(s,s)$ *iff* $E^{\mathcal{T}}(t,t)$, *and*
- *for every equivalence class $C \in PDTREE/R$:*
  $\left|\{ s' : E^{\mathcal{S}}(s,s'),\ not\ E^{\mathcal{S}}(s',s)\ and\ subtree(s') \in C \}\right| =_k$
  $\left|\{ t' : E^{\mathcal{T}}(t,t'),\ not\ E^{\mathcal{T}}(t',t)\ and\ subtree(t') \in C \}\right|$, *and*

  $\left|\{ s' : E^{\mathcal{S}}(s',s),\ not\ E^{\mathcal{S}}(s,s')\ and\ subtree(s') \in C \}\right| =_k$
  $\left|\{ t' : E^{\mathcal{T}}(t',t),\ not\ E^{\mathcal{T}}(t,t')\ and\ subtree(t') \in C \}\right|$, *and*

  $\left|\{ s' : E^{\mathcal{S}}(s,s')\ and\ E^{\mathcal{S}}(s',s), subtree(s') \in C \}\right| =_k$
  $\left|\{ t' : E^{\mathcal{T}}(t,t')\ and\ E^{\mathcal{T}}(t',t), subtree(t') \in C \}\right|$

For any set X let $Equiv(X)$ be the set of equivalence relations on X.

Definition 2 *For $k \in \mathcal{N}$ let $\Gamma_k : Equiv(PDTREE) \to Equiv(PDTREE)$ be the following mapping:*

$$\Gamma_k(R) = \{(\mathcal{S}, \mathcal{T}) \ :\ \Delta_k(R, \mathcal{S}, \mathcal{T})\}$$

The definition is correct ($\Gamma_k(R)$ is an equivalence relation for all k and R) since $=_k$ is an equivalence relation for all k.

Lemma 1 *For all $k \in \mathcal{N}$ the mapping Γ_k is monotonic.*

Proof.
Let $R_1, R_2 \in Equiv(PDTREE)$ and $\mathcal{S}, \mathcal{T} \in PDTREE$. Assume $R_1 \subseteq R_2$ and $(\mathcal{S}, \mathcal{T}) \in \Gamma_k(R_1)$. Let $X \in PDTREE/R_2$. Since $R_1 \subseteq R_2$, then $X = \bigcup Y$ for some $Y \subseteq PDTREE/R_1$. Since $\Delta_k(R_1, \mathcal{S}, \mathcal{T})$, then for all $Z \in Y$

$$\left|\{ s : E^{\mathcal{S}}(r^{\mathcal{S}}, s),\ \text{not}\ E^{\mathcal{S}}(s, r^{\mathcal{S}})\ \text{and}\ subtree(s) \in Z \}\right| =_k$$

$$\left|\{ t : E^{\mathcal{T}}(r^{\mathcal{T}}, t),\ \text{not}\ E^{\mathcal{T}}(t, r^{\mathcal{T}})\ \text{and}\ subtree(t) \in Z \}\right|$$

and obviously

$$\left|\{ s : E^{\mathcal{S}}(r^{\mathcal{S}}, s),\ \text{not}\ E^{\mathcal{S}}(s, r^{\mathcal{S}})\ \text{and}\ subtree(s) \in X \}\right| =_k$$

$$\left|\{ t : E^{\mathcal{T}}(r^{\mathcal{T}}, t),\ \text{not}\ E^{\mathcal{T}}(t, r^{\mathcal{T}})\ \text{and}\ subtree(t) \in X \}\right|$$

Similar conclusion we get for two other equalities from definition of Δ_k. Finally $\Delta_k(R_2, \mathcal{S}, \mathcal{T})$ and $(\mathcal{S}, \mathcal{T}) \in \Gamma_k(R_2)$. □

Lemma 2 *For any set A, the set $Equiv(A)$ of equivalence relations on A ordered by the inclusion relation is a complete lattice.*

Proof. It is enough to show that any subset of $Equiv(A)$ has a greatest lower bound. Let $X \subseteq Equiv(A)$, and $R = \bigcap X$. Obviously R is an equivalence relation. We show that R is a greatest lower bound of X. Obviously R is an lower bound, since $R \subseteq Y$ for all $Y \in X$. Let $R' \in Equiv(A)$ be an lower bound of X. Then for all $Y \in X$, $R' \subseteq Y$. Then $R' \subseteq R$. □

Note that a least upper bound in $Equiv(A)$ need not coincide with the set-theoretic union.

By Knaster-Tarski fixed point theorem we get:

Lemma 3 *For all $k \in \mathcal{N}$, Γ_k has a least and a greatest fixed point.* □

Let $k \in \mathcal{N}$ be fixed. Let $L_0, L_1, ...$ and $U_0, U_1, ...$ be sequences of equivalence relations on $PDTREE$ defined by:

- $L_0 := Id(PDTREE)$, $U_0 = PDTREE^2$
- $L_{i+1} = \Gamma_k(L_i)$, $U_{i+1} = \Gamma_k(U_i)$ for $i \in \mathcal{N}$

Lemma 4 *$L_0, L_1, ...$ is increasing and $U_0, U_1, ...$ is decreasing.* □

Denote $\bigcup\{L_i : i \in \mathcal{N}\}$ by L_∞ and $\bigcap\{U_i : i \in \mathcal{N}\}$ by U_∞.

Lemma 5 *Let R be a fixed point of Γ_k. Then $L_\infty \subseteq R$ and $R \subseteq U_\infty$.* □

Lemma 6 *$L_\infty = U_\infty$, consequently Γ_k has exactly one fixed point.*

Proof. Let $PDTREE_h$ be a set of pointed trees of height at most h. By induction w.r.t. i we show that $L_{i+1} \cap (PDTREE_i)^2 = U_{i+1} \cap (PDTREE_i)^2$:

- $L_1 \cap (PDTREE_0)^2 = Id(PDTREE_0) = U_1 \cap (PDTREE_0)^2$.
  (Warning $L_0 \cap (PDTREE_0)^2 = \{(a,a),(b,b)\} \neq \{(a,a),(b,b),(a,b),(b,a)\} = U_0 \cap (PDTREE_0)^2$, where $a = \langle\{0\}, E^a =, c^a = 0\rangle$ and $b = \langle\{0\}, E^b = \{(0,0)\}, c^b = 0\rangle$.)
- If $L_{i+1} \cap (PDTREE_i)^2 = U_{i+1} \cap (PDTREE_i)^2$, then for all $\mathcal{S}, \mathcal{T} \in PDTREE_{i+1}$, $\Delta_k(L_{i+1}, \mathcal{S}, \mathcal{T})$ iff $\Delta_k(L_{i+1}, \mathcal{S}, \mathcal{T})$, and then $L_{i+2}(\mathcal{S}, \mathcal{T})$ iff $U_{i+2}(\mathcal{S}, \mathcal{T})$, so $L_{i+2} \cap (PDTREE_{i+1})^2 = U_{i+2} \cap (PDTREE_{i+1})^2$

Since $L_i \subseteq U_i$ for all $i \in \mathcal{N}$ and by lemma 4 we get that $L_{i+1} \cap (PDTREE_i)^2 = L_\infty \cap (PDTREE_i)^2$ and $U_{i+1} \cap (PDTREE_i)^2 = U_\infty \cap (PDTREE_i)^2$ for all i i.e. $L_\infty \cap (PDTREE_i)^2 = U_\infty \cap (PDTREE_i)^2$ for all i. Since for every $\mathcal{S} \in PDTREE$ there is $i \in \mathcal{N}$ such that $\mathcal{S} \in PDTREE_i$ we get $L_\infty = U_\infty$. □

By lemma 3, 5 and 6 we get:

Lemma 7 *For all $k \in \mathcal{N}$, Γ_k has exactly one fixed point.* □

Let us denote this unique fixed point by $\equiv_k$.

Lemma 8 *Let $k \geq 3$, $\mathcal{S}, \mathcal{T} \in DTREE$ and $x_1, y_1 \in S, x_2, y_2 \in T$ such that $E^{\mathcal{S}}(x_1, y_1)$ and $E^{\mathcal{T}}(x_2, y_2)$.*
Assume that in the play of $Pebble_k(\mathcal{S}, \mathcal{T})$ the vertices x_1 and x_2 are marked by pebble p_i and y_1, y_2 by p_j. Assume also that the length of a longest path which begins with the vertices x_1, y_1 is different from the length of a longest path which begins with x_2, y_2. Then Spoiler wins the play.

Proof. By three pebbles - p_i, p_j and any third - Spoiler can "pass" across the longer path. □

Theorem 1 *Let $\mathcal{S}, \mathcal{T} \in DTREE$ and $k \geq 3$. Duplicator wins $Pebble_k(\mathcal{S}, \mathcal{T})$ iff there are $s \in S, t \in T$ such that $\mathcal{S}(s) \equiv_k \mathcal{T}(t)$.*

Proof.

$\Leftarrow$ Suppose that for some $s \in S$ and $t \in T$ $\mathcal{S}(s) \equiv_k \mathcal{T}(t)$.

Lemma 9 *Suppose that in the play of $Pebble_k(\mathcal{S}, \mathcal{T})$ the following condition is satisfied:*
($\star$) There is an isomorphism $f : spread^{\mathcal{S}(s)} \to spread^{\mathcal{T}(t)}$ such that $f(p_i^{\mathcal{S}}) = p_i^{\mathcal{T}}$ for all $i \in Peb$ and $subtree^{\mathcal{S}(s)}(x) \equiv_k subtree^{\mathcal{T}(t)}(f(x))$ for all $x \in spread^{\mathcal{S}(s)}$.
Then for every move of Spoiler there exists a move of Duplicator such that the requirement ($\star$) is preserved.

Proof. Assume $|Peb| < k$ (if not, Spoiler removes one pebble and then requirement ($\star$) is also preserved). Assume that Spoiler places a pebble p_i for some $i \notin Peb$ on $x \in S$. If $x \in spread^{\mathcal{S}(s)}$ then Duplicator answers $(p_i, f(x))$. Obviously the requirement ($\star$) is preserved (by the same isomorphism f). Assume then that $x \notin spread^{\mathcal{S}(s)}$. Let $a \in S$ be the last vertex in the path from s to x that belongs to $spread^{\mathcal{S}(s)}$.
Let $w \in T$ be a vertex computed by this algorithm:
1. $v := a$, $w := f(a)$
2. while $v \neq x$ do
   $v := c_v$, where $c_v \in children^{\mathcal{S}(s)}(v)$ and c_v is in the path from a to x.
   $w := c_w$, where $c_w \in children^{\mathcal{T}(t)}(w)$ such that:
   a. $E^{\mathcal{S}}(v, c_v)$ iff $E^{\mathcal{T}}(w, c_w)$ and $E^{\mathcal{S}}(c_v, v)$ iff $E^{\mathcal{T}}(c_w, w)$, and
   b. $subtree^{\mathcal{S}(s)}(c_v) \equiv_k subtree^{\mathcal{T}(t)}(c_w)$, and
   c. $c_w \notin spread^{\mathcal{T}(t)}$.

There exists a vertex $c_w \in children^{\mathcal{T}(t)}(w)$ in a loop which satisfy **a.** and **b.** because $subtree^{\mathcal{S}(s)}(v) \equiv_k subtree^{\mathcal{T}(t)}(w)$ is an invariant of this algorithm (and because $\equiv_k$ is a fixed point of Γ_k). Additionally we can choose a vertex satisfying **a**, **b.** and **c.** because the invariants of the algorithm are:
- $c_v \notin spread^{\mathcal{S}(s)}$
- every vertex in T has at most $k-1$ children in$spread^{\mathcal{T}(t)}$ (since $spread^{\mathcal{T}(t)}$ is spread by at most $k-1$ pebbles)

If Duplicator answers w, then the requirement ($\star$) will be preserved ($spread^{\mathcal{S}(s)}$ and $spread^{\mathcal{T}(t)}$ will be extended by all values of c_v and c_w from the algorithm and f will be extended by: $f(c_v) = c_w$ for every c_v and c_w computed by the algorithm in the same turn of the loop). □
Initial configuration of $Pebble_k(\mathcal{S}, \mathcal{T})$ obviously satisfies the requirement ($\star$). When the requirement ($\star$) is satisfied then the mapping $p_i^{\mathcal{S}} \mapsto p_i^{\mathcal{T}}$ is a partial isomorphism. So, by lemma 9 Duplicator wins the game.

$\Rightarrow$ Let $s \in S$ be a middle vertex of the longest path in S (any longest path; if its length is odd, then let it be any of the two middle vertices). For $x \in S$ we write *children*(x), *height*(x) and *subtree*(x) instead of $children^{S(s)}(x)$, $height^{S(s)}(x)$ and $subtree^{S(s)}(x)$ respectively. Let $h := height(s)$. Notice that there exist $s_1, s_2 \in children(s)$, $s_1 \neq s_2$ such that $height(s_1) \geq h-2 \leq height(s_2)$.
Let Spoiler place a pebble p_1 on s. Let $t \in \mathcal{T}$ be an answer of Duplicator. Similarly, for $x \in \mathcal{T}$ we write *children*(x), *height*(x) and *subtree*(x) instead of $children^{\mathcal{T}(t)}(x)$, $height^{\mathcal{T}(t)}(x)$ and $subtree^{\mathcal{T}(t)}(x)$ respectively.
If $height(t) < h$ then Spoiler places a pebble p_2 on $s' \in children(s)$ such that $height(s') = h - 1$ and then he wins by lemma 8. If $height(t) > h$, Spoiler plays similarly. Assume then that $height(t) = h$. If there is no $t_1, t_2 \in children(t)$, $t_1 \neq t_2$ such that $height(t_1) \geq h - 2 \leq height(t_2)$ then Spoiler places pebbles p_2 and p_3 on s_1 and s_2 respectively, and then he also wins by lemma 8. Assume then that such a vertices exist (i.e. t is a middle vertex of the longest path in $\mathcal{T}$).
Assume that $S(s) \not\equiv_k \mathcal{T}(t)$.
Let Spoiler play according to this interactive algorithm:
1. $v := s$, $w := t$.
2. If not $E^S(v,v)$ iff $E^{\mathcal{T}}(w,w)$ then Spoiler wins
3. Let $X \subseteq S$, $Y \subseteq \mathcal{T}$ be such that:
   - $X = \{ x \in children(v) \ : \ cond^S(v,x) \text{ and } subtree(x) \in C \}$, $Y = \{ y \in children(w) \ : \ cond^{\mathcal{T}}(w,y) \text{ and } subtree(y) \in C \}$ for some $C \in PDTREE/\equiv_k$ and $cond^A(a,b)$ – one of the following conditions: 1. $E^A(a,b)$ and not $E^A(b,a)$, 2. $E^A(b,a)$ and not $E^A(a,b)$, 3. $E^A(a,b)$ and $E^A(b,a)$.
   - $|X| \neq_k |Y|$

   Without loss of generality we can assume that $|X| < |Y|$. Then obviously $|X| < k$. Now Spoiler makes a sequence of moves $(w_1,q_1),..,(w_{|X|+1},q_{|X|+1})$, where $w_1,..,w_{|X|+1} \in Y$, $w_i \neq w_j$ for $i \neq j$ and $q_1,..,q_{|X|+1} \in \{p_1,..,p_k\}$ is any sequence of different pebbles such that neither q_1 nor q_2 are placed on w.
   Let $(v_1,q_1),..,(v_{|X|+1},q_{|X|+1})$ be an answer of Duplicator. If any of v_j is not a child of v or $v_i = v_j$ for $i \neq j$ then Duplicator loses the game.
   Let $j \in \{1,..,|X|+1\}$ be such that $v_j \notin X$ (such j obviously exists).
   Let $v := v_j$ and $w := w_j$ and repeat this step of algorithm.

Notice that ever at the beginning of step 2. of the algorithm we have:
- Vertices $v \in S$ and $w \in \mathcal{T}$ are marked by the same pebble.
- $subtree(v) \not\equiv_k subtree(w)$

and by that Spoiler wins the game when he plays according to above algorithm (since a relation of being a child in a tree is well founded). □

3.2 Complexity of Pebble Problem on Directed Trees

Theorem 2 *The problem Pebble[DTREE] is in PTIME. More precisely, there exists algorithm that for $k \in \mathcal{N}$ and $S, \mathcal{T} \in DTREE$ solves the problem Pebble$_k(S,\mathcal{T})$ in time $O(log(k) * (|S| + |\mathcal{T}|)^3)$.*

Proof.
The algorithm of deciding whether Duplicator wins $Pebble_k(\mathcal{S},\mathcal{T})$:

- For every middle vertex s of the longest path in $\mathcal{S}$ and every middle vertex t of the longest path in $\mathcal{T}$ check whether $\mathcal{S}(s) \equiv_k \mathcal{T}(t)$.

The algorithm of checking if $\mathcal{S}(s) \equiv_k \mathcal{T}(t)$:

1. For every $v \in children(s) \cup children(t)$, $l_v^{\mathcal{S}(s)} := 0$ and $l_v^{\mathcal{T}(t)} := 0$.
2. For every $v \in children^X(x)$ and $w \in children^Y(y)$, where $(x,X),(y,Y) \in \{(s,\mathcal{S}(s)),(t,\mathcal{T}(t))\}$ such that $E^X(x,v)$ iff $E^Y(y,w)$ and $E^X(v,x)$ iff $E^Y(w,y)$ check recursively whether $subtree(v) \equiv_k subtree(w)$ and if so then increase the counters l_v^Y and l_w^X by 1 (any counter increase only if it is less then k).
3. If for every $v \in children(s) \cup children(t)$, $l_v^{\mathcal{S}(s)} =_k l_v^{\mathcal{T}(t)}$ then $\mathcal{S} \equiv_k \mathcal{T}$. Otherwise $\mathcal{S}(s) \not\equiv_k \mathcal{T}(t)$.

In any directed tree there may be at most two middle vertices of the longest path. Finding these vertices in $\mathcal{S}$ and $\mathcal{T}$ has complexity $O((|S|+|T|)^2)$. Complexity of checking if $\mathcal{S}(s) \equiv_k \mathcal{T}(t)$ is estimated by $O((|S|+|T|)^3 * log(k))$, since for every pair of vertices in $S \cup T$ algorithm is executed at most once (number of such pairs is $(|S|+|T|)^2$) and complexity of one execution is estimated by $O((|S|+|T|) * log(k))$ (without calculating of recursive executions; the factor $log(k)$ appears because the counters l_v^X are less then $k+1$ and can be represented by $O(log(k))$ bits).

Let $R \subseteq PDTREE^2$ be a relation computed by the algorithm. From the structure of the algorithm after step 2. we have:

$$l_v^X = |\{w \in children^X(x) \ : \ R(subtree(v), subtree(w)) \text{ and }$$

$$E^Y(y,v) \text{ iff } E^X(x,w) \text{ and } E^Y(v,y) \text{ iff } E^X(w,x)\}|$$

for all $v \in children(y)$, where $(x,X),(y,Y) \in \{(s,\mathcal{S}(s)),(t,\mathcal{T}(t))\}$.
So R is a fixed point of Γ_k and then $R = \equiv_k$. □

4 Pebble Problem on Rooted Trees

4.1 Characterisation of Pebble Problem on Rooted Trees

For a rooted tree $\mathcal{S}$ and $v \in S$ we denote by $childtrees^{\mathcal{S}}(v)$ (or shortly by $childtrees(v)$) the set $\{subtree^{\mathcal{S}}(w) : w \in children^{\mathcal{S}}(v)\}$.

Definition 3 *Let $\mathcal{S},\mathcal{T} \in RTREE$ and $k \in \mathcal{N}$. Let $R \subseteq RTREE^2$ be an equivalence relation. We denote by $\Delta_k(R,\mathcal{S},\mathcal{T})$ the following condition:*

- *For all $X \in RTREE/R$:*
  - $|childtrees(r^{\mathcal{S}}) \cap X| =_k |childtrees(r^{\mathcal{T}}) \cap X|$

Definition 4 *For $k \in \mathcal{N}$ let $\Gamma_k : Equiv(RTREE) \to Equiv(RTREE)$ be the following mapping:*

$$\Gamma_k(R) = \{(\mathcal{S}, \mathcal{T}) \ : \ \Delta_k(R, \mathcal{S}, \mathcal{T})\}$$

The definition is correct ($\Gamma_k(R)$ is an equivalence relation for all k and R) since $=_k$ is an equivalence relation for all k.

Lemma 10 *For all $k \in \mathcal{N}$ the mapping Γ_k is monotonic.*

Proof.
Let $R_1, R_2 \in Equiv(RTREE)$ and $\mathcal{S}, \mathcal{T} \in RTREE$. Assume $R_1 \subseteq R_2$ and $(\mathcal{S}, \mathcal{T}) \in \Gamma_k(R_1)$. Let $X \in RTREE/R_2$. Since $R_1 \subseteq R_2$, then $X = \bigcup Y$ for some $Y \subseteq RTREE/R_1$. Since $\Delta_k(R_1, \mathcal{S}, \mathcal{T})$, then for all $Z \in Y$

$$|childtrees(r^{\mathcal{S}}) \cap Z| =_k |childtrees(r^{\mathcal{T}}) \cap Z|$$

and obviously

$$|childtrees(r^{\mathcal{S}}) \cap X| =_k |childtrees(r^{\mathcal{T}}) \cap X|$$

Finally $\Delta_k(R_2, \mathcal{S}, \mathcal{T})$ and $(\mathcal{S}, \mathcal{T}) \in \Gamma_k(R_2)$. □

By Knaster-Tarski fixed point theorem, for all $k \in \mathcal{N}$, Γ_k has a greatest fixed point. Let us denote it by $\equiv_k$.

Theorem 3 *Let $k \geq 3$ and $\mathcal{S}, \mathcal{T} \in RTREE$. Duplicator wins $Pebble_k(\mathcal{S}, \mathcal{T})$ iff $\mathcal{S} \equiv_k \mathcal{T}$.*

Proof. Since $\equiv_k$ is the greatest fixed point of Γ_k then by Knaster-Tarski fixed point theorem, $\equiv_k = \bigvee\{\alpha \in Equiv(RTREE) : \alpha \subseteq \Gamma_k(\alpha)\}$. So, to prove that $R = \equiv_k$ for some equivalence relation R, it is sufficient to show that $\equiv_k \subseteq R$ and $R \subseteq \Gamma_k(R)$.

Let $R = \{(\mathcal{S}, \mathcal{T}) :$ Duplicator wins $Pebble_k(\mathcal{S}, \mathcal{T})\}$. Obviously R is an equivalence relation.

- $\mathbf{\equiv_k \subseteq R}$
  Suppose that $\mathcal{S} \equiv_k \mathcal{T}$.

  Lemma 11 *Suppose that in the play of $Pebble_k(\mathcal{S}, \mathcal{T})$ the following condition is satisfied:*

  ($\star$) *There is an isomorphism $f : spread^{\mathcal{S}} \to spread^{\mathcal{T}}$ such that $f(p_i^{\mathcal{S}}) = p_i^{\mathcal{T}}$ for all $i \in Peb$ and $subtree^{\mathcal{S}}(x) \equiv_k subtree^{\mathcal{T}}(f(x))$ for all $x \in spread^{\mathcal{S}}$.*

  Then for every move of Spoiler there exists a move of Duplicator such that the requirement ($\star$) is preserved.

Proof. Assume $|Peb| < k$ (if not, Spoiler removes one pebble and then requirement $(\star)$ is also preserved). Assume that Spoiler places a pebble p_i for some $i \notin Peb$ on $x \in S$. If $x \in spread^{\mathcal{S}}$ then Duplicator answers $(p_i, f(x))$. Obviously the requirement $(\star)$ is preserved (by the same isomorphism f). Assume then that $x \notin spread^{\mathcal{S}}$. Let $a \in S$ be the last vertex in the path from $r^{\mathcal{S}}$ to x that belongs to $spread^{\mathcal{S}}$.

Let $w \in T$ be a vertex computed by this algorithm:

1. $v := a$, $w := f(a)$
2. while $v \neq x$ do
   $v := c_v$, where $c_v \in children^{\mathcal{S}}(v)$ and c_v is in the path from a to x.
   $w := c_w$, where $c_w \in children^{\mathcal{T}}(w)$ such that:
   a. $subtree^{\mathcal{S}}(c_v) \equiv_k subtree^{\mathcal{T}}(c_w)$, and
   b. $c_w \notin spread^{\mathcal{T}}$.

There exists a vertex $c_w \in children^{\mathcal{T}}(w)$ in a loop which satisfies **a.** because $subtree^{\mathcal{S}}(v) \equiv_k subtree^{\mathcal{T}}(w)$ is an invariant of this algorithm (and because $\equiv_k$ is a fixed point of Γ_k). Additionally we can choose a vertex satisfying both **a** and **b.** because the invariants of the algorithm are:

- $c_v \notin spread^{\mathcal{S}}$
- every vertex in T has at most $k-1$ children in $spread^{\mathcal{T}}$ (since $spread^{\mathcal{T}}$ is spread by at most $k-1$ pebbles)

If Duplicator answers w, then the requirement $(\star)$ will be preserved ($spread^{\mathcal{S}}$ and $spread^{\mathcal{T}}$ will be extended by all values of c_v and c_w from the algorithm and f will be extended by: $f(c_v) = c_w$ for every c_v and c_w computed by the algorithm in the same turn of the loop). □

Initial configuration of $Pebble_k(\mathcal{S}, \mathcal{T})$ obviously satisfies the requirement $(\star)$. When requirement $(\star)$ is satisfied then the mapping $p_i^{\mathcal{S}} \mapsto p_i^{\mathcal{T}}$ is a partial isomorphism. So, by lemma 11 Duplicator wins the game.

- $\mathbf{R \subseteq \Gamma_k(R)}$

  Assume that Duplicator wins $Pebble_k(\mathcal{S}, \mathcal{T})$. Moreover suppose that $(\mathcal{S}, \mathcal{T}) \notin \Gamma_k(R)$, i.e. there is $X \in RTREE/R$ such that $|childtrees(r^{\mathcal{S}}) \cap X| \neq_k |childtrees(r^{\mathcal{T}}) \cap X|$. Let us denote $\{v \in children(r^{\mathcal{S}}) : subtree(v) \in X\}$ by A and $\{w \in children(r^{\mathcal{T}}) : subtree(w) \in X\}$ by B. Without loss of generality we can assume that $|A| < |B|$, hence $|A| < k$. Suppose Spoiler puts $|A|+1$ different pebbles on different vertices from B. Then Duplicator has to answer by putting $|A|+1$ pebbles on different children of $r^{\mathcal{S}}$. Let $v \in children(r^{\mathcal{S}}), w \in children(r^{\mathcal{T}})$ be the vertices marked by the same pebble, such that $(subtree(v), subtree(w)) \notin R$. It means that Spoiler has a winning strategy in $Pebble_k(subtree(v), subtree(w))$. We can assume that in this strategy always some pebble remains present in both trees and that the game starts from positions v and w marked by the same pebble. Then this strategy expanded to $(\mathcal{S}, \mathcal{T})$ is a winning strategy of Spoiler – contradiction. It shows that $(\mathcal{S}, \mathcal{T}) \in \Gamma_k(R)$, that is $R \subseteq \Gamma_k(R)$. □

4.2 Complexity of Pebble Problem on Regular Trees

Let $k \in \mathcal{N}$ be fixed. Let $R_0, R_1, \ldots$ be a sequence of equivalence relations on $REGTREE$ defined by:

- $R_0 = REGTREE^2$
- $R_{i+1} = \Gamma_k(R_i)$

Lemma 12 *$R_0, R_1, \ldots$ is decreasing.*

Proof. By induction w.r.t. i:

- $R_1 \subseteq REGTREE^2 = R_0$
- If $R_i \subseteq R_{i-1}$ then $R_{i+1} = \Gamma_k(R_i) \subseteq \Gamma_k(R_{i-1}) = R_i$ (since Γ_k is monotonic).

□

Let us denote $\bigcap\{R_i : i \in \mathcal{N}\}$ by R_∞

Lemma 13 *Let R be a fixed point of Γ_k. Then $R \subseteq R_\infty$.*

Proof. By induction w.r.t. i we show that $R \subseteq R_i$:

- $R \subseteq REGTREE^2 = R_0$
- If $R \subseteq R_i$ then $R = \Gamma_k(R) \subseteq \Gamma_k(R_i) = R_{i+1}$ (since Γ_k is monotonic). □

Lemma 14 *R_∞ is a fixed point of Γ_k.*

Proof. Since Γ_k is monotonic and $R_\infty \subseteq R_i$ for all $i \in \mathcal{N}$, then $\Gamma_k(R_\infty) \subseteq \Gamma_k(R_i) = R_{i+1}$ i.e. for all $i \geq 1$ $\Gamma_k(R_\infty) \subseteq R_i$. Since $\Gamma_k(R_\infty) \subseteq REGTREE^2 = R_0$ then $\Gamma_k(R_\infty) \subseteq R_i$ for all $i \in \mathcal{N}$ and then $\Gamma_k(R_\infty) \subseteq R_\infty$.

Let $(\mathcal{S}, \mathcal{T}) \in R_\infty$. Then for all $i \in \mathcal{N}$ $(\mathcal{S}, \mathcal{T}) \in R_i$ and then for all $i \in \mathcal{N}$ $(\mathcal{S}, \mathcal{T}) \in R_{i+1} = \Gamma_k(R_i)$ i.e. $\Delta_k(R_i, \mathcal{S}, \mathcal{T})$. Let $i \in \mathcal{N}$ be such that $R_i \cap (Dom(\mathcal{S}) \cup Dom(\mathcal{T}))^2 = R_\infty \cap (Dom(\mathcal{S}) \cup Dom(\mathcal{T}))^2$. Then obviously $\Delta_k(R_\infty, \mathcal{S}, \mathcal{T})$ and then $(\mathcal{S}, \mathcal{T}) \in \Gamma_k(R_\infty)$. □

By lemmas 13 and 14 we obtain:

Lemma 15 *$\equiv_k = R_\infty$.* □

Let $\mathcal{S}, \mathcal{T} \in REGTREE$ be fixed. We denote $Dom(\mathcal{S}) \cup Dom(\mathcal{T})$ shortly by Dom.

Definition 5 *Let $\mathcal{A}, \mathcal{B} \in Dom$ and $R \in Equiv(Dom)$. We denote by $\Delta'_k(R, \mathcal{A}, \mathcal{B})$ the following condition:*

- *For all $X \in Dom/R$:*
  - $|childtrees(r^{\mathcal{A}}) \cap X| =_k |childtrees(r^{\mathcal{B}}) \cap X|$

Definition 6 *Let $\Gamma'_k : Equiv(Dom) \to Equiv(Dom)$ be the following mapping:*

$$\Gamma'_k(R) = \{(\mathcal{A}, \mathcal{B}) \ : \ \Delta'_k(R, \mathcal{A}, \mathcal{B})\}$$

The definition is correct ($\Gamma'_k(R)$ is an equivalence relation for all k and R) since $=_k$ is an equivalence relation for all k.

Let $R'_0, R'_1, \ldots$ be a sequence of equivalence relations on *Dom* defined by:

- $R'_0 = Dom^2$
- $R'_{i+1} = \Gamma'_k(R_i)$

Lemma 16 *For all $i \in \mathcal{N}$, $R'_i = R_i \cap Dom$.*

Proof. By induction w.r.t. i:

- $R'_0 = Dom = R_0 \cap Dom$.
- Assume $R'_i = R_i \cap Dom$. Then obviously for all $\mathcal{S}, \mathcal{T} \in Dom$, $\Delta_k(R_i, \mathcal{S}, \mathcal{T})$ iff $\Delta'_k(R'_i, \mathcal{S}, \mathcal{T})$ i.e. $(\mathcal{S}, \mathcal{T}) \in R'_{i+1}$ iff $(\mathcal{S}, \mathcal{T}) \in R_{i+1}$. So $R'_{i+1} = R_{i+1}$. □

We represent a regular tree $\mathcal{T}$ by the finite automaton $\mathcal{S}$. A state of $\mathcal{S}$ is a pair (t, i) such that $t \in Dom(\mathcal{T})$ and $1 \leq i \leq n$, where n is the maximal number of siblings in $\mathcal{T}$ with subtrees t. We denote the set of states in $\mathcal{S}$ by S. We let $(t_1, i_1) \to (t_2, i_2)$ if the root of t_1 has at least i_2 children with subtrees t_2. The initial state in $\mathcal{S}$ is $(\mathcal{T}, 1)$ and we denote it by $p_{\mathcal{S}}$.

In the next theorem we consider the regular trees as representing them automata. In particular the size $|S|$ of regular tree $\mathcal{S}$ is the number of states in the automaton representing $\mathcal{S}$.

Theorem 4 *The problem Pebble[REGTREE] is in PTIME. More precisely, there exists algorithm that for $k \in \mathcal{N}$ and $\mathcal{S}, \mathcal{T} \in REGTREE$ solves problem $Pebble_k(\mathcal{S}, \mathcal{T})$ in time $O(log(k) * (|S| + |T|)^6)$.*

Proof.
Algorithm of deciding whether $\mathcal{S} \equiv_k \mathcal{T}$:

1. For every $p, q \in S \cup T$ do $R(p, q) := true$
2. While $\Gamma'_k(R) \neq R$ do $R := \Gamma'_k(R)$
3. If $R(p_{\mathcal{S}}, p_{\mathcal{T}})$ then $\mathcal{S} \equiv_k \mathcal{T}$, otherwise $\mathcal{S} \not\equiv_k \mathcal{T}$

Algorithm of computing $\Gamma'_k(R)$ in step 2:

- For every $p, q \in S \cup T$ do
  a. For every r such that $p \to r$ or $q \to r$ do $l^p_r := 0$ and $l^q_r := 0$
  b. For every inordered pair (r_0, r_1) such that $x_0 \to r_0$ and $x_1 \to r_1$ where $x_0, x_1 \in \{p, q\}$, if $R(r_0, r_1)$ then increase $l^{x_1}_{r_0}$ and $l^{x_0}_{r_1}$ by 1 (increase $l^{x_{1-i}}_{r_i}$ only if it is less then k; if $x_0 = x_1$ and $r_0 = r_1$ then increase the counter only once).
  c. If for all r such that $p \to r$ or $q \to r$, $l^p_r = l^q_r$ then $\Gamma'_k(R)(p, q) := true$, else $\Gamma'_k(R)(p, q) := false$.

Remark that after step b. the value of counter l^p_r is the number of children of p which are in the relation R with r.
Correctness of the algorithm follows from lemmas 15 and 16.

Complexity of the algorithm:

- Initialization in step 1. has complexity $O(|S| + |T|)^2$.
- In step 2. there is at most $(|S| + |T|)^2$ turns of the loop, since $\Gamma'_k(R) \subseteq R$.
- Computing of $\Gamma'_k(R)$ has complexity $O(log(k)*(|S|+|T|)^4)$. The factor $log(k)$ appears because the counters l^p_r are less than $k + 1$ and can be represented by $O(log(k))$ bits.
- Checking whether $\Gamma'_k(R) = R$ has complexity $O(|S| + |T|)^2$.
- Substitution $R := \Gamma'_k(R)$ has complexity $O(|S| + |T|)^2$.

Total complexity of the algorithm is then $O(log(k) * (|S| + |T|)^6)$. □

5 Conclusion

We have shown that given a number k and two finite directed trees, or two regular trees, one can decide in polynomial time if Duplicator wins the k-pebbles Barwise game on these structures. Directed trees can be viewed as a most general class of tree-shaped directed graphs. (It generalizes undirected trees as well as rooted trees.) We hope that the techniques developed in this paper can be used for extending the result to determine the complexity of pebble problem on wider classes of tree-like structures e.g. graphs of bounded tree-width. A step in this direction was made in [6] where we show that one can decide in polynomial time if Duplicator wins the k-pebbles Barwise game (for given number k) on given two graphs with explicit tree decompositions (represented by binary relations).

References

1. Barwise, J.: *On Moschovakis closure ordinals.* Journal of Symbolic Logic 42 (1977), 292–296
2. Grohe, M.: *Equivalence in finite-variable logics is complete for polynomial time.* Proceeding FOCS (1996)
3. Ebbinghaus, H.-D., Flum, J.: *Finite Model Theory.* Perspectives in Mathematical Logic. Springer, Berlin (1995)
4. Pezzoli, E.: *Computational Complexity of Ehrenfeucht-Fraïsse Games on Finite Structures.* Computer Science Logic, LNCS 1584 (1998), 159–170.
5. Pezzoli, E.: *On the computational complexity of type-two functionals and logical games on finite structures.* Ph.D thesis. Stanford University. June 1998
6. Krzeszczakowski, Ł.: *On the complexity of the problem of finding a winning strategy in model-theoretic games.* (in Polish) M.Sc thesis. Warsaw University. September 2001

Bistability: An Extensional Characterization of Sequentiality

James Laird

COGS, University of Sussex, UK
jiml@cogs.susx.ac.uk

Abstract. We give a simple order-theoretic construction of a cartesian closed category of sequential functions. It is based on biordered sets analogous to Berry's bidomains, except that the stable order is replaced with a new notion, the bistable order, and instead of preserving stably bounded greatest lower bounds, functions are required to preserve bistably bounded least upper bounds and greatest lower bounds. We show that bistable cpos and bistable and continuous functions form a CCC, yielding models of functional languages such as the simply-typed λ-calculus and SPCF. We show that these models are strongly sequential and use this fact to prove universality and full abstraction results.

1 Introduction

A longstanding problem in domain theory has been to find a simple characterization of higher-order sequential functions which is wholly extensional in character. Typically, what is sought is some form of mathematical structure, such that set-theoretic functions which preserve this structure are sequential *and* can be used to construct a cartesian closed category. Clearly, any solution to this problem is dependant on what one means by sequential. It has been closely associated with the full abstraction problem for PCF, although it is now known that PCF sequentiality cannot be characterized effectively in this sense [8, 11]. As an alternative, we have the strongly stable, or sequentially realizable functionals [2, 12]. Here the difficulty is perhaps to understand precisely the sense in which these are sequential at higher types, in the absence of direct connections to a sequential programming language or an explicitly sequential model.

Another notion of sequentiality — the observably sequential functionals — was discovered by Cartwright and Felleisen [3]. They observed that if one or more errors are added to a functional language, then the order of evaluation of programs becomes observable by varying their inputs. Thus each function corresponds to a unique evaluation tree or sequential algorithm [4], which can be reconstructed from its graph. The observably sequential functionals do form a cartesian closed category, which contains a fully abstract model of SPCF — PCF with errors and a simple control operator. However, the definitions of observably sequential functions and sequential algorithms are based implicitly or explicitly on (Kahn-Plotkin) sequentiality, and hence they cannot offer a characterization

M. Baaz and J.A. Makowsky (Eds.): CSL 2003, LNCS 2803, pp. 372–383, 2003.

of it in the above sense. So we may refine our original problem to ask whether there is a simple, order-theoretic characterization of observable sequentiality, as a preservation property for set-theoretic functions.

This paper proposes a solution to this problem. We will construct a cartesian closed category of biordered sets, and show that it contains universal models of the simply-typed λ-calculus and finitary SPCF, and fully abstract models of SPCF. The basis of our model is a kind of biorder analogous to Berry's bidomains [1], except that the stable order used in the latter has been replaced with a "bistable order". There are several reasons for this step. Although the bidomain model of PCF is not sequential, even at first order types, the bidomain model of *unary* PCF is sequential, and universal [9]. The proof of this result relies on the fact that there is a "top" element at each type. The connection with observably sequential functions is made by viewing top as an error element. Under this interpretation, the monotone and stable functions on bidomains are not observably sequential, because they are not "error-propagating" (i.e. sequential with respect to $\top$ as well as $\bot$). However, the duality between $\bot$ and $\top$ suggests that we "symmetrize" the stable order, which leads to the bistable order, and the notion of bistable function. The latter preserve both the least upper bounds and greatest lower bounds of bistably bounded finite sets and are observably sequential, as we establish by universality and full abstraction results.

2 Bistable Biorders

We shall introduce bistability in the context of a form of bidomain. Although the bistable order can be defined independently of the extensional order, it is the interaction between the extensional order and bistable order in biorders that will be used to characterize sequentiality

Definition 1. *A* bistable biorder *is a tuple $\langle |D|, \sqsubseteq_D, \leq_D \rangle$ consisting of a set $|D|$ with two partial orders (an* extensional *order $\sqsubseteq$, and a* bistable *order $\leq$), such that:*

- *$(|D|, \sqsubseteq)$ has a least element $\bot$ and a greatest element $\top$, such that $\bot \leq \top$, although $\bot$ and $\top$ are not in general least and greatest in $(|D|, \leq)$.*
- *$a, b \in (|D|, \leq)$ are bounded above if and only if they are bounded below (if a and b are so bounded, we shall write $a \updownarrow b$),*
- *if $a \updownarrow b$ then there are elements $a \wedge b, a \vee b \in |D|$ which are (respectively) the greatest lower bound and least upper bound of a and b with respect to both orders,*
- *if $a \updownarrow b$, $a \updownarrow c$ and $b \updownarrow c$, then $a \vee (b \wedge c) = (a \vee b) \wedge (a \vee c)$ (and so $a \wedge (b \vee c) = (a \vee b) \wedge (a \vee c)$.*

Note that $\leq$ is included in $\sqsubseteq$ since if $a \leq b$ then $a \wedge b = a$ is a glb for a, b with respect to $\sqsubseteq$.

Thus a bistable biorder may be presented as a tuple $(|D|, \sqsubseteq_D, \updownarrow_D)$ where $(|D|, \sqsubseteq)$ is a partial order with least and greatest elements, $\updownarrow$ is an equivalence relation, and for each $x \in |D|$, the equivalence class $([x], \sqsubseteq\restriction [x])$ is a *distributive lattice* (having greatest and least elements, in many cases, but not in general). This is equivalent to the above definition, under the interpretation $x \leq y$ if $x \updownarrow y$ and $x \sqsubseteq y$.

Simple examples of bistable biorders include the one-point order, **1** (in which $\top = \bot$), and the *Sierpinski Space* Σ, which contains distinct $\top$ and $\bot$ elements such that $\bot \leq \top$ (and so $\bot \sqsubseteq \top$). The product and separated sum operations on bistable orders are straightforward (although note that the latter adds *two* new points, $\top$ and $\bot$).

Definition 2. *Let $\{A_i \mid i \in I\}$ be a family of bistable biorders. The product $\Pi_{i \in I} A_i$ is defined pointwise:*

- $|\Pi_{i \in I} A_i| = \Pi_{i \in I} |A_i|$,
- $x \sqsubseteq_{\Pi_{i \in I} A_i} y$ *if* $\pi_i(x) \sqsubseteq \pi_i(y)$ *for all* $i \in I$,
- $x \leq_{\Pi_{i \in I} A_i} y$ *if* $\pi_i(x) \leq \pi_i(y)$ *for all* $i \in I$.

The separated sum $\Sigma_{i \in I} A_i$ is defined:

- $|\Sigma_{i \in I} A_i| = (\coprod_{i \in I} |A_i|) \cup \{\bot, \top\}$,
- $x \sqsubseteq_{\Sigma_{i \in I} A_i} y$ *if* $x = \bot$ *or* $y = \top$ *or* $x = \mathtt{in}_i(z_1)$ *and* $y = \mathtt{in}_i(z_2)$, *where* $z_1 \sqsubseteq_i z_2$,
- $x \leq_{\Sigma_{i \in I} A_i} y$ *if* $x = \bot$ and $y = \top$, *or* $x = \mathtt{in}_i(z_1)$ *and* $y = \mathtt{in}_i(z_2)$, *where* $z_1 \leq_i z_2$.

2.1 Bistable Functions

We shall now construct a CCC of bistable biorders and monotone and bistable functions.

Definition 3. *A function $f : D \to E$ is* monotone *if for all $x, y \in |D|$, $x \sqsubseteq_D y$ implies $f(x) \sqsubseteq_E f(y)$.*
f is bistable *if it is monotone with respect to the bistable order, and for all $x, y \in |D|$, if $x \updownarrow y$ then $f(x \wedge y) = f(x) \wedge f(y)$ and $f(x \vee y) = f(x) \vee f(y)$.*

Thus a monotone function f is bistable if $x \updownarrow y$ implies $f(x) \updownarrow f(y)$ and for each x, $f\restriction [x]_\updownarrow$ is a lattice homomorphism.

Definition 4. *We shall say that a bistable and monotone function f is* strict *if f prserves the least upper bound and greatest lower bound of the empty set — i.e. $f(\top) = \top$ and $f(\bot) = \bot$.*

We now construct a bistable biorder of functions, in which the extensional order is standard, and the bistable order is a symmetric version of the stable order.

Definition 5. *Given bistable biorders D, E, we define the function-space $D \Rightarrow E$ as follows:*

- $|D \Rightarrow E|$ *is the set of monotone and bistable functions from* D *to* E,
- $f \sqsubseteq_{D \Rightarrow E} g$ *if for all* $x, y \in |D|$, $x \sqsubseteq_D y$ *implies* $f(x) \sqsubseteq_E f(y)$,
- $f \leq_{D \Rightarrow E} g$ *if for all* $x, y \in |D|$ *such that* $x \leq_D y$, $f(y) \updownarrow g(x)$ *and* $f(x) = f(y) \wedge g(x)$ *and* $g(y) = f(y) \vee g(x)$.

Lemma 1. *$D \Rightarrow E$ is a bistable biorder.*

Proof. If $f \updownarrow g$ we can define $f \wedge g$ and $f \vee g$ by:
$(f \wedge g)(a) = f(a) \wedge g(a)$ and $(f \vee g)(a) = f(a) \vee g(a)$ for all a.
$f \wedge g$ and $f \vee g$ are monotone and bistable — e.g. to show that if $a \updownarrow b$ then $(f \wedge g)(a) \vee (f \wedge g)(b) = (f \wedge g)(a \vee b)$ suppose $f, g \leq h$. Then since $a, b \leq a \vee b$, $f(a) = f(a \vee b) \wedge h(a), f(b) = f(a \vee b) \wedge h(b), g(b) = g(a \vee b) \wedge h(a), g(b) = g(a \vee b) \wedge h(b)$ so:
$(f \wedge g)(a) \vee (f \wedge g)(b) = (f(a) \wedge g(a)) \vee (f(b) \wedge g(b)) = (f(a \vee b) \wedge h(a) \wedge g(a \vee b) \wedge h(a)) \vee (f(a \vee b) \wedge h(b) \wedge g(a \vee b) \wedge h(b)) = (h(a) \wedge ((f \wedge g)(a \vee b))) \vee (h(b) \wedge ((f \wedge g)(a \vee b))) = (h(a) \vee h(b)) \wedge ((f \wedge g)(a \vee b)) = h(a \vee b) \wedge ((f \wedge g)(a \vee b)) = (f \wedge g)(a \vee b)$, since $(f \wedge g)(a \vee b) \leq h(a \vee b)$.

$f \wedge g$ and $f \vee b$ are glbs and lubs with respect to $\sqsubseteq$ and $\leq$: e.g. to show $f \wedge g \leq f$, suppose $a \leq b$, then $f(a) = f(b) \wedge h(a)$, $g(a) = g(b) \wedge h(a)$ and $h(a) = f(b) \vee h(a) = g(b) \vee h(a)$ and so:
$(f \wedge g)(a) = f(a) \wedge g(a) = (h(a) \wedge f(b)) \wedge (h(a) \wedge g(b)) = (h(a) \wedge f(b)) \wedge g(b) = f(a) \wedge g(b) = f(a) \wedge ((f \wedge g)(b))$ and $f(a) \vee (f \wedge g)(b) = f(a) \vee (f(b) \wedge g(b)) = f(b) \wedge (f(a) \vee g(b)) = f(b) \wedge ((h(a) \wedge f(b)) \vee g(b)) = f(b) \wedge ((h(a) \vee g(b)) \wedge (f(b) \vee g(b))) = f(b) \wedge (h(b) \wedge (f(b) \vee g(b))) = f(b) \wedge (f(b) \vee g(b)) = f(b)$ as required.

Thus if $\mathcal{BBO}$ is the category of bistable biorders and monotone and bistable functions, it is now straightforward to show the following.

Proposition 1. *$(\mathcal{BBO}, \mathbf{1}, \times, \Rightarrow)$ is cartesian closed.*

The separated sum is a co-product in the subcategory of $\mathcal{BBO}$ consisting of bistable biorders and strict functions.

2.2 Bicpos

We shall now extend our notion of bistable biorder with a notions of completeness and continuity.

Definition 6. *A bistable bicpo is a bistable biorder D such that $(|D|, \sqsubseteq)$ is a cpo (i.e. every $\sqsubseteq$-directed set has a $\sqsubseteq$-least upper bound), and the following properties hold:*

i *If $X, Y \subseteq |D|$ are $\sqsubseteq$-directed sets such that for all $x \in X$ there exists $y \in Y$ such that $x \leq y$, and if $y \in Y$ there exists $x \in X$ such that $x \leq y$ then $\bigsqcup X \leq \bigsqcup Y$.*

ii *For any $X \subseteq |D| \times |D|$ such that $\pi_1(X), \pi_2(X)$ are directed and for all $\langle y, z \rangle \in X$, $y \updownarrow z$:*
$(\bigsqcup_{y \in \pi_1(X)} y) \wedge (\bigsqcup_{z \in \pi_2(X)} z) = \bigsqcup_{\langle y, z \rangle \in X} (y \wedge z))$, *and*
$(\bigsqcup_{y \in \pi_1(X)} y) \vee (\bigsqcup_{z \in \pi_2(X)} z) = \bigsqcup_{\langle y, z \rangle \in X} (y \vee z))$.

Let $\mathcal{BBC}$ be the category of bistable bicpos and (extensionally) continuous and bistable functions.

Proposition 2. *$\mathcal{BBC}$ is a cartesian closed category.*

Proof. We use the additional properties to show that for any directed set F of functions from A to B, a bistable and continuous least upper bound can be defined pointwise — $(\bigsqcup F)(a) = \bigsqcup\{f(a) \mid f \in F\}$.
E.g. if $a \leq b$ then $(\bigsqcup F)(a) \leq (\bigsqcup F)(b)$ by property (i) since for all $f \in F$, $f(a) \leq f(b)$.
$\bigsqcup F$ is bistable, since if $a \updownarrow b$ then for each $f \in F$, $f(a) \updownarrow f(b)$ and hence applying property (ii) to the set $\{\langle f(a), f(b)\rangle \mid f \in F\}$, we have:
$(\bigsqcup F)(a \sqcap b) = \bigsqcup_{f \in F} f(a \sqcap b) = \bigsqcup_{f \in F}(f(a) \sqcap f(b)) = (\bigsqcup_{f \in F} f(a)) \sqcap (\bigsqcup_{f \in F} f(b))$, etcetera.

We must also check that the exponential preserves properties (i) and (ii): e.g. given directed sets of functions F, G such that for all $f \in F$ there exists $g \in G$ such that $f \leq g$ and for all $g \in G$ there exists f such that $f \leq g$, then $\bigsqcup F \leq \bigsqcup G$. Firstly, for all a, $(\bigsqcup F)(a) \leq (\bigsqcup G)(a)$ by property (i). Now suppose $a \leq b$. Let $X = \{\langle f(b), g(a)\rangle \mid f \in F \wedge g \in G \wedge f \leq g\}$. Then $\pi_1(X) = \{f(b) \mid f \in F\}$ and $\pi_2(X) = \{g(a) \mid g \in G\}$ and hence by property (ii), $(\bigsqcup F)(a) = \bigsqcup_{\langle f,g\rangle \in X}(f(b) \wedge g(a)) = (\bigsqcup F)(b) \wedge (\bigsqcup G)(a)$, and so on.

2.3 Bistability and Bisequentiality

We may now give a simple proof that the monotone and bistable functions into Σ are *bisequential* (i.e. sequential with respect to both $\bot$ and $\top$ elements), and also *strongly sequential* (having a unique sequentiality index).

Definition 7. *Let $\{A_i \mid i \in I\}$ and B be bistable biorders. A function $f : \Pi_{i \in I} A_i \to B$ is i-strict if $f(x) = \bot$ whenever $\pi_i(x) = \bot$ and $f(x) = \top$ whenever $\pi_i(x) = \bot$.*

Lemma 2. *Let $\{A_i \mid i \in I\}$ be a finite family of bistable biorders. If $f : \Pi_{i \in I} A_i \to \Sigma$ is a strict, monotone and bistable function, then f is i-strict for some $i \in I$ (and providing $\Pi_{i \in I} A_i$ has at least two points, i is unique).*

Proof. Given $j \in I$, let $\bot[\top]_j = \langle x_i \mid i \in I\rangle$, where $x_i = \top$ if $i = j$, and $x_i = \bot$ otherwise. Similarly $\top[\bot]_j = \langle x_i \mid i \in I\rangle$, where $x_i = \bot$ if $i = j$, and $x_i = \top$ otherwise.

Then if $\pi_i(x) = \bot$, $x \sqsubseteq \top[\bot]_i$, and if $\pi_i(x) = \top$, $\bot[\top]_i \sqsubseteq x$, and so f is i-strict if $f(\top[\bot]_i) = \bot$, and $f(\bot[\top]_i) = \top$. Since $\top[\bot]_j \updownarrow \top[\bot]_k$ for all j, k, and $\bigwedge_{i \in I} \top[\bot]_i = \bot$, we have $f(\bigwedge_{i \in I} \top[\bot]_i) = f(\bot) = \bot$, and so $f(\top[\bot]_i) = \bot$ for some i, and similarly $f(\bigvee_{i \in I}(\bot[\top]_i) = \top$, and so $f(\bot[\top]_j) = \top$ for some j. Moreover, if $i \neq j$, then $\bot[\top]_j \sqsubseteq \top[\bot]_i$, and so $i = j$ as required.

In the case where A_i is the Sierpinski order, Σ, there is exactly one i-strict function $f : \Pi_{i \in I} A_i \to \Sigma$ — the ith projection. Thus for each indexing set i, the "biflat" domain $\Sigma_{i \in I}\mathbf{1}$ is isomorphic to $\Sigma^I \Rightarrow I$. (i.e. $(\Pi_{i \in I}\Sigma) \to \Sigma$.

Lemma 3. *For $I \subseteq \omega$, $\Sigma^I \Rightarrow \Sigma \cong \Sigma_{i \in I}\mathbf{1}$.*

Proof. The map which associates the ith projection with $\mathtt{in}_i(\top)$ for each i is an isomorphism of bistable biorders. (The proof extends to $I = \omega$ by continuity.)

The extensional and bistable orderings on $\Sigma \times \Sigma \Rightarrow \Sigma$ are depicted in Figure 1.

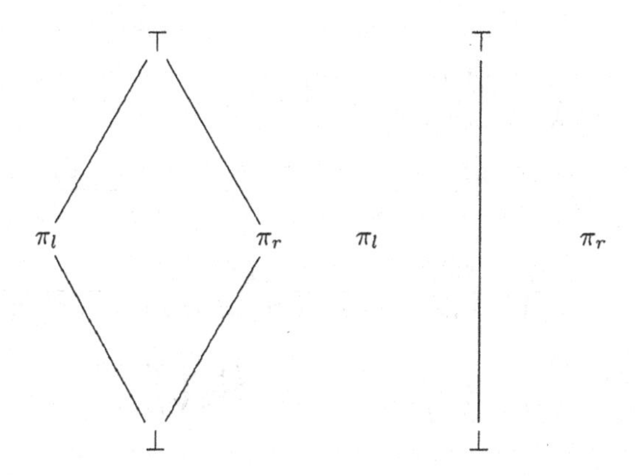

Fig. 1. Extensional and bistable orderings on $\Sigma \times \Sigma \Rightarrow \Sigma$

3 Universality and Full Abstraction for SPCF+

We shall now describe a bistable biorder model of SPCF+, which is a version of Cartwright and Felleisen's SPCF [3] with (infinitary) lifted sums and products, and a single error, $\top$, at each type. This may be thought of either as a simple functional programming language with non-local control, or as a term language for certain bistable biorders. We will show that the finitary fragment of the model is *universal* — i.e. every element is the denotation of a term. Full abstraction follows from universality for the finite fragment, as we have a complete and order-extensional model. The presence of separated sums is significant for our proof of universality, as we shall show that every finite type-object in the bistable biorder model is a (definable) retract of a type-object built from sums and products. We can, however, prove universality for restricted fragments of SPCF+ using straightforward syntactic arguments.

Let SPCF+ be the language with types generated by the following grammar (where $I \leq \omega$, and $T_i \ : \ i < I$ is a sequence of types):

$$T ::= \Pi_{i<I}T_i \mid \Sigma_{i<I}T_i \mid T_1 \Rightarrow T_2$$

We will sometimes write $\mathbf{1}$ for the empty product ($I = 0$), Σ for the empty sum, and $\overline{I}$ for $\Sigma_{i<I}\mathbf{1}$ (the "flat" domain containing I values). The types of finitary SPCF+ are those generated from finite products and sums ($I < \omega$). Terms are generated from the λ-calulus with products, lifted sums, and fixpoint operators, plus a catch operator from $\Sigma^I \Rightarrow \Sigma$ to $\overline{I}$ which sends the ith projection to the numeral $\bar{i} = \mathtt{in}_i(\top)$. From this we can derive a more general form of catch of type $(\Pi_{i\leq I} A_i \Rightarrow \Sigma) \Rightarrow \bar{I}$ which sends any i-strict function to $\bar{i}$, as follows: catch M = catch $\lambda x.(M \langle(\mathsf{case}_0\, \pi_i(x))\, \top \mid i < I\rangle)$.

Table 1. Typing judgements for SPCF+

$$\frac{}{\Gamma\vdash\top{:}T,\bot{:}T} \qquad \frac{}{\Gamma,x{:}T\vdash x{:}T}$$

$$\frac{\Gamma,x{:}S\vdash M{:}T}{\Gamma\vdash\lambda x.M{:}S\Rightarrow T} \qquad \frac{\Gamma\vdash M{:}S\Rightarrow T \quad \Gamma\vdash N{:}S}{\Gamma\vdash M\,N{:}T}$$

$$\frac{\forall i{<}I.\Gamma\vdash M_i{:}T_i}{\Gamma\vdash\langle M_i \mid i{<}I\rangle{:}\Pi_{i<I}T_i} \qquad \frac{\Gamma\vdash M{:}\Pi_{i<I}T_i \quad j{<}I}{\Gamma\vdash\pi_j(M){:}T_j}$$

$$\frac{\Gamma\vdash M{:}T_i \quad i{<}I}{\Gamma\vdash\mathtt{in}_i(M){:}\Sigma_{j<I}T_j} \qquad \frac{\Gamma\vdash M{:}\Sigma_{j<I}T_j}{\Gamma\vdash\mathsf{case}_I M{:}\Pi_{i<I}(T_i{\Rightarrow}S){\Rightarrow}S}$$

$$\frac{\Gamma\vdash(\Pi_{i<I}\Sigma){\Rightarrow}\Sigma}{\Gamma\vdash\mathrm{catch}_I M{:}\bar{I}} \qquad \frac{\Gamma\vdash M{:}T{\Rightarrow}T}{\Gamma\vdash\mathbf{Y}M{:}T}$$

We may give a simple operational semantics for SPCF programs — closed terms of type Σ — using *evaluation contexts.*

Definition 8. *Evaluation contexts of SPCF are given by the following grammar:*

$$E[\cdot] ::= [\cdot] \mid E[\cdot]\,M \mid \mathsf{case}\,E[\cdot] \mid \pi_i E[\cdot] \mid \mathbf{Y}E[\cdot]$$

The small-step operational semantics of SPCF programs is as follows:

$$\begin{gathered}
E[\top] \longrightarrow \top\\
E[\lambda x.M\,N] \longrightarrow E[M[N/x]]\\
E[\pi_j(\langle M_i \mid i < I\rangle)] \longrightarrow E[M_j]\\
E[(\mathsf{case}\,\mathtt{in}_i(M))\,N] \longrightarrow E[\pi_i(N)\,M]\\
E[\mathrm{catch}_I\,M] \longrightarrow M\,\langle E[\bar{i}] \mid i < I\rangle\\
E[\mathbf{Y}\,M] \longrightarrow E[M\,(\mathbf{Y}M)]
\end{gathered}$$

For a program M we write $M \Downarrow$ if $M \twoheadrightarrow \top$.

The interpretation of types as bistable bicpos and terms as bistable and continuous maps is wholly straightforward since we have a cartesian closed category with lifted sums, least fixed points of directed sets with which to interpret the fixed-point combinator, and the isomorphism from $\Pi_{i\leq I}(\Sigma) \Rightarrow \Sigma$ to $\Sigma_{i<n}\mathbf{1}$ (Lemma 3) with which to interpret catch.

Proposition 3. *$M \Downarrow$ if and only if $[\![M]\!] = \top$.*

Proof. Standard: Soundness is proved by induction, using the fact that each evaluation context represents a strict map. Adequacy is proved using a computability predicate argument.

To show that the bistable biorder model is universal for finitary types, we shall show that each of the latter is the retract of a *ΠΣ type* — i.e a type generated by the grammar:

$$P ::= \Sigma_{i<I} P_i \mid \Pi_{i<I} P_i$$

Lemma 4. *The bistable biorder model of finitary SPCF+ is universal for the ΠΣ types (i.e. element of a finite ΠΣ type is the denotation of a term).*

Proof. This is a straightforward induction on type structure.

We now use the bisequentiality of bistable functions to prove the following key lemma.

Lemma 5. *For any finite family of bistable biorders* $\{A_{ij} \mid i \in I, j \in J_i\}$*, there is an isomorphism*

$$\Pi_{i\in I}\Sigma_{j\in J_i} A_{ij} \Rightarrow \Sigma \cong \Sigma_{k\in I}\Pi_{l\in J_k}((A_{kl} \times \Pi_{i\in I-\{k\}}\Sigma_{j\in J_i} A_{ij}) \Rightarrow \Sigma)$$

Proof. Any monotone and bistable function $f : \Pi_{i\in I}\Sigma_{j\in J_i} A_{ij} \to \Sigma$ is either constant or strict. So by Lemma 2, f is either $\top$, $\bot$, or k-strict for some unique $k \in I$, and thus corresponds by currying to a unique strict function $f_k : \Sigma_{j\in J_k} A_{kj} \to (\Pi_{i\in I-\{k\}}\Sigma_{j\in J_i} A_{ij}) \Rightarrow \Sigma$. Because the separated sum is a co-product in the category of strict maps, $f_i = [\mathtt{in}_l; f_k \mid l \in J_k]$. Finally, $\mathtt{in}_l; f_k : A_{kl} \to (\Pi_{i\in I-\{k\}}\Sigma_{j\in J_i} A_{ij}) \Rightarrow \Sigma$ uncurries to $g_l : (A_{kl} \times \Pi_{i\in I-\{k\}}\Sigma_{j\in J_i} A_{ij}) \to \Sigma$.

Thus there is a bijection from $\Pi_{i\in I}\Sigma_{j\in J_i} A_{ij} \Rightarrow \Sigma$ to $\Sigma_{k\in I}\Pi_{l\in J_k}((A_{kl} \times \Pi_{i\in I-\{k\}}\Sigma_{j\in J_i} A_{ij}) \Rightarrow \Sigma)$ which sends $\top$ to $\top$, $\bot$ to $\bot$, and each k-strict f to $\mathtt{in}_k(\langle g_l \mid l \in j_k\rangle)$, where g_l is such that if $\pi_k(x) = \mathtt{in}_l(y_{kl})$, then $f(x) = g_l(\langle y_{kl}, \langle \pi_i(x) \mid i \in I - \{k\}\rangle\rangle)$. Moreover, this is an isomorphism of bistable biorders.

Given two SPCF+ types S, T, we shall say that there is a definable retraction from S to T if $[\![S]\!] \trianglelefteq [\![T]\!]$, and this retraction is the denotation of SPCF+ terms of type $S \Rightarrow T$ and $T \Rightarrow S$ (if it is an isomorphism, we say that S and T are definably isomorphic).

Lemma 6. *If SPCF+ is universal at type* T*, and there is a definable retraction from* S *to* T*, then SPCF+ is universal at type* S*.*

Proof. Suppose $\mathtt{in} : [\![S]\!] \trianglelefteq [\![T]\!] : \mathtt{proj}$. Given an element $e \in [\![S]\!]$, we have a term $M_{\mathtt{in}(e)} : T$ such that $[\![M_{\mathtt{in}(e)}]\!] = \mathtt{in}(e)$ and thus if $[\![\mathsf{OUT} : S \Rightarrow T]\!] = \mathtt{proj}$ then $e = [\![\mathsf{OUT}\ M_{\mathtt{in}(e)}]\!]$.

We shall now show that the isomorphism described in Lemma 5 is definable.

Definition 9. *Given a finitary SPCF+ type* $T = \Pi_{i<n}\Sigma_{j<m_i} T_{ij} \Rightarrow \Sigma$*, and* $k < n$*,* $l < m_k$*, let* $S_{kl} = \Pi_{i<k}\Sigma_{j<m_i} T_{ij} \times T_{kl} \times \Pi_{i<n-k}\Sigma_{j<m_{i+k}} T_{(i+k)j}$.

Lemma 7. T *is definably isomorphic to* $\Sigma_{k<n}\Pi_{l<m_k}(S_{kl} \Rightarrow \Sigma)$.

Proof. The defining terms for the isomorphism of Lemma 5 are, from T to S:

$$\lambda f.(\mathsf{case}(\mathsf{catch}(f))\ \langle \mathtt{in}_i(\langle M_{i,j} \mid j < m_i\rangle) \mid i < n\rangle)$$

where $M_{i,j} = \lambda x.f\ \langle \pi_0(x), \ldots, \pi_{i-1}(x), \mathtt{in}_j(\pi_i(x)), \pi_{i+1}(x), \ldots, \pi_{n-1}(x)\rangle$, and from S to T:

$$\lambda xy.(\mathsf{case}(x)\ \langle \lambda u.\mathsf{case}(\pi_i(y))\ \langle N_{i,j} \mid j < m_i\rangle \mid i < n\rangle)$$

where $N_{i,j} = \lambda v.(\pi_j(u)\ \langle \pi_0(y), \ldots, \pi_{i-1}(y), v, \pi_{i+1}(y), \ldots, \pi_{n-1}(y)\rangle)$.

Lemma 8. *If T is a finitary $\Pi\Sigma$ type, then $T \Rightarrow \Sigma$ is definably isomorphic to a $\Pi\Sigma$ type.*

Proof. This is by induction on the size of T (as a syntax tree), for which the base case is trivial. For the induction step we may assume T has the form $\Pi_{i<n}\Sigma_{j<m_i}T_{ij}$. Then $T \Rightarrow \Sigma$ is definably isomorphic to $\Sigma_{k<n}\Pi_{l<m_k}(S_{kl} \Rightarrow \Sigma)$ by Lemma 7. Since the size of each S_{kl} is less than that of T, the inductive hypothesis now applies.

Now let the SPCF types (i.e. types in which the only sums are the atomic types) be those generated by the grammar:

$$S, T ::= \overline{I} \mid S \Rightarrow T$$

Proposition 4. *Every SPCF type is definably isomorphic to a $\Pi\Sigma$ type.*

Proof. By induction on type-structure. For the induction step, suppose $T = S_0 \Rightarrow \ldots S_{n-1} \Rightarrow \overline{I}$. Then T is definably isomorphic to $((\Pi_{i<n}S_i) \times \Sigma^I) \Rightarrow \Sigma$. By induction each S_i is definably isomorphic to a $\Pi\Sigma$ type, and by Lemma 8, so is T.

We now use a definable retraction to extend universality to all finitary SPCF+ types.

Lemma 9. *For any family of types $\{T_i \mid i < I\}$, there is a definable retraction $\Sigma_{i<I}T_i \trianglelefteq \overline{I} \times \Pi_{i<I}T_i$*

Proof. The defining terms are, from $\Sigma_{i<I}T_i$ to $\overline{I} \times \Pi_{i<I}T_i$:

$$\lambda x.((\mathsf{case}(x))\ \langle \lambda y.\langle \overline{i}, \bot, \ldots \bot, y, \bot, \ldots, \bot\rangle \mid i < I\rangle)$$

and from $\overline{i} \times \Pi_{i<I}T_i$ to $\Sigma_{i<I}T_i$:

$$\lambda x.((\mathsf{case}\ \pi_1(x))\ \langle \mathtt{in}_i(\pi_{i+1}(x)) \mid i < I\rangle)$$

So every finitary $\Pi\Sigma$ is the retract of a product of ground types.

Proposition 5. *Every finitary SPCF+ type is a definable retract of a $\Pi\Sigma$ type.*

Proof. We use the retraction of Lemma 9 to show that every finitary SPCF+ type is the definable retract of a product of SPCF types, which is definably isomorphic to a $\Pi\Sigma$ type.

(In fact, using a more general version of Lemma 5, we can show that every finitary SPCF+ type is isomorphic to a $\Pi\Sigma$ type.) By Lemma 6 we have the following.

Corollary 1. *The bistable biorder model of finitary SPCF+ is universal.*

We may prove further universality results for fragments of SPCF+, such as (finitary) SPCF, or the simply-typed λ-calculus over the single ground type Σ (the bistable model of the latter is equivalent to the "minimal model" characterized syntactically by Padovani [13]). To do so, it suffices to show that every element of a type-object of the restricted language is definable as a term *of which every subterm is in the language.* For SPCF and the minimal model, this can be achieved using straightforward Tait-Girard style reducibility arguments.

We could extend universality to all types either by constructing an effectively presented model, or by allowing infinitely deep types and terms. However, universality for the finite fragment is sufficient to prove full abstraction for SPCF+ by showing that each type-object is the limit of a chain of retractions which are SPCF definable.

Lemma 10. *For each SPCF+ type S there is a sequence of finitary types S_i : $i \in \omega$ and SPCF-definable retractions:* $\mathtt{in}_i : [\![S^i]\!] \trianglelefteq [\![S]\!] : \mathtt{proj}_i$ *such that if $e \in [\![S]\!]$ then $e = \bigsqcup_{i \in \omega} \mathtt{in}_i(\mathtt{proj}_i(e))$.*

Proof. We define:

- $(\Sigma_{j<n} T_j)^i = \Sigma_{j<i}(T_j)^i$, if $i < n$, $(\Sigma_{j<n} T_j)^i = \Sigma_{j<n}(T_j)^i$, otherwise.
- $(\Pi_{j<n} T_j)^i = \Pi_{j<i}(T_j)^i$, if $i < n$, $(\Pi_{j<n} T_j)^i = \Pi_{j<n}(T_j)^i$, otherwise,
- $(S \Rightarrow T)^i = S^i \Rightarrow T^i$.

Recall that, given terms $M, N : T$, we say that M observationally approximates N ($M \lesssim N$) if for all compatible program contexts $C[\cdot]$, $C[M] \Downarrow$ implies $C[N] \Downarrow$.

Proposition 6. *The bistable bicpo model of SPCF+ is fully abstract — i.e. for all terms M, N, $M \lesssim N$ if and only if $[\![M]\!] \sqsubseteq [\![N]\!]$.*

Proof. Equational soundness follows from adequacy in the standard way. To prove completeness — that for all $M, N : S$, $N \lesssim M$ implies $[\![N]\!] \sqsubseteq [\![M]\!]$, suppose $[\![N]\!] \not\sqsubseteq [\![M]\!]$. Then $[\![N]\!] = \bigsqcup_{i \in \omega} \mathtt{in}_i(\mathtt{proj}_i([\![N]\!])) \not\sqsubseteq \bigsqcup_{i \in \omega} \mathtt{in}_i(\mathtt{proj}_i([\![M]\!])) = [\![M]\!]$. So for some i, $\mathtt{proj}_i([\![N]\!]) = [\![\mathsf{OUT}_i\ N]\!] \not\sqsubseteq \mathtt{proj}_i([\![M]\!])) = [\![\mathsf{OUT}_i\ M]\!]$, and so by universality of the finite fragment, there is a term $L : S^i \Rightarrow \Sigma$ such that $[\![L\ (\mathsf{OUT}_i\ N)]\!] \neq \bot$ and $[\![L\ (\mathsf{OUT}_i\ M)]\!] = \bot$ and so $L\ (\mathsf{OUT}_i\ N) \Downarrow$ and $L\ (\mathsf{OUT}_i\ M) \not\Downarrow$ as required.

4 Conclusions

A natural question concerns the connections between bistable bidomains and more intensional characterizations of the same type structure, such as observably sequential domains, sequential algorithms, and Hyland and Schalk's "graph games" model [7]. It is fairly straightforward to show that sequential algorithms over sequential data structures with a single error act as bistable functions on their arguments, and Curien [6] has shown that the category of sequential data structures and sequential algorithms embeds *fully* in the category of bistable biorders and bistable functions. However, this leaves open the question of how the correspondence works in the opposite direction; what is the image of the embedding, and given an object in that image, can we construct the corresponding sequential data structures and sequential algorithms? The underlying linear structure of the sequential algorithms model also suggests a further question; it is equivalent to the co-Kleisli category of a certain co-monad [10, 5] on a symmetric monoidal category of games and strategies. Can we provide a parallel "linear decomposition" of a CCC of bistable domains, and describe a purely extensional model of intuitionistic linear logic equivalent to the games model?

To answer to these questions, we have investigated a notion of "locally boolean" domain — a partial order (the extensional order) with an involutive negation, which can be used to give simple definitions of the stable and bistable orders. Our fundamental representation result for these domains is that they can all be generated (up to isomorphism) by taking products and co-products, lifting, and limits of ω-chains. In particular, each pointed domain is isomorphic to one of the form $\Sigma_{i\in I}\Pi_{j\in J_i}A_{ij}$, which is a more general version of Proposition 4 (every type-object of SPCF has this form).

References

1. G. Berry. Stable models of typed λ-calculi. In *Proceedings of the 5th International Colloquium on Automata, Languages and Programming*, number 62 in LNCS, pages 72–89. Springer, 1978.
2. A. Bucciarelli and T. Ehrhard. A theory of sequentiality. *Theoretical Computer Science*, 113:273–292, 1993.
3. R. Cartwright and M. Felleisen. Observable sequentiality and full abstraction. In *Proceedings of POPL '92*, 1992.
4. R. Cartwright, P.-L. Curien and M. Felleisen. Fully abstract semantics for observably sequential languages. *Information and Computation*, 1994.
5. P.-L. Curien. On the symmetry of sequentiality. In *Mathematical Foundations of Computer Science*, number 802 in LNCS. Springer, 1993.
6. P.-L. Curien. Sequential algorithms as bistable maps. Unpublished note, 2002.
7. Martin Hyland and Andrea Schalk. Games on graphs and sequentially realizable functionals. In *Proceedings of LICS '02*. IEEE Press, 2002.
8. G. Kahn, and G. Plotkin. Concrete domains. *Theoretical Computer Science*, Böhm Festschrift special issue, 1993. First appeared as technical report 338 of INRIA-LABORIA, 1978.

9. J. Laird. A fully abstract and effectively presentable model of unary FPC. To appear in the proceedings of TLCA '03, 2003.
10. F. Lamarche. Sequentiality, games and linear logic. In *Proceedings, CLICS workshop, Aarhus University*. DAIMI-397–II, 1992.
11. R. Loader. Finitary PCF is undecidable. *Annals of Pure and Applied Logic*, 2000.
12. J. Longley. The sequentially realizable functionals. Technical Report ECS-LFCS-98-402, LFCS, Univ. of Edinburgh, 1998.
13. V. Padovani. Decidability of all minimal models. In M. Coppo and S. Berardi, editor, *Types for proofs and programs*, volume 1158 of *LNCS*. Springer, 1996.

Automata on Lempel-Ziv Compressed Strings

Hans Leiß[1] and Michel de Rougemont[2]

[1] Universität München, CIS, D-80538 München, Germany
leiss@cis.uni-muenchen.de
[2] Université Paris-II & LRI Bâtiment 490, F-91405 Orsay Cedex, France
mdr@lri.fr

Abstract. Using the Lempel-Ziv-78 compression algorithm to compress a string yields a dictionary of substrings, i.e. an edge-labelled tree with an order-compatible enumeration, here called an *LZ*-trie. Queries about strings translate to queries about *LZ*-tries and hence can in principle be answered without decompression. We compare notions of automata accepting *LZ*-tries and consider the relation between acceptable and MSO-definable classes of *LZ*-tries. It turns out that regular properties of strings can be checked efficiently on compressed strings by *LZ*-trie automata.

1 Introduction

We are interested in the *compressed model checking problem*: which properties of strings can be checked given the compressed strings? The challenge is to beat the decompress-and-then-check method. We restrict ourselves to the classical Ziv-Lempel[8] string compression algorithm *LZ*-78. It compresses a string $w \in \Sigma^+$ to a generally much shorter sequence $LZ(w) \in (\mathbb{N} \times \Sigma)^+$, where the numbers point to previous elements of the sequence.

As usual, we view a string w as a colored finite linear order $\mathcal{S}_w = (D, <, U_a)_{a\in\Sigma}$, where $D = \{0, \ldots, |w| - 1\}$ is the set of positions, ordered by $<$ as usual, and $U_a(i)$ means that a occurs at position i of w. Properties of strings are expressed in first-order (FO) or second-order (SO) logic of colored linear orders. In a similar way, we give two representations of *LZ*-78-compressed strings α by relational structures, one by node-labelled *LZ-graphs* $\mathcal{G}_\alpha$ and another one by *LZ-tries* $\mathcal{T}_\alpha$, which are a kind of edge-labelled trees.

A natural approach to the compressed model checking problem is to translate properties of strings to properties of compressed strings. In fact, monadic second-order (MSO) formulas φ in the language of strings can be translated to dyadic second-order (DSO) formulas φ^{LZ} in the language of *LZ*-graphs $\mathcal{G}_\alpha$. One can therefore answer queries φ about $\mathcal{S}_w$ by evaluating φ^{LZ} on the smaller structure $\mathcal{G}_{LZ(w)}$. However, since the translation doubles the arity of relation variables, this is not guaranteed to provide an efficient solution.

By Büchi's well-known theorem (cf. [4], Theorem 5.2.3), MSO for strings is equally expressive as regular expressions, or finite automata, are. This raises the question whether MSO for *LZ*-graphs leads to a notion of *LZ*-automaton that provides an efficient method for checking a reasonably rich class of properties of compressed strings.

M. Baaz and J.A. Makowsky (Eds.): CSL 2003, LNCS 2803, pp. 384–396, 2003.

We study this question in the slightly more suitable format of LZ-tries. These come with an enumeration of their nodes and can be viewed as simple acyclic directed graphs. We introduce notions of *LZ-trie automata* by modifying corresponding notions of tree-automata. We show that deterministic bottom-up LZ-trie-automata are less powerful than non-deterministic ones, and that the latter capture ∃-MSO for LZ-tries, where ∃-MSO is the set of MSO-formulas of the form $\exists \boldsymbol{X} \psi$ where $\boldsymbol{X}$ are set variables and ψ is an FO-formula. Finally, we show that MSO-properties of strings can be checked efficiently on LZ-compressed strings by deterministic top-down LZ-automata. Thus, regular expression search in strings can be done on the LZ-compressed strings, without decompression.

2 Lempel-Ziv-78 Compression

We fix a finite alphabet Σ and, to avoid structures with empty universe, only consider non-empty strings $w \in \Sigma^+$. The classical Lempel-Ziv-78 compression algorithm has many variations (cf. [2], [3]). It decomposes a string $w \in \Sigma^+$ into a sequence of substrings or *blocks* $B_i \in \Sigma^+$, so that $w = B_0 \cdots B_{m-1}$. The first block B_0 consists of the first letter of w. Suppose for some $n > 0$, we have constructed blocks $B_0, \ldots, B_{n-1}$ such that $w = B_0 \cdots B_{n-1}v$ for some $v \in \Sigma^+$. Then B_n is the shortest non-empty prefix of v that is not among $\{B_0, \ldots, B_{n-1}\}$, if this exists[1], otherwise B_n is v. The *LZ-compression* $LZ(w)$ of w is the sequence $p_0 \cdots p_{m-1}$ of pairs $p_n = (k, a)$ such that $B_n = B_{k-1}a$ (where $B_{-1} := \epsilon$) and $w = B_0 \cdots B_{m-1}$. The decompression is given by $B_k = \mathit{decode}(p_k)$ where $\mathit{decode}((0,a)) = a$ and $\mathit{decode}((n+1,a)) = \mathit{decode}(p_n)a$.

Example 1. The blocks of $w = abbbaabbabbb$ are $a.b.bb.aa.bba.bbb$, and its compression is $LZ(w) = (0,a)(0,b)(2,b)(1,a)(3,a)(3,b)$.

2.1 LZ-Graphs

Definition 1. *A compressed string* $\alpha = p_0 \cdots p_{m-1}$ *is represented as a* finite labelled ordered graph

$$\mathcal{G}_\alpha := (D_m, <, U_a, E)_{a \in \Sigma},$$

where $m = |\alpha|$ *is the number of blocks,* $D_m = \{0, \ldots, m-1\}$, $<$ *the natural order on* D_m, $U_a(i)$ *is true iff the last letter of block* B_i *is* a. *The binary relation* E *describes the reference to previous pairs: if* $p_i = (k, a)$ *for some* $a \in \Sigma$ *and* $k > 0$, *(i.e. the* k*-th block* B_{k-1} *is the longest strict prefix of* B_i*), then there is an edge* $E(i, k-1)$ *from node* i *to* $k-1$.

Example 1 (Cont.). For $w = a.b.bb.aa.bba.bbb. = B_0B_1B_2B_3B_4B_5$, the graph $\mathcal{G}_{LZ(w)}$ is

[1] and then B_n extends one of $B_0, \ldots, B_{n-1}$ by exactly one letter.

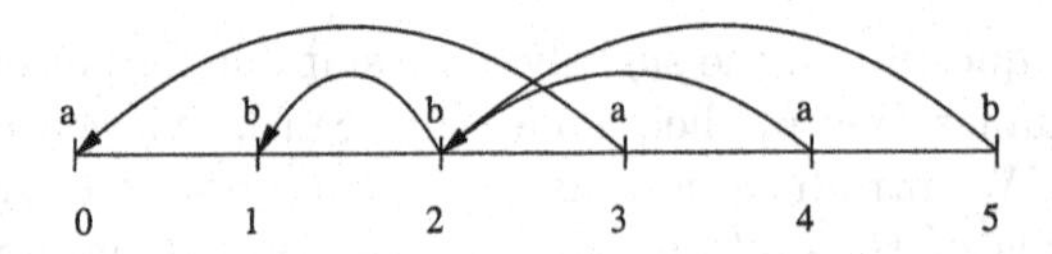

Observe that $\mathcal{S}_w$ can be interpreted in $\mathcal{G}_{LZ(w)}$ as a binary relation: a position i in w is mapped to the pair $h(i) = (k, j)$ in $LZ(w)$ iff i lies in block B_k and B_j is the nonempty prefix of B_k ending in i. Note that in $\mathcal{G}_{LZ(w)}$, node j can be reached from k by a path of E-edges.

Example 2. If $w = a.aa.ab.aba.aa$, the positions 5,6,7 occurring in block $B_3 = aba$ are represented by $(3,0), (3,2), (3,3)$, because the nonempty prefixes of B_3 are $a = B_0$, $ab = B_2$, and $aba = B_3$.

Theorem 1 ([1]). *For every MSO-formula $\varphi(x_1, \ldots, x_n, \boldsymbol{X}^{(1)})$ about strings there is a DSO formula $\varphi^{LZ}(x_1, y_1, \ldots, x_n, y_n, \boldsymbol{X}^{(2)})$ about LZ-graphs, such that for each $\mathcal{S}_w$ and all $i_1, \ldots, i_n \in \mathcal{S}_w$ and $\boldsymbol{S} \subseteq \mathcal{S}_w$,*

$$\mathcal{S}_w \models \varphi[\boldsymbol{i}, \boldsymbol{S}] \iff \mathcal{G}_{LZ(w)} \models \varphi^{LZ}[h(\boldsymbol{i}), h(\boldsymbol{S})].$$

Proof. (Sketch) In DSO, we can define E^*, the reflexive transitive closure of E, and then define φ^{LZ} inductively using

$$\begin{aligned}
(U_a(x_i))^{LZ} &:= U_a(y_i), \\
(x_i \leq x_j)^{LZ} &:= x_i < x_j \vee (x_i = x_j \wedge y_i \leq y_j), \\
(\exists x_{n+1} \varphi)^{LZ} &:= \exists x_{n+1} \exists y_{n+1} \, (E^*(x_{n+1}, y_{n+1}) \wedge \varphi^{LZ}), \\
(\exists X^1 \varphi)^{LZ} &:= \exists X^2 (\forall x \forall y (X^2(x,y) \rightarrow E^*(x,y)) \wedge \varphi^{LZ}).
\end{aligned} \tag{1}$$

For the atomic cases, note that in φ^{LZ} a variable x_i stands for a block of w and y_i for a relative position in this block.

A property L of non-empty strings is *definable on strings* (resp. *on compressed strings* or *LZ-graphs*), if for some formula φ of the appropriate language, $L = \{w \mid \mathcal{S}_w \models \varphi\}$ (resp. $L = \{w \mid \mathcal{G}_{LZ(w)} \models \varphi\}$).

Remark 1. There are properties of strings that are FO-definable on strings, but not on LZ-graphs, like $\exists x(U_a(x) \wedge U_b(x+1))$. There are properties of strings that are FO-definable on LZ-graphs, but not even MSO-definable on strings (cf. [1]).

2.2 *LZ*-Tries

While compressing $w = B_0 \cdots B_{n-1} v$, the blocks $B_0, \ldots, B_{n-1}$ found are maintained as a *dictionary* of subwords of w and stored as a tree by sharing common prefixes. The linear order of the blocks in $LZ(w)$ amounts to an enumeration of the nodes of the tree.

Definition 2. *A (finite) Σ-tree $(T, \leq, \xleftarrow{a}, 0)_{a\in\Sigma}$ is a (finite) tree $(T, \leq, 0)$ with root 0, where $\{\xleftarrow{a} \subseteq T \times T \mid a \in \Sigma\}$ are pairwise disjoint minimal relations such that $\leq$ is the reflexive transitive closure of their union.*

A Σ-tree is a Σ-trie if to each node $n \in T$ and each $a \in \Sigma$ there is at most one $n' \in T$ such that $n \xleftarrow{a} n'$. A (finite) enumerated Σ-trie

$$\mathcal{T} = (T, \leq, \xleftarrow{a}, 0, Succ)_{a\in\Sigma},$$

*or an LZ-*trie *for short, is a Σ-trie $(T, \leq, \xleftarrow{a}, 0)_{a\in\Sigma}$ with a successor relation*[2] *Succ on T that is compatible with the partial order $\leq$. We assume that $T = \{0, 1, 2, \ldots, m\}$ and $Succ(i,j)$ iff $i+1 = j$ in $\mathbb{N}$.*

Example 1 (Cont.). Enumerating the pairs of $LZ(w)$ by 1,2, etc. in a third component, we obtain a sequence $(0,a,1)(0,b,2)(2,b,3)$ $(1,a,4)(3,a,5)(3,b,6)$ of triples. These represent a tree in which block B_k labels the path from the root 0 to node $k+1$:

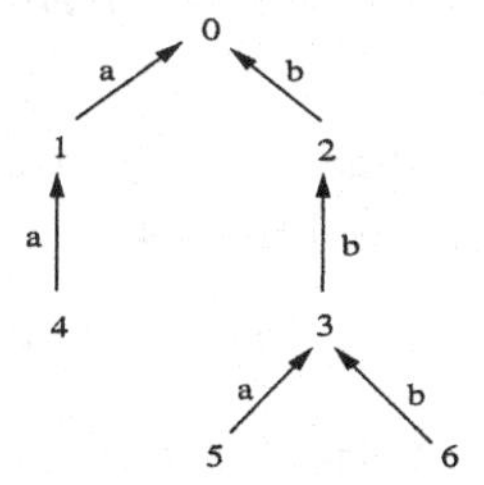

A tuple $p_n = (k,a)$ of $LZ(w)$ is drawn as an edge $k \xleftarrow{a} n+1$.

We write $\mathcal{T}_\alpha$ for the enumerated trie representing the compressed word α. We always assume that our strings w have a distinguished end symbol; then the final block of $LZ(w)$ is different from the previous ones and the tree of blocks indeed is a trie.

Remark 2. What differs in choosing $\mathcal{G}_\alpha$ or $\mathcal{T}_\alpha$ is the logical language used to talk about LZ-compessed strings α. Basically, we have

$$\mathcal{G}_\alpha \models E(i,j) \wedge U_a(i) \iff \mathcal{T}_\alpha \models (j+1) \xleftarrow{a} (i+1).$$

Modulo the additional root node in the trie, $\geq$ in the trie amounts to E^* in the graph, and $\leq$ in the graph to $Succ^*$ in the trie. Since E^* resp. $Succ^*$ is $\exists$- and $\forall$-MSO-definable in the language of LZ-graphs resp. LZ-tries, $\exists$-MSO-properties of LZ-graphs translate to $\exists$-MSO-properties of LZ-tries and vice versa.

Using (1), quantifiers Qx and QX about strings translate to *bounded* quantifiers $Q(x,y) \in E^*$ and $QX \subseteq E^*$. So when translating to the language of LZ-tries we only quantify over tuples (and relations of such) whose components lie on a *path* of $\mathcal{T}_{LZ(w)}$.

[2] i.e. a minimal binary relation *Succ* whose transitive reflexive closure $Succ^*$ is a total ordering of T.

3 MSO-Equivalence for *LZ*-Graphs

For relational structures $\mathcal{A}, \mathcal{B}$ of the same signature, $\mathcal{A} \equiv_r^{MSO} \mathcal{B}$ says that $\mathcal{A}$ and $\mathcal{B}$ satisfy the same MSO-sentences of quantifier rank $\leq r$.

Two facts about MSO for strings imply the existence of finite automata that can check MSO-properties of strings (cf. [4]):

a) for each r, there are only finitely many $\equiv_r^{MSO}$-equivalence classes for word structures $\mathcal{S}_w$, and
b) for compound strings wa, the $\equiv_r^{MSO}$-class of $\mathcal{S}_{wa}$ depends only on the $\equiv_r^{MSO}$-classes of $\mathcal{S}_w$ and $\mathcal{S}_a$.

(The analogous situatation holds for trees over Σ.) *LZ*-compressed words $\alpha = p_0 \cdots p_{m-1}$ are words over the infinite alphabet $\mathbb{N} \times \Sigma$. Can we check MSO-properties of compressed strings by a kind of finite automaton for *LZ*-graphs? Since we deal with a finite relational language, we still have a):

Proposition 1. *For each r, the equivalence relation $\equiv_r^{MSO}$ between LZ-graphs has finite index.*

But what about b)? By extending winning strategies for duplicator in the Ehrenfeucht-Fraisse-game $G_r(\mathcal{G}_\alpha, \mathcal{G}_\beta)$, we can show:

Lemma 1. *(i) For LZ-compressed words $\alpha(0,a), \beta(0,a')$ over Σ,*

$$\mathcal{G}_\alpha \equiv_r^{MSO} \mathcal{G}_\beta \ \wedge a = a' \implies \mathcal{G}_{\alpha(0,a)} \equiv_r^{MSO} \mathcal{G}_{\beta(0,a')}.$$

(ii) For LZ-compressed words $\alpha(k+1,a)$ and $\beta(k'+1,a')$ over Σ,

$$(\mathcal{G}_\alpha, k) \equiv_r^{MSO} (\mathcal{G}_\beta, k') \ \wedge \ a = a' \implies \mathcal{G}_{\alpha(k+1,a)} \equiv_r^{MSO} \mathcal{G}_{\beta(k'+1,a')}.$$

However, in (ii) one has to assume that the elements k, k' pointed to from the new maximal elements share the same properties. Instead, one would need the stronger claim

$$\mathcal{G}_\alpha \equiv \mathcal{G}_\beta \wedge \mathcal{G}_\alpha{\restriction}k \equiv \mathcal{G}_\beta{\restriction}k' \wedge a = a' \implies \mathcal{G}_{\alpha(k+1,a)} \equiv \mathcal{G}_{\beta(k'+1,a')},$$

where $\mathcal{G}_\alpha{\restriction}k$ is the restriction of $\mathcal{G}_\alpha$ with k as its maximal element. The equivalence class of $\mathcal{G}_\alpha{\restriction}k$ would be the automaton state assigned to k. (Notice that $\mathcal{G}_\alpha{\restriction}k$ is a *LZ*-graph, but $(k+1,a)$ is not a *LZ*-compressed word.) The problem is that duplicator's winning strategies for $G_r(\mathcal{G}_\alpha, \mathcal{G}_\beta)$ and $G_r(\mathcal{G}_\alpha{\restriction}k, \mathcal{G}_\beta{\restriction}k')$ may pick *different* elements to answer spoilers playing of some element of, say, $\mathcal{G}_\alpha{\restriction}k$.

Thus, unlike in the case of strings or trees, for compound compressed words $\alpha(k,a)$ we have component *LZ*-graphs $\mathcal{G}_\alpha$ and $\mathcal{G}_\alpha{\restriction}k$ that are not disjoint, and winning strategies in games for these do not combine to winning strategies for composed *LZ*-graphs.

From this we conclude that we cannot use a Büchi-Myhill-Nerode construction to obtain from the $\equiv_r^{MSO}$-classes a finite sequential automaton for *LZ*-graphs, and likewise for *LZ*-tries.

4 *LZ*-Trie-Automata

If we view *LZ*-tries as trees with an additional edge *Succ* between nodes, we obtain directed acyclic graphs of a special kind: the successor child may be equal to some decendant with respect to the $\xleftarrow{a}$-child-relations. Since these are still very close to trees, it is natural to use a variation of tree-automata as an approximative notion of *LZ*-automaton for checking properties of *LZ*-tries.

Definition 3. *Let $n \in G$ be a node in a graph (G, E). For $m \in \mathbb{N}$, the* sphere of radius m around n, *$s_m(n)$, is the set of nodes $k \in G$ such that there is a E-path of length $\leq m$ from n to k or vice versa. The* hemisphere of radius m around n, *$hs_m(n)$, is the set of nodes k such that there is an E-path of length $\leq m$ from n to k.*

Definition 4. *Let $n \in T$ be a node in the LZ-trie $\mathcal{T}$. The* bottom-up *LZ*-hemisphere of radius m around n, *$bu\text{-}hs_m^{\mathcal{T}}(n)$, is the restriction of $\mathcal{T}$ to the m-hemisphere around n in the graph (T, E), where*

$$E := \bigcup\{\xleftarrow{a} \mid a \in \Sigma\} \cup \{Succ\}.$$

The top-down *LZ*-hemisphere of radius m around n, *$td\text{-}hs_m^{\mathcal{T}}(n)$, is the restriction of $\mathcal{T}$ to the m-hemisphere around n in the graph $(T, \breve{E})$, where $\breve{E}$ is the converse of E. An* *LZ*-hemisphere *is an LZ-hemisphere of some radius around some node in some LZ-trie $\mathcal{T}$.*

Example 1 (Cont.). An *LZ*-trie and the bottom-up resp. top-down 2-hemispheres of node 3 (with dashed edges for *Succ* resp. *Pred*):

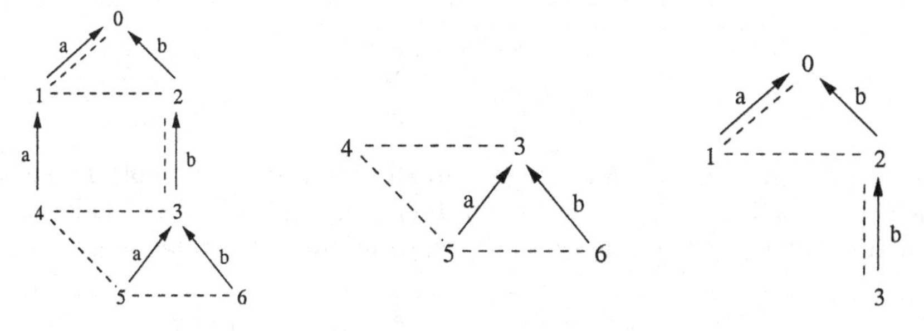

Definition 5. *A finite* bottom-up (resp. top-down) *m*-*LZ*-automaton $\mathcal{A} = (Q, \Sigma, \delta, q_{in}, F)$ *consists of a finite set Q of* states, *sets $I, F \subseteq Q$ of* initial *and* final *states, a finite* alphabet Σ, *a finite* transition relation δ *consisting of pairs (P, q), written $P \to q$, where $q \in Q$ and P is a bottom-up (resp. top-down) LZ-hemisphere of radius m whose nodes except the root are labelled by elements of Q.*

A run *of $\mathcal{A}$ on an LZ-trie $\mathcal{T}$ is a function $r : T \to Q$ where $r(\max) \in I$ (resp. $r(0) \in I$) and for each $n \in T$ there is some $(P, q) \in \delta$ such that $bu\text{-}hs_m^{\mathcal{T}}(n)$ (resp. $td\text{-}hs_m^{\mathcal{T}}(n)$), expanded by the labelling of nodes given by r, is isomorphic to P with label q at its root.* $\mathcal{A}$ accepts *$\mathcal{T}$ if there is a run r of $\mathcal{A}$ on $\mathcal{T}$ such* *that* $r(0) \in F$ *(resp.* $r(\max) \in F$*). Let* $L(\mathcal{A}) := \{\mathcal{T} \mid \mathcal{A}$ *accepts* $\mathcal{T}\}$ *be* the class of *LZ*-tries accepted by $\mathcal{A}$.

$\mathcal{A}$ *is* deterministic *if* $|I| = 1$ *and* $q = q'$ *when* $(P, q), (P, q') \in \delta$.

While an *m*-*LZ*-automaton $\mathcal{A}$ sequentially follows the enumeration of a trie $\mathcal{T}_\alpha$, it can access some states reached at suffixes (resp. prefixes) of α. Strictly speaking, it does not have a 'finite memory'.

4.1 Bottom-Up *LZ*-Trie-Automata

A 1-*LZ*-automaton working bottom-up the *LZ*-trie towards the root has transitions that determine the state at a node from the states at the node's Σ-children and successor. But it also has to distinguish which of the Σ-children is the successor of the node, if any.

Example 2. Consider the class $\mathcal{K}$ of *LZ*-tries over $\Sigma = \{a, b\}$ which have a node whose successor and *a*-child agree, i.e. which satisfy the sentence $\varphi := \exists x \exists y\, [y = x + 1 \wedge (x \overset{a}{\longleftarrow} y)]$. We give a bottom-up 1-*LZ*-automaton $\mathcal{A}$ accepting $\mathcal{K}$. We write a transition in the form

$$(q_a, q_b, q_{succ}, i) \rightarrow p,$$

where q_a, q_b, q_{succ} are the states of the *a*-, *b*- and *Succ*-child or $\perp$, when there is no such child, and $i \in \{1, 2, 3, \perp\}$ says which of the children is equal to the successor node, if any. Thus, $(p, q, p, 1) \rightarrow q'$ corresponds to the transition $P \rightarrow q'$ which can be shown as

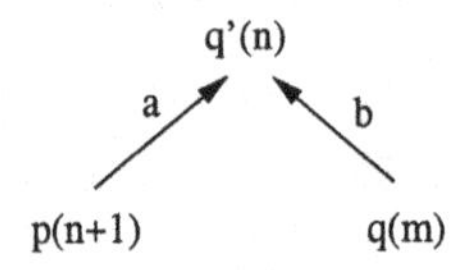

$\mathcal{A}$ has a final state q_1, which is assigned to all ancestors of the root of a subtrie satisfying φ, and an initial state q_0, which is assigned to all other nodes of the input trie. Letting q, p, q' range over $\{q_0, q_1\}$, the transition table is

a) $(\perp, \perp, \perp, \perp) \rightarrow q_0$	*e)* $(q, p, q', 3) \rightarrow q'$
b) $(q, \perp, q, 1) \rightarrow q_1$	*f)* $(q, \perp, q', 3) \rightarrow q'$
c) $(q, p, q, 1) \rightarrow q_1$	*g)* $(\perp, q, q', 3) \rightarrow q'$
d) $(q, p, q', 2) \rightarrow q'$	*h)* $(\perp, \perp, q', 3) \rightarrow q'$.

Rule *a)* means that if there is no successor-node, $\mathcal{A}$ is in state q_0. Rules *b)* and *c)* say that if the successor-node is the *a*-child, then $\mathcal{A}$ goes to q_1 as we just saw the pattern φ. Rules *d)* – *f)* say that if the successor node differs from the *a*-child, $\mathcal{A}$ remains in the state of the successor node. Similar for *g)* and *h)*, when there is no *a*-child.

Theorem 2. *There is a property of LZ-tries that is recognized by a non-deterministic bottom-up 1-LZ-automaton but not by any deterministic bottom-up m-LZ-automaton.*

Proof. (Sketch) Let $\Sigma = \{a, b, c\}$ and consider the following property φ of enumerated Σ-tries:

> There are two subsequent nodes $i-1$ and i such that some node j is both the a-predecessor of i and a b-ancestor of $i-1$.

An LZ-trie satisfies φ iff it contains nodes linked as follows (in the trie and in the LZ-graph, respectively), where $j < i-1$:

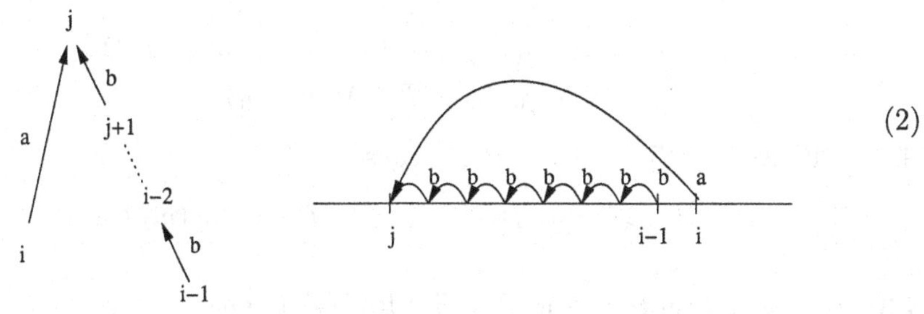

(2)

When finding nodes connected like $i-1$ and i, a bottom-up 1-LZ-automaton can (non-deterministically) guess that a b-ancestor of $i-1$ will be the a-parent of i and check this while proceeding. But no deterministic bottom-up m-LZ-automaton can check the property φ because, intuitively speaking, the m-hemisphere of a node j is not big enough to see if the predecessor of its a-child is one of its b-descendants. The proof details are not quite obvious but have to be omitted for lack of space.

4.2 Top-Down LZ-Trie-Automata

G.Navarro [5] has shown how to do regular expression search on LZ-78-compressed texts by simulating an automaton reading the original text, beating the decompression-and-search approach by a factor of 2. Actually, this simulation is a deterministic top-down 1-LZ-automaton on the LZ-compressed text:

Theorem 3. *The LZ-compression $\{\mathcal{T}_{LZ(w)} \mid w \in R\}$ of any regular set $R \subseteq \Sigma^+$ is accepted by a deterministic top-down 1-LZ-automaton.*

Proof. Let $\mathcal{A} = (Q, \Sigma, q_0, \delta, F)$ be a deterministic finite automaton accepting R. Define a deterministic top-down 1-LZ-automaton $\mathcal{A}' = (Q', \Sigma, \delta', q'_{in}, F')$ by $Q' := Q \times (Q \to Q)$, $q'_0 := (q_0, \lambda q.q)$, $F' := \{(q, f) \mid q \in F\}$ and δ' according to the following transitions:

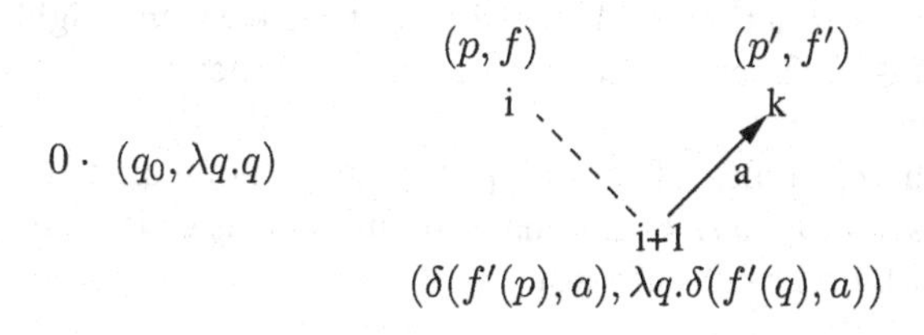

for each $a \in \Sigma$ (including the case $i = k$).

Suppose r' is a run of $\mathcal{A}'$ on $\mathcal{T}_{LZ(w)}$ for a compressed word $LZ(w) = p_0 \cdots p_{m-1}$ where p_k represents block B_k of $w = B_0 \cdots B_{m-1}$. By induction we see that for each $k < m$,

$$r'(k) = (\delta(q_0, B_0 \cdots B_{k-1}), \lambda p.\delta(p, B_{k-1})). \quad (3)$$

For $k > 0$, suppose $r'(i) = (p, f) = (\delta(q_0, B_0 \cdots B_{i-1}), f)$ and $p_i = (k, a)$. Then $B_i = B_{k-1}a$, and with $r'(k) = (p', f') = (p', \lambda p.\delta(p, B_{k-1}))$

$$\begin{aligned} r'(i+1) &= (\delta(f'(p), a), \lambda q.\delta(f'(q), a)) \\ &= (\delta(p, B_{k-1}a), \lambda q.\delta(q, B_{k-1}a)) \\ &= (\delta(\delta(q_0, B_0 \cdots B_{i-1}), B_i), \lambda q.\delta(q, B_i)) \\ &= (\delta(q_0, B_0 \cdots B_i), \lambda q.\delta(q, B_i)). \end{aligned}$$

Hence $\mathcal{A}'$ accepts $\{\mathcal{T}_{LZ(w)} \mid w \in R\}$, because

$$w \in R \iff \delta(q_0, B_0 \cdots B_{m-1}) \in F \iff r'(m) \in F'.$$

4.3 Graph Acceptors and ∃-MSO for *LZ*-Tries

If we consider the states q of an automaton $\mathcal{A}$ as monadic predicates on nodes of tries, the condition "there is an accepting $\mathcal{A}$-run" can be expressed by an ∃-MSO-sentence $\exists \boldsymbol{q}\psi$ in the language of *LZ*-tries:

Proposition 2. *For every m-LZ-automaton $\mathcal{A}$ there is an ∃-MSO-sentence $\varphi_{\mathcal{A}}$ defining the class of LZ-tries accepted by $\mathcal{A}$.*

For sentences, this improves on the translation of Theorem 1:

Corollary 1. *Every property of strings which is definable in MSO on strings is definable in ∃-MSO on LZ-tries.*

Proof. By Büchi's theorem, Theorem 3 and Proposition 2.

The converse of Corollary 1 is wrong: $\{a.b. \cdots .b^n a.b^n b. \mid n \in \mathbb{N}\}$ is not a regular language, but has an FO-definable class of *LZ*-tries.

To show the converse of Proposition 2, we use the graph acceptors for directed acyclic graphs presented by W. Thomas [7].

On graphs $G = (V, E)$, a *tile* δ over a set Q is an m-neighbourhood of a node with a labelling in Q, i.e. a finite node-labelled graph. A *graph acceptor* is a triple $\mathcal{A} = (Q, \Delta, Occ)$ where Δ is a finite set of tiles over the finite set Q and Occ is a boolean combination of conditions $\chi_{p,\delta}$:= *there are* $\geq p$ *occurences of tile* δ. A *run* of $\mathcal{A}$ on G is a function $r : V \to Q$ such that each m-neighbourhood of G becomes a tile $\delta \in \Delta$. Then $\mathcal{A}$ *accepts* a graph G if there exists a run of $\mathcal{A}$ on G whose tiles satisfy Occ.

Note that the existence of an accepting run is non-constructive. The main result of [7] says that a class of graphs of bounded degree is definable in ∃-MSO iff it is accepted by a graph acceptor.

Theorem 4. *A class $\mathcal{K}$ of LZ-tries is $\exists$-MSO-definable iff for some m, $\mathcal{K}$ is accepted by an m-LZ-automaton.*

Proof. Note that each LZ-trie $\mathcal{T}$ satisfies the following conditions:

(i) each node of $\mathcal{T}$ has a degree $\leq |\Sigma| + 1$,
(ii) $\mathcal{T}$ is an acyclic (directed) graph with a designated out-edge (the successor) for each node,
(iii) $\mathcal{T}$ has a node that is reachable from any node by a path.

Using (i), by Theorem 3 of [7], $\mathcal{K}$ is $\exists$-MSO-definable iff $\mathcal{K}$ is recognizable by a graph acceptor. Moreover, from (i)-(iii) and Proposition 6 of [7], it follows that $\mathcal{K}$ is recognizable by a graph acceptor iff it is recognizable by a graph acceptor without occurrence constraints.

$\Leftarrow$: Suppose $\mathcal{K}$ is accepted by some m-LZ-automaton $\mathcal{A}$. We may assume that final states of $\mathcal{A}$ do not occur in the m-hemispheres P of transitions (P, q) of $\mathcal{A}$. Consider the transitions of $\mathcal{A}$ as a tiling system. Then an accepting run of $\mathcal{A}$ is a tiling that uses at least one of the tiles (P, q) where $q \in F$. Thus, $\mathcal{K}$ has a graph acceptor.

$\Rightarrow$: By the above remarks, $\mathcal{K}$ is recognizable by a graph acceptor without occurrence constraints. On the LZ-tries, the tiles have a root node, so we can view each tile as a transition rule saying that the automaton enters a state at the root depending on the m-sphere of the root and the states at the descendants in the sphere. Let each state be final; then the automaton accepts iff there is a tiling.

A graph $G = (V, E)$ is *k-colorable* if there is a partitioning of the set V of nodes into at most k classes (colors) such that any two adjacent nodes belong to different classes.

Clearly, k-colorability can be expressed by an $\exists$-MSO sentence, and hence has a graph acceptor. While 3-colorability on arbitrary graphs is an NP-complete problem, it is trivial on LZ-tries, because the maximal node m is connected to at most two nodes:

Proposition 3. *Every LZ-trie is 3-colorable.*

Potthoff e.a. [6] have shown that on the class of directed acyclic graphs whose edges are uniquely labelled with a bounded number of labels, 2-colorability is *not* recognizable by a deterministic graph acceptor, i.e. one with at most one accepting run on each graph. For the subclass of LZ-tries, however, it is (actually by a 4-state deterministic bottom-up-LZ-automaton):

Proposition 4. *On the class of LZ-tries, 2-colorability is recognizable by a deterministic graph acceptor.*

4.4 Strictness of the Automata Hierarchy

The bottom-up resp. top-down hierarchies of m-LZ-automata are strict, and no level exhausts the MSO-properties of tries:

Theorem 5. *For every m, there is an MSO-sentence defining a class of LZ-tries that is not accepted by any m-LZ-automaton.*

Proof. For $m = 1$, let $L = \{b^1 b^2 \cdots b^n a b^1 a b^2 a \cdots b^{n-1} a \mid n \in \mathbb{N}\} \subseteq \{a, b\}^+$. Each $w \in L$ has a LZ-block decomposition as indicated by

$$w_n = b^1.b^2.\cdots.b^n.a.b^1a.b^2a.\cdots.b^{n-1}a$$

and a compression

$$LZ(w_n) = (0, b)(1, b) \cdots (n-1, b)(0, a)(1, a) \cdots (n-1, a).$$

The corresponding tries $\mathcal{T}_{LZ(w_n)}$ look like

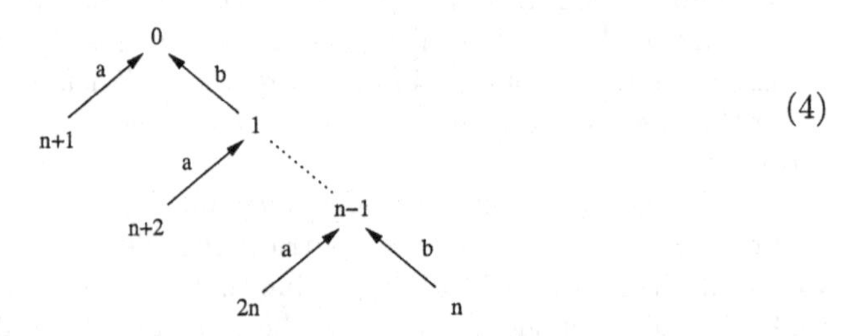

(4)

where the successor relation is given by the node numbers.

The class of enumerated tries in $LZ(L)$ can be defined by the MSO-sentence φ saying that the set B of nodes that are the root 0 or a b-child, has the following properties (of which (i) is neither $\exists$-MSO nor $\forall$-MSO):

(i) $B \supseteq \{0\}$ is the smallest set being closed under b-children,
(ii) a node is not in B iff it is an a-child of a node in B,
(iii) for all nodes x, y, we have $y = x + 1$ iff one of the following holds:
  (a) y is the b-child of x,
  (b) y is the a-child of the b-child y' of some x' whose a-child is x,
  (c) y is the a-child of 0 and x the member of B that has no b-child.

Claim. $LZ(L)$ is not accepted by a 1-bottom-up-LZ-automaton.

Suppose that $\mathcal{A}$ is a 1-bottom-up-LZ-automaton that accepts the class defined by φ. Let $m > |Q|$ and $w = w_{2m}$. An accepting run of $\mathcal{A}$ on $\mathcal{T}_{LZ(w)}$ assigns the same state, say q, to at least two different a-childs, say nodes $n + k + 2$ and $n + 2$. The state of their predecessors is determined by a rule

$$(\bot, \bot, q, 3) \rightarrow p.$$

Let $\mathcal{T}'$ be like $\mathcal{T}_{LZ(w)}$, except that nodes $n+1$ and $n+k+1$ are switched in the ordering. Then $\mathcal{A}$ will also accept the modified structure, as indicated by the states assigned to nodes as follows:

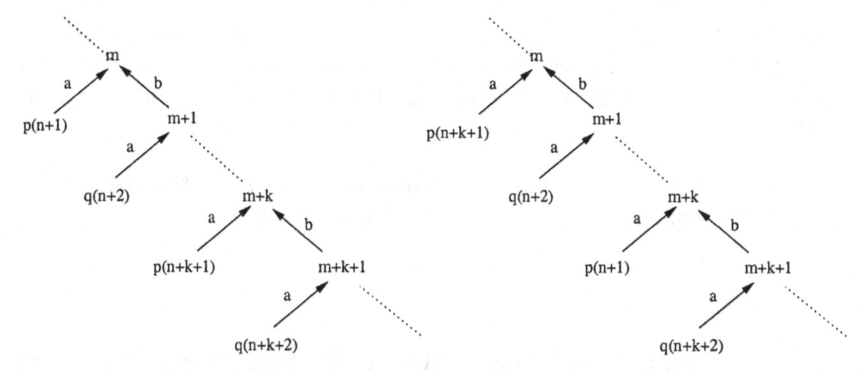

But the enumeration of a-children in $\mathcal{T}'$ does not conform to φ, so $\mathcal{T}' \in T(\mathcal{A}) \setminus LZ(L)$.

For the case $m > 1$, we modify the example as follows: along the B-part, between two nodes that have both an a-child and a b-child, we add $m-1$ nodes that have no a-child, i.e. the subgraphs

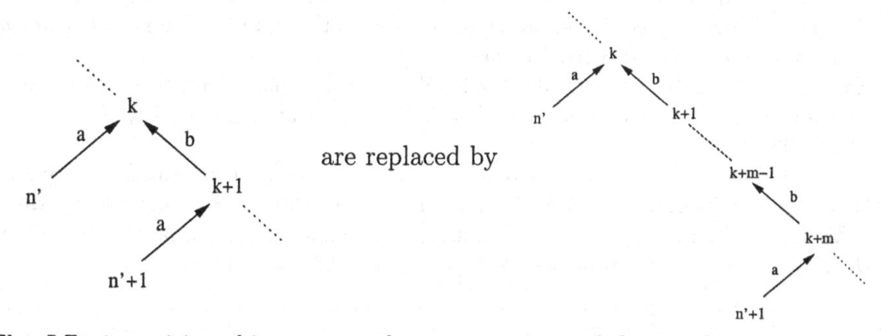

The LZ-tries arising this way are the compressions of the words

$$w_{n,k} = b^1.b^2.b^3.\cdots.b^{nm+k}.a.b^m a.b^{2m}a.\cdots.b^{nm}a.$$

Then $\{\mathcal{T}_{LZ(w_{n,k})} \mid n,k \in \mathbb{N}, k \leq m\}$ is accepted by a $(m+1)$-bottom-up-LZ-automaton, hence MSO-definable. But it is not accepted by any m-bottom-up-LZ-automaton: intuitively, the m-hemisphere of a node k does not tell whether the path from k following a and successor and the path following $b^m a$ end at the same node.

A similar argument shows that the class defined by φ is also not accepted by any m-top-down-LZ-automaton with the same m.

5 Conclusion

We have shown some relations between properties of strings, properties of their Lempel-Ziv-78-compressions, and automata accepting compressed strings, but the picture is not complete yet. For example:

Conjecture 1. On the class of all LZ-tries, not every MSO-formula is equivalent to an $\exists$-MSO-formula.

It seems likely that the class of acceptable LZ-tries is not closed under complement (cf. Theorem 2). We also believe that deterministic bottom-up-LZ-automata are much weaker than deterministic top-down-LZ-automata (cf. Theorem 3):

Conjecture 2. There is no deterministic bottom-up m-LZ-automaton that can check on $LZ(w)$ whether $w \in \{a, b\}^+$ has a subword ab.

References

1. Foto Afrati, Hans Leiß, and Michel de Rougemont. Definability and compression. In *15th Annual IEEE Symposium on Logic In Computer Science, LICS'2000. Santa Barbara, CA, June 22-26, 2000*, pages 151–172. Computer Society Press, 2000.
2. Timothy C. Bell, John G. Cleary, and Ian H. Witten. *Text Compression.* Prentice Hall, Englewood Cliffs, NJ, 1990.
3. T. Cover and J. Thomas. *Elements of Information Theory.* John Wiley, 1991.
4. H.-D. Ebbinghaus and J. Flum. *Finite Model Theory.* Springer-Verlag, 1991.
5. Gonzalo Navarro. Regular expression searching on compressed text. *Journal of Discrete Algorithms*, 2003 (to appear).
6. Andreas Potthoff, Sebastian Seibert, and Wolfgang Thomas. Nondeterminism versus determinism of finite automata over directed acyclic graphs. *Bull. Belg. Math. Soc.*, 1:285–298, 1994.
7. Wolfgang Thomas. Automata theory on trees and partial orders. In eds. M. Bidoit, M. Dauchet, editor, *TAPSOFT'97*, LNCS 1214, pages 20–38. Springer Verlag, 1997.
8. J. Ziv and A. Lempel. Compression of individual sequences via variable-rate coding. *IEEE Transactions on Information Theory*, pages 530–536, 1978.

Complexity of Some Problems in Modal and Intuitionistic Calculi

Larisa Maksimova[1] and Andrei Voronkov[2]

[1] Institute of Mathematics, Novosibirsk
[2] University of Manchester

Abstract. Complexity of provability and satisfiability problems in many non-classical logics, for instance, in intuitionistic logic, various systems of modal logic, temporal and dynamic logics was studied in [3, 5, 9, 22–24]. Ladner [9] proved that the provability problem is PSPACE-complete for modal logics K, T and S4 and coNP-complete for S5. Statman [24] proved that the problem of determining if an arbitrary implicational formula is intuitionistically valid is PSPACE-complete.
We consider the complexity of some properties for non-classical logics which are known to be decidable. The complexity of tabularity, pre-tabularity, and interpolation problems in extensions of the intuitionistic logic and of the modal logic S4 is studied, as well as the complexity of amalgamation problems in varieties of Heyting algebras and closure algebras.

1 Main Results

A *superintuitionistic logic* is a set of modality-free formulas containing the set Int of all intuitionistically valid formulas and closed under substitutions and modus ponens. We consider the families $E(\text{Int})$ of all superintuitionistic logics and $NE(\text{S4})$ of the normal extensions of S4. Logics in $NE(\text{S4})$ are, in addition, closed under the necessitation rule $A/\Box A$ and contain all formulas valid in S4. For any logic L, we denote by $L + A$ the extension L by an extra axiom scheme A. In particular,

$$\begin{aligned}
\text{Cl} &= \text{Int} + (p \vee \neg p),\\
For &= \text{Int} + \bot,\\
\text{KC} &= \text{Int} + (\neg p \vee \neg\neg p),\\
\text{LC} &= \text{Int} + (p \to q) \vee (q \to p),\\
\text{S4} &= \text{K} + (\Box p \to p) + (\Box p \to \Box\Box p),\\
\text{S4.1} &= \text{S4} + (\Box\Diamond p \to \Diamond\Box p),\\
\text{S4.2} &= \text{S4} + (\Diamond\Box p \to \Box\Diamond p),\\
\text{S4.3} &= \text{S4} + \Box(\Box p \to q) \vee \Box(\Box q \to p),\\
\text{S5} &= \text{S4} + (\Diamond p \to \Box\Diamond p),\\
\text{Grz} &= \text{S4} + (\Box(\Box(p \to \Box p) \to p) \to p),\\
\text{Grz.2} &= \text{Grz} + \text{S4.2},\\
\text{Grz.3} &= \text{Grz} + \text{S4.3}.
\end{aligned}$$

A logic is called *tabular* if it can be characterized by finitely many finite models; and *pretabular* if it is maximal among the non-tabular logics. A logic L is called *locally*

M. Baaz and J.A. Makowsky (Eds.): CSL 2003, LNCS 2803, pp. 397–412, 2003.

tabular if for any finite set P of propositional variables there exist only finitely many formulas of variables in P non-equivalent in L. It is known that each tabular logic is locally tabular.

A logic L is said to have *Craig's interpolation property*, denoted *CIP*, if for every formula $(A \to B) \in L$ there exists a formula C such that (i) both $(A \to C) \in L$ and $(C \to B) \in L$, and (ii) every variable of C occurs in both A and B. A logic L is said to have the *interpolation property*, denoted *IPD*, if $A \vdash_L B$ implies that exists a formula C such that (i) both $A \vdash_L C$ and $C \vdash_L B$, and (ii) every variable of C occurs in both A and B.

We assume that the language contains the propositional constants $\top$ or $\bot$. A *size* of a formula A, denoted $|A|$, is the number of occurrences of variables and logical symbols in A.

Let L be Int or some modal logic. By the *tabularity (pretabularity etc.) problem over* L we mean the problem of determining for an arbitrary formula A, whether $L + A$ is tabular (pretabular, etc.), and consider its complexity with respect to the size of A.

One can find the necessary definitions from Complexity Theory in [7,21]. In Section 3 we prove the following result.

THEOREM 1 *(i) The tabularity problems over both* Int *and* S4 *are* NP*-complete.*
(ii) The pretabularity problems over both Int *and* S4 *are both* NP*-hard and* coNP*-hard, and belong to in* Δ_2^p.
(iii) The local tabularity problem over S4 *is* NP*-complete.* ❑

It is not known whether the local tabularity problem over Int is decidable.

The interpolation problem is considered in Section 4. We prove the following results.

THEOREM 2 *(i) The interpolation problems over both* Int *and* Grz *are* PSPACE*-complete.*
(ii) The problem of determining whether Int $+ A$ *is a tabular logic with CIP is* NP*-complete.*
(iii) Both CIP and IPD problems over S4 *are in* coNEXP *and* PSPACE*-hard.*
(iv) The problems of determining whether S4 $+ A$ *is a tabular (pretabular or locally tabular) logic with CIP (or IPD) are both* NP*-hard and* coNP*-hard, and belong to* Δ_2^p. ❑

There exists a well-known duality between $E(\text{Int})$ and the family of varieties of Heyting algebras, and also between $NE(\text{S4})$ and the family of varieties of closure algebras. If Ax is a set of formulas and $L = \text{Int} + Ax$ then $\{A = \top \mid A \in Ax\}$ forms a base of identities for the variety $V(L)$ associated with L. The definitions for $L \in NE(\text{S4})$ are analogous. On the other hand, with any identity $A = B$ one can associate a formula $A \leftrightarrow B$ which is valid in an algebra if and only if the identity $A = B$ holds in this algebra. By rewriting Theorem 2 in the algebraic language, we obtain the following result.

THEOREM 3 *(i) The amalgamability problem for finitely based varieties of Heyting algebras is* PSPACE*-complete.*

(ii) The amalgamability and super-amalgamability problems for finitely-based varieties of closure algebras are PSPACE*-hard and belong to* coNEXP. ❑

2 Complexity and Decision Problems

We assume the knowledge of the standard material in complexity theory which can be found, for example, in [21]. In this paper we will use complexity classes P, NP, EXP, NEXP, PSPACE, coNP, coNEXP, DP, and Δ_2^p.

If we take two superintuitionistic logics L_1 and L_2 then $L_1 + L_2$ and $L_1 \cap L_2$ are superintuitionistic logics, too. The former logic can be axiomatized by all extra axioms of L_1 and L_2. For an axiomatization of the intersection one can use Miura's theorem [20]. To formulate it, we recall some definitions.

If A and B are two formulas, we denote by $A \vee' B$ a disjunction $A \vee B'$, where B' is obtained from B by renaming variables so that A and B' have no variables in common.

It was proved by Miura (see [20]) that the intersection $L_1 \cap L_2$ of two logics $L_1, L_2 \in E(\text{Int})$ is axiomatizable by the set of formulas $A \vee' B$, where A ranges over the set of extra axioms of L_1 and B over the set of extra axioms of L_2. For logics in $NE(\text{S4})$ one should take the set of formulas $\Box A \vee' \Box B$ for all extra axioms A of L_1 and B of L_2 [19].

With each logic L one can associate the following decision problems.

1. Given a formula A, is this formula provable in L (L-provability);
2. The dual problem of unprovability in L;

In addition, with each pair of logic L_0, L one can associate the following decision problem.

3. Given a formula A, is the logic $L_0 + A$ equivalent to L (*equivalence to L over* L_0);

The equivalence problem between logics will play an important role in this paper. In general, the equivalence problem is undecidable. When we restrict ourself to considering particular families of logics, for instance, superintuitionistic logics, then the problem of equivalence to a particular logic L may be undecidable too. On the other hand, this problem is decidable if we take into consideration only superintuitionistic logics and deal with their equivalence to one of logics Cl, KC, or LC.

For arbitrary fixed $L_1 \in E(\text{Int})$ we consider two inclusions $\text{Int} + A \subseteq L_1$ and $L_1 \subseteq \text{Int} + A$. The former inclusion is the same as provability of A in L_1, and the latter is the same as provability of extra axioms of L_1 in $\text{Int} + A$. It is evident that for any logic $L_1 \in E(\text{Int})$, the logic $\text{Int} + A$ is equivalent to L_1 if and only if both inclusions $\text{Int} + A \subseteq L_1$ and $L_1 \subseteq \text{Int} + A$ hold. On the other hand, we have

PROPOSITION 4 *Let L_0 be a logic in $E(\text{Int})$ or in $NE(\text{S4})$ and L_1 be a finitely axiomatizable extension of L_0. Then each of the inclusion problems $L_0 + A \subseteq L_1$ and $L_1 \subseteq L_0 + A$ is linearly reducible to equivalence to L_1 over L_0.*

PROOF. Let us consider the case $L_0 \in E(\text{Int})$. Let B be a single axiom of L_1 w.r.t L_0, i.e. $L_1 = L_0 + B$. Evidently, $L_0 + A \subseteq L_1$ if and only if $L_0 + (A \& B)$ is equivalent to L_1, so the first inclusion problem is polynomial-time reducible to equivalence to L_1 over L_0.

Further, $L_1 \subseteq L_0 + A$ if and only if $L_1 = L_1 \cap (L_0 + A)$. By Miura's theorem, $L_1 \cap (L_0 + A)$ is $L_0 + (B \vee' A)$, so $L_1 \subseteq L_0 + A$ if and only if $L_1 = L_0 + (B \vee' A)$, which gives us a reduction of the second inclusion.

The proof for the case $L_0 \in NE(\mathrm{S4})$ is analogous, but one should use the formula $\Box B \vee' \Box A$ instead of $B \vee' A$. ❑

If S is a set of extensions of a given logic L_0, we can consider the problem of membership in S (over L_0), i.e. the problem of recognizing whether $L_0 + A$ is in S.

We say that a logic L has the *Hallden Property*, denoted *HP* if for any formulas A and B without common variables, $A \vee B \in L$ implies $A \in L$ or $B \in L$.

PROPOSITION 5 *Let L_0 be* Int *or a logic in $NE(\mathrm{S4})$ and S be a family of extensions of L_0 containing a finitely axiomatizable logic $L_1 = L_0 + A$ satisfying the conditions:*

1. $(\exists L' \supset L_1)(\forall L)((L_1 \subseteq L \subseteq L'$ *and* $L \in S) \Rightarrow L = L_1)$,
2. *L_1 has the Hallden property.*

Then the problem of L_1-provability is linearly reducible to the problem of membership in S over L_0.

PROOF. Suppose that $L_1 = \mathrm{Int} + A$, A' is any formula in $L' - L_1$, and B is an arbitrary formula. Consider the formula $A\&(A' \vee' B)$.

If $L_1 \vdash B$, then $L_1 \vdash A' \vee' B$, so $\mathrm{Int} + A\&(A' \vee' B) = L_1 + (A' \vee' B) = L_1 \in S$.

If $L_1 \nvdash B$, then $L_1 \nvdash A' \vee' B$ by the Hallden property, so $L_1 + (A' \vee' B) \neq L_1$. On the other hand, $L_1 + (A' \vee' B) \subseteq L'$. By condition (1) we get $\mathrm{Int} + A_1\&(A' \vee' B) = L_1 + (A' \vee' B) \notin S$, so we have

$$L_1 \vdash B \text{ iff } (\mathrm{Int} + A_1\&(A' \vee' B)) \in S.$$

The proof for modal logics is carried out analogously, by inserting $\Box$ in formulas.

Let us now prove some lower bounds on complexity.

PROPOSITION 6 *Let L_0 be a logic in $E(\mathrm{Int})$ or in $NE(\mathrm{S4})$, S be a family of extensions of L_0, and L_1, L_2 be two finitely axiomatizable extensions of L_0 such that $L_1 \subset L_2$.*

1. *If $L_1 \notin S$ and $L_2 \in S$ then the problem of membership in S over L_0 is* NP*-hard.*
2. *If $L_1 \in S$ and $L_2 \notin S$ then the problem of membership in S over L_0 is* coNP*-hard.*

PROOF.

1. Take, for example, $L_0 = \mathrm{Int}$. Let $L_1 = \mathrm{Int} + A_1$, $L_2 = \mathrm{Int} + A_2$. For an arbitrary formula B we have (by Glivenko's theorem)

$$\begin{array}{ll} B \text{ is satisfiable} & \Rightarrow \mathrm{Int} + \neg B \text{ is inconsistent, i.e. } \mathrm{Int} + \neg B = For; \\ B \text{ is unsatisfiable} & \Rightarrow Cl \vdash \neg B \Rightarrow \mathrm{Int} \vdash \neg B, \text{ i.e., } \mathrm{Int} + \neg B = \mathrm{Int}. \end{array}$$

Hence, if B is satisfiable then

$$\text{Int} + (A_1 \& (A_2 \vee' \neg B)) = L_1 + (A_2 \vee' \neg B) = (L_1 + A_2) \cap For = L_2 \in S.$$

If B is unsatisfiable, then

$$\text{Int} + (A_1 \& (A_2 \vee' \neg B)) = (L_1 + A_2) \cap (\text{Int} + A_1 + \neg B) = L_1 \notin S.$$

If L_0 is a modal logic in $NE(\text{S4})$, then for arbitrary modality-free B we have $Cl \vdash B$ if and only if $L_0 \vdash B$, so the prove for $L_0 = \text{Int}$, we only must replace $A_2 \vee' \neg B$ by $\Box A_2 \vee' \Box \neg B$.

2. Similar to the previous case, but replace S by its complement. ❑

We immediately obtain

COROLLARY 7 *Let L_0 be* Int *or* S4, *P a property of logic non-trivial on the class of finitely axiomatizable extensions of L_0. Then the problem of determining whether $L_0 + A$ has the property P is either* NP*-hard or* coNP*-hard.*

PROPOSITION 8 *Let L_0 be* Int *or* S4 *and S be a family of extensions of L_0. Let there exist finitely axiomatizable extensions L, L', L'' of L_0 such that $L' \subset L \subset L''$, $L \in S$, and L', $L'' \notin S$. Then the problem of membership in S over L_0 is* DP*-hard.*

PROOF. Assume $L_0 = \text{Int}$, $L = \text{Int} + A$, $L' = \text{Int} + A'$, and $L'' = \text{Int} + A''$. For each pair φ, ψ of boolean formulas define

$$f(\varphi, \psi) = A' \& (A'' \vee' \neg\varphi \vee' (A \& \neg\psi)).$$

We will prove that φ is satisfiable and ψ is unsatisfiable if and only if $(\text{Int} + f(\varphi, \psi)) \in S$. To this end consider three cases.

1. *φ is satisfiable and ψ is unsatisfiable.* Then there exists a substitution s of $\bot$ or $\top$ for the variables of φ such that $\text{Int} \vdash s(\varphi) \leftrightarrow \top$. So in the logic $\text{Int} + f(\varphi, \psi)$ one can derive $A' \& (A'' \vee' \neg\top \vee' (A \& \neg\psi))$, that is $A' \& (A'' \vee' (A \& \neg\psi))$. It is clear that $\text{Int} + A' \& (A'' \vee' (A \& \neg\psi))$ contains $\text{Int} + f(\varphi, \psi))$, hence these two calculi are equivalent. Further, $Cl \vdash \neg\psi$, so $\text{Int} \vdash \neg\psi$ by Glivenko's theorem. It follows that $\text{Int} + A' \& (A'' \vee' (A \& \neg\psi))$ is equivalent to $\text{Int} + A' \& (A'' \vee' A)$, hence also to $\text{Int} + A = L$ by the condition $L' \subset L \subset L''$. Therefore, $\text{Int} + f(\varphi, \psi) \in S$.
2. *Both ψ and φ are satisfiable.* Since instances of φ and ψ in $f(\varphi, \psi)$ have no variables in common, there exists a substitution s of $\top$ or $\bot$ for the variables of φ and ψ in $f(\varphi, \psi)$ such that $\text{Int} \vdash s(\varphi) \leftrightarrow \top$ and $\text{Int} \vdash s(\psi) \leftrightarrow \top$. As a consequence, $\text{Int} + f(\varphi, \psi)$ is equivalent to $\text{Int} + (A' \& A'')$, which in turn is equivalent to $\text{Int} + A'' = L''$, and $\text{Int} + f(\varphi, \psi) \notin S$.
3. *φ is unsatisfiable.* We have $Cl \vdash \neg\varphi$. By Glivenko's theorem $\text{Int} \vdash \neg\varphi$. Then $\text{Int} + f(\varphi, \psi)$ is equivalent to $\text{Int} + A' = L'$, and $\text{Int} + f(\varphi, \psi) \notin S$.

The proof for normal extensions of S4 is analogous. We only should redefine

$$f(\varphi, \psi) = A'\&(\Box A'' \vee' \neg\varphi \vee' (\Box A\&\neg\psi)). \qquad \square$$

PROPOSITION 9 *Let L_0 be in $E(\mathrm{Int})$ or in $NE(\mathrm{S4})$ and L be any finitely axiomatizable extension of L_0.*

(a) If L is consistent, then the problem of provability in L is coNP*-hard and the problem of unprovability* NP*-hard.*

(b) If $L \neq L_0$, then the problem of equivalence to L over L_0 is NP*-hard.*

(c) If L is consistent, then the problem of equivalence to L over L_0 is coNP*-hard.*

(d) If $L \neq L_0$ and L is consistent, then the problem of equivalence to L over L_0 is DP*-hard.*

PROOF. (a) B is a satisfiable boolean formula iff $\neg B$ is not provable in L_0.

(b) and (c) follow from Proposition 6 by $S = \{L\}$.

(d) In Proposition 8 take $S = \{L\}$, $L' = L_0$, $L'' = For = L_0 + \bot$, $L = L_0 + A$. Then φ is satisfiable and ψ is unsatisfiable if and only if $(L_0 + (\neg\varphi \vee' (A\&\neg\psi)) = L \in S$. $\square$

3 Tabularity and Related Properties

To find out the complexity of the tabularity and related problems we need some computational characteristics of particular logics and their models. Although there exist logics in $E(\mathrm{Int})$ and $NE(\mathrm{S4})$ which are not Kripke-complete, our main results can be proved in terms of Kripke structures. We recall some definitions.

A *frame* $\mathbf{W} = (W, R)$ is a non-empty set with a binary relation R. As we will consider only normal extensions of S4, all our frames are supposed to be reflexive and transitive. A *Kripke structure* $\mathbf{M} = (W, R, \models)$ is a frame with a truth-relation $\models$, where the boolean operations are defined in the usual way, and $x \models \Box A$ iff $\forall y(xRy \Rightarrow y \models A)$.

For extensions of Int we need a different definition. An *intuitionistic Kripke structure* is a tuple $\mathbf{M} = (W, \leq, \models_i)$ which is (a) partially ordered by $\leq$, (b) satisfies the *monotonicity condition*: for each variable p,

$$(x \models_i p \text{ and } x \leq y) \Rightarrow y \models_i p$$

and (c)

$$\begin{aligned} x \models_i A \rightarrow B &\text{ iff } \forall y(x \leq y \Rightarrow (y \models_i A \Rightarrow y \models_i B)),\\ x \models_i \neg A &\text{ iff } \forall y(x \leq y \Rightarrow y \not\models_i A). \end{aligned}$$

The relations $x \models_i (A\&B)$ and $x \models_i (A \vee B)$ are defined as usual.

If $\mathbf{M} = (W, \leq, \models)$ is a (modal or intuitionistic) Kripke structure, we say that a formula A is *true in* $\mathbf{M}$ if $x \models A$ for all $x \in W$; A is *satisfiable in* $\mathbf{M}$ if $x \models A$ for some $x \in W$; A is *refutable* in $\mathbf{M}$ if $x \not\models A$ for some $x \in W$. We say that a formula A is *valid in a frame* $\mathbf{W}$ and $\mathbf{W}$ *validates* A if A is true in any Kripke structure $\mathbf{M} = (W, \leq, \models)$ based on $\mathbf{W}$; A is *refutable in* $\mathbf{W}$ if it is not valid in $\mathbf{W}$.

Let L be a modal logic. A frame $\mathbf{W}$ is called an *L-frame* if all formulas provable in L are true in each Kripke structure based on $\mathbf{W}$. If L is a superintuitionistic logic, a frame $\mathbf{W}$ is called an *L-frame* if all formulas provable in L are true in each intuitionistic Kripke structure based on $\mathbf{W}$. For intuitionistic or modal logic L, we say that a formula A is *L-valid* if A is valid in all L-frames; A is *L-refutable* if it is not L-valid.

A logic L is called *Kripke-complete* if provability in L is equivalent to L-validity. A logic L is said to have the *finite model property (FMP)* if provability in L is equivalent to validity in all finite L-frames. It is clear that any logic with FMP is Kripke-complete. Each tabular logic L can be characterized by a suitable finite L-frame. A logic L is *polynomially (exponentially) approximable* if any formula A unprovable in L is refutable in some finite L-frame whose cardinality is bounded by a polynomial (respectively, exponential) function of the size of A.

Remark. For the aim of this paper, the notion of refutability is more important than satisfiability. In the case of modal logics one can easily see that a formula A is satisfiable in a structure if and only if $\neg A$ is refutable in the same structure. There is no such a straightforward direct reduction for intuitionistic structures.

In our calculation we use a lemma whose proof is given in Proposition 3.1 of [5]:

LEMMA 10 *(See [5].) Given a structure M and a formula A, there is an algorithm for calculating the value of A in M that runs in time $\mathcal{O}(\|M\| \times |A|)$, where $\|M\|$ is the sum of the number of elements in the frame and the number of pairs in R.*

As a consequence, we easily obtain the following result.

LEMMA 11 *For each finite (intuitionistic or modal) frame* $\mathbf{W}$, *refutability in* $\mathbf{W}$ *is* NP-*complete and validity in* $\mathbf{W}$ *is* coNP-*complete.*

It is known that any property of frames expressible in first order logic with one binary relation is recognizable on a finite frame in polynomial time w.r.t. the size of the frame (e.g., [21], Theorem 5.1). By Lemma 10 we obtain the following result.

LEMMA 12 *If a logic L in E(Int) or in NE(K) is polynomially (or exponentially) approximable by a class of frames definable by finitely many first order formulas, then the L-refutability problem is in* NP *(resp. in* NEXP*) and the L-provability problem is in* coNP *(respectively, in* coNEXP*).*

By Proposition 9(a) we conclude the following.

LEMMA 13 *If a consistent logic L in E(Int) or in NE(K) is polynomially approximable by a class of frames definable by finitely many first order formulas, then the L-refutability problem is* NP-*complete and the L-provability problem is* coNP-*complete.*

The NP-completeness of refutability problem for many superintuitionistic and modal logics was proved in [2].

PROPOSITION 14 *(a) For each consistent tabular logic L in E(Int) or in NE(S4) the problem of equivalence to L over* Int *(respectively, over* S4*) is* DP-*complete.*

(b) The consistency problems over both Int *and* S4 *are* coNP-*complete.*

PROOF. (a) Kuznetsov [8] proved that for every tabular logic $L \in E(\text{Int})$ there exist finitely many tabular logics $L_1, \ldots, L_k$ in $E(\text{Int})$ such that for every $L' \in E(\text{Int})$,

$$L' \supseteq L \text{ iff } L' \not\subseteq L_i \text{ for } i = 1, \ldots, k.$$

It means that for every tabular $L \in E(\text{Int})$ there exist finitely many finite frames $\mathbf{W}_1, \ldots, \mathbf{W}_k$ such that for each formula A

$$\text{Int} + A \supseteq L \text{ iff } A \text{ is refutable in all } \mathbf{W}_1, \ldots, \mathbf{W}_k.$$

We see from Lemma 11 that the inclusion problem $\text{Int} + A \supseteq L$ is in NP. By the same Lemma, the problem $\text{Int} + A \subseteq L$ (which is the problem of L-validity of A) is coNP-complete, so the problem of equivalence to L over Int is in DP. By Proposition 9 this problem is DP-hard, so it is DP-complete. The proof for modal logics in $NE(\text{S4})$ is analogous. An analog of mentioned Kuznetsov's result for extensions of S4 was proved in [19].

(b) It is known that $\text{Int}+A$ is consistent iff A is a two-valued tautology. Also $\text{S4}+A$ is consistent iff A is valid in the one-element reflexive frame. ❑

Let us define the following sequences of frames for $n \geq 1$:

S_n is the set $\{1, \ldots, n\}$ with the natural ordering relation;
U_{n+1} is the set $\{0, 1, \ldots, n+1\}$, where $0RxR(n+1)$ for all x and $\neg xRy$ for $1 \leq x, y \leq n$, $x \neq y$;
V_n is the subframe of U_{n+1} obtained by deleting $(n+1)$;
X_n is $\{1, \ldots, n\}$, where xRy for all x, y;
Y_{n+1} is the set $\{1, \ldots, n+1\}$, where xRy for all $x, y \leq n$ and $xR(n+1)$ for all x.

In order to prove Theorem 1, we remind [10] that there are exactly three pretabular extensions of Int, namely, the logics

$$\begin{aligned} \text{LC} &= \text{Int} + (p \to q) \vee (q \to p), \\ LP_2 &= \text{Int} + (p \vee (p \to q \vee \neg q)), \\ LQ_3 &= \text{Int} + (r \vee (r \to (p \vee (p \to q \vee \neg q))) + \neg p \vee \neg\neg p. \end{aligned}$$

The logic LC is characterized by the finite linearly ordered intuitionistic frames, LP_2 by the finite frames satisfying the condition $\forall x \forall y \forall z (x \leq y \leq z \Rightarrow (x = y \text{ or } y = z))$. At last, LQ_3 is characterized by the finite frames with the least and the greatest elements and with the chains of length not more than 3.

The family $NE(\text{S4})$ contains exactly five pretabular logics ([12]):

$$\begin{aligned} PM1 &= \text{Grz.3}, \\ PM2 &= \text{Grz.2} + (\Box p \vee \Box(\Box p \to \Box q \vee \Box(\Box q \to \Box r \vee \Box\Diamond\neg r))), \\ PM3 &= \text{S4} + (\Box(p \to \Box p) \vee \Box(\Diamond p \to p)), \\ PM4 &= \text{S4.1} + (\Box q \vee \Box(\Box q \to \Box r \vee \Box\Diamond\neg r)), \\ PM5 &= \text{S5}. \end{aligned}$$

They are characterized, respectively, by the classes of frames $\{S_n | n > 0\}$, $\{U_n | n > 0\}$, $\{V_{n+1} | n > 0\}$, $\{Y_{n+1} | n > 0\}$, $\{X_n | n > 0\}$.

It means that provability in L is equivalent to L-validity for all these logics, and unprovability is equivalent to L-refutability.

LEMMA 15 *Each pretabular logic in $E(\text{Int})$ or $E(\text{S4})$ is linearly approximable.*

PROOF. One can show that if a modal formula A is refutable in a frame S_n (respectively, V_n, U_n, Y_n, X_n), then it is refutable in S_m (respectively, V_m, U_m, Y_m, X_m), where $m \leq |A| + 3$. Also if a modal-free formula A is refutable in an intuitionistic structure based on S_n, V_n or U_n then it is refutable in, respectively, S_m, V_m or U_m for $m \leq |A| + 3$. ❑

NP-completeness of the satisfiability problem was proved for S5 in [9] and for Grz.3 in [23]. We state

LEMMA 16 *For each pretabular logic L in $E(\text{Int})$ or in $NE(\text{S4})$, refutability problem of L is* NP*-complete, and provability problem of L is* coNP*-complete.*

PROOF. By Lemmas 15 and 12 L-refutability problem is in NP for each logic under consideration. On the other hand, refutability problem is NP-hard for any pretabular logic by Proposition 9(a). ❑

To find out the complexity of pretabularity problem, we need one more proposition.

PROPOSITION 17 *Let L_0 be* Int *or* S4 *and L any its pretabular extension. Then the inclusion problem $L_0 + A \supseteq L$ over L_0 is* NP*-complete, and the problem of equivalence to L over L_0 is* DP*-complete.*

PROOF. For the inclusion problem we use the following well-known criteria [6, 13, 14] (refutability is considered in intuitionistic structures based on the above-mentioned frames):

$$\begin{array}{lll} \text{Int} + A \supseteq \text{LC} & \text{iff} & A \text{ is refutable in both } V_2 \text{ and } U_3, \\ \text{Int} + A \supseteq LP_2 & \text{iff} & A \text{ is refutable in } S_3, \\ \text{Int} + A \supseteq LQ_3 & \text{iff} & A \text{ is refutable in both } V_2 \text{ and } S_4. \end{array}$$

Below we mean refutability in modal Kripke structures.

$$\begin{array}{lll} \text{S4} + A \supseteq PM1 & \text{iff} & A \text{ is refutable in all } V_2, U_3, X_2, Y_3, \\ \text{S4} + A \supseteq PM2 & \text{iff} & A \text{ is refutable in all } V_2, S_4, X_2, Y_3, \\ \text{S4} + A \supseteq PM3 & \text{iff} & A \text{ is refutable in all } S_3, X_2, Y_3, \\ \text{S4} + A \supseteq PM4 & \text{iff} & A \text{ is refutable in all } S_3, V_2 \text{ and } X_2, \\ \text{S4} + A \supseteq PM5 & \text{iff} & A \text{ is refutable in } S_2. \end{array}$$

We see that for every pretabular logic L we only should verify refutability of A in finitely many fixed finite frames, so our inclusion problem for L is in NP by Lemma 11. Further, $L_0 + A$ is equivalent to L if and only if A is valid in L and $L_0 + A \supseteq L$. The problem of L-validity is coNP-complete by Lemma 16. It follows that the problem of equivalence to L over L_0 is in DP, so it is DP-complete by Proposition 9(d). ❑

Now we are in a position to prove Theorem 1:
PROOF (of Theorem 1). (i) *The tabularity problems over both* Int *and* S4 *are* NP*-complete.* Kuznetsov [8] proved that a logic is tabular if and only if it is not contained in L_1 for any pretabular logic L_1. So a logic Int $+ A$ is tabular if and only if A is

refutable in each of LC, LP_2 and LQ_3. Thus the tabularity problem over Int is in NP by Lemma 16. On the other hand, this problem is NP-hard by Proposition 6(a) (one can take $L_1 = \text{Int}$ and $L_2 = For$).

The family $NE(\text{S4})$ contains five pretabular logics, and one can prove the statement by analogy.

(ii) *The pretabularity problems over both* Int *and* S4 *are both* NP*-hard and* coNP*-hard, and belong to in* Δ_2^p. The pretabularity problem over Int is the problem of membership in the set $\{\text{LC}, LP_2, LQ_3\}$ which is in Δ_2^p by Proposition 17. On the other hand, it is DP-hard by Proposition 8. The proof for modal logics is analogous.

(iii) *The local tabularity problem over* S4 *is* NP*-complete.* A logic in $NE(\text{S4})$ is locally tabular if and only if it is not contained in Grz.3 [11]. So local tabularity of $\text{S4} + A$ is equivalent to refutability of A in Grz.3 which is NP-complete by Lemma 16.

❑

4 Interpolation and Amalgamation

In this section we prove Theorems 2 and 3. Theorem 2(i) is an immediate corollary of Propositions 18(i) and 19. Theorem 2(ii) is proved in Proposition 18(ii) and the rest of Theorem 2 in Proposition 24. Theorem 3 follows from Theorem 2 and Proposition 25.

We start with

PROPOSITION 18 *(i) The interpolation problem over* Int *is* PSPACE*-complete;*
(ii) The problem of determining whether $\text{Int} + A$ *is a tabular logic with CIP is* NP*-complete.*

PROOF. In [13] the first author proved that in $E(\text{Int})$ there are exactly eight logics with Craig's interpolation property CIP, which is equivalent to IPD in the extensions of Int. The list of these logics consists of Int, KC, LC, LP_2, Cl, *For*, $\text{LC} + LP_2$ and $LP_2 + (p \to q) \vee (q \to p) \vee (p \leftrightarrow \neg q)$. Note that all these logics have the Hallden Property.

The logics Cl, $\text{LC} + LP_2$, $LP_2 + (p \to q) \vee (q \to p) \vee (p \leftrightarrow \neg q)$ are characterized, respectively, by the frames S_1, S_2 and V_2; LC by $\{S_n | n > 0\}$, LP_2 by $\{V_n | n > 0\}$ and KC by finite partially ordered frames having the greatest element. In addition, we have the following criterion for inclusion ([13]):

$\text{Int} + A \supseteq For$	iff A is refutable in S_1,
$\text{Int} + A \supseteq \text{Cl}$	iff A is refutable in S_2,
$\text{Int} + A \supseteq (\text{LC} + LP_2)$	iff A is refutable in both V_2 and S_3,
$\text{Int} + A \supseteq LP_2 +$ $(p \to q) \vee (q \to p) \vee (p \leftrightarrow \neg q)$	iff A is refutable in both V_3 and S_3,
$\text{Int} + A \supseteq \text{LC}$	iff A is refutable in both V_2 and U_3,
$\text{Int} + A \supseteq LP_2$	iff A is refutable in S_3,
$\text{Int} + A \supseteq \text{KC}$	iff A is refutable in V_2.

So for all proper extensions of Int with CIP, the inclusion problem $\text{Int} + A \supseteq L$ is equivalent to refutability of A in one or two finite frames which is in NP by Lemma 11.

The provability problem is coNP-complete for tabular logics Cl, LC$+LP_2$, $LP_2+(p\to q)\vee(q\to p)\vee(p\leftrightarrow\neg q)$, and also for pretabular logics LC and LP_2. It is clear that the equivalence problem is in DP for each of these logics.

Further, since the validity problem in Int is PSPACE-complete [24], so is the problem of equivalence to Int over Int. To find the complexity bounds for validity in KC, we note that

$$\mathrm{KC}\vdash A \text{ iff } \mathrm{Int}\vdash(\neg P_1\vee\neg\neg P_1)\&\dots\&(\neg P_n\vee\neg\neg P_n)\to A,$$

where $P_1,\dots,P_n$ are all variables of A, so validity in KC is in PSPACE. On the other hand, KC contains the implicational fragment of Int [25] which is PSPACE-hard [24], so validity in KC is PSPACE-complete. It follows that equivalence to KC is in PSPACE, and it is PSPACE-complete by Proposition 5. Thus equivalence to L is in PSPACE for each logic L with CIP in $E(\mathrm{Int})$, so the interpolation problem over Int is in PSPACE. To use Proposition 5, we take, for instance, $L_1=\mathrm{Int}$ and $L_2=\mathrm{KC}\cap LP_2$. Therefore the interpolation problem over Int is PSPACE-complete.

To prove (ii), we note that Int$+A$ is a tabular logic with CIP if and only if Int$+A\supseteq$ Int $+(p\to q)\vee(q\to p)\vee(p\leftrightarrow\neg q)$, and we already know that this inclusion problem is NP-complete. ❑

Note. PSPACE-completeness of the provability problem in KC was proved in [2]. Tableau calculi and cut-free sequent calculi for superintuitionistic logics with CIP are constructed in [1].

Boolos [4] proved Craig's Interpolation Property CIP for the Grzegorchyk logic Grz. The complete list of normal extensions of Grz with the interpolation property IPD contains seven logics (see [18]), namely, Grz, Grz.2, $PM3=\mathrm{Grz}+(\Box(p\to\Box p)\vee\Box(\Diamond p\to p))$, $Gr_4=PM3+(\Diamond\Box p\&\Diamond\Box q\&\Diamond\Box r\to\Diamond\Box(p\&q)\vee\Diamond\Box(p\&r)\vee\Diamond\Box(q\&r)$, $Gr_5=\mathrm{Grz.2}+Gr_4$, $Gr_6=\mathrm{Grz}+(p\to\Box p)$, $Gr_7=Fm$. All of them have CIP, so IPD and CIP are equivalent in $NE(\mathrm{Grz})$.

It is known [2] that provability problem for Grz is PSPACE-complete. Using this result, we assert

PROPOSITION 19 *The interpolation problem over* Grz *is* PSPACE-*complete.*

PROOF. We note that the logics Gr_4-Gr_7 are tabular and $PM3$ is pretabular, so for each of these logics the problem of equivalence is in DP by Propositions 14 and 17.

Further, we note that

$$\mathrm{Grz.2}\vdash B \text{ iff } \mathrm{Grz}\vdash A(p_1)\&\dots\&A(p_n)\to B,$$

where $A(p)=\Box\Diamond p\to\Diamond\Box p$ and $p_1,\dots,p_n$ are all variables of B, so the provability problem of Grz.2 is linearly reducible to that of Grz, and is in PSPACE.

In addition, we have ([14]):

$$\mathrm{S4}+B\supseteq\mathrm{Grz.2} \text{ iff } B \text{ is refutable in all } X_2, Y_3 \text{ and } V_2,$$

so the inclusion problem S4 $+B\supseteq$ Grz.2 is in NP, and the problem of equivalence to Grz.2 is in PSPACE.

Taking into account that the problem of equivalence to Grz over Grz is PSPACE-complete, we conclude that the interpolation problem over Grz is in PSPACE. On the other hand, the problem of provability in Grz is reducible to the interpolation problem by Proposition 5 (it is sufficient to take $L_1 = \mathrm{Grz}$ and $L_2 = \mathrm{Grz.2} \cap PM3$). ❑

Note. One can reduce the provability problem of KC to the provability problem of Grz.2, so the latter problem as well as the problem of equivalence to Grz.2 is PSPACE-complete.

Let us now consider the interpolation properties CIP and IPD over S4. It is known that CIP implies IPD in extensions of S4. The first author proved in [14, 17] that there exist not more than forty nine logics with IPD in $NE(\mathrm{S4})$, among them at most thirty seven have CIP, and at least twelve logics have IPD and do not possess CIP. At the present time, it is known that at least thirty one logics in $NE(\mathrm{S4})$ really have CIP and at least forty three logics have IPD ([16, 18]). Although the interpolation problem in $NE(\mathrm{S4})$ is not yet completely solved, nevertheless, it is decidable since for each logic L in the mentioned list of forty nine logics, the problem of equivalence to L is decidable over S4 [14].

The list $\mathcal{L}$ of the forty nine logics containing all logics with IPD in $NE(\mathrm{S4})$ consists of thirty one locally tabular logics (for which the interpolation problem is completely solved) and of eighteen logics which are not locally tabular. All logics in $\mathcal{L}$ are finitely axiomatizable, and have FMP and Hallden property. Each logic in $\mathcal{L}$ was described in detail in [14]. Using this description, we can state two lemmas:

LEMMA 20 *Let L be a logic with IPD in $NE(\mathrm{S4})$, $L \neq \mathrm{S4}$. Then the inclusion problem* $\mathrm{S4} + A \supseteq L$ *is* NP*-complete.*

LEMMA 21 *Let L be a consistent and locally tabular logic with IPD in $NE(\mathrm{S4})$. Then L is polynomially approximable, the L-provability problem is* coNP*-complete, and the problem of equivalence to L over* S4 *is* DP*-complete.*

In order to find complexity bounds for logics in $\mathcal{L}$ which are not locally tabular, we recall some definitions.

In [12] the first author introduced two characteristics μ_1 and μ_2 of frames and logics in $NE(\mathrm{S4})$ as follows. With any element a of a frame $\mathbf{W}$ we associate its *cluster*, i.e., the set $C(a) = \{x \in \mathbf{W} | aRxRa\}$. A cluster of a is *internal* if there is y such that aRy and $\neg yRa$, and *external* otherwise. We define $\mu_1(\mathbf{W})$ and $\mu_2(\mathbf{W})$ as the suprema of cardinalities of external and of internal clusters in $\mathbf{W}$ respectively. If L is a calculus, we define

$$\mu_i(L) = sup\{\mu_i(\mathbf{W}) | \mathbf{W} \text{ is a finite } L\text{-frame}\}.$$

It was proved in [12] that $\mu_1(L) = \omega$ iff $L \subseteq \mathrm{S5}$, and $\mu_2(L) = \omega$ iff $L \subseteq PM4$, where $PM4$ is one of five pretabular logics in $NE(\mathrm{S4})$.

Also we need two sequences of formulas ξ_n and η_{n+1}, defined as follows:

$\xi_n = \varphi_n \to \neg p_1$, where φ_n is the conjunction of formulas
$\Box(p_i \to \neg p_j)$ for all $i \neq j$;
$\Box(p_i \to \Diamond p_j)$ for all $i, j \leq n$;
$\Box(p_1 \vee \ldots \vee p_n)$;

$\eta_{n+1} = \psi_n \to \neg p_1$, where ψ_n is the conjunction of formulas
$\Box(p_i \to \neg p_j)$ for all $i \neq j$;
$\Box(p_i \to \Diamond p_j)$ for all $i \leq n$ and $j \leq n+1$;
$\Box(p_{n+1} \to \neg\Diamond p_j)$ for all $j \leq n$;
$\Box(p_1 \vee \ldots \vee p_{n+1})$.

The following Lemma was proved in [15].

LEMMA 22 *Let* $L = \mathrm{S4} + A$. *Then for all* $n > 0$
(a) $L \vdash \xi_{n+1}$ *iff* $\mu_1(L) \leq n$;
(b) $L \vdash \eta_{n+2}$ *iff* $\mu_2(L) \leq n$.

Now we turn to logics in the list $\mathcal{L}$ which are not locally tabular.

LEMMA 23 *Let L be a non-locally-tabular logic with IPD in NE*(S4). *Then both problems of L-provability and of equivalence to L over* S4 *are in* coNEXP *and* PSPACE-*hard.*

PROOF. Our list $\mathcal{L}$ contains eighteen logics which are not locally tabular. They are of the form L, $L + \xi_{i+1}$, $L + \eta_{j+2}$, $L + \xi_{i+1} + \eta_{j+2}$ for $L \in \{\mathrm{S4}, \mathrm{S4.2}\}$ and $1 \leq i, j \leq 2$. It was proved in [15, 14] that all logics $\mathrm{S4} + \xi_{i+1}$, $\mathrm{S4} + \eta_{j+2}$, $\mathrm{S4} + \xi_{i+1} + \eta_{j+2}$ have the finite model property. They are characterized, respectively, by all finite frames $\mathbf{W}$ with $\mu_1(\mathbf{W}) \leq i$ and/or $\mu_2(\mathbf{W}) \leq j$. The classes of finite frames for S4.2, $\mathrm{S4.2} + \xi_{i+1}$, $\mathrm{S4.2} + \eta_{j+2}$, $\mathrm{S4.2} + \xi_{i+1} + \eta_{j+2}$ should, in addition, satisfy the condition: $(xRy$ and $xRz) \Rightarrow \exists u(yRu$ and $zRu)$.

Moreover, if a formula A is unprovable in some logic L of this list, then there exists a finite frame of a size not more than $2^{14|A|}$ which validates L and refutes A [15, 14]. It follows by Lemma 12 that the refutability problem of L is in NEXP, and L-provability problem is in coNEXP. The inclusion problem $\mathrm{S4} + A \supseteq L$ is in NP for all these logics by Lemma 20, so the problem of equivalence to L over S4 is in coNEXP.

On the other hand, there exists the well-known Gödel-Tarski translation T from Int to S4 defined as follows:

$$T(p) = \Box p \text{ for any variable } p;$$
$$T(A\&B) = T(A)\&T(B);$$
$$T(A \vee B) = T(A) \vee T(B);$$
$$T(A \to B) = \Box(T(A) \to T(B));$$
$$T(\neg A) = \Box\neg T(A).$$

This function T also reduces Int to the logics $\mathrm{S4}+\xi_{i+1}$, $\mathrm{S4}+\eta_{j+2}$, $\mathrm{S4}+\xi_{i+1}+\eta_{j+2}$, and KC to each of logics S4.2, $\mathrm{S4.2} + \xi_{i+1}$, $\mathrm{S4.2} + \eta_{j+2}$, $\mathrm{S4.2} + \xi_{i+1} + \eta_{j+2}$. It follows that L-provability problem is PSPACE-hard for all logics of this lemma. ❑

PROPOSITION 24 *(i) Both CIP and IPD problems over* S4 *are in* coNEXP *and* PSPACE-*hard.*
(ii) The problems of determining whether S4 + A *is a tabular (pretabular or locally tabular) logic with CIP (or IPD) are in* Δ_2^p *and* DP-*hard.*

PROOF. It was proved in [16, 18] that the family of logics with IPD in $NE(\mathrm{S4})$ contains the thirty one locally tabular logics (of which one is inconsistent and thirty are consistent) and also the logics L, $L+\xi_{i+1}$, $L+\eta_3$, $L+\xi_{i+1}+\eta_3$, where $L \in \{\mathrm{S4}, \mathrm{S4.2}\}$ and $1 \leq i \leq 2$, which are not locally tabular. These non-locally-tabular logics and, in addition, nineteen locally tabular logics in $NE(\mathrm{S4})$ have CIP. We do not know whether the six logics $L+\eta_4$, $L+\xi_2+\eta_4$, $L+\xi_3+\eta_4$ have IPD (or CIP) or not.

The problem of equivalence over S4 to the inconsistent logic Fm is the same as the inconsistency problem, so it is NP-complete by Proposition 14(b). For each consistent and locally tabular logic L with IPD the problem of equivalence to L over S4 is DP-complete by Lemma 21. It follows from Lemma 23 that both CIP and IPD problems over S4 are in coNEXP.

On the other hand, the logic S4 has both CIP and IPD, and no proper extension of S4 contained in $(\mathrm{S4}+\xi_4)$ has IPD, so both CIP and IPD problems over S4 are PSPACE-hard by Proposition 5. Thus, we proved (i).

Pretabular logics with IPD are $PM2-PM5$, each of them has CIP and is locally tabular. The problem of membership in the set $\{PM2,\ldots,PM5\}$ is in Δ_2^p and DP-hard by Propositions 17 and 8.

The problem of determining whether $\mathrm{S4}+A$ is a tabular (or locally tabular) logic with IPD (or CIP) is in Δ_2^p by Lemma 21. On the other hand, this problem is DP-hard by Proposition 8: it is sufficient to note that the logic L determined by the three-element frame $\{0,1,2\}$, where $0R1R2R1$, is tabular and has CIP, $L' = \mathrm{S4}$ is not locally tabular, and a tabular logic L'' determined by two two-element frames S_2 and X_2 (see Section 4) includes L and does not possess IPD. ❑

It is well-known that there is a duality between $E(\mathrm{Int})$ and the family of varieties of Heyting algebras, and also between $NE(\mathrm{S4})$ and the family of varieties of closure algebras. If Ax is a set of formulas and $L = \mathrm{Int} + Ax$ then $\{A = \top | A \in Ax\}$ forms a base of identities for the variety $V(L)$ associated with L. The definitions for the logics $L \in NE(\mathrm{S4})$ are analogous. On the other hand, with any identity $A = B$ one can associate a formula $A \leftrightarrow B$ which is valid in some algebra if and only if the identity $A = B$ holds in this algebra.

We remind that a class $\mathfrak{T}$ is *amalgamable* if it satisfies the following condition for all algebras $\mathbf{A}, \mathbf{B}, \mathbf{C}$ in $\mathfrak{T}$.

(AP) For all monomorphisms $\alpha : \mathbf{A} \to \mathbf{B}$, $\beta : \mathbf{A} \to \mathbf{C}$ there exist an algebra $\mathbf{D}$ in $\mathfrak{T}$ and monomorphisms $\gamma : \mathbf{B} \to \mathbf{D}$, $\delta : \mathbf{C} \to \mathbf{D}$ such that $\gamma\alpha = \delta\beta$.

We say that $\mathfrak{T}$ is *super-amalgamable*, if for any algebras $\mathbf{A}, \mathbf{B}, \mathbf{C}$ in $\mathfrak{T}$, the condition AP is satisfied and, in addition, for each $x \in \mathbf{B}$, $y \in \mathbf{C}$ the following equivalences hold:

$$\gamma(x) \leq \delta(y) \Leftrightarrow (\exists z \in \mathbf{A})(x \leq \alpha(z) \& \beta(z) \leq y) \text{ and}$$
$$\gamma(x) \geq \delta(y) \Leftrightarrow (\exists z \in \mathbf{A})(x \geq \alpha(z) \& \beta(z) \geq y).$$

We will describe the complexity of the amalgamability problem for a variety with respect to the sum of the lengths of its base of identities.

The following Proposition is an immediate corollary of Theorem 1 in [13] and of Theorems 1 and 2 in [14].

PROPOSITION 25 *Let L be in $E(\mathrm{Int})$ or in $NE(\mathrm{S4})$. Then*
(i) L has CIP iff $V(L)$ is super-amalgamable;
(ii) L has IPD iff $V(L)$ is amalgamable.

Using Theorem 2 and Proposition 25, we immediately obtain

Theorem 3 *(i) The amalgamability problem for finitely based varieties of Heyting algebras is* PSPACE*-complete.*
(ii) The amalgamability and super-amalgamability problems for finitely based varieties of closure algebras are in coNEXP *and* PSPACE*-hard.*

References

1. A.Avellone, M.Ferrari, and P.Miglioli. Duplication-free tableau calculi and related cut-free sequent calculi for the interpolable propositional intermediate logics. *Logic Journal of the IGPL*, 7(4):447–480, 1999.
2. A. Chagrov and M. Zakharyaschev. *Modal Logic*. Clarendon Press, Oxford, 1997.
3. M.J. Fisher and R.E. Ladner. Propositional dynamic logic of regular programs. *Journal of Computer and System Sciences*, 18(2):194–211, 1979.
4. G.Boolos. On systems of modal logic with provability interpretations. *Theoria*, 45(1):7–18, 1980.
5. J.Y. Halpern and Y. Moses. A guide to completeness and complexity for modal logics of knowledge and belief. *Artificial Intelligence*, 54:319–379, 1992.
6. T. Hosoi and H. Ono. Intermediate propositional logics. a survey. *J. of Tsuda College*, 5:67–82, 1963.
7. D. S. Johnson. A catalog of complexity classes. In J. van Leeuwen, editor, *Handbook of Theoretical Computer Science*, volume A, chapter 8, pages 67–161. Elsevier Science, 1990.
8. A.V. Kuznetsov. Some properties of the lattice of varieties of pseudo-boolean algebras. In *11th Sovjet Algebraic Colloquium, Abstracts*, pages 255–256, Kishinev, 1971.
9. R.E. Ladner. The computational complexity of provability in systems of modal propositional logic. *SIAM Journal of Computing*, 6(3):467–480, 1977.
10. L.L. Maksimova. Pretabular superintuitionistic logics. *Algebra and Logic*, 11(5):558–570, January 1972.
11. L.L. Maksimova. Modal logics of finite slices. *Algebra and Logic*, 14(3):304–319, 1975.
12. L.L. Maksimova. Pretabular extensions of Lewis' logic S4. *Algebra and Logic*, 14(1):28–55, 1975.
13. L.L. Maksimova. Craig's theorem in superintuitionistic logics and amalgamable varieties of pseudoboolean algebras. *Algebra and Logic*, 16(6):643–681, 1977.
14. L.L. Maksimova. Interpolation theorems in modal logics and amalgamable varieties of topoboolean algebras. *Algebra and Logic*, 18(5):556–586, 1979.
15. L.L. Maksimova. On a classification of modal logics. *Algebra and Logic*, 18(3):328–340, 1979.
16. L.L. Maksimova. Interpolation theorems in modal logic: Sufficient conditions. *Algebra and Logic*, 19(2):194–213, 1980.
17. L.L. Maksimova. Absence of interpolation in modal companions of Dummett's logic. *Algebra and Logic*, 21:690–694, 1982.
18. L.L. Maksimova. Amalgamation and interpolation in normal modal logics. *Studia Logica*, 50(3/4):457–471, 1991.

19. L.L. Maksimova and V.V.Rybakov. On a lattice of normal modal logics. *Algebra and Logic*, 13:188–216, 1974.
20. S. Miura. A remark on the intersection of two logics. *Nagoya Math. Journal*, 26:167–171, 1966.
21. C.H. Papadimitriou. *Computational Complexity*. Addison-Wesley, 1994.
22. V.R. Pratt. Models of program logics. In *Proceedings 20th IEEE Symposium on Foundations of Computer Science*, pages 115–122, 1979.
23. E. Spaan. *Complexity of Modal Logics*. PhD thesis, Institute for Logic, Language and Computation, University of Amsterdam, 1992.
24. R. Statman. Intuitionistic propositional logic is polynomial-space complete. *Theoretical Computer Science*, 9:67–72, 1979.
25. V.A. Yankov. On the calculus of weak low of excluded middle. *Izvestia Academii Nauk SSSR, Ser. math.*, 32(5):1044–1051, 1968.

Goal-Directed Calculi for Gödel-Dummett Logics

George Metcalfe[1], Nicola Olivetti[2], and Dov Gabbay[1]

[1] Department of Computer Science, King's College London, Strand, London WC2R 2LS, UK
{metcalfe,dg}@dcs.kcl.ac.uk

[2] Department of Computer Science, University of Turin, Corso Svizzera 185, 10149 Turin, Italy
olivetti@di.unito.it

Abstract. In this work we present *goal-directed calculi* for the Gödel-Dummett logic **LC** and its finite-valued counterparts, $\mathbf{LC_n}$ ($n \geq 2$). We introduce a terminating hypersequent calculus for the implicational fragment of **LC** with local rules and a single identity axiom. We also give a *labelled* goal-directed calculus with invertible rules and show that it is co-NP. Finally we derive labelled goal-directed calculi for $\mathbf{LC_n}$.

Keywords: Gödel Logics, Intermediate Logics, Fuzzy Logics, Hypersequents, Goal-Directed Calculi.

1 Introduction

Matrices for the finite-valued Gödel logics $\mathbf{LC_n}$ ($n = 2, 3, \ldots$) were introduced by Gödel in [14]; an infinite-valued version was presented and axiomatised by Dummett in [7] and is known as the Gödel-Dummett logic **LC**.

LC is important as a logic *intermediate* between intuitionistic logic **LJ** and classical logic **LK**. It has been used in investigations of relevance logics [8] and the provability logic of Heyting algebras [22], and has attracted renewed interest recently as one of the fundamental *t-norm based fuzzy logics* [16]. Moreover, the three-valued Gödel logic $\mathbf{LC_3}$ is of interest as the logic of "here and there" [18] used to check the strong equivalence of logic programs.

A proof calculus for **LC** extending Gentzen's calculus for **LJ** has been provided by Avron [3] using *hypersequents*, a generalisation of Gentzen-style sequents. Although very natural, this calculus is not particularly suitable for proof search since its rules are both non-invertible and non-terminating. The first *single* sequent calculus for **LC** was given by Sonobe in [21]; terminating and contraction-free versions with invertible rules were subsequently developed by Avellone et al. [2], Dykhoff [9] and Fiorino [10]. These calculi all possess rules with an arbitary number of premises however. Responding to this problem, Avron and Koniskowa [4] and Larchey-Wendling [17] have presented systems where all rules are *local*. Indeed Avron and Konsiskowa have introduced three new systems; two single sequent calculi and a hypersequent calculus, the latter turning out to be identical under translation to the *sequents of relations* calculus given by Baaz and Fermüller in [5]. Systems for the *finite-valued* Gödel logics can be obtained

M. Baaz and J.A. Makowsky (Eds.): CSL 2003, LNCS 2803, pp. 413–426, 2003.

from these calculi by adding extra axioms. More general approaches to automated deduction in finite-valued logics are described in [15, 6]. We also mention here the calculus of Aguzzoli and Gerla [1] which uses the fact that a formula is valid in **LC** iff it is valid in $\mathbf{LC_n}$ where n is a function of the size of the formula.

In this work we develop *goal-directed calculi* for the implicational fragments[1] of **LC** and $\mathbf{LC_n}$ following the methodology of Gabbay and Olivetti [12]. The goal-directed approach is a generalisation of the logic programming style of deduction particularly suitable for proof-search. It has been applied to a wide range of logics including classical and intuitionistic logics, intermediate logics, modal logics and substructural logics [12, 20]. Goal-directed methods have also been developed for **LC** [12] but the systems obtained are not particularly suitable for proof search.

We proceed as follows. In Sections 2 and 3 we introduce Gödel-Dummett logics and goal-directed methods respectively. Then in Sections 4 and 5 we give goal-directed hypersequent and labelled goal-directed calculi for **LC**. Finally in Section 6 we present goal-directed labelled calculi for $\mathbf{LC_n}$.

2 Gödel-Dummett Logics

The language is built inductively in the usual way from propositional variables $p_1, p_2, \ldots$, binary connectives $\rightarrow, \wedge$ and $\vee$, and the constants $\bot$ and t.

Definition 1 (Gödel-Dummett Logics). $\mathbf{LC_n} = \langle [0, 1/(n-1), \ldots, (n-2)/(n-1), 1], \rightarrow, \wedge, \vee, \bot, t \rangle$ *and* $\mathbf{LC} = \langle [0,1]_{\mathbb{R}}, \rightarrow, \wedge, \vee, \bot, t \rangle$ *where* $A \wedge B = min(A, B)$, $A \vee B = max(A, B)$, $\bot = 0$, $t = 1$ *and* $A \rightarrow B = 1$ *if* $A \leq B$, B *otherwise. We write* $\models A$ ($\models_n A$) *iff* $1 \leq v(A)$ *for every valuation* v *in* **LC** ($\mathbf{LC_n}$).

LC is characterised by *linearly ordered* Kripke models and $\mathbf{LC_n}$ by linearly ordered Kripke models with $n-1$ elements (see eg [11] for details).

Definition 2 (Axiomatisations for $\mathbf{LC}^{\rightarrow}$ and $\mathbf{LC_n}^{\rightarrow}$). *The axiomatisation of* $\mathbf{LC}^{\rightarrow}$ *consists of modus ponens and the following axioms [7], adopting the convention of writing* $A_1 \rightarrow \ldots \rightarrow A_n \rightarrow B$ *for* $A_1 \rightarrow (A_2 \rightarrow (\ldots (A_n \rightarrow B) \ldots)$*:*

1. $A \rightarrow B \rightarrow A$
2. $(A \rightarrow B \rightarrow C) \rightarrow (A \rightarrow B) \rightarrow A \rightarrow C$
3. $((A \rightarrow B) \rightarrow C) \rightarrow ((B \rightarrow A) \rightarrow C) \rightarrow C$

$\mathbf{LC_n}^{\rightarrow}$ *is obtained for* $n \geq 2$ *by adding the axiom:*

$$(A_1 \rightarrow B) \rightarrow ((A_1 \rightarrow A_2) \rightarrow B) \rightarrow \ldots \rightarrow ((A_{n-1} \rightarrow A_n) \rightarrow B) \rightarrow B$$

3 Goal-Directed Methods

In this paper we adopt the following "logic programming style" goal-directed paradigm of deduction. For a given logic let us denote by $\Gamma \vdash^? A$ the query

[1] Goal-directed calculi for the full propositional language have been developed but are not presented here due to lack of space.

"does A follow from Γ?" where Γ is a database (collection) of formulae and A is a goal formula. The deduction is *goal-directed* in the sense that the next step in a proof is determined by the form of the current goal: a complex goal is decomposed until its atomic constituents are reached, an atomic goal q is matched (if possible) with the "head" of a formula $G \to q$ in the database, and its "body" G asked in turn. This can be viewed as a sort of resolution step, or generalised Modus Tollens.

This model of deduction can be refined in several ways: (1) by putting constraints/labels on database formulae, restricting those available to match an atomic goal, (2) by adding more control to ensure termination, either by loop-checking or by "diminishing resources" ie removing formulae "used" to match an atomic goal, (3) by re-asking goals previously occurring in the deduction using *restart rules.* Note however that for applications such as deductive databases and logic programming, a terminating proof procedure is not always essential. We might want to get a proof of the goal from the database quickly if one exists, but be willing to have no answer otherwise (and preempt termination externally).

Goal-directed procedures have been proposed for a variety of logics, in some cases as refinements of sequent calculi called "Uniform proof systems" [20]. Here we illustrate the goal-directed methodology by presenting the diminishing resources with bounded restart algorithm given in [12] for the implicational fragment of intuitionistic logic[2].

Definition 3 (GDLJ$^{\to}$). *Queries have the form $\Gamma \vdash^? G, H$ where Γ is a multiset of formulae and H (called the history) is a sequence of atomic goals. The rules are as follows (where $H * (q)$ is q appended to H):*

(*success*)	$\Gamma \vdash^? q, H$ *succeeds if* $q \in \Gamma$
(*implication*)	*From* $\Gamma \vdash^? A \to B, H$ *step to* $\Gamma, A \vdash^? B, H$
(*reduction*)	*From* $\Gamma, A_1 \to \ldots \to A_n \to q \vdash^? q, H$ *step to* $\Gamma \vdash^? A_i, H * (q)$ *for* $i = 1, \ldots, n$
(*bounded restart*)	*From* $\Gamma \vdash^? q, H$ *step to* $\Gamma \vdash^? q_1, H * (q)$ *if* q_1 *follows* q *in the sequence* H

Example 1. Consider the following proof, observing that (bounded restart) is needed at (2) to compensate for the removal of $p \to q$ at (1):

$$\vdash^? ((p \to q) \to p) \to (p \to q) \to q,\ \emptyset$$
$$(p \to q) \to p \vdash^? (p \to q) \to q,\ \emptyset$$
$$(p \to q) \to p, p \to q \vdash^? q,\ \emptyset \qquad (1)$$
$$(p \to q) \to p \vdash^? p,\ (q)$$
$$\vdash^? p \to q,\ (q, p)$$
$$p \vdash^? q,\ (q, p) \qquad (2)$$
$$p \vdash^? p,\ (q, p, q) \text{ by bounded restart}$$
$$\textit{success}$$

[2] In fact the calculus is an $O(n \log n)$-space decision procedure for this fragment [13].

4 A Hypersequent Calculus for $\mathbf{LC}^{\rightarrow}$

We begin by generalising the goal-directed approach used for intuitionistic logic in Section 3 analagously to Avron's generalisation of Gentzen sequents, here defining a hypersequent as a set of sequents with *restarts.*

Definition 4 (Hypersequent). *A component is a structure $S = \Gamma \vdash B, R$ where Γ is a set of formulae, B is a formula and $R = ()$ or $R = (q)$ for some atom q. A hypersequent is a set of components written $S_1 | \ldots | S_n$.*

We interpret a hypersequent as a *disjunction*, each restart being interpreted as an extra formula on the left hand side of the appropriate component.

Definition 5 (Interpretation of Hypersequents for LC). *The interpretation of $S = A_1, \ldots, A_n \vdash B, R$ is $\phi^S = A_1 \rightarrow \ldots \rightarrow A_n \rightarrow (B \rightarrow \phi^R) \rightarrow B$ where $\phi^{()} = t$ and $\phi^{(C)} = C$. The interpretation of a hypersequent $G = S_1 | \ldots | S_n$ is $\phi^{S_1} \vee \ldots \vee \phi^{S_n}$ and we write $\models G$ iff $\models \phi^G$.*

We present the following hypersequent calculus[3] for $\mathbf{LC}^{\rightarrow}$.

Definition 6 ($\mathbf{GDLC}_{\mathbf{h}}^{\rightarrow}$). $\mathbf{GDLC}_{\mathbf{h}}^{\rightarrow}$ *has the following rules:*

(success)
$$G|\Gamma, q \vdash q, R$$

(implication)
$$\frac{G|\Gamma, A \vdash B, R}{G|\Gamma \vdash A \rightarrow B, R}$$

(reduction1)
$$\frac{\overbrace{G|\Gamma \vdash q, R| \vdash A_i, (q)}^{\textit{for } i = 1 \ldots n \ (n > 0)}}{G|\Gamma, A_1 \rightarrow \ldots \rightarrow A_n \rightarrow q \vdash q, R}$$

(reduction2)
$$\frac{G|\Gamma_1, q \vdash p, R_1|\Gamma_2 \vdash q, R_2 \quad \overbrace{G|\Gamma_1 \vdash p, R_1| \vdash A_i, (q)|\Gamma_2 \vdash q, R_2}^{\textit{for } i = 1 \ldots n \ (n > 0)}}{G|\Gamma_1, A_1 \rightarrow \ldots \rightarrow A_n \rightarrow q \vdash p, R_1|\Gamma_2 \vdash q, R_2}$$

(mingle)
$$\frac{G|\Gamma_1, \Gamma_2 \vdash p, R_2}{G|\Gamma_1 \vdash q, R_1|\Gamma_2, q \vdash p, R_2}$$

(restart)
$$\frac{G|\Gamma_1, \Gamma_2 \vdash p, R}{G|\Gamma_1 \vdash q, R|\Gamma_2 \vdash p, (q)}$$

Example 2. The prelinearity axiom for $\mathbf{LC}^{\rightarrow}$ is proved as follows.

$$\cfrac{\cfrac{\cfrac{\cfrac{\cfrac{\cfrac{\cfrac{\vdash r, ()|p \vdash p, (r)}{\vdash r, ()|p \vdash q, (r)|q \vdash p, (r)}\ (\textit{mingle})}{\vdash r, ()| \vdash p \rightarrow q, (r)|q \vdash p, (r)}\ (\textit{implication})}{(p \rightarrow q) \rightarrow r \vdash r, ()|q \vdash p, (r)}\ (\textit{reduction1})}{(p \rightarrow q) \rightarrow r \vdash r, ()| \vdash q \rightarrow p, (r)}\ (\textit{implication})}{(p \rightarrow q) \rightarrow r, (q \rightarrow p) \rightarrow r \vdash r, ()}\ (\textit{reduction1})}{(p \rightarrow q) \rightarrow r \vdash ((q \rightarrow p) \rightarrow r) \rightarrow r), ()}\ (\textit{implication})}{\vdash ((p \rightarrow q) \rightarrow r) \rightarrow ((q \rightarrow p) \rightarrow r) \rightarrow r, ()}\ (\textit{implication})$$

[3] It is easy to transform this calculus into a more goal-directed style of presentation as in Section 3; here the intention is rather to facilitate comparison with other hypersequent calculi developed for **LC**.

Read upwards $\mathbf{GDLC}_{\mathbf{h}}^{\rightarrow}$ is *goal-directed* for each component. Both reduction rules look for a formula on the left hand side of a component S_1, with a head that matches an atomic formula on the right hand side of a component S_2. (*reduction*1) deals with the case where $S_1 = S_2$ and (*reduction*2) with the case where $S_1 \neq S_2$. The (*mingle*) and (*restart*) rules are also goal-directed, combining the left hand sides of two components into a single new component. Only the *choice* of components is not goal-directed.

Theorem 1. *If a hypersequent G is derivable in* $\mathbf{GDLC}_{\mathbf{h}}^{\rightarrow}$ *then* $\models G$.

Termination of $\mathbf{GDLC}_{\mathbf{h}}^{\rightarrow}$ is guaranteed since the set of subformulae occurring in each premise of the rules is strictly smaller than the set of subformulae occurring in the conclusion.

Theorem 2. $\mathbf{GDLC}_{\mathbf{h}}^{\rightarrow}$ *is terminating.*

To prove completeness we distinguish two functions of $\mathbf{GDLC}_{\mathbf{h}}^{\rightarrow}$: (1) to decompose complex formulae using invertible rules, and (2) to determine the validity of atomic hypersequents. We show that (1) allows us to reduce the provability of a hypersequent to the provability of hypersequents containing only atoms and irrelevant formulae (in the sense that they do not affect the validity of the hypersequent), and that (2) allows us to prove all valid atomic hypersequents (and thus also valid atomic hypersequents with irrelevant formulae).

Definition 7 (Invertible). *A rule is invertible if whenever the conclusion of the rule is valid then all its premises are valid.*

Proposition 1. (*implication*), (*reduction*1) *and* (*reduction*2) *are invertible.*

Applying the (*implication*), (*reduction*1) and (*reduction*2) rules exhaustively to a hypersequent results in hypersequents where for each component the right hand side is an atom that fails to match the head of a formula on the left hand side of *any* component.

Definition 8 (Irreducible). *Let* $Head(\Gamma) = \{q | A_1 \rightarrow \ldots \rightarrow A_n \rightarrow q \in \Gamma\}$. *A hypersequent* $\Gamma_1 \vdash A_1, R_1 | \ldots | \Gamma_n \vdash A_n, R_n$ *is irreducible if A_i is atomic for* $i = 1 \ldots n$ *and there exist* $\Pi_1, \ldots, \Pi_n$ *and* $\Delta_1, \ldots, \Delta_n$ *such that* $\Gamma_i = \Delta_i \cup \Pi_i$ *and Δ_i is atomic for* $i = 1 \ldots n$ *and* $Head(\Pi_1 \cup \ldots \cup \Pi_n) \cap \{A_1, \ldots, A_n\} = \emptyset$.

Proposition 2. *For a valid hypersequent G we can apply the rules of* $\mathbf{GDLC}_{\mathbf{h}}^{\rightarrow}$ *to obtain a multiset of valid irreducible hypersequents.*

If an irreducible hypersequent is valid then by removing non-atomic formulae we obtain an atomic hypersequent that is also valid.

Proposition 3. *If* $\models \Pi_1, \Delta_1 \vdash q_1, R_1 | \ldots | \Pi_n, \Delta_n \vdash q_n, R_n$ *where Δ_i is atomic for* $i = 1 \ldots n$ *and* $Head(\Pi_1 \cup \ldots \cup \Pi_n) \cap \{q_1, \ldots, q_n\} = \emptyset$ *then* $\models \Delta_1 \vdash q_1, R_1 | \ldots | \Delta_n \vdash q_n, R_n$.

We now turn our attention to proving that all valid atomic hypersequents are derivable in $\mathbf{GDLC}_{\mathbf{h}}^{\rightarrow}$. We introduce for this purpose the notion of a (strict) G-cycle for a hypersequent G and show that if a G-cycle exists then G is provable in $\mathbf{GDLC}_{\mathbf{h}}^{\rightarrow}$. We then make use of a proposition by Avron and Koniskowa [4] which states that every valid atomic hypersequent G contains a G-cycle.

Definition 9 ($\leq_G, <_G, \lhd_G$). *For a hypersequent G and atomic p, q we define $\leq_G$, $<_G$ and $\lhd_G$ as follows: (1) $p <_G q$ iff $\Gamma, q \vdash p, R$ is a component of G, (2) $p \leq_G q$ iff $\Gamma \vdash p, (q)$ is a component of G, (3) $p \lhd_G q$ iff $p <_G q$ or $p \leq_G q$.*

Definition 10 ((Strict) G-cycle). *A G-cycle is a sequence of atoms $p_1, \ldots, p_n$ such that $p_i \lhd_G p_{i+1}$ for $i = 1 \ldots n-1$, $p_i <_G p_{i+1}$ for some $1 \leq i \leq n-1$ and $p_1 = p_n$. A strict G-cycle is a G-cycle where $p_i \neq p_j$ for $1 \leq i < j \leq n-1$.*

Proposition 4 (Avron and Koniskowa [4]). *An atomic hypersequent G is valid iff there exists a G-cycle.*

Proposition 5. *If there is a G-cycle then there is a strict G-cycle.*

Proof. Let C be a G-cycle $p_1, \ldots, p_n$. We proceed by induction on $r_C = |\{p_i | p_i = p_j,\ 1 \leq i < j \leq n-1\}|$. If $r_C = 0$ then C is strict and we are done. For $r_C > 0$, $p_i = p_j$ for some $i < j$; if $C' = p_i, \ldots, p_j$ is a G-cycle (ie if there exists $i \leq k \leq j-1$ such that $p_k <_G p_{k+1}$) then $r_{C'} < r_C$ and we are done, otherwise $C' = p_1, \ldots, p_i, p_{j+1}, \ldots, p_n$ must be a G-cycle where $r_{C'} < r_C$. □

Proposition 6. *For a hypersequent G if there exists a strict G-cycle then G is provable in* $\mathbf{GDLC}_{\mathbf{h}}^{\rightarrow}$.

Proof. By induction on the length l of a strict G-cycle. If $l = 2$ then we have $p <_G p$ which means that $\Gamma, p \vdash p, R$ is a component of G so G is derivable by (*success*). For a strict G-cycle $C = p_1, \ldots, p_l$ where $l > 2$ we have a subsequence q_1, q_2, q_3 which must have one of the following patterns: (1) $q_1 <_G q_2 <_G q_3$, (2) $q_1 \leq_G q_2 <_G q_3$, (3) $q_1 <_G q_2 \leq_G q_3$.

For (1) we have that $S_1 = \Gamma_1, q_2 \vdash q_1, R_1$ and $S_2 = \Gamma_2, q_3 \vdash q_2, R_2$ are components of G. If they are the same component then $q_1 = q_2$ contradicts the strictness of C. By applying (*mingle*) to G we obtain G' where S_1 and S_2 have been replaced by $S_3 = \Gamma_1, \Gamma_2, q_3 \vdash q_1, R_1$. We claim that by removing q_2 from C we obtain a G'-cycle of length $l-1$ whence by the induction hypothesis G' is derivable and we are done. We have that $q_1 <_{G'} q_3$ since S_3 is a component of G', and that the only inequalities in G not in G' are of the form $q_2 \lhd_G p$ or $p \lhd_G q_2$ which cannot occur since C is strict. (2) follows similarly to (1).

For (3) we first consider the subcase where $l = 3$ so we have $q_1 = q_3$. Here we have $S_1 = \Gamma_1, q_2 \vdash q_1, R$ and $S_2 = \Gamma_2 \vdash q_2, (q_1)$ are distinct components of C; by applying (*restart*) to G we obtain G' where S_1 and S_2 have been replaced by $S_3 = \Gamma_1, \Gamma_2, q_2 \vdash q_2, R$ which is derivable by (*success*). We now consider the possible patterns for $l > 3$: (3a) $q_1 <_G q_2 \leq_G q_3 \leq_G p$, (3b) $q_1 <_G q_2 \leq_G q_3 <_G p$, (3c) $p \leq_G q_1 <_G q_2 \leq_G q_3$, (3d) $p <_G q_1 <_G q_2 \leq_G q_3$. (3b) and (3c) are covered by pattern (2) and (3d) is covered by pattern (1) so we only need to consider (3a). $S_1 = \Gamma_1 \vdash q_2, (q_3)$ and $S_2 = \Gamma_2 \vdash q_3, (p)$ are components of G. Applying (*restart*) to G we obtain G' where S_1 and S_2 have been replaced by $S_3 = \Gamma_1, \Gamma_2 \vdash q_2, (p)$. Now removing q_3 from C we get a strict

G'-cycle of length $l-1$, and by the induction hypothesis G' is derivable. □

Since it follows from Propositions 2 and 3 that the rules of $\mathbf{GDLC}_{\mathbf{h}}^{\rightarrow}$ can be applied to a valid hypersequent to obtain a set of valid atomic hypersequents, and from Propositions 4, 5 and 6 that all valid atomic hypersequents are provable in $\mathbf{GDLC}_{\mathbf{h}}^{\rightarrow}$, we get that $\mathbf{GDLC}_{\mathbf{h}}^{\rightarrow}$ is complete.

Theorem 3. *For a hypersequent G, if $\models G$ then G is derivable in* $\mathbf{GDLC}_{\mathbf{h}}^{\rightarrow}$.

5 A Labelled Goal-Directed Calculus for $\mathbf{LC}^{\rightarrow}$

$\mathbf{GDLC}_{\mathbf{h}}^{\rightarrow}$ has some nice properties for proof search; it is terminating, contraction-free and has invertible reduction rules that operate on the principal connective of one formula at a time. As a *goal-directed* calculus it is not quite satisfactory however. For a start the (*mingle*) and (*restart*) rules are not invertible so backtracking may be required to complete a proof. Alternative rules are:

$$(mingle')\ \frac{G|\Gamma_1,\Gamma_2 \vdash p, R_2|\Gamma_1 \vdash q, R_1}{G|\Gamma_1 \vdash q, R_1|\Gamma_2, q \vdash p, R_2} \qquad (restart')\ \frac{G|\Gamma_1,\Gamma_2 \vdash p,(q)|\Gamma_2 \vdash q, R}{G|\Gamma_1 \vdash q, R|\Gamma_2 \vdash p,(q)}$$

Replacing (*mingle*) and (*restart*) with (*mingle'*) and (*restart'*) gives a sound and complete system with invertible rules but to keep termination the condition $\Gamma_1 \not\subseteq \Gamma_2$ is needed for both cases, thereby losing the logical purity of the calculus[4].

A further problem for the goal-directedness of $\mathbf{GDLC}_{\mathbf{h}}^{\rightarrow}$ is the creation of components with just goals and restarts. Consider the following derivation:

$$\cfrac{\cfrac{\cfrac{p,q \vdash q,()}{p,q \vdash r,()| \vdash q,(r)}\ (restart) \qquad \cfrac{\cfrac{p \vdash p}{p \vdash q,()| \vdash p,(q)}\ (restart)}{p \vdash r,()| \vdash p,(q)| \vdash q,(r)}\ (restart)}{p, p \rightarrow q \vdash r,()| \vdash q,(r)}\ (reduction2)}{p, p \rightarrow q, q \rightarrow r \vdash r,()}\ (reduction1)$$

Multiple uses of (*restart*) seem unwarranted for such a simple theorem. The problem is that when a reduction rule is applied to a formula $p \rightarrow q$ a new component is created with goal p, restart q and no antecedent formulae; hence a formula with head p can only be reduced using (*reduction*2), which adds an extra branch, and (*restart*). Alternative rules for (*reduction*1) and (*reduction*2) can be given that avoid this problem by keeping the same database formulae for the new component. However this loses the nice property of having the size of hypersequents bounded linearly by the size of the original formula and may require an exponential number of copies of the same formula to be reduced.

Our solution here is to maintain a *global* database using *labels*[5] to keep track of which formulae are accessible to which goals. We can then give rules that

[4] In practice the goal-directedness of rules can reduce the amount of backtracking required dramatically; hence for proof search the non-invertible rules without subset conditions might be preferable.

[5] Labels are also used in [12] to give a non-terminating goal-directed system for $\mathbf{LC}^{\rightarrow}$.

reduce copies of a formula occurring in different components *at the same time.* To this end we define and interpret goal-directed queries as follows:

Definition 11 (Goal-Directed Query). *A goal-directed query (query) is a structure of the form* $\Gamma \vdash^? \delta_1 : G_1, R_1, H$ *where:*

Γ *is a set of labelled formulae called the database.*
$H = \{(\delta_2 : G_2, R_2), \ldots, (\delta_n : G_n, R_n)\}$ *is called the history.*
$\delta_1, \ldots, \delta_n$ *are sets of labels.*
$G_1, \ldots, G_n$ *are formulae called goals.*
$R_i = ()$ *or* $R_i = (q)$ *(q atomic) for* $i = 1 \ldots n$ *are called restarts.*

Definition 12 (Interpretation of Goal-Directed Queries). *Let* $Acc(\Gamma, \delta) = \{A|x : A \in \Gamma \text{ and } x \in \delta\}$ *and for a query* $Q = \Gamma \vdash^? \delta_1 : G_1, R_1, H$ *where* $H = \{(\delta_2 : G_2, R_2), \ldots, (\delta_n : G_n, R_n)\}$*, define* $G^Q = Acc(\Gamma, \delta_1) \vdash G_1, R_1 | \ldots | Acc(\Gamma, \delta_n) \vdash G_n, R_n$ *and write* $\models Q$ *iff* $\models G^Q$.

We introduce the following labelled goal-directed calculus for $\mathbf{LC}^{\rightarrow}$.

Definition 13 ($\mathbf{GDLC}_l^{\rightarrow}$). $\mathbf{GDLC}_l^{\rightarrow}$ *has the following rules:*

(*success*) $\Gamma \vdash^? \delta : q, R, H$ *succeeds if* $x : q \in \Gamma$ *and* $x \in \delta$

(*implication*) *From* $\Gamma \vdash^? \delta : A \rightarrow B, R, H$ *step to:*
$\Gamma, x : A \vdash^? \delta \cup \{x\} : B, R, H$ *(x new)*

(*reduction1*) *From* $\Gamma, x : A_1 \rightarrow \ldots \rightarrow A_n \rightarrow q \vdash^? \delta : q, R, H$
where: $x \in \delta$ *and* $n > 0$ *step to:*
$\Gamma \vdash^? \delta : A_i, (q), H \cup \{(\delta : q, R)\}$ *for* $i = 1 \ldots n$

(*reduction2*) *From* $\Gamma, x : A_1 \rightarrow \ldots \rightarrow A_n \rightarrow q \vdash^? \delta : q, R_1, H$
where: $H = H' \cup \{(\gamma : p, R_2)\}$, $n > 0$ *and* $x \in \gamma$ *step to:*
$\Gamma, x : q \vdash^? \gamma \cup \delta : p, R_2, H' \cup \{(\delta : q, R_1)\}$ *and:*
$\Gamma \vdash^? \delta : A_i, (q), H \cup \{(\delta : q, R_1)\}$ *for* $i = 1 \ldots n$

(*mingle*) *From* $\Gamma, x : q \vdash^? \delta : q, R_1, H$
where: $H = H' \cup \{(\gamma : p, R_2)\}$ *and* $x \in \gamma$ *step to:*
$\Gamma, x : q \vdash^? \delta \cup \gamma : p, R_2, H' \cup \{(\delta : q, R_1)\}$

(*restart*) *From* $\Gamma \vdash^? \delta : q, R, H$
where: $H = H' \cup \{(\gamma : p, (q))\}$ *step to:*
$\Gamma \vdash^? \delta \cup \gamma : p, (q), H' \cup \{(\delta : q, R)\}$

(*switch*) *From* $\Gamma \vdash^? \delta : q, R_1, H$
where: $H = H' \cup \{(\gamma : p, R_2)\}$ *step to:*
$\Gamma \vdash^? \gamma : p, R_2, H' \cup \{(\delta : q, R_1)\}$

The (*success*) and (*implication*) rules are straightforward translations of the corresponding hypersequent rules. (*reduction1*) and (*reduction2*) match the current goal with a labelled formula in the database; the former for a label accessible from the current goal, the latter for a label accessible from a goal in the history. Both remove the formula from the database so that the formula is no longer accessible to *any* goal in the query. The (*mingle*) and (*restart*) rules perform similar functions to the revised (*mingle*′) and (*restart*′) rules in Section 4. (*switch*) which has no counterpart in the hypersequent calculus, allows the computation to move from the current goal to one in the history.

Example 3. We show below how $\mathbf{GDLC}_{\mathrm{l}}^{\rightarrow}$ avoids the repeated applications of (*restart*) characteristic of $\mathbf{GDLC}_{\mathrm{h}}^{\rightarrow}$:

$x : p, y : p \to q, z : q \to r \vdash^? \{x, y, z\} : r, (), \emptyset$

$x : p, y : p \to q \quad \vdash^? \{x, y, z\} : q, (r), \{H1\} \quad H1 = (\{x, y, z\} : r, ())$

$x : p \quad \vdash^? \{x, y, z\} : p, (q), \{H1, H2\} \quad H2 = (\{x, y, z\} : q, (r))$

success

Example 4. A proof of the formula $((p_1 \to p_2) \to q) \to ((p_2 \to p_3) \to q) \to \ldots \to ((p_{n-1} \to p_n) \to q) \to ((p_n \to p_1) \to q) \to q$ requires an exponential number of steps in Dyckhoff's calculus [9] but can be proved in $\mathbf{GDLC}_{\mathrm{l}}^{\rightarrow}$ in a linear number of steps. Below we give the proof for the case where $n = 3$ using labels to record the contents of the database where appropriate:

$\vdash^? \emptyset : ((p_1 \to p_2) \to q) \to ((p_2 \to p_3) \to q) \to ((p_3 \to p_1) \to q) \to q, (), \emptyset$

$x : (p_1 \to p_2) \to q, y : (p_2 \to p_3) \to q, z : (p_3 \to p_1) \to q \vdash^? \{x, y, z\} : q, (), \emptyset$

	y, z	$\vdash^? \{x, y, z\} : p_1 \to p_2, (q), \{H_1\}$	$H_1 = (\{x, y, z\} : q, ())$
(1)	$y, z, u : p_1$	$\vdash^? \{x, y, z, u\} : p_2, (q), \{H_1\}$	
	y, z, u	$\vdash^? \{x, y, z\} : q, (), \{H_2\}$	$H_2 = (\{x, y, z, u\} : p_2, (q))$
	z, u	$\vdash^? \{x, y, z\} : p_2 \to p_3, (q), \{H_1, H_2\}$	
(2)	$z, u, v : p_2$	$\vdash^? \{x, y, z, v\} : p_3, (q), \{H_1, H_2\}$	
	z, u, v	$\vdash^? \{x, y, z\} : q, (), \{H_2, H_3\}$	$H_3 = (\{x, y, z, v\} : p_3, (q))$
	u, v	$\vdash^? \{x, y, z\} : p_3 \to p_1, (q), \{H_1, H_2, H_3\}$	
(3)	$u, v, w : p_3$	$\vdash^? \{x, y, z, w\} : p_1, (q), \{H_1, H_2, H_3\}$	
(4)	u, v, w	$\vdash^? \{x, y, z, u, w\} : p_2, (q), \{H_1, H_3, H_4\}$	$H_4 = (\{x, y, z, w\} : p_1, (q))$
	u, v, w	$\vdash^? \{x, y, z, u, v, w\} : p_3, (q), \{H_1, H_2', H_4\}$	$H_2' = (\{x, y, z, u, w\} : p_2, (q))$
	success		

At (1) and (2) q fails to match anything in the database so (*switch*) is applied. At (3) and (4) the current goal matches an atom in the database accessible to a goal in the history, so (*mingle*) is applied.

Theorem 4. *If a query Q succeeds in* $\mathbf{GDLC}_{\mathrm{l}}^{\rightarrow}$ *then* $\models Q$.

We now turn our attention to the termination of $\mathbf{GDLC}_{\mathrm{l}}^{\rightarrow}$. Unlike hypersequents, queries focus on *one goal at a time* and require the rule (*switch*) to move to a goal recorded in the history. This can cause unwanted looping, eg:

$\emptyset \vdash^? \emptyset : p_1, (), \{(\emptyset : p_2, ())\}$ use (*switch*) to step to:

$\emptyset \vdash^? \emptyset : p_2, (), \{\emptyset : p_1, ())\}$ then use (*switch*) again to step to:

$\emptyset \vdash^? \emptyset : p_1, (), \{(\emptyset : p_2, ())\}$ and so on ...

Our solution is to divide the history into *two parts*, the left and right history. Goal-directed queries are written as $\Gamma \vdash^? \delta_1 : G_1, R_1, LH, RH$ with the same properties as Definition 11 except that *both* LH and RH are sets of goals with sets of labels and restarts. Definition 12 is revised accordingly. The calculus $\mathbf{GDLC}_\mathbf{I}^{\rightarrow}$ remains almost the same with the union of the left history and the right history replacing the history. Goal-directed queries resulting from applications of all rules except (*switch*) move everything from the right history into the left history, leaving the former empty. The (*switch*) rule is changed as follows:

(*switch*) From $\Gamma \vdash^? \delta : q, LH, RH$ where: $LH = LH' \cup \{(\gamma : p, R_2)\}$ step to:
$\Gamma \vdash^? \gamma : p, R_2, LH', RH \cup \{(\delta : q, R_1)\}$

Clearly (*switch*) can be applied successively only as many times as there are items in the left history.

Looping may also occur as a result of (*restart*) and (*mingle*) which both add labels to the set available for a particular goal. If the labels added already occur in the set then clearly nothing has been gained. We therefore add the subset condition ($\delta \not\subseteq \gamma$) to both rules. These changes do not affect the soundness of the calculus and allow us to prove the following termination theorem for $\mathbf{GDLC}_\mathbf{I}^{\rightarrow}$.

Theorem 5. $\mathbf{GDLC}_\mathbf{I}^{\rightarrow}$ *is terminating.*

The completeness of $\mathbf{GDLC}_\mathbf{I}^{\rightarrow}$ is proved similarly to Theorem 3, noting that dividing the history only causes the calculus to terminate when no rule is applicable in *any* component of the corresponding hypersequent.

Theorem 6. *For a query* Q, *if* $\models Q$ *then* Q *succeeds in* $\mathbf{GDLC}_\mathbf{I}^{\rightarrow}$.

To show that $\mathbf{GDLC}_\mathbf{I}^{\rightarrow}$ is co-NP we give the following *algorithm* to decide whether a query fails or not for $\mathbf{GDLC}_\mathbf{I}^{\rightarrow}$ and show that it is NP.

Definition 14 (fail(Q)). *For a rule* R *and query* Q, *applies*(R, Q) *is true if* R *can be applied to* Q, *result*(R, Q) *is the set of queries produced by applying* R *to* Q *and choose*(S) *for a set* S *is a member of* S *chosen non-deterministically*[6].

```
fail(query Q){
do Q = reduce(Q);
  while (Q != "Irreducible" && Q != "Success");
if (Q == "Success") return REJECT;
else {
  do Q = expand(Q);
    while (Q != "Fail" && Q != "Success");
  if (Q == "Fail") return ACCEPT;
  else return REJECT; }}
```

[6] Only (*reduction*1) and (*reduction*2) can result in more than one query to be chosen.

```
reduce(query Q){
if (applies(success, Q))
  Q = "Success";
else if (applies(implication, Q))
  Q = choose(result(implication, Q));
else if (applies(reduction1, Q))
  Q = choose(result(reduction1, Q));
else if (applies(reduction2, Q))
  Q = choose(result(reduction2, Q));
else if (applies(switch, Q))
  Q = choose(result(switch, Q));
else Q = "Irreducible";
return Q}
```

```
expand(query Q){
if (applies(success, Q))
  Q = "Success";
else if (applies(mingle, Q))
  Q = choose(result(mingle, Q));
else if (applies(restart, Q))
  Q = choose(result(restart, Q));
else if (applies(switch, Q))
  Q = choose(result(switch, Q));
else Q = "Fail";
return Q}
```

Theorem 7. $fail(Q) \in NP$

Proof. We show that every execution of $fail(Q)$ is polynomially bounded in the size s of Q. $fail$ has two distinct stages; applying *reduce* until Q succeeds or is irreducible, then for the latter case applying *expand* until Q succeeds or fails. We examine the length of these two steps separately, starting with *reduce*. Since each application of (*implication*), (*reduction*1) and (*reduction*2) reduces the multiset complexity of Q we have that there can be at most $O(s)$ applications of these rules. However, before each application of (*implication*), (*reduction*1) and (*reduction*2) there may be a number of applications of (*switch*). Now since each application of (*switch*) reduces the size of the right history of Q which is of order $O(s)$ we can limit this number to $O(s)$. Hence we get that there will be $O(s.s) = O(s^2)$ *reduce* steps. To apply *reduce* requires checking the applicability of (*success*), (*implication*), (*reduction*1), (*reduction*2) and (*switch*) which is $O(s^2)$ so our first step is $O(s^2.s^2) = O(s^4)$. We turn our attention now to *expand*. Since each (*mingle*) and (*restart*) step decreases the measure $\Sigma_{i=1}^{n}(| \cup_{j=1}^{n} \delta_j| - |\delta_i|)$ there can only be $O(s^2)$ applications of these rules. Again however we may require $O(s)$ applications of (*switch*) giving $O(s^2.s) = O(s^3)$ *expand* steps. Checking the applicability of (*success*), (*implication*) and (*switch*) is $O(s^2)$ but checking the applicability of (*mingle*) and (*restart*) is $O(s^4)$ due to the subset conditions so each *expand* step is $O(s^4)$. Hence our second stage is $O(s^4.s^3) = O(s^7)$ and we have that $fail$ is $O(s^7)$. $\square$[7]

6 Finite-Valued Gödel-Dummett Logics

$\mathbf{LC_n}$ is obtained from **LC** by restricting the number of truth values to n. This can be achieved for our calculi by adding further axioms eg to $\mathbf{GDLC_h^{\rightarrow}}$:

$$\text{(n-success) } G|\Gamma_1 \vdash q_1, R_1|\Gamma_2, q_1 \vdash q_2, R_2| \ldots |\Gamma_n, q_{n-1} \vdash q_n, R_n$$

A more elegant approach however is to check such a restriction *as a proof progresses.* Each application of the (*mingle*) rule restricts the set of truth values

[7] Note that the bound $O(s^7)$ is worst-case. In practice there are many improvements that could be made to reduce the complexity of the algorithm.

taken by the new goal, thus for example in $\mathbf{LC_3}$ we have that $q \leq p$ or $p \leq r$ iff $q \leq p$ or $p \leq r$ *or* $r > 0$. Hence we can add a *counter* for each goal (current and in the history) interpreted as restrictions on the truth values for that goal. Validity for goal-directed queries in $\mathbf{LC_n}$ is then defined as follows:

Definition 15 (Validity of Goal-Directed Queries for $\mathbf{LC_n}$). *$\Gamma \vdash^?_n \delta_1 : G_1, a_1, R_1, \{(\delta_2 : G_2, a_2, R_2), \ldots, (\delta_m : G_m, a_m, R_m)\}$ is valid in $\mathbf{LC_n}$ if for all valuations v for $\mathbf{LC_n}$ there exists i, $1 \leq i \leq m$, such that either $v(G_i) \geq 1 - (a_i/(n-1))$ or $v(G_i) \geq v(\wedge Acc(\Gamma, \delta_i) \wedge (G_i \to \phi^{R_i}))$ where $\phi^{()} = t$ and $\phi^{(C)} = C$.*

We present the following goal-directed calculus for $\mathbf{LC_n}$:

Definition 16 ($\mathbf{GDLC_I^{\to}(n)}$). *$\mathbf{GDLC_I^{\to}(n)}$ consists of the rules (implication), (reduction1), (reduction2) and (switch) of $\mathbf{GDLC_I^{\to}}$ (with the same counters in both premises and conclusions) and also:*

(*success*) *$\Gamma \vdash^?_n \delta : q, a, R, H$ succeeds if either $x : q \in \Gamma$ and $x \in \delta$ or $a = n - 1$*

(*mingle*) *From $\Gamma, x : q \vdash^?_n \delta : q, a, R_1, H$*
where: $H = H' \cup \{(\gamma : p, b, R_2)\}$ and $x \in \gamma$ step to:
$\Gamma, x : q \vdash^?_n \delta \cup \gamma : p, max(a+1, b), R_2, H' \cup \{(\delta : q, a, R_1)\}$

(*restart*) *From $\Gamma \vdash^?_n \delta : q, a, R, H$*
where: $H = H' \cup \{(\gamma : p, b, (q))\}$ step to:
$\Gamma \vdash^?_n \delta \cup \gamma : p, max(a, b), (q), H' \cup \{(\delta : q, a, R)\}$

Example 5. Consider the following proof of the characteristic axiom for $\mathbf{LC_3}$:

$\vdash^?_3 \emptyset : (p_1 \to q) \to ((p_2 \to p_1) \to q) \to ((p_2 \to p_3) \to q) \to q, 0, (), \emptyset$

$x : p_1 \to q, y : (p_1 \to p_2) \to q, z : (p_2 \to p_3) \to q \vdash^?_3 \{x, y, z\} : q, 0, (), \emptyset$

$y, z \vdash^?_3 \{x, y, z\} : p_1, 0, (q), \{H_1\}$ $\quad H_1 = (\{x, y, z\} : q, 0, ())$

$y, z \vdash^?_3 \{x, y, z\} : q, 0, (), \{H_2\}$ $\quad H_2 = (\{x, y, z\} : p_1, 0, (q))$

$z \vdash^?_3 \{x, y, z\} : p_1 \to p_1, 0, (q), \{H_1, H_2\}$

$x, u : p_1 \vdash^?_3 \{x, y, z, u\} : p_2, 0, (q), \{H_1, H_2\}$

$z, u \vdash^?_3 \{x, y, z\} : q, 0, (), \{H_2, H_3\}$ $\quad H_3 = (\{x, y, z, u\} : p_2, 0, (q))$

$u \vdash^?_3 \{x, y, z\} : p_2 \to p_3, 0, (), \{H_1, H_2, H_3\}$

$u, v : p_2 \vdash^?_3 \{x, y, z, v\} : p_3, 0, (), \{H_1, H_2, H_3\}$

$u, v \vdash^?_3 \{x, y, z\} : p_1, 0, (q), \{H_1, H_3, H_4\}$ $\quad H_4 = (\{x, y, z, v\} : p_3, 0, ())$

$v \vdash^?_3 \{x, y, z, u\} : p_2, 1, (q), \{H_1, H_2, H_4\}$

$\vdash^?_3 \{x, y, z, u, v\} : p_3, 2, (q), \{H_1, H_2, H_3\}$

success

Theorem 8. *A query Q succeeds in $\mathbf{GDLC_I^{\to}(n)}$ iff $\models_n Q$.*

7 Discussion

In this paper we have presented goal-directed systems for the implicational fragments of the finite-valued and infinite-valued Gödel-Dummett logics, $\mathbf{LC_n}$ and

LC, including terminating hypersequent calculi with simple axioms and local rules, and co-NP labelled calculi with invertible rules.

The first achievement of this work is to have extended the goal-directed approach to Gödel-Dummett logics. Goal-directed methods have been developed for a wide range of logics including classical, intuitionistic, substructural, intermediate and basic modal logics [12], but not as yet for many-valued or fuzzy logics[8]. This work provides a starting point for an investigation of goal-directed methods for these logics. In particular we intend to look at Łukasiewicz logic **Ł**, recently given first sequent and hypersequent calculi [19].

Goal-directed systems are particularly suitable for proof search since they focus on a goal formula and do not decompose unnecessary database formulae. Our calculi have other advantages over others given in the literature however. Avron's original hypersequent calculus [3] is clearly the most natural proof system for **LC** but is unsuitable for proof search, while Sonobe's calculus and related approaches [21, 2, 9, 10] suffer from the problem of having non-local rules. In our opinion the state of the art is provided by Avron and Koniskowa's calculi $\mathbf{GLC_{RS}}$, $\mathbf{GLC^*_{RS}}$ and $\mathbf{GLC^*}$ [4], and also Baaz and Fermüller's sequents of relations calculus $\mathbf{RG_\infty}$ [5]. One immediate advantage that $\mathbf{GDLC_h^\rightarrow}$ possesses over all these systems is that it has just one basic axiom and no expansion rules. Moreover, $\mathbf{GDLC_h^\rightarrow}$, unlike Avron and Koniskowas's calculi, has rules that operate only on principal connectives. $\mathbf{RG_\infty}$ also has this property but uses two types of components. A more important advantage of $\mathbf{GDLC_h^\rightarrow}$ over these calculi however is the fact that the size of hypersequents is *linear* in the size of the original formula. Even better from the point of view of proof search, is the labelled calculus $\mathbf{GDLC_l^\rightarrow}$, which by maintaining a *global* database, ensures that subformulae are reduced *only once* in each branch of a proof, without removing accessibility to goals. Moreover, by imposing subset conditions $\mathbf{GDLC_l^\rightarrow}$ becomes co-NP, although we note that using rules which rely on removing formulae from the database may prove to be more efficient.

Finally we comment on some proposed extensions. First, we note that it is straightforward to extend our calculi to cope with *negation* and *conjunction*, the former by introducing a constant $\perp$, and the latter by defining a simple normal form. *Disjunction* may be defined as $A \vee B = ((A \rightarrow B) \rightarrow B) \wedge ((B \rightarrow A) \rightarrow A)$ but at the cost of an exponential blow-up in the size of formulae. A better approach is to define a normal form and refine the rules of our calculi accordingly. To extend goal-directed methods for **LC** and $\mathbf{LC_n}$ to the *first-order* case, a good starting point is the universal fragment, a goal-directed calculus having been obtained for this fragment of intuitionistic logic using unification and run-time skolemisation [12].

[8] Goal-directed calculi for **LC** have been presented in [12] but they are non-terminating and significantly inferior to those given here.

References

1. S. Aguzzoli and B. Gerla. Finite-valued reductions of infinite-valued logics. *Archive for Mathematical Logic*, 41(4):361–399, 2002.
2. A. Avellone, M. Ferrari, and P. Miglioli. Duplication-free tableau calculi and related cut-free sequent calculi for the interpolable propositional intermediate logics. *Logic Journal of the IGPL*, 7(4):447–480, 1999.
3. A. Avron. Hypersequents, logical consequence and intermediate logics for concurrency. *Annals of Mathematics and Artificial Intelligence*, 4(3–4):225–248, 1991.
4. A. Avron and B. Konikowska. Decomposition Proof Systems for Gödel-Dummett Logics. *Studia Logica*, 69(2):197–219, 2001.
5. M. Baaz and C. G. Fermüller. Analytic calculi for projective logics. In *Proc. TABLEAUX '99*, volume 1617, pages 36–50, 1999.
6. M. Baaz, C. G. Fermüller, and G. Salzer. Automated deduction for many-valued logics. In A. Robinson and A. Voronkov, editors, *Handbook of Automated Reasoning*, volume II, chapter 20, pages 1355–1402. Elsevier Science B.V., 2001.
7. M. Dummett. A propositional calculus with denumerable matrix. *Journal of Symbolic Logic*, 24:97–106, 1959.
8. J. M. Dunn and R. K. Meyer. Algebraic completeness results for Dummett's LC and its extensions. *Zeitschrift für Mathematische Logik und Grundlagen der Mathematik*, 17:225–230, 1971.
9. R. Dyckhoff. A deterministic terminating sequent calculus for Gödel-Dummett logic. *Logic Journal of the IGPL*, 7(3):319–326, 1999.
10. G. Fiorino. An o(nlog n)-space decision procedure for the propositional Dummett logic. *Journal of Automated Reasoning*, 27(3):297–311, 2001.
11. D. Gabbay. *Semantical Investigations in Heyting's Intuitionistic Logic.* Reidel, Dordrecht, 1981.
12. D. Gabbay and N. Olivetti. *Goal-directed Proof Theory.* Kluwer Academic Publishers, 2000.
13. D. Gabbay and N. Olivetti. Goal oriented deductions. In D. Gabbay and F. Guenthner, editors, *Handbook of Philosopophical Logic*, volume 9, pages 199–285. Kluwer Academic Publishers, second edition, 2002.
14. K. Gödel. Zum intuitionisticschen Aussagenkalkül. *Anzeiger Akademie der Wissenschaften Wien, mathematisch-naturwiss. Klasse*, 32:65–66, 1932.
15. R. Hähnle. *Automated Deduction in Multiple-Valued Logics.* Oxford University Press, 1993.
16. P. Hájek. *Metamathematics of Fuzzy Logic.* Kluwer Academic Publishers, Dordrecht, 1998.
17. D. Larchey-Wendling. Combining proof-search and counter-model construction for deciding gödel-dummett logic. *Lecture Notes in Computer Science*, 2392, 2002.
18. V. Lifschitz, D. Pearce, and A. Valverde. Strongly equivalent logic programs. *ACM Transactions on Computational Logic*, 2(4):526–541, October 2001.
19. G. Metcalfe, N. Olivetti, and D. Gabbay. Sequent and hypersequent calculi for abelian and Łukasiewicz logics. Submitted.
20. D. Miller, G. Nadathur, F. Pfenning, and A. Scedrov. Uniform proofs as a foundation for logic programming. *Annals of Pure and Applied Logic*, 51:125–157, 1991.
21. O. Sonobe. A Gentzen-type formulation of some intermediate propositional logics. *Journal of Tsuda College*, 7:7–14, 1975.
22. A. Visser. On the completeness principle: a study of provability in heyting's arithmetic. *Annals of Mathematical Logic*, 22:263–295, 1982.

A Logic for Probability in Quantum Systems

Ron van der Meyden and Manas Patra*

School of Computer Science and Engineering
University of New South Wales, Sydney 2052, Australia
{meyden,manasp}@cse.unsw.edu.au

Abstract. Quantum computation deals with projective measurements and unitary transformations in finite dimensional Hilbert spaces. The paper presents a propositional logic designed to describe quantum computation at an operational level by supporting reasoning about the probabilities associated to such measurements: measurement probabilities, and transition probabilities (a quantum analogue of conditional probabilities). We present two axiomatizations, one for the logic as a whole and one for the fragment dealing just with measurement probabilities. These axiomatizations are proved to be sound and complete. The logic is also shown to be decidable, and we provide results characterizing its complexity in a number of cases.

1 Introduction

Quantum computing promises to open fresh vistas for computer science – almost 100 years after quantum mechanics revolutionized physics. We expect that logical methods will play a key role in the development of quantum computer systems, much like the role they have played in classical computing. (Pratt [Pra92] has also argued that logical support for reasoning about quantum effects may become essential in the next ten years to designers of classical hardware.) In this paper, we study what is potentially one of the building blocks of modal logics for quantum computing, by developing a logic for expressing properties of quantum states, in the same way that propositional logic, which expresses properties of classical states, provides the basis for modal logics of computation.

A new propositional basis is required for a logic of quantum computation because quantum mechanics introduced a new paradigm for the desciption of microscopic phenomena, radically revising the classical notions of state and observable. Both the notion of quantum state and the predictions quantum mechanics makes about measurements in the laboratory are inherently probabilistic in nature. This suggests that a logic for reasoning about quantum computations should support reasoning about the values of measurement probabilities.

There already exists an extensive body of literature on logics for reasoning about probability [AH94,Bac90,Car50,FHM90,Nil86]. Of these works, relatively few concern explicit statements about probability values; the most expressive

* On leave from Department of Physics, S. V. College, Univrsity of Delhi, Delhi-110021, India.

M. Baaz and J.A. Makowsky (Eds.): CSL 2003, LNCS 2803, pp. 427–440, 2003.

and most fully analyzed framework that does admit such statements is that of Fagin et al. [FHM90], which permits statements such as $\Sigma_{i=1\ldots k} a_i P(\phi_i) \leq a_{k+1}$, where the ϕ_i are propositional formulas and the a_i are integer constants.

However, these existing results cannot be directly applied to reasoning about probabilities in the context of quantum computation. One of the obstacles is that the space of propositions in quantum mechanics has a significantly more complicated structure than the space of classical propositions. Classical physics is based on a notion of measurement of observables that leaves the state of the system invariant, so that the order in which measurements are performed is irrelevant. By contrast, measurement in quantum mechanics changes the state, and the order of measurements matters.

Two measurements that commute are said to correspond to compatible observables, and it can be shown that such observables are co-measurable, i.e. their values may be inferred from a single measurement of another observable. While compatible observables interact classically, incompatible observables do not. Moreover, the probabilities related to incompatible observables are connected by a non-linear equation over the complex number field. The subject of "quantum logic" [BvN36] studies the algebraic structure of the subspaces of the quantum state space that are associated with measurements. This algebra turns out to have a non-boolean structure in which the the distributive law fails. "Quantum probability" theory [Gud89] goes on to define a revised notion of probability space based on these non-boolean algebraic spaces. Quantum logic has been motivated by concerns for the philosophical foundations of quantum mechanics, seeking in particular to justify the use of Hilbert spaces in quantum mechanics.

Our approach in this paper will be somewhat more pragmatic: we will take the Hilbert space formalization for granted, and focus on the operationally observable probabilities of quantum measurement outcomes. We take *finite dimensional* Hilbert spaces as our semantic basis; the restriction to finite dimensions is precisely that used in quantum computation. We consider two probability operators: a monadic operator P and a dyadic operator T. The operator P resembles a standard probability operator. Unlike quantum logic, which permits boolean combinations of incompatible propositions, we restrict the application of this probability operator to boolean combinations of a set of compatible propositions. A consequence of this is that the logic of this probability operator is similar to the logic of the classical probability operator. The second type of probability operator T that we consider resembles a classical conditional probability operator: in the quantum case, this notion corresponds to the probablity of a transition between subspaces taking place when a measurement is performed.

The main results of the paper are completeness theorems for two axiomatizations. The first deals with the fragment of the logic containing only the operator P, and permitting the expression of linear inequalities over probability values. The complete axiomatization in this case is wholly propositional (i.e. does not require quantification), and consists of a set of axioms concerning the interaction of the probability operators and boolean logic, together with a set of axioms for reasoning about linear inequalities. The second axiomatization is for the full language, with both operators P and T. When axiomatizing reasoning about classical conditional probabilities, one needs an axiomatization of real closed fields [Tar51]. For our logic, we need an extended theory of algebraically closed fields with that of real closed fields embedded in it.

Further, we show that the satisfiability problem is decidable for both languages we consider, and we characterize its complexity. The complexity classes we identify for the satisfiability problem turn out to be the same classes as in the complexity results of Fagin et al. [FHM90] for the classicalcase, viz., NP complete for the language with just the operator P and linear constraints, and in exponential space for the language with both operators P and T and with polynomial equations and quantification over numbers.

The structure of the paper is as follows.

Section 2 provides an overview of the basic definitions of quantum mechanics. We introduce the syntax of the language in section 3.1, and provide its semantics in Section 3.2. Section 4 provides a few illustrations of what can be expressed in the logic. The axiomatizations and completeness results are presented in Section 5.1 (for the logic based on P), and Section 5.2 (for the logic based on both P and T.) Section 6 concludes by discussing some questions raised by these results and sketching future work.

2 Quantum Theory

The most complete description of a physical system is given by its state. According to quantum theory the state of physical system S is a unit vector in a Hilbert space. Associated with S is a Hilbert space, that is complex vector space with an inner product. We shall restrict ourselves to finite-dimensional spaces, which are adequate for quantum computing and our logic. We assume that all states are pure states, as described below. However, we could) easily extend the syntax and semantics to "mixed" states which are positive, normalised combination of pure states.

Postulate 1 *Associated to S is a finite-dimensional complex Hilbert space $\{H, \langle\rangle\}$ called the state space. The dimension n is determined by the system.*

We identify H with C^n with standard inner product; if $|\alpha\rangle = (x_1 \ldots x_n)$, $|\beta\rangle = (y_1 \ldots y_n)$ then $\langle\alpha|\beta\rangle = \sum \overline{x_i} y_i$ where $\langle\alpha|$ is the conjugate vector $(\overline{x_1}, \cdots, \overline{x_n})$. The inner product satisfies $\langle\alpha|\beta_1 + \beta_2\rangle = \langle\alpha|\beta_1\rangle + \langle\alpha|\beta_2\rangle$ and $\langle\alpha|\beta\rangle = \overline{\langle\beta|\alpha\rangle}$ and $\langle\alpha|\alpha\rangle \geq 0$. The quantity $\||\alpha\rangle\| = \sqrt{\langle\alpha|\alpha\rangle}$ is called the length of $|\alpha\rangle$. If it is 1 then α is called a unit vector. Note that the length or more generally $|\langle\alpha|\beta\rangle|$ is invariant w.r.t. multiplication of $|\alpha\rangle$ and $|\beta\rangle$ by arbitrary complex constants of modulus 1. With this notation we extend Postulate 1.

Postulate 2 *The state of the system is given by a unit vector, determined up to a scalar multiple of modulus 1. Moreover, each such vector is realisable as a state.*

Thus $|\alpha\rangle$ and $|\beta\rangle$ represent the same state iff $|\alpha\rangle = e^{ic}|\beta\rangle$ c real.

A basis $\mathbf{b} = \{\alpha_1, \cdots, \alpha_n\}$ of H is a linearly independent set of vectors such that every vector in H is a (unique) linear combination of the $|\alpha_i\rangle$'s. It is orthonormal iff $\langle\alpha_i|\alpha_j\rangle = \delta_{ij}$ where $\delta_{ij} = 1$ if $i = j$ and 0 otherwise. From any set of n linearly independent vectors we can construct an orthonormal basis. It is the set of orthonormal vectors which correspond to the classical notion of states. Their occurence in any test can be considered as mutually exclusive events. Henceforth basis will mean an orthonormal one.

Postulate 3 *Any orthornormal basis represents a realisable maximal test.*

Let n be the maximum number of different outcomes possible in a given system for any test. For example we may test for the value of the z-component of spin of an electron or polarisation of a photon. We imagine we have large number of similarly prepared systems called an ensemble and we test for the values of different measurable quantities like spin etc.

For a spin-1/2 system we always get a maximum of 2 outcomes ('up' and 'down') for any test. So $n = 2$. This number is a property of the system and according to the postulate equals the dimension of the state space. In general, we postulate that for an ensemble in an arbitrary state, it is always possible to devise a test that yields the n outcomes corresponding to an orthonormal basis with definite probability.

Postulate 4 *If the system is prepared in state $|\alpha\rangle$ and a maximal test corresponding to a basis $\mathbf{b} = \{|\beta_i\rangle|\ i = 1, \cdots, n\}$ is performed, the probability that the outcome will correspond to $|\beta_i\rangle$ is given by $p_i = |\langle\alpha|\beta_i\rangle|^2$.*

Since one of the outcomes must occur, $\sum p_i = 1$. In the relative frequency interpretation of probability this means that if we have an ensemble of N systems and perform a maximal test corresponding to $\{|\beta_j\rangle\}$ then if the frequency of outcome corresponding to $|\beta_i\rangle$ is n_i, we have $p_i = \lim_{N\to\infty} \frac{n_i}{N}$. If the system is known to be in one of the states in a basis $\{|\beta_i\rangle\}$, say $|\beta_1\rangle$, then $p_1 = 1$ and $p_i = 0$ for $i \neq 1$. That is, we can predict the outcome with certainty for this maximal test. This is the case that corresponds to the classical theory. However, if we choose a different maximal test corresponding to a different basis then the outcomes become random.

We note that if H is a Hilbert space with inner product $\langle\ \rangle$, then H is isomorphic to the dual space of linear functionals (i.e. complex valued functions) on H. Thus for each $|\alpha\rangle$ let $\langle\alpha|$ denote its image under this isomorphism, called the dual. Then $(\langle\alpha|)(|\beta\rangle) = \langle\alpha|\beta\rangle$ by definition. Further$(|\alpha\rangle\langle\beta|)(|\gamma\rangle) \stackrel{\text{def}}{=} (\langle\beta|\gamma\rangle)|\alpha\rangle$ is a linear operator on H. In particular, $|\alpha\rangle\langle alpha|$ is identified as the projection operator on $|\alpha\rangle$. Let $\mathcal{L}(\mathcal{H})$ denote the space of linear operators on H. An operator $A \in \mathcal{L}(\mathcal{H})$ is hermitian if $\langle\alpha|A\beta\rangle = \langle A\alpha|\beta\rangle$ for all $|\alpha\rangle$ and $|\beta\rangle$. An operator U is called unitary if $\langle U\alpha|U\beta\rangle = \langle\alpha|\beta\rangle$ for all $|\alpha\rangle$ and $|\beta\rangle$. In matrix notation let $B^\dagger$ denote the transposed conjugate of a square matrix B. Then B is hermitian if $B = B^\dagger$ and U is unitary if $U^{-1} = U^\dagger$. Hermitian and unitary matrices play a crucial role in quantum theory.

Let the system be in a state $|\alpha_i\rangle$ where $\{|\alpha_j\rangle \mid j = 1, \cdots, n\}$ is an orthonormal basis. Let $\{|\beta_j\rangle \mid j = 1 \cdots, n\}$ be another orthonormal basis. Then if we do a maximal test with respect to the later basis the probability of obtaining result $|\beta_k\rangle$ is $p_{ik} = |\langle\alpha_i|\beta_k\rangle|^2$. p_{ik} can also be written as $\mathtt{Tr}((|\alpha_i\rangle\langle\alpha_i|)(|\beta_k\rangle\langle\beta_k|))$ where the *trace* Tr is the sum of the diagonal elements of a square matrix, which is independent of the representation. The p_{ij} are called the transition probabilities. Let $U = (u_{ij} = \langle\alpha_i|\beta_j\rangle)$ be a matrix. Then U is unitary. It is the matrix which expresses the change of basis and $p_{ij} = |u_{ij}|^2$. We thus see that the transition probability matrix is doubly stochastic, i.e., $\sum_i p_{ij} = \sum_j p_{ij} = 1$. But an arbitrary doubly stochastic matrix (for example appearing in classical Markhov processes) may not correspond to transition probability matrix in quantum theory because it may not satisfy $p_{ij} = |u_{ij}|^2$ for some unitary $U = (u_{ij})$. Such matrices (p_{ij}) are called orthostochastic. Thus the p_{ij} must satisfy some relations. We thus see an important difference with classical probability theory. We can not make arbitrary probability assignments (satisfying of course the usual probability constraints) but the probabilities must satisfy certain nonlinear inequalities. This is also true of the probabilities p_i introduced earlier.

3 Syntax and Semantics

We now present the syntax and semantics of our logic of quantum probabilities.

3.1 Syntax

For each dimension $n \geq 1$ we define two languages interpreted with respect to n-dimensional Hilbert Space: $\mathcal{L}_n(P)$ and $\mathcal{L}_n(P,T)$. (We write $\mathcal{L}_n(X)$ when making assertions that apply to both languages.) We begin by describing the language $\mathcal{L}_n(P)$, which is based on the probability operator P.

Maximal measurements correspond to orthonormal bases of the Hilbert space, which are related by unitary transformations, as discussed above. Bases are represented in both languages by means of *basis variables* $\mathtt{b}, \mathtt{c}, \mathtt{d}, \ldots$. A *basis component* of $\mathcal{L}_n(X)$ is an expression of the form $\mathtt{b}_i$ where $1 \leq i \leq n$ and $\mathtt{b}$ is a basis variable[1]. Semantically, basis components correspond to the elements of an orthonormal basis of the n-dimensional Hilbert space.

The probability operators in our language will apply to formulas expressing properties of the outcome of a maximal measurement. We capture these properties by b-*formulas*, where b is a basis variable representing the measurement. A b-formula is a Boolean combination of b-components, i.e., an expression of the form $\mathtt{b}_i$, or $\alpha \wedge \alpha'$, or $\neg\alpha$, where α and α' are b-formulas. We define $\alpha \vee \alpha'$ as $\neg(\neg\alpha \wedge \neg\alpha')$. Note that all the basis components in a b-formula must be constructed from the same basis variable b; if t and t' are distinct basis formulas then

[1] Note that we do not use subscripting to distinguish basis variables: $\mathtt{b}_1$ and $\mathtt{b}_2$ always denote components of the same basis b, rather than two distinct basis variables. We use superscripting to denote distinct basis variables when use of distinct letters for the basis variables do not suffice.

$\mathsf{b}_1 \wedge \mathsf{b}_1'$ is not a b-formula. Intuitively, this restriction ensures that b-formulas describe the outcomes of measurements compatible with the basis b, and prevents construction of formulas combining results of incompatible measurements.

A *probability term* is an expression of the form $P(\alpha)$ where α is a b-formula for some basis variable b. A *linear probability atom* is an expression of the form $a_1 \cdot P(\alpha_1) + \ldots + a_k \cdot P(\alpha_k) \leq c$, where each a_i is an integer, c is an integer, and each α_i is a b^i-formula for some basis constant b^i. The formulas of the language $\mathcal{L}_n(P)$ are all the boolean combinations of linear probability atoms i.e., each linear probability atom is a formula of $\mathcal{L}_n(P)$, and if ϕ_1 and ϕ_2 are formulas of $\mathcal{L}_n(P)$ then so are $\neg\phi_1$, and $\phi_1 \wedge \phi_2$. The constructions $\phi 690_1 \vee \phi_2$, $\phi_1 \Rightarrow \phi_2$, $\phi_1 \Leftrightarrow \phi_2$ may be defined in this language as usual. Expressions such as $X < Y$, $X = Y$, where X, Y are linear combinations of probability terms are also definable in this language.

For the language $\mathcal{L}_n(P,T)$, we add *transition probability terms*, which are expressions of the form $T(\alpha, \beta)$, where α is a b-formula for some basis variable b and β is a c-formula for some basis variable c. Intuitively, these are a kind of conditional probability, expressing the probability that a measurement in basis c will have outcome satisfying β, given that the current state satisfies α. A *transition probability atom* is an expression of the form $p \leq 0$, where p is a polynomial expression with integer coefficients over probability terms and transition probability terms. Note that linear probability atoms are a special case of transition probability atoms. As above, comparison operators other than '$\leq$' are definable. Note that quantum probabilities are inherently quadratic. Define a *simple transition probability formula* to be a boolean combination of transition probability atoms. We need quantification over both real and complex numbers and a new type of term, denoting a complex number, to represent the entries of unitary matrices.

The proof theory for $\mathcal{L}(P,T)$ presented below will make use of results concerning the first order language $\mathcal{L}_{\{+,\cdot,=\}}$ with equality the only predicate symbol and function symbols representing addition and multiplication. The language $\mathcal{L}_{\{+,\cdot,=\}}$ leads to a recursively axiomatizable (in fact, decidable) theory both when interpreted with respect to the real field $\mathbf{R}$ and the complex field $\mathbf{C}$ [Tar51]. For our logic, it is convenient to first define the sorted first order language $\mathcal{L}_{\mathbf{RC}}$ with two sorts $\mathbf{R}$ and $\mathbf{C}$, with $\mathbf{R}$ a subsort of $\mathbf{C}$. The language $\mathcal{L}_{\mathbf{RC}}$ has equality as its only predicate symbol, a constant symbol c (of sort $\mathbf{R}$) for each integer, and the (infix) binary functions $+$ and $\cdot$ as its only function symbols. Both function symbols are overloaded: if t_1, t_2 are terms of sort $\mathbf{R}$ (respectively, $\mathbf{C}$), then so are $t_1 + t_2$ and $t_1 \cdot t_2$. We write $\forall x : \mathbf{R}.(\phi)$ and $\forall x : \mathbf{C}.(\phi)$ for universal quantification over the reals and complex numbers respectively. In this language, we may define operations such as complex conjugation $\overline{x + iy} = x - iy$, and the modulus of a complex number $|x + iy| = \sqrt{x^2 + y^2}$ (where x and y are real), so we use such operations freely. The inequality $x \leq y$ on real terms x, y is also definable (by $\exists z : R.(y = x + z^2)$).

The language $\mathcal{L}_n(P,T)$ is defined to be the extension of $\mathcal{L}_{\mathbf{RC}}$ in which we add probability terms and transition probability terms as terms of sort $\mathbf{R}$, as well as the terms $m_{ij}(\mathtt{b},\mathtt{c})$, of sort **C**, for $1 \le i, j \le n$ and basis variables $\mathtt{b}, \mathtt{c}$. Intuitively, $m_{ij}(\mathtt{b},\mathtt{c})$ denotes the ij-th entry of the unitary matrix that transforms the basis denoted by $\mathtt{b}$ into the basis denoted by $\mathtt{c}$.

3.2 Semantics

We now present the semantics for the language $\mathcal{L}_n(P,T)$ (and consequently for the sublanguage $\mathcal{L}_n(P)$). As remarked above, the language has been designed to allow it to serve as the propositional fragment of a modal logic. Thus, although there are no explicit modal operators, the semantics has some resemblences to Kripke semantics. In particular, we interpret formulas at a state within a collection of states, with respect to an interpretation of the atomic symbols.

A *structure* for $\mathcal{L}_n(P,T)$ is an n-dimensional Hilbert space H. For notational convenience we use plain Greek letters (ψ not $|\psi\rangle$ of 2) in this section. A *state* within this structure is a unit vector ψ in H. An *interpretation* of $\mathcal{L}_n(P,T)$ in a structure H is function π, such that

1. for each basis variable $\mathtt{b}$, $\pi(\mathtt{b})$ is an orthonormal basis $\psi_1, \ldots, \psi_n$ of H; (we write $\pi(\mathtt{b})_i$ for ψ_i)
2. for each real variable x, $\pi(x)$ is a real number;
3. for each complex variable X, $\pi(X)$ is a complex number.

If $M = (m_{ij})$ is an $n \times n$ unitary matrix and $B = \psi_1, \ldots, \psi_n$ is a sequence of vectors of H, we write MB for the sequence of vectors $\psi'_1, \ldots, \psi'_n$, where $\psi'_i = \Sigma_{i=1}^n m_{ik}\psi_i$. If B is an orthonormal basis of H then so is MB.

We extend the interpretation π to terms $\mathbf{t}$ of various sorts as follows. Given the term $\mathbf{t}$, a state ψ and an interpretation π, we define the interpretation $[\![\mathbf{t}]\!]_{\pi,\psi}$ of X with respect to π and ψ as follows. Basis variables are interpreted as bases:

1. $[\![\mathtt{b}]\!]_{\pi,\psi} = \pi(\mathtt{b})$, when $\mathtt{b}$ is a basis variable.

When $\mathtt{b}$ is a basis variable, we interpret $\mathtt{b}$-formulas as projection operators on H (these may also be understood as representing the subspaces of H onto which they project):

2. $[\![\mathtt{b}_i]\!]_{\pi,|\psi\rangle} = |\psi'\rangle\langle\psi'|$, where $\psi' = \pi(\mathtt{b})_i$;
3. $[\![\alpha_1 \wedge \alpha_2]\!]_{\pi,\psi} = [\![\alpha_1]\!]_{\pi,\psi} \cdot [\![\alpha_2]\!]_{\pi,\psi}$ (this is the projection operator projecting onto the intersection of the subspaces of H that are the images of the projection operators $[\![\alpha_1]\!]_{\pi,\psi}$ and $[\![\alpha_1]\!]_{\pi,\psi}$) which could be written as the product of these projectors.;
4. $[\![\neg\alpha]\!]_{\pi,\psi} = [\![\alpha]\!]^{\perp}_{\pi,\psi}$ is the projection operator projecting onto the orthogonal complement of the image of H under $[\![\alpha]\!]_{\pi,\psi}$.

(The reason we have taken $\pi(\mathtt{b})$ to be a basis rather than sequence of projection operators is to provide semantics for the terms $m_{ij}(\mathtt{b},\mathtt{c})$.) Terms of sort **R**, including probability terms and transition probability terms, are interpreted as real numbers, and terms of sort **C**, including the unitary matrix entry terms $m_{ij}(\mathtt{b},\mathtt{c})$, are interpreted as complex numbers:

5. $[\![x]\!]_{\pi,\psi} = \pi(x)$, when x is real or complex variable;
6. $[\![k]\!]_{\pi,\psi} = k$, when k is an integer;
7. $[\![P(\alpha)]\!]_{\pi,\psi} = \|[\![\alpha]\!]_{\pi,\psi}(\psi)\|^2$;
8. $[\![T(\alpha,\beta)]\!]_{\pi,\psi} = \mathtt{Tr}([\![\beta]\!]_{\pi,\psi}[\![\alpha]\!]_{\pi,\psi})$
9. $[\![m_{ij}(\mathtt{b},\mathtt{c})]\!]_{\pi,\psi} = c_{ij}$, where $M = (c_{ij})$ is the $n \times n$ (unitary) complex array such that $M\pi(\mathtt{b}) = \pi(\mathtt{c})$;
10. $[\![X \cdot Y]\!]_{\pi,\psi} = [\![X]\!]_{\pi,\psi} \cdot [\![X]\!]_{\pi,\psi}$
11. $[\![X + Y]\!]_{\pi,\psi} = [\![X]\!]_{\pi,\psi} + [\![Y]\!]_{\pi,\psi}$

To give semantics to formulas of $\mathcal{L}_n(P,T)$, we define a relation of satisfaction of a formula ϕ at a state ψ in a structure H, with respect to an interpretation π, denoted by $H,\pi,\psi \models \phi$. The definition is by the following induction:

1. $H,\pi,\psi \models X = Y$ if $[\![X]\!]_{\pi,\psi} = [\![Y]\!]_{\pi,\psi}$ (in case of $\mathcal{L}_n(P)$, this clause is replaced by $H,\pi,\psi \models X \leq c$ if $[\![X]\!]_{\pi,\psi} \leq c$);
2. $H,\pi,\psi \models \neg\phi$ if not $H,\pi,\psi \models \phi$;
3. $H,\pi,\psi \models \phi_1 \wedge \phi_2$ if $H,\pi,\psi \models \phi_1$ and $H,\pi,\psi \models \phi_2$;
4. $H,\pi,\psi \models \exists x : \mathbf{R}.(\phi)$ if there is a real number r such that $H,\pi[r/x],\psi \models \phi$;
5. $H,\pi,\psi \models \exists x : \mathbf{C}.(\phi)$ if there is a complex number c such that $H,\pi[c/x],\psi \models \phi$;

A formula ϕ of $\mathcal{L}_n(P,T)$ is *satisfiable* (in the n-dimensional Hilbert space H) if there exists an interpretation π and a state ψ such that $H,\pi,\psi \models \phi$. A formula ϕ is *valid* (in H) if $H,\pi,\psi \models \phi$ for all interpretations π and states ψ.

4 Examples

We give some examples of formulas in our language which express important concepts of quantum mechanics.

Superposition: A vector $|\alpha\rangle$ is a *superposition* of two vectors $|\beta_1\rangle$ and $|\beta_2\rangle$ if it is a linear combination of the two, i.e., $|\alpha\rangle = c_1|\beta_1\rangle + c_2|\beta_2\rangle$ for some complex numbers c_1, c_2. Consider the formula $T(\mathtt{b}_1, \mathtt{b}'_1 \vee \mathtt{b}'_2) = 1$. If $\pi(\mathtt{b}_1) = |\alpha\rangle$, $\pi(\mathtt{b}'_1) = |\beta_1\rangle$ and $\pi(\mathtt{b}'_2) = |\beta_2\rangle$ then $H,\pi,\psi \models T(\mathtt{b}_1, \mathtt{b}'_1 \vee \mathtt{b}'_2) = 1$ iff $\mathbf{Tr}(|\alpha\rangle\langle\alpha|(|\beta_1\rangle\langle\beta_1| + |\beta_2\rangle\langle\beta_2|)) = 1$. This is equivalent to $|\langle\beta_1|\alpha\rangle|^2 + |\langle\beta_2|\alpha\rangle|^2 = 1$, which is true iff the state $|\alpha\rangle$ is a superposition of the states $|\beta_1\rangle$ and $|\beta_2\rangle$. That is, the formula expresses that "$\mathtt{b}_1$ is a superposition of $\mathtt{b}'_1$ and $\mathtt{b}'_2$".

Phase Relations: Let $\mathtt{b}^0, \ldots, \mathtt{b}^k$ be $k+1$ bases. Then the following formula states a relation between $\mathtt{b}^0$ and the $\mathtt{b}^j$, for $j = 1 \ldots k$.

$$\begin{array}{ll} \mathbf{MP}_k & \forall x_1 \ldots x_n : \mathbf{R}.(\bigwedge_{i=1}^n P(\mathtt{b}_i^0) = x_i^2 \Rightarrow \\ & \quad \exists z_1 \ldots z_n : \mathbf{C}(\bigwedge_{i=1}^n |z_i| = 1 \wedge \\ & \quad \bigwedge_{j=1}^k \bigwedge_{i=1}^n P(\mathtt{b}_i^j) = |\Sigma_{r=1}^n m_{ir}(\mathtt{b}^0, \mathtt{b}^j) x_r z_r|^2)) \end{array}$$

Proposition 1. *The formula* $\mathbf{MP}_k$ *is valid for all* $k \geq 1$.

Proof. Let π be any interpretation and $|\psi\rangle$ any vector in H_n. If $[\![P(\mathtt{b}_i^0)]\!]_{\pi,|\psi\rangle} = \pi(x_i)^2$, then we may write $|\psi\rangle = \sum_{i=1}^n c_i \pi(x_i)\pi(\mathtt{b}^0)_i$, where the c_i are complex numbers with $|c_i| = 1$. Define $\pi(z_i) = c_i$. We can then calculate the probabilities with respect to other bases as follows:

$$\begin{aligned}
[\![P(\mathtt{b}_i^j)]\!]_{\pi,|\psi\rangle} &= |\langle \pi(\mathtt{b}^j)_i \mid \psi\rangle|^2 \\
&= |\langle \textstyle\sum_{r=1}^n [\![m_{ir}(\mathtt{b}^0,\mathtt{b}^j)]\!]_{\pi,|\psi\rangle} \cdot \pi(\mathtt{b}^0)_r \mid \psi\rangle|^2 \\
&= |\textstyle\sum_{r=1}^n [\![m_{ir}(\mathtt{b}^0,\mathtt{b}^j)]\!]_{\pi,|\psi\rangle} \cdot \pi(x_r) \cdot \pi(\mathtt{b}^0)_r|^2
\end{aligned}$$

from which it can be seen that $\mathbf{MP}_k$ holds.

Quantum State Tomography: Suppose we are given a collection of identically prepared systems (an ensemble), which corresponds to an unknown quantum state. Quantum state tomography (QST) addresses the problem of determining this unknown state. By measuring the ensemble in a single basis, we may determine a probability distribution over the outcomes associated to the basis elements. This distribution does not suffice to determine the state of the system. However, we may also divide the original collection of systems into subcollections and subject each subcollection to maximal measurement corresponding to an appropriately chosen orthonormal bases. We get sets of probability distributions. For an appropriately chosen set of measurements, this set of distributions suffices to compute the state. The following formula expresses this fact.

Let $\mathbf{u}$ be the sequence of variables u_{ij}^k where $1 \leq i,j,k \leq n$.

$$\begin{aligned}
&\exists \mathbf{u} : \mathbf{C}[\ \textstyle\bigwedge_{1\leq i,j,k\leq n} m_{ij}(\mathtt{b},\mathtt{c}^k) = u_{ij}^k \\
&\quad \Rightarrow \forall z_1 \ldots z_n : \mathbf{C}(\\
&\qquad \textstyle\bigwedge_{1\leq i,k\leq n} P(\mathtt{c}_i{}^k) = |\sum_j u_{ij}^k z_j \sqrt{P(\mathtt{b}_j)}|^2 \\
&\qquad \Rightarrow \textstyle\bigwedge_i P(\mathtt{b}_i{}') = |\sum_j m_{ij}(\mathtt{b},\mathtt{b}')z_j\sqrt{P(\mathtt{b}_j)}|^2\ \ )\ ].
\end{aligned}$$

This formula is valid. It expresses the fact that there is a "pattern of interrelation" between a set of bases $\mathtt{b}$ and $\mathtt{c}^1, \mathtt{c}^2, \ldots \mathtt{c}^n$, captured by the values $\mathbf{u}$, such that for any vector ψ, the probabilities of measurements associated to a set of bases related in this pattern provide sufficient information to calculate the probabilities of measurements with respect to any other basis. Note that by the discussion of the formula MP_k above, it is always possible to find values for $z_1, \ldots z_k$ such that

$$\bigwedge_{1\leq i,k\leq n} P(\mathtt{c}_i{}^k) = |\sum_j u_{ij}^k z_j \sqrt{P(\mathtt{b}_j)}|^2 \tag{1}$$

is satisfied. Thus the universally quantified formula in the conclusion is never true vacuously. The formula therefore expresses that given an appropriately related set of bases $\mathtt{b}$ and $\mathtt{c}^1, \mathtt{c}^2, \ldots \mathtt{c}^n$, it is possible, given any basis $\mathtt{b}'$, to calculate the values $P(\mathtt{b}_i')$ from the values u_{ij}^k, $P(\mathtt{b}_i)$, $P(\mathtt{c}_i^k)$ and $m_{ij}(\mathtt{b},\mathtt{b}')$. We do this by first solving the equation (1) for the phase values $z_1, \ldots, z_n$, and then computing $P(\mathtt{b}_i')$ as $|\sum_j m_{ij}(\mathtt{b},\mathtt{b}')z_j\sqrt{P(\mathtt{b}_j)}|^2$.

5 Axiomatization

We now present axiomatizations and state the completeness results for the languages $\mathcal{L}_n(P)$ and $\mathcal{L}_n(P,T)$.

5.1 Axiomatizing $\mathcal{L}_n(P)$

The axiomatization of $\mathcal{L}_n(P)$ consists of a number of parts, each dealing with one of the syntactic constructs of the language.

The first fragment of our axiomatization deals with the boolean logic of the b-formulas. As this is slightly richer than propositional logic, we identify for each basis variable b a fragment of the proof theory that deals only with b-formulas. The axioms of this fragment can be taken to be any complete axiomatization of propositional logic over the atomic formulas $\mathtt{b}_1, \ldots, \mathtt{b}_n$, with, e.g., Modus Ponens as the proof rule, plus the following axioms that capture the fact that we are dealing with an n-dimensional Hilbert space:

B1 $\mathtt{b}_1 \vee \ldots \vee \mathtt{b}_n$

B2 $\neg(\mathtt{b}_i \wedge \mathtt{b}_j)$ for $i \neq j$

We say that a b-formula ϕ is a b-tautology, and write $\vdash_{\mathtt{b}} \phi$ if it can be derived from these axioms alone. Note that these definitions isolate reasoning about b-formulas from reasoning about c-formulas when b and c are distinct.

Next, we have some axioms capturing the properties of the probability operator. The following axioms correspond very closely to the axioms W1-W4 of Fagin et al. [FHM90], but with the difference that we need to be careful to respect the syntactic constraints on probability terms. In the following, we require that there exists a basis term b such that ϕ, ϕ_1 and ϕ_2 are b-formulas:

P1 $0 \leq P(\phi) \leq 1$

P2 $P(\phi) = 1$ if ϕ is a b-tautology

P3 $P(\phi_1 \wedge \phi_2) + P(\phi_1 \wedge \neg\phi_2) = P(\phi_1)$

P4 $P(\phi_1) = P(\phi_2)$ if $\phi_1 \Leftrightarrow \phi_2$ is a b-tautology

Note that in P2 and P4, we deal with b-tautologies where Fagin et al have tautologies of propositional logic. Fagin et al note that in their axiomatization, there is no axiom corresponding to the fact that probability measures are countably additive. This does not make their axiomatization incomplete, because the countable additivity is not expressible in the logic. In our logic, we also do not have a countable additivity axiom, but the reason is somewhat simpler: the only measurable properties with respect to a basis denoted by b are those corresponding to the 2^n inequivalent b-formulas. Semantically, these formulas correspond to the linear subspaces generated by subsets of the n basis vectors.

In addition to the above axioms, we also need a set of axioms that capture reasoning about linear inequalities. That is, we need to be able to derive formulas such as $(2P(\phi_1) \leq 3 \wedge 4P(\phi_2) \leq 1) \Rightarrow P(\phi_1) + 2P(\phi_2) \leq 2$, the validity of which follows just from the meaning of these operations on real numbers, rather than the meaning of the probability terms. We refer the reader to Fagin et al. [FHM90] for such an axiomatization $\mathtt{AX}_{\mathtt{INEQ}}$ of such reasoning, and leave it as an exercise to construct an axiomatization $\mathtt{AX_{INEQ}}'$ suited to $\mathcal{L}_n(P)$ (the difference is to replace all variable occurences by probability terms).

Let $\mathtt{AX_n(P)}$ be the abovementioned axioms and rules of inference. Then we have the following:

Theorem 1. $\mathtt{AX_n(P)}$ *is a sound and complete axiomatization of* $\mathcal{L}_n(P)$.

The axiomatization $\mathtt{AX_n(P)}$ is almost identical to Fagin et al's axiomatization of for linear inequalities over classical probabilities — the main difference is the syntactic restrictions relating to b-formulas. Thus, from the point of view of the language $\mathcal{L}_n(P)$, quantum probabilities behave similarly to classical probabilities. The similarity is also reflected in the complexity of the logic:

Theorem 2. *Satisfiability of a formula of* $\mathcal{L}_n(P)$ *in n-dimensional Hilbert space (with* $n \geq 2$*) is NP-complete.*

The same complexity was obtained by FHM for their logic of classical probabilities. Interestingly, the proof of Theorem 1 shows that probabilities with respect to different bases act independently. More precisely, the completeness proof can be used to establish the following:

Proposition 2. *Let* $\phi_1, \ldots, \phi_m$ *be formulas and* $\mathtt{b}^1, \ldots, \mathtt{b}^m$ *be distinct basis variables such that for each* $j = 1, \ldots, n$*, the only probability terms occuring in* ϕ_j *are* $\mathtt{b}^j$ *probability terms. Then* $\phi_1 \wedge \ldots \wedge \phi_m$ *is satisfiable iff each* ϕ_j *is satisfiable.*

Thus, unlike quantum logic, $\mathcal{L}_n(P)$ is unable to express constraints on the ways that incompatible propositions are "pasted together". To capture such constraints, we need to turn to our richer logic, dealing with transition probabilities.

5.2 Axiomatizing $\mathcal{L}_n(P,T)$

We now present the axiomatization of $\mathcal{L}_n(P,T)$.

To capture the properties of the probability operator P, the axiomatization contains the axiomatization of b-formulas used above, and the probability axioms P1-P4. We need a similar set of axioms for the transition probabilities. In the following, ϕ, ϕ_1, ϕ_2 are b-formulas for some basis variable b, and ϕ', ϕ'_1, ϕ'_2 are b$'$-formulas for some basis variable b$'$. For a fixed b-formula ϕ, the probabilities of a transition to a b$'$-formula satisfy four properties directly analogous to P1-P4 above:

T1 $T(\phi, \phi') \geq 0$

T2 $T(\phi, \phi') = 1$ when ϕ' is a b$'$-tautology

T3 $T(\phi, \phi'_1 \wedge \phi'_2) + T(\phi, \phi'_1 \wedge \neg\phi'_2) = T(\phi, \phi'_1)$

T4 $T(\phi, \phi'_1) = T(\phi, \phi'_2)$ if $\phi'_1 \Leftrightarrow \phi'_2$ is a b$'$-tautology

These properties allow us to decompose a transition probability term into an equivalent expression of transition probability terms in which the second argument contains only atomic basis formulas. A similar decomposition with respect to the first argument can be obtained using the following property:

T5 $T(\phi, \phi') = T(\phi', \phi)$

Additionally, we have the following properties concerning a number of special cases of transitions. The first concerns transitions within a given basis:

T6a $T(\mathtt{b}_i, \mathtt{b}_j) = 0$ when $i \neq j$

T6b $T(\mathtt{b}_i, \mathtt{b}_i) = 1$

Next, note that the formula $P(\mathtt{b}_i) = 1$ can be understood as saying that the state ψ at which the formula is being evaluated is equal to the i-th vector in the basis $\mathtt{b}$. The following property can be understood as stating that transition probabilities for transitions from the current state reduce to simple probabilities:

T7 $P(\mathtt{b}_i) = 1 \Rightarrow T(\mathtt{b}_i, \phi) = P(\phi)$

For $\mathcal{L}_n(P)$ our axiomatization used a set of axioms for reasoning about linear inequalities. In the case of $\mathcal{L}_n(P,T)$, we need to reason about non-linear polynomials. Moreover, to capture the quantum nature of the probabilities, we need to reason about complex numbers. This leads us to include in the axiomatization a set of axioms for reasoning about real and complex numbers. The following proposition follows from the fact that the complex numbers may be represented as pairs of real numbers, that the operations of complex addition and multiplication may be defined as operations on these pairs, and that the language $\mathcal{L}_{\{+,\cdot,=\}}$ has an axiomatizable theory with respect to the standard model $\mathbf{R}$ [Tar51].

Proposition 3. *The set of valid formulas of $\mathcal{L}_{\mathbf{RC}}$ has a recursive axiomatization* $\mathtt{AX_{RC}}$.

The rules of inference of $\mathtt{AX_{RC}}$ are the usual rules for first order logic with equality. For reasons of space we do not list the axioms of $\mathtt{AX_{RC}}$ here. We include these axioms and rules of inference in our axiomatization. The language $\mathcal{L}_n(P,T)$ contains the terms $m_{ij}(\mathtt{b}, \mathtt{c})$, of complex number sort, to represent the unitary operators associated with basis transformations. The following properties are direct from the definition of these terms:

M1 $m_{ij}(\mathtt{b}, \mathtt{c}) = \overline{m_{ji}(\mathtt{c}, \mathtt{b})}$

The fact that the transformation from a basis to itself corresponds to the identity matrix, and that consecutive basis transformations correspond to matrix multiplication, are captured by the next two properties:

M2a $m_{ij}(\mathtt{b}, \mathtt{b}) = 1$ if $i = j$

M2b $m_{ij}(\mathtt{b}, \mathtt{b}) = 0$ if $i \neq j$

M3 $m_{ij}(\mathtt{b}, \mathtt{d}) = \Sigma_{k=1}^{n} m_{ik}(\mathtt{b}, \mathtt{c}) m_{kj}(\mathtt{c}, \mathtt{d})$

We note that the following property, expressing unitarity of the transformation, follows from M1-M3:

M4 $\Sigma_{k=1}^{n} m_{ik}(\mathtt{b}, \mathtt{c}) \overline{m_{jk}(\mathtt{b}, \mathtt{c})} = 1$

These matrices are connected to probabilities by the following axiom:

MT $T(\mathtt{b}_i, \mathtt{c}_j) = |m_{ij}(\mathtt{b}, \mathtt{c})|^2$

We note that M1-M3 and MT imply some of the properties of transition probabilities noted above. In particular, T1, T2 and T6 become derivable, as does the case $T(\mathtt{b}_i, \mathtt{b}_j) = T(\mathtt{b}_j, \mathtt{b}_i)$ of T5.

Let the axiomatization $\mathtt{AX_n(P,T)}$ consist of the propositional component with **B1-B2** for the b-formulas, the axioms **P1-P4**, **T1-T7**, the axioms and rules of $\mathtt{AX_{RC}}$ (including the usual rules of inference for first order logic), **M1-M3**, **MT** and the axiom $\mathbf{MP}_k$ (see Section 4) for all $k \leq n^2 - n + 1$. Then we have the following:

Theorem 3. $\mathtt{AX_n(P,T)}$ *is a sound and complete axiomatization for the language* $\mathcal{L}_n(P,T)$.

Note that although we have shown that $\mathbf{MP}_k$ is sound for all $k \geq 1$, we have only included its instances for $k \leq n^2 - n + 1$ in the set of axioms. Indeed, as part of the completeness proof we show that it is not necessary to include $\mathbf{MP}_k$ for larger k since it already follows:

Theorem 4. *For every number* $k \geq 1$*, the formula* $\mathbf{MP}_k$ *is a theorem of* $\mathtt{AX_n(P,T)}$.

We have stated this result as a theorem because we feel that it is of significance for physics as well as the logic of quantum probabilities. It shows that to determine whether a set of numbers can have arisen as the probabilities associated to a set of k bases in n-dimensional Hilbert space, it suffices to check the probabilities associated to every subset of size $n^2 - n + 1$.

As for the language $\mathcal{L}_n(P)$, we also can also obtain from the completeness proof some complexity bounds for $\mathcal{L}_n(P,T)$.

Theorem 5. *Satisfiability of a formula in* $\mathcal{L}_n(P,T)$ *can be decided in exponential space. If the formula is quantifier free, then its satisfiability can be decided in polynomial space.*

The proof of this result makes use of results of Ben-Or, Kozen and Reif [BKR86] for the full language, and of Canny [Can88] for the quantifier free case.

6 Conclusion

A topic that has been of some interest in the quantum mechanics literature is the extent to which it is possible to eliminate the use of complex numbers, and to reason about quantum probabilities purely as real numbers. This requires the characterization of the relationships between the quantum probabilities that follow from their Hilbert space definition. These relationships have been characterized in some low dimensions [Per95], but their charaterization in general remains an open problem. Our work may provide an avenue to address this problem, by applying quantifier elimination to our axiomatization.

We have shown our logic to be to be decidable. An interesting topic for further research is to determine the extent to which it is possible to further enrich the logic while retaining decidability/axiomatizability. Extensions that suggest themselves are temporal logic, dynamic logic and the logic of knowledge. Even before embarking on a study of such modal extensions, a variety of constructs dealing only with a single state are worthy of study. Constructs such as quantification over bases and unitary transformations, can also be added while keeping the language decidable. One construct that is of critical significance for quantum computing is the tensor product. We can already handle this to some extent simply by applying our language to the case where the dimension n is a product $n_1 \cdot n_2$, but it is desirable to have the tensor product as a more integral part of the language. We plan to study such extensions in future work.

References

[AH94] M. Abadi and J.Y. Halpern. Decidability and expressiveness for first-order logics of probability. *Information and Computation*, 112(1):1–36, 1994.

[Bac90] F. Bacchus. *Representing and Reasoning with Probabilistic Knowledge*. MIT Press, Cambridge, Mass., 1990.

[BKR86] M. Ben-Or, D. Kozen, and J. H. Reif. The complexity of elementary algebra and geometry. *Journal of Computer and System Sciences*, 32(1):251–264, 1986.

[BvN36] G. Birkhoff and J. von Neumann. The logic of quantum mechanics. *Annals of Mathematics*, 37:823–843, 1936.

[Can88] J. F. Canny. Some algebraic and geometric computations in PSPACE. In *Proc. 20th ACM Symp. on Theory of Computing*, pages 460–467, 1988.

[Car50] R. Carnap. *Logical Foundations of Probability*. University of Chicago Press, Chicago, 1950.

[FHM90] R. Fagin, J. Y. Halpern, and N. Megiddo. A logic for reasoning about probabilities. *Information and Computation*, 87(1/2):78–128, 1990.

[Gud89] S. Gudder. *Quantum Probability Theory*. Academic Press, San Diego, 1989.

[Nil86] N. Nilsson. Probabilistic logic. *Artificial Intelligence*, 28:71–87, 1986.

[Per95] A. Peres. *Quantum Theory: Concepts and Methods*. Kluwer Academic Publishers, Dordrecht, 1995.

[Pra92] V.R. Pratt. Linear logic for generalized quantum mechanics. In *Proc. of Workshop on Physics and Computation (PhysComp'92)*, pages 166–180, Dallas, Oct 1992. IEEE.

[Tar51] A. Tarski. *A Decision Method for Elementary Algebra and Geometry*. Univ. of California Press, 2nd edition, 1951.

A Strongly Normalising Curry-Howard Correspondence for IZF Set Theory

Alexandre Miquel

Laboratoire de Recherche en Informatique
Université Paris-Sud, 91405 Orsay Cedex, France
Alexandre.Miquel@lri.fr

Abstract. We propose a method for realising the proofs of Intuitionistic Zermelo-Fraenkel set theory (IZF) by strongly normalising λ-terms. This method relies on the introduction of a Curry-style type theory extended with specific subtyping principles, which is then used as a low-level language to interpret IZF via a representation of sets as pointed graphs inspired by Aczel's hyperset theory.

As a consequence, we refine a classical result of Myhill and Friedman by showing how a strongly normalising λ-term that computes a function of type $\mathbb{N} \to \mathbb{N}$ can be extracted from the proof of its existence in IZF.

1 Introduction

In this paper, we revisit the work of Myhill [12], Friedman [4], McCarty [9] and Krivine [6, 7] related to the study of the computational contents of the proofs of set theory. Unlike the former approaches that are based on variants of Kleene's realisability, we propose a different framework that combines ideas coming from hyperset theory to the theory of type systems [2, 5, 8].

Technically, our work relies on the introduction of a type system which is used as an assembly language to interpret the axioms of intuitionisitic set theory in terms of pointed graphs. Conceptually, this translation is the natural reformulation of Aczel's model of hyperset theory [1] (originally achieved in set theory) in a type-theoretical framework. Of course, building this model in type theory allows us to benefit from the representation of proofs as λ-terms. Moreover, since the type system we are using enjoys the strong normalisation property, any proof of IZF is thus realised by a strongly normalising λ-term via our translation.

Apart from the strong normalisation property for IZF—which is the main contribution of the paper—there are several interests in using a type-theoretical framework as an intermediate language for studying set theory.

On the practical side, one may benefit from the use of many proof-assistants based on type theory to build the corresponding λ-terms explicitly: since the proofs of the axioms of IZF tend to be quite large, writing them down would be almost impossible by hand. By using the method described in this paper, the

M. Baaz and J.A. Makowsky (Eds.): CSL 2003, LNCS 2803, pp. 441–454, 2003.

author could explicitely construct (and check) all the proof-terms of IZF axioms with the help of the Coq proof-assistant [3][1].

On a more theoretical side, this decomposition gives new insights about the relationship between set theories and type theories. As far as we know, there is currently no type system whose proof theoretical strength reaches the one of ZF. Despite its non-standard subtyping rules, the extended type system we will present in section 5 deserves the credit of enjoying this property.

2 The Logical Framework

In this section, we introduce the core part of the logical framework we will use in this paper. The initial system, called $F\omega.2$, will be first extended in paragraph 3.2 and then in section 5 to reach the proof theoretical strength of IZF.

2.1 System $F\omega.2$

System $F\omega.2$—or *system $F\omega$ with one universe*[2]—is organised in two syntactic categories: a syntactic category of *object-terms* in order to represent mathematical objects, and a syntactic category of *proof-terms* in order to represent mathematical proofs.

Object terms actually form an autonomous type system (the 'higher part' of system $F\omega.2$) which is completely independent from proof-terms. (Unlike the calculus of constructions, system $F\omega.2$ has no proof-dependent types.) This type system can be seen as an extension of Martin-Löf's logical framework

Object terms

$$\begin{array}{rcl} M,N,T,U,A,B & ::= & \mathsf{Type}_1 \ \mid \ \mathsf{Type}_2 \ \mid \ \Pi x:T.U \\ & \mid & x \ \mid \ \lambda x:T.U \ \mid \ M(N) \\ & \mid & \mathsf{Prop} \ \mid \ A \Rightarrow B \\ & \mid & \forall x:T.B \ \mid \ \exists x:T.B \end{array}$$

with a primitive type Prop of propositions (below the first universe Type_1), plus extra constructions to represent implication and universal/existential quantification. In this presentation, the letters M, N denote arbitrary object terms, whereas the letters T, U are reserved for types (i.e. the terms of type Type_1 or Type_2) and the letters A, B for propositions (i.e. the terms of type Prop).

As in Martin-Löf's logical framework [8], we make a distinction between a universe Type_1 of *small data-types* and a (top) universe Type_2 of *large data-types*. For convenience, we also consider a cumulativity rule $\mathsf{Type}_1 \subset \mathsf{Type}_2$ [3]. Notice that at the level of object terms, Prop is not a sort, and propositions are not

[1] The corresponding proof-scripts can be downloaded on the author's web page at `http://pauillac.inria.fr/~miquel`.

[2] "$F\omega.2$" means: "system F with higher-order twice".

[3] To preserve simplicity, we do not address here the problem of propagating cumulativity through dependent products.

Formation of signatures: $\Sigma \vdash$

$$\frac{}{[] \vdash} \qquad \frac{\Sigma \vdash T : \mathsf{Type}_i}{\Sigma; [x : T] \vdash} \; x \notin DV(\Sigma)$$

Typing rules of object terms: $\Sigma \vdash M : T$

$$\frac{\Sigma \vdash}{\Sigma \vdash \mathsf{Prop} : \mathsf{Type}_1} \qquad \frac{\Sigma \vdash}{\Sigma \vdash \mathsf{Type}_1 : \mathsf{Type}_2} \qquad \frac{\Sigma \vdash M : \mathsf{Type}_1}{\Sigma \vdash M : \mathsf{Type}_2}$$

$$\frac{\Sigma \vdash}{\Sigma \vdash x : T} \; (x:T) \in \Sigma \qquad \frac{\Sigma \vdash T : \mathsf{Type}_i \qquad \Sigma; [x : T] \vdash U : \mathsf{Type}_i}{\Sigma \vdash \Pi x : T . U : \mathsf{Type}_i} \; i \in \{1;2\}$$

$$\frac{\Sigma; [x : T] \vdash M : U}{\Sigma \vdash \lambda x : T . M : \Pi x : T . U} \qquad \frac{\Sigma \vdash M : \Pi x : T . U \qquad \Sigma \vdash N : T}{\Sigma \vdash M(N) : U\{x := N\}}$$

$$\frac{\Sigma \vdash A : \mathsf{Prop} \qquad \Sigma \vdash B : \mathsf{Prop}}{\Sigma \vdash A \Rightarrow B : \mathsf{Prop}} \qquad \frac{\Sigma; [x : T] \vdash B : \mathsf{Prop}}{\Sigma \vdash Q x : T . B : \mathsf{Prop}} \; Q \in \{\forall; \exists\}$$

$$\frac{\Sigma \vdash M : T}{\Sigma \vdash M : T'} \; T =_\beta T'$$

Formation of logical contexts: $\Sigma \vdash \Gamma \; \mathsf{ctx}$

$$\frac{\Sigma \vdash}{\Sigma \vdash [] \; \mathsf{ctx}} \qquad \frac{\Sigma \vdash \Gamma \; \mathsf{ctx} \qquad \Sigma \vdash A : \mathsf{Prop}}{\Sigma \vdash \Gamma; [\xi : A] \; \mathsf{ctx}} \; \xi \notin DV(\Gamma)$$

Typing rules of proof-terms: $\langle \Sigma \rangle \Gamma \vdash t : A$

$$\frac{\Sigma \vdash \Gamma \; \mathsf{ctx}}{\langle \Sigma \rangle \Gamma \vdash \xi : A} \; (\xi : A) \in \Gamma \qquad \frac{\langle \Sigma \rangle \Gamma \vdash t : A}{\langle \Sigma \rangle \Gamma \vdash t : A'} \; A =_\beta A'$$

$$\frac{\langle \Sigma \rangle \Gamma; [\xi : A] \vdash t : B}{\langle \Sigma \rangle \Gamma \vdash \lambda \xi . t : A \Rightarrow B} \qquad \frac{\langle \Sigma \rangle \Gamma \vdash t : A \Rightarrow B \qquad \langle \Sigma \rangle \Gamma \vdash u : A}{\langle \Sigma \rangle \Gamma \vdash tu : B}$$

$$\frac{\langle \Sigma; [x : T] \rangle \Gamma \vdash t : B}{\langle \Sigma \rangle \Gamma \vdash t : \forall x : T . B} \; x \notin FV(\Gamma) \qquad \frac{\langle \Sigma \rangle \Gamma \vdash t : \forall x : T . B \qquad \Sigma \vdash N : T}{\langle \Sigma \rangle \Gamma \vdash t : B\{x := N\}}$$

$$\frac{\Sigma; [x : T] \vdash B : \mathsf{Prop} \quad \Sigma \vdash N : T \qquad \langle \Sigma \rangle \Gamma \vdash t : B\{x := N\}}{\langle \Sigma \rangle \Gamma \vdash t : \exists x : T . B}$$

$$\frac{\Sigma; [x : T] \vdash A : \mathsf{Prop} \quad \Sigma \vdash B : \mathsf{Prop} \qquad \langle \Sigma \rangle \Gamma \vdash t : \forall x : T . (A \Rightarrow B)}{\langle \Sigma \rangle \Gamma \vdash t : (\exists x : T . A) \Rightarrow B}$$

Fig. 1. Typing rules of system $F\omega.2$

data-types. (Of course, propositions will be considered as data-types, but only at the level of proof-terms.)

Formally, the type system of object terms is based on a judgement $\Sigma \vdash M : T$ which expresses that the term M has type T under the assumptions in Σ. The corresponding typing assumptions are regrouped in *signatures*

Signatures $\Sigma \quad ::= \quad [x_1 : T_1; \ldots ; x_n : T_n]$

that are finite ordered lists of declarations of the form $(x : T)$. In order to ensure that the terms T involved in such declarations are well-formed types, we also need a judgement $\Sigma \vdash$ which expresses that the signature Σ is well-formed. The typing rules for both judgements are then defined by mutual recursion thanks to the rules given in Fig. 1.

The proof system (i.e. the 'lower part') of system $F\omega.2$ follows the typing discipline *à la* Curry, so that proofs terms are actually pure λ-terms:

Proof-terms $t, u \quad ::= \quad \xi \mid \lambda\xi\,.\,t \mid tu$

By this, we mean that the universal and existential quantifications are treated as infinitary intersection and union types respectively, so that in practice the corresponding introduction and elimination rules have no impact on proof-terms. On the other hand, implication is introduced and eliminated as usual, by the means of λ-abstraction and application.

Proof-terms depend on assumptions that are declared in *logical contexts*:

Logical contexts $\Gamma \quad ::= \quad [\xi_1 : A_1; \ldots ; \xi_k : A_k]\,.$

Since the object-terms A_i involved in such declarations cannot depend on proof-variables, the order of these declarations is irrelevant. To ensure that each A_i is a well-formed proposition (in a given signature), we introduce a judgement $\Sigma \vdash \Gamma$ ctx which expresses that the logical context Γ is well-formed in the signature Σ. Finally, the last judgement $\langle\Sigma\rangle\Gamma \vdash t : A$ expresses that in the signature Σ, the proof-term t is a proof of the sequent $\Gamma \vdash A$. The rules of inference for both judgements are given in Fig. 1.

Although the logic of system $F\omega.2$ is ultimately built on implication and quantifiers, the connectives $\wedge$, $\vee$ and units $\top$, $\bot$ are definable by the means of standard second-order encodings, as well as Leibniz equality.

The Primitive Existential Quantifier. Although the existential quantifier could have been defined purely in terms of $\forall$ and $\Rightarrow$ (by the mean of the standard second-order encoding of the existential quantifier), system $F\omega.2$ introduces a primitive form of existential quantification that behaves exactly as an infinitary union type constructor. (See the last two proof-typing rules of Fig. 1.)

In practice, the primitive form of existential quantifier—which restores a symmetry between quantifiers in a spirit which is very close to the one of standard realisability—tends to produce more compact proof-terms than the second-order encoding. But the main reason for using the primitive form of existential quantification is that denotations of propositions become much more readable in the normalisation model, a point which is important when studying intuitionistic forms of the axiom of choice as we will do in paragraphs 4.3 and 4.4.

From the point of view of provability of course, both forms of existential quantification are logically equivalent, and the full system with primitive form of existential quantification is nothing else but a conservative extension of the system restricted to $\forall$ and $\Rightarrow$ only.

3 Interpreting Zermelo's Set Theory

In this section, we briefly explain how to encode Intuitionistic Zermelo's set theory in system $F\omega.2$. We only give the basics of the encoding whose implementation details can be found in the author's PhD thesis [11].

Formally, Intuitionistic Zermelo's set theory (IZ) is the intuitionisitic first-order theory based on two binary predicates '=' (equality) and '$\in$' (membership) whose axioms are given in Fig. 2.

Equality & Compatibility axioms

(REFL)	$\forall x$	$x = x$
(SYM)	$\forall x, y$	$x = y \Rightarrow y = x$
(TRANS)	$\forall x, y, z$	$x = y \Rightarrow y = z \Rightarrow x = z$
(COMPAT-L)	$\forall x, y, z$	$x = y \Rightarrow y \in z \Rightarrow x \in z$
(COMPAT-R)	$\forall x, y, z$	$x \in y \Rightarrow y = z \Rightarrow x \in z$

Zermelo's axioms

(EXT)	$\forall a, b \ (\forall x \ x \in a \Leftrightarrow x \in b) \Rightarrow a = b$
(PAIR)	$\forall a, b \, \exists c \, \forall x \quad x \in c \Leftrightarrow x = a \vee x = b$
(SELECT)	$\forall a \, \exists b \, \forall x \quad x \in b \Leftrightarrow x \in a \wedge \phi$ where ϕ is any formula such that $b \notin FV(\phi)$
(POWER)	$\forall a \, \exists b \, \forall x \quad x \in b \Leftrightarrow (\forall y \ y \in x \Rightarrow y \in a)$
(UNION)	$\forall a \, \exists b \, \forall x \quad x \in b \Leftrightarrow (\exists y \ y \in a \wedge x \in y)$
(INFINITY)	$\exists a \quad \varnothing \in a \wedge (\forall x \ x \in a \Rightarrow x \cup \{x\} \in a)$

Fig. 2. Axioms of Zermelo's set theory

3.1 Sets as Pointed Graphs

A set is represented in system $F\omega.2$ as a *pointed graph*, that is, as a triple (X, A, a) where:

1. $X : \mathsf{Type}_1$ is the *carrier* (i.e. the type of *vertices*)
2. $A : X \to X \to \mathsf{Prop}$ is the *edge relation* (or *local membership*)
3. $a : X$ is the *root* of the pointed graph.

Intuitively, a pointed graph (X, A, a) can be thought as a kind of transitive closure of a set whose structure is given by the relation $A(x, y)$ (that can be read as: 'x is a local element of y') and by the root a (i.e. the entry point in the transitive closure).

Notice that unlike the terms X, A and a, the triple (X, A, a) is not a real object of system $F\omega.2$ (which does not provide any pairing mechanism) but only an informal notation to group related components. Similarly, we also introduce the following shorthands

$$\begin{array}{lcl} \forall(X,A,a).\phi & \stackrel{\Delta}{\equiv} & \forall X:\mathsf{Type}_1 \,.\, \forall A:(X\to X\to\mathsf{Prop}) \,.\, \forall a:X \,.\, \phi \\ \exists(X,A,a).\phi & \stackrel{\Delta}{\equiv} & \exists X:\mathsf{Type}_1 \,.\, \exists A:(X\to X\to\mathsf{Prop}) \,.\, \exists a:X \,.\, \phi \\ \lambda(X,A,a).M & \stackrel{\Delta}{\equiv} & \lambda X:\mathsf{Type}_1 \,.\, \lambda A:(X\to X\to\mathsf{Prop}) \,.\, \lambda a:X \,.\, M \end{array}$$

to denote the universal and existential quantifications as well as the λ-abstraction over the class of pointed graphs.

Extensional equality of set theory is interpreted as *bisimilarity* in the class of pointed graphs. This relation denoted by $(X, A, a) \approx (Y, B, b)$ is defined by:

$$\begin{array}{l} (X,A,a) \approx (Y,B,b) \quad \stackrel{\Delta}{\equiv} \\ \quad \exists R:(X\to Y\to\mathsf{Prop}) \,. \\ \quad\quad (\forall x,x':X \,.\, \forall y:Y \,.\quad A(x',x) \wedge R(x,y) \Rightarrow \exists y':Y \,.\; B(y',y) \wedge R(x',y')) \;\wedge \\ \quad\quad (\forall y,y':Y \,.\, \forall x:X \,.\quad B(y',y) \wedge R(x,y) \Rightarrow \exists x':X \,.\; A(x',x) \wedge R(x',y')) \;\wedge \\ \quad\quad R(a,b) \,. \end{array}$$

Membership is then interpreted as *shifted bisimilarity*, namely:

$$(X,A,a) \in (Y,B,b) \quad \stackrel{\Delta}{\equiv} \quad \exists b':B \,.\quad B(b',b) \;\wedge\; (X,A,a) \approx (Y,B,b') \,.$$

Once the interpretation of equality and membership has been defined, it is straightforward to translate any formula ϕ of set theory as a proposition ϕ^* in system $F\omega.2$. For that, we consider a fixed mapping that associates three distinct variables $X_i : \mathsf{Type}_1$, $A_i : X_i \to X_i \to \mathsf{Prop}$ and $a_i : X_i$ of the type system $F\omega.2$ to each variable x_i of the first-order language of set theory, and we set:

$$\begin{array}{rcll} (x_i = x_j)^* & \stackrel{\Delta}{\equiv} & (X_i, A_i, a_i) \approx (X_j, A_j, a_j) & \\ (x_i \in x_j)^* & \stackrel{\Delta}{\equiv} & (X_i, A_i, a_i) \in (X_j, A_j, a_j) & \\ (\phi \diamond \psi)^* & \stackrel{\Delta}{\equiv} & \phi^* \diamond \psi^* & (\diamond \in \{\Rightarrow;\ \wedge;\ \vee\}) \\ (Qx\ \phi)^* & \stackrel{\Delta}{\equiv} & Q(X,A,a) \,.\, \phi^* & (Q \in \{\forall;\ \exists\}) \end{array}$$

It is straightforward to check that this translation validates the equality and compatibility axioms of Fig. 2.

3.2 Soundness of Zermelo's Axioms

The main interest of interpreting sets as pointed graphs and equality as bisimilarity is that it automatically validates the axiom of extensionality:

Proposition 1 (Extensionality) — *The translation of the extensionality axiom is provable in system $F\omega.2$.*

As pointed out by [6, 9, 12], the interpretation of the extensionality axiom is the cornerstone of any computational interpretation of (the proofs of) set theory. On the other hand, proving that the bisimilarity relation is extensional w.r.t. the shifted bisimilarity relation is quite easy, and the formalisation of this proof in type-theory gives the corresponding λ-term for free.

The other axioms of Zermelo express the possibility of constructing new sets from other sets by several means. Except for the axiom of infinity, all the corresponding constructions have a natural translation in terms of pointed graphs that can be formalised in system $F\omega.2$, so that:

Proposition 2 (Finitary Zermelo axioms) — *The translation of* (PAIR), (POWER), (UNION) *and of each instance of* (SELECT) *is provable in $F\omega.2$.*

However, the axiom of infinity poses another problem, since its interpretation in terms of pointed graphs requires an infinite small data-type whose existence cannot be proved in the logical framework we presented in section 2[4].

As proposed in [11], a way to solve this problem is to add an extra universe below the universe Type_1 (i.e. a universe Type_0) from which one easily reconstructs a type of numerals by suitable encodings (so that the full construction of a model of IZ actually takes place in system $F\omega.3$).

In this paper, we consider a simpler solution by extending our logical framework with primitive numerals. For that, we introduce a small type $\mathsf{Nat} : \mathsf{Type}_1$ with two constructors $\mathsf{0} : \mathsf{Nat}$ and $\mathsf{S} : \mathsf{Nat}\to\mathsf{Nat}$, as well as two primitive functions $\mathsf{pred} : \mathsf{Nat}\to\mathsf{Nat}$ and $\mathsf{null} : \mathsf{Nat}\to\mathsf{Prop}$ with the computational rules

$$\mathsf{pred}(\mathsf{0}) \to_\beta \mathsf{0} \qquad \mathsf{pred}(\mathsf{S}(M)) \to_\beta M \qquad \mathsf{null}(\mathsf{0}) \to_\beta \top \qquad \mathsf{null}(\mathsf{S}(M)) \to_\beta \bot$$

from which we easily derive that S is injective, and that $\mathsf{S}(n) \neq \mathsf{0}$ for any $n : \mathsf{Nat}$ (where $\neq$ denotes the negation of Leibniz equality). In this framework, the induction principle comes for free provided we restrict all the quantifications with the predicate $\mathsf{wf\_nat}$ defined by

$$\mathsf{wf\_nat}(n) \ \stackrel{\Delta}{\equiv}\ \forall P : (\mathsf{Nat}\to\mathsf{Prop})\,.\ P(\mathsf{0}) \Rightarrow (\forall p : \mathsf{Nat}\,.\, P(p) \Rightarrow P(\mathsf{S}(p))) \Rightarrow P(n)\,.$$

Using this, it is then easy to build a pointed graph that represents the set ω of von Neumann numerals, so that:

---

[4] A simple counter-model is the obvious extension of the finitary boolean model of Church's theory of simple types to system $F\omega.2$, in which small types are interpreted by hereditarily finite sets whereas large types are interpreted by the elements of a fixed set-theoretical universe.

Proposition 3 (Infinity) — *The translation of the axiom of infinity is provable in system $F\omega.2$ extended with primitive numerals.*

3.3 Beyond Zermelo

It is natural to ask whether the former soundness result can be extended to IZF, which is obtained by adding the *collection scheme* to IZ:

$$(\textsc{Coll}) \qquad \forall a \quad (\forall x{\in}a\ \exists y\ \phi) \ \Rightarrow\ \exists b\ \forall x{\in}a\ \exists y{\in}b\ \phi$$

(where ϕ is an arbitrary formula such that $b \notin FV(\phi)$).

Unfortunately, the answer is negative, since the soundness of (Coll) would entail the relative consistency of ZF w.r.t. $F\omega.2$ with primitive numerals (via Gödel's negation translation, which maps provable formulas of ZF to provable formulas of IZF). On the other hand, our type system can be seen as a subsystem of the calculus of constructions with universes whose strong normalisation property (and logical consistency) has been proved in ZF [10], so that its proof-theoretical strength is actually less than the one of ZF[5].

4 The Normalisation Model $\mathcal{M}$

This section is devoted to the construction of a strong normalisation model $\mathcal{M}$ of $F\omega.2$. The main interest of such a model is not only that it constitutes the main device for proving the strong normalisation property of system $F\omega.2$ but that it naturally validates more propositions than the syntax, and thus suggests extensions of it.

4.1 Interpreting Object-Terms

The model $\mathcal{M}$ is defined in classical set theory with axiom of choice (ZFC). To interpret type-theoretical universes, we also assume the following axiom:

Axiom 4 — *There exists two nested ZF-universes.*

By ZF-universe (or set-theoretical universe), we mean any transitive set $\mathfrak{U}$ that fulfils the following conditions:

1. $A \in \mathfrak{U} \ \Rightarrow\ \mathfrak{P}(A) \in \mathfrak{U}$
2. $A \in \mathfrak{U}\ \wedge\ (B_x)_{x\in A} \in \mathfrak{U}^A \ \Rightarrow\ (\bigcup_{x\in A} B_x) \in \mathfrak{U}$
3. $\omega \in \mathfrak{U}$

ZF-universes are closed under all the operations that can be defined in ZFC, among which the generalised cartesian product that will be used to interpret dependent products:

$$A \in \mathfrak{U} \wedge (B_x)_{x\in A} \in \mathfrak{U}^A \ \Rightarrow\ \Big(\prod_{x\in A} B_x\Big) \in \mathfrak{U}.$$

[5] We conjecture that system $F\omega.2$ with primitive numerals has the same proof-theoretical strength as higher-order Zermelo's set theory.

In the following, we assume that $\mathfrak{U}_1$ and $\mathfrak{U}_2$ are two ZF-universes such that $\mathfrak{U}_1 \in \mathfrak{U}_2$. Since we want to interpret any signature, we must prohibit empty types by setting $[\![\mathsf{Type}_i]\!] = \mathfrak{U}_i \setminus \{\varnothing\}$ for $i \in \{1;2\}$. (Notice that these sets are still closed under generalised cartesian products, thanks to the axiom of choice.)

Propositions are interpreted as saturated sets [10]. The set of all saturated sets, denoted by $\mathfrak{SAT}$, is closed under arbitrary (non-empty) unions and intersections, and forms a complete lattice whose top element is SN, and whose bottom element is the set $Neut$ of neutral terms. Moreover, $\mathfrak{SAT}$ is closed under the construction $S \to T$ defined by

$$S \to T \quad \triangleq \quad \{t \in \Lambda; \quad \forall u \in S \quad tu \in T\} \quad \in \quad \mathfrak{SAT}.$$

Let $\mathcal{M} = \mathfrak{U}_2$. A valuation is a function $\rho : \mathcal{V} \to \mathcal{M}$, where $\mathcal{V}$ denotes the set of object-variables. The interpretation $(M, \rho) \mapsto [\![M]\!]_\rho$ is defined by

$$[\![\mathsf{Type}_i]\!]_\rho = \mathfrak{U}_i \setminus \{\varnothing\} \qquad [\![\mathsf{Prop}]\!]_\rho = \mathfrak{SAT} \qquad [\![x]\!]_\rho = \rho(x)$$

$$[\![\Pi x : T \,.\, U]\!]_\rho = \prod_{v \in [\![T]\!]_\rho} [\![U]\!]_{(\rho; x \leftarrow v)} \qquad [\![\lambda x : T \,.\, M]\!]_\rho = \left(v_{\in [\![T]\!]_\rho} \mapsto [\![M]\!]_{(\rho; x \leftarrow v)}\right)$$

$$[\![M(N)]\!]_\rho = [\![M]\!]_\rho([\![N]\!]_\rho) \qquad [\![A \Rightarrow B]\!]_\rho = [\![A]\!]_\rho \to [\![B]\!]_\rho$$

$$[\![\forall x : T \,.\, A]\!]_\rho = \bigcap_{v \in [\![T]\!]_\rho} [\![A]\!]_{(\rho; x \leftarrow v)} \qquad [\![\exists x : T \,.\, A]\!]_\rho = \bigcup_{v \in [\![T]\!]_\rho} [\![A]\!]_{(\rho; x \leftarrow v)}$$

whereas the constants Nat, 0, S, pred and null are interpreted in the obvious way. Notice that the right-hand side of the equation which gives the interpretation of the application may be undefined, so that the function $\rho \mapsto [\![M]\!]_\rho$ is partial.

Each signature Σ is interpreted as the set $[\![\Sigma]\!]$ of all valuations ρ such that $\rho(x) \in [\![T]\!]$ for each declaration $(x : T) \in \Sigma$ (i.e. the set of *adapted valuations*). We finally get by a straightforward induction:

Proposition 5 (Soundness of typing) — *If $\Sigma \vdash M : T$, then for all $\rho \in [\![\Sigma]\!]$, the denotations $[\![M]\!]_\rho$ and $[\![T]\!]_\rho$ are well-defined, and $[\![M]\!]_\rho \in [\![T]\!]_\rho$.*

As for any strong normalisation model, our model enjoys the crucial property that the denotation of a (well-formed) type is always inhabited. For this reason, any well-formed signature admits at least an adapted valuation.

4.2 The Normalisation Invariant

As well as we interpreted signatures as sets of adapted valuations, each logical context $\Gamma = [\xi_1 : A_1; \ldots; \xi_n : A_n]$ is now interpreted as a set of adapted substitutions. Formally, $[\![\Gamma]\!]_\rho$ is defined (for a given valuation ρ) as the set of all substitutions $\sigma = [\xi_1 := u_1; \ldots; \xi_n := u_n]$ such that $u_i \in [\![A_i]\!]_\rho$ for all $i = 1..n$.

We can now express the strong normalisation invariant of (Curry-style) system $F\omega.2$ as follows:

Proposition 6 (Strong normalisation invariant) — *If $\langle\Sigma\rangle\Gamma\vdash t:A$ is derivable, then for all $\rho\in[\![\Sigma]\!]$ and for all $\sigma\in[\![\Gamma]\!]_\rho$ one has $t[\sigma]\in[\![A]\!]_\rho$.*

By instantiating this result to an arbitrary adapted valuation $\rho\in[\![\Sigma]\!]$ and to the identity substitution (adapted to Γ) we conclude that:

Corollary 7 (Strong normalisation) — *In $F\omega.2$, all the typable proof-terms are strongly normalising λ-terms.*

Truth in the Model. Although the strong normalisation model $\mathcal{M}$ is quite close of a realisability model (think of the interpretation of $\forall$, $\exists$ and $\Rightarrow$), an important difference with the standard realisability approach is that in $\mathcal{M}$, the denotation of a proposition is never empty, since a saturated set contains at least all the neutral terms. To define a suitable notion of truth in the model, one has to exclude such 'paraproofs' by only considering *closed* terms.

Formally, we will say that a proposition A is true in the model $\mathcal{M}$ if its denotation $[\![A]\!]\in\mathfrak{SAT}$ contains at least a closed proof-term. Notice that the proposition $\bot=\forall p:\mathsf{Prop}\,.\,p$ (whose denotation is *Neut*) is false in the model, which shows that system $F\omega.2$ is logically consistent.

4.3 Axiom of Choice and Uniform Collection

An example of a true but non-provable proposition is the following formulation of the axiom of choice in $F\omega.2$:

$$(\forall x:T\,.\,\exists x:U\,.\,A(x,y))\ \Rightarrow\ \exists f:T\to U\,.\,\forall x:T\,.\,A(x,f(x))$$

where T and U are arbitrary data-types. Although this proposition is not provable in $F\omega.2$, it is straightforward to check that in the model $\mathcal{M}$, its denotation contains the closed proof-term $\lambda\xi\,.\,\xi$ since both members of the corresponding implication have the very same denotation, namely:

$$\bigcap_{x\in T}\bigcup_{y\in U}A(x,y)\quad=\quad\bigcup_{f\in U^T}\bigcap_{x\in T}A(x,f(x))\,.$$

(In the model of course, the left-to-right inclusion relies on the axiom of choice.)

This fact suggests that we can add the following proof principle

$$\frac{\langle\Sigma\rangle\Gamma\vdash t:\forall x:T\,.\,\exists y:U\,.\,A(x,y)}{\langle\Sigma\rangle\Gamma\vdash t:\exists f:(T\to U)\,.\,\forall x:T\,.\,A(x,f(x))}$$

to our system without breaking the normalisation invariant.

The Uniform Collection Scheme. Coming back to our translation of set theory, it is interesting to notice that this additional rule is actually sufficient for proving a weak form of collection scheme—that will be called here the *uniform collection scheme*—whose statement is the following:

$$(\forall x\ \exists y\ \phi(x,y))\quad\Rightarrow\quad\forall a\ \exists b\ \ \forall x{\in}a\ \exists y{\in}b\ \ \phi(x,y)$$

This statement is weaker than the collection scheme since it relies on a totality assumption which is stronger than that of collection, by requiring that the binary relation $\phi(x, y)$ should be total not only on a, but on the whole universe. (Of course, the uniform collection is classically equivalent to the collection scheme.)

4.4 An Intuitionistic Choice Operator

The difficulty in realising the full collection scheme is that for any $x \in a$, the implicit witness y given by the proof of $\exists y\ \phi(x, y)$ does not only depend on x, but also on the proof of the relativisation $x \in a$. To overcome this difficulty, we propose to express this dependency by using a trick inspired by Krivine's interpretation of the denumerable axiom of choice [6, 7].

For that, we extend system $F\omega.2$ with an *intuitionistic choice operator* that associates to any predicate $A(x)$ a sequence of objects $\epsilon x : T . A(x) : \mathsf{Nat} \to T$ whose intended meaning is: if A holds for some $x : T$, then A holds for some element of the sequence $\epsilon x : T . A(x)$.

Formally, the interpretation of this new construction in the model $\mathcal{M}$ relies on a fixed enumeration $(t_n)_{n \in \omega}$ of all the λ-terms. The construction $\epsilon x : T . A(x)$ is then interpreted as a function $f \in T^\omega$ (defined by using the axiom of choice) such that for all $n \in \omega$, either:

1. $f(n)$ is some $x \in T$ such that $t_n \in A(x)$ if such an element exists; or
2. $f(n)$ is an arbitrary element of T otherwise.

Once the function $f \in T^\omega$ that interprets the construction $\epsilon x : T . A(x)$ has been defined, it is straightforward to check that $\bigcup_{x \in T} A(x) \subset \bigcup_{n \in \omega} A(f(n))$. (The converse inclusion also holds, but is not of interest.) Coming back to our logical framework, this inclusion means that the model validates the typing rule

$$\frac{\langle\Sigma\rangle\Gamma \vdash t : \exists x : T . A(x)}{\langle\Sigma\rangle\Gamma \vdash t : \exists n : \mathsf{Nat} . A((\epsilon x : T . A(x))(n))}$$

which can thus be added to our type system without harm for the strong normalisation property.

5 The Extended Framework $F\omega.2^{++}$

Using the material we presented above, we can now define an extended Curry-style framework called $F\omega.2^{++}$ that actually contains enough proof principles to allow the (translation of the) collection scheme to be derived.

5.1 Syntax and Typing Rules

Object-Terms and Signatures. The syntax and typing rules of object-terms and signatures of system $F\omega.2^{++}$ are the one of system $F\omega.2$ extended with primitive numerals and the intuitionistic choice operator discussed in paragraph 4.4 (see Fig. 3). The corresponding reduction rules are the usual β-rule and the reduction rules of the constants pred and null that we gave in paragraph 3.2.

Judgements $\Sigma \vdash$, $\Sigma \vdash M : T$ and $\Sigma \vdash \Gamma$ ctx

Rules of Fig. 1 p. 443 + constants Nat, 0, S, pred, null

+

$$\frac{\Sigma; [x : T] \vdash A : \mathsf{Prop}}{\Sigma \vdash \epsilon x : T . A : \mathsf{Nat} \to T}$$

Propositional subtyping: $\Sigma \vdash A \leq B$

$$\frac{\Sigma \vdash A : \mathsf{Prop} \qquad \Sigma \vdash A' : \mathsf{Prop}}{\Sigma \vdash A \leq A'} {\scriptstyle A =_\beta A'} \qquad \frac{\Sigma \vdash A \leq B \qquad \Sigma \vdash B \leq C}{\Sigma \vdash A \leq C}$$

$$\frac{\Sigma \vdash A' \leq A \qquad \Sigma \vdash B \leq B'}{\Sigma \vdash \ A \Rightarrow B \ \leq \ A' \Rightarrow B'}$$

$$\frac{\Sigma; [x : T] \vdash A \leq B}{\Sigma \vdash \ A \ \leq \ \forall x : T . B} {\scriptstyle x \notin FV(A)} \qquad \frac{\Sigma; [x : T] \vdash A : \mathsf{Prop} \qquad \Sigma \vdash N : T}{\Sigma \vdash \ \forall x : T . A \ \leq \ A\{x := N\}}$$

$$\frac{\Sigma; [x : T] \vdash A : \mathsf{Prop} \qquad \Sigma \vdash N : T}{\Sigma \vdash \ A\{x := N\} \ \leq \ \exists x : T . A} \qquad \frac{\Sigma; [x : T] \vdash A \leq B}{\Sigma \vdash \exists x : T . A \leq B} {\scriptstyle x \notin FV(B)}$$

$$\frac{\Sigma; [x : T] \vdash A : \mathsf{Prop} \qquad \Sigma \vdash B : \mathsf{Prop}}{\Sigma \vdash \ \forall x{:}T . (A \Rightarrow B) \ \leq \ (\exists x{:}T . A) \Rightarrow B}$$

$$\frac{\Sigma; [x : T; \ y : U] \vdash A : \mathsf{Prop}}{\Sigma \vdash \ \forall x{:}T . \exists y{:}U . A \ \leq \ \exists f : (\Pi x{:}T.U) . \forall x{:}T . A\{y{:=}f(x)\}}$$

$$\frac{\Sigma; [x : T] \vdash A : \mathsf{Prop}}{\Sigma \vdash \ \exists x : T . A \ \leq \ \exists n : \mathsf{Nat} . A\{x := (\epsilon x : T . A)(n)\}}$$

Typing rules of proofs-terms: $\langle\Sigma\rangle\Gamma \vdash t : A$

$$\frac{\Sigma \vdash \Gamma \ \mathsf{ctx}}{\langle\Sigma\rangle\Gamma \vdash \xi : A} {\scriptstyle (\xi : A) \in \Gamma} \qquad \frac{\langle\Sigma\rangle\Gamma; [\xi : A] \vdash t : B}{\langle\Sigma\rangle\Gamma \vdash \lambda\xi . t : A \Rightarrow B}$$

$$\frac{\langle\Sigma\rangle\Gamma \vdash t : A \Rightarrow B \qquad \langle\Sigma\rangle\Gamma \vdash u : A}{\langle\Sigma\rangle\Gamma \vdash tu : B}$$

$$\frac{\langle\Sigma; [x : T]\rangle\Gamma \vdash t : A}{\langle\Sigma\rangle\Gamma \vdash t : \forall x : T . A} {\scriptstyle x \notin FV(\Gamma)} \qquad \frac{\langle\Sigma\rangle\Gamma \vdash t : A \qquad \Sigma \vdash A \leq A'}{\langle\Sigma\rangle\Gamma \vdash t : A'}$$

Fig. 3. Rules of inference of system $F\omega.2^{++}$

Logical Contexts and Proof-Terms. Logical contexts have the same syntax and formation rules as in system $F\omega.2$, and proof-terms are still pure λ-terms.

The main novelty of system $F\omega.2^{++}$ is the introduction of a new form of judgement called *propositional subtyping* and written $\Sigma \vdash A \leq B$. This judgement—whose logical meaning is a *direct implication*—is intended to capture the different inclusions between saturated sets that we pointed out throughout section 4. In particular, this judgement (whose rules of inference are given in Fig. 3) now incorporates the introduction and elimination rules for both quantifiers, as well as a very natural rule that expresses the contravariance of the domain of implication (and the covariance of its codomain) w.r.t. subtyping. The last two subtyping rules express both intuitionistic forms of axiom of choice discussed in paragraphs 4.3 and 4.4. (Notice that the latter is actually sufficient to derive the collection scheme.)

In this framework, the proof-typing rules become simpler than in system $F\omega.2$, for that many logical rules have been incorporated in the propositional subtyping judgement, and are now accessed via a standard subsumption rule.

5.2 Strong Normalisation

From the results of section 4, it is clear that $\mathcal{M}$ is still a strong normalisation model for $F\omega.2^{++}$. The proofs of propositions 5 and 6 are easily adapted to system $F\omega.2^{++}$ by interpreting the new construction $\epsilon x : T \, . \, A$ as explained in 4.4. Notice that in order to prove the strong normalisation invariant for system $F\omega.2^{++}$, we first have to check the soundness of propositional subtyping:

Proposition 8 — *If $\Sigma \vdash A \leq B$, then for all $\rho \in [\![\Sigma]\!]$ one has $[\![A]\!]_\rho \subset [\![B]\!]_\rho$.*

From this, we easily deduce that all the well-formed proof-terms of system $F\omega.2^{++}$ are strongly normalising, and that the system is logically consistent.

5.3 Deriving the Collection Scheme

Of course, the main interest of using system $F\omega.2^{++}$ is that we can now realise the collection scheme, by the means of the intuitionistic choice operator:

Proposition 9 (Collection scheme) — *The translation of each instance of* (COLL) *is provable in system $F\omega.2^{++}$.*

Notice that this result entails that the proof-theoretical strength of system $F\omega.2^{++}$ is at least the one of IZF/ZF.

5.4 Extracting Functions from Proofs

In this paragraph, we aim to show that from any proof (in IZF) of a statement ϕ of the form

$$\phi \quad \equiv \quad \forall x{\in}\omega \;\; \exists y{\in}\omega \;\; \psi(x,y)$$

(where ψ is an arbitrary formula s.t. $FV(\psi) = \{x; y\}$) we can extract a strongly normalising λ-term that computes the corresponding function.

The extraction process relies on the fact that, internally, the set ω of von Neumann numerals is implemented [11] as a pointed graph (X, A, a) equipped with an injection $i : \mathsf{Nat} \to X$ that associates to any object $n : \mathsf{Nat}$ such that $\mathsf{wf\_nat}(n)$ the vertex $i(n) : X$ which represents the corresponding von Neumann numeral in the graph (X, A). Using this, it is easy to derive from a proof of ϕ (in IZF) a proof-term (in $F\omega.2^{++}$) of the statement:

$$\forall n : \mathsf{Nat}.\ \mathsf{wf\_nat}(n) \Rightarrow \exists p : \mathsf{Nat}.\ \mathsf{wf\_nat}(p) \wedge \phi^*(X, A, i(n), X, A, i(p))$$

(where ψ^* denotes the translation of the binary relation ψ in $F\omega.2^{++}$). By dropping the second component of the conjunction, we thus get a proof-term

$$\lambda\xi\,.\,t'\xi(\lambda\xi_1, \xi_2\,.\,\xi_1) \quad : \quad (\exists n : \mathsf{Nat}\,.\,\mathsf{wf\_nat}(n)) \Rightarrow (\exists p : \mathsf{Nat}\,.\,\mathsf{wf\_nat}(p))$$

that obviously computes the desired function, since:

Fact 10 — *The closed inhabitants of the saturated set* $[\![\mathsf{wf\_nat}(x)]\!]_{x \leftarrow n}$ *(for a given* $n \in \omega$*) are the SN-terms whose normal form is Church numeral* $\lceil n \rceil$.

References

1. P. Aczel. Non well-founded sets. *Center for the Study of Language and Information*, 1988.
2. H. Barendregt. Introduction to generalized type systems. Technical Report 90-8, University of Nijmegen, Department of Informatics, May 1990.
3. B. Barras, S. Boutin, C. Cornes, J. Courant, J.C. Filliâtre, E. Giménez, H. Herbelin, G. Huet, C. Muñoz, C. Murthy, C. Parent, C. Paulin, A. Saïbi, and B. Werner. The Coq Proof Assistant Reference Manual – Version V6.1. Technical Report 0203, INRIA, August 1997.
4. H. Friedman. Some applications of Kleene's methods for intuitionistic systems. In *Cambridge Summer School in Mathematical Logic*, volume 337 of *Springer Lecture Notes in Mathematics*, pages 113–170. Springer-Verlag, 1973.
5. J.H. Geuvers and M.J. Nederhof. A modular proof of strong normalization for the calculus of constructions. In *Journal of Functional Programming*, volume 1,2(1991), pages 155–189, 1991.
6. J.-L. Krivine. Typed lambda-calculus in classical Zermelo-Fraenkel set theory. *Archive for Mathematical Logic*, 40(3):189–205, 2001.
7. J.-L. Krivine. Dependent choice, 'quote' and the clock. *Theoretical Computer Science*, 2003.
8. P. Martin-Löf. *Intuitionistic Type Theory*. Bibliopolis, Napoli, 1984.
9. D. McCarty. *Realizability and Recursive Mathematics*. PhD thesis, Ohio State University, 1984.
10. P.-A. Melliès and B. Werner. A generic normalization proof for pure type systems. In *Proceedings of TYPES'96*, 1997.
11. A. Miquel. *Le calcul des constructions implicite: syntaxe et sémantique*. PhD thesis, Université Paris VII, 2001.
12. J. Myhill. Some properties of intuitionistic Zermelo-Fraenkel set theory. In *Cambridge Summer School in Mathematical Logic*, volume 337 of *Springer Lecture Notes in Mathematics*, pages 206–231. Springer-Verlag, 1973.

The Epsilon Calculus

Georg Moser[1,*] and Richard Zach[2]

[1] WWU Münster, Institut für Mathematische Logik und Grundlagenforschung, D-48149 Münster, Germany
moserg@math.uni-muenster.de

[2] University of Calgary, Department of Philosophy University of Calgary, Calgary, AB T2N 1N4, Canada
rzach@ucalgary.ca

Hilbert's ε-calculus [1, 2] is based on an extension of the language of predicate logic by a term-forming operator ε_x. This operator is governed by the *critical axiom*

$$A(t) \to A(\epsilon_x A(x)) \quad ,$$

where t is an arbitrary term. Within the ε-calculus, quantifiers become definable by $\exists x A(x) \Leftrightarrow A(\epsilon_x A(x))$ and $\forall x A(x) \Leftrightarrow A(\epsilon_x \neg A(x))$. (The expression $\epsilon_x A(x)$ is called an ε-*term*.)

The ε-calculus was developed in the context of Hilbert's program of consistency proofs. It is of independent interest, however, and a study from a computational and proof-theoretic point of view is worthwhile. As a simple example of the advantages of the ε-calculus note that by encoding quantifiers on the term-level, formalizing (informal) proofs is sometimes easier in the ε-calculus, compared to formalizing proofs in, e.g., sequent calculi.

In this tutorial we will restrict our attention to predicate logic. After formally introducing the relevant concepts we present some *conservativity results* for the ε-calculus. In particular we will prove the first and second ε-theorem.

Let T denote a finitely axiomatized open theory with axioms $P_1, \ldots, P_t$ based on the ε-calculus. We suppose some usual formalization $\mathcal{S}$ of predicate logic, e.g. Gentzen's LK. Then the first ε-theorem expresses that any formula (without quantified variables) deducible in T is also deducible from $P_1, \ldots, P_t$ in the quantifier-free fragment of $\mathcal{S}$. The second ε-theorem expresses that any formula (free of ε-terms) deducible in T is also deducible from $P_1, \ldots, P_t$ in $\mathcal{S}$.

As an easy consequence of the first ε-theorem one obtains: Any consequence F of a finitely axiomatized open theory T, where F is quantifier-free is provable from the axioms of T by the quantifier-free fragment of predicate logic. Another consequence is Herbrand's theorem, while the converse of Herbrand's theorem follows from the second ε-theorem.

References

1. D. Hilbert and P. Bernays. *Grundlagen der Mathematik*, volume I,II. Springer Verlag, Second Edition, 1970.
2. A.C. Leisenring. *Mathematical Logic and Hilbert's ϵ-symbol.* MacDonald Technical and Scientific, London, 1969.

* Supported by a Marie Curie fellowship, grant number HPMF-CT-2002-015777

M. Baaz and J.A. Makowsky (Eds.): CSL 2003, LNCS 2803, p. 455, 2003.

Modular Semantics and Logics of Classes

Bernhard Reus*

School of Cognitive and Computing Sciences
University of Sussex
bernhard@cogs.susx.ac.uk

Abstract. The semantics of class-based languages can be defined in terms of objects only [1, 7, 8] if classes are viewed as objects with a constructor method. One obtains a store in which method closures are held together with field values. Such a store is also called "higher-order" and does not come for free [13]. It is much harder to prove properties of such stores and as a consequence (soundness of) programming logics can become rather contrived (see [2]).
A simpler semantics separates methods from the object store [4, 12]. But again, there is a drawback. Once the semantics of a package of classes is computed it is impossible to add other classes in a compositional way. Modular reasoning principles are therefore not obtainable either.
In this paper we improve a simple class-based semantics to deal with extensions compositionally and derive modular reasoning principles for a logic of classes. The domain theoretic reasoning principle behind this is fixpoint induction.
Modularity is obtained by endowing the denotations of classes with an additional parameter that accounts for those classes added "later at linkage time."
Local class definitions (inner classes) are possible but for dynamic class-loading one cannot do without higher-order store.

1 Introduction and Motivation

In our quest for a complete denotational understanding of the semantics *and* logics of object-oriented programming languages we have treated object-based languages [13] in terms of Abadi and Cardelli's object calculus [1] and class-based languages [12] in terms of a simple sublanguage of sequential Java.

Class-based languages use the *class* concept to describe a collection of objects that are the *instances* of the class. In [8] two denotational models for various languages (including classes and inheritance) are given. One uses fixpoint closures, the other one self-application but in both models objects contain fields and methods (method closures). This leads to what is called "higher-order store", the domain of which is defined by a mixed variant equation similar to the one below:

$$\mathsf{St} = \mathsf{Rec}_{\mathsf{Loc}}(\mathsf{Rec}_{\mathcal{F}}\mathsf{Val}) \;\times\; \mathsf{Rec}_{\mathcal{M}}(\mathsf{St} \rightharpoonup \mathsf{Val} \times \mathsf{St}) \;.$$

* partially supported by the EPSRC under grant GR/R65190/01 and by the Nuffield Foundation under grant NAL/00244/A

M. Baaz and J.A. Makowsky (Eds.): CSL 2003, LNCS 2803, pp. 456–469, 2003.

For such stores invariance properties are hard to show.

A simpler approach aims at separating methods from objects. Methods do *not* have to be stored in the heap but are gathered in an environment that contains a method suite for each class. The class name acts as a link between the objects in the store and the method suites and is used for method dispatch. There are several variations of such a denotational semantics in the literature, [3,4,12], but in all of them only a *fixed* set of classes can be interpreted. Logics based on *operational semantics* which address these problems were presented in [11,14]. A *denotational semantics*, however, provides easier handling of mutual recursive modules as well as better tools for analysing and extending the language and logic. It also allows us to study the differences between logics for languages with higher-order store (object-based languages) and class-based languages more thoroughly.

In this paper we introduce a *denotational* semantics that is "open" with respect to extensions of classes and thus allows for a modular programming logic. The main contributions are:

- modular (compositional) denotational semantics for classes without higher-order store
- modular logic for classes (in terms of denotations)
- semantics for local classes
- explanation of modularity in class-based languages and relationship w.r.t. object-based languages.

Correctness of programs is usually shown using Hoare-like calculi. But it is the design, analysis, and soundness proof of these calculi that can be simplified using denotational semantics.

In the next section we briefly present the syntax of a simple class-based language. Then a (standard) denotational semantics is provided. Section 4 improves this semantics to obtain modularity and compositionality. Finally, modular proof rules are discussed. The paper end with a short summary and outlook (Section 6).

2 Syntax

The syntax and semantics of the class-based language is a refinement of the one in [12]. Let $\mathcal{F}$ be the set of field names, $\mathcal{M}$ the set of method names, and $\mathcal{C}$ the set of class names. Usually, f stands for a field name, m, n stand for method names, and C, D for class names.

$a, b ::=$	x	variables
|	`new C()`	object creation
|	$a.\mathsf{f}$	field selection
|	$a.\mathsf{f} = b$	field update
|	$a.\mathsf{m}()$	method call
|	$\mathtt{let}\ x{=}a\ \mathtt{in}\ b$	local def.

The "self" reference `this` is just considered a special (predefined) variable name, therefore $\mathtt{this} \in \mathsf{Var}$. Note that we do not distinguish between expressions and statements (like in the object calculus). There is no extra operator for sequential composition as it can be expressed using `let`, i.e. $a; b \equiv \mathtt{let}\ x = a\ \mathtt{in}\ b$ where x does not occur freely in a and b. Methods always have to return a result.

Additionally, there is a syntax for class definitions:

$$
\begin{aligned}
c &::= \mathtt{class}\ \mathsf{C}\ \{[\mathtt{inherits}\ \mathsf{D}]\ \mathsf{f}_i\ ^{i=1,\ldots,n} \\
&\qquad\qquad\qquad\qquad\qquad \mathsf{m}_j = b_j\ ^{j=1,\ldots,m} \\
&\qquad\qquad\qquad\qquad\qquad \mathsf{C}\,() = b \\
&\qquad\qquad\qquad\qquad\qquad \} \\
cl &::= \epsilon \mid c\ cl
\end{aligned}
$$

A class definition is similar to one in Java with all type declarations omitted. It contains a list of field declarations, a list of method declarations and a constructor. For the sake of simplicity all functions and the constructor are nullary and shadowing of field variables by inheritance is disallowed, ie. it is not possible to declare a field in a subclass that was already declared in a superclass[1].

3 Denotational Semantics

The denotational semantics is supposed to live in the category of predomains – ie. complete partially ordered sets – and continuous maps. Equivalently one can work in a category with domains (that have a least element) and strict, continuous maps. Note that the fixpoint operator fix is only defined on domains with least element.

Indexed collections, like stores and method suites, can be nicely modelled as records.

3.1 Record Types

Let $\mathbb{L}$ be a (countable) set of labels and A a predomain then the type of records with entries from A and labels from $\mathbb{L}$ is defined as follows:

$$\mathsf{Rec}_{\mathbb{L}}(A) = \Sigma_{L \in \mathcal{P}_{\mathsf{fin}}(\mathbb{L})} A^L$$

where A^L is the set of all total functions from L to A. It is easily seen that $\mathsf{Rec}_{\mathbb{L}}$ is a locally continuous functor on PreDom. A record with labels l_i and corresponding entries v_i $(1 \leq i \leq k)$ is written $\{\!|l_1 = v_1, \ldots l_k = v_k|\!\}$. Notice that $\mathsf{Rec}_{\mathbb{L}}(A)$ is always non–empty as it contains the element $\langle \emptyset, \emptyset \rangle$.

Records are ordered as follows:

$$\langle L_1, f_1 \rangle \sqsubseteq \langle L_2, f_2 \rangle \Leftrightarrow L_1 = L_2\ \wedge\ \forall \ell \in L_1.\, f_1(\ell) \sqsubseteq_A f_2(\ell)$$

[1] A more sophisticated treatment shadowing the duplicate fields of the superclass as used in Java can be modelled by adding the duplicate fields with appropriately changed field names.

Therefore, a record and its extension are *in*comparable. Note that records do not have a least element, only minimal ones. This can be remedied by lifting the record type to attach a new least element, usually called $\bot$: $\mathsf{Rec}_{\mathbb{L}}(A)$ is a complete poset if A is so, thus the lifted $\mathsf{Rec}_{\mathbb{L}}(A)_{\bot}$ is always a domain.

Basic record operations like selection, update, extension, and deletion are defined below.

Definition 1. *Let $r \in \mathsf{Rec}_{\mathbb{L}}(A)$ such that $r = \langle L, f\rangle$ with $L \subseteq \mathbb{L}$ and $f \in A^L$. Definedness of label l in record r is as follows: $l \in \mathsf{dom}\, r \Leftrightarrow l \in L$.*
Selection of a label $l \in \mathbb{L}$ in record r, short $r.l$, is defined if $l \in \mathsf{dom}\, r$ and yields $f(l) \in A$.

An update function for records is define in Table 1. It is undefined for labels which do not appear in the argument record.

Table 1. Definition of record update

$$\{\!|l_i{=}f_i|\!\}^{i=1\ldots n}[l{:=}f] = \begin{cases} \text{undefined} & \text{if } l \notin \mathsf{dom}\{\!|l_i{=}f_i|\!\}^{i=1\ldots n} \\ \{\!|l_1{=}f_1,\ldots,l_i{=}f,\ldots,l_n{=}f_n|\!\} & \text{if } l = l_i \end{cases}$$

If $r_1 = \langle L_1, f_1\rangle$ and $r_2 = \langle L_2, f_2\rangle$ are two records in $\mathsf{Rec}_{\mathbb{L}}(A)$ then the extension $r_1 + r_2$ is defined as follows:

$$\langle L_1, f_1\rangle + \langle L_2, f_2\rangle = \langle L_1 \cup L_2, \lambda l.\ \textit{if } l \in L_1 \ \textit{then } f_1(l) \ \textit{else } f_2(l)\rangle$$

Note that in $r_1 + r_2$ the values of the fields in r_2 which are also used in r_1 are lost.

Finally, if $r = \langle L, f\rangle$ and $l \in \mathbb{L}$ then the deletion operation which erases l in r, $r \setminus l$, is defined as follows:

$$r \setminus l = \begin{cases} r & \textit{if } l \notin \mathsf{dom}\, r \\ \langle L \setminus l, f\rangle & \textit{otherwise} \end{cases}$$

For a nested record $r \in \mathsf{Rec}_{\mathbb{L}_1}(\mathsf{Rec}_{\mathbb{L}_2}(A))$ we abbreviate $r[l_1 := r.l_1[l_2 := a]]$ by the simpler and more intuitive $r[l_1.l_2 := a]$.

3.2 Semantic Domains

We assume flat predomains for basic values (like booleans or integers) Val and a flat predomain of locations Loc.

The class-based language introduced above finds its interpretation within the following *non-recursive* system of domains:

Definition 2. *Let the semantic counterparts of objects (1), stores (2), closures (3), method suites (4), and class descriptors (5) be defined as below.*

$$\mathsf{Ob} = \mathsf{Rec}_{\mathcal{F}}(\mathsf{Val}) \times \mathcal{C} \tag{1}$$

$$\mathsf{St} = \mathsf{Rec}_{\mathsf{Loc}}(\mathsf{Ob}) \tag{2}$$

$$\mathsf{Cl} = \mathsf{Loc} \times \mathsf{St} \rightharpoonup \mathsf{Val} \times \mathsf{St} \tag{3}$$

$$\mathsf{Ms} = \mathsf{Rec}_{\mathcal{M}}(\mathsf{Cl}) \times (\mathsf{Loc} \times \mathsf{St} \rightharpoonup \mathsf{St}) \times \mathcal{C} \tag{4}$$

$$\mathsf{Mss} = \mathsf{Rec}_{\mathcal{C}}(\mathsf{Ms}) \tag{5}$$

An *object* consists of a record of field values plus the name of its class type. This information is stored in order to be able to do the dynamic dispatch for method calls (due to subtype polymorphism). A *store* is a record of objects indexed by locations. A *closure* is a partial function mapping a location, representing the callee object, and (the old) store to a value, the result of the method, and a new store which accounts for the side effects during execution of the method in question. A *method suite* (Ms) is a record $\mathsf{Rec}_{\mathcal{M}}(\mathsf{Cl}) \times (\mathsf{Loc} \times \mathsf{St} \rightharpoonup \mathsf{St}) \times \mathcal{C}$ which contains all methods of a class, the initialisation code for the constructor, and the name of the superclass. We assume that there is always a superclass, ie. there is a root class `Root` that all classes inherit from[2]. The constructor takes a location in which to create the object. This reference is created freshly by the semantics of `new` and then passed to the constructors of the superclass. A *class descriptor* (Mss) is the collection of method suites for all declared classes. The function $\mathsf{type} : \mathsf{Ob} \to \mathcal{C}$ returns an object's class, ie. $\mathsf{type}(o) = o.2$, which is the object's dynamic type. For interpreting variables we need an environment Env which maps variables to values, ie. $\mathsf{Env} = \mathsf{Var} \rightharpoonup \mathsf{Val}$.

3.3 Semantic Equations

The interpretation of the syntax is given by the following semantic equations:

Definition 3. *Given an environment* $\rho \in \mathsf{Env}$, *an environment of method suites* $\mu \in \mathsf{Mss}$ *and an expression* a, *its interpretation* $[\![a]\!]\rho : \mathsf{Mss} \to \mathsf{St} \rightharpoonup \mathsf{Val} \times \mathsf{St}$ *is defined in Table 2 (page 461). Note that the semantic* **let** *on the right hand side of the definitions is strict in its first argument.*
Given an environment of method suites $\mu \in \mathsf{Mss}$ *a class definition* c *is defined via* $[\![c]\!] : \mathsf{Mss} \to \mathsf{Ms}$ *in Table 3 (page 461).*

The methods of the superclass are copied into the method suite of the subclass. It is more memory efficient not to copy them and use a more sophisticated method dispatch instead that searches in super classes. Finally, since we work in an untyped setting, it is legitimate to initialize any field with zero. In a typed setting zero would be replaced by a default value for the type of the field. It is no contradiction to the untypedness of our semantics to have classes (class names) stored in objects as these classes represent *runtime* types needed for dynamic dispatch. There are no static types in use.

[2] In Java this would be `Object`.

Table 2. Denotational semantics of expressions

$$
\begin{array}{ll}
[\![x]\!]\,\rho\,\mu\,\sigma & = \langle \rho(x), \sigma\rangle \\
[\![\texttt{this}]\!]\,\rho\,\mu\,\sigma & = \langle \rho(\text{this}), \sigma\rangle \\
[\![\texttt{new}\ \mathsf{C}()]\!]\,\rho\,\mu\,\sigma & = \langle \ell, \mu.\mathsf{C}.2(\ell,\sigma)[\ell.2 = \mathsf{C}]\rangle \qquad \text{where } \ell \notin \mathsf{dom}\sigma \\
[\![a.\mathsf{f}]\!]\,\rho\,\mu\,\sigma & = \textbf{let}\ \langle \ell, \sigma'\rangle = [\![a]\!]\,\rho\,\mu\,\sigma\ \textbf{in}\ \langle \sigma'.\ell.1.\mathsf{f}, \sigma'\rangle \\
[\![a.\mathsf{f}{=}b]\!]\,\rho\,\mu\,\sigma & = \textbf{let}\ \langle \ell, \sigma'\rangle = [\![a]\!]\,\rho\,\mu\,\sigma\ \textbf{in} \\
 & \qquad \textbf{let}\ \langle v, \sigma''\rangle = [\![b]\!]\,\rho\,\mu\,\sigma'\ \textbf{in}\ \langle v, \sigma''[\ell.1.\mathsf{f} := v]\rangle \\
[\![a.\mathsf{m}()]\!]\,\rho\,\mu\,\sigma & = \textbf{let}\ \langle \ell, \sigma'\rangle = [\![a]\!]\,\rho\,\mu\,\sigma\ \textbf{in}\ \mu.\mathsf{type}(\sigma'.\ell).1.\mathsf{m}(\ell,\sigma') \\
[\![\texttt{let}\ x{=}a\ \texttt{in}\ b]\!]\,\rho\,\mu\,\sigma & = \textbf{let}\ \langle \ell, \sigma'\rangle = [\![a]\!]\,\rho\,\mu\,\sigma\ \textbf{in}\ [\![b]\!]\,\rho[\ell/x]\,\mu\,\sigma' \,.
\end{array}
$$

Table 3. Denotational semantics of class definitions

$$
\left[\!\!\left[\begin{array}{l}
\texttt{class C inherits D}\ \{ \\
\quad \mathsf{f}_i\ ^{i=1..n} \\
\quad \mathsf{m}_j{=}b_j\ ^{j=1..m} \\
\quad \mathsf{C}()\ {=}b \\
\quad \}
\end{array}\right]\!\!\right] \mu = \left\langle \begin{array}{l}
\{\!|\mathsf{m}_j{=}\lambda\langle\ell,\sigma\rangle.\,[\![b_j]\!]\,(\text{this}{\mapsto}\ell)\,\mu\,\sigma|\!\}^{j=1..m} + \mu.\mathsf{D}.1, \\
\lambda\langle\ell,\sigma\rangle.\,([\![b]\!]\,(\text{this}{\mapsto}\ell)\,\mu\,(\mu.\mathsf{D}.2(\ell,\sigma))).2[\mathsf{f}_i := 0], \\
\mathsf{D}
\end{array}\right\rangle
$$

The semantics of a *module* `class C` ... could be defined as

$$
\mathsf{fix}\ \mu : \mathsf{Mss}_\perp.\ \{\!|\mathsf{C} = [\![\texttt{class C}\ldots]\!]\,\mu|\!\} \quad .
$$

Fixpoint induction would then provide the corresponding proof principle. But this definition does not give the desired result if classes depend on other classes. An example is used below to emphasize the problem:

```
class Top = {
 f
 m() = this.f.n()
 n() = 0
 Top() = this.f = this
}

class C inherits Top {
 n() = 1
 C() = this = this  /* skip */
}
```

The semantics of these two classes according to the above are:

$$
\begin{array}{l}
[\![\texttt{class Top}\ldots]\!] = \mathsf{fix}\ \mu.\ \{\!|\mathsf{Top} = \langle \{\!|\ \mathsf{m} = \lambda\langle\ell,\sigma\rangle.\mu.\mathsf{type}(\sigma.(\sigma.\ell.\mathsf{f})).1.\mathsf{n}(\sigma.\ell.\mathsf{f},\sigma), \\
\qquad\qquad\qquad\qquad\qquad\qquad \mathsf{n} = \lambda\langle\ell,\sigma\rangle.\,0|\!\}, \\
\qquad\qquad\qquad\qquad\qquad\qquad \lambda\langle\ell,\sigma\rangle.\,\sigma[\ell := \langle\{\!|\mathsf{f} = \ell|\!\},\mathsf{Top}\rangle],\mathsf{Root}\rangle\ |\!\}
\end{array}
$$

$$
\begin{array}{l}
[\![\texttt{class C inherits Top}\ldots]\!] = \\
\quad \mathsf{fix}\ \mu.\ \{\!|\mathsf{C} = \langle \qquad \{\!|\mathsf{m} = \lambda\langle\ell,\sigma\rangle.\,\mu.\mathsf{type}(\sigma.(\sigma.\ell.\mathsf{f})).1.\mathsf{n}(\sigma.\ell.\mathsf{f},\sigma), \\
\qquad\qquad\qquad\qquad \mathsf{n} = \lambda\langle\ell,\sigma\rangle.\,1|\!\}, \\
\qquad\qquad\qquad\qquad \lambda\langle\ell,\sigma\rangle.\,\sigma[\ell := \langle\{\!|\mathsf{f} = \ell|\!\},\mathsf{C}\rangle],\mathsf{Top}\ \rangle\ |\!\}
\end{array}
$$

If we join the two classes into one class descriptor μ defined as follows

$$\mu = \{\!|\texttt{Top} = [\![\texttt{class Top}\ldots]\!]|\!\} + \{\!|\texttt{C} = [\![\texttt{class C inherits Top}\ldots]\!]|\!\}$$

and interpret the following expression t

```
let o = new Top()
in let _ = o.f = new C()
   in  o.m()
```

w.r.t. μ we obtain:

$$\begin{aligned}[\![t]\!]\,\emptyset\,\mu\,\sigma = \ & \textbf{let } \sigma_1 = \sigma[\ell := \langle\{\!|\mathsf{f} = \ell|\!\}, \mathtt{Top}\rangle] \textbf{ in} \\ & \textbf{let } \sigma_2 = \mu.\mathsf{Top}.2(\ell', \sigma_1)[\ell'.2 := \mathtt{C}] \textbf{ in} \\ & \textbf{let } \sigma_3 = \sigma_2[\ell.\mathsf{f} := \ell'] \textbf{ in} \\ & \mu.\mathsf{type}(\sigma_3.\ell).1.\mathsf{m}(\ell, \sigma_3)\end{aligned}$$

where ℓ is fresh in σ and ℓ' is fresh in σ_1. This simplifies to

$$\textbf{let } \sigma_4 = \sigma[\ell := \langle\{\!|\mathsf{f} = \ell'|\!\}, \mathtt{Top}\rangle, \ell' := \langle\{\!|\mathsf{f} := \ell'|\!\}, \mathtt{C}\rangle] \textbf{ in } \mu.\mathsf{Top}.1.\mathsf{m}(\ell, \sigma_4)$$

which by the semantics of class Top simplifies to

$$\mu.\mathsf{type}(\sigma_4.(\sigma_4.\ell.\mathsf{f})).1.\mathsf{n}(\sigma_4.\ell.\mathsf{f}, \sigma_4)$$

which in turn evaluates to

$$\mu.\mathsf{C}.1.\mathsf{n}(\ell', \sigma_4) \quad .$$

Obviously, the result depends on the semantics of the method n in C, ie. μ.C.1.n. But the fixpoint operation yields no defined result for μ.C, just one for μ.Top and thus the result is undefined whereas it should be 1. The reason for this unexpected behaviour is that the fixpoint has closed the method suites available. It cannot be updated to contain the methods of other classes added later.

To avoid this problem we define the semantics of classes slightly differently, in a parameterized way.

4 A Modular Semantics for Classes

Modules depend on a context of class declarations unknown at the time of definition. Later they may be linked together with this context changing the semantics of the original modules. This effect will be modelled by (mutual) fixpoints. Recall that records defined by fixpoints can not get bigger (because of their invariance under extensions) but the values of their fields can become more defined. This is enough for our purposes, since in module or package definitions the number of methods and fields is fixed. Nevertheless, the record extension operator is needed to express a temporarily growing number of classes.

Definition 4. *We define a polymorphic operator maprec that applies a function to all components of a record:* $maprec_A$: $(A \to B) \to \mathsf{Rec}_{\mathbb{L}}(A) \to \mathsf{Rec}_{\mathbb{L}}(B)$ *as follows:*

$$\text{maprec}\, f\, \langle D, g \rangle = \langle D, f \circ g \rangle$$

which is in analogy with map for lists.

A modular class definition has the following semantics taking into account further packaging with other classes:

Definition 5. *The semantics* $[\![\_]\!]^{\mathrm{m}} : \mathsf{Mss} \to \mathsf{Ms}$ *is defined as*

$$[\![\texttt{class}\, \mathsf{C} \ldots]\!]^{\mathrm{m}} = \lambda\mu : \mathsf{Mss}.\, \mathsf{fix}\, \tau : \mathsf{Ms}_{\perp}.\, [\![\texttt{class}\, \mathsf{C} \ldots]\!]\ (\{\!|\mathsf{C} = \tau|\!\} + \mu)$$

where the parameter μ *represents the possible additional classes* C *may depend on.*

Even if C is not intended to refer to any other class, in an open world it must reserve the right to refer to future *subclasses* of itself.

A "package" or "linkage" operator is needed that links a module with a given package, ie. a list of linked (or packed) modules. Like modules packages have an additional parameter in order to be open to future extensions. In order to accomplish this, we define the type

$$\mathsf{CTs} = \mathsf{Rec}_{\mathcal{C}}(\mathsf{Mss} \to \mathsf{Ms})$$

which represents collections of class transformers of type $\mathsf{Mss} \to \mathsf{Ms}$ that describe the semantics of a class depending on a class descriptor for the context.

Definition 6. *The package operator*

$$\text{pack}_{\mathsf{C}} : (\mathsf{Mss} \to \mathsf{Ms}) \times \mathsf{CTs} \to \mathsf{CTs}_{\perp}$$

is defined as follows:

$$\begin{aligned} \text{pack}_{\mathsf{C}}(M, P) = \mathsf{fix}\ \delta : \mathsf{CTs}_{\perp}.\ & \\ & \{\!|\mathsf{C} = \lambda\mu : \mathsf{Mss}.\, M(\delta(\mu) \setminus \mathsf{C} + \mu)|\!\} + \\ & \text{maprec}\, (\lambda x : \mathsf{Mss} \to \mathsf{Ms}.\, \lambda\mu : \mathsf{Mss}.\, x(\delta(\mu).\mathsf{C} + \mu))\, P\ . \end{aligned}$$

Definition 7. *The semantics of a package of class definitions can be defined as follows:*

$$[\![\texttt{class C\{...\}}\ cl]\!] = \text{pack}_{\mathsf{C}}([\![\texttt{class C\{...\}}]\!]^{\mathrm{m}}, [\![cl]\!])$$

and $[\![\epsilon]\!] = \{\!|\,|\!\}$.

Lemma 1. *Packaging with nothing does not have any effect, ie.*

$$\text{pack}([\![\texttt{class}\ \mathsf{C} = \ldots]\!]^{\mathrm{m}}, \emptyset) = \{\!|\mathsf{C} = [\![\texttt{class}\ \mathsf{C} = \ldots]\!]^{\mathrm{m}}|\!\}$$

Proof. Let $M = [\![\texttt{class C} = \ldots]\!]^{\mathrm{m}}$ and assume $\delta = \{\!|\mathsf{C} = M|\!\}$ (1). If we can show that

$$\{\!|\mathsf{C} = \lambda\mu : \mathsf{Mss}.\ [\![\texttt{class C} = \ldots]\!]^{\mathrm{m}}(\delta(\mu) \setminus \mathsf{C} + \mu)|\!\} = \{\!|\mathsf{C} = M|\!\}$$

then by fixpoint induction we have shown

$$\mathsf{fix}\,\delta : \mathsf{CTs}_{\bot}.\ \{\!|\mathsf{C} = \lambda\mu : \mathsf{Mss}.\ [\![\texttt{class C} = \ldots]\!]^{\mathrm{m}}(\delta(\mu) \setminus \mathsf{C} + \mu)|\!\} = \{\!|\mathsf{C} = M|\!\}$$

which implies the claim.

$$\begin{aligned}
&\{\!|\mathsf{C} = \lambda\mu : \mathsf{Mss}.\ [\![\texttt{class C} = \ldots]\!]^{\mathrm{m}}(\delta(\mu) \setminus \mathsf{C} + \mu)|\!\} =_{(1)} \\
&= \{\!|\mathsf{C} = \lambda\mu : \mathsf{Mss}.\ [\![\texttt{class C} = \ldots]\!]^{\mathrm{m}}(\{\!|\mathsf{C} = M(\mu)|\!\} \setminus \mathsf{C} + \mu)|\!\} = \\
&= \{\!|\mathsf{C} = \lambda\mu : \mathsf{Mss}.\ [\![\texttt{class C} = \ldots]\!]^{\mathrm{m}}\ \mu|\!\} = \\
&= \{\!|\mathsf{C} = M|\!\} \quad .
\end{aligned}$$

One can always "close" a package or module by just applying it to the empty environment.

4.1 Inner Classes

Inner classes are classes defined "on-the-fly" inside classes or methods. Simple class descriptors are not sufficient as argument to the interpretation function since packaging needs transformers. The type of the interpretation function thus changes to

$$[\![]\!]\rho : \mathsf{CTs} \to \mathsf{St} \rightharpoonup \mathsf{Val} \times \mathsf{St} \quad .$$

If "`let class C` ... `in` s" denotes the syntax for local inner class declarations the interpretation is as follows:

$$[\![\texttt{let class C} \ \ldots \ \texttt{in}\ s]\!]\ \rho\,\delta\,\sigma = [\![s]\!]\ \rho\ \mathrm{pack}_{\mathsf{C}}([\![\texttt{class C} \ \ldots]\!]^{\mathrm{m}}, \delta)\,\sigma$$

Whenever the interpretation really needs the method environment – ie. when evaluating a method call or creating an object – the environment can be closed for this moment of time to find the right closure:

$$\begin{aligned}
[\![\texttt{new}\ \mathsf{C}()]\!]\,\rho\,\delta\,\sigma &= \langle \ell, (\delta.\mathsf{C}\,\{\!||\!\}).2(\ell,\sigma)[\ell.2 := \mathsf{C}]\rangle \\
&\quad\ \text{where } \ell \notin \mathsf{dom}\,\sigma \\
[\![a.\mathsf{m}()]\!]\,\rho\,\delta\,\sigma &= \textbf{let}\ \langle \ell, \sigma' \rangle = [\![a]\!]\,\rho\,\delta\,\sigma\ \textbf{in}\ (\delta.\mathsf{type}(\sigma'.\ell)\ \{\!||\!\}).1.\mathsf{m}(\ell,\sigma')
\end{aligned}$$

The corresponding change to method store transformers means that we have to change the semantics of the class modules slightly:

$$\begin{aligned}
[\![\texttt{class}\,C \ \ldots]\!]^{\mathrm{m}} = \lambda\mu : \mathsf{Mss}.\,\mathsf{fix}\,\tau : \mathsf{Ms}_{\bot}.\,[\![\texttt{class}\,C \ \ldots]\!] \\
\{\!|C = \lambda_ : \mathsf{Mss}.\,\tau|\!\} + \mathrm{maprec}\,(\lambda x : \mathsf{Ms}.\ \lambda_ : \mathsf{Mss}.\,x)\,\mu
\end{aligned}$$

5 Logics of Programs

Having fixed the denotational semantics of classes we can start reasoning about denotations. This is in analogy with the LCF (Logic of Computable Functions) project (see e.g. [10]) for the functional paradigm.

5.1 Specifications

First we have to define an appropriate notion of *specification* for classes following [6, 11, 14]. Compare also with [12].

As every class contains a number of methods, some common structures in specifications can be singled out: for every method there is a result specification and a transition specification.

$B_{\mathsf{m}} \in \wp(\mathsf{Val} \times \mathsf{St})$ result specification for $\mathsf{m} \in \mathcal{M}$
$T_{\mathsf{m}} \in \wp(\mathsf{Loc} \times \mathsf{St} \times \mathsf{Val} \times \mathsf{St})$ transition specification for $\mathsf{m} \in \mathcal{M}$

Taking into account that we are only interested in partial correctness (at least for the considerations of this paper) the meaning of the specifications can be described informally as follows:

B_{m} : "If method m terminates its *result* fulfils the result specification."
T_{m} : "If method m terminates its *effect* fulfils the transition specification."

A modular specification may depend on the specification of some other classes. The existence of other classes in the environment may indeed be part of the specification itself.

Definition 8. *Specification building operators for methods (*mth*), classes (*cls*), and packages (*pck*) are defined below. For the sake of readability we abbreviate the type of result specifications* $\mathsf{RSpec} = \wp(\mathsf{Val} \times \mathsf{St})$ *and transition specifications* $\mathsf{TSpec} = \wp(\mathsf{Loc} \times \mathsf{St} \times \mathsf{Val} \times \mathsf{St})$.

$$\begin{aligned}
\mathrm{mth} &: \mathsf{RSpec} \times \mathsf{TSpec} \to \wp(\mathcal{C} \times \mathsf{Cl}) \\
\mathrm{cls} &: \mathsf{Rec}_{\mathcal{M}}(\mathsf{RSpec} \times \mathsf{TSpec}) \to \wp(\mathcal{C} \times \mathsf{Ms}) \\
\mathrm{pck} &: \mathsf{Rec}_{\mathcal{C}}(\mathsf{Rec}_{\mathcal{M}}(\mathsf{RSpec} \times \mathsf{TSpec})) \to \wp(\mathsf{Mss})
\end{aligned}$$

$$\begin{aligned}
\langle \mathsf{C}, f \rangle \in \mathrm{mth}(B,T) \text{ iff } & \forall \ell \in \mathsf{Loc}.\, \forall v \in \mathsf{Val}.\, \forall \sigma, \sigma' \in \mathsf{St}. \\
& (\mathsf{type}(\sigma.\ell) = \mathsf{C} \wedge f(\ell,\sigma) = \langle v, \sigma' \rangle) \\
& \Rightarrow B(v,\sigma') \wedge T(\ell,\sigma,v,\sigma')
\end{aligned}$$

$$\langle \mathsf{C}, \mu \rangle \in \mathrm{cls}(R) \text{ iff } \forall \mathsf{m} \in \mathcal{M}.\, \mathsf{m} \in \mathsf{dom}\, R \Rightarrow \mathsf{m} \in \mathsf{dom}\, \mu \wedge \langle \mathsf{C}, \mu.\mathsf{m} \rangle \in \mathrm{mth}(R.\mathsf{m})$$

$$\delta \in \mathrm{pck}(\boldsymbol{R}) \text{ iff } \forall \mathsf{C} \in \mathcal{C}.\, \mathsf{C} \in \mathsf{dom}\, \boldsymbol{R} \Rightarrow \mathsf{C} \in \mathsf{dom}\, \delta \wedge \langle \mathsf{C}, \delta.\mathsf{C} \rangle \in \mathrm{cls}(\boldsymbol{R}.\mathsf{C})$$

The following abbreviation will be used repeatedly:

$$\gamma \in \mathsf{C} \mapsto R \text{ iff } \mathsf{C} \in \mathsf{dom}\, \gamma \wedge \langle \mathsf{C}, \gamma.\mathsf{C} \rangle \in \mathrm{cls}(R)$$

Proposition 1. *The specifications above are admissible predicates.*

Proof. For example,

$$\begin{aligned}
\langle \mathsf{C}, f \rangle \in \mathrm{mth}(B,T) \text{ iff } & \forall \ell \in \mathsf{Loc}.\, \forall v \in \mathsf{Val}.\, \forall \sigma, \sigma' \in \mathsf{St}. \\
& (\mathsf{type}(\sigma.\ell) \neq \mathsf{C} \vee f(\ell,\sigma)\uparrow \vee (B(f(\ell,\sigma)) \wedge T(\ell,\sigma,f(\ell,\sigma)))
\end{aligned}$$

As $\mathcal{C}$ is a flat predomain admissibility must only be shown with respect to the second component f. Since admissible predicates are closed under universal quantification, disjunction, conjunction, and composition with continuous maps (and function application is continuous) it only remains to show that $\uparrow$, B, and T are admissible. The former is by definition, the latter are because they are predicates on a flat predomain. One can show similarly that the other predicates are admissible. Note that $\mathsf{C} \in \mathsf{dom}\,\gamma$ is admissible in γ due to the non-extension order on records.

Specifications for class description transformers can also be defined in the logical relations style:

Definition 9. *Given predicates $P \in \wp(A)$ and $Q \in \wp(B)$, a predicate $P \to Q \in \wp(A \to B)$ is defined as follows:*

$$f \in P \to Q \textit{ iff } \forall a \in A.\ a \in P \Rightarrow f(a) \in Q\ .$$

Corollary 1. *If Q is admissible so is $P \to Q$.*

5.2 Modular Proof Rules

As the semantics of classes and packages is defined via fixpoints it seems adequate to use fixpoint induction. Fortuitously, as shown above, the specifications in use are admissible such that fixpoint induction is applicable.

A spatial conjunction operator, $*$, for records (of method suites) will be used below. It is defined in analogy with separation logic [5, 9] where the operator works on heaps. It is more appropriate than normal conjunction $\wedge$ due to the modularity of class definitions.

Definition 10. *Let $P, Q \in \wp(\mathsf{Mss})$, then $P * Q \in \wp(\mathsf{Mss})$ is defined as follows:*

$$\gamma \in P * Q \textit{ iff } \exists \gamma_1, \gamma_2 : \mathsf{Mss}.\ \mathsf{dom}\,\gamma_1 \cap \mathsf{dom}\,\gamma_2 = \emptyset \wedge \gamma = \gamma_1 + \gamma_2 \wedge \gamma_1 {\in} P \wedge \gamma_2 {\in} Q\ .$$

We assume that $*$ has higher precedence than $\to$.

Admissibility of $P * Q$ is a more delicate matter but fortunately it is not required for the rules below since $*$ always appears on the left hand side of an implication $\to$.

The rules below show which assumptions can be used when in order to prove packages which are broken down into classes. The correctness of classes under certain context assumptions can the be proved using standard techniques not discussed in this paper.

Theorem 1. *The following proof rules are correct:*

$$(1)\ \frac{[\![\texttt{class C }\ldots]\!] \in (\mathsf{C} \mapsto R) * \Gamma \to \mathrm{cls}(R)}{[\![\texttt{class C }\ldots]\!]^{\mathrm{m}} \in \Gamma \to \mathrm{cls}(R)}$$

$$(2)\ \frac{M \in \Gamma * \mathrm{pck}(\boldsymbol{R'}) \to \mathrm{cls}(R) \qquad P \in (\mathsf{C} \mapsto R) * \Gamma\ \to \mathrm{pck}(\boldsymbol{R'})}{\mathrm{pack}_{\mathsf{C}}(M, P) \in \Gamma \to \mathrm{pck}(\{\!|\mathsf{C} = R|\!\} + \boldsymbol{R'})}$$

$$(3)\ \frac{[\![c]\!] \in (\mathsf{C} \mapsto R) * \mathrm{pck}(\boldsymbol{R'}) * \Gamma \to \mathrm{cls}(R) \qquad [\![cl]\!] \in (\mathsf{C} \mapsto R) * \Gamma \to \mathrm{pck}(\boldsymbol{R'})}{[\![c\ cl]\!] \in \Gamma \to \mathrm{pck}(\{\!|\mathsf{C} = R|\!\} + \boldsymbol{R'})}$$

Proof. Since the predicates in use are admissible one obtains correctness by fixpoint induction.
For rule (1) assume (i) that $\mu \in \Gamma$ and (induction hyp) that $\tau \in \mathrm{pck}(R)$. We have to show that $[\![\texttt{class C...}]\!](\{\!|\mathsf{C} = \tau|\!\} + \mu) \in \mathrm{pck}(R)$. But this holds by the premise of rule (1) since $\{\!|\mathsf{C} = \tau|\!\} + \mu \in C \mapsto R * \Gamma$ by (i) and (ind.hyp.).
Rules (2) and (3) can be shown similarly with (3) using (2).

The following predicate states that a certain class in a class descriptor is a subclass of another. The predicate is well-defined since the subtype hierarchy is not circular.

Definition 11. *Let* $\mathsf{C}, \mathsf{D} \in \mathcal{C}$ *and* $\gamma \in \mathsf{Mss}$*:*

$$\mathsf{D} \leq_\gamma \mathsf{C} \textit{ iff } \gamma.\mathsf{D}.3 = \mathsf{C} \vee (\gamma.\mathsf{D}.3 = \mathsf{D}' \wedge \mathsf{D}' \leq_\gamma \mathsf{C}) .$$

5.3 Method Invocation and Inheritance

The semantics of method invocation depends on the (dynamic) class type of the callee object. The class is only known, however, for `this`, it is not known for arbitrary objects on the heap. To verify a property of the method it could be necessary to stipulate that the method behaves identically in all classes. This is unrealistic. It is more convenient to assume that it behaves identically for a certain branch of the class hierarchy. If it is known that the maximal class type of an object is C, then the possible methods to be taken into account can be reduced to those in the hierarchy below C. For all of those one could stipulate that they share some behaviour sufficient to prove the given specification. This can be done using the spatial implication operator $\triangleright$ (see also [9]):

Definition 12. *For any* $\gamma \in \mathsf{Mss}$ *and* $P, Q \in \wp(\mathsf{Mss})$ *we have*

$$\gamma \in P \triangleright Q \ \textit{ iff } \ \forall \beta : \mathsf{Mss}.\ \beta \in P \Rightarrow \gamma + \beta \in Q .$$

With the help of $\triangleright$ one can now express that no extension of a context may contain a subclass of C that does not fulfill ϕ, briefly $Sub(C, \phi) \triangleright \mathit{false}$ where $Sub(C, \phi)$ is defined as shown below:

$$\gamma \in Sub(\mathsf{C}, \phi) \text{ iff } \exists \mathsf{D} : \mathcal{C}.\, \mathsf{D} \in \mathsf{dom}\,\gamma \ \wedge \ \gamma.\mathsf{D} \notin \phi \ \wedge \mathsf{D} <_\gamma \mathsf{C} .$$

Thus, for a proof of $\delta \in \Gamma * Sub(C, \phi) \triangleright \mathit{false} \to \Delta$ one can assume that the input $\mu \in \Gamma * Sub(C, \phi) \triangleright \mathit{false}$ does not contain any subclasses of C that do not fulfill ϕ. In order to employ behavioural subtyping for C one chooses ϕ to be the specification of C.

6 Conclusion

We have presented a modular denotational semantics for a class-based language and sound proof rules for managing the modules. No recursive domains are necessary as long as persistent classes are not loaded at run time. To achieve the latter one could resort to the approach presented in [13] for object-based languages, paying however the price of a much more involved semantics and logics with more severe restrictions.

The advantage of the semantics presented in this paper is that fixpoint induction is sufficient to derive useful proof rules for modular specifications, the only restriction being that specifications are admissible. Furthermore, the denotational approach does not commit itself to a particular logic so the techniques are applicable to all kinds of languages.

Future work comprises the integration of data abstraction such that accessibility restrictions can be modelled. The applicability of the presented technique to separation logic needs to be investigated as much as the benefit of using separation logic for the records of class definitions.

Acknowledgements

The idea of this paper emerged in discussion with Peter O'Hearn and Uday Reddy about the alleged limitations of a simple denotational semantics for classes. Thanks to Thomas Streicher and Hubert Baumeister for useful remarks and pointers.

References

1. M. Abadi and L. Cardelli. *A Theory of Objects.* Springer Verlag, 1996.
2. M. Abadi and K.R.M. Leino. A logic of object-oriented programs. In Michel Bidoit and Max Dauchet, editors, *Theory and Practice of Software Development: Proceedings / TAPSOFT '97, 7th International Joint Conference CAAP/FASE*, volume 1214 of *Lecture Notes in Computer Science*, pages 682–696. Springer-Verlag, 1997.
3. P. America and F.S. de Boer. A proof theory for a sequential version of POOL. Technical Report http://www.cs.uu.nl/people/frankb/Available-papers/spool.dvi, University of Utrecht, 1999.
4. Anindya Banerjee and David Naumann. Representation independence, confinement, and access control. In *Proceedings of ACM Principles of Programming Languages POPL*, volume 164, pages 166–177. ACM press, 2002.
5. Cristiano Calcagno and Peter W. O'Hearn. On garbage and program logic. In *FoSSaCS*, volume 2030 of *LNCS*, pages 137–151, Berlin, 2001. Springer.
6. F.S. de Boer. A WP-calculus for OO. In W. Thomas, editor, *Foundations of Software Science and Computations Structures*, volume 1578 of *Lecture Notes in Computer Science.* Springer-Verlag, 1999.
7. Andreas V. Hense. Wrapper semantics of an object-oriented programming language with state. In *Proceedings Theoretical Aspects of Computer Software*, volume 526 of *Lecture Notes in Computer Science*, pages 548–568. Springer-Verlag, 1991.

8. S.N. Kamin and U.S. Reddy. Two semantic models of object-oriented languages. In Carl A. Gunter and John C. Mitchell, editors, *Theoretical Aspects of Object-Oriented Programming: Types, Semantics, and Language Design*, pages 464–495. The MIT Press, 1994.
9. Peter W. O'Hearn, John C. Reynolds, and Hongseok Yang. Local reasoning about programs that alter data structures. In *CSL*, volume 2142 of *LNCS*, pages 1–19, Berlin, 2001. Springer.
10. L.C. Paulson. *Logic and Computation*, volume 2 of *Cambridge Tracts in Theoretical Computer Science.* Cambridge University Press, 1987.
11. A. Poetzsch-Heffter and P. Müller. Logical foundations for typed object-oriented languages. In D. Gries and W. De Roever, editors, *Programming Concepts and Methods*, 1998.
12. B. Reus. Class based vs. object based: A denotational comparison. In *Algebraic Methodology And Software Technology*, volume 2422 of *Lecture Notes in Computer Science*, pages 473–488, Berlin, 2002. Springer Verlag.
13. B. Reus and Th. Streicher. Semantics and logics of objects. In *Proceedings of the 17th Symp. Logic in Computer Science*, pages 113–122, 2002.
14. B. Reus, M. Wirsing, and R. Hennicker. A Hoare-Calculus for Verifying Java Realizations of OCL-Constrained Design Models. In *FASE 2001*, volume 2029 of *Lecture Notes in Computer Science*, pages 300–317, Berlin, 2001. Springer.

Validity of CTL Queries Revisited*

Marko Samer and Helmut Veith

Institute of Information Systems
Database and Artificial Intelligence Group
Vienna University of Technology, Austria
{samer,veith}@dbai.tuwien.ac.at

Abstract. We systematically investigate temporal logic queries in model checking, adding to the seminal paper by William Chan at CAV 2000. Chan's temporal logic queries are CTL specifications where one unspecified subformula is to be filled in by the model checker in such a way that the specification becomes true. Chan defined a fragment of CTL queries called CTL^v which guarantees the existence of a unique strongest solution. The starting point of our paper is a counterexample to this claim. We then show how the research agenda of Chan can be realized by modifying his fragment appropriately. To this aim, we investigate the criteria required by Chan, and define two new fragments CTL^v_{new} and CTL^d where the first is the one originally intended; the latter fragment also provides unique strongest solutions where possible but admits also cases where the set of solutions is empty.

1 Introduction

Temporal logic queries are a generalization of model checking [7,6,12] which allows system properties not only to be verified, but to be computed in a systematic manner. A temporal logic query is an incomplete CTL specification containing a special placeholder symbol "?". Intuitively, the query asks for those system properties which yield a correct specification when inserted into the query. For example, the query $\mathbf{AF}\,\mathbf{AG}\,?$ asks for the invariants the system must eventually satisfy, the query $\mathbf{AG}(\mathit{shutdown} \rightarrow \mathbf{AG}\,?)$ asks for the invariants after system shutdown, and the query $\mathbf{AG}(? \rightarrow \neg\mathbf{AF}\,\mathit{ack})$ asks for properties which prevent that an acknowledgment is eventually sent. Query solving is the task of finding solutions, i.e., appropriate instantiations, to temporal logic queries in a given model. Query solving is more powerful than simple model checking because solving a query amounts to reasoning about a large number of specifications simultaneously. In particular, query solving can be used for diagnostic tasks and for understanding the behavior of an unknown system.

Temporal logic queries were first proposed by William Chan at CAV 2000 [4]. He realized that the number of solutions to CTL queries usually is very large for

* This work was supported by the European Community Research Training Network "Games and Automata for Synthesis and Validation" (GAMES) and by the Austrian Science Fund Project Z29-N04.

M. Baaz and J.A. Makowsky (Eds.): CSL 2003, LNCS 2803, pp. 470–483, 2003.

practically interesting systems. Thus, he focused on queries with unique strongest solutions – called *exact solutions* – from which all other solutions can be inferred. Unfortunately, not every CTL query has an exact solution, and determining whether a given query has an exact solution in every model is EXPTIME-complete [4]. Consequently, Chan defined an expressive yet easily recognizable syntactic fragment CTL^v, and claimed that each query in CTL^v has an exact solution in every model. Moreover, Chan proposed an efficient symbolic query solving algorithm for CTL^v with the same complexity as CTL model checking.

Temporal logic queries have been well received in the model checking community. In the following, we shortly describe the most important developments initiated by Chan's paper. Their main focus, however, is on generalizing Chan's work, i.e., they investigate temporal logic queries in a broader framework and do not use CTL^v as basis for their work. As a result of these generalizations, their approaches have in general higher complexity, i.e., double exponential in the number of atomic propositions. However, they have the same cost as model checking when applied to queries with a single strongest solution.

Bruns and Godefroid [2] have investigated temporal logic queries where multiple strongest solutions are allowed. Based on an automata-theoretic approach, they have shown how to obtain a query solving algorithm for any logic having a translation to alternating automata. Gurfinkel, Devereux, and Chechik [9, 5], on the other hand, have proposed an extension of Chan's approach to multiple placeholders in the framework of their multi-valued model checker XChek.

In the framework of CTL^*, Hornus and Schnoebelen [10] have shown that determining whether there exists an exact solution in a given model and computing this solution can be reduced to a linear number of model checking problems. Moreover, they have shown that if there exist multiple strongest solutions, the second one can be computed with a quadratic number of model checking calls, the third one with a cubic number of model checking calls, and so on. Their results can be seen as further evidence that it is important to have a good understanding of when there exists a unique strongest solution.

Current work announced by Ramakrishnan [13] in the area of logic programming based model checkers is investigating the connection between temporal logic queries and value-passing μ-calculus with free variables.

Our contribution. In this paper, we present a CTL^v query along with a suitable Kripke structure, such that the query does not have an exact solution. This contradicts Chan's main claim. The counterexample appeared when we attempted to prove formally the results in [4]. Since we consider Chan's paper a highly interesting and important contribution, and the paper has been frequently cited, we believe that it is important to give a sound foundation to get a better understanding of temporal logic queries. Based on the insight obtained by the counterexample we define two new languages:

- CTL^d generalizes Chan's approach in that it guarantees an exact solution, but only in those cases where the set of solutions is not empty.
- CTL^v_{new} is the subset of CTL^d where all queries necessarily have a solution in every Kripke structure. Thus, CTL^v_{new} fulfills Chan's original criteria.

We will show that the queries in CTL^d are distributive over conjunction, i.e., that any two solutions to a query can be conjoined ("collected") into a stronger common solution, and that a strong compound solution can be broken down ("separated") into individual solutions. It is then a natural step to restrict CTL^d to CTL^v_{new}. We also discuss a syntactic characterization of validity and provide partial results towards this aim. All our results are accompanied by exact proofs. The full version of this paper is available as [14].

This paper is organized as follows. In Section 2, we introduce the formalisms used in our work. Section 3 summarizes Chan's results on temporal logic queries and introduces CTL^v. After presenting the counterexample in Section 4, we investigate CTL^d and CTL^v_{new} in Sections 5 and 6 respectively. Section 7 provides partial results towards a syntactic characterization of validity. Finally, in Section 8, we summarize our results and indicate future research.

2 Preliminaries

We assume the reader is familiar with Kripke structures and the computation tree logic CTL based on the temporal operators $\mathbf{X}$ ("next"), $\mathbf{G}$ ("global"), and $\mathbf{U}$ ("until"). Let $\mathcal{K}$ be a Kripke structure, $\pi = s_0, s_1, s_2, \ldots$ a path, and φ a CTL formula. We write π^n to denote the suffix of π that begins at state s_n, and we write $\pi^{(n)}$ to denote the state s_n on π. If it is clear from the context, we write for simplicity π^n also to denote the state $\pi^{(n)}$. Furthermore, paths(s) denotes the set of paths starting at state s. We write $\mathcal{K} \models \varphi$ to denote $\mathcal{K}, s_0 \models \varphi$, where s_0 is the initial state in $\mathcal{K}$. Note that $\mathbf{F}\,\varphi \Leftrightarrow \mathit{true}\,\mathbf{U}\,\varphi$ and $\varphi \rightarrow \psi \Leftrightarrow \neg\varphi \vee \psi$ are as usual. Since we will consider formulas in negation normal form only, all other operators are defined directly and not as abbreviations.

Following Chan [4], we use some additional CTL operators. In particular, we use the weak until operator $\varphi\,\mathbf{W}\,\psi \Leftrightarrow (\mathbf{G}\,\varphi) \vee (\varphi\,\mathbf{U}\,\psi)$. The other operators are variants of the strong until operator $\mathbf{U}$ and the weak until operator $\mathbf{W}$:

$$\varphi\,\mathring{\mathbf{U}}\,\psi \Leftrightarrow \varphi\,\mathbf{U}\,(\varphi \wedge \psi) \qquad \varphi\,\mathring{\mathbf{W}}\,\psi \Leftrightarrow \varphi\,\mathbf{W}\,(\varphi \wedge \psi)$$

$$\varphi\,\bar{\mathbf{U}}\,\psi \Leftrightarrow \varphi\,\mathbf{U}\,(\neg\varphi \wedge \psi) \qquad \varphi\,\bar{\mathbf{W}}\,\psi \Leftrightarrow \varphi\,\mathbf{W}\,(\neg\varphi \wedge \psi)$$

Of course, the additional operators do not increase the expressive power of CTL, but they will give rise to stronger temporal logic query languages. Therefore, we will use the additional CTL operators throughout this paper.

3 CTL Queries

In this section, we summarize Chan's results on temporal logic queries [4]. Throughout the paper we will write γ, $\bar{\gamma}$, etc. to denote queries, and use all other Greek letters to denote formulas.

Definition 1 (CTL Query). *A* CTL query γ *is a CTL formula where exactly one subformula is replaced by the special symbol* ? *called* placeholder. *A query is* positive, *if the placeholder appears under an even number of negations; otherwise, it is* negative.

Table 1. Dual Operators

	γ pos.	γ neg.
$\varphi \Rightarrow^\gamma \psi$	$\varphi \Rightarrow \psi$	$\psi \Rightarrow \varphi$
$\varphi \wedge^\gamma \psi$	$\varphi \wedge \psi$	$\varphi \vee \psi$
$\gamma[\top]$	$\gamma[true]$	$\gamma[false]$
$\gamma[\bot]$	$\gamma[false]$	$\gamma[true]$

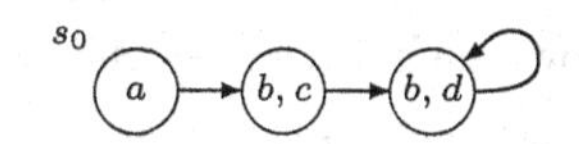

Fig. 1. CTL Query Example

For example, $\mathbf{A}(p\,\mathbf{U}\,\mathbf{AG}\,?)$ is a positive CTL query, whereas the CTL query $\mathbf{AG}(? \rightarrow \mathbf{AF}\,p)$ is negative. To avoid tedious case distinctions depending on the polarity of CTL queries in the remainder of this paper, we will use the dual operators defined in Table 1.

Definition 2 (Solution). *Let γ be a CTL query, $\mathcal{K}$ a Kripke structure, and φ a propositional formula. We write $\gamma[\varphi]$ to denote the result of substituting φ for the placeholder in the query γ. If $\mathcal{K} \models \gamma[\varphi]$, then we say that φ is a* solution *to γ in $\mathcal{K}$. We denote the set of all solutions to a query γ in a Kripke structure $\mathcal{K}$ by $sol(\mathcal{K}, \gamma) = \{\varphi \mid \mathcal{K} \models \gamma[\varphi]\}$.*

For example, consider the Kripke structure $\mathcal{K}$ shown in Fig. 1 and the CTL query $\gamma = \mathbf{A}((a \vee c)\,\mathbf{U}\,\mathbf{AG}\,?)$. It is easy to see that b, d, and $b \wedge d$ are solutions to γ in $\mathcal{K}$. So we have $\mathcal{K} \models \gamma[b]$, $\mathcal{K} \models \gamma[d]$, and $\mathcal{K} \models \gamma[b \wedge d]$.

To obtain the maximum information a query provides, it is necessary to consider all solutions. Since the number of solutions is likely to be very large, it is desirable to have a strongest solution that subsumes all other solutions. In the remainder of this section we present Chan's approach how to obtain such a strongest solution.

3.1 Monotone CTL Queries

The first important property we need to consider is monotonicity. A CTL query γ can be viewed as a function $\gamma : \varphi \mapsto \gamma[\varphi]$ that maps propositional formulas to CTL formulas. Depending on the polarity of a query, the corresponding function is either monotonically increasing or monotonically decreasing.

Definition 3 (Monotonicity). *A CTL query γ is* monotone *iff $\varphi \Rightarrow \psi$ implies $\gamma[\varphi] \Rightarrow^\gamma \gamma[\psi]$ for all propositional formulas φ and ψ.*

It is easy to define a class of CTL queries that are guaranteed to be monotone.

Definition 4 ($\mathbf{CTL}^m$). *The language CTL^m is the largest set of CTL queries that do not contain a subquery of the form $\gamma\,\bar{\mathbf{U}}\,\varphi$ or $\gamma\,\bar{\mathbf{W}}\,\varphi$, where γ is a query.*

For example, $\mathbf{EX}\,\mathbf{A}(\varphi\,\bar{\mathbf{U}}\,\mathbf{AG}\,?)$ is in CTL^m, whereas $\mathbf{EX}\,\mathbf{A}((\varphi \vee ?)\,\bar{\mathbf{U}}\,\psi)$ is not in CTL^m. The following lemma justifies our claim from above.

Lemma 1 ([4]). *Each query γ in CTL^m is monotone.*

Proof. Structural induction on γ. See [14] for details. □

The following characterizes the cases when a monotone CTL query has a solution.

Lemma 2 ([4]). *Let γ be a monotone CTL query, and $\mathcal{K}$ any Kripke structure.*

1. *γ has a solution in $\mathcal{K}$ iff $\mathcal{K} \models \gamma[\top]$.*
2. *Every proposition is a solution to γ in $\mathcal{K}$ iff $\mathcal{K} \models \gamma[\bot]$.*

3.2 Exact Solutions and Valid Queries

Recall that our aim is to obtain a strongest solution that subsumes all other solutions to a given query. We call such a strongest solution an *exact solution.*

Definition 5 (Exact Solution). *A solution ξ to a CTL query γ in a Kripke structure $\mathcal{K}$ is* exact *iff for every solution φ, it holds that $\xi \Rightarrow^\gamma \varphi$.*

It is easy to see that ξ is an exact solution to γ in $\mathcal{K}$ iff $\text{sol}(\mathcal{K}, \gamma) \subseteq \{\varphi \mid \xi \Rightarrow^\gamma \varphi\}$. The following proposition shows that for monotone CTL queries also any implication of an exact solution is a solution. In other words, for an exact solution ξ to a monotone CTL query γ in $\mathcal{K}$ we have $\text{sol}(\mathcal{K}, \gamma) = \{\varphi \mid \xi \Rightarrow^\gamma \varphi\}$.

Proposition 1 ([4]). *Let ξ be an exact solution to a monotone CTL query γ in a Kripke structure $\mathcal{K}$. The propositional formula φ is a solution to γ in $\mathcal{K}$ iff $\xi \Rightarrow^\gamma \varphi$.*

Note that not every monotone CTL query has an exact solution. Therefore, we are interested in queries that have an exact solution in every Kripke structure. For example, consider the Kripke structure $\mathcal{K}$ shown in Fig. 1 and the query $\gamma = \mathbf{AF}\,?$. It is easy to see that a and c are solutions to γ in $\mathcal{K}$, i.e., $\mathcal{K} \models \gamma[a] \wedge \gamma[c]$. Now, assume that there exists an exact solution ξ to γ in $\mathcal{K}$. Then, we know that $\xi \Rightarrow a$ and $\xi \Rightarrow c$, which is trivially equivalent to $\xi \Rightarrow a \wedge c$. Thus, by Proposition 1, we obtain that $a \wedge c$ must also be a solution to γ in $\mathcal{K}$. However, it is easy to see that $\mathcal{K} \not\models \gamma[a \wedge c]$.

Definition 6 (Valid Query). *A CTL query is* valid *iff it has an exact solution in every Kripke structure. We say a set of queries is valid if each query in this set is valid.*

As we will see, validity of CTL queries is closely related to the following property.

Definition 7 (Distributivity over Conjunction). *Let γ be a CTL query. γ is* distributive over conjunction *iff for all propositional formulas φ and ψ, $\gamma[\varphi] \wedge \gamma[\psi] \Leftrightarrow \gamma[\varphi \wedge^\gamma \psi]$.*

The following lemma shows the relation between validity and distributivity.

Lemma 3 ([4]). *A monotone CTL query is valid iff it has a solution in every Kripke structure and is distributive over conjunction.*

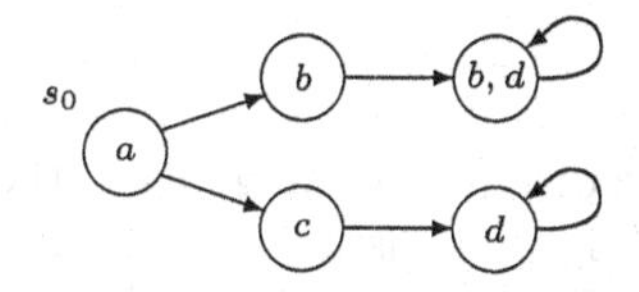

Fig. 2. CTL^v Counterexample

3.3 Definition of CTL^v

We have seen in Section 3.2 that not all CTL queries are valid. Notwithstanding the EXPTIME-completeness of deciding the validity of a query, Chan defined the following fragment CTL^v intended to comprise only valid queries.

Definition 8 (CTL^v). *CTL^v is the smallest set of queries satisfying:*

1. *? and ¬? are CTL^v queries.*
2. *If φ is a CTL formula and γ is a CTL^v query, then $\varphi \vee \gamma$, $\mathbf{AX}\,\gamma$, $\mathbf{A}(\gamma\,\mathring{\mathbf{W}}\,\varphi)$, and $\mathbf{A}(\varphi\,\bar{\mathbf{W}}\,\gamma)$ are also CTL^v queries.*
3. *A persistence query (CTL^p) is also a CTL^v query.*

The language CTL^p of persistence queries *is the smallest set of queries satisfying:*

4. *If γ is a CTL^v query, then $\mathbf{AG}\,\gamma$ is a persistence query.*
5. *If φ is a CTL formula and γ is a persistence query, then $\varphi \vee \gamma$, $\mathbf{AX}\,\gamma$, $\mathbf{A}(\varphi\,\mathbf{U}\,\gamma)$, and $\mathbf{A}(\varphi\,\mathbf{W}\,\gamma)$ are also persistence queries.*

4 The Counterexample

Chan's paper [4] did not contain a proof for its main claim, the validity of CTL^v. We present a counterexample to this claim.

Proposition 2. *The language CTL^v contains invalid queries.*

Proof. Consider the CTL query

$$\gamma = \mathbf{A}(a\,\mathbf{U}\,\mathbf{AX}(c \vee \mathbf{AG}\,?))$$

and the Kripke structure $\mathcal{K}$ shown in Fig. 2. It is easy to see that $\gamma \in$ CTL^v, and that b and d are solutions to γ in $\mathcal{K}$, i.e., $\mathcal{K} \models \gamma[b] \wedge \gamma[d]$. However, $b \wedge d$ is not a solution to γ in $\mathcal{K}$, i.e., $\mathcal{K} \not\models \gamma[b \wedge d]$. Thus, γ is not distributive over conjunction. Hence, by Lemma 3, γ is invalid. □

In the remainder of this paper, we will construct a language of valid CTL queries following Chan's approach. Instead of simply restricting CTL^v in such a way that the above counterexample is not possible, we take a more systematic approach. From Lemma 3 we know that validity means distributivity plus the existence of a solution in every Kripke structure. Among these properties, distributivity is both the much harder and the much more important one; we will therefore first concentrate on a possibly broad language for distributive queries, and then restrict it to valid queries.

5 CTL^d Queries

We will now present CTL^d, a distributive language without the ptoblem of CTL^v. The intuition behind CTL^d is that disjunction is not allowed to appear in the scope of the next operator **AX** in the definition of the persistence queries, unless there appears the global operator **AG** between them. To incorporate this insight into the definition of CTL^d, it is natural to split the language CTL^p of persistence queries into two sublanguages; the proofs to be shown below will provide further evidence for this choice.

Definition 9 (CTL^d). *CTL^d is the smallest set of queries satisfying:*

1. *? and ¬? are CTL^d queries.*
2. *If φ is a CTL formula and γ is a CTL^d query, then $\varphi \wedge \gamma$, $\varphi \vee \gamma$, $\mathbf{AX}\,\gamma$, $\mathbf{A}(\gamma\,\mathbf{U}\,\varphi)$, $\mathbf{A}(\gamma\,\mathring{\mathbf{U}}\,\varphi)$, $\mathbf{A}(\varphi\,\bar{\mathbf{U}}\,\gamma)$, $\mathbf{A}(\gamma\,\mathbf{W}\,\varphi)$, $\mathbf{A}(\gamma\,\mathring{\mathbf{W}}\,\varphi)$, and $\mathbf{A}(\varphi\,\bar{\mathbf{W}}\,\gamma)$ are also CTL^d queries.*
3. *An outer persistence query (CTL^{p_o}) is also a CTL^d query.*

The language CTL^{p_o} of outer persistence queries *is the smallest set of queries satisfying:*

4. *An inner persistence query (CTL^{p_i}) is also an outer persistence query.*
5. *If φ is a CTL formula and γ is an outer persistence query, then $\varphi \vee \gamma$, $\mathbf{A}(\gamma\,\mathbf{U}\,\varphi)$, $\mathbf{A}(\varphi\,\mathbf{U}\,\gamma)$, $\mathbf{A}(\gamma\,\mathbf{W}\,\varphi)$, and $\mathbf{A}(\varphi\,\mathbf{W}\,\gamma)$ are also outer persistence queries.*

The language CTL^{p_i} of inner persistence queries *is the smallest set of queries satisfying:*

6. *If γ is a CTL^d query, then $\mathbf{AG}\,\gamma$ is an inner persistence query.*
7. *If φ is a CTL formula and γ is an inner persistence query, then $\varphi \wedge \gamma$, $\mathbf{AX}\,\gamma$, $\mathbf{A}(\gamma\,\mathring{\mathbf{U}}\,\varphi)$, $\mathbf{A}(\varphi\,\mathbf{U}\,\gamma)$, $\mathbf{A}(\varphi\,\mathring{\mathbf{U}}\,\gamma)$, $\mathbf{A}(\varphi\,\bar{\mathbf{U}}\,\gamma)$, $\mathbf{A}(\gamma\,\mathring{\mathbf{W}}\,\varphi)$, $\mathbf{A}(\varphi\,\mathbf{W}\,\gamma)$, $\mathbf{A}(\varphi\,\mathring{\mathbf{W}}\,\gamma)$, and $\mathbf{A}(\varphi\,\bar{\mathbf{W}}\,\gamma)$ are also inner persistence queries.*

It can be easily seen that CTL^d is a subset of CTL^m.

5.1 Distributivity of CTL^d

This section is devoted to the nested inductive proof of our main result.

Theorem 1. *Each query in CTL^d is distributive over conjunction.*

According to Definition 7, we have to show that for each query $\gamma \in$ CTL^d and all formulas φ and ψ it holds that $\gamma[\varphi] \wedge \gamma[\psi] \Leftrightarrow \gamma[\varphi \wedge^\gamma \psi]$. We split this equivalence into two implications:

Definition 10 (Separating, Collecting). *Let γ be a CTL query. γ is* separating *iff it satisfies $\gamma[\varphi \wedge^\gamma \psi] \Rightarrow \gamma[\varphi] \wedge \gamma[\psi]$, and* collecting *iff it satisfies $\gamma[\varphi] \wedge \gamma[\psi] \Rightarrow \gamma[\varphi \wedge^\gamma \psi]$. γ is* **AG**-collecting *iff for each subquery of the form $\mathbf{AG}\,\bar{\gamma}$, $\bar{\gamma}$ is collecting.*

The separation property is shown very easily.

Lemma 4. *Each query in CTL^d is separating.*

Proof. Choose any query $\gamma \in \mathrm{CTL}^d \subset \mathrm{CTL}^m$. Consider the two implications $\varphi \wedge^\gamma \psi \Rightarrow^\gamma \varphi$ and $\varphi \wedge^\gamma \psi \Rightarrow^\gamma \psi$, which are both tautologies. By Lemma 1, we obtain both $\gamma[\varphi \wedge^\gamma \psi] \Rightarrow \gamma[\varphi]$ and $\gamma[\varphi \wedge^\gamma \psi] \Rightarrow \gamma[\psi]$. This is trivially equivalent to $\gamma[\varphi \wedge^\gamma \psi] \Rightarrow \gamma[\varphi] \wedge \gamma[\psi]$. □

Note that the separation property in fact holds for all monotone CTL queries. For the collection property, however, we have to do more work. By a series of lemmas we prove variations of the collection property for each language CTL^d, CTL^{p_i}, and CTL^{p_o}. These lemmas will then be combined in an inductive proof.

Definition 11 (Reachability). *Let s_1 and s_2 be two states in a Kripke structure. We say that state s_2 is* reachable *from state s_1, in symbols $s_1 \rightsquigarrow s_2$, iff there exists a path $\pi \in paths(s_1)$ and an $n \in \mathbb{N}$ such that $\pi^n \in paths(s_2)$.*

The following lemma shows that solving collecting queries preceded by the **AG** operator is similar to finding invariants.

Lemma 5. *Let γ be a collecting CTL query, and let s_1 and s_2 be two states such that $s_1 \rightsquigarrow s_2$. If $s_1 \models \mathbf{AG}\,\gamma[\varphi]$ and $s_2 \models \mathbf{AG}\,\gamma[\psi]$, then $s_2 \models \mathbf{AG}\,\gamma[\varphi \wedge^\gamma \psi]$.*

Proof. Since $s_2 \models \mathbf{AG}\,\gamma[\psi]$, we know that for each state s_3 with $s_2 \rightsquigarrow s_3$ it holds that $s_3 \models \gamma[\psi]$. In addition, since $s_1 \models \mathbf{AG}\,\gamma[\varphi]$ and $s_1 \rightsquigarrow s_2$, we know that for each state s_3 with $s_2 \rightsquigarrow s_3$ it holds that $s_3 \models \gamma[\varphi]$. Now, since γ is collecting, we obtain $s_3 \models \gamma[\varphi \wedge^\gamma \psi]$ for each state s_3 with $s_2 \rightsquigarrow s_3$. Hence, we have $s_2 \models \mathbf{AG}\,\gamma[\varphi \wedge^\gamma \psi]$. □

Note that each query in CTL^{p_i} contains a subquery of the form $\mathbf{AG}\,\gamma$. Thus, we can use Lemma 5 as the induction start in Lemma 6.

Lemma 6. *Let $\gamma \in CTL^{p_i}$ be* **AG***-collecting, and let s_1 and s_2 be two states such that $s_1 \rightsquigarrow s_2$. If $s_1 \models \gamma[\varphi]$ and $s_2 \models \gamma[\psi]$, then $s_2 \models \gamma[\varphi \wedge^\gamma \psi]$.*

Proof. Structural induction on γ.
Induction start: If $\gamma = \mathbf{AG}\,\bar{\gamma}$, then, by Lemma 5, we obtain $s_2 \models \gamma[\varphi \wedge^\gamma \psi]$.
Induction step: Let $\gamma = \mathbf{A}(\theta\,\mathbf{U}\,\bar{\gamma})$ and assume $s_1 \models \gamma[\varphi]$ as well as $s_2 \models \gamma[\psi]$. W.l.o.g., we choose a path $\pi \in \mathrm{paths}(s_2)$, and a path $\sigma \in \mathrm{paths}(s_1)$ s.t. for some $n \in \mathbb{N}$ we have $\sigma^n = \pi$. Therefore, we know that there exists a smallest k and a smallest l such that $\sigma^k \models \bar{\gamma}[\varphi]$ and $\sigma^{n+l} \models \bar{\gamma}[\psi]$. Now, by induction hypothesis, we obtain $\sigma^{\max(k,n+l)} \models \bar{\gamma}[\varphi \wedge^{\bar{\gamma}} \psi]$, where $\max(k, n+l) \geq n$. In addition, it is easy to see that $\sigma^i \models \theta$ for all $n \leq i < \max(k,\, n+l)$. Since π was chosen w.l.o.g., we obtain $s_2 \models \mathbf{A}(\theta\,\mathbf{U}\,\bar{\gamma}[\varphi \wedge^{\bar{\gamma}} \psi])$. Hence, we obtain $s_2 \models \gamma[\varphi \wedge^\gamma \psi]$.
See [14] for the remaining cases. □

Note that in CTL^m, all subqueries of a given query have the same polarity; the only exception is the subquery ? which has a different polarity than ¬?. Similarly as above, the case of inner persistence queries CTL^{p_i} shown in Lemma 6 constitutes the induction start for the outer persistence queries in Lemma 7.

Lemma 7. *Let $\gamma \in CTL^{P_o}$ be* **AG***-collecting, s_1, s_2 be states and $\rho \in paths(s_1)$ be a path where $\rho^n \in paths(s_2)$ for some suitable $n \in \mathbb{N}$. If $s_1 \models \gamma[\varphi]$ and $s_2 \models \gamma[\psi]$, then there exists an $r \leq n$ such that $\rho^r \models \gamma[\varphi \wedge^\gamma \psi]$.*

Proof. Structural induction on γ.
Induction start: If $\gamma \in \text{CTL}^{P_i}$, then, by Lemma 6, we obtain $s_2 \models \gamma[\varphi \wedge^\gamma \psi]$.
Induction step: Let $\gamma = \theta \vee \bar{\gamma}$ and assume $s_1 \models \gamma[\varphi]$ as well as $s_2 \models \gamma[\psi]$. If $s_1 \not\models \theta$ and $s_2 \not\models \theta$, we know that $s_1 \models \bar{\gamma}[\varphi]$ and $s_2 \models \bar{\gamma}[\psi]$. Now, by induction hypothesis, we obtain that there exists $i_0 \leq n$ such that $\rho^{i_0} \models \bar{\gamma}[\varphi \wedge^{\bar{\gamma}} \psi]$. Otherwise, if $s_1 \models \theta$ or $s_2 \models \theta$, we have $\rho^0 \models \theta$ or $\rho^n \models \theta$ respectively. Thus, in all cases there exists $r \leq n$, s.t. $\rho^r \models \theta \vee \bar{\gamma}[\varphi \wedge^{\bar{\gamma}} \psi]$ which implies the assertion.

Let $\gamma = \mathbf{A}(\theta \,\mathbf{U}\, \bar{\gamma})$ and assume $s_1 \models \gamma[\varphi]$ as well as $s_2 \models \gamma[\psi]$. We define $P = \{\pi \in \text{paths}(s_1) \mid \forall i \leq n.\ \pi^{(i)} = \rho^{(i)}\}$ and choose w.l.o.g. any path $\sigma \in P$. Thus, we know that there exists a smallest k and a smallest l such that $\sigma^k \models \bar{\gamma}[\varphi]$ and $\sigma^{n+l} \models \bar{\gamma}[\psi]$. Now, by induction hypothesis, we obtain that there exists $k \leq i_0 \leq n + l$ such that $\sigma^{i_0} \models \bar{\gamma}[\varphi \wedge^{\bar{\gamma}} \psi]$. We distinguish two cases: If $i_0 \leq n$, we trivially obtain $\rho^{i_0} \models \mathbf{A}(\theta \,\mathbf{U}\, \bar{\gamma}[\varphi \wedge^{\bar{\gamma}} \psi])$. Note that in this case we do not have to consider the paths paths(ρ^{i_0}) in detail, because the truth of the formula depends only on the first state $\rho^{(i_0)}$. Otherwise, if $i_0 > n$, we know for all $n \leq i < i_0$ that $\sigma^i \models \theta$ and $\sigma^{i_0} \models \bar{\gamma}[\varphi \wedge^{\bar{\gamma}} \psi]$. Since σ was chosen w.l.o.g., we obtain $\rho^n \models \mathbf{A}(\theta \,\mathbf{U}\, \bar{\gamma}[\varphi \wedge^{\bar{\gamma}} \psi])$. Note that the paths in P contain all elements of paths(s_2) as suffixes.

See [14] for the remaining cases. □

Remark. The case of $\gamma = \theta \vee \bar{\gamma}$ is important because it is this case which made us split up the persistence queries into inner and outer persistence queries. More precisely, the inductive proof for the more restrictive property of Lemma 6 does not go through for $\gamma = \theta \vee \bar{\gamma}$. The chosen less restrictive property of Lemma 7 works for $\gamma = \theta \vee \bar{\gamma}$, but its weakness in turn forces us to restrict the allowed operators in CTL^{P_o}. A posteriori, it is not surprising that disjunction plays a crucial role in the counterexample of Section 4.

The following property of CTL^{P_o} queries is directly implied by the special case of $s_1 = s_2$ in the previous lemma.

Corollary 1. *If $\gamma \in CTL^{P_o}$ is* **AG***-collecting, then γ is collecting.*

Finally, Corollary 1 serves as the induction start for Lemma 8.

Lemma 8. *If $\gamma \in CTL^d$ is* **AG***-collecting, then γ is collecting.*

Proof. Structural induction on γ.
Induction start: If γ is the (negated) placeholder[1], then the assertion is trivially true, and if $\gamma \in \text{CTL}^{P_o}$, we obtain the assertion by Corollary 1.
Induction step: Let $\gamma = \mathbf{A}(\bar{\gamma} \,\mathbf{U}\, \theta)$ and assume $s \models \gamma[\varphi] \wedge \gamma[\psi]$. W.l.o.g., we choose any path $\sigma \in \text{paths}(s)$. Therefore, we know that there exists a smallest k such

[1] This refers to the case where no subquery $\mathbf{AG}\,\bar{\gamma}$ exists.

that $\sigma^k \models \theta$. If $k > 0$, we know that $\sigma^{i_0} \models \bar{\gamma}[\varphi] \wedge \bar{\gamma}[\psi]$ for all $i_0 < k$. Now, by induction hypothesis, we obtain $\sigma^{i_0} \models \bar{\gamma}[\varphi \wedge^{\bar{\gamma}} \psi]$ for all $i_0 < k$. The case $k = 0$ is trivial. Since σ was chosen w.l.o.g., we obtain the assertion.

See [14] for the remaining cases. □

The following lemma brings our auxiliary results together such that we obtain the collecting property for distributivity over conjunction.

Lemma 9. *Each query in CTL^d is collecting.*

Proof. The proof is by induction over the number of subqueries of the form $\mathbf{AG}\,\bar{\gamma}$ of γ. It is important to note that according to the definition of CTL^d, whenever we have two subqueries $\mathbf{AG}\,\bar{\gamma}$ and $\mathbf{AG}\,\bar{\bar{\gamma}}$ of a query γ, then either $\mathbf{AG}\,\bar{\bar{\gamma}}$ is a subquery of $\mathbf{AG}\,\bar{\gamma}$ or vice versa. Thus, all subqueries starting with $\mathbf{AG}$ are linearly ordered in a natural way.

Induction start: If γ contains no subquery of the form $\mathbf{AG}\,\bar{\gamma}$, then the assumptions of Lemma 8 are trivially satisfied. Therefore, we obtain the assertion by Lemma 8. *Induction hypothesis:* Assume that the assertion holds for each query γ that contains at most n subqueries of the form $\mathbf{AG}\,\bar{\gamma}$. *Induction step:* Assume γ contains $n+1$ subqueries of the form $\mathbf{AG}\,\bar{\gamma}$. Thus, for each subquery $\mathbf{AG}\,\bar{\gamma}$ of γ, we trivially know that $\bar{\gamma}$ contains at most n subqueries of the form $\mathbf{AG}\,\bar{\bar{\gamma}}$. Hence, by induction hypothesis, we obtain that γ is $\mathbf{AG}$-collecting. Now, by Lemma 8, we obtain the assertion. □

Now, Theorem 1 follows trivially by Lemma 4 and Lemma 9.

6 Valid Queries in CTL^d

If we are primarily concerned with finding an exact solution to CTL queries while accepting that there may not exist a solution, we can use queries in CTL^d. However, following the original motivation of Chan we are also interested in queries for which there certainly exists an exact solution in *every* Kripke structure. This leads us to the language CTL$^v_{new}$ of valid queries.

Definition 12 (CTL$^v_{new}$). *CTL$^v_{new}$ is the smallest set of queries satisfying:*

1. *? and ¬? are CTL$^v_{new}$ queries.*
2. *If φ is a CTL formula and γ is a CTL$^v_{new}$ query, then $\varphi \vee \gamma$, $\mathbf{AX}\,\gamma$, $\mathbf{A}(\gamma\,\mathbf{W}\,\varphi)$, $\mathbf{A}(\gamma\,\mathring{\mathbf{W}}\,\varphi)$, and $\mathbf{A}(\varphi\,\bar{\mathbf{W}}\,\gamma)$ are also CTL$^v_{new}$ queries.*
3. *A valid outer persistence query is also a CTL$^v_{new}$ query.*

The language CTL$^{p_o}_v$ of valid outer persistence queries *is the smallest set of queries satisfying:*

4. *A valid inner persistence query is also a valid outer persistence query.*
5. *If φ is a CTL formula and γ is a valid outer persistence query, then $\varphi \vee \gamma$, $\mathbf{A}(\varphi\,\mathbf{U}\,\gamma)$, $\mathbf{A}(\gamma\,\mathbf{W}\,\varphi)$, and $\mathbf{A}(\varphi\,\mathbf{W}\,\gamma)$ are also valid outer persistence queries.*

The language $CTL_v^{p_i}$ *of* valid inner persistence queries *is the smallest set of queries satisfying:*

6. *If* γ *is a* CTL_{new}^v *query, then* $\mathbf{AG}\,\gamma$ *is a valid inner persistence query.*
7. *If* φ *is a CTL formula and* γ *is a valid inner persistence query, then* $\mathbf{AX}\,\gamma$, $\mathbf{A}(\varphi\,\mathbf{U}\,\gamma)$, $\mathbf{A}(\varphi\,\mathbf{W}\,\gamma)$, $\mathbf{A}(\gamma\,\mathring{\mathbf{W}}\,\varphi)$, *and* $\mathbf{A}(\varphi\,\bar{\mathbf{W}}\,\gamma)$ *are also valid inner persistence queries.*

It can be easily seen that CTL_{new}^v is a subset of CTL^d. For the validity of CTL_{new}^v it only remains to show that each query in CTL_{new}^v has a solution in *every* Kripke structure.

Lemma 10. *If* $\gamma \in CTL_{new}^v$, *then* $\mathcal{K} \models \gamma[\top]$ *for any Kripke structure* $\mathcal{K}$.

Proof. Structural induction on γ. See [14] for details. □

Now, by Theorem 1, Lemma 10, Lemma 2, and Lemma 3, we have reached our final result in the following theorem.

Theorem 2. *Each query in* CTL_{new}^v *is valid.*

7 Towards a Syntactic Characterization

In this section we will justify the restrictions on temporal operators in the definition of CTL^d and CTL_{new}^v. In addition, we will shortly discuss a syntactic characterization of distributivity and validity.

In order to do this, we first show that releasing the constraints on the occurrences of temporal operators in CTL^d necessarily violates distributivity over conjunction. Recall that the definition of CTL^m excludes only the operators $\mathbf{A}(\gamma\,\bar{\mathbf{U}}\,\varphi)$ and $\mathbf{A}(\gamma\,\bar{\mathbf{W}}\,\varphi)$. Thus, these operators are the only ones that could violate monotonicity and hence separability. Counterexamples to monotonicity resp. separability can be easily found when allowing these operators in Definition 4 [14]. Moreover, counterexamples to the collection property concerning the CTL^d operators when adding $\mathbf{A}(\varphi\,\mathbf{U}\,\gamma)$, $\mathbf{A}(\varphi\,\mathring{\mathbf{U}}\,\gamma)$, $\mathbf{A}(\varphi\,\mathbf{W}\,\gamma)$, or $\mathbf{A}(\varphi\,\mathring{\mathbf{W}}\,\gamma)$ to line 2 in Definition 9 can also be found very easily [14]. Now, it remains to consider the more complicated counterexamples to extensions of CTL^{p_i} and CTL^{p_o} as follows in detail.

CTL^{p_i} Collection Property. Consider the Kripke structure shown in Fig. 3 together with the following queries:

- $\mathbf{A}(a\,\mathbf{U}\,\mathbf{AX}\,\mathbf{A}(\mathbf{AG}\,?\,\mathbf{U}\,(c \vee (b \wedge d))))$
- $\mathbf{A}(a\,\mathbf{U}\,\mathbf{AX}\,\mathbf{A}(\mathbf{AG}\,?\,\mathbf{W}\,(c \vee (b \wedge d))))$

If we added $\mathbf{A}(\gamma\,\mathbf{U}\,\varphi)$ or $\mathbf{A}(\gamma\,\mathbf{W}\,\varphi)$ to line 7 in Definition 9, then the corresponding query above would be in this extension of CTL^d. For both queries γ, however, we have $s_0 \models \gamma[b] \wedge \gamma[d]$, but $s_0 \not\models \gamma[b \wedge^\gamma d]$. Hence, both operators must indeed be excluded from the definition of CTL^{p_i}.

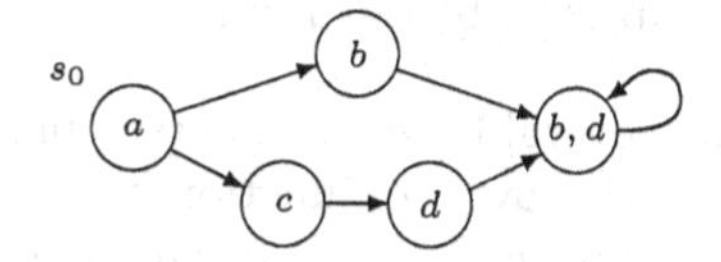

Fig. 3. CTL^{p_i} Demarcation

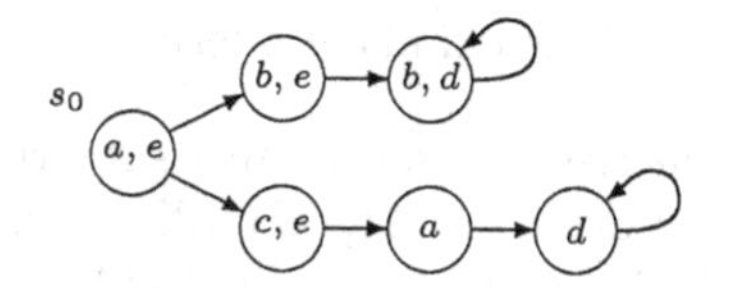

Fig. 4. CTL^{p_o} Demarcation

CTL^{p_o} Collection Property. Consider the Kripke structure shown in Fig. 4 together with the following queries:

- $\mathbf{A}(e\,\mathbf{U}\,(\neg c \wedge \mathbf{A}(a\,\mathbf{U}\,(c \vee \mathbf{AG}\,?))))$
- $\mathbf{A}(e\,\mathbf{U}\,\mathbf{A}(\neg c\,\mathring{\mathbf{U}}\,\mathbf{A}(a\,\mathbf{U}\,(c \vee \mathbf{AG}\,?))))$
- $\mathbf{A}(e\,\mathbf{U}\,\mathbf{A}(\mathbf{A}(a\,\mathbf{U}\,(c \vee \mathbf{AG}\,?))\,\mathring{\mathbf{W}}\,\neg c))$
- $\mathbf{A}(e\,\mathbf{U}\,\mathbf{A}((a \vee b)\,\mathring{\mathbf{W}}\,\mathbf{A}(a\,\mathbf{U}\,(c \vee \mathbf{AG}\,?))))$
- $\mathbf{A}(e\,\mathbf{U}\,\mathbf{A}(\mathbf{A}(a\,\mathbf{U}\,(c \vee \mathbf{AG}\,?))\,\mathring{\mathbf{U}}\,\neg c))$
- $\mathbf{A}(e\,\mathbf{U}\,\mathbf{A}(c\,\bar{\mathbf{U}}\,\mathbf{A}(a\,\mathbf{U}\,(c \vee \mathbf{AG}\,?))))$
- $\mathbf{A}(e\,\mathbf{U}\,\mathbf{A}(c\,\bar{\mathbf{W}}\,\mathbf{A}(a\,\mathbf{U}\,(c \vee \mathbf{AG}\,?))))$

If we added $\varphi \wedge \gamma$, $\mathbf{A}(\gamma\,\mathring{\mathbf{U}}\,\varphi)$, $\mathbf{A}(\gamma\,\mathring{\mathbf{W}}\,\varphi)$, $\mathbf{A}(\varphi\,\mathring{\mathbf{U}}\,\gamma)$, $\mathbf{A}(\varphi\,\bar{\mathbf{U}}\,\gamma)$, $\mathbf{A}(\varphi\,\mathring{\mathbf{W}}\,\gamma)$, or $\mathbf{A}(\varphi\,\bar{\mathbf{W}}\,\gamma)$ to line 5 in Definition 9, then the corresponding query above would be in this extension of CTL^d. Again, for all of these queries γ above we have $s_0 \models \gamma[b] \wedge \gamma[d]$, but $s_0 \not\models \gamma[b \wedge^\gamma d]$. Hence, these operators must indeed be excluded from the definition of CTL^{p_o}. The counterexample concerning the next operator has already been shown in Section 4. Hence, the operator $\mathbf{AX}\,\gamma$ also must be excluded from the definition of CTL^{p_o}.

Finally, the restrictions of CTL^d to CTL$^v_{new}$ can also be justified by counterexamples to validity. Due to space restrictions, the reader is referred to [14] for these simple examples.

The above results demonstrate that CTL^d and CTL$^v_{new}$ are maximal in the sense that the clauses in Definition 9 and Definition 12 cannot be extended by releasing the restrictions on the temporal operators.

Of course, it would be interesting and important to obtain an exact syntactic characterization of distributivity and validity. We believe, however, that this is a highly non-trivial question, and cannot be obtained by trivial modifications of CTL^d and CTL$^v_{new}$. As mentioned above, deciding both validity and distributivity is EXPTIME-complete [4, 14]. Therefore, it is not possible to define the natural syntactic fragment in terms of a *simple* grammar such as a context-free grammar (as the complexity of deciding membership in the language would be low). It may, however, in principle be possible to capture distributivity and validity by simple syntactic fragments such that every distributive resp. valid query is *equivalent* to a query in these fragments[2]. Alternatively, a characterization with respect to a special kind of grammar similar to [3] could also be of interest.

[2] Compare Maidl's characterization of ACTL ∩ LTL [11] which seemingly contradicts the PSPACE-completeness of deciding whether a given ACTL formula is in LTL.

There are two kinds of queries which are *not* captured by our fragments:

1. Queries where the same subformula occurs several times at different places. For example, $\mathbf{A}(\varphi\, \mathbf{U}\, (\mathbf{AX}\, \neg\varphi) \wedge ?)$ is distributive over conjunction but not in CTL^d, and $(\mathbf{AG}\, \neg\varphi) \vee \mathbf{A}(?\, \mathbf{U}\, \varphi)$ is valid but not in CTL^v_{new}. It can be easily seen that our first example is very similar to the definition of the disjoint strong until operator $\bar{\mathbf{U}}$, i.e., we could define a new operator equivalent to $\mathbf{A}(\varphi\, \mathbf{U}\, (\mathbf{AX}\, \neg\varphi) \wedge ?)$ and extend our fragment in this way. However, there is an unlimited number of possible extensions in this way, e.g., consider the queries $\mathbf{A}(\varphi\, \mathbf{U}\, (\mathbf{AX}\, \mathbf{AX}\, \neg\varphi) \wedge ?)$, $\mathbf{A}(\varphi\, \mathbf{U}\, (\mathbf{AX}\, \mathbf{AX}\, \mathbf{AX}\, \neg\varphi) \wedge ?)$, ...
2. Queries where all subformulas are independent from each other. For example, $\mathbf{A}(a\, \mathbf{U}\, (d \wedge \mathbf{A}((c \vee \mathbf{AG}\, ?)\, \mathbf{U}\, b)))$ is distributive over conjunction but not in CTL^d, and $\mathbf{A}(a\, \mathbf{U}\, \mathbf{A}((c \vee \mathbf{AG}\, ?)\, \mathring{\mathbf{W}}\, b))$ is valid but not in CTL^v_{new}.

While we believe that the queries of type 2 can be dealt with by an extension of the approach in this paper, type 1 raises a much harder problem.

8 Conclusion and Future Work

In this paper we have systematically investigated Chan's program on temporal logic query languages, and corrected some flaws as to put a thorough foundation for future work. Among these, we have defined the two fragments CTL^d and CTL^v_{new}. CTL^d has a more systematic appearance; in particular, the behavior of the variants of the until operator is completely analogous to the behavior of the corresponding variants of the weak until operator. This does obviously not hold true for CTL^v_{new}. The most challenging among the theoretical issues left is to capture the distributive and the valid queries by simple syntactic fragments as discussed in Section 7.

A natural different extension already mentioned in [4] is to consider LTL. While the work in [10] covers CTL*, and hence LTL, it does not characterize the existence of exact solutions. We are currently working on a formal foundation of distributive resp. valid LTL queries.

Our original interest in temporal logic queries was triggered by practical applications on which we are currently working. The first one is *invariant discovery* [8], i.e., the automatic discovery of program invariants. The second application area is counterexample analysis. Recent work on software verification [1] is dealing with the question how to extract error causes from multiple counterexamples generated by a model checker. Indeed, temporal logic queries can be used in a natural way to determine error causes. Consider for example a safety property where the good states are described by a formula φ, and define the queries $\gamma_1 = \mathbf{AG}(? \to \mathbf{AF}\, \neg\varphi)$ and $\gamma_2 = \mathbf{AG}(? \to \mathbf{EG}\, \varphi)$. The solution to γ_1 describes the set of states which possibly violate φ, and the solution to γ_2 describes the set of states which are on a correct trace. From these sets, the error causes of [1] can be obtained easily.

References

1. Thomas Ball, Mayur Naik, and Sriram K. Rajamani. From symptom to cause: Localizing errors in counterexample traces. In *Proceedings of the 30th Annual Symposium on Principles of Programming Languages*, pages 97–105. ACM Press, 2003.
2. Glenn Bruns and Patrice Godefroid. Temporal logic query checking. In *Proceedings of the 16th Annual IEEE Symposium on Logic in Computer Science*, pages 409–417. IEEE Computer Society Press, 2001.
3. Francesco Buccafurri, Thomas Eiter, Georg Gottlob, and Nicola Leone. On ACTL formulas having linear counterexamples. *Journal of Computer and System Sciences*, 62(3):463–515, 2001.
4. William Chan. Temporal-logic queries. In *Proceedings of the 12th International Conference on Computer Aided Verification*, volume 1855 of *Lecture Notes in Computer Science*, pages 450–463. Springer-Verlag, 2000.
5. Marsha Chechik and Arie Gurfinkel. TLQSolver: A temporal logic query checker. In *Proceedings of the 15th International Conference on Computer Aided Verification*, Lecture Notes in Computer Science. Springer-Verlag, 2003. To appear.
6. Edmund M. Clarke and E. Allen Emerson. Design and synthesis of synchronization skeletons using branching time temporal logic. In *Proceedings of the Workshop on Logics of Programs*, volume 131 of *Lecture Notes in Computer Science*, pages 52–71. Springer-Verlag, 1981.
7. Edmund M. Clarke, Orna Grumberg, and Doron A. Peled. *Model Checking*. MIT Press, 1999.
8. Michael D. Ernst, Jake Cockrell, William G. Griswold, and David Notkin. Dynamically discovering likely program invariants to support program evolution. In *Proceedings of the 21st International Conference on Software Engineering*, pages 213–224. IEEE Computer Society Press, 1999.
9. Arie Gurfinkel, Benet Devereux, and Marsha Chechik. Model exploration with temporal logic query checking. In *Proceedings of the 10th ACM SIGSOFT Symposium on Foundations of Software Engineering*, pages 139–148. ACM Press, 2002.
10. Samuel Hornus and Philippe Schnoebelen. On solving temporal logic queries. In *Proceedings of the 9th International Conference on Algebraic Methodology and Software Technology*, volume 2422 of *Lecture Notes in Computer Science*, pages 163–177. Springer-Verlag, 2002.
11. Monika Maidl. The common fragment of CTL and LTL. In *Proceedings of the 41st Annual Symposium on Foundations of Computer Science*, pages 643–652. IEEE Computer Society Press, 2000.
12. Jean-Pierre Queille and Joseph Sifakis. Specification and verification of concurrent systems in CESAR. In *Proceedings of the 5th International Symposium on Programming*, volume 137 of *Lecture Notes in Computer Science*, pages 337–350. Springer-Verlag, 1982.
13. C. R. Ramakrishnan. A model checker for value-passing mu-calculus using logic programming. In *Proceedings of the 3rd International Symposium on Practical Aspects of Declarative Languages*, volume 1990 of *Lecture Notes in Computer Science*, pages 1–13. Springer-Verlag, 2001.
14. Marko Samer. Temporal logic queries in model checking. Diploma thesis, Vienna University of Technology, May 2002.

Calculi of Meta-variables

Masahiko Sato[1], Takafumi Sakurai[2],
Yukiyoshi Kameyama[3], and Atsushi Igarashi[1]

[1] Graduate School of Informatics, Kyoto University
{masahiko,igarashi}@kuis.kyoto-u.ac.jp
[2] Department of Mathematics and Informatics, Chiba University
sakurai@math.s.chiba-u.ac.jp
[3] Institute of Information Sciences and Electronics, University of Tsukuba, and JST
kam@is.tsukuba.ac.jp

Abstract. The notion of meta-variable plays a fundamental role when we define formal systems such as logical and computational calculi. Yet it has been usually understood only informally as is seen in most textbooks of logic. Based on our observations of the usages of meta-variables in textbooks, we propose two formal systems that have the notion of meta-variable.

In both calculi, each variable is given a level (non-negative integer), which classifies variables into object variables (level 0), meta-variables (level 1), metameta-variables (level 2) and so on. Then, simple arity systems are used to exclude meaningless terms like a meta-level function operating on the metameta-level. A main difference of the two calculi lies in the definitions of substitution. The first calculus uses *textual* substitution, which can often be found in definitions of quantified formulae: when a term is substituted for a meta-variable, free object-level variables in the term may be captured. The second calculus is based on the observation that predicates can be regarded as meta-level functions on object-level terms, hence uses *capture-avoiding* substitution.

We show both calculi enjoy a number of properties including Church-Rosser and Strong Normalization, which are indispensable when we use them as frameworks to define logical systems.

Keywords: Meta-variable, logical framework, context, λ-calculus

1 Introduction

The notion of *meta-variable* is a fundamental notion both in logic and computer science. It is because both logic and computer science mainly deal with linguistic objects such as formulas, proofs, programs etc., and whenever we make a *general* statement about these objects we use meta-variables to refer to these objects. For example, if we look at any book of logic, most variables we see are meta-variables. Meta-variables are also known as *metamathematical variables* [7] and *syntactical variables* [12]. However, it seems that, so far, only a very few attempts have been made to formalize the notion of meta-variable. One reason for this may be that we have to go to metameta-level to do so.

M. Baaz and J.A. Makowsky (Eds.): CSL 2003, LNCS 2803, pp. 484–497, 2003.

In this paper, we present two new formalizations of the concept of the meta-variable. These formalizations are based on our observations of the usages of meta-variables in text books and technical papers on logic.

The first observation is from Shoenfield [12] and Kleene [7]. In these books, we find the following sentence as one of the inductive clauses which define formulas. (We have slightly modified notations from the originals.)

If x is a variable and A is a formula, then $\exists x\ A$ is a formula.

In the above sentence, both 'x' and 'A' are meta-variables, and when we use the sentence as a rule which is used to construct a concrete formula, we must *instantiate* these meta-variables by concrete linguistic objects of the object language. Thus, for example, we may instantiate x by a concrete variable $\mathtt{x}$ and A by a concrete formula $\mathtt{x} = \mathtt{x}$. Then, by applying the instantiated rule, we have that $\exists \mathtt{x}\ \mathtt{x} = \mathtt{x}$ is a formula. Here, it is important to remark that the process of instantiation we just described is a form of *substitution*, but, unlike ordinary substitutions, the variables being substituted are meta-variables. There is another subtle point in this substitution process. To see this, we analyze the above instantiation process in two steps. Namely, we assume that the meta-variable 'x' is first instantiated and then, in the second step, the meta-variable 'A' is instantiated. Then, after the first step, we get the following sentence.

If $\mathtt{x}$ is a variable and A is a formula, then $\exists \mathtt{x}\ A$ is a formula.

In the second step, we substitute $\mathtt{x} = \mathtt{x}$ for A in the above sentence, and we get the fully instantiated sentence:

If $\mathtt{x}$ is a variable and $\mathtt{x} = \mathtt{x}$ is a formula, then $\exists \mathtt{x}\ \mathtt{x} = \mathtt{x}$ is a formula.

We note here that, unlike ordinary substitution, we have substituted $\mathtt{x} = \mathtt{x}$ for A in $\exists \mathtt{x}\ A$ without following the usual convention of renaming the name of the binding variable ($\mathtt{x}$ in this case) to avoid the capture of free variables.

The second observation we now make is also a very common one. Often in the literature, notation like $A(x)$ is used to indicate a formula where free occurrences of x in $A(x)$ is implicitly understood. Thus, if t is a term, then $A(t)$ stands for a formula which is obtained from $A(x)$ by substituting t for x in $A(x)$. In this usage, 'x', 't' and 'A' are all meta-variables and the first two meta-variables range over variables and terms in the object language. As for the third meta-variable 'A' it is possible to interpret its range in two ways.

The first interpretation is to regard 'A' as ranging over functions which, when applied to terms, will yield formulas. In this interpretation $A(t)$ denotes the result of applying the meta-level function A to a term t. So, in this interpretation, A is a *metameta*-variable, since its denotation is not a linguistic object of the object language but it is a function in the meta language.

The second interpretation is to regard 'A' as ranging over *abstracts* of the object language which can be instantiated to formulas by supplying terms of the object language. This interpretation is possible only if the object language contains such abstracts as its formal entities. Higher-order abstract syntax employed by, e.g., Edinburgh LF [5] is based on this interpretation.

In this paper, we will introduce two typed calculi $\lambda\mathcal{M}$ and λm, which are respectively designed based on the above two observations. In $\lambda\mathcal{M}$ and λm, with each variable a non-negative integer, which we call the level of the variable, is associated. We consider level 0 as the object-level, level 1 as the meta-level, level 2 as the metameta-level and so on.

In these formalizations, we believe that we can take virtually any formal system as the object-level of our systems. However, for the sake of concrete presentation, we take as the object-level a system of symbolic expressions we introduced in Sato [10] which is simple but powerful enough to represent syntax of many of the commonly used formal systems such as the λ-calculus and predicate calculus. Both $\lambda\mathcal{M}$ and λm will be constructed on top of this object-level system by adding higher-level structures, and we will show that these calculi enjoy nice properties such as confluence and strong normalizability. We will also show that the calculus $\lambda\mathcal{M}$ can represent the notion of context naturally since a context, which is an expression containing some holes in it, is inherently a linguistic object in the meta-level and not in the object-level.

Due to lack of space, we have omitted some lemmas and details of proofs. A full version of this paper with proofs is accessible at

`http://www.sato.kuis.kyoto-u.ac.jp/~masahiko/index-e.html`.

2 Informal Introduction to the Calculi

In this section we informally explain the two calculi $\lambda\mathcal{M}$ and λm which we propose in this paper. We assume our object language contains constants, abstraction ($(x)\,[M]$), and pair ($\langle M, N\rangle$). Abstraction and application in the meta-level are denoted by $\lambda X.M$ and MN. (We often use capital letters for meta-variables in examples, although both object- and meta-level variables belong to the same syntactic category in the formal definition.)

2.1 The Calculus $\lambda\mathcal{M}$

The first calculus $\lambda\mathcal{M}$ is based on the first observation in Section 1. Let us consider the first observation again and assume that we have just completed the first step. Then, we have the expression $\exists \mathtt{x}\ A$. We can represent this expression by $\langle '\exists', (x^0)\,[A^1]\rangle$ using a constant $'\exists'$. On the shoulder of a variable, we write a natural number to indicate its level, although we often omit the level if it is clear from the context. So, we simply write x for x^0 and it corresponds to the concrete object-level variable $\mathtt{x}$. In $\lambda\mathcal{M}$, the instantiation process of the meta-variable A by the object-level formula $x = x$ can be represented as the reduction process of the following β-redex:

$$(\lambda A.\, \langle '\exists', (x)\,[A]\rangle)\langle '=', \langle x, x\rangle\rangle.$$

In the reduction, as pointed out in Section 1, *non-standard* substitution is performed and we get:

$$\langle '\exists', (x)\,[\langle '=', \langle x, x\rangle\rangle]\rangle$$

which represents the formula $\exists \mathtt{x}\ \mathtt{x} = \mathtt{x}$ as expected. Note that the object-level variable x is captured through the substitution.

The non-standard (textual) substitution we have just introduced gives rise to the following two technical problems.

The first one is the non-confluence of the calculus. As argued in the literature on context calculi [8, 6, 11], calculi that have textual substitution cannot be confluent unless we restrict the evaluation order. For instance, let M be the term $(\lambda X^2.\, (\lambda x^1.\, X^2)y^0)x^1$. Depending on the evaluation-order, we will get different results y^0 and x^1. Our solution to this problem is (roughly) that, a redex may not be reduced if it contains variables of higher levels than the level of the redex[1]. In this example, the inner redex has level-1, so its reduction is postponed until the variable X^2 disappears.

The second problem in formulating $\lambda\mathcal{M}$ is that, since we restrict the evaluation-order, some reductions may get stuck. Consider the terms $(\lambda X^2.\, \lambda x^1.\, X^2)$ $y^0 z^0$, and $(\lambda x^1.\, \lambda X^2.\, X^2)y^0 z^0$. The first term reduces to y^0, while the second term cannot be reduced. Since we do not consider terms like the second one meaningful, we introduce *arities* to rule out such terms. For instance, $\lambda x^1.\, X^2 : 0 \to_1 0$ signifies that this term denotes a level-1 function from objects to objects. Similarly we have $\lambda X^2.\, \lambda x^1.\, X^2 : 0 \to_2 (0 \to_1 0)$. On the other hand, $\lambda x^1.\, \lambda X^2.\, X^2$ would have arity $0 \to_1 (0 \to_2 0)$, and it would denote a level-1 function which returns a level-2 function. We will exclude such a term by defining arity properly, and show that the evaluation in $\lambda\mathcal{M}$ does not get stuck (Theorem 4).

Although $\lambda\mathcal{M}$ has non-standard substitution, we need the standard capture-avoiding substitution as well, when the variable being substituted for and one being captured are of the same level. Let us see the following reduction:

$$(\lambda X^2.\, \lambda Y^2.\, \lambda z^1.\, X^2)(Y^2 z^1) \to \lambda W^2.\, \lambda z^1.\, Y^2 z^1$$

in which the variable z^1 is captured, while the variable Y^2 is not captured since its level is the same as that of the variable X^2.

2.2 The Calculus λm

The second calculus λm formalizes the first interpretation[2] of the second observation in Section 1.

The formula $A(t)$ can be represented as $A^2(t^1)$ using the level-2 variable A^2 of arity $0 \to_1 0$, and the level-1 variable t of arity 0 in λm. The existential formula $\exists x\ A(x)$ is represented as:

$$\langle {}'\exists', (x^0)\, [A^2(x^0)] \rangle,$$

and the substitution of $\lambda x^1 : 0.\, \langle {}'=', \langle x^1, x^1 \rangle \rangle$ for A^2 is realized by β-reduction, but this time we do *not* use the non-standard substitution. Hence, the term:

---

[1] The level of the redex $(\lambda X^i.\, M)N$ is i.

[2] In this paper, we do not formalize the second interpretation in which A in $A(t)$ ranges over abstracts of the object language. It should be a straightforward extension of this work, but details are left for future work.

$$(\lambda A^2 : 0 \to_1 0. \langle '\exists', (x^0)\, [A^2(x^0)] \rangle)(\lambda x^1 : 0. \langle '=', \langle x^1, x^1 \rangle \rangle)$$

reduces (using *standard* substitution) to

$$\langle '\exists', (x^0)\, [\langle '=', \langle x^0, x^0 \rangle \rangle] \rangle$$

as expected.

3 The Calculus $\lambda\mathcal{M}$

In this section, we give a formal definition of the first calculus $\lambda\mathcal{M}$. The second calculus λm will be introduced in the next section.

3.1 Arities and Terms

We define *arity* (α) and its *level* ($|\alpha|$) as follows:

1. 0 is an arity and $|0| = 0$.
2. If α and β are arities, $0 < i$, $|\alpha| < i$, and $|\beta| \leq i$, then $\alpha \to_i \beta$ is an arity and $|\alpha \to_i \beta| = i$.

Note that the side condition of the second clause reflects the intended notion of level introduced in Section 1. Intuitively, the arity $\alpha \to_i \beta$ is for level-i functions from terms of arity α to terms of arity β, thus $|\alpha| < i$ and $|\beta| < i$ must be satisfied. The restriction on β is relaxed to $|\beta| \leq i$ to allow currying.

We assume that, for each natural number i and arity α, there are infinitely many variables, and the sets of variables of each level and arity are mutually disjoint. For a variable x of level i and arity α, we sometimes write it as x^i. The set of all variables is denoted by V. A *(variable) declaration* is an expression of the form $x^i : \alpha$, where α is an arity and either $|\alpha| < i$ or $|\alpha| = i = 0$. We say the *level* of this declaration is i. A *hypothesis sequence* is a finite sequence of declarations. A *judgment* is an expression of the form $\Gamma \vdash M : \alpha$ where Γ is a hypothesis sequence and α is an arity.

We have the following rules that are used to derive judgments.

The first rule introduces variables for each i and α, where we assume x is a variable of level i and arity α.

$$\frac{x^i : \alpha \in \Gamma}{\Gamma \vdash x^i : \alpha}\ (\mathsf{var})$$

The next two rules introduce abstraction and application for level-i ($i > 0$).

$$\frac{\Gamma, x^i : \alpha \vdash M : \beta \quad |\beta| \leq i}{\Gamma \vdash \lambda x^i{:}\alpha.\, M : \alpha \to_i \beta}\ (\mathsf{abs}) \qquad \frac{\Gamma \vdash M : \alpha \to_i \beta \quad \Gamma \vdash N : \alpha}{\Gamma \vdash MN : \beta}\ (\mathsf{app})$$

Note that, in the rule (abs), the level of the variable x^i and that of the arity $\alpha \to_i \beta$ should agree, and the side-condition $|\beta| \leq i$ is needed to form a (well-formed) arity $\alpha \to_i \beta$. Note also that we may not construct a term like $\lambda x^1 : 0. \lambda X^2 : 0. X^2$.

The last group of rules are those for the level-0, the object language.

$$\frac{c \text{ is a constant}}{\Gamma \vdash c : 0}\ (\mathsf{const0}) \qquad \frac{\Gamma, x^0 : 0 \vdash M : 0}{\Gamma \vdash (x^0)\,[M] : 0}\ (\mathsf{abs0})$$

$$\frac{\Gamma \vdash M : 0 \quad \Gamma \vdash N : 0}{\Gamma \vdash \langle M, N \rangle : 0}\ (\mathsf{pair0})$$

An expression M is said to be a *term* if a judgment of the form $\Gamma \vdash M : \alpha$ is derivable for some Γ and α. We sometimes just write $\lambda x^i.\, M$ for $\lambda x^i{:}\alpha.\, M$ when the arity α of x^i is irrelevant.

The scope of $\lambda x.$ and free occurrences of variables in a term are defined as usual. For a term M, the set of free variables in M is denoted by $\mathrm{FV}(M)$. The *level of a term* M, denoted by $|M|$, is the maximum level of variables in M, or 0 if there is no variable in M. Note that we take all variables (even variable occurrences in the scope of λ) into account—for instance, $|(\lambda x^2{:}\alpha.\, x^2)y^1| = 2$. Note also that $|M|$ is not necessarily equal to the level of its arity. The *level of a hypothesis sequence* Γ, denoted by $|\Gamma|$, is the maximum level of variables in Γ, or 0 if Γ is the empty sequence.

3.2 α-Equivalence

In the calculus $\lambda\mathcal{M}$, we need special care to define α-equivalence: occurrences of a variable x in the scope of $\lambda x.$ – usually called bound occurrences – may or may not be subject to renaming since textual substitution does not commute with naive variable renaming. For instance, we may identify $\lambda x^1.\, y^2(\lambda z^1.\, x^1 z^1)$ with $\lambda x^1.\, y^2(\lambda u^1.\, x^1 u^1)$, but not with $\lambda w^1.\, y^2(\lambda z^1.\, w^1 z^1)$. To see its reason, let us substitute x^1 for y^2 in these terms. Since the level of y^2 is higher than x^1 and w^1, the textual substitution is used, and the first and the third terms become $\lambda x^1.\, x^1(\lambda z^1.\, x^1 z^1)$, and $\lambda w^1.\, x^1(\lambda z^1.\, w^1 z^1)$, resp., which do not have the same denotational meaning. Hence, we let an abstraction $\lambda x^i.\, e$ be α-convertible only when no variable at a level higher than i occurs in its scope.

We define the α-equivalence after a few auxiliary definitions. A *renaming* of variables is a partial function from V to V which is injective and satisfies $|f(x^i)| = i$ for all x^i in the domain of the partial function. For a partial function f, its domain is denoted by $\mathrm{dom}(f)$. For a renaming f and a variable x^i (which may not be in $\mathrm{dom}(f)$), $f \downarrow x^i$ is a renaming of variables such that $\mathrm{dom}(f \downarrow x^i) = \mathrm{dom}(f) - \{x^i\}$, and $f \downarrow x^i$ agrees with f on its domain.

For a renaming f and terms M and N in $\lambda\mathcal{M}$, we derive a judgment of the form $f \vdash M \simeq N$ by the following inference rules:

$$\frac{f(x^i) = y^i}{f \vdash x^i \simeq y^i} \qquad \frac{f \vdash M \simeq M' \quad f \vdash N \simeq N'}{f \vdash MN \simeq M'N'}$$

$$\frac{f \vdash M \simeq M'}{f \vdash \lambda x^i : \alpha.\, M \simeq \lambda x^i : \alpha.\, M'}\ (f(x^i) = x^i)$$

$$\frac{f \vdash M \simeq M'}{g \vdash \lambda x^i : \alpha.\, M \simeq \lambda y^i.\, \alpha.M'} \left(\begin{array}{c} |M| \le i,\ |M'| \le i, \\ f(x^i) = y^i,\ g \downarrow x^i = f \downarrow x^i \end{array} \right)$$

The last rule can be applied only when both M and M' are of level less than or equal to i. Otherwise, the term $\lambda x^i. M$ contains (not necessarily free) occurrences of meta-variables which have higher levels than i, and we cannot rename the bound variable x^i. In this case we can still apply the second last rule, since it does not rename the bound variable x^i. For brevity, we omit the inference rules for terms constructed by the rules (const0), (abs0), and (pair0), which are similarly defined. For instance, a term $(x^0)\,\texttt{[}M\texttt{]}$ has the same rules as the term $\lambda x^i : \alpha. M$.

Let $\texttt{id}$ be the identity function on V. If $\texttt{id} \vdash M \simeq N$ is derived by the rules above, we say M *is α-equivalent to* N (written by $M \equiv_\alpha N$). It is easy to show the relation $\equiv_\alpha$ is a congruence on terms.

3.3 Substitution and Reduction

The notion of reduction in the calculus $\lambda\mathcal{M}$ is the union of those in the object language (which we do not specify) and the following β-reduction:

$$(\beta)\ (\lambda x^i. M)N \to [x^i := N]M \qquad \text{if } |M| \leq i \text{ and } |N| \leq i$$

in which $[x^i := N]M$ denotes the (non-standard) substitution defined below. We write $\xrightarrow{*}$ for the reflexive and transitive closure of $\to$.

For a level $i > 0$, a level-i variable x^i, and terms M and N such that $|M| \leq i$ and $|N| \leq i$, we define $[x^i := N]M$ as follows:

1. $[x^i := N]x^i \triangleq N$
2. $[x^i := N]y^j \triangleq y^j \qquad \text{if } y^j \not\equiv x^i$
3. $[x^i := N](M_1 M_2) \triangleq ([x^i := N]M_1)([x^i := N]M_2)$
4. $[x^i := N](\lambda y^j. M) \triangleq \lambda y^j. [x^i := N]M \quad \text{if } j < i$
5. $[x^i := N](\lambda y^i. M) \triangleq \lambda y^i. [x^i := N]M \quad \text{if } y^i \notin \mathrm{FV}(N) \text{ and } x^i \not\equiv y^i$
6. $[x^i := N]c \triangleq c$
7. $[x^i := N](y^0)\,\texttt{[}M\texttt{]} \triangleq (y^0)\,\texttt{[}[x^i := N]M\texttt{]}$
8. $[x^i := N]\langle M_1, M_2\rangle \triangleq \langle [x^i := N]M_1, [x^i := N]M_2\rangle$

The first three clauses are standard. For the fourth line, if the level of y^j is strictly less than that of x^i, the substitution behaves like textual replacement, that is, free variables may get captured through this substitution. For the fifth line, the bound variable y^i has the same level as x^i that is being substituted, in which case the substitution is the standard capture-avoiding one, hence the side-conditions $y^i \notin \mathrm{FV}(N)$ and $x^i \not\equiv y^i$ must be satisfied. (The second side-condition $x \not\equiv y$ is not needed for the fourth clause because variables at different levels are assumed to be different.) This side-condition can be always satisfied by taking an α-equivalent term. The last three clauses deal with the level 0 in a straightforward manner, except textual substitution.

For brevity, we identify α-equivalent terms in the following. Under this convention, we simply have that $[x^i := N]M$ is defined if and only if $\max(|M|, |N|) \leq i$.

The object-level may contain other redices than the β-redices in the above form, and to distinguish these two, we say that a redex $(\lambda x^i : \alpha.\, M)N$ is a *meta-redex*. A term is *meta-normal* if it does not have meta-redices and a reduction is called a *meta-reduction* if its redex is a meta-redex.

3.4 Replacing Level-0 Languages

Our level-0 language here is a simplest possible language, which has no notion of types or even computation. We have adopted such a language because it is simple but expressible enough to represent expressions that appear in typical logical systems. It would be possible, however, to adopt other languages such as untyped and simply typed λ-calculi as the level-0 language.

For example, for untyped λ-calculus, we could introduce another term constructor $M_1 \cdot M_2$ by the rule

$$\frac{\Gamma \vdash M : 0 \quad \Gamma \vdash N : 0}{\Gamma \vdash M \cdot N : 0} \; (\mathsf{app0})$$

and a reduction rule $(x^0)\,[M] \cdot N \to [x^0 := N]M$ (if $|M| = |N| = 0$) where the level-0 substitution would be defined as expected. It would be also straightforward to substitute the simply typed λ-calculus for the level-0 language, by extending the base arities from 0 to the set of simple types.

4 The Calculus λm

As discussed in the previous sections, the second calculus λm is based on the second observation discussed in the introduction. It is obtained from $\lambda\mathcal{M}$ by replacing the definition of substitution with the standard capture-avoiding one. Also, we need to use the standard definition of α-equivalence to identify $\lambda x^1.y^2(\lambda z^1.x^1 z^1)$ with $\lambda w^1.y^2(\lambda z^1.w^1 z^1)$, which are not α-equivalent in $\lambda\mathcal{M}$. Since the other definitions of arities, rules to derive judgments, β-reduction remain the same, we avoid repeating definitions and just show changes to be made.

For α-equivalence, we replace the third and fourth rules to derive $f \vdash M \simeq N$ with the following one:

$$\frac{f \vdash M \simeq M'}{g \vdash \lambda x^i : \alpha.M \simeq \lambda y^i : \alpha.M'} \; (f(x^i) = y^i,\, g \downarrow x^i = f \downarrow x^i)$$

Notice that there is no restriction on the levels of the bodies M and M' of λ-abstraction. Similarly, the definition of substitution will be as follows.

For a level $i > 0$, a level-i variable x^i, and terms M and N such that $|M| \leq i$ and $|N| \leq i$, $[x^i := N]M$ is defined by:

1. $[x^i := N]x^i \triangleq N$
2. $[x^i := N]y^j \triangleq y^j \quad$ if $y^j \not\equiv x^i$
3. $[x^i := N](M_1 M_2) \triangleq ([x^i := N]M_1)([x^i := N]M_2)$
4. $[x^i := N](\lambda y^j.\, M) \triangleq \lambda y^j.\, [x^i := N]M \quad$ if $y^j \notin \mathrm{FV}(N)$ and $x^i \not\equiv y^j$
5. $[x^i := N]c \triangleq c$

6. $[x^i := N](y^0)\,[M] \triangleq (y^0)\,[[x^i := N]M] \quad \text{if } y^0 \notin \mathrm{FV}(N)$
7. $[x^i := N]\langle M_1, M_2\rangle \triangleq \langle [x^i := N]M_1, [x^i := N]M_2\rangle$

Notice that the first three clauses are the same as before and the fourth clause is now a familiar one that avoids variable capturing.

5 Examples

In this section, we show a few examples of $\lambda\mathcal{M}$ and λm.

5.1 Representing Formulas

In earlier sections, we informally explained how the formula $\mathrm{x} = \mathrm{x}$ is substituted for the meta-variable A in $\exists \mathrm{x}\ A$. Here we consider this process formally in $\lambda\mathcal{M}$.

The arity of the corresponding term is inferred in $\lambda\mathcal{M}$ as follows:

$$\dfrac{\dfrac{\dfrac{\dfrac{}{x^0:0, A^1:0 \vdash {}'\exists' : 0} \quad \dfrac{\dfrac{}{x^0:0, A^1:0, x^0:0 \vdash A^1 : 0}}{x^0:0, A^1:0 \vdash (x^0)\,[A^1] : 0}}{x^0:0, A^1:0 \vdash \langle {}'\exists', (x^0)\,[A^1]\rangle : 0}}{x^0:0 \vdash \lambda A^1:0.\,\langle {}'\exists', (x^0)\,[A^1]\rangle : 0 \to_1 0} \qquad \dfrac{\vdots}{x^0:0 \vdash \langle {}'=', \langle x^0, x^0\rangle\rangle : 0}}{x^0:0 \vdash (\lambda A^1:0.\,\langle {}'\exists', (x^0)\,[A^1]\rangle)(\langle {}'=', \langle x^0, x^0\rangle\rangle) : 0}$$

Note that, the variable x^0 occurs free in the term of the conclusion, as indicated by the hypothesis sequence.

We can compute this term as:

$$\begin{aligned}
&(\lambda A^1:0.\,\langle {}'\exists', (x^0)\,[A^1]\rangle)(\langle {}'=', \langle x^0, x^0\rangle\rangle)\\
\to\ &[A^1 := \langle {}'=', \langle x^0, x^0\rangle\rangle]\langle {}'\exists', (x^0)\,[A^1]\rangle\\
\equiv\ &\langle {}'\exists', [A^1 := \langle {}'=', \langle x^0, x^0\rangle\rangle](x^0)\,[A^1]\rangle\\
\equiv\ &\langle {}'\exists', (x^0)\,[\langle {}'=', \langle x^0, x^0\rangle\rangle]\rangle
\end{aligned}$$

Note that non-standard (textual) substitution is applied in the last step, since the level of A^1 is higher than that of x^0. As a result, the free occurrences of x^0 get bound, and x^0 does not occur free in the resulting term, which represents the formula $\exists \mathrm{x}\ \mathrm{x} = \mathrm{x}$.

We also informally explained how the same example can be written in λm. Its formal counterpart can be written as follows[3]:

$$\dfrac{\dfrac{\dfrac{\dfrac{}{A^2:00 \vdash {}'\exists' : 0} \quad \dfrac{\dfrac{\dfrac{}{\Gamma \vdash A^2 : 00} \quad \dfrac{}{\Gamma \vdash x^0 : 0}}{\Gamma \vdash A^2(x^0) : 0}}{A^2:00 \vdash (x^0)\,[A^2(x^0)] : 0}}{A^2:00 \vdash \langle {}'\exists', (x^0)\,[A^2(x^0)]\rangle : 0}}{\vdash \lambda A^2:00.\,\langle {}'\exists', (x^0)\,[A^2(x^0)]\rangle : 00 \to_2 0} \qquad \dfrac{\dfrac{\vdots}{y^1:0 \vdash \langle {}'=', \langle y^1, y^1\rangle\rangle : 0}}{\vdash \lambda y^1:0.\,\langle {}'=', \langle y^1, y^1\rangle\rangle : 00}}{\vdash (\lambda A^2:00.\,\langle {}'\exists', (x^0)\,[A^2(x^0)]\rangle)(\lambda y^1:0.\,\langle {}'=', \langle y^1, y^1\rangle\rangle) : 0}$$

[3] We changed the level-1 variable x^1 to y^1 for readability. Formally this renaming is justified by the α-equivalence.

where we put an arity $00 = 0 \to_1 0$ and $\Gamma = A^2 : 0 \to_1 0, x^0 : 0$ Note that we replaced the meta-variable A^1 in $\lambda\mathcal{M}$ by an application term $A(x)$, and to ensure this application is resolved in a meta-level, the arity of A should be $0 \to_1 0$, hence the level of the variable A must be 2 (or higher).

We omit the computation of the term here, since it is essentially the same as the standard β-reduction.

5.2 Representing Contexts

Contexts can be represented in $\lambda\mathcal{M}$ using meta-variables naturally. Let M be a term, C be a context in the object language, i.e., a term with a hole $[\,]$ in it, and $C[M]$ be the result of the hole-filling operation. In $\lambda\mathcal{M}$, the context C is represented as a term $\overline{C} \equiv (\lambda X^1{:}0.\, C[X^1])$, and $C[M]$ as an application $(\overline{C}M)$, which reduces to $C[M]$ by the textual substitution in $\lambda\mathcal{M}$.

Let us take an example from Hashimoto and Ohori's context calculus [6]. Consider the context $C \equiv (\lambda u.\, (\lambda x.\, [\,])u + y)3$ and the term $M \equiv (\lambda z.\, C[x+z])x$ in lambda calculus. They can be written in $\lambda\mathcal{M}$ (using our notation for object language) as:

$$\overline{C} \equiv \lambda X^1{:}0.\ (u^0)\,[((x^0)\,[X^1]\cdot u^0) + y^0]\cdot 3$$
$$\overline{M} \equiv (z^0)\,[\overline{C}\ (x^0 + z^0)]\cdot x^0$$

We can reduce $\overline{M}$ as:

$$\begin{aligned}\overline{M} &\to (z^0)\,[[X^1 := x^0 + z^0]((u^0)\,[((x^0)\,[X^1]\cdot u^0) + y^0]\cdot 3)]\cdot x^0\\ &\equiv (z^0)\,[(u^0)\,[((x^0)\,[x^0 + z^0]\cdot u^0) + y^0]\cdot 3]\cdot x^0\\ &\xrightarrow{*} 3 + x^0 + y^0\end{aligned}$$

Note that, by the side-condition of the β-reduction, we cannot reduce the outermost level-0 redex first. If it would be reduced first, we would get $3+x^0+x^0$ as a result, so the Church-Rosser property would be broken. Hashimoto and Ohori's context calculus has a similar restriction that the β-reduction is prohibited when the redex contains a hole.

A good point of our representation of contexts is that, since contexts are functions, they can be *composed*, in other words, we can fill a context in another context. As a simple example, let C and D be the contexts $\lambda x.\, [\,]$ and $\lambda y.\, [\,]$ in lambda calculus, and consider hole-filling of D in C, i.e. $C[D]$. The contexts C and D are represented in $\lambda\mathcal{M}$ as $\overline{C} \equiv \lambda X^1.\, (x^0)\,[X^1]$ and $\overline{D} \equiv \lambda X^1.\, (y^0)\,[X^1]$, then we can compose them in the same way as composition of two functions:

$$\overline{C}\circ\overline{D} \equiv \lambda X^1.\overline{C}\ (\overline{D}\ X^1)$$

$\overline{C}\circ\overline{D}$ reduces to $\lambda X^1.\, (x^0)\,[(y^0)\,[X^1]]$ which represents the context $\lambda x.\, \lambda y.\, [\,]$. It should be noted that, in several existing context calculi including Hashimoto-Ohori's and our previous work [11], contexts cannot be composed, since these calculi keep track of possible bound variables in a hole.

6 Properties of $\lambda\mathcal{M}$ and λm

We have a number of desirable properties for the calculi $\lambda\mathcal{M}$ and λm, that is, they enjoy subject reduction, confluence, and strong normalization properties.

In the following, we focus on the properties of $\lambda\mathcal{M}$, but the modification for λm is straightforward.

We can prove the subject reduction property in the standard way.

Theorem 1 (Subject Reduction Property). *If $\Gamma \vdash M : \alpha$ is derived and $M \xrightarrow{*} N$, then $\Gamma \vdash N : \alpha$ can be derived.*

Note that even if $\Gamma \vdash (\lambda x^i. M)N : \alpha$ is derived, $[x^i := N]M$ is not necessarily defined. But we can prove from Theorem 2 and 3 that if $|\Gamma| \leq i$ then there exist terms M' and N' such that $M \xrightarrow{*} M'$, $N \xrightarrow{*} N'$, and $[x^i := N']M'$ is defined.

Lemma 1. *Suppose $\Gamma \vdash M : \alpha$ is derived and the highest level of the redices of M is i. Then, any reduction sequence that reduces only level-i redices leads to a term that does not have redices of level-i or higher.*

We can prove this lemma by reducing it to the strong normalizability of the simply typed λ-calculus. We translate $\lambda\mathcal{M}$ to the simply typed λ-calculus by mapping a level-i abstraction $\lambda x^i. M$ to an abstract of the simply typed λ-calculus and a level-j ($j < i$) abstraction $\lambda x^j. M$ to a pair of x^j and M. Since we prove a stronger property in Theorem 6, we omit the details here.

Note that, we cannot simply map λ-abstractions of $\lambda\mathcal{M}$ to those of the simply typed λ-calculus, since the textual substitution cannot be simulated by the capture-avoiding substitution in the standard calculi.

Theorem 2. *If $\Gamma \vdash M : \alpha$ is derived, then M has a meta-normal form.*

Proof. By repeatedly using Lemma 1. □

Theorem 3. *If $\Gamma \vdash M : \alpha$ is derived and M is meta-normal, then $|M| \leq \max(|\Gamma|, |\alpha|)$.*

Proof. Suppose M is a meta-normal term. We prove the theorem by induction on M.

If M is $(\lambda x^i{:}\beta_1. N_0)N_1 \cdots N_n$ for some $i \geq 0, x^i, \beta_1, n \geq 0, N_0, \cdots, N_n$, then the arity of N_0 should be $\gamma \equiv \beta_2 \rightarrow_{j_2} \cdots \rightarrow_{j_n} \beta_n \rightarrow_i \alpha$ where β_k is the arity of N_k for $k = 2, \cdots, n$. Since N_0 is meta-normal, we have $|N_0| \leq \max(|\Gamma|, i, |\gamma|)$ by induction hypothesis, and $|\gamma| \leq i$ by the side-condition of rule (abs). Therefore, we have $|\lambda x^i{:}\beta_1. N_0| = \max(i, |N_0|) \leq \max(i, |\Gamma|)$. Now, we have two cases.

1. $n = 0$. $|M| \leq \max(i, |\Gamma|) = \max(|\beta_1 \rightarrow_i \gamma|, |\Gamma|)$.
2. $n > 0$. We have, by induction hypothesis, $|N_k| \leq \max(|\Gamma|, |\beta_k|)$ for $1 \leq k \leq n$. Hence $|N_1| \leq \max(|\Gamma|, i)$. Since M is meta-normal, $\max(|N_0|, |N_1|) > i$. Hence $|\Gamma| > i$. Since $|\gamma| \geq |\beta_k|$, we have $|N_k| \leq |\Gamma|$ for $1 \leq k \leq n$. Hence $|M| \leq |\Gamma|$.

We omit the proof of the case where M is of the form $x^i N_1 \cdots N_n$ and we can easily prove the claim in other cases (M is a constant c, $(x^0)\,[N_0]$, or $\langle N_0, N_1 \rangle$) by induction hypothesis. □

The following theorem ensures that meta-level evaluation in $\lambda\mathcal{M}$ does not get stuck.

Theorem 4 (Normal Form Property). *Suppose* $\Gamma \vdash M : \alpha$ *is derived and* M *is meta-normal. (1) If* $|\Gamma| = |\alpha| = 0$*, then* $|M| = 0$*. (2) If* $|\Gamma| < |\alpha|$*, then* M *is a* λ*-abstraction.*

Proof. (1) Clear from Theorem 3. (2) If M is not a λ-abstraction, then it must be of the form $x^i N_1 N_2 \cdots N_n$ or $(\lambda x^i{:}\gamma.\, N_0)N_1 N_2 \cdots N_n$ for $n \geq 0$. In the former case, $|\alpha| \leq |x^i| \leq |\Gamma|$. Contradiction. In the latter case, let β be the arity of N_0. From Theorem 3, $|(\lambda x^i : \gamma.N_0)N_1| \leq |\beta|$. By the side-condition of rule (abs), $|\beta| \leq i$. Hence $|N_0| \leq i$ and $|N_1| \leq i$, which implies $(\lambda x^i{:}\gamma.\, N_0)N_1$ is a redex. Contradiction. □

We can prove the confluence and strong normalizability of $\lambda\mathcal{M}$. The confluence can be proved using the standard technique (parallel reduction), but the strong normalizability is not trivial. Since space is limited, we just give the idea of the proof here. For the detailed proof, see the full version of this paper.

Lemma 2 (Substitution Lemma). *Let* x^i *and* y^i *be distinct variables,* $|M_0| \leq i$*,* $|M_1| \leq i$*, and* $|N| \leq i$*. Then we have*

$$[x^i := N][y^i := M_1]M_0 \equiv [y^i := [x^i := N]M_1][x^i := N]M_0$$

Theorem 5 (Confluence). *The meta-reduction is confluent.*

Theorem 6 (Strong Normalizability). *If* $\Gamma \vdash M : \alpha$ *is derived, then* M *is strongly normalizable with respect to the meta-reduction.*

Proof. (idea) We can prove this theorem by using the reducibility method, but we cannot follow the standard way because substitution operations can be applied only when the level restriction is satisfied. In other words, we cannot express the lemma, which is usually claimed in proof by the reducibility method, that

> If $\Gamma \vdash M : \alpha$ is derived, then a term that is obtained by substituting reducible terms for free variables in M is reducible.

To deal with this difficulty, we extend $\lambda\mathcal{M}$ so that it has 'postponed substitutions', that is, we define an *extended judgment* of the form $\Gamma_1, x^i := N : \alpha, \Gamma_2 \vdash M$ and give a reduction rule that substitutes N for x in Γ_2 or M under some conditions. Then, we define reducibility sets that consist of extended judgments and prove lemmas similar to the ones in the case of simply typed lambda calculus. □

We remark that, for many object languages such as the simply typed lambda calculus, these theorems still hold if we replace the meta-reduction by the union of the meta-reduction and the reductions in the object language.

7 Related Work and Conclusion

We have proposed two formal systems $\lambda\mathcal{M}$ and λm that formalize the notion of meta-variable, motivated by the observations how meta-variables are used in textbooks of logic.

Edinburgh Logical Framework (LF) [5] gives a typed framework by which we can define various logical systems as object calculi. Unlike LF and its descendants, we clearly distinguish the meta-levels from the object-level. The textual substitution in $\lambda\mathcal{M}$ is another characteristic feature.

Geuvers and Jogjov [3] have introduced the notion of meta-variable into the proof assistant system, so that they can describe open proofs. Their motivation is similar to ours, so they have also encountered the problem that free variables are captured when a meta-variable is instantiated. Their solution to this problem is similar to the one in our work on the calculus of context [11].

Recently much effort has been devoted to formalize the notion of context in lambda calculi, and various calculi of context have been proposed. In the work of Talcott [13] and Mason [8], the notion of contexts are formalized outside of the object language, because contexts are meta-level objects in nature and should be characterized independently of the object language. In other work such as Dami [1], Hashimoto-Ohori [6], Sands [9], and Sato-Sakurai-Kameyama [11], contexts are built into the system, so that context manipulations and object-level calculations can be carried out in the same framework. By representing contexts in $\lambda\mathcal{M}$, we have integrated the two approaches.

It turns out that λm is similar to the calculi for binding-time analysis in off-line partial evaluation with multiple computation stages [4, 2]. In those calculi, types are stratified like ours but, there, levels represent binding time—i.e., when certain expression can be computed. Reduction is similar to ours in that the order of computation is also determined by levels. Aside from the base language (for which they use typed or untyped λ-calculus), one subtle difference is that, in those calculi, variables can range over expressions of the *same* level, resulting in a relaxed condition on function types: $\alpha \to_i \beta$ is a type if $|\alpha| \leq i$ (not $|\alpha| < i$), and $|\beta| \leq i$.

Davies [2] also presents a reformulation $\lambda^{\bigcirc}$ of the calculus for binding-time analysis and shows that it corresponds to a proof system of linear-time temporal logic with a modal operator. Actually, our earlier attempt to formalize meta-variables [14] was done in a close style. In these calculi, terms are annotated with information that indicates levels of subexpressions, hence λm is a simpler calculus to formalize meta-variables.

As far as we know, the calculus $\lambda\mathcal{M}$ is the first one which formalizes non-standard (textual) substitution and is confluent and strongly normalizing. We believe that our calculi can be a basis of formalizing logical and computation systems naturally and directly, but their further development is left for future work.

Acknowledgements

We thank anonymous referees for helpful comments. The first author thanks Henk Barendregt and Jan Willem Klop for having discussions on meta-variables with him. This work is supported in part by Grant-in-Aid for Scientific Research on Priority Areas Research No. 15017247 (Sato, Sakurai) and Grant-in-Aid for Young Scientists (B) No. 15700011 (Igarashi), from MEXT of Japan.

References

1. L. Dami. A Lambda-Calculus for Dynamic Binding. *Theoretical Computer Science*, 192:201–231, 1998.
2. R. Davies. A Temporal-Logic Approach to Binding-Time Analysis. In *11th Annual IEEE Symposium on Logic in Computer Science (LICS'96)*, pages 184–195, 1996.
3. H. Geuvers and G. Jojgov. Open Proofs and Open Terms: A Basis for Interactive Logic. In *CSL 2002*, LNCS 2471, pages 537–552, 2002.
4. R. Glück and J. Jørgensen. Efficient multi-level generating extensions for program specialization. In *Programming Languages, Implementations, Logics and Programs (PLILP'95)*, LNCS 982, pages 259–278, 1995.
5. R. Harper, F. Honsell, and G. Plotkin. A Framework for Defining Logics. *Journal of the Association for Computing Machinery*, 40(1):143–194, 1993.
6. M. Hashimoto and A. Ohori. A Typed Context Calculus. *Theoretical Computer Science*, 266(1-2):249–272, 2001.
7. S. C. Kleene. *Introduction to Metamathematics*. North-Holland, 1952.
8. I. Mason. Computing with Contexts. *Higher-Order and Symbolic Computation*, 12:171–201, 1999.
9. D. Sands. Computing with Contexts - a Simple Approach. *Electronic Notes in Theoretical Computer Science*, 10, 1998.
10. M. Sato. Theory of Judgments and Derivations. In *Discovery Science*, LNAI 2281, pages 78–122, 2001.
11. M. Sato, T. Sakurai, and Y. Kameyama. A Simply Typed Context Calculus with First-Class Environments. *Journal of Functional and Logic Programming*, 2002(4):1–41, 2002.
12. J. R. Shoenfield. *Mathematical Logic*. Addison-Wesley, 1967.
13. C. Talcott. A Theory of Binding Structures and Applications to Rewriting. *Theoretical Computer Science*, 112(1):99–143, 1993.
14. K. Yamamoto, A. Okamoto, M. Sato, and A. Igarashi. A Typed Lambda Calculus with Quasi-quotation (in Japanese). In *Informal Proceedings of the 4th JSSST Workshop on Programming and Programming Languages*, pages 87–102, 2003.

Henkin Models of the Partial λ-Calculus

Lutz Schröder*

BISS, Department of Computer Science, University of Bremen

Abstract. We define (set-theoretic) notions of intensional Henkin model and syntactic λ-algebra for Moggi's partial λ-calculus. These models are shown to be equivalent to the originally described categorical models via the global element construction; the proof makes use of a previously introduced construction of classifying categories. The set-theoretic semantics thus obtained is the foundation of the higher order algebraic specification language HasCasl, which combines specification and functional programming.

Introduction

The partial λ-calculus has been introduced in [14, 15, 17] as a natural generalization of the simply typed λ-calculus that encompasses abstraction for partial functions. Partial functions generally serve to model both non-termination and irregular termination of programs. Consequently, they feature in several specification languages such as RSL [7], SPECTRUM [4], and Casl [2]. The language HasCasl [20], which offers a setting for both specification and implementation of higher order functional programs, is based on the partial λ-calculus. The work presented here forms the centerpiece of the semantical underpinnings of HasCasl.

The semantics of the partial λ-calculus as presented in [14] is given in terms of models in partial cartesian closed categories (pcccs). HasCasl is designed as an extension of the first order algebraic specification language Casl, which comes with a set-theoretic semantics; this necessitates the development of a set-theoretic semantics for the partial λ-calculus, in the spirit of Henkin's semantics for classical higher order logic [8]. The notion of model chosen for HasCasl is that of *intensional Henkin models*, where partial function types are interpreted by sets that need not contain all set-theoretic partial functions and, moreover, may contain several elements representing the same set-theoretic function. This choice avoids problems such as incompleteness of deduction or non-existence of initial models, and moreover allows a more detailed analysis of the function space [20].

The central result presented here states that pccc models and intensional Henkin models are in a suitable sense equivalent, in accordance with promises made in [20]. This result generalizes (and elucidates) a corresponding statement made in [3] for the *total* λ-calculus. One direction of the equivalence works by

* Research supported by the DFG project HasCASL (KR 1191/7-1)

M. Baaz and J.A. Makowsky (Eds.): CSL 2003, LNCS 2803, pp. 498–512, 2003.

passing from a pccc model to the Henkin model determined by its global elements (i.e. the elements of a type T in the Henkin model are the morphisms $1 \to T$ in the pccc model); it is non-trivial to prove that this process can indeed be reversed.

The required background w.r.t. the partial λ-calculus and ppcccs is recalled in Sections 1 and 2; Section 2 also introduces the novel notion of pre-pccc. In Section 3, results on classifying categories proved in [19, 18] are summarized and adapted to deal with pre-ppcccs. The main novel results are presented in Sections 4 and 5. In Section 4, Henkin models are defined as first-order set-valued functors and shown to be equivalent to a notion of *syntactic λ-algebra* that generalizes the corresponding definition for the total case given in [3]. Moreover, we define internal languages of Henkin models, which play a crucial role in the proof of the equivalence with pccc-models carried out in Section 5. As general references for categorical concepts, we recommend [1, 12].

1 The Partial λ-Calculus

The natural generalization of the simply typed λ-calculus to the setting of partial functions is the partial λ-calculus as introduced in [14, 15, 17]. The basic idea is that function types are replaced by partial function types, and λ-abstractions denote partial functions instead of total ones. Due to the fact that terms need not denote, equational reasoning requires some care. We will focus on *existential* equations here, to be read 'both sides are defined and equal'; this will be discussed further below (Remark 8).

A *signature* consists of a set of *sorts* and a set of partial *operators* with given *profiles* (or arities) written $f : \bar{s} \to t$, where t is a *type* and $\bar{s}$ is a *multi-type*, i.e. a (possibly empty) list of types. A type is either a sort or a *partial function type*

$$\bar{s} \rightharpoonup t,$$

with $\bar{s}$ and t as above (one cannot resort to currying for multi-argument partial functions [14]). Following [14], we assume application operators in the signature, so that application does not require extra typing or deduction rules. For $\bar{t} = (t_1, \ldots, t_m)$, $\bar{s} \rightharpoonup \bar{t}$ denotes the multi-type $(\bar{s} \rightharpoonup t_1, \ldots, \bar{s} \rightharpoonup t_m)$, not to be confused with the (non-existent) 'type' $\bar{s} \rightharpoonup t_1 \times \cdots \times t_m$. A *morphism* between two signatures is a pair of maps between the corresponding sets of sorts and operators, respectively, that is compatible with operator profiles.

A signature gives rise to a notion of typed terms in context according to the typing rules given in Figure 1, where a context Γ is a list $(x_1 : s_1, \ldots, x_n : s_n)$, shortly $(\bar{x} : \bar{s})$, of type assignments for distinct variables. More precisely, we speak simultaneously about terms and *multi-terms*, i.e. lists of terms. The judgement $\Gamma \rhd \alpha : t$ reads '(multi-)term α has (multi-)type t in context Γ'. The empty multi-term (), also denoted $*$, doubles as a term of 'type' (), also denoted 1. When convenient, we use a context to denote the associated multi-type, as e.g. in $\Gamma \rightharpoonup t$; moreover, we write λ-abstraction in the form $\lambda\Gamma.\, \alpha$ where suitable.

$$\frac{x : s \text{ in } \Gamma}{\Gamma \rhd x : s} \qquad \frac{\Gamma \rhd \alpha : \bar{t} \quad f : \bar{t} \to u}{\Gamma \rhd f(\alpha) : u} \qquad \frac{\Gamma, \Delta \rhd \alpha : u}{\Gamma \rhd \lambda \Delta .\, \alpha : \Delta \twoheadrightarrow u}$$

Fig. 1. Typing rules for the partial λ-calculus

A *partial λ-theory* $\mathcal{T}$ is a signature Σ together with a set $\mathcal{A}$ of *axioms* that take the form of existentially conditioned equations: an (existential) *equation* $\alpha_1 \stackrel{e}{=} \alpha_2$ is read 'α_1 and α_2 are defined and equal'. Equations $\alpha \stackrel{e}{=} \alpha$ are abbreviated as $\operatorname{def} \alpha$ and called *definedness judgements.* An *existentially conditioned equation (ECE)* is a sentence of the form $\operatorname{def} \alpha \Rightarrow_\Gamma \phi$, where α is a multi-term and ϕ is an equation in context Γ, to be read 'ϕ holds on the domain of α'. By equations between multi-terms, we can express conjunction of equations (e.g. $\operatorname{def}(\alpha, \beta) \equiv \operatorname{def} \alpha \wedge \operatorname{def} \beta$); true will denote $\operatorname{def} *$.

In Figure 2, we present a set of proof rules for existential equality in a partial λ-theory. The rules are parametrized over a fixed context Γ. We write $\operatorname{def} \alpha \vdash_\Gamma \phi$ if an equation ϕ can be deduced from $\operatorname{def} \alpha$ in context Γ by means of these rules; in this case, $\operatorname{def} \alpha \Rightarrow_\Gamma \phi$ is a *theorem.* The rules are essentially a version of the calculus presented in [14], adapted for existential (rather than strong) equations. Of course, there is no reflexive law, since $\alpha \stackrel{e}{=} \alpha$ is false if α is undefined. For conciseness, subderivations are denoted in the form $\operatorname{def} \alpha \vdash_\Delta \phi$, where the context Δ and the assumption $\operatorname{def} \alpha$ are to be understood as *extending* the ambient context and assumptions. *Strong equations* $\Delta \rhd \alpha \stackrel{s}{=} \beta$, or just $\alpha \stackrel{s}{=} \beta$, are abbreviations for '$\operatorname{def} \alpha \vdash_\Delta \operatorname{def} \beta$ and $\operatorname{def} \beta \vdash_\Delta \alpha \stackrel{e}{=} \beta$'; in particular, rule ($\beta$) is really two rules. Rule (ξ) implies that all λ-terms are defined.

The higher order rules (ξ) and (β) show a slight preference for strong equations. Note, however, that the usual form of the η-equation, $\lambda \bar{y} : \bar{t}.\, \alpha(\bar{y}) = \alpha$, is an ECE, not a strong equation.

A *translation* between partial λ-theories is a signature morphism which transforms axioms into theorems.

Remark 1. An essential feature of partial equational logic are *conditioned terms* $\alpha \restriction \beta$, which denote the restriction of a multi-term α to the domain of a multi-term β [5, 15]. In our setting, conditioned terms can be coded using projection operators. Conditioned terms, in turn, provide a coding for λ-abstraction of multi-terms.

Remark 2. A notion of *predicates* is provided in the shape of terms $\Gamma \rhd \alpha : 1$, for which we write α in place of $\operatorname{def} \alpha$. The sentence $\operatorname{def} \beta$ can be coded as the predicate $(\lambda x.\, *)(\beta)$.

The expressive power of ECEs is greatly increased in the presence of an equality predicate (see also [6, 14]):

$$\textbf{(var)}\ \frac{x : s \text{ in } \Gamma}{\operatorname{def} x} \quad \textbf{(st)}\ \frac{\operatorname{def} f(\alpha)}{\operatorname{def} \alpha} \quad \textbf{(unit)}\ \frac{x : 1 \text{ in } \Gamma}{x \stackrel{e}{=} *} \quad \textbf{(sym)}\ \frac{\alpha \stackrel{e}{=} \beta}{\beta \stackrel{e}{=} \alpha}$$

$$\textbf{(tr)}\ \frac{\alpha \stackrel{e}{=} \beta \quad \beta \stackrel{e}{=} \gamma}{\alpha \stackrel{e}{=} \gamma} \quad \textbf{(cg)}\ \frac{\alpha \stackrel{e}{=} \beta \quad \operatorname{def} f(\alpha)}{f(\alpha) \stackrel{e}{=} f(\beta)} \quad \textbf{(ax)}\ \frac{\operatorname{def} \alpha \Rightarrow_{\bar{y}:\bar{t}} \phi \in \mathcal{A} \quad \bar{y} : \bar{t} \text{ in } \Gamma \quad \operatorname{def} \alpha}{\phi} \quad \textbf{(sub)}\ \frac{\operatorname{def} \alpha \vdash_{\bar{y}:\bar{t}} \phi \quad \operatorname{def}(\beta, \alpha[\bar{y}/\beta])}{\phi[\bar{y}/\beta]}$$

$$(\eta)\ \frac{x : \bar{t} \rightharpoonup u \text{ in } \Gamma}{\lambda \bar{y} : \bar{t}.\, x(\bar{y}) \stackrel{e}{=} x} \quad (\beta)\ \frac{\bar{y} : \bar{t} \text{ in } \Gamma}{(\lambda \bar{y} : \bar{t}.\, \alpha)(\bar{y}) \stackrel{s}{=} \alpha} \quad (\xi)\ \frac{\Delta \rhd \alpha \stackrel{s}{=} \beta}{\lambda \Delta.\, \alpha \stackrel{e}{=} \lambda \Delta.\, \beta}$$

Fig. 2. Deduction rules for existential equality in context Γ

Definition 3. A partial λ-theory has *internal equality* if there exists, for each type s, a binary predicate (cf. Remark 2) eq_s such that

$$eq_s(x, y) \Rightarrow_{x,y:s} x \stackrel{e}{=} y \quad \text{and} \quad \mathsf{true} \Rightarrow_{x:s} eq_s(x, x)$$

Such a predicate allows coding conditional equations as ECEs. In combination with λ-abstraction, it gives rise to a full-fledged intuitionistic logic [6, 11, 20].

2 Partial Cartesian Closed Categories

As suggested in [14], the categorical correlate for partial λ-theories are partial cartesian closed categories (pcccs). The crucial feature of a pccc is that, in analogy to (total) morphisms in a cartesian closed category, partial morphisms from $A \times B$ to C are in one-to-one correspondence to total morphisms from A into a partial function space $B \rightharpoonup C$. Here, a partial morphism is taken to be a span

$$\begin{array}{ccc} \bullet & \xrightarrow{f} & B \\ {\scriptstyle m}\downarrow & & \\ A & & \end{array},$$

where m is a monomorphism (of a restricted class) to be thought of as the domain of definition. Alternative approaches to the concept of 'partial maps' in categories regard partial maps as morphisms in their own right; see [6, 16] for in-depth comparisons of the two concepts. The reason for choosing the former approach here is that the transition from pccc models to Henkin models via representable functors would not work in this straightforward manner with the latter approach.

Definition 4. Following [17], we call a class of monomorphisms in a category $\mathbf{C}$ a *dominion* if it contains all identities and is closed under composition and pullbacks, i.e. pullbacks of $\mathcal{M}$-morphisms exist and are in $\mathcal{M}$. A *dominional category* is a pair $(\mathbf{C}, \mathcal{M})$, where $\mathcal{M}$ is a dominion on $\mathbf{C}$; an *admissible* subobject is an element of $\mathcal{M}$. A functor F between dominional categories is called *dominional* if it preserves admissible subobjects and their pullbacks. F is a *dominional equivalence* if it is an equivalence of categories (which, of course, already implies pullback preservation) and reflects admissible subobjects.

A dominional category $(\mathbf{A}, \mathcal{N})$ is called a *dominional subcategory* of $(\mathbf{C}, \mathcal{M})$ if $\mathbf{A}$ is a subcategory of $\mathbf{C}$ and the inclusion $\mathbf{A} \hookrightarrow \mathbf{C}$ is a dominional functor, i.e. if $\mathcal{N} \subset \mathcal{M}$, and $\mathbf{A}$ is closed under pullbacks of $\mathcal{N}$-subobjects in $\mathbf{C}$. $(\mathbf{A}, \mathcal{N})$ is called *(dominionally) full* if, moreover, $\mathbf{A}$ is full in $\mathbf{C}$, and, for $A \in \operatorname{Ob} \mathbf{A}$, each $\mathcal{M}$-subobject of A in $\mathbf{C}$ is in $\mathcal{N}$.

A dominion $\mathcal{M}$ is closed under intersections. If m is a monomorphism and $mg \in \mathcal{M}$, then $g \in \mathcal{M}$. In particular, $\mathcal{M}$ contains all isomorphisms.

A *partial morphism* is a span (m, f) as above with $m \in \mathcal{M}$. Partial morphisms (m_1, f_1) and (m_2, f_2) are regarded as equal if there exists an isomorphism h such that $f_1 h = f_2$ and $m_1 h = m_2$. The composite of (m, f) and a partial morphism (g, n) from B to C is defined as $(mf^{-1}(n), gf^*)$, where

$$\begin{array}{ccc} \bullet & \xrightarrow{f^*} & \bullet \\ {\scriptstyle f^{-1}(n)}\downarrow & & \downarrow{\scriptstyle n} \\ \bullet & \xrightarrow[f]{} & B \end{array}$$

is a pullback.The partial morphisms in $(\mathbf{C}, \mathcal{M})$ form a category $\mathbf{P}(\mathbf{C}, \mathcal{M})$ which contains $\mathbf{C}$ as a (non-full) subcategory [16, 17]. Note that a dominional subcategory $(\mathbf{A}, \mathcal{N})$ of $(\mathbf{C}, \mathcal{M})$ is full iff $\mathbf{P}(\mathbf{A}, \mathcal{N})$ is a full subcategory of $\mathbf{P}(\mathbf{C}, \mathcal{M})$.

As usual, we call a category (functor, subcategory) *cartesian* if it has (preserves, is closed under) finite products; the terminal object is denoted by 1. In a cartesian dominional category $(\mathbf{C}, \mathcal{M})$, $\mathcal{M}$ is closed under products (but not under pairing). We say that $(\mathbf{C}, \mathcal{M})$ has *internal equality* if $\mathcal{M}$ contains all diagonals $A \to A \times A$. It then follows that $\mathbf{C}$ has equalizers (hence is finitely complete, shortly: left exact or *lex*), and that $\mathcal{M}$ contains all regular monomorphisms and is closed under pairing. Cartesian dominional categories are equivalent to first order partial equational theories [19].

Definition 5. A cartesian dominional category $(\mathbf{C}, \mathcal{M})$ is called a *partial cartesian closed category (pccc)* if the composite functor

$$\mathbf{C} \xrightarrow{- \times A} \mathbf{C} \hookrightarrow \mathbf{P}(\mathbf{C}, \mathcal{M})$$

is left adjoint for each object A in $\mathbf{C}$

(This definition is slightly weaker than the one given in [14] in that we do not require finite completeness.) In a pccc we have, for each pair of objects A, B, a *partial function space* $A \rightharpoonup B$ and a co-universal partial evaluation morphism *ev* from $(A \rightharpoonup B) \times A$ to B. Explicitly, every partial morphism $C \times A \overset{m}{\hookleftarrow} \bullet \overset{f}{\longrightarrow} B$ factors uniquely as $ev \circ (\hat{f} \times A)$ in $\mathbf{P}(\mathbf{C}, \mathcal{M})$ by a total morphism $\hat{f}$ called its *abstraction.* As a special case, we have a *classifier* $\top : 1 \rightarrow \Omega$ for $\mathcal{M}$-subobjects, where $\Omega = 1 \rightharpoonup 1$; i.e. $\top \in \mathcal{M}$, and each $\mathcal{M}$-subobject m of A has a unique *characteristic map* $\chi_m : A \rightarrow \Omega$ such that

$$\begin{array}{ccc} \bullet & \xrightarrow{m} & A \\ \downarrow & & \downarrow \chi_m \\ 1 & \xrightarrow[\top]{} & \Omega \end{array}$$

is a pullback. This implies that $\mathcal{M}$ consists of regular monomorphisms, so that $\mathcal{M} = \text{RegMono}(\mathbf{C})$ in case $(\mathbf{C}, \mathcal{M})$ has internal equality.

More generally, we will need to consider the case that partial function spaces exist only for certain objects, the relevant example being the syntactic category of a partial λ-theory without internal equality (cf. Section 3):

Definition 6. A *pre-pccc* is a cartesian dominional category $(\mathbf{C}, \mathcal{M})$, equipped with a class $\mathfrak{A} \subset \text{Ob}\,\mathbf{C}$ of objects called *type objects* such that $\mathfrak{A}$ contains 1 and is closed under partial function spaces in the sense that $(A_1 \times \cdots \times A_n \rightharpoonup B)$ exists and is in $\mathfrak{A}$ for all $A_1, \ldots, A_n, B \in \mathfrak{A}$, $n \geq 0$, and such that $\mathfrak{A}$ generates $(\mathbf{C}, \mathcal{M})$ as a cartesian dominional category. Pcccs are regarded as pre-pcccs with $\mathfrak{A} = \text{Ob}\,\mathbf{C}$. A cartesian dominional functor between two pre-pcccs is called *partial cartesian closed (pcc)* if it preserves type objects and their partial function spaces and evaluation morphisms.

(Compare this to the treatment of the simply typed λ-calculus in [10].) It will turn out that, for model theoretic purposes, the class of type objects is really a property of $(\mathbf{C}, \mathcal{M})$ rather than extra structure on it (cf. Remark 18). Note that in the total case, i.e. when $\mathcal{M} = \text{Iso}(\mathbf{C})$, every pre-pccc is already a pccc, i.e. in this case a cartesian closed category.

Example 7. Every (quasi-)topos is a pccc with internal equality (but not conversely). A typical example of a pccc without internal equality is the category of cpos and continuous functions with Scott open sets [23] as admissible subobjects.

Remark 8. In a higher-order setting, existential equations and strong equations are equivalent in expressiveness (e.g., the strong equation $\alpha = \beta$ is equivalent to the existential equation $\lambda x : 1.\, \alpha \overset{e}{=} \lambda x : 1.\, \beta$). However, in the context of first order, which remains important even for higher-order languages as illustrated below, existential equations have stronger ties to the semantics (see also [19]). In particular, existential equations can, unlike strong equations, be interpreted as subobjects in any lex dominional category. Moreover, internal equality is necessarily existential. Finally, existential equations are suitable as premises in conditional equations, while implications with strong equations as premises fail to be in Horn form.

3 Internal Languages and Classifying Categories

In [18], we show that partial λ-theories and pcccs are equivalent in the same sense as established for λ-theories and cartesian closed categories by Lambek and Scott [11]. Here, we need a variant of this result that makes use of pre-pcccs instead, the benefit being that the resulting classifying categories are more easily manageable. Note that throughout the exposition, the classifying pccc of a partial λ-theory $\mathcal{T}$ is denoted by $\mathsf{Cl}(\mathcal{T})$, and the classifying pre-pccc by $\mathsf{Sy}(\mathcal{T})$.

The correspondence between categories and theories works by assigning to each pre-pccc $(\mathbf{C}, \mathcal{M}, \mathfrak{A})$ a partial λ-theory $\mathsf{L}(\mathbf{C}, \mathcal{M}, \mathfrak{A})$, its *internal language*, and, conversely, to each partial λ-theory $\mathcal{T}$ a classifying pre-pccc $\mathsf{Sy}(\mathcal{T})$, also called the *syntactic category*. In [18], $\mathsf{Cl}(\mathcal{T})$ is constructed as the *free pccc* over $\mathsf{Sy}(\mathcal{T})$; moreover, it is shown that $\mathsf{Sy}(\mathcal{T})$ already is a pccc in case $\mathcal{T}$ has internal equality, and that the two constructions induce an equivalence of partial λ-theories and pcccs, respectively, with internal equality. The latter result is easily modified to obtain an *equivalence between partial λ-theories and pre-pcccs.* In particular, models of a partial λ-theory $\mathcal{T}$ in a pre-pccc $(\mathbf{C}, \mathcal{M}, \mathfrak{A})$, i.e. translations $\mathcal{T} \to \mathsf{L}(\mathbf{C}, \mathcal{M}, \mathfrak{A})$, are in bijection with pcc functors $\mathsf{Sy}(\mathcal{T}) \to (\mathbf{C}, \mathcal{M}, \mathfrak{A})$; soundness of the deduction system of Figure 2 for such models is proved along the way. Thanks to the generatedness condition in the definition of pre-pccc, $\mathsf{Sy}(\mathsf{L}(\mathbf{C}, \mathcal{M}, \mathfrak{A}))$ is equivalent to $(\mathbf{C}, \mathcal{M}, \mathfrak{A})$; we will thus liberally identify the two categories when reasoning about $(\mathbf{C}, \mathcal{M}, \mathfrak{A})$. The relevant constructions are summarized below; the proofs are analogous to the ones in [18].

To begin, we define, for a given pre-pccc $(\mathbf{C}, \mathcal{M}, \mathfrak{A})$, the signature Σ of $\mathsf{L}(\mathbf{C}, \mathcal{M}, \mathfrak{A})$. The sorts in Σ are just the objects in $\mathfrak{A}$. We recursively define an interpretation $[\![\_]\!]$ in $\mathbf{C}$ for (multi-)types using the pre-pccc structure. The partial operators of profile $\bar{s} \to t$ in Σ are the partial morphisms from $[\![\bar{s}]\!]$ to $[\![t]\!]$ in $(\mathbf{C}, \mathcal{M})$, with evaluation morphisms (cf. Section 2) as application operators.

The interpretation $[\![\_]\!]$ is then extended to contexts, terms, multi-terms, and definedness jugdements: for a context $\Gamma = (\bar{x} : \bar{s})$, $[\![\Gamma]\!] = [\![\bar{s}]\!]$. Given a term or multi-term $\Gamma \rhd \alpha : \bar{t}$, we define a partial morphism

$$\begin{array}{ccc} [\![\Gamma.\ \mathrm{def}\,\alpha]\!] & \xrightarrow{[\![\Gamma.\,\alpha]\!]} & [\![\bar{t}]\!] \\ \downarrow & & \\ [\![\Gamma]\!] & & \end{array}$$

by recursion over the term structure, using projections, partial composition, intersection/tupling, and abstractions, respectively. We will denote any existing domain-codomain restrictions of $[\![\Gamma.\,\alpha]\!]$ by $[\![\Gamma.\,\alpha]\!]$ as well.

This interpretation leads to a notion of satisfaction in $(\mathbf{C}, \mathcal{M})$:

Definition 9. An ECE $\phi \Rightarrow_\Gamma \alpha \overset{e}{=} \beta$ in Σ *holds* in $(\mathbf{C}, \mathcal{M})$ if the subobject $[\![\Gamma.\,\phi]\!]$ is contained in $[\![\Gamma.\ \mathrm{def}(\alpha, \beta)]\!]$ and the morphisms $[\![\Gamma.\,\alpha]\!]$ and $[\![\Gamma.\,\beta]\!]$ coincide when restricted to $[\![\Gamma.\,\phi]\!]$.

This allows us to complete the definition of $\mathsf{L}(\mathbf{C}, \mathcal{M}, \mathfrak{A})$: the axioms of $\mathsf{L}(\mathbf{C}, \mathcal{M}, \mathfrak{A})$ are the ECEs that hold in $(\mathbf{C}, \mathcal{M})$.

Conversely, given a partial λ-theory $\mathcal{T}$, the syntactic category $\mathsf{Sy}(\mathcal{T})$ has objects $(\Gamma.\,\phi)$, where ϕ is a definedness judgement in context Γ (with $(\Gamma.\,\mathsf{true})$ abbreviated as (Γ)); morphisms $(\Gamma.\,\phi) \to (\Delta.\,\psi)$, where $\Gamma = (\bar{x} : \bar{s})$ and $\Delta = (\bar{y} : \bar{t})$, are multi-terms $\Gamma \rhd \alpha : \bar{t}$ such that

$$\phi \vdash_\Gamma \psi[\bar{y}/\alpha] \wedge \operatorname{def} \alpha,$$

taken modulo equality deducible from ϕ. The identity on $(\Gamma.\,\phi)$ is represented by $\bar{x}$; composition is substitution. A subobject in $\mathsf{Sy}(\mathcal{T})$ is admissible iff it has a representative of the form

$$\bar{x} : (\Gamma.\,\phi) \hookrightarrow (\Gamma.\,\psi).$$

Finally, the class $\mathfrak{A}$ of type objects in $\mathsf{Sy}(\mathcal{T})$ consists of all objects of the form $(x : t)$. In the absence of internal equality, objects other than the type objects $(x : t)$ need not have partial function spaces, so that the pre-pccc $\mathsf{Sy}(\mathcal{T})$ may fail to be a pccc.

4 Henkin Models and Their Internal Language

In [3], it is stated that intensional Henkin models of a (typed) total λ-theory are equivalent to models in cartesian closed categories; here, a Henkin model is a set-theoretic model where function types need not be interpreted by full function sets (this is the original idea of [8]), and the word intensional is meant to indicate that two inhabitants of a function type may be distinct although they produce the same output on all inputs. We now proceed to establish a corresponding result for the partial λ-calculus.

In the total case, the correspondence can be made precise as follows. Recall from [11] that a λ-theory $\mathcal{T}$ generates a cartesian closed category (ccc) called its *classifying category.*

Definition 10. Given a ccc $\mathbf{C}$, a *ccc model* of $\mathbf{C}$ is a cartesian closed (cc) functor $F : \mathbf{C} \to \mathbf{A}$ into some ccc $\mathbf{A}$. An (intensional) *Henkin model* of $\mathbf{C}$, or of a (total) λ-theory $\mathcal{T}$ of which $\mathbf{C}$ is the classifying category, is a cartesian functor $\mathbf{C} \to \mathbf{Set}$.

The use of the term 'Henkin model' is justified by the fact that such models are easily seen to be the same as Henkin models in the original sense (called *syntactic λ-algebras* in [3]). This is in tune with the intuition that Henkin models demote higher order to first order structure. Every ccc model gives rise to a Henkin model by composition with hom(1, _), since hom(1, _) is itself a cartesian functor:

$$\begin{array}{ccc} \mathbf{C} & \xrightarrow{F} & \mathbf{A} \\ & \searrow & \downarrow \hom(1,\_) \\ & & \mathbf{Set} \end{array}.$$

Conversely, every Henkin model $G : \mathbf{C} \to \mathbf{Set}$ arises in this way: one can construct from G a ccc $\mathbf{A}$ by taking the same objects as in $\mathbf{C}$, and the elements of $G(A \to B)$, where $A \to B$ denotes the function space, as morphisms from A to B in $\mathbf{A}$. Moreover, one has a functor $G^* : \mathbf{C} \to \mathbf{A}$ that maps a morphism $f : A \to B$ to the element of $G(A \to B)$ determined by its *name* $1 \to (A \to B)$. The functor G^* is a ccc model because the ccc structure is in fact equationally internalized in $\mathbf{C}$. This model is determined uniquely in the sense that, for any cc model F such that $\hom(1, \_) \circ F \cong G$, the codomain restriction of F to the full subcategory spanned by its image is equivalent to G^*.

We have thus outlined a novel proof of the equivalence of ccc-models and Henkin models as stated in [3], emphasizing the fact that the correspondence is really via postcomposition with $\hom(1, \_)$. This will also be the program for the partial case. The analogy to the total case suggests the following definitions:

Definition 11. A *pccc model* of a pre-pccc $(\mathbf{C}, \mathcal{M}, \mathfrak{A})$ is a pcc functor from $(\mathbf{C}, \mathcal{M}, \mathfrak{A})$ into a pccc $(\mathbf{A}, \mathcal{N})$. An *(intensional) Henkin model* of $(\mathbf{C}, \mathcal{M}, \mathfrak{A})$ (or of a partial λ-theory $\mathcal{T}$ with $\mathsf{Sy}(\mathcal{T}) = (\mathbf{C}, \mathcal{M}, \mathfrak{A})$) is a cartesian dominional functor $(\mathbf{C}, \mathcal{M}) \to \mathbf{Set}$.

(Note that the definition of Henkin model is independent of $\mathfrak{A}$! For pccc models, see Remark 18.)

Remark 12. In order to construct a pccc model, it suffices to provide a pcc functor into a pre-pccc. This follows from the fact that for every cartesian dominional category, the Yoneda embedding provides a full dominional cartesian extension to a pccc which preserves existing partial function spaces [17, 18].

Henkin models may be described as syntactic λ-algebras modeled on the corresponding notion defined for the total λ-calculus in [3]:

Definition 13. A *syntactic λ-algebra* for a partial λ-theory $\mathcal{T}$ is a family of sets $[\![s]\!]$, indexed over all *types* of $\mathcal{T}$, together with partial interpretation functions

$$[\![\Gamma.\, \alpha]\!] : [\![\Gamma]\!] \to [\![t]\!]$$

for each term $\Gamma \rhd \alpha : t$ in $\mathcal{T}$, where $[\![\Gamma]\!]$ denotes the extension of the interpretation to contexts via the cartesian product. This interpretation is subject to the following conditions:

(i) $[\![\Gamma.\, x_i]\!]$, where $\Gamma = (\bar{x} : \bar{s})$, is the i-th projection;
(ii) $[\![\bar{y} : \bar{t}.\, \gamma]\!] \circ [\![\Gamma.\, \beta]\!] = [\![\Gamma.\, \gamma[\bar{y}/\beta]]\!]$, where $\Gamma \rhd \beta : \bar{t}$ is a multi-term, with the interpretation extended to multi-terms in the obvious way;
(iii) whenever $\phi \vdash_\Gamma \alpha \stackrel{e}{=} \beta$ in $\mathcal{T}$ and $[\![\Gamma.\, \phi]\!](x)$ holds (i.e. is defined), then $[\![\Gamma.\, \alpha]\!](x) = [\![\Gamma.\, \beta]\!](x)$ are defined.

This is the definition of model used in the semantics of HasCasl given in [20].

Remark 14. It is implicit in [15] that syntactic λ-algebras are equivalent to the combinatorically defined *λ_p-algebras* considered there.

Both Henkin models and syntactic λ-algebras can be provided with the obvious notions of morphism. Using results of [19], one easily proves the following correspondence theorem.

Theorem 15. *The category of Henkin models of a partial λ-theory $\mathcal{T}$ is equivalent to the category of syntactic λ-algebras for $\mathcal{T}$.*

A pccc model $F : (\mathbf{C}, \mathcal{M}, \mathfrak{A}) \to (\mathbf{A}, \mathcal{N})$ gives rise to a Henkin model in the same manner as in the total case: since $\hom(1, \_) : (\mathbf{A}, \mathcal{N}) \to \mathbf{Set}$ is a cartesian dominional functor, so is $\hom(1, \_) \circ F : (\mathbf{C}, \mathcal{M}) \to \mathbf{Set}$. (It is in particular at this point that the representation of partial morphisms as spans pays off.)

The central point in the proof that, conversely, Henkin models give rise to pccc models is the fact that a Henkin model $F : (\mathbf{C}, \mathcal{M}) \to \mathbf{Set}$ has a theory $\mathsf{L}(F)$, its *internal language*, which can be soundly interpreted in F. We shall use the following notational convention: a provably defined term $() \rhd \alpha : A$ in $\mathsf{L}(\mathbf{C}, \mathcal{M}, \mathfrak{A})$ determines a morphism $f : 1 \to A$; the map Ff defines an element of FA which we denote by $F(\alpha)$. Now $\mathsf{L}(F)$ is constructed as follows:

The sorts of $\mathsf{L}(F)$ are the elements of $\mathfrak{A}$; the operators of profile $\bar{A} \to B$ are the elements of $F(\bar{A} \rightharpoonup B)$, where we abuse multi-types $\bar{A}$ in $\mathsf{L}(F)$ to denote their obvious interpretations in $(\mathbf{C}, \mathcal{M})$. The application operators are the elements $F(\lambda f : \bar{A} \rightharpoonup B, \bar{x} : \bar{A}.\, f(\bar{x}))$. Note that $F\bar{A} \cong FA_1 \times \cdots \times FA_n$.

We have the following structure on F, transferred from $\mathbf{C}$:

- for $\Gamma = (\bar{x} : \bar{A})$, we have an element $p_i = F(\lambda\Gamma.\, x_i) \in F(\Gamma \rightharpoonup A_i)$ representing the *i-th projection* for each i.
- For (multi-)types $\bar{A}$, $\bar{B}$, C, the morphism

$$\lambda\bar{x} : \bar{A}.\, f(\bar{g}(\bar{x})) : (f : \bar{B} \rightharpoonup C, \bar{g} : \bar{A} \rightharpoonup \bar{B}) \to (\bar{A} \rightharpoonup C)$$

  in $\mathbf{C}$ is mapped by F to a *multi-composition* map

$$F(\bar{B} \rightharpoonup C) \times F(\bar{A} \rightharpoonup \bar{B}) \to F(\bar{A} \rightharpoonup C)$$

  written $(f, \bar{g}) \mapsto f \circ_p \bar{g}$.
- For (multi-)types $\bar{A}$, $\bar{B}$, C, the morphism

$$\lambda\bar{x} : \bar{A}.\, \lambda\bar{y} : \bar{B}.\, f(\bar{x}, \bar{y}) : (f : \bar{A}\bar{B} \rightharpoonup C) \to (\bar{A} \rightharpoonup \bar{B} \rightharpoonup C)$$

  induces an *abstraction* map

$$\Lambda : F(\bar{A}\bar{B} \rightharpoonup C) \to F(\bar{A} \rightharpoonup \bar{B} \rightharpoonup C).$$

A term $\Gamma \rhd \alpha : \bar{C}$ in $\mathsf{L}(F)$, where $\Gamma = (\bar{x} : \bar{A})$, can be given an interpretation

$$\langle\!\langle \Gamma.\, \alpha \rangle\!\rangle \in F(\Gamma \rightharpoonup \bar{C})$$

in F, defined by recursion over the term structure using the operations introduced above:

$$\begin{aligned}
\langle\!\langle \Gamma.\, x_i \rangle\!\rangle &= p_i \\
\langle\!\langle \Gamma.\, (\alpha_1, \ldots, \alpha_n) \rangle\!\rangle &= (\langle\!\langle \Gamma.\, \alpha_1 \rangle\!\rangle, \ldots \langle\!\langle \Gamma.\, \alpha_n \rangle\!\rangle) \\
\langle\!\langle \Gamma.\, f(\alpha) \rangle\!\rangle &= f \circ_p \langle\!\langle \Gamma.\, \alpha \rangle\!\rangle \\
\langle\!\langle \Gamma.\, \lambda\bar{y} : \bar{B}.\, \alpha \rangle\!\rangle &= \Lambda(\langle\!\langle \Gamma, \bar{y} : \bar{B}.\, \alpha \rangle\!\rangle)
\end{aligned}$$

We say that a sentence $\phi \Rightarrow_\Gamma \alpha \stackrel{e}{=} \beta$ in $\mathsf{L}(F)$ *holds* in F if

$$\langle\!\langle \Gamma.\, \mathrm{def}\, \alpha \wedge \phi \rangle\!\rangle = \langle\!\langle \Gamma.\, \phi \rangle\!\rangle \quad \text{and} \quad \langle\!\langle \Gamma.\, \alpha \restriction \phi \rangle\!\rangle = \langle\!\langle \Gamma.\, \beta \restriction \phi \rangle\!\rangle.$$

The axioms of $\mathsf{L}(F)$ are the sentences that hold in F.

As announced, the interpretation of $\mathsf{L}(F)$ in F is sound:

Theorem 16. *If $\phi \vdash_\Gamma \psi$, then $\phi \Rightarrow_\Gamma \psi$ holds in F.*

5 The Equivalence Theorem

We now proceed to prove that every Henkin model arises essentially uniquely from a pccc model by postcomposition with $\hom(1, \_)$. More formally:

Theorem 17. *Let $F : (\mathbf{C}, \mathcal{M}, \mathfrak{A}) \to \mathbf{Set}$ be a Henkin model of a pre-pccc. Then there exists a pccc model $F^* : (\mathbf{C}, \mathcal{M}, \mathfrak{A}) \to (\mathbf{A}, \mathcal{N})$ such that*

$$\hom(1, \_) \circ F^* \cong F.$$

The codomain restriction of F^* to the full dominional subcategory of $(\mathbf{A}, \mathcal{N})$ spanned by its image is determined uniquely up to equivalence.

This includes the case that $(\mathbf{C}, \mathcal{M})$ is a pccc. Even then, the mentioned span $(\mathbf{B}, \mathcal{P})$ of the image of F^* in general need not be a pccc [18]. Of course, it has a pccc hull in $(\mathbf{A}, \mathcal{N})$, but in the light of an example given in [18], it seems unlikely that this pccc is uniquely determined by F. However, $(\mathbf{B}, \mathcal{P})$ *is* a pccc in two important special cases, namely if $(\mathbf{C}, \mathcal{M})$ has internal equality, or if $\mathbf{C}$ is a cartesian closed category, with $\mathcal{M} = \mathrm{Iso}(\mathbf{C})$. The latter case is just the abovementioned result of [3] for the total λ-calculus.

Remark 18. It is an immediate (and pleasing) consequence of Theorem 17 that the notion of pccc model of a pre-pccc $(\mathbf{C}, \mathcal{M}, \mathfrak{A})$ depends only on $(\mathbf{C}, \mathcal{M})$.

Proof (Theorem 17). The uniqueness statement follows from the fact that, for *(multi-)types* $\bar{A}$, B in $(\mathbf{C}, \mathcal{M}, \mathfrak{A})$, the set $\hom(1, F(\bar{A} \multimap B))$ determines the partial morphisms from $F^*\bar{A}$ to F^*B. This, in turn, determines the full dominional subcategory spanned by the image of $\mathfrak{A}$ under F, which is identical to the subcategory mentioned in the theorem by the generatedness condition for pre-pcccs.

The strategy for the existence proof is as follows: we shall define a translation

$$\sigma : \mathsf{L}(\mathbf{C}, \mathcal{M}, \mathfrak{A}) \to \mathsf{L}(F)$$

which gives rise to a functor $(\mathbf{C}, \mathcal{M}, \mathfrak{A}) \cong \mathsf{Sy}(\mathsf{L}(\mathbf{C}, \mathcal{M}, \mathfrak{A})) \to \mathsf{Sy}(\mathsf{L}(F))$ that preserves the pccc structure, since the latter is given syntactically. $\mathsf{Sy}(\mathsf{L}(F))$ is a pre-pccc, which by Remark 12 has a full dominional extension to a pccc $(\mathbf{A}, \mathcal{N})$, so that we obtain a pccc model $F^* : (\mathbf{C}, \mathcal{M}) \to \mathsf{Sy}(\mathsf{L}(F)) \hookrightarrow (\mathbf{A}, \mathcal{N})$. The proof is finished by establishing that $\hom(1, F^*A) \cong FA$.

The translation σ is the identity on sorts (i.e. type objects). An operator $f : \bar{A} \to B$ in $\mathsf{L}(\mathbf{C}, \mathcal{M}, \mathfrak{A})$ is mapped to $F(\lambda \bar{x} : \bar{A}. f(\bar{x})) \in F(\bar{A} \twoheadrightarrow B)$. Thus defined, σ preserves application operators. We have to show that σ maps axioms of $\mathsf{L}(\mathbf{C}, \mathcal{M}, \mathfrak{A})$ to theorems (in fact, axioms) in $\mathsf{L}(F)$. This is easily proved using

Lemma 19. *For a term $\Gamma \rhd \alpha : A$ in $\mathsf{L}(\mathbf{C}, \mathcal{M}, \mathfrak{A})$,*

$$\langle\!\langle \Gamma. \sigma(\alpha) \rangle\!\rangle = F(\lambda \Gamma. \alpha) \in F(\Gamma \twoheadrightarrow A).$$

It remains to be shown that $\hom(1, \_) \circ F^*$ is isomorphic to F. We first treat the case that $(\mathbf{C}, \mathcal{M})$ is a pccc. In this case, we can define a functor $\tilde{F} : \mathbf{C} \to \mathbf{Set}$ by

$$\tilde{F}A = \{x \in F(1 \twoheadrightarrow A) \mid \mathsf{true} \Rightarrow \operatorname{def} x(*) \text{ holds in } F\}.$$

Lemma 20. *We have natural isomorphisms*

$$\hom(1, \_) \circ F^* \cong \tilde{F} \cong F.$$

This lemma completes the proof of Theorem 17 for the case that $(\mathbf{C}, \mathcal{M})$ is a pccc. In the general case, one shows that the restrictions of $\hom(1, \_) \circ F^*$ and F, respectively, to $\mathfrak{A}$ regarded as a full subcategory of $\mathbf{C}$ are isomorphic via the same intermediate step as above. This isomorphism can then be extended to all of $\mathbf{C}$ thanks to the observation that $\mathfrak{A}$ contains the classifier and hence in fact generates $\mathbf{C}$ via finite products and pullbacks. □

Remark 21. Since $\mathsf{Cl}(\mathcal{T})$ is the free pccc over $\mathsf{Sy}(\mathcal{T})$, pccc models of $\mathsf{Cl}(\mathcal{T})$ are equivalent to pccc models of $\mathsf{Sy}(\mathcal{T})$. This means that every Henkin model of $\mathsf{Sy}(\mathcal{T})$ extends to a Henkin model of $\mathsf{Cl}(\mathcal{T})$; whether or not this extension is unique is the subject of further research.

Remark 22. Some care is required concerning the notion of satisfaction in Henkin models/syntactic λ-algebras. The set of sentences $\phi \Rightarrow_\Gamma \alpha \stackrel{e}{=} \beta$ such that $[\![\Gamma. \phi]\!](x)$ entails $[\![\Gamma. \alpha]\!](x) \stackrel{e}{=} [\![\Gamma. \beta]\!](x)$ in a given syntactic λ-algebra $\mathcal{A}$ fails to be closed under (ξ) (see also [3, 13]), so that this notion of satisfaction would not coincide with the one given by the associated pccc model. We remedy this by using instead the notion of satisfaction in Henkin models as defined in Section 4. Then an ECE holds in $\mathcal{A}$ iff it holds in the associated pccc model; in particular, *the rules of Figure 2 are sound and complete for Henkin models.*

6 HasCasl

HasCasl is a higher order extension of the first order specification language Casl based on the partial λ-calculus and Haskell-style type-class polymorphism. It has been introduced in [20] and used in order to formalize monad-independent computation logics in [22, 21], as well as in a case study concerned with higher order Petri nets [9].

Basic HasCasl does not have internal equality. However, internal equality, the internal logic defined on top of it, and HOLCF-style general recursion are added as optional features in a bootstrap fashion and are provided with built-in syntactical sugar. This allows the combination of programming and specification within a single language. A simple example is the implementation of a function that iterates a partial transition function until a given predicate fails, subject to the requirement that transitions are possible while the predicate holds:

spec REPEAT = RECURSION **and** BOOL **then**
type *State*
ops $step : \quad State \rightarrow ?\ State$
$\quad cont : \quad State \rightarrow Bool$
$\quad repeat : State \rightarrow ?\ State$
internal $\{\forall s : State \bullet cont\ s \implies def\ step\ s\ \}$
program $repeat\ s = if\ cont\ s\ then\ repeat\ (step\ s)\ else\ s$

Here, the keywords **internal** and **program** invoke syntactical sugar for the internal logic and general recursion, respectively. Note that, thanks to the axiom, undefinedness of *repeat* occurs only for states that admit infinitely many steps.

The benefit of the results above w.r.t. the semantics of HASCASL is that the latter now profits from the best of both worlds. As stated above, the semantics of HASCASL is defined in terms of syntactic λ-algebras, or, equivalently, Henkin models. This makes for a tight connection to the set-theoretic semantics of first-order CASL. For instance, thanks to the fact that both CASL models and HASCASL models can be understood as set-valued cartesian dominional functors, one rather easily proves

Theorem 23. *Given a* CASL *signature* Σ, *every model of* Σ *embeds universally into a Henkin-model of* Σ, *where* Σ *is reinterpreted as a* HASCASL *signature.*

On the other hand, arguments concerning the existence of models for certain specifications are often easier when the notion of pccc models is used. For example, the models of the HASCASL type definition

$$\textbf{type}\ L ::= abs(rep : L \rightarrow ?\ L)$$

that axiomatizes the untyped partial λ-calculus are essentially the solutions of the corresponding domain equation in arbitrary pcccs, e.g. in categories of domains. (More precisely, the specification only requires that the partial function type, denoted $L \rightarrow ?\ L$ in HASCASL, is a retraction of L.) Pcccs where such a type exists necessarily fail to have internal equality, since the arising internal logic would lead to a Russell-type paradox. Consider, however, the HASCASL specification

$$\text{INTERNALLOGIC}\ \textbf{then type}\ L ::= abs(rep : L \rightarrow L)$$

that axiomatizes the untyped *total* λ-calculus ($L \rightarrow L$ denotes the total function type) in the presence of internal equality, introduced by referencing a named specification of the internal logic as laid out in [20]. This specification is consistent: take a cartesian closed category **C** with reflexive object U. The functor category $\mathbf{Set}^{\mathbf{C}^{op}}$ is a topos and as such gives rise to a Henkin model with internal equality. Moreover, the Yoneda embedding $Y : \mathbf{C} \hookrightarrow \mathbf{Set}^{\mathbf{C}^{op}}$ preserves the cartesian closed structure [24], so that $Y(U)$ is a reflexive object and hence provides us with a model of the above type.

7 Conclusion and Future Work

We have established equivalence results for notions of model for the partial λ-calculus, to wit, syntactic λ-algebras in the spirit of [3], first order (i.e. cartesian dominional) functors from the classifying category into **Set**, and higher order models in arbitrary partial cartesian closed categories (pccc models), thus completing the semantical foundations of HasCasl. These results generalize known results for the total λ-calculus [3], and illuminate them by demonstrating that the correspondence works simply by postcomposing pccc models with the global element functor $\hom(1, \_)$. We have emphasized some of the immediate benefits of this result for the model theory of HasCasl. More generally, the equivalence result reconciles the mostly set-theoretic viewpoint of mainstream logic (or algebraic specification) and the concepts of categorical logic and semantics. The connection to Henkin's seminal work, which concerned models of higher order *logic*, is tightened by the fact that the partial λ-calculus *with internal equality* in fact incorporates full (intuitionistic) higher order logic [20].

One of the upshots of this is that the known deduction system for pccc models is sound and complete for Henkin models. A pressing task in the further development of HasCasl is to provide proof support using this deduction system via an encoding into an existing theorem prover.

It is hard to resist the temptation to generalize the correspondence between first order and higher order models to base categories other than **Set**. The proof of the conjecture that higher order models in $\mathcal{V}$-enriched categories are equivalent to first order models in $\mathcal{V}$ is the subject of further research.

Acknowledgements

The author wishes to thank Till Mossakowski, Christian Maeder, and Kathrin Hoffmann for collaboration on HasCasl, and Christoph Lüth for useful comments and discussions.

References

1. J. Adámek, H. Herrlich, and G. E. Strecker, *Abstract and concrete categories*, Wiley Interscience, 1990.
2. E. Astesiano, M. Bidoit, H. Kirchner, B. Krieg-Brückner, P. D. Mosses, D. Sannella, and A. Tarlecki, *Casl: the Common Algebraic Specification Language*, Theoret. Comput. Sci. **286** (2002), 153–196.
3. V. Breazu-Tannen and A. R. Meyer, *Lambda calculus with constrained types*, Logic of Programs, LNCS, vol. 193, Springer, 1985, pp. 23–40.

4. M. Broy, C. Facchi, R. Grosu, R. Hettler, H. Hussmann, D. Nazareth, F. Regensburger, and K. Stølen, *The requirement and design specification language SPECTRUM, an informal introduction, version 1.0*, Tech. report, Department of Informatics, Technical University of Munich, 1993.
5. P. Burmeister, *Partial algebras — an introductory survey*, Algebras and Orders (I. G. Rosenberg and G. Sabidussi, eds.), NATO ASI Series C, Kluwer, 1993, pp. 1–70.
6. P.-L. Curien and A. Obtułowicz, *Partiality, cartesian closedness and toposes*, Inform. and Comput. **80** (1989), 50–95.
7. C. George, P. Haff, K. Havelund, A. E. Haxthausen, R. Milne, C. Bendix Nielson, S. Prehn, and K. R. Wagner, *The Raise Specification Language*, Prentice Hall, 1992.
8. L. Henkin, *Completeness in the theory of types*, J. Symbolic Logic **15** (1950), 81–91.
9. K. Hoffmann and T. Mossakowski, *Algebraic higher order nets: Graphs and Petri nets as tokens*, 16th Workshop on Algebraic Development Techniques (WADT 02), LNCS, 2003, to appear.
10. B. Jacobs, *Categorical logic and type theory*, Elsevier, Amsterdam, 1999.
11. J. Lambek and P. J. Scott, *Introduction to higher order categorical logic*, Cambridge University Press, 1986.
12. S. Mac Lane, *Categories for the working mathematician*, 2nd ed., Springer, 1998.
13. J. C. Mitchell and P. J. Scott, *Typed lambda models and cartesian closed categories*, Categories in Computer Science and Logic, Contemp. Math., vol. 92, AMS, 1989, pp. 301–316.
14. E. Moggi, *Categories of partial morphisms and the λ_p-calculus*, Category Theory and Computer Programming, LNCS, vol. 240, Springer, 1986, pp. 242–251.
15. ______, *The partial lambda calculus*, Ph.D. thesis, University of Edinburgh, 1988.
16. E. Robinson and G. Rosolini, *Categories of partial maps*, Inform. and Comput. **79** (1988), 95–130.
17. G. Rosolini, *Continuity and effectiveness in topoi*, Ph.D. thesis, Merton College, Oxford, 1986.
18. L. Schröder, *Classifying categories for partial higher order logic*, available as `http://www.tzi.de/~lschrode/hascasl/hoclasscat.ps`
19. ______, *Classifying categories for partial equational logic*, Category Theory and Computer Science (CTCS 02), ENTCS, vol. 69, 2003.
20. L. Schröder and T. Mossakowski, HASCASL*: Towards integrated specification and development of functional programs*, Algebraic Methodology And Software Technology, LNCS, vol. 2422, Springer, 2002, pp. 99–116.
21. L. Schröder and T. Mossakowski, *Monad-independent dynamic logic in* HASCASL, 16th Workshop on Algebraic Development Techniques (WADT 02), LNCS, 2003, to appear.
22. L. Schröder and T. Mossakowski, *Monad-independent Hoare logic in* HASCASL, Fundamental Approaches to Software Engineering, LNCS, vol. 2621, Springer, 2003, pp. 261–277.
23. D. Scott, *Continuous lattices*, Toposes, Algebraic Geometry and Logic, Lect. Notes Math., vol. 274, Springer, 1972, pp. 97–136.
24. ______, *Relating theories of the λ-calculus*, To H.B. Curry: Essays in Combinatory Logic, Lambda Calculus and Formalisms, Academic Press, 1980, pp. 403–450.

Nominal Unification

Christian Urban, Andrew Pitts, and Murdoch Gabbay

University of Cambridge, Cambridge, UK

Abstract. We present a generalisation of first-order unification to the practically important case of equations between terms involving *binding operations*. A substitution of terms for variables solves such an equation if it makes the equated terms *α-equivalent*, i.e. equal up to renaming bound names. For the applications we have in mind, we must consider the simple, textual form of substitution in which names occurring in terms may be captured within the scope of binders upon substitution. We are able to take a 'nominal' approach to binding in which bound entities are explicitly named (rather than using nameless, de Bruijn-style representations) and yet get a version of this form of substitution that respects α-equivalence and possesses good algorithmic properties. We achieve this by adapting an existing idea and introducing a key new idea. The existing idea is terms involving explicit substitutions of names for names, except that here we only use *explicit permutations* (bijective substitutions). The key new idea is that the unification algorithm should solve not only equational problems, but also problems about the *freshness* of names for terms. There is a simple generalisation of the classical first-order unification algorithm to this setting which retains the latter's pleasant properties: unification problems involving α-equivalence and freshness are decidable; and solvable problems possess most general solutions.

1 Introduction

Decidability of unification for equations between first-order terms and algorithms for computing most general unifiers form a fundamental tool of computational logic with many applications to programming languages and computer-aided reasoning. However, very many potential applications fall outside the scope of first-order unification, because they involve term languages with binding operations where at the very least we do not wish to distinguish terms differing up to the renaming of bound names.

There is a large body of work studying languages with binders through the use of various λ-calculi as term representation languages, leading to *higher-order unification* algorithms for solving equations between λ-terms modulo $\alpha\beta\eta$-equivalence. However, higher-order unification is technically complicated without being completely satisfactory from a pragmatic point of view. The reason lies in the difference between substitution for first-order terms and for λ-terms. The former is a simple operation of textual replacement (sometimes called *grafting* [7], or *context substitution* [12, Sect. 2.1]), whereas the latter also involves renamings to avoid capture. Capture-avoidance ensures that substitution respects α-equivalence, but it complicates higher-order unification algorithms. Furthermore it is the simple textual form of substitution rather than the more complicated capture-avoiding form which occurs in many informal applications of 'unification modulo α-equivalence'. For example, consider the following schematic rule

M. Baaz and J.A. Makowsky (Eds.): CSL 2003, LNCS 2803, pp. 513–527, 2003.

which might form part of the inductive definition of a binary evaluation relation $\Downarrow$ for the expressions of an imaginary functional programming language:

$$\frac{\texttt{app}(\texttt{fn}\, a.Y, X) \Downarrow V}{\texttt{let}\, a = X\, \texttt{in}\, Y \Downarrow V}. \tag{1}$$

Here X, Y and V are metavariables standing for unknown programming language expressions. The binders $\texttt{fn}\, a.(-)$ and $\texttt{let}\, a = X\, \texttt{in}\, (-)$ may very well capture free occurrences of a when we instantiate the schematic rule by replacing Y with an expression. For instance, using the rule scheme in a bottom-up search for a proof of

$$\texttt{let}\, a = 1\, \texttt{in}\, a \Downarrow 1 \tag{2}$$

we would use a substitution that involves capture, namely $[X := 1, Y := a, V := 1]$, to unify the goal with the conclusion of (1)—generating the new goal $\texttt{app}(\texttt{fn}\, a.a, 1) \Downarrow 1$ from the hypothesis of the rule. The problem with this is that in informal practice we usually identify terms up to α-equivalence, whereas textual substitution does not respect α-equivalence. For example, up to α-equivalence, the goal

$$\texttt{let}\, b = 1\, \texttt{in}\, b \Downarrow 1 \tag{3}$$

is the same as (2). We might think (erroneously!) that the conclusion of (1) is the same as $\texttt{let}\, b = X\, \texttt{in}\, Y \Downarrow V$ without changing the rule's hypothesis—after all, if we are trying to make α-equivalence disappear into the infrastructure, then we must be able to replace any *part* of what we have with an equivalent part. So we might be tempted to unify the conclusion with (3) via the textual substitution $[X := 1, Y := b, V := 1]$, and then apply this substitution to the hypothesis to obtain a wrong goal, $\texttt{app}(\texttt{fn}\, a.b, 1) \Downarrow 1$. Using λ-calculus and higher-order unification saves us from such sloppy thinking, but at the expense of having to make explicit the dependence of metavariables on bindable names via the use of function variables and application. For example, (1) would be replaced by

$$\frac{\texttt{app}\,(\texttt{fn}\, \lambda a.F\, a)\, X \Downarrow V}{\texttt{let}\, X\, (\lambda a.F\, a) \Downarrow V} \quad \text{or, modulo } \eta\text{-equivalence,} \quad \frac{\texttt{app}\,(\texttt{fn}\, F)\, X \Downarrow V}{\texttt{let}\, X\, F \Downarrow V}. \tag{4}$$

Now goal (3) becomes $\texttt{let}\, 1\, \lambda b.b \Downarrow 1$ and there is no problem unifying it with the conclusion of (4) via a capture-avoiding substitution of 1 for X, $\lambda c.c$ for F and 1 for V.

This is all very fine, but the situation is not as pleasant as for first-order terms: higher-order unification problems can be undecidable, decidable but lack most general unifiers, or have such unifiers only by imposing some restrictions [20]; see [6] for a survey of higher-order unification. We started out wanting to compute with binders modulo α-equivalence, and somehow the process of making possibly-capturing substitution respectable has led to function variables, application, capture-avoiding substitution and $\beta\eta$-equivalence. Does it have to be so? No!

For one thing, several authors have already noted that one can make sense of possibly-capturing substitution modulo α-equivalence by using *explicit substitutions* in the term representation language: see [7, 13, 15, 26]. Compared with those works, we make a number of simplifications. First, we find that we do *not* need to use function variables, application or $\beta\eta$-equivalence in our representation language—leaving just binders and α-equivalence. Secondly, instead of using explicit substitutions of names for names, we use *explicit permutations* of names. The idea of using name-permutations, and in particular name-swappings, when dealing with α-conversion dates back to [10] and there is growing evidence of its usefulness (see [3, 4], for example). When a name substitution is actually a permutation, the function it induces from terms to terms is a bijection; this bijectivity gives the operation of permuting names very good logical properties compared with name substitution. Consider for example the α-equivalent terms $\mathtt{fn}\,a.b$ and $\mathtt{fn}\,c.b$, where a, b and c are distinct. If we apply the substitution $[b \mapsto a]$ (renaming all free occurrences of b to be a) to them we get $\mathtt{fn}\,a.a$ and $\mathtt{fn}\,c.a$, which are no longer α-equivalent. Thus renaming substitutions do not respect α-equivalence in general, and any unification algorithm using them needs to take extra precautions to not inadvertently change the intended meaning of terms. The traditional solution for this problem is to introduce a more complicated form of renaming substitution that avoids capture of names by binders. In contrast, the simple operation of name-permutation respects α-equivalence; for example, applying the name-permutation $(a\,b)$ that swaps all occurrences of a and b (be they free, bound or binding) to the terms above gives $\mathtt{fn}\,b.a$ and $\mathtt{fn}\,c.a$, which are still α-equivalent. We exploit such good properties of name-permutations to give a conceptually simple unification algorithm.

In addition to the use of explicit name-permutations, we also compute symbolically with predicates expressing *freshness* of names for terms. This seems to be the key novelty of our approach. Although it arises naturally from the work reported in [11, 23], it is easy to see directly why there is a need for computing with freshness, given that we take a 'nominal' approach to binders. (In other words we stick with concrete versions of binding and α-equivalence in which the bound entity is named explicitly, rather than using de Bruijn-style representations, as for example in [7, 26].) A basic instance of our generalised form of α-equivalence identifies $\mathtt{fn}\,a.X$ with $\mathtt{fn}\,b.(a\,b)\cdot X$ provided *b is fresh for X*, where the subterm $(a\,b)\cdot X$ indicates an explicit permutation—namely the swapping of a and b—waiting to be applied to X. We write 'b is fresh for X' symbolically as $b \# X$; the intended meaning of this relation is that b does not occur free in any (ground) term that may be substituted for X. If we know more about X we may be able to eliminate the explicit permutation in $(a\,b)\cdot X$; for example, if we knew that $a \# X$ holds as well as $b \# X$, then $(a\,b)\cdot X$ can be replaced by X. It should already be clear from these simple examples that in our setting the appropriate notion of term-equality is not a bare equation, $t \approx t'$, but rather a hypothetical judgement of the form

$$\nabla \vdash t \approx t' \tag{5}$$

where ∇ is a *freshness environment*—a finite set $\{a_1 \# X_1, \ldots, a_n \# X_n\}$ of freshness assumptions. For example $\{a \# X, b \# X\} \vdash \mathtt{fn}\,a.X \approx \mathtt{fn}\,b.X$ is a valid judgement of our *nominal equational logic*. Similarly, judgements about freshness itself will take the form

$$\nabla \vdash a \# t\,. \tag{6}$$

To summarise: We will represent languages involving binders using the usual notion of first-order terms over a many-sorted signature. These give us terms with: distinguished constants naming bindable entities, that we call *atoms*; terms $a.t$ expressing a generic form of *binding* of an atom a in a term t; and terms $\pi\cdot X$ representing an explicit *permutation* of atoms π waiting to be applied to whatever term is substituted for the variable X. Section 2 presents this term-language together with a syntax-directed inductive definition of the provable judgements of the form (5) and (6) which for *ground* terms (i.e. ones with no variables) agrees with the usual notions of α-equivalence and 'not a free variable of'. However, on open terms our judgements *differ* from these standard notions and appear to be an extension that has not yet been studied in the literature (including [11,24,23]). Section 3 considers unification in this setting. Solving equalities between abstractions $a.t \approx ?\ a'.t'$ entails solving both equalities $t \approx ?\ (a\, a')\cdot t'$ and freshness problems $a \ \#?\ t'$. Therefore our general form of *nominal unification problem* is a finite collection of individual equality and freshness problems. Such a problem P is solved by providing not only a substitution σ (of terms for variables), but also a freshness environment ∇ (as above), which together have the property that $\nabla \vdash \sigma(t) \approx \sigma(t')$ and $\nabla \vdash a \# \sigma(t'')$ hold for each individual equality $t \approx ?\ t'$ and freshness $a \ \#?\ t''$ in the problem P. Our main result with respect to unification is that *solvability is decidable and that solvable problems possess most general solutions* (for a reasonably obvious notion of 'most general'). The proof is via a unification algorithm that is very similar to the first-order algorithm given in the now-common transformational style [18]. (See [17, Sect. 2.6] or [1, Sect. 4.6] for expositions of this.) Section 4 considers the relationship of our version of 'unification modulo α-equivalence' to existing approaches. Section 5 assesses what has been achieved and the prospects for applications. To appreciate the kind of problem that nominal unification solves, you might like to try the following quiz about the λ-calculus [2] before we apply our algorithm to solve it at the end of Section 3.

Quiz *Assuming that a and b are distinct variables,* is it possible to find λ-terms $M_1, \ldots, M_7$ that make the following pairs of terms α-equivalent?

1. $\lambda a.\lambda b.(M_1\, b)$ and $\lambda b.\lambda a.(a\, M_1)$
2. $\lambda a.\lambda b.(M_2\, b)$ and $\lambda b.\lambda a.(a\, M_3)$
3. $\lambda a.\lambda b.(b\, M_4)$ and $\lambda b.\lambda a.(a\, M_5)$
4. $\lambda a.\lambda b.(b\, M_6)$ and $\lambda a.\lambda a.(a\, M_7)$.

If it is possible to find a solution for any of these four problems, can you describe what all possible solutions for that problem are like?

Answers: see Example 2.

2 Nominal Equational Logic

We take a concrete approach to the syntax of binders in which bound entities are explicitly named. Furthermore we separate the names of bound entities from the names of variables, which is inspired for example by the π-calculus [21], in which the restriction operator binds channel names and these are quite different from names of unknown processes. Names of bound entities will be called *atoms*. This is partly for historical reasons (stemming from the work by the second two authors [11]) and partly to indicate that the internal structure of such names is irrelevant to us: all we care about is their identity (i.e. whether or not one atom is the same as another) and that the supply of atoms is inexhaustible.

Although there are several general frameworks in the literature for specifying languages with binders, not all of them meet the requirements mentioned in the previous paragraph. Use of the simply typed λ-calculus for this purpose is common; but as discussed in the Introduction, it leads to a problematic unification theory. Among *first-order* frameworks, Plotkin's notion of *binding signature* [25,9], being unsorted, equates names used in binding with names of variables standing for unknown terms; so it is not sufficiently general for us. A first-order framework that does meet our requirements is the notion of *nominal algebras* in [16]. The *nominal signatures* that we use in this paper are a mild (but practically useful) generalisation of nominal algebras in which name-abstraction and pairing can be mixed freely in arities (rather than insisting as in [16] that the argument sort of a function symbol be normalised to a tuple of abstractions).

Definition 1. *A **nominal signature** is specified by: a set of **sorts of atoms** (typical symbol ν); a disjoint set of **sorts of data** (typical symbol δ); and a set of **function symbols** (typical symbol f), each of which has an **arity** of the form $\tau \to \delta$. Here τ ranges over (compound) **sorts** given by the grammar $\tau ::= \nu \mid \delta \mid 1 \mid \tau \times \tau \mid \langle\nu\rangle\tau$. Terms of sort $\langle\nu\rangle\tau$ are binding abstractions of atoms of sort ν over terms of sort τ. We will explain the syntax and properties of such terms in a moment.*

Example 1. Here is a nominal signature for expressions in a small fragment of ML:

sort of atoms: *vid* function symbols: $\mathtt{vr} : \mathit{vid} \to \mathit{exp}$
sort of data: *exp*
$\mathtt{app} : \mathit{exp} \times \mathit{exp} \to \mathit{exp}$
$\mathtt{fn} : \langle\mathit{vid}\rangle\mathit{exp} \to \mathit{exp}$
$\mathtt{lv} : \mathit{exp} \times \langle\mathit{vid}\rangle\mathit{exp} \to \mathit{exp}$
$\mathtt{lf} : \langle\mathit{vid}\rangle((\langle\mathit{vid}\rangle\mathit{exp}) \times \mathit{exp}) \to \mathit{exp}$.

The function symbol `vr` constructs terms of sort *exp* representing value identifiers (named by atoms of sort *vid*); `app` constructs application expressions from pairs of expressions; `fn`, `lv` and `lf` construct terms representing respectively function abstractions (`fn x => e`), local value declarations (`let val x = e1 in e2 end`) and local recursive function declarations (`let fun f x = e1 in e2 end`). The arities of the function symbols specify which are binders and in which way their arguments are bound. This kind of specification of binding scopes is of course a feature of *higher-order abstract syntax* [22], using function types $\nu{\to}\tau$ in simply typed λ-calculus where we use abstraction sorts $\langle\nu\rangle\tau$. We shall see that the latter have much more elementary (indeed, first-order) properties compared with the former. To make this point clear we deliberately use a first-order syntax for terms, and *not* higher-order abstract syntax, although we often refer to abstractions, binders and free atoms by analogy with the λ-calculus.

Definition 2. *Given a nominal signature, we assume that there are countably infinite and pairwise disjoint sets of **atoms** (typical symbol a) for each sort of atoms ν, and **variables** (typical symbol X) for each sort τ. The **terms** over a nominal signature and their sorts are inductively defined as follows, where we write $t : \tau$ to indicate that a term t has sort τ.*

Unit value $\langle\rangle : 1$.
Pairs $\langle t_1, t_2\rangle : \tau_1 \times \tau_2$, *if* $t_1 : \tau_1$ *and* $t_2 : \tau_2$.
Data $f\,t : \delta$, *if* f *is a function symbol of arity* $\tau \to \delta$ *and* $t : \tau$.
Atoms $a : \nu$, *if* a *is an atom of sort* ν.
Atom-abstraction $a.t : \langle\nu\rangle\tau$, *if* a *is an atom of sort* ν *and* $t : \tau$.
Suspension $\pi{\cdot}X : \tau$, *if* $\pi = (a_1\,b_1)(a_2\,b_2)\cdots(a_n\,b_n)$ *is a finite list whose elements* $(a_i\,b_i)$ *are pairs of atoms, with* a_i *and* b_i *of the same sort, and* X *is a variable of sort* τ. *In the case that* π *is the empty list* [], *we just write* X *for* $\pi{\cdot}X$.

Recall that every finite permutation can be expressed as a composition of swappings $(a_i\,b_i)$; the list π of pairs of atoms occurring in a suspension term $\pi{\cdot}X$ specifies a finite permutation of atoms waiting to be applied once we know more about the variable

$$[]\cdot a \stackrel{\text{def}}{=} a \qquad ((a_1\,a_2) :: \pi)\cdot a \stackrel{\text{def}}{=} \begin{cases} a_1 & \text{if } \pi\cdot a = a_2 \\ a_2 & \text{if } \pi\cdot a = a_1 \\ \pi\cdot a & \text{otherwise} \end{cases}$$

$$\begin{aligned} \pi\cdot\langle\rangle &\stackrel{\text{def}}{=} \langle\rangle \\ \pi\cdot\langle t_1, t_2\rangle &\stackrel{\text{def}}{=} \langle \pi\cdot t_1, \pi\cdot t_2\rangle \\ \pi\cdot(f\,t) &\stackrel{\text{def}}{=} f\,(\pi\cdot t) \\ \pi\cdot(a.t) &\stackrel{\text{def}}{=} (\pi\cdot a).(\pi\cdot t) \\ \pi\cdot(\pi'\cdot X) &\stackrel{\text{def}}{=} (\pi @ \pi')\cdot X\,. \end{aligned}$$

Fig. 1. Permutation action on terms, $\pi\cdot t$.

$$\frac{}{\nabla \vdash \langle\rangle \approx \langle\rangle}\,(\approx\text{-unit}) \qquad \frac{\nabla \vdash t_1 \approx t_1' \quad \nabla \vdash t_2 \approx t_2'}{\nabla \vdash \langle t_1, t_2\rangle \approx \langle t_1', t_2'\rangle}\,(\approx\text{-pair}) \qquad \frac{\nabla \vdash t \approx t'}{\nabla \vdash f\,t \approx f\,t'}\,(\approx\text{-function symbol})$$

$$\frac{\nabla \vdash t \approx t'}{\nabla \vdash a.t \approx a.t'}\,(\approx\text{-abstraction-1}) \qquad \frac{a \neq a' \quad \nabla \vdash t \approx (a\,a')\cdot t' \quad \nabla \vdash a \# t'}{\nabla \vdash a.t \approx a'.t'}\,(\approx\text{-abstraction-2})$$

$$\frac{}{\nabla \vdash a \approx a}\,(\approx\text{-atom}) \qquad \frac{(a \# X) \in \nabla \text{ for all } a \in ds(\pi, \pi')}{\nabla \vdash \pi\cdot X \approx \pi'\cdot X}\,(\approx\text{-suspension})$$

$$\frac{}{\nabla \vdash a \# \langle\rangle}\,(\#\text{-unit}) \qquad \frac{\nabla \vdash a \# t_1 \quad \nabla \vdash a \# t_2}{\nabla \vdash a \# \langle t_1, t_2\rangle}\,(\#\text{-pair}) \qquad \frac{\nabla \vdash a \# t}{\nabla \vdash a \# f\,t}\,(\#\text{-function symbol})$$

$$\frac{}{\nabla \vdash a \# a.t}\,(\#\text{-abstraction-1}) \qquad \frac{a \neq a' \quad \nabla \vdash a \# t}{\nabla \vdash a \# a'.t}\,(\#\text{-abstraction-2})$$

$$\frac{a \neq a'}{\nabla \vdash a \# a'}\,(\#\text{-atom}) \qquad \frac{(\pi^{-1}\cdot a \# X) \in \nabla}{\nabla \vdash a \# \pi\cdot X}\,(\#\text{-suspension})$$

Fig. 2. Inductive definition of $\approx$ and $\#$.

X (by substituting for it, for example). We represent finite permutations in this way because it is really the operation of swapping which plays a fundamental role in the theory. Since, semantically speaking (see Remark 1 below about semantics), swapping commutes with all term-forming operations, we can normalise terms involving an explicit swapping operation by pushing the swap in as far as it will go, until it reaches a variable (cf. Fig. 1); the terms in Def. 2 are all normalised in this way, with explicit swappings 'piled up' in front of variables giving what we have called *suspensions*.

We wish to give a definition of α-equivalence for terms over a nominal signature that is respected by substitution of terms for variables, even though the latter may involve capture of atoms by binders. To do so we will need to make use of an auxiliary relation of *freshness* between atoms and terms, whose intended meaning is that the atom does not occur free in any substitution instance of the term. As discussed in the Introduction, our judgements about term equivalence ($t \approx t'$) need to contain hypotheses about the freshness of atoms with respect to variables ($a \# X$); and the same goes for our judgements about freshness itself ($a \# t$). Figure 2 gives a syntax-directed inductive definition of equivalence and freshness using judgements of the form

$$\nabla \vdash t \approx t' \quad \text{and} \quad \nabla \vdash a \# t$$

where t and t' are terms of the same sort over a given nominal signature, a is an atom, and the **freshness environment** ∇ is a finite set of **freshness constraints** $a \# X$, each specified by an atom and a variable. Rules (≈-abstraction-2), (≈-suspension) and (#-suspension) in Fig. 2 make use of the following definitions.

Definition 3. *Recall from Def. 2 that we specify* ***finite permutations of atoms*** *by finite lists* $(a_1\, b_1,)(a_2\, b_2)\cdots(a_n\, b_n)$ *representing the composition of finitely many swappings* $(a_i\, b_i)$, *with* a_i *and* b_i *of the same sort. Since we will apply permutations to terms on the left, the order of the composition is from right to left. So with this representation, the composition of a permutation* π *followed by a swap* $(a\, b)$ *is given by list-cons, written* $(a\, b) :: \pi$; *the composition of* π *followed by another permutation* π' *is given by list-concatenation, written as* $\pi' @ \pi$; *the* ***identity permutation*** *is given by the empty list* $[]$; *and the* ***inverse*** *of a permutation is given by list reversal, written as* π^{-1}. *The* ***permutation action****,* $\pi{\cdot}t$, *of a finite permutation of atoms* π *on a term* t *is defined as in Fig. 1; it pushes the list* π *into the structure of the term* t *until it 'piles up' in front of suspensions (applying the actual permutation that* π *represents to atoms that it meets on the way). The* ***disagreement set*** *of two permutations* π *and* π' *(used in rule (≈-suspension) in Fig. 2) is defined by*

$$ds(\pi, \pi') \stackrel{def}{=} \{a \mid \pi{\cdot}a \neq \pi'{\cdot}a\}\,. \tag{7}$$

Note that every disagreement set of the lists π and π' is a subset of the *finite* set of atoms occurring in either of the lists, because if a does not occur in those lists, then from Fig. 1 we get $\pi{\cdot}a = a = \pi'{\cdot}a$. To illustrate the use of disagreement sets, consider

$$\{a \# X, c \# X\} \vdash (a\, c)(a\, b){\cdot}X \approx (b\, c){\cdot}X$$

which holds by (≈-suspension), since the disagreement set of $(a\, c)(a\, b)$ and $(b\, c)$ is $\{a,c\}$.

Lemma 1. $\nabla \vdash - \approx -$ *is an equivalence relation; it is preserved by all of the term-forming operations in Def. 2; and it respects the freshness relation (i.e. if* $\nabla \vdash a \# t$ *and* $\nabla \vdash t \approx t'$, *then* $\nabla \vdash a \# t'$*). Both* $\approx$ *and* $\#$ *are preserved by the permutation action given in Fig. 1 in the following sense: if* $\nabla \vdash t \approx t'$, *then* $\nabla \vdash \pi{\cdot}t \approx \pi{\cdot}t'$; *and if* $\nabla \vdash a \# t$, *then* $\nabla \vdash \pi{\cdot}a \# \pi{\cdot}t$.

Proof. Although reasoning about $\approx$ and $\#$ is rather pleasant once the above facts are proved, establishing them first is rather tricky—mainly because of the large number of cases, but also because the facts in the lemma are inter-dependent; in addition some further properties of the permutation action and disagreement sets need to be established first (statements omitted)[1]. □

The main reason for using suspensions in the syntax of terms is to enable a definition of *substitution of terms for variables* which allows capture of free atoms by atom-abstractions while still respecting α-equivalence. The following lemma establishes this. First we give some terminology and notation for term-substitution.

Definition 4. *A* ***substitution*** σ *is a sort-respecting function from variables to terms with the property that* $\sigma(X) = X$ *for all but finitely many variables* X. *We shall write*

[1] A machine-checked proof of all the results using the theorem prover Isabelle can be found at `http://www.cl.cam.ac.uk/~cu200/Unification`.

$dom(\sigma)$ for the finite set of variables X satisfying $\sigma(X) \neq X$. If $dom(\sigma)$ consists of variables $X_1, \ldots, X_n$ and $\sigma(X_i) = t_i$ for $i = 1..n$, we shall sometimes write σ as

$$\sigma = [X_1 := t_1, \ldots, X_n := t_n]. \tag{8}$$

We write $\sigma(t)$ for the result of **applying a substitution** *σ to a term t; this is the term obtained from t by replacing each suspension $\pi{\cdot}X$ in t (as X ranges over $dom(\sigma)$) by the term $\pi{\cdot}\sigma(X)$ got by letting π act on the term $\sigma(X)$ using the definition in Fig. 1. For example, if $\sigma = [X := \langle b, Y\rangle]$ and $t = a.(a\,b){\cdot}X$, then $\sigma(t) = a.\langle a, (a\,b){\cdot}Y\rangle$. Given substitutions σ and σ', and freshness environments ∇ and ∇', we write*

$$(a) \quad \nabla' \vdash \sigma(\nabla) \qquad \textit{and} \qquad (b) \quad \nabla \vdash \sigma \approx \sigma' \tag{9}$$

to mean that (for a) $\nabla' \vdash a \# \sigma(X)$ holds for each $(a \# X) \in \nabla$ and (for b) $\nabla \vdash \sigma(X) \approx \sigma'(X)$ holds for all $X \in dom(\sigma) \cup dom(\sigma')$.

Lemma 2 (Substitution). *Substitution commutes with the permutation action: $\sigma(\pi{\cdot}t) = \pi{\cdot}(\sigma(t))$. Substitution preserves $\approx$ and $\#$ in the following sense:*

- *if $\nabla' \vdash \sigma(\nabla)$ and $\nabla \vdash t \approx t'$, then $\nabla' \vdash \sigma(t) \approx \sigma(t')$;*
- *if $\nabla' \vdash \sigma(\nabla)$ and $\nabla \vdash a \# t$, then $\nabla' \vdash a \# \sigma(t)$.*

Proof. The first sentence follows by induction on the structure of t. The second follows by induction on the proofs of $\nabla \vdash t \approx t'$ and $\nabla \vdash a \# t$ from the rules in Fig. 2, using the first sentence and the (proof of) Lemma 1. □

We claim that the relation $\approx$ defined in Fig. 2 gives the correct notion of α-equivalence for terms over a nominal signature. This is reasonable, given Lemma 1 and the fact that, by definition, it satisfies rules ($\approx$-abstraction-1) and ($\approx$-abstraction-2). Further evidence is provided by the following theorem, which shows that for ground terms $\approx$ agrees with the following more traditional definition of α-equivalence.

Definition 5 (Naïve α-equivalence). *Define the binary relation $t =_\alpha t'$ between terms over a nominal signature to be the least sort-respecting congruence relation satisfying $a.t =_\alpha b.[a \mapsto b]t$ whenever b is an atom (of the same sort as a) not occurring at all in t. Here $[a \mapsto b]t$ indicates the result of replacing all free occurrences of a with b in t.*

Theorem 1 (Adequacy). *If t and t' are* **ground terms** (i.e. terms with no variables and hence no suspensions) *over a nominal signature, then the relation $t =_\alpha t'$ of Def. 5 holds if and only if $\varnothing \vdash t \approx t'$ is provable from the rules in Fig. 2. Furthermore, $\varnothing \vdash a \# t$ is provable if and only if a is not in the set $FA(t)$ of free atoms of t.*
Proof. The proof is similar to the proof of [11, Proposition 2.2]. □

For non-ground terms, the relations $=_\alpha$ and $\approx$ *differ*. For example $a.X =_\alpha b.X$ always holds, whereas $\varnothing \vdash a.X \approx b.X$ is not provable unless $a = b$. This disagreement is to be expected, since we noted in the Introduction that $=_\alpha$ is *not* preserved by substitution, whereas from Lemma 2 we know that $\approx$ is.

Remark 1. Further evidence for the status of $\approx$ and $\#$ is provided by a natural interpretation of judgements provable from the rules in Fig. 2 in the universe of FM-sets [11]. The details will appear in the full version of this paper.

3 Unification

Given terms t and t' of the same sort over a nominal signature, can we decide whether or not there is a substitution of terms for the variables in t and t' that makes them equal in the sense of the relation $\approx$ introduced in the previous section? Since instances of $\approx$ in general are established modulo freshness constraints, it makes more sense to ask whether or not there is both a substitution σ and a freshness environment ∇ for which $\nabla \vdash \sigma(t) \approx \sigma(t')$ holds. As for ordinary first-order unification, solving such an equational problem may throw up *several* equational subproblems; but an added complication here is that because of rule ($\approx$-abstraction-2) in Fig. 2, equational problems may generate *freshness* problems, i.e. ones involving the relation $\#$. We are thus led to the following definition of unification problems for nominal equational logic.

Definition 6. *A* ***unification problem*** *P over a nominal signature is a finite set of atomic problems, each of which is either an* ***equational problem*** *$t \approx ?\ t'$ where t and t' are terms of the same sort over the signature, or a* ***freshness problem*** *$a \ \#?\ t$ where a is an atom and t a term over the signature. A* ***solution*** *for P consists of a pair (∇, σ) where ∇ is a freshness environment and σ is a substitution satisfying*

- $\nabla \vdash a \# \sigma(t)$ *for each* $(a \ \#?\ t) \in P$ *and*
- $\nabla \vdash \sigma(t) \approx \sigma(t')$ *for each* $(t \approx ?\ t') \in P$.

Such a pair is a ***most general solution*** *for P if given any other solution (∇', σ'), then there is a substitution σ'' satisfying $\nabla' \vdash \sigma''(\nabla)$ and $\nabla' \vdash \sigma'' \circ \sigma \approx \sigma'$. (Here we have used the notation of (9); and $\sigma'' \circ \sigma$ denotes the* ***substitution composition*** *of σ followed by σ'', given by $(\sigma'' \circ \sigma)(X) \stackrel{def}{=} \sigma''(\sigma(X))$.)*

Theorem 2 (Nominal unification). *There is an algorithm which, given any nominal unification problem, decides whether or not it has a solution and if it does, returns a most general solution.*

Proof. We describe an algorithm using labelled transformations directly generalising the presentation of first-order unification in [17, Sect. 2.6], which in turn is based upon the approach in [18]. (See also [1, Sect. 4.6] for a detailed exposition, but not using labels.) We use two types of labelled transformation between unification problems, namely

$$P \stackrel{\sigma}{\Longrightarrow} P' \quad \text{and} \quad P \stackrel{\nabla}{\Longrightarrow} P'$$

where the substitution σ is either the identity ε, or a single replacement $[X := t]$; and where the freshness environment ∇ is either empty $\varnothing$, or a singleton $\{a \# X\}$. The legal transformations are given in Fig. 3. This figure uses the notation $P \uplus P'$ to indicate *disjoint union* of problem sets; and the notation σP to indicate the problem resulting from applying the substitution σ to all the terms occurring in the problem P.

Given a unification problem P, the algorithm proceeds in two phases. In the first phase it applies as many $\stackrel{\sigma}{\Longrightarrow}$ transformations as possible (non-deterministically). If this results in a problem containing no equational subproblems then it proceeds to the second phase; otherwise it halts with failure. In the second phase it applies as many $\stackrel{\nabla}{\Longrightarrow}$ transformations as possible (non-deterministically). If this does not result in the empty

(≈?-unit)	$\{\langle\rangle \approx? \langle\rangle\} \uplus P \overset{\varepsilon}{\Longrightarrow} P$
(≈?-pair)	$\{\langle t_1, t_2\rangle \approx? \langle t_1', t_2'\rangle\} \uplus P \overset{\varepsilon}{\Longrightarrow} \{t_1 \approx? t_1', t_2 \approx? t_2'\} \cup P$
(≈?-function symbol)	$\{f\,t \approx? f\,t'\} \uplus P \overset{\varepsilon}{\Longrightarrow} \{t \approx? t'\} \cup P$
(≈?-abstraction-1)	$\{a.t \approx? a.t'\} \uplus P \overset{\varepsilon}{\Longrightarrow} \{t \approx? t'\} \cup P$
(≈?-abstraction-2)	$\{a.t \approx? a'.t'\} \uplus P \overset{\varepsilon}{\Longrightarrow} \{t \approx? (a\,a')\cdot t', a \,\#?\, t'\} \cup P$ provided $a \neq a'$
(≈?-atom)	$\{a \approx? a\} \uplus P \overset{\varepsilon}{\Longrightarrow} P$
(≈?-suspension)	$\{\pi\cdot X \approx? \pi'\cdot X\} \uplus P \overset{\varepsilon}{\Longrightarrow} \{a \,\#?\, X \mid a \in ds(\pi, \pi')\} \cup P$
(≈?-variable)	$\left.\begin{array}{l}\{t \approx? \pi\cdot X\} \uplus P\\ \{\pi\cdot X \approx? t\} \uplus P\end{array}\right\} \overset{\sigma}{\Longrightarrow} \sigma P$ with $\sigma = [X := \pi^{-1}\cdot t]$ provided X does not occur in t
(#?-unit)	$\{a \,\#?\, \langle\rangle\} \uplus P \overset{\varnothing}{\Longrightarrow} P$
(#?-pair)	$\{a \,\#?\, \langle t_1, t_2\rangle\} \uplus P \overset{\varnothing}{\Longrightarrow} \{a \,\#?\, t_1, a \,\#?\, t_2\} \cup P$
(#?-function symbol)	$\{a \,\#?\, f\,t\} \uplus P \overset{\varnothing}{\Longrightarrow} \{a \,\#?\, t\} \cup P$
(#?-abstraction-1)	$\{a \,\#?\, a.t\} \uplus P \overset{\varnothing}{\Longrightarrow} P$
(#?-abstraction-2)	$\{a \,\#?\, a'.t\} \uplus P \overset{\varnothing}{\Longrightarrow} \{a \,\#?\, t\} \cup P$ provided $a \neq a'$
(#?-atom)	$\{a \,\#?\, a'\} \uplus P \overset{\varnothing}{\Longrightarrow} P$ provided $a \neq a'$
(#?-suspension)	$\{a \,\#?\, \pi\cdot X\} \uplus P \overset{\nabla}{\Longrightarrow} P$ with $\nabla = \{\pi^{-1}\cdot a \,\#\, X\}$

Fig. 3. Labelled transformations.

problem, then it halts with failure; otherwise overall it has constructed a transformation sequence of the form

$$P \overset{\sigma_1}{\Longrightarrow} \cdots \overset{\sigma_n}{\Longrightarrow} P' \overset{\nabla_1}{\Longrightarrow} \cdots \overset{\nabla_m}{\Longrightarrow} \varnothing \tag{10}$$

(where P' does not contain any equational subproblems) and the algorithm returns the solution $(\nabla_1 \cup \cdots \cup \nabla_m, \sigma_n \circ \cdots \circ \sigma_1)$.

It is not hard to devise a well-founded ordering on nominal unification problems to show that each phase of the algorithm must terminate. So one just has to show that

(a) if the algorithm fails on P, then P has no solution; and
(b) if the algorithm succeeds on P, then the result it produces is a most general solution.

When failure happens it is because of certain subproblems that manifestly have no solution (e.g. in the first phase, $a \approx? a'$ with $a \neq a'$, and $\pi\cdot X \approx? f\,t$ or $f\,t \approx? \pi\cdot X$ with X occurring in t; in the second phase, $a \,\#?\, a$). Therefore part (a) is a consequence of the following two properties of transformations, where we write $\mathcal{U}(P)$ for the set of all solutions for a problem P:

$$\text{if } (\nabla', \sigma') \in \mathcal{U}(P) \text{ and } P \overset{\sigma}{\Longrightarrow} P', \text{ then } (\nabla', \sigma') \in \mathcal{U}(P') \text{ and } \nabla' \vdash \sigma' \circ \sigma \approx \sigma' \tag{11}$$

$$\text{if } (\nabla', \sigma') \in \mathcal{U}(P) \text{ and } P \overset{\nabla}{\Longrightarrow} P', \text{ then } (\nabla', \sigma') \in \mathcal{U}(P') \text{ and } \nabla' \vdash \sigma'(\nabla). \tag{12}$$

For part (b), one first shows

$$\text{if } (\nabla', \sigma') \in \mathcal{U}(P') \text{ and } P \overset{\sigma}{\Longrightarrow} P', \text{ then } (\nabla', \sigma' \circ \sigma) \in \mathcal{U}(P) \tag{13}$$

$$\text{if } (\nabla', \sigma') \in \mathcal{U}(P'),\ P \overset{\nabla}{\Longrightarrow} P' \text{ and } \nabla'' \vdash \sigma'(\nabla), \text{ then } (\nabla' \cup \nabla'', \sigma') \in \mathcal{U}(P). \tag{14}$$

From these and the fact that $(\varnothing, \varepsilon) \in \mathcal{U}(\varnothing)$, one gets that if a sequence like (10) exists, then $(\nabla, \sigma) \overset{\text{def}}{=} (\nabla_1 \cup \cdots \cup \nabla_m, \sigma_n \circ \cdots \circ \sigma_1)$ is in $\mathcal{U}(P)$. Furthermore from (11) and (12), we get that any other solution $(\nabla', \sigma') \in \mathcal{U}(P)$ satisfies $\nabla' \vdash \sigma'(\nabla)$ and $\nabla' \vdash \sigma' \circ \sigma \approx \sigma'$, so that (∇, σ) is indeed a most general solution. $\square$

Example 2. Using the first three function symbols of the nominal signature of Example 1 to represent λ-terms, the Quiz at the end of the Introduction translates into the following four unification problems over that signature, where a and b are distinct atoms of sort vid and $X_1, \ldots, X_7$ are distinct variables of sort exp:

$$\begin{aligned}
P_1 &\stackrel{\text{def}}{=} \{\mathtt{fn}\, a.\mathtt{fn}\, b.\mathtt{app}\langle X_1, \mathtt{vr}\, b\rangle \approx ?\ \mathtt{fn}\, b.\mathtt{fn}\, a.\mathtt{app}\langle \mathtt{vr}\, a, X_1\rangle\},\\
P_2 &\stackrel{\text{def}}{=} \{\mathtt{fn}\, a.\mathtt{fn}\, b.\mathtt{app}\langle X_2, \mathtt{vr}\, b\rangle \approx ?\ \mathtt{fn}\, b.\mathtt{fn}\, a.\mathtt{app}\langle \mathtt{vr}\, a, X_3\rangle\},\\
P_3 &\stackrel{\text{def}}{=} \{\mathtt{fn}\, a.\mathtt{fn}\, b.\mathtt{app}\langle \mathtt{vr}\, b, X_4\rangle \approx ?\ \mathtt{fn}\, b.\mathtt{fn}\, a.\mathtt{app}\langle \mathtt{vr}\, a, X_5\rangle\},\\
P_4 &\stackrel{\text{def}}{=} \{\mathtt{fn}\, a.\mathtt{fn}\, b.\mathtt{app}\langle \mathtt{vr}\, b, X_6\rangle \approx ?\ \mathtt{fn}\, a.\mathtt{fn}\, a.\mathtt{app}\langle \mathtt{vr}\, a, X_7\rangle\}.
\end{aligned}$$

Applying the nominal unification algorithm described above, we find that

- P_1 has no solution;
- P_2 has a most general solution given by $\nabla_2 = \varnothing$ and $\sigma_2 = [X_2 := \mathtt{vr}\, b, X_3 := \mathtt{vr}\, a]$;
- P_3 has a most general solution given by $\nabla_3 = \varnothing$ and $\sigma_3 = [X_4 := (a\, b) \cdot X_5]$;
- P_4 has a most general solution given by $\nabla_4 = \{b \# X_7\}$ and $\sigma_3 = [X_6 := (b\, a) \cdot X_7]$.

Derivations for P_1 and P_4 are sketched in Fig. 4. Using the Adequacy Theorem 1, one can interpret these solutions as the following statements about the λ-terms from the quiz.

Quiz answers

1. There is no λ-term M_1 making the first pair of terms α-equivalent.
2. The only solution for the second problem is to take $M_2 = b$ and $M_3 = a$.
3. For the third problem we can take M_5 to be any λ-term, so long as we take M_4 to be the result of swapping all occurrences of a and b throughout M_5.
4. For the last problem, we can take M_7 to be any λ-term that *does not contain free occurrences of* b, so long as we take M_6 to be the result of swapping all occurrences of b and a throughout M_7, or equivalently (since b is not free in M_7), taking M_6 to be the result of replacing all free occurrences of a in M_7 with b.

Remark 2 (Atoms are not variables). Nominal unification unifies variables, but it does not unify atoms. Indeed the operation of identifying two atoms by renaming does not necessarily preserve the validity of the judgements in Fig. 2. For example, $\varnothing \vdash a.b \approx c.b$ holds if $b \neq a, c$; but renaming b to be a in this judgement we get $\varnothing \vdash a.a \approx c.a$, which does not hold so long as $a \neq c$. Referring to Def. 2, you will see that we do allow variables ranging over sorts of atoms; and such variables can be unified like any other variables. However, if A is such a variable, then it cannot appear in abstraction position, i.e. as $A.t$. This is because we specifically restricted abstraction to range over atoms, rather than over arbitrary terms of atom sort. Such a restriction seems necessary to obtain single, most general, solutions to nominal unification problems. For without such a restriction, because of rule ($\approx$-abstraction-2) we would also have to allow variables to appear on the left-hand side of freshness relations and in suspended permutations. So then we would get unification problems like $\{(A\, B) \cdot C \approx ?\ C\}$, where A, B and C are variables of atom sort; this has two incomparable solutions, namely $(\varnothing, [A := B])$ and $(\{A \# C, B \# C\}, \varepsilon)$.

$P_1 \stackrel{\varepsilon}{\Longrightarrow} \{\mathtt{fn}\, b.\mathtt{app}\langle X_1, \mathtt{vr}\, b\rangle \approx?\, \mathtt{fn}\, b.\mathtt{app}\langle \mathtt{vr}\, b, (a\, b)\cdot X_1\rangle,\ a \,\#?\, \mathtt{fn}\, a.\mathtt{app}\langle \mathtt{vr}\, a, X_1\rangle\}$	($\approx$?-abstraction-2)
$\stackrel{\varepsilon}{\Longrightarrow} \{\mathtt{app}\langle X_1, \mathtt{vr}\, b\rangle \approx?\, \mathtt{app}\langle \mathtt{vr}\, b, (a\, b)\cdot X_1\rangle,\ a \,\#?\, \mathtt{fn}\, a.\mathtt{app}\langle \mathtt{vr}\, a, X_1\rangle\}$	($\approx$?-abstraction-1)
......	...
$\stackrel{\varepsilon}{\Longrightarrow} \{X_1 \approx?\, \mathtt{vr}\, b,\ \mathtt{vr}\, b \approx?\, (a\, b)\cdot X_1,\ a \,\#?\, \mathtt{fn}\, a.\mathtt{app}\langle \mathtt{vr}\, a, X_1\rangle\}$	($\approx$?-pair)
$\stackrel{\sigma}{\Longrightarrow} \{\mathtt{vr}\, b \approx?\, \mathtt{vr}\, a,\ a \,\#?\, \mathtt{fn}\, a.\mathtt{app}\langle \mathtt{vr}\, a, \mathtt{vr}\, b\rangle\}$ with $\sigma = [X_1 := \mathtt{vr}\, b]$	($\approx$?-variable)
$\stackrel{\varepsilon}{\Longrightarrow} \{b \approx?\, a,\ a \,\#?\, \mathtt{fn}\, a.\mathtt{app}\langle \mathtt{vr}\, a, \mathtt{vr}\, b\rangle\}$	($\approx$?-fnctn symbol)
FAIL	
$P_4 \stackrel{\varepsilon}{\Longrightarrow} \{\mathtt{fn}\, b.\mathtt{app}\langle \mathtt{vr}\, b, X_6\rangle \approx?\, \mathtt{fn}\, a.\mathtt{app}\langle \mathtt{vr}\, a, X_7\rangle\}$	($\approx$?-abstraction-1)
$\stackrel{\varepsilon}{\Longrightarrow} \{\mathtt{app}\langle \mathtt{vr}\, b, X_6\rangle \approx?\, \mathtt{app}\langle \mathtt{vr}\, b, (b\, a)\cdot X_7\rangle,\ b \,\#?\, \mathtt{app}\langle \mathtt{vr}\, a, X_7\rangle\}$	($\approx$?-abstraction-2)
......	...
$\stackrel{\varepsilon}{\Longrightarrow} \{b \approx?\, b,\ X_6 \approx?\, (b\, a)\cdot X_7,\ b \,\#?\, \mathtt{app}\langle \mathtt{vr}\, a, X_7\rangle\}$	($\approx$?-fnctn symbol)
$\stackrel{\varepsilon}{\Longrightarrow} \{X_6 \approx?\, (b\, a)\cdot X_7,\ b \,\#?\, \mathtt{app}\langle \mathtt{vr}\, a, X_7\rangle\}$	($\approx$?-atom)
$\stackrel{\sigma}{\Longrightarrow} \{b \,\#?\, \mathtt{app}\langle \mathtt{vr}\, a, X_7\rangle\}$ with $\sigma = [X_6 := (b\, a)\cdot X_7]$	($\approx$?-variable)
$\stackrel{\varnothing}{\Longrightarrow} \{b \,\#?\, \langle \mathtt{vr}\, a, X_7\rangle\}$	($\#$?-fnctn symbol)
......	...
$\stackrel{\varnothing}{\Longrightarrow} \{b \,\#?\, a,\ b \,\#?\, X_7\}$	($\#$?-fnctn symbol)
$\stackrel{\varnothing}{\Longrightarrow} \{b \,\#?\, X_7\}$	($\#$?-atom)
$\stackrel{\nabla}{\Longrightarrow} \varnothing$ with $\nabla = \{b \,\#\, X_7\}$	($\#$?-suspension)

Fig. 4. Example derivations.

4 Related Work

Most previous work on unification for languages with binders is based on forms of higher-order unification, i.e. solving equations between λ-terms modulo $\alpha\beta\eta$-equivalence by capture-avoiding substitution of terms for function variables. Notable among that work is Miller's *higher-order pattern unification* used in his L_λ logic programming language [20]. This kind of unification retains the good properties of first-order unification: a linear-time decision procedure and existence of most general unifiers. However it imposes a restriction on the form of λ-terms to be unified; namely that function variables may only be applied to distinct bound variables. An empirical study by Michaylov and Pfenning [19] suggests that most unifications arising dynamically in higher-order logic programming satisfy Miller's restriction, but that it rules out some useful programming idioms. For us, the main disadvantage of L_λ is one common to most approaches based on higher-order abstract syntax: one cannot *directly* express the common idiom of possibly-capturing substitution of terms for metavariables. Instead one has to replace metavariables, X, with function variables applied to distinct lists of (bound) variables, $F\, x_1 \ldots x_n$, and use capture-avoiding substitution.

Dowek *et al* [8] present a version of higher-order pattern unification for $\lambda\sigma$ (a λ-calculus with de-Bruijn indices and explicit substitutions) in which unification problems are solved, like in nominal unification, by textual replacements of terms for variables; however a 'pre-cooking' operation ensures that the textual replacements can be (faithfully) related to capture-avoiding substitutions. It seems that nominal unification problems can be encoded into higher-order pattern unification problems using a *non-trivial* translation (the details of this encoding still remain to be investigated). But even if it turns out that it is possible to simulate nominal unification by higher-order pattern unification, the calculations involved in translating our terms into higher-order pattern terms and then using higher-order pattern unification are far more intricate than our simple algorithm that solves nominal unification problems directly.

Hamana [13, 14] manages to add possibly-capturing substitution to a language like Miller's L_λ. This is achieved by adding syntax for explicit renaming operations and by recording implicit dependencies of variables upon bindable names in a typing context. The mathematical foundation for Hamana's system is the model of binding syntax of Fiore *et al* [9]. The mathematical foundation for our work appeared concurrently [10] and is in a sense complementary. For in Hamana's system the typing context restricts which terms may be substituted for a variable by giving a finite set of names that *must contain* the free names of such a term; whereas we give a finite set of names which the term's free variables *must avoid*. Since α-conversion is phrased in terms of avoidance, i.e. freshness of names, our approach seems more natural if one wants to compute α-equivalences concretely. On top of that, our use of name permutations, rather than arbitrary renaming functions, leads to technical simplifications. In any case, the bottom line is that Hamana's system seems more complicated than the one presented here and does not possess most general unifiers.

5 Conclusion

In this paper we have proposed a solution to the problem of finding possibly-capturing substitutions that unify terms involving binders up to α-conversion. To do so we considered a many-sorted first-order term language with distinguished collections of constants called *atoms* and with *atom-abstraction* operations for binding atoms in terms. This provides a simple, but flexible, framework for specifying binding operations and their scopes, in which the bound entities are explicitly named. By using variables prefixed with suspended permutations, one can have substitution of terms for variables both allow capture of atoms by binders and respect α-equivalence (renaming of bound atoms). The definition of α-equivalence for the term language makes use of an auxiliary *freshness* relation between atoms and terms which generalises the 'not a free atom of' relation from ground terms to terms with variables; furthermore, because variables stand for unknown terms, hence with unknown free atoms, it is necessary to make hypotheses about the freshness of atoms for variables in judgements about term equivalence and freshness. This reliance on 'freshness' is the main novelty—it arises from the work reported in [11, 23]. It leads to a new notion of unification problem in which instances of both equivalence and freshness have to be solved by giving term-substitutions and (possibly) freshness conditions on variables in the solution. We showed that this unification problem is decidable and unitary.

Cheney and Urban [5] are investigating the extent to which nominal unification can be used in resolution-based proof search for a form of first-order logic programming for languages with binders (with a view to providing better machine-assistance for structural operational semantics). Such a logic programming language should permit a concrete, 'nominal' approach to bound entities in programs while ensuring that computation (which in this case is the computation of answers to queries) respects α-equivalence between terms. This is illustrated with the Prolog-like program in Fig. 5, which implements a simple typing algorithm for λ-terms. Interesting is the third clause. First, note the term `(lam x.M)`, which unifies with any λ-abstraction. The binder `x`,

```
type Gamma (var X) A :- mem (pair X A) Gamma.
type Gamma (app M N) B :- type Gamma M (arrow A B), type Gamma N A.
type Gamma (lam x.M) (arrow A B) / x#Gamma :- type (pair x A)::Gamma M B.

mem A A::Tail.
mem A B::Tail :- mem A Tail.
```

Fig. 5. A Prolog-like program implementing the typing rules for the simply-typed λ-calculus.

roughly speaking, has in the 'nominal' approach a value which can be used in the body of the clause, for example for adding `(pair x A)` to the context Gamma. Second, the freshness constraint x#Gamma ensures that Gamma cannot be replaced by a term that contains x freely. Since this clause is intended to implement the usual rule for typing λ-abstractions

$$\frac{\{x : A\} \cup \Gamma \rhd M : B}{\Gamma \rhd \lambda x.M : A \supset B}$$

its operational behaviour is given by: choose fresh names for Gamma, x, M, A and B (this is standard in Prolog-like languages), unify the head of the clause with the goal formula, apply the resulting unifier to the body of the clause and make sure that Gamma is not replaced by a term that contains freely the fresh name we have chosen for x. Similar facilities for *functional programming* already exist in the FreshML language, built upon the same foundations: see [27] and `www.freshml.org`. We are also interested in the special case of 'nominal matching' and its application to term-rewriting modulo α-equivalence.

If these applications show that nominal unification is practically useful, then it becomes important to study its complexity. The presentations of the term language in Section 2 and of the algorithm in Section 3 were chosen for clarity and to make the proof of correctness easier[2] rather than for efficiency. In any case, it remains to be investigated whether the swapping and freshness computations that we have added to ordinary, first-order unification result in greater than linear-time complexity.

Acknowledgements: We thank Gilles Dowek, Roy Dyckhoff, Dale Miller, Francois Pottier and Helmut Schwichtenberg for comments on this work. This research was supported by UK EPSRC grants GR/R29697 (Urban) and GR/R07615 (Pitts and Gabbay).

References

1. F. Baader and T. Nipkow. *Term Rewriting and All That.* Cambridge University Press, 1998.
2. H. P. Barendregt. *The Lambda Calculus: Its Syntax and Semantics*. North-Holland, 1984.
3. L. Caires and L. Cardelli. A spatial logic for concurrency II. In *Proc. of CONCUR 2002*, volume 2421 of *LNCS*, pages 209–225. Springer-Verlag, 2002.
4. L. Cardelli, P. Gardner, and G. Ghelli. Manipulating trees with hidden labels. In *Proc. of FOSSACS*, volume 2620 of *LNCS*, pages 216–232. Springer-Verlag, 2003.

[2] See `http://www.cl.cam.ac.uk/~cu200/Unification` for the Isabelle proof scripts.

5. J. Cheney and C. Urban. αProlog, a fresh approach to logic programming modulo α-equivalence. In *Proc. of UNIF 2003*, number DSIC-II/12/03 in Departamento de Sistemas Informáticos y Computación Technical Report Series. Universidad Politécnica de Valencia, 2003.
6. G. Dowek. Higher-order unification and matching. In A. Robinson and A. Voronkov, editors, *Handbook of Automated Reasoning*, chapter 16, pages 1009–1062. Elsevier, 2001.
7. G. Dowek, T. Hardin, and C. Kirchner. Higher-order unification via explicit substitutions. In *10th Symposium of LICS*, pages 366–374. IEEE Computer Society Press, 1995.
8. G. Dowek, T. Hardin, C. Kirchner, and F. Pfenning. Higher-order unification via explicit substitutions: the case of higher-order patterns. In *Proc. of JICSLP*, pages 259–273, 1996.
9. M. P. Fiore, G. D. Plotkin, and D. Turi. Abstract syntax and variable binding. In *14th Symposium of LICS*, pages 193–202. IEEE Computer Society Press, 1999.
10. M. J. Gabbay and A. M. Pitts. A new approach to abstract syntax involving binders. In *14th Symposium of LICS*, pages 214–224. IEEE Computer Society Press, 1999.
11. M. J. Gabbay and A. M. Pitts. A new approach to abstract syntax with variable binding. *Formal Aspects of Computing*, 13:341–363, 2002.
12. C. A. Gunter. *Semantics of Programming Languages: Structures and Techniques*. Foundations of Computing. MIT Press, 1992.
13. M. Hamana. A logic programming language based on binding algebras. In *Proc. of TACS 2001*, volume 2215 of *LNCS*, pages 243–262. Springer-Verlag, 2001.
14. M. Hamana. Simple β_0-unification for terms with context holes. In *Proc. of UNIF 2002*, 2002. Unpublished proceedings.
15. M. Hashimoto and A. Ohori. A typed context calculus. *TCS*, 266:249–271, 2001.
16. F. Honsell, M. Miculan, and I. Scagnetto. An axiomatic approach to metareasoning on nominal algebras in HOAS. In *Proc. of ICALP 2001*, volume 2076 of *LNCS*, pages 963–978. Springer-Verlag, 2001.
17. J. W. Klop. Term rewriting systems. In S. Abramsky, D. M. Gabbay, and T. S. E. Maibaum, editors, *Handbook of Logic in Computer Science, Volume 2*, pages 1–116. OUP, 1992.
18. A. Martelli and U. Montanari. An efficient unification algorithm. *ACM Trans. Programming Languages and Systems*, 4(2):258–282, 1982.
19. S. Michaylov and F. Pfenning. An empirical study of the runtime behaviour of higher-order logic programs. In *Proc. Workshop on the λProlog Programming Language*, pages 257–271, 1992. CIS Technical Report MS-CIS-92-86.
20. D. Miller. A logic programming language with lambda-abstraction, function variables, and simple unification. *Journal of Logic and Computation*, 1:497–536, 1991.
21. R. Milner, J. Parrow, and D. Walker. A calculus of mobile processes (parts I and II). *Information and Computation*, 100:1–77, 1992.
22. F. Pfenning and C. Elliott. Higher-order abstract syntax. In *Proc. ACM-SIGPLAN Conference on Programming Language Design and Implementation*, pages 199–208. ACM Press, 1988.
23. A. M. Pitts. Nominal logic: A first order theory of names and binding. In *Proc. of TACS 2001*, volume 2215 of *LNCS*, pages 219–242. Springer-Verlag, 2001.
24. A. M. Pitts and M. J. Gabbay. A metalanguage for programming with bound names modulo renaming. In *Proc. of MPC2000*, volume 1837 of *LNCS*, pages 230–255. Springer-Verlag, 2000.
25. G. D. Plotkin. An illative theory of relations. In *Situation Theory and its Applications*, volume 22 of *CSLI Lecure Notes*, pages 133–146. Stanford University, 1990.
26. M. Sato, T. Sakurai, and Y. Kameyama. A simply typed context calculus with first-class environments. *Journal of Functional and Logic Programming*, 2002(4), 2002.
27. M. R. Shinwell, A. M. Pitts, and M. J. Gabbay. FreshML: Programming with binders made simple. In *Proc. of ICFP 2003*. ACM Press, 2003.

Friends or Foes? Communities in Software Verification*

Helmut Veith

Abt. f. Datenbanken und Artificial Intelligence
Institut für Informationsysteme, Technische Universität Wien
Favoritenstrasse 9-11, 1040 Wien, Austria
veith@dbai.tuwien.ac.at

Abstract. In contrast to hardware which is finite-state and based on relatively few ample principles, software systems generally give rise to infinite state spaces, and are described in terms of programming languages involving rich semantical concepts. The challenges of software verification can be addressed only by a combined effort of different communities including, most notably, model checking, theorem proving, symbolic computation, static analysis, compilers, and abstract interpretation. We focus on a recent family of tools which use predicate abstraction and theorem proving to extract a finite state system amenable to model checking.

The last fifteen years have seen model checking mature from a purely academic pursuit to an important tool for hardware developers in industry which complements more traditional techniques such as simulation and testing. The main practical problem in model checking is the state explosion problem, i.e., the enormous size of the finite-state system. In addition to information-preserving symbolic methods using BDDs or SAT procedures, *abstraction* is the most important method to combat state explosion. The critical problem is how to establish an appropriate level of abstraction of the system to make the analysis tractable: Too fine a level of granularity leads to a large state space, while a coarse abstraction may result in false negatives. As manual abstraction is tedious and error-prone, *counterexample-guided abstraction methods* [4] are used to automatically improve an initial abstraction mapping until a good abstraction is found.

Simultaneously, (semi)manual proof assistants have also been used successfully for hardware verification tasks involving intricate mathematical reasoning, for example, to assert the correctness of floating point arithmetic [6].

As hardware verification has matured, and the number of critical software applications is constantly increasing, software verification has attracted renewed interest during the last years. Software however abounds in features not present in hardware, and programs do in general not yield finite state models. Specific

* This research was supported by the Austrian Science Fund Project N Z29-N04, the EU Research and Training Network GAMES, and the Office of Naval Research (ONR). The views and conclusions contained in this document are those of the author.

M. Baaz and J.A. Makowsky (Eds.): CSL 2003, LNCS 2803, pp. 528–529, 2003.

concepts distinguishing hardware from software include complex base types such as strings and high precision floating point variables, dynamic data structures, object orientation, user-defined types and classes, procedure calls, recursion, function variables, dynamic method lookup, overloading, templates, include files, exceptions, callbacks, use of files and databases, absent source code for libraries and system calls, as well as embedded, mobile and self-modifying code. It is thus evident that software verification is a highly interdisciplinary enterprize which has to draw on the experience and methods from diverse fields including theorem proving, symbolic computation, compilers, static analysis, and abstract interpretation. Despite differences in tradition, mathematical schools, and engineering aptitude, the collaboration between these fields is steadily increasing.

Predicate abstraction [5] is a paradigmatic approach to combine theorem proving, abstraction, and model checking. In predicate abstraction, the abstract model is described in terms of logical formulas ("predicates") ranging over the concrete model; the relationships between the predicates are computed using a theorem prover, and the abstract model is then verified by a model checker. Several model checkers for C based on predicate abstraction have been developed, in particular SLAM [1, 2] which is targeted at device drivers, BLAST [7] for verifying embedded systems using a lazy abstraction technique, and MAGIC [3] whose focus is on modular verification and security applications. The verification tools use automated theorem provers, efficient decision procedures, and static analysis to extract an abstract model, and employ counterexample-guided refinement techniques for improving the abstraction.

References

1. T. Ball and S. K. Rajamani. Boolean programs: A model and process for software analysis, 2000. MSR Technical Report 2000-14.
2. T. Ball and S. K. Rajamani. The SLAM project: debugging system software via static analysis. In *Proc. Principles of Programming Languages (POPL)*, pages 1–3, 2002.
3. S. Chaki, E. M. Clarke, A. Groce, S. Jha, and H. Veith. Modular verification of software components in C. In *Proc. Int. Conference on Software Engineering (ICSE)*, pages 385–395, 2003.
4. E. Clarke, O. Grumberg, S. Jha, Y. Lu, and H. Veith. Counterexample-guided abstraction refinement. In *Proc. Computer Aided Verification (CAV)*, volume 1855 of *Lecture Notes in Computer Science*, pages 154–169, 2000. Extended Version to appear in J.ACM.
5. S. Graf and H. Saidi. Construction of abstract state graphs with PVS. In *Proc. Computer Aided Verification (CAV)*, volume 1254 of *Lecture Notes in Computer Science*, pages 72–83, June 1997.
6. J. Harrison. Formal verification of floating point trigonometric functions. In *Proc. 3rd International Conference on Formal Methods in Computer-Aided Design (FMCAD)*, volume 1954 of *Lecture Notes in Computer Science*, pages 217–233, 2000.
7. T. A. Henzinger, R. Jhala, R. Majumdar, and G. Sutre. Lazy abstraction. In *Proc. Principles of Programming Languages (POPL)*, pages 58–70, 2002.

More Computation Power for a Denotational Semantics for First Order Logic

Kess F.M. Vermeulen

CWI, P.O. Box 94079, 1090 GB Amsterdam, The Netherlands
kees.vermeulen@cwi.nl

Abstract. This paper starts from a denotational semantics for first order logic proposed by Apt. He interprets first order logic as a programming language within the declarative paradigm, but different from the usual approach of logic programming. For his proposal Apt obtains soundness with respect to the standard interpretation of first order logic. Here we consider the *expressivity* (or: *computation power*) of the system proposed, i.e., the programs for which no *error* messages are produced. To gain computational realism a decrease in expressivity compared to first order logic is inevitable. In [1] the expressivity was compared with logic programming with both positive and negative results. Here we consider three ways of improving the situation: replacing the original interpretation of conjunction-as-sequential-composition by a symmetric interpretation of conjunction; extending the framework with the option of recursive definitions by Clark completions; replacing the interpretation of disjunction-as-choice by an interpretation which is in the style of 'backtracking': only produce an *error* if all options lead to *error*.
For each improvement we obtain a soundness result. The three improvements are developed independently in a way that makes sure that also their combination preserves soundness.
The combination gives a considerable improvement of the expressivity of the denotational semantics that Apt proposed, both in the comparison with first order logic and in the comparison with logic programming.

Keywords: declarative programming, denotational semantics, expressivity of programming languages

1 Introduction

The declarative programming paradigm has a lot to offer. It treats simple declarative statements as instructions to investigate how to make the statement true: programs make you dreams come true. Logic programming (LP) has shown that such promises are not always beyond the scope of reason. Still, as was pointed out in Apt's paper [1], there are drawbacks inherent in logic programming. In particular, the Horn clause fragment, the fragment initially considered in Kowalski [7], is too small for serious programming purposes. Extensions of the fragment to arbitrary first order formulas — see, for example [8] and [2] — turned out to complicate matters considerably and the attempts did not converge to a commonly accepted solution. Second, in logic programming computation relies heavily on

M. Baaz and J.A. Makowsky (Eds.): CSL 2003, LNCS 2803, pp. 530–543, 2003.

the term model. Or, to put it differently, it is confined to the Herbrand universe. This makes it hard for programmers to benefit from specific properties of the model for which their programs are written. The enhancement of logic programming to constraint programming (CLP) improves the situation[1], but it does so by introducing an oracle: the constraint solver.

In Apt [1] these points are the motivation for a new attempt for the interpretation of first order logic as a programming language. These theoretical developments go hand in hand with the development of Alma_0, as described in [3]. In fact, [1] tries to give a denotational semantics for the declarative fragment of Alma_0.

In the proposed set up ordinary terms are replaced by $\mathcal{J}$-terms. The second important ingredient of [1] is its treatment of equations: they sometimes work as an instruction for unification, sometimes as a test for 'real' equality. This way the spirit of computation of logic programming is preserved and at the same time the gap between the level of terms and the intended model is narrowed. The interpretation of equations is then extended in a natural way to the full first order language, where computational realism is ensured by the introduction of a special value for *error*. The result is a natural interpretation of first order logic as a programming language which is sound with respect to the standard interpretation. But, as was already pointed out in [1], the comparison with LP is not decisive: there are clear examples where Apt's proposal works better, but also examples where LP outperforms the denotational semantics.

Hence it is clearly desirable to further extend the computation power of the proposal, i.e., the set of programs that does not give rise to an error statement. The expressivity question was taken up by Smaus [10] for equations. Here we discuss three other ways to improve the expressivity of the interpretation. First we reconsider the interpretation of conjunction. Apt interprets $\wedge$ as sequential composition. This introduces an order sensitivity into the system that seriously restricts its expressivity. Hence we propose an alternative interpretation for $\wedge$ which is symmetric, i.e., not sensitive to order. Next we include the option of definition by recursion for predicates. We use ideas from [4] and use Clark completions of logic programs. Finally, we consider the interpretation of $\vee$. Apt [1] interprets $\vee$ as choice. Hence, if one of the disjuncts produces an error message the overall system reproduces this message for the disjunction as a whole. Here we prevent this propagation of *errors* and replace the treatment of $\vee$ with an interpretation reminiscent of backtracking: *error* is reported only if none of the choices gives a proper outcome.

This gives us three ways of improving the system *qua* expressivity. They are considered independently and each improvement is shown to respect *soundness* with respect to first order logic. It will be clear that they can be combined to gain a considerable overall improvement of expressivity in a sound way.

In what follows we start by summarizing the proposal by Apt (section 2). Then, the three improvements are presented (in sections 3, 4 and 5). Finally, we sum up the results in the conclusion (section 6) where we also present additional

[1] Consult [9] and [6] for more information.

discussion and options for further work. The overall conclusion will be that by the proposed improvements we obtain a considerable gain in expressivity for the approach of Apt [1] and hence achieve an improvement in both the comparison with LP and the standard interpretation of the first order language. Thereby the denotational semantics becomes an even more attractive way of working out the idea of declarative programming and starts to look like a serious alternative to LP.

2 The Semantic Toolkit

We start by recalling the semantics provided by Apt in [1]. The idea of this semantics is to provide a uniform computational meaning for the first-order formulas independent of the underlying interpretation and without a constraint store. This yields a limited way of processing formulas in the sense that occasionally an *error* may arise. Let us recall the relevant definitions:

Definition 1. *Consider a language of terms L and an algebra $\mathcal{J}$ for it.*

- *Consider a term of L in which we replace some of the variables by the elements of the domain of $\mathcal{J}$. We call the resulting object a* generalized term.
- *Given a generalized term t we define its $\mathcal{J}$-*evaluation *as follows. Each ground term of L evaluates to a unique value in $\mathcal{J}$. In a generalized term t we replace each maximal ground subterm of t by its value in $\mathcal{J}$ (maximal partial evaluation). The resulting generalized term is called a $\mathcal{J}$-*term *and denoted $[\![t]\!]_{\mathcal{J}}$.*
- *A $\mathcal{J}$-*substitution *is a finite mapping from variables to $\mathcal{J}$-terms that assigns to each variable x in its domain a $\mathcal{J}$-term different from x. The notation is: $\{x_1/h_1, \ldots, x_n/h_n\}$. We write ϵ for the $\mathcal{J}$-*substitution with the empty domain. Application of a $\mathcal{J}$-substitution θ to a generalized term t is defined in the standard way and denoted by $t\theta$. The set of $\mathcal{J}$-substitutions is called *Subs*.
- The *composition* of two $\mathcal{J}$-substitutions θ and η, *is written as $\theta\eta$ and is defined as the unique $\mathcal{J}$-substitution γ such that for a variable x*

$$x\gamma = [\![(x\theta)\eta]\!]_{\mathcal{J}}.$$

$\mathcal{J}$-substitutions generalize both the usual notion of substitution and the notion of a valuation, i.e., the assignment of domain values to variables. Now we use $\mathcal{J}$-substitutions to interpret equations as follows: (Here and elsewhere we do not indicate the dependency of the semantics on the underlying interpretation or algebra.)

$$[\![s = t]\!](\theta) := \begin{cases} \{\theta\{s\theta/[\![t\theta]\!]_{\mathcal{J}}\}\} & \text{if } s\theta \text{ is a variable that does not occur in } t\theta, \\ \{\theta\{t\theta/[\![s\theta]\!]_{\mathcal{J}}\}\} & \text{if } t\theta \text{ is a variable that does not occur in } s\theta \\ & \text{and } s\theta \text{ is not a variable,} \\ \{\theta\} & \text{if } [\![s\theta]\!]_{\mathcal{J}} \text{ and } [\![t\theta]\!]_{\mathcal{J}} \text{ are identical,} \\ \emptyset & \text{if } s\theta \text{ and } t\theta \text{ are ground and } [\![s\theta]\!]_{\mathcal{J}} \neq [\![t\theta]\!]_{\mathcal{J}}, \\ \{\mathit{error}\} & \text{otherwise.} \end{cases}$$

Now consider an interpretation $\mathcal{I}$ based on the algebra $\mathcal{J}$. Given an atomic formula $A(t_1,\ldots,t_n)$ and a $\mathcal{J}$-substitution θ we denote the interpretation of A in $\mathcal{I}$ by $A_\mathcal{I}$. We say that:

- $A(t_1,\ldots,t_n)\theta$ is *true* if $A(t_1,\ldots,t_n)\theta$ is ground and $([\![t_1\theta]\!]_\mathcal{J},\ldots,[\![t_n\theta]\!]_\mathcal{J}) \in A_\mathcal{I}$,
- $A(t_1,\ldots,t_n)\theta$ is *false* if $A(t_1,\ldots,t_n)\theta$ is ground and $([\![t_1\theta]\!]_\mathcal{J},\ldots,[\![t_n\theta]\!]_\mathcal{J}) \notin A_\mathcal{I}$.

Note that this way the interpretation of A is fixed by $\mathcal{I}$. In LP terminology: it is a 'built-in' predicate. We do not (yet) have the option of defining predicates by recursion. To deal with existential quantification we use the $DROP_x$ operation which is defined as follows:

$$\begin{aligned} &xDROP_x(\theta) := x \\ &yDROP_x(\theta) := y\theta \qquad \text{for } y \neq x \end{aligned}$$

$DROP_x$ is extended pointwise to subsets of $Subs \cup \{error\}$. This way we add *error* as a kind of 'special substitution.' Now $[\![\cdot]\!]$ is defined for a $\mathcal{J}$-substitution θ by structural induction[2]:

- $[\![A(\mathbf{t})]\!](\theta) := \begin{cases} \{\theta\} & \text{if } A(\mathbf{t})\theta \text{ is true,} \\ \emptyset & \text{if } A(\mathbf{t})\theta \text{ is false,} \\ \{error\} & \text{otherwise, that is if } A(\mathbf{t})\theta \text{ is not ground,} \end{cases}$
- $[\![\phi_1 \wedge \phi_2]\!](\theta) := [\![\phi_2]\!]([\![\phi_1]\!](\theta))$,
- $[\![\phi_1 \vee \phi_2]\!](\theta) := [\![\phi_1]\!](\theta) \cup [\![\phi_2]\!](\theta)$,
- $[\![\neg\phi]\!](\theta) := \begin{cases} \{\theta\} & \text{if } [\![\phi]\!](\theta) = \emptyset, \\ \emptyset & \text{if } \theta \in [\![\phi]\!](\theta), \\ \{error\} & \text{otherwise,} \end{cases}$
- $[\![\exists x\, \phi]\!](\theta) := DROP_u([\![\phi\{x/u\}]\!](\theta))$, where u is a fresh variable.

For *error* we set: $[\![\phi]\!](error) := \{error\}$. The following example clarifies the way we interpret atoms and conjunction.

Example 1. Consider the standard algebra for the language of arithmetic with the set of integers as domain. We denote its elements by $\ldots, \mathbf{-2}, \mathbf{-1}, \mathbf{0}, \mathbf{1}, \mathbf{2}, \ldots$. Each constant i evaluates to the element $\mathbf{i}$. Then we have:

1. $[\![y = z - 1 \wedge z = x + 2]\!](\{x/\mathbf{1}\}) = [\![z = x + 2]\!](\{x/\mathbf{1}, y/z - \mathbf{1}\}) = \{\{x/\mathbf{1}, y/\mathbf{2}, z/\mathbf{3}\}\}$,
2. $[\![y = 1 \wedge z = 1 \wedge y - 1 = z - 1]\!](\epsilon) = \{\{y/\mathbf{1}, z/\mathbf{1}\}\}$,
3. $[\![y = 1 \wedge z = 2 \wedge y < z]\!](\epsilon) = \{\{y/\mathbf{1}, z/\mathbf{2}\}\}$,
4. $[\![x = 0 \wedge \neg(x = 1)]\!](\epsilon) = \{\{x/\mathbf{0}\}\}$,
5. $[\![y - 1 = z - 1 \wedge y = 1 \wedge z = 1]\!](\epsilon) = \{error\}$,
6. $[\![y < z \wedge y = 1 \wedge z = 2]\!](\epsilon) = \{error\}$,
7. $[\![\neg(x = 1) \wedge x = 0]\!](\epsilon) = \{error\}$.

[2] Here and elsewhere we write $\mathbf{t}$ for a sequence $t_1,\ldots,t_n$.

In this semantics conjunction is not commutative: it is sequential composition. Consequently it is important in which order the formulas are processed. Also, all *error* assignments are propagated, without exception.

In [1] (and also in [5]) the semantics is shown to be *sound* w.r.t. first order logic, i.e.:

- $\eta \in [\![\phi]\!](\theta) \;\Rightarrow\; \models_{\mathcal{I}} \phi\eta$
- $[\![\phi]\!](\theta) = \emptyset \;\Rightarrow\; \models_{\mathcal{I}} \neg\phi\theta$

This is a crucial result: it shows that the answers that the programs produce are reliable, as long as there are no *error* messages. But it does not show how often we get an *error* message. In this paper we take up this question and investigate how to get more successful runs for more programs.

3 Symmetric Conjunction

We refer to the full first order language as $\mathcal{L}_\wedge$. We compare it to $\mathcal{L}_{\overline{\wedge}}$, the language in which the binary connective $\overline{\wedge}$ replaces $\wedge$. $\overline{\wedge}$ is a symmetric conjunction and is interpreted as follows:

$$[\![\phi_1 \overline{\wedge} \phi_2]\!](\theta) \;=\; [\![\phi_1]\!]([\![\phi_2]\!](\theta)) \;\cup\; [\![\phi_2]\!]([\![\phi_1]\!](\theta)).$$

Hence $\phi_1 \overline{\wedge} \phi_2$ tells us to compute the ϕ_i $(i = 1, 2)$ in arbitrary order. This is one way to get out of the problem noted in the example: we have simple local instructions to consider both orders for the execution of the ϕ_i.

Note that $\mathcal{L}_{\overline{\wedge}}$ inherits the soundness of $\mathcal{L}_\wedge$: $\phi_1 \overline{\wedge} \phi_2 \in \mathcal{L}_{\overline{\wedge}}$ is interpreted as the union of the interpretations of the formulas $\phi_1 \wedge \phi_2$ and $\phi_2 \wedge \phi_1 \in \mathcal{L}_\wedge$, both of which have the same truth value in first order logic. As the semantics of $\wedge$ respects this truth value in a sound way, also the semantics of $\overline{\wedge}$ has to be sound. More formally we note that:

$$\begin{aligned}
&\eta \in [\![\phi_1 \overline{\wedge} \phi_2]\!](\theta) \Rightarrow \\
&\eta \in [\![\phi_1 \wedge \phi_2]\!](\theta) \text{ or } \eta \in [\![\phi_2 \wedge \phi_1]\!](\theta) \Rightarrow \text{(by soundness of } \mathcal{L}_\wedge) \\
&\mathcal{I} \models \phi_i\eta \text{ for both } i = 1, 2 \Rightarrow \\
&\mathcal{I} \models (\phi_1 \overline{\wedge} \phi_2)\eta
\end{aligned}$$

and a similar argument works for the case where $[\![\phi_1 \overline{\wedge} \phi_2]\!](\theta) = \emptyset$.

We want to compare executions of one formula $\phi \in \mathcal{L}_{\overline{\wedge}}$ with computations of all the formulas in $\mathcal{L}_\wedge$ that can be obtained by taking permutations of the conjuncts that occur in ϕ. For this purpose we introduce the following translation:

Definition 2. *The permutation translation* $\pi(\cdot) : \mathcal{L}_{\overline{\wedge}} \to \mathcal{L}_\wedge$ *is given by the following table:*

$\phi \in \mathcal{L}_{\overline{\wedge}}$	$\pi(\phi) \in \mathcal{L}_\wedge$
$s = t$	$s = t$
$A(t_1, \ldots, t_n)$	$A(t_1, \ldots, t_n)$
$\phi_1 \vee \phi_2$	$\pi(\phi_1) \vee \pi(\phi_2)$
$\neg\phi$	$\neg\pi(\phi)$
$\exists x\, \phi$	$\exists x\, \pi(\phi)$
$\phi_1 \overline{\wedge} \phi_2$	$(\pi(\phi_1) \wedge \pi(\phi_2)) \;\vee\; (\pi(\phi_2) \wedge \pi(\phi_1))$

If we want to distinguish the semantics for $\mathcal{L}_\wedge$ from the semantics for $\mathcal{L}_{\overline{\wedge}}$, we use the distinguishing connective as subscript: $[\![\cdot]\!]_\wedge$ vs $[\![\cdot]\!]_{\overline{\wedge}}$. The definition of $\pi(\phi)$ basically replaces $\phi_1 \overline{\wedge} \phi_2$ by $(\phi_1 \wedge \phi_2) \vee (\phi_2 \wedge \phi_1)$, but it repeats this 'all the way down' in the subformulas ϕ_i $(i = 1, 2)$. This way of translating has its limits: it respects the bracket structure of larger conjunctions. This shows up as soon as we consider conjunctions of three formulas. Consider, for given ϕ_1, ϕ_2 and ϕ_3: $\psi_1 = \phi_1 \wedge (\phi_2 \wedge \phi_3)$ and $\psi_2 = (\phi_1 \wedge \phi_2) \wedge \phi_3$. Then we find:

- $\pi(\psi_1) =$
  $(\phi_1 \wedge (\phi_2 \wedge \phi_3)) \vee (\phi_1 \wedge (\phi_3 \wedge \phi_2)) \vee ((\phi_2 \wedge \phi_3) \wedge \phi_1) \vee ((\phi_3 \wedge \phi_2) \wedge \phi_1)$
- $\pi(\psi_2) =$
  $((\phi_1 \wedge \phi_2) \wedge \phi_3) \vee ((\phi_2 \wedge \phi_1) \wedge \phi_3) \vee (\phi_3 \wedge (\phi_1 \wedge \phi_2)) \vee (\phi_3 \wedge (\phi_2 \wedge \phi_1))$

Both translations generate four orders of evaluation for the three ϕ_i, where there are, of course, six orders of three formulas to consider. A binary connective will always generate a bracketing of the three formulas and then it cannot generate more than four orders of computation in the semantics. This is a limit on what one can do with one binary connective in the current set up.

The translation provides a good way to test the behavior of $\overline{\wedge}$. We can check whether the permutation translation is correct: does the systematic replacement of $\wedge$ by $\overline{\wedge}$ in a formula ϕ, lead to the evaluation of all the orders that turn up in $\pi(\phi)$? In other words, does the following property hold for all θ:

Proposition 1. *For every $\phi \in \mathcal{L}_{\overline{\wedge}}$ we have:*

$$[\![\phi]\!]_{\overline{\wedge}} = [\![\pi(\phi)]\!]_\wedge \qquad (\sharp)$$

Proof: The proof is by induction on the complexity of $\phi \in \mathcal{L}_{\overline{\wedge}}$. We only give the crucial case for $\overline{\wedge}$: $[\![\phi_1 \overline{\wedge} \phi_2]\!]_{\overline{\wedge}}(\theta) = [\![\phi_1]\!]_{\overline{\wedge}}([\![\phi_2]\!]_{\overline{\wedge}}(\theta)) \cup [\![\phi_2]\!]_{\overline{\wedge}}([\![\phi_1]\!]_{\overline{\wedge}}(\theta)) = [\![(\pi(\phi_1) \wedge \pi(\phi_2)) \vee (\pi(\phi_2) \wedge \pi(\phi_1))]\!]_\wedge(\theta) = [\![\pi(\phi_1 \overline{\wedge} \phi_2)]\!]_\wedge(\theta)$ □

To conclude this section, we present an interesting trick: we show how, at the cost of more *error* messages, order of computation can be regulated in $\mathcal{L}_{\overline{\wedge}}$. Hence the trick shows that replacing $\wedge$ by $\overline{\wedge}$ does not lead to a significant loss of control over the order of computation in programs.

In items 2 and 5 of the list on page 533 we saw that:

$$[\![(y = 1 \wedge z = 1) \wedge y - 1 = z - 1]\!](\epsilon) = \{\{y/1,\ z/1\}\}$$
$$[\![(y - 1 = z - 1 \wedge (y = 1 \wedge z = 1)]\!](\epsilon) = \{\mathit{error}\}$$

because $[\![y - 1 = z - 1]\!](\epsilon) = \{\mathit{error}\}$. Now:

$$[\![(y - 1 = z - 1 \,\overline{\wedge}\, (y = 1 \wedge z = 1)]\!](\epsilon) = \{\{y/1,\ z/1\}, \mathit{error}\}$$

This shows that $\overline{\wedge}$ does improve the situation of item 5: we find the obvious outcome $\{y/1,\ z/1\}$. But it also shows that it does not improve the situation for item 2: $\overline{\wedge}$ automatically considers *both* orders, so *error* gets included.

This observation generalizes to a trick for fixing the order of computation at the cost of including *error* messages.

Example 2. Consider six distinct variables x, y, u, v, p, q and set:

- $\rho_1 = x = y$
- $\rho_2 = (x+1 = y+1 \barwedge u = v)$
- $\rho_3 = (x+1 = y+1 \barwedge u+1 = v+1) \barwedge p = q$

Call $r_i = [\![\rho_i]\!]$ (for $i = 1, 2, 3$). It is easy to check that, starting from the empty state ϵ, $r_1; r_2; r_3$ is the only successful order of execution of the r_i. It gives: $\langle \epsilon, \{x/y,\ u/v,\ p/q\} \rangle$.

The trick[3] shows that the semantics of equations allows control over the order of execution of a program, simply by adding equations of the type ρ_i at the appropriate points in the program. In particular, this allows us to simulate sequential $\wedge$ with symmetric $\barwedge$. For example: $A(\mathbf{t}) \wedge B(\mathbf{s})$ compares to $(A(\mathbf{t}) \barwedge \rho_1) \barwedge (B(\mathbf{s}) \barwedge \rho_2)$. The example can easily be extended into a kind of 'backwards translation,' but there is one *proviso*: the control of order obtained by adding equations of variables generates extra *error* messages in many cases! Setting this *proviso* aside, from the point of view of expressivity it does not really matter whether we *add* $\barwedge$ to the language with $\wedge$ or whether we *replace* $\wedge$ by $\barwedge$ (as long as enough variables are available).

By combination of the symmetric conjunction with the interpretation of disjunction discussed in section 5 we can also remove the extra *error* messages, to get an even more elegant way of regulating the order of execution of programs, in a setting where conjunction is symmetric.

4 Recursive Procedures

Apt and Smaus [4] discuss the incorporation of recursive procedures in the framework. They show how this makes the connection with logic programs precise. However, they give an *operational* rather then a denotational semantics for first order logic. Here we present the appropriate denotational analogue. Then we compare the way the two conjunctions work out in the extension with recursive procedures.

First we introduce new atomic predicates $p_i(x^i_1, \ldots, x^i_{m_i})$ $(1 \leq i \leq n)$ for the recursive procedures. For each of these predicates a *definition*: $\psi_i(x^i_1, \ldots, x^i_{m_i})$ is available. Such a definition is a *positive* formula[4] in which the new predicates can occur. Let's agree to refer to the positive fragment of a language $\mathcal{L}$ by adding the superscript $+$: $\mathcal{L}^+$. The free variables of the definition of a procedure are always a subset of the variables of the procedure itself. Definitions like this are collected in sets for which we use notation $\mathcal{D}$.

[3] Note that the successful order of execution is generated by the formula $\rho_1 \barwedge (\rho_2 \barwedge \rho_3)$, but not by the formula $(\rho_1 \barwedge \rho_2) \barwedge \rho_3$. Hence $\barwedge$ is not associative. To put this in a more computational perspective: there are limits to the options for delay of computation in a setting with just one binary connective $\barwedge$.

[4] This means that $\neg$ does not occur in ψ_i. As we regard $\phi \rightarrow \psi$ as shorthand for $\neg(\phi \wedge \neg\psi)$ and $\forall x\ \psi$ as shorthand for $\neg\exists x\ \neg\phi$, implications and universal quantifications do not occur in positive formulas either.

Apt and Smaus [4] have shown that the definition sets $\mathcal{D}$ correspond to Clark completions of logic programs. Because of this correspondence it makes sense to think of $p_i(x_1^i,\ldots,x_{m_i}^i)\{x_1^i/t_1,\ldots,x_{m_i}^i/t_{m_i}\}$ as a *procedure call*. Below we write $\mathbf{x^i}$ for the sequence $(x_1^i,\ldots,x_{m_i}^i)$ and $p_i(\mathbf{t^i})$ for a procedure call. The extension of $\mathcal{L}_\wedge$ with the procedures defined by $\mathcal{D}$ is called $\mathcal{L}_{\wedge\mathcal{D}}$. We can do exactly the same thing for $\mathcal{L}_{\bar{\wedge}}$, of course, to obtain $\mathcal{L}_{\bar{\wedge}\mathcal{D}}$.

We do not just want to consider definitions by recursion using the sequential conjunction, but also for the symmetric conjunction. Then, we will also investigate the correctness of the permutation translation for the extended languages. In order to do this, we first extend the definition of the permutation translation to recursive procedures by setting: $\pi(p_i(\mathbf{t^i})) = p_i(\mathbf{t^i})$. The second thing we have to take care of is the proper transfer of definitions from $\mathcal{L}_{\bar{\wedge}}^+$ to $\mathcal{L}_\wedge^+$. For a sensible comparison it is necessary that we translate the definitions. As notation we use: $\chi_i(\mathbf{x^i}) = \pi(\psi_i(\mathbf{x^i}))$. This is how the translation is extended. Next we extend the semantics to include interpretations of procedure calls by using *approximating formulas* for the definitions in $\mathcal{D}$ [5].

Definition 3. *Let a definition set* $\mathcal{D} = \{\psi_1(\mathbf{x^1}),\ldots,\psi_n(\mathbf{x^n})\}$ *be given for procedures* $p_1(\mathbf{x^1}),\ldots,p_n(\mathbf{x^n})$*. Then we set for each* $1 \le i \le n$ *and each natural number* $\alpha \ge 0$*:*

- $\psi_i^0 = \bot$
- $\psi_i^{\alpha+1} = \psi_i\{p_1/\psi_1^\alpha,\ldots,p_n/\psi_n^\alpha\}$

A similar definition works for the translated definitions, χ_i in $\mathcal{L}_\wedge$.

Let's consider a simple example to see how it works. In the example we see first the definitions of two recursive procedures $p_i(x,y)$ $(i = 1,2)$. Then we see the approximating formulas for the definitions.

Example:

- $\psi_1(x,y) = r(x,y) \vee \exists z\,(r(x,z) \wedge p_2(z,y))$
- $\psi_2(x,y) = s(x,y) \vee \exists z\,(s(x,z) \wedge p_1(z,y))$

- $\psi_1(x,y)^0 = \bot$
- $\psi_2(x,y)^0 = \bot$
- $\psi_1(x,y)^1 = r(x,y)$
- $\psi_2(x,y)^1 = s(x,y)$
- $\psi_1(x,y)^2 = r(x,y) \vee \exists z\,(r(x,z) \wedge s(z,y))$
- $\psi_2(x,y)^2 = s(x,y) \vee \exists z\,(s(x,z) \wedge r(z,y))$
- $\psi_1(x,y)^3 = r(x,y) \vee \exists z\,(r(x,z) \wedge (s(z,y) \vee \exists u\,(s(z,u) \wedge r(u,y))))$
- $\psi_2(x,y)^3 = s(x,y) \vee \exists z\,(s(x,z) \wedge (r(z,y) \vee \exists u\,(r(z,u) \wedge s(u,y))))$
- ...

[5] Here we do not mention the variables to improve readability. $\bot$ is used as a shorthand for some convenient contradiction.

Note that formulas never contain procedure calls. So, we already have interpretations $[\![\psi_i(\mathbf{t^i})^\alpha]\!]_{\overline{\wedge}}(\theta)$ and $[\![\chi_i(\mathbf{t^i})^\alpha]\!]_{\wedge}(\theta)$ (for $\alpha \geq 0$). Hence the following definition makes sense:

Definition 4. *We define for each $\theta \in Subs_{\mathcal{J}} \cup \{error\}$ and $1 \leq i \leq n$:*

$$\circ\ [\![p_i(\mathbf{t^i})]\!]_{\overline{\wedge}}(\theta) = \bigcup\{[\![\psi_i(\mathbf{t^i})^\alpha]\!]_{\overline{\wedge}}(\theta) : \alpha \geq 0\}$$
$$\circ\ [\![p_i(\mathbf{t^i})]\!]_{\wedge}(\theta) = \bigcup\{[\![\chi_i(\mathbf{t^i})^\alpha]\!]_{\wedge}(\theta) : \alpha \geq 0\}$$

This gives a perfectly consistent definition of a denotational semantics for the procedure calls, but is it the one we want? After all, the idea was that the $p_i(\mathbf{x^i})$ would correspond to recursively defined procedures. Hence we are after a semantics in terms of least fixed points of appropriate operators. So, we have to find the appropriate operators and check that the approximating formulas indeed give us the least fixed point of these operators.

Here it is convenient to regard $[\![\phi]\!]$ as a binary relation on $Subs_{\mathcal{J}} \cup \{error\}$. This works by considering $\eta[\![\phi]\!]\theta$ instead of $\eta \in [\![\phi]\!](\theta)$. We can read $\mathcal{D} = \{\psi_1(\mathbf{x^1}), \ldots, \psi_n(\mathbf{x^n})\}$ as a set of formula *schemes* in $\mathcal{L}^{+}_{\overline{\wedge}}$ with *parameters* $p_i(\mathbf{x^i})$ ($1 \leq i \leq n$) and each $\psi_i(\mathbf{x^i})$ can be seen as an n-ary operator that assigns to denotations: $[\![\phi_1]\!], \ldots, [\![\phi_n]\!]$, the value: $[\![\psi_i\{p_1/\phi_1, \ldots, p_n/\phi_n\}]\!]$.

The operator $[\![\psi_i(\mathbf{x^i})]\!]$ is continuous, because $\psi_i(\mathbf{x^i})$ only contains the connectives: $\vee$, $\overline{\wedge}$, $\exists x$. And the denotational semantics of these connectives only uses the operations in $\{\cup, ;, DROP_u\}$, each of which is continuous. Hence the definitions $\psi_i(\mathbf{x^i})$ generate continuous operators on binary relations and the theory of fixed points for continuous operators applies: the operators have a least fixed point and we can approximate these fixed points from below, in the standard way.

The exact same argument[6] works for the $\chi_i(\mathbf{x^i})$ in the language with the sequential conjunction. This gives the following result:

Proposition 2. *For each $1 \leq i \leq n$:*

$$\circ\ [\![p_i(\mathbf{t^i})]\!]_{\overline{\wedge}} = lfp([\![\psi_i(\mathbf{t^i})]\!]_{\overline{\wedge}})$$
$$\circ\ [\![p_i(\mathbf{t^i})]\!]_{\wedge} = lfp([\![\chi_i(\mathbf{t^i})]\!]_{\wedge})$$

At this point we have a sensible way of extending $\mathcal{L}_{\overline{\wedge}}$ and $\mathcal{L}_{\wedge}$ with recursive procedures and we have a proper denotational semantics for the extended languages. This is the second significant improvement of the expressivity of the denotational format proposed by Apt in [1].

Now all that remains to be done is to check that the two ways of extending expressivity can be combined. We do this by checking the correctness of the translation π with the added clause for procedure calls. Fortunately this is easy to check, as the following lemma holds:

[6] Note that in the argument details about substitutions (of terms $\mathbf{t^i}$ for variables $\mathbf{x^i}$ and of formulas ϕ_i for parameters p_i) have been omitted.

Lemma 1. $[\![p_i(\mathbf{t^i})]\!]_{\overline{\wedge}} = [\![p_i(\mathbf{t^i})]\!]_{\wedge}$.

Proof:

$[\![p_i(\mathbf{t^i})]\!]_{\overline{\wedge}} =$ (by the extended definition of $[\![\cdot]\!]_{\overline{\wedge}}$)

$\bigcup\{[\![\psi_i(\mathbf{t^i})^\alpha]\!]_{\overline{\wedge}} : \alpha \geq 0\} =$ ($\psi_i(\mathbf{t^i})^\alpha$ does not contain parameters p_j, so it enjoys ($\sharp$))

$\bigcup\{[\![\pi(\psi_i(\mathbf{t^i})^\alpha)]\!]_{\wedge} : \alpha \geq 0\} =$ (re-ordering disjuncts gives the χ_i^α from the permutations of ψ_i^α and vice versa)

$\bigcup\{[\![\chi_i(\mathbf{t^i})^\alpha]\!]_{\wedge} : \alpha \geq 0\} =$ (by the extended definition of $[\![\cdot]\!]_{\wedge}$)

$[\![p_i(\mathbf{t^i})]\!]_{\wedge}$ □

As we have set $\pi(p_i(\mathbf{t^i})) = p_i(\mathbf{t^i})$, this indeed gives the correctness of $\pi(\cdot)$ for the language extended with recursive procedures.

Proposition 3. *The extension of the permutation translation $\pi(\cdot)$ to $\mathcal{L}_{\overline{\wedge}\mathcal{D}}$ preserves correctness: for each $\phi \in \mathcal{L}_{\overline{\wedge}\mathcal{D}} : [\![\phi]\!]_{\overline{\wedge}} = [\![\pi(\phi)]\!]_{\wedge}$.* □

By now we have considered two ways of extending the expressivity of the proposal in [1] and we have shown that they can be combined. In the next step we reconsider the semantics of $\vee$, in particular the amount of *error* messages that $\vee$ produces.

5 Choice without *error*

In order to produce even more sound computations without *error* messages, we alter the semantics of disjunction. The inspiration for this alternative notion is *backtracking* in LP: search along alternative branches of the computation tree and only report *error* if no successful computation arises.

The altered semantics is presented with a slightly modified notion of computation state: instead of working with $\text{POW}(Subs \cup \{error\})$, we now work with $\text{POW}(Subs) \cup \{error\}$. This means, informally, that we now consider *error* as an additional computation state, where before we considered *error* as an additional substitution. We do this because we do not want computation states in which both *error* and real substitutions are present, i.e., we want to avoid the situation $\{error, \theta, \ldots\}$. Instead of checking all the time that we succeed in doing so, we prefer to exclude such combinations from the outset, by working in $\text{POW}(Subs) \cup \{error\}$.

As an example where we now get an improved result we consider $[\![(y = x + 1 \ \wedge \ x = 0) \ \vee \ (x = 0 \ \wedge \ y = x + 1)]\!](\epsilon)$. In the original set up this is interpreted as an instruction to either compute $[\![(y = x + 1 \ \wedge \ x = 0)]\!](\epsilon)$ or $[\![(x = 0 \ \wedge \ y = x + 1)]\!](\epsilon)$. (So, it corresponds to the symmetric conjunction $y = x + 1 \mathbin{\overline{\wedge}} x = 0$ introduced in section 3.) The result is $\{error, \{x/0, y/1\}\}$. In the semantics proposed in this section we only produce $\{\{x/0, y/1\}\}$. This way only the 'sensible options' that the original semantics produces, remain. *error* messages only arise if there are no sensible solutions.

On $\text{POW}(Subs) \cup \{error\}$ we consider an ordering $\sqsubseteq$ instead of $\subseteq$. This is necessary, because we now have a special computation state that is not a subset of *Subs*. So, we have to explain how it compares to subsets of *Subs*. The definition is as follows ($\Theta, \Theta' \subseteq Subs$) :

$\Theta \sqsubseteq \Theta'$ iff $\Theta \subseteq \Theta'$ and
$\emptyset \sqsubseteq error \sqsubseteq \Theta$ (for $\Theta \neq \emptyset$)

In a picture:

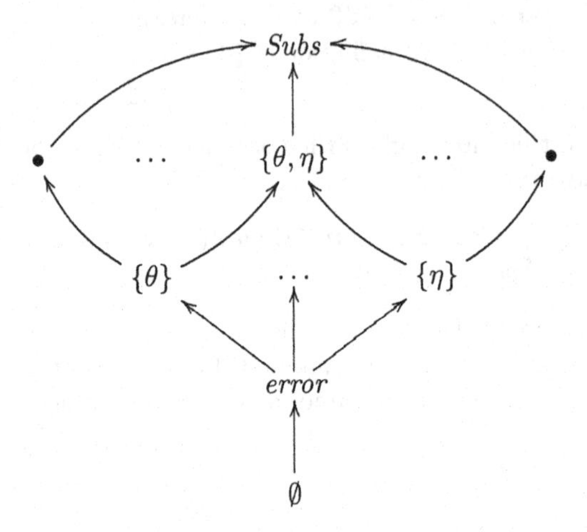

It is easy to check that this is a lattice. We write $\sqcup$ (and $\sqcap$) for the join (and meet, resp.) of the lattice. Hence, the truth table for $\sqcup$ is as follows:

$\sqcup$	$\emptyset$	*error*	Θ
$\emptyset$	$\emptyset$	*error*	Θ
error	*error*	*error*	Θ
Θ'	Θ'	Θ'	$\Theta \cup \Theta'$

Here and later on Θ and Θ' are subsets of *Subs*. Note that the new ingredient of the semantics, $\sqcup$, avoids the co-occurrence of *error* with real substitutions: nowhere in the table a co-occurrence of substitutions with *error* messages turns up. This is crucial for the correctness of the definition of the semantics below.

We also have to define $DROP_x$ for the new *error* state. It will be no surprise that we choose:

$DROP_x(error) := error.$

The semantics basically remains 'as before', i.e., as in section 2 and [1]. But the definitions now have to be read as definitions in the new semantic universe. So,

$$[\![\phi_1 \vee \phi_2]\!](\theta) = [\![\phi_1]\!](\theta) \sqcup [\![\phi_2]\!](\theta)$$

and *error* replaces $\{error\}$ in the other clauses of the definition. For example, $[\![\phi]\!](error) = error.$

We obtain a soundness result for the new set up of the semantics by comparing it to the semantics from [1], for which we write $[\![\cdot]\!]_o$: the old semantics. The proposition below makes explicit what we promised above: in the new version of the semantics we have simply removed *error* messages from the old version. We give the proposition as a result about $[\![\phi]\!](\theta)$ (for $\theta \in \mathit{Subs}$). Then it is obvious how to generalize it to arbitrary computation states.

Proposition 4.

1. $\theta \in [\![\phi]\!](\eta) \;\Rightarrow\; \theta \in [\![\phi]\!]_o(\eta)$
2. $[\![\phi]\!](\eta) = \emptyset \;\Rightarrow\; [\![\phi]\!]_o(\eta) = \emptyset$
3. $[\![\phi]\!]_o(\theta) = \{\mathit{error}\} \;\Rightarrow\; [\![\phi]\!](\theta) = \mathit{error}$

Proof:

The third clause follows from the first two. The proof of the first two cases is a simultaneous induction on the complexity of ϕ. For the atomic cases, we just note that the two versions of the semantics co-incide. Now the induction proceeds without problems. We just present two cases. Of course we should give the crucial case: $\vee$. And for purpose of illustration we also include another case: $\wedge$.

$\phi_1 \wedge \phi_2$: 1. $\theta \in [\![\phi_1 \wedge \phi_2]\!](\eta) \Rightarrow$
$\exists\zeta : \zeta \in [\![\phi_1]\!](\eta) \;\&\; \theta \in [\![\phi_2]\!](\zeta) \Rightarrow$ (by induction hypothesis)
$\exists\zeta : \zeta \in [\![\phi_1]\!]_o(\eta) \;\&\; \theta \in [\![\phi_2]\!]_o(\zeta) \Rightarrow$
$\theta \in [\![\phi_1 \wedge \phi_2]\!]_o(\eta)$

2. $[\![\phi_1 \wedge \phi_2]\!](\eta) = \emptyset \Rightarrow$
either: $[\![\phi_1]\!](\eta) = \emptyset$. Now the induction hypothesis (for 2) gives $[\![\phi_1]\!]_o(\eta) = \emptyset$, from which $[\![\phi_1 \wedge \phi_2]\!]_o(\eta) = \emptyset$ follows.
else: $[\![\phi_1]\!](\eta) \neq \emptyset$. But in this case we know that *error* has not turned up in $[\![\phi_1]\!](\eta)$, as it does not appear in $[\![\phi_2]\!]([\![\phi_1]\!](\eta))$. So we know, by induction hypothesis (for 1), that for each of the $\zeta \in [\![\phi_1]\!](\eta)$, also $\zeta \in [\![\phi_1]\!]_o(\eta)$, and that for each of these ζ: $[\![\phi_2]\!](\zeta) = \emptyset$. Now we apply the induction hypothesis (for 2) to each $\zeta \in [\![\phi_1]\!](\eta)$.

$\phi_1 \vee \phi_2$: 1. $\theta \in [\![\phi_1 \vee \phi_2]\!](\eta) \Rightarrow$
$\theta \in [\![\phi_i]\!](\eta)$ for some $i \in \{1,2\} \Rightarrow$ (by induction hypothesis)
$\theta \in [\![\phi_i]\!]_o(\eta)$ for some $i \in \{1,2\} \Rightarrow$
$\theta \in [\![\phi_1 \vee \phi_2]\!]_o(\eta)$

2. $[\![\phi_1 \vee \phi_2]\!](\eta) = \emptyset \Rightarrow$
$[\![\phi_i]\!](\eta) = \emptyset$ for both $i \in \{1,2\} \Rightarrow$ (by induction hypothesis)
$[\![\phi_i]\!]_o(\eta) = \emptyset$ for both $i \in \{1,2\} \Rightarrow$
$[\![\phi_1 \vee \phi_2]\!]_o(\eta) = \emptyset$

From this proposition the soundness of the semantics follows as a corollary from the soundness of the old semantics, $[\![\cdot]\!]_o$.

Corollary 1. *The semantics is sound with respect to first order logic, i.e.:*

- *For each $\mathcal{J}$-substitution $\eta \in [\![\phi]\!](\theta)$:*
  $\mathcal{J} \models \phi\eta$
- *If $[\![\phi]\!](\theta) = \emptyset$, then:*
  $\mathcal{J} \models \neg\phi\theta$

So, we have succeeded in yet another extension of the expressivity of the denotational semantics in [1]: disjunction-as-backtracking. For the combination of this extension with the extensions of the previous two sections, simply note that the definitions we gave there still make perfect sense with the new notion of computation state and disjunction: also when we treat *error* as a special computation state rather than a special substitution, we can define a symmetric conjunction and use Clark completions for the incorporation of recursion. Then the relevant propositions and their proofs only require minimal alterations.

6 Conclusion and Prospects

In this paper we have considered ways of extending the expressivity of the denotational semantics for first order logic proposed by Apt in [1]. Such extensions of expressivity correspond to extensions in computation power: we can write more '*error*-free' programs.

The extensions of expressivity considered are obtained by: deminishing order sensitivity of the semantics; adding options for definition by recursion; interpreting choice in the style of backtracking.

We were able to show that the proposed extensions are *sound*: the extensions do not give rise to a conflict with respect to the standard semantics of the first order language. Also, we have seen that the three ways of improving expressivity can be combined and that their combination also preserves soundness. Hence, we now have an interpretation of first order logic as a programming language with a computation power that is significantly larger than in the original proposal in [1].

The increase of expressivity is also relevant for the comparison with LP. In LP conjunction is symmetric, as witnessed by the strong completeness result for LP. So, by removing the order sensitivity of conjunction already one important disadvantage is removed from the approach of [1]. Recursion is at the heart of LP. Hence adding the option of definition by recursion here clearly is another crucial way of regaining ground on LP. And also the removal of *error* messages in the interpretation of $\vee$ is relevant: it reminds us of backtracking in LP.

So, we conclude that the view of first order logic as a programming language proposed in [1] can be made to work for a considerable fragment of the first order language. Furthermore, this fragment compares favorably with the expressive power of LP.

As further topics of immediate interest we note the combination of the semantics of this paper with tools from *constraint programming*. LP has been able to gain computational attractiveness by combinations with such tools, resulting in *constraint logic programming* (CLP). For the semantics of [1] this combination has been achieved in [5].

References

1. K.R. Apt. A denotational semantics for first-order logic. In *Proc. of the computational logic conference (CL2000)*, Lecture Notes in Artificial Intelligence 1861, pages 53–69. Springer Verlag, 2000.
2. K.R. Apt and R. Bol. Logic programming and negation: a survey. *Journal of Logic Programming*, pages 9–71, 1994.
3. K.R. Apt, J. Brunekreef, V. Partington, and A. Schaerf. Alma-0: An imperative language that supports declarative programming. *ACM Toplas*, 20(5):1014–1066, 1998.
4. K.R. Apt and J.-G. Smaus. Rule-based versus procedure-based view on programming. In *Joint Bulletin of the Novosibirsk Computing Center and Institute of Informatics Systems*, volume 16 of *Computer Science*, pages 75–97. , 2001.
5. K.R. Apt and C.F.M. Vermeulen. First-order logic viewed as a constraint programming language. In A. Voronkov and M. Baaz, editors, *Proceedings of LPAR02*, LNAI2514, pages 19–35. Springer Verlag, 2002.
6. J. Jaffar and J.M. Maher. Constraint logic programming: a survey. *Journal of Logic Programming*, 19/20, 1994.
7. R. Kowalski. Predicate logic as a programming language. In *Proc. of the IFIP Congress 1974*, pages 569–574. North Holland, 1974.
8. J.W. Lloyd. *Foundations of Logic Programming.* Springer Verlag, 1987. second edition.
9. K. Marriott and P. Stuckey. *Programming with Constraints: an introduction.* The MIT Press, 1998.
10. J.-G. Smaus. The optimality of apt's generic semantics for first-order equations. work in progress, 2002.

Effective Model Completeness of the Theory of Restricted Pfaffian Functions
(Abstract)

Nicolai Vorobjov

Dept. of Computer Science
University of Bath

Pfaffian functions, introduced by Khovanskii in late 70s, are analytic functions satisfying triangular systems of first order partial differential equations with polynomial coefficients. They include for instance algebraic and elementary transcendental functions in the appropriate domains, iterated exponentials, and *fewnomials*. A simple example, due to Osgood, shows that the first order theory of reals expanded by restricted Pfaffian functions does not admit quantifier elimination. On the other hand, Gabrielov and Wilkie proved (non-constructively) that this theory is model complete, i.e., *one type* of quantifiers can be eliminated. The talk will explain some ideas behind recent algorithms for this *quantifier simplification* which are based on effective cylindrical cell decompositions of sub-Pfaffian sets. Complexities of these algorithms are essentially the same as the ones which existed for a particular case of semialgebraic sets.

M. Baaz and J.A. Makowsky (Eds.): CSL 2003, LNCS 2803, p. 544, 2003.

Effective Quantifier Elimination over Real Closed Fields
(Abstract)

Nicolai Vorobjov

Dept. of Computer Science
University of Bath

In early 30s A. Tarski, motivated by a problem of automatic theorem proving in elementary algebra and geometry, suggested an algorithm for quantifier elimination in the first order theory of the reals. The complexity of Tarski's algorithm is a non-elementary function of the format of the input formula. In mid-70s a group of algorithms appeared based on the idea of a cylindrical cell decomposition and having an elementary albeit doubly-exponential complexity, even for deciding closed existential formulae. The tutorial will explain some ideas behind a new generation of algorithms which were designed during 80s and 90s and have, in a certain sense, optimal (singly-exponential) complexity. In a useful particular case of closed existential formulae (i.e., deciding feasibility of systems of polynomial equations and inequalities) these new algorithms are theoretically superior to procedures known before in numerical analysis and computer algebra.

M. Baaz and J.A. Makowsky (Eds.): CSL 2003, LNCS 2803, p. 545, 2003.

Fast Infinite-State Model Checking in Integer-Based Systems

Tatiana Rybina and Andrei Voronkov

University of Manchester
{rybina,voronkov}@cs.man.ac.uk

Abstract. In this paper we discuss the use of logic for reachability analysis for infinite-state systems. Infinite-state systems are formalized using transition systems over a first-order structure. We establish a common ground relating a large class of algorithms by analyzing the connections between the symbolic representation of transition systems and formulas used in various reachability algorithms. We consider in detail the so-called *guarded assignment systems* and *local reachability algorithms*. We show how an implementation of local reachability algorithms and a new incremental algorithm for finding Hilbert's base in the system BRAIN resulted in much faster reachability checking than in systems using constraint libraries and decision procedures for Presburger's arithmetic. Experimental results demonstrate that problems in protocol verification which are beyond the reach of other existing systems can be solved completely automatically.

1 Introduction

Reachability properties arise is many applications of verification. In this paper we discuss reachability algorithms in infinite-state systems. Infinite-state systems are formalized using transition systems over a first-order structure. We analyze the connections between the symbolic representation of transition systems and formulas which are used in various reachability algorithms. Our main results are related to the so-called *guarded assignment systems*.

This paper serves two main purposes. First, it formalizes infinite-state systems using model-theoretic notions and discusses reachability algorithms based on this formalization. Though many results and observations of this paper form part of folklore circulating in the infinite-state model checking community, we believe that our formalization is useful since it gives a common model-theoretic approach to otherwise quite diverse formalizations. Second, we observe that for a large class of systems, called guarded assignment systems (GAS), the reachability analysis is simpler since only formulas of a special form are used for satisfiability- and entailment-checking. Many known formalizations of broadcast, cache coherence, and other protocols belong to the simplest kind of GAS, called simple GAS. It follows from our results that the so-called local backward reachability algorithms can be used for simple GAS over structures in which satisfiability-checking is decidable for conjunctions of atomic formulas. This, for example, allows one to extend the existing reachability algorithms to some theories of queues.

M. Baaz and J.A. Makowsky (Eds.): CSL 2003, LNCS 2803, pp. 546–573, 2003.

We also discuss how this general framework applies to a special case of integer-based transition systems. We show how the main operations required for reachability analysis in such systems can be implemented using an incremental algorithm for building Hilbert's bases of systems of linear equations and inequations over integers.

This paper is organized as follows. In Section 2 we briefly overview relevant parts of model theory and define several classes of first-order formulas. In Section 3 we give a model theory-based formalization of transition systems and their symbolic representations and introduce basic forward and backward reachability algorithms. We also discuss requirements on the underlying first-order structures. In Section 5 we define guarded assignment systems and reachability algorithms for them. In Section 6 we introduce local reachability algorithms for guarded assignment systems. Finally, in Section 10 we discuss some other issues related to our formalization of infinite-state systems and reachability.

2 Preliminaries

In this section we define notation and several classes of formulas which will be used in the rest of the paper. We assume knowledge of standard model-theoretic definitions, such as first-order formulas, structure and truth, which can be found in any standard textbook on logic or model theory.

Unless stated otherwise, we will deal with a fixed first-order structure $\mathbb{M}$ with a domain $\mathcal{D}$ and assume that all formulas are formulas of the signature of this structure. A *valuation* for a set of variables V in $\mathbb{M}$ is any mapping $s : V \to \mathcal{D}$. We will use the standard model-theoretical notation $\mathbb{M}, s \models A$ to denote that the formula A is true in the structure $\mathbb{M}$ under a valuation s. When we use this notation, we assume that s is defined on all free variables of A.

A formula A with free variables V is said to be *satisfiable* (respectively, *valid*) in $\mathbb{M}$ if there exists a valuation s for V in $\mathbb{M}$ such that $\mathbb{M}, s \models A$ (respectively, for every valuation s we have $\mathbb{M}, s \models A$).

A formula A is called

- ▷ *quantifier-free* if A contains no quantifiers.
- ▷ *positive existential* if A can be built from atomic formulas using just $\exists, \wedge, \vee$;
- ▷ a *conjunctive constraint* if A can be built from atomic formulas using just $\exists, \wedge$;
- ▷ a *simple constraint* if A is a conjunction of atomic formulas.

3 Transition Systems and Reachability: A Formalization in Logic

In this section we introduce a logical formalization of transition systems.

3.1 Infinite-State Transition Systems and Their Symbolic Representation

Our formalization of transition systems is as follows. A transition system has a finite number of variables with values in a possibly infinite domain $\mathcal{D}$. A state is a mapping from variables to values. Transitions may change values of variables. A symbolic representation of such a system uses first-order formulas interpreted in a structure with the domain $\mathcal{D}$.

DEFINITION 1 (Transition System) A *transition system* is a tuple $\mathbb{S} = (\mathcal{V}, \mathcal{D}, \mathcal{T})$, where

1. $\mathcal{V}$ is a finite set of *state variables*.
2. $\mathcal{D}$ is a non-empty set, called the *domain*. Elements of $\mathcal{D}$ are called *values*.

A *state* of the transition system $\mathbb{S}$ is a function $s : \mathcal{V} \rightarrow \mathcal{D}$.

3. $\mathcal{T}$ is a set of pairs of states, called the *transition relation* of $\mathbb{S}$.

A transition system $\mathbb{S}$ is *finite-state* if $\mathcal{D}$ is finite, and *infinite-state* otherwise. ❑

We call any set of pairs of states a *transition*. Transition systems arising in practical applications often have variables ranging over different domains. For example, some state variables may range over the natural numbers, while others over the boolean values or other finite domains. We introduce a single domain for simplicity. It is not hard to generalize our formalization to several domains using many-sorted first order logic. In the sequel we assume a fixed transition system $\mathbb{S} = (\mathcal{V}, \mathcal{D}, \mathcal{T})$. Suppose that $\mathbb{M}$ is a first-order structure whose domain is $\mathcal{D}$. For example, if $\mathcal{D}$ is the set of natural numbers, then the structure $\mathbb{M}$ can be the set of natural numbers together with the order $<$, operation $+$ and constants $0, 1$. In addition to the set of state variables $\mathcal{V}$, we also introduce a set $\mathcal{V}'$ of *next state variables* of the same cardinality as $\mathcal{V}$. We fix a bijection $' : \mathcal{V} \rightarrow \mathcal{V}'$ such that $v \in \mathcal{V}$ we have $v' \in \mathcal{V}'$.

We can treat the variables in $\mathcal{V} \cup \mathcal{V}'$ also as logical variables. Then any mapping $s : \mathcal{V} \rightarrow \mathcal{D}$ can be considered as both a state of the transition system $\mathbb{S}$ and a valuation for $\mathcal{V}$ in the structure $\mathbb{M}$, and similarly for $s' : \mathcal{V}' \rightarrow \mathcal{D}$.

DEFINITION 2 (Symbolic Representation) Let S be a set of states and A be a formula with free variables in $\mathcal{V}$. We say that A *symbolically represents* S *in* $\mathbb{M}$, or simply *represents* S if for every valuation s for $\mathcal{V}$ in $\mathbb{M}$ we have

$$s \in S \Leftrightarrow \mathbb{M}, s \models A.$$

Likewise, we say that a formula B with free variables in $\mathcal{V} \cup \mathcal{V}'$ *(symbolically) represents a transition* T *in* $\mathbb{M}$ if for every pair of valuations s, s' in $\mathbb{M}$ for $\mathcal{V}$ and $\mathcal{V}'$ respectively we have

$$(s, s') \in T \Leftrightarrow \mathbb{M}, s, s' \models B.$$

If a formula Tr *symbolically represents* the transition relation T of a transition system $\mathbb{S}$, we simply say that Tr symbolically represents $\mathbb{S}$. ❑

In the sequel we will follow the following convention. We will often identify a symbolic representation of a transition with the transition itself. For example, when T is a formula with free variables in $\mathcal{V} \cup \mathcal{V}'$ we can refer to T as a transition.

DEFINITION 3 We say that a state s_n is *forward reachable* from a state s_0 w.r.t. $\mathcal{T}$ if there exists a sequence of states $s_1, \ldots, s_{n-1}$ such that for all $i \in \{0, \ldots, n-1\}$ we have $(s_i, s_{i+1}) \in \mathcal{T}$. In this case we also say that s_n is reachable from s_0 *in n steps* and that s_0 is *backward reachable* from s_n in n steps. ❑

Instead of "forward reachable" we will say "reachable". When we speak about reachability with respect to $\mathcal{T}$, and $\mathcal{T}$ is clear from the context, we will simply say "reachable".

The reachability problem can now be defined as follows.

DEFINITION 4 (Reachability Problem) The *reachability problem for* $\mathbb{M}$ is the following decision problem. Given formulas In, Fin, and Tr such that

1. In represents a set of states I, called the set of *initial states*;
2. Fin represents a set of states F, called the set of *final states*;
3. Tr represents the transition relation of a transition system $\mathbb{S}$,

do there exist states $s_1 \in I$, $s_2 \in F$ such that s_2 is reachable from s_1 w.r.t. Tr? ❑

In fact, reachability is a family of decision problems parametrized by the structure $\mathbb{M}$.

When we discuss instances of the reachability problem, we will call the formulas In and Fin the *initial* and *final conditions*, respectively, and Tr the *transition formula*.

The reachability problem for infinite-state systems is, in general, undecidable. Various results on reachability are discussed in many papers, including [13, 1, 20, 2].

Let us give an example instance of the reachability problem. Before defining it formally, we will introduce a convenient syntax for representing transitions. Suppose that $T = T_1 \cup \ldots \cup T_n$ is a transition such that every T_i can be symbolically represented by $P_i(\mathcal{V}) \wedge Q_i(\mathcal{V}, \mathcal{V}')$ such that the free variables of $P_i(\mathcal{V})$ are in $\mathcal{V}$. Then we will write the transition T as

$$\begin{array}{c} P_1(\mathcal{V}) \Rightarrow Q_1(\mathcal{V}, \mathcal{V}'), \\ \cdots \\ P_n(\mathcal{V}) \Rightarrow Q_n(\mathcal{V}, \mathcal{V}'). \end{array} \tag{1}$$

We will call the formula $P_i(\mathcal{V})$ the *guard* or the *precondition*, and the formula $Q_i(\mathcal{V}')$ the *postcondition* of this transition. Note that the notions of pre- and post- conditions refer to a particular representation of the transition, since the same transition may have representations using different pre- and post-conditions. Guarded representation (1) thus denotes the transition

$$(P_1(\mathcal{V}) \wedge Q_1(\mathcal{V}, \mathcal{V}')) \vee \ldots \vee (P_n(\mathcal{V}) \wedge Q_n(\mathcal{V}, \mathcal{V}')),$$

that is, the union of all transitions $P_i(\mathcal{V}) \Rightarrow Q_i(\mathcal{V}, \mathcal{V}')$.

An example of a transition system presented using the guarded representation is given in Figure 1.

3.2 Algorithms for Checking Reachability

The algorithms for checking reachability are based on the idea of building a symbolic representation of the set of reachable states. There are two main kinds of reachability

$$\begin{array}{rl} drinks > 0 \wedge customers > 0 \Rightarrow drinks' = drinks - 1 & \texttt{(*dispense-drink*)} \\ true \Rightarrow drinks' = drinks + 64 & \texttt{(*recharge*)} \\ true \Rightarrow customers' = customers + 1 & \texttt{(*customer-coming*)} \\ customers > 0 \Rightarrow customers' = customers - 1 & \texttt{(*customer-going*)} \end{array}$$

Fig. 1. Guarded representation of transitions

algorithms: *forward reachability* and *backward reachability*. Forward reachability algorithms try to build the set of states reachable from the initial states and check whether this set of states contains any final state. Backward reachability algorithms try to build the set of states backward reachable from the final states and check if this set of states contains any initial state.

Before defining these algorithms, let us discuss symbolic representations of forward and backward reachable states. We assume fixed initial and final conditions $In(\mathcal{V})$, $Fin(\mathcal{V})$ and the transition formula $Tr(\mathcal{V}, \mathcal{V}')$.

Let $A(\mathcal{V})$ be a formula which represents a set of states S. It is not hard to argue that the set of states reachable in one step from a state in S can be represented by the formula

$$\exists \mathcal{V}_1(A(\mathcal{V}_1) \wedge Tr(\mathcal{V}_1, \mathcal{V})).$$

Likewise, the set of states backward reachable in one step from a state in S is represented by the formula

$$\exists \mathcal{V}_1(A(\mathcal{V}_1) \wedge Tr(\mathcal{V}, \mathcal{V}_1)).$$

This observation implies the following facts about forward and backward reachability.

LEMMA 5 (Forward Reachability) *Consider the sequence of formulas FR_i defined as follows:*

$$FR_0(\mathcal{V}) = In(\mathcal{V});\ FR_{i+1}(\mathcal{V}) = FR_i(\mathcal{V}) \vee \exists \mathcal{V}_1(FR_i(\mathcal{V}_1) \wedge Tr(\mathcal{V}_1, \mathcal{V})).$$

Then each FR_i symbolically represents the set of states reachable from I in at most i steps. ❑

LEMMA 6 (Backward Reachability) *Consider the sequence of formulas BR_i defined as follows:*

$$\begin{aligned} BR_0(\mathcal{V}) &= Fin(\mathcal{V}); \\ BR_{i+1}(\mathcal{V}) &= BR_i(\mathcal{V}) \vee \exists \mathcal{V}_1(BR_i(\mathcal{V}_1) \wedge Tr(\mathcal{V}, \mathcal{V}_1)). \end{aligned}$$

Then each BR_i symbolically represents the set of states backward reachable from F in at most i steps. ❑

Using these lemmas, one can prove two theorems which form the basis of reachability algorithms.

THEOREM 7 (Reachability) *(i) There exists a final state reachable from an initial state if and only if there exists a number $i \geq 0$ such that $\mathbb{M} \models \exists \mathcal{V}(FR_i(\mathcal{V}) \wedge Fin(\mathcal{V}))$. (ii) If there exists a number $i \geq 0$ such that $\mathbb{M} \not\models \exists \mathcal{V}(FR_i(\mathcal{V}) \wedge Fin(\mathcal{V}))$ and $\mathbb{M} \models \forall \mathcal{V}(FR_{i+1}(V) \rightarrow FR_i(\mathcal{V}))$, then there exists no final state reachable from an initial state. The same statements hold if one replaces FR by BR and Fin by In.*

```
procedure Forward
input: formulas In, Fin, Tr
output: "reachable" or "unreachable"
begin
  current(V) := In(V)
  while M ⊭ ∃V(current(V) ∧ Fin(V))
    next(V) :=
      simp(current(V) ∨
           ∃V1(current(V1) ∧ Tr(V1, V)) )
    if M ⊨ ∀V(next(V) → current(V))
      then return "unreachable"
    current(V) := next(V)
  return "reachable"
end
```

Fig. 2. Forward reachability algorithm

```
procedure Backward
input: formulas In, Fin, Tr
output: "reachable" or "unreachable"
begin
  current(V) := Fin(V)
  while M ⊭ ∃V(current(V) ∧ In(V))
    prev(V) :=
      simp(current(V) ∨
           ∃V1(current(V1) ∧ Tr(V, V1)) )
    if M ⊨ ∀V(prev(V) → current(V))
      then return "unreachable"
    current(V) := prev(V)
  return "reachable"
end
```

Fig. 3. Backward reachability algorithm

This theorem reduces, in some sense, reachability-checking to checking whether formulas of a special form are true in the structure $\mathbb{M}$. The form of these formulas depends on the symbolic representations In, Fin, Tr of the sets of initial and finite states and the transition relation.

The basic reachability algorithms are based on explicit computation of the formulas FR_i and BR_i and the use of Theorem 7.

DEFINITION 8 The *forward reachability algorithm* Forward and the *backward reachability algorithm* Backward are shown in Figures 2 and 3. They are parametrized by a function simp, called a *simplification function.* ❑

Usually, this formula transforms formulas into equivalent ones, i.e., $\mathbb{M} \models \forall \mathcal{V}(A \leftrightarrow \mathsf{simp}(A))$ for all formulas A.

In different reachability algorithms, the simplification function simp may make a simple transformation, for example into a normal form, but may also perform more complex ones, such as quantifier elimination. We will sometimes need the simplification function simp to preserve some class of formulas. To this end, we introduce the following definition. We call a function simp on formulas *stable* w.r.t. a class of formulas $\mathcal{C}$, if for every formula $A \in \mathcal{C}$ we have $\mathsf{simp}(A) \in \mathcal{C}$.

3.3 First-Order Theories of Infinite Structures

In this section we analyze first-order structures which are suitable for reachability algorithms. In both forward and backward reachability algorithms one has to solve repeatedly the following problems:

1. *Satisfiability:* whether the formula $\mathsf{current}(\mathcal{V}) \wedge Fin(\mathcal{V})$ or $\mathsf{current}(\mathcal{V}) \wedge In(\mathcal{V})$ is satisfiable in $\mathbb{M}$.
2. *Entailment:* whether the formula $\mathsf{next}(\mathcal{V}) \rightarrow \mathsf{current}(\mathcal{V})$ or $\mathsf{prev}(\mathcal{V}) \rightarrow \mathsf{current}(\mathcal{V})$ is valid in $\mathbb{M}$.

Of course the reachability algorithms of Figures 2 and 3 can only be considered as algorithms modulo the assumption that satisfiability- and entailment-checking are decidable.

It is not hard to argue that the formulas current($\mathcal{V}$) at the ith iteration of the while-loop are equivalent to the formulas FR_i for forward reachability and BR_i for backward reachability, provided that simp is equivalence-preserving. If simp(A) $=$ A for every formula A, then the formulas current($\mathcal{V}$) are exactly the formulas FR_i for the forward reachability algorithm and BR_i for the backward reachability algorithm. This implies that the reachability algorithms have the following properties.

THEOREM 9 (Soundness and Semi-Completeness)
The algorithms Forward *and* Backward *have the following properties:*

1. *there is a final state reachable from an initial state if and only if the algorithm returns "reachable";*
2. *if the algorithm returns "unreachable", then there is no final state reachable from an initial state.* ❑

On some inputs the algorithms do not terminate. In this case we say that the reachability analysis *diverges*.

Complexity of the satisfiability and entailment problems depends on the structure $\mathbb{M}$ and the form of the formulas current used in the algorithm.

If the first-order theory of $\mathbb{M}$ is decidable, one can implement satisfiability- and entailment-checking using a decision procedure for the first-order theory of $\mathbb{M}$. In some cases, one can even use off-the-shelf decision procedures or libraries. For example, [8] use the Omega Library [25] for deciding Presburger arithmetic. The use of general-purpose algorithms to decide specific classes of formulas may be inefficient. Then specialized tools may be needed to decide these classes of formulas more efficiently.

It is not hard to see that in many important special cases the formulas FR_i and BR_i have a special form.

LEMMA 10 *Let In, Fin, and Tr be positive existential formulas and* simp *be stable w.r.t. positive existential formulas. Then the formulas* current *in the algorithms* Forward *and* Backward *are positive existential too.* ❑

It follows from this lemma that it is enough to solve satisfiability and entailment for positive existential formulas only, as soon as the initial and final conditions and the transition formula are positive existential formulas, which they often are. As we shall see below, for some important special cases even quantifier-free formulas suffice.

If the first-order theory of $\mathbb{M}$ admits quantifier elimination, i.e., every formula is effectively equivalent to a quantifier-free formula, then it is enough to check satisfiability and entailment of quantifier-free formulas only. This can be achieved by applying quantifier elimination, i.e., replacing quantified formulas by equivalent quantifier-free formulas, whenever quantified formulas appear, for example, when transitions are applied. However, quantifier elimination may be expensive.

4 Integer-Based Systems

Denote by I the set of integers. Consider the structure

$$\mathbb{I} = (I, >, <, \geq, \leq, +, -, 0, 1, 2, \ldots),$$

where all the function and predicate symbols (for example $>$) have their standard interpretation over integers. The first-order theory of $\mathbb{I}$ is decidable, which means that one can use the reachability algorithms Forward and Backward even when the initial and final conditions and the transition formula are arbitrary first-order formulas. In addition, infinite-state transition systems (of a very simple form) over the domain of integers can be used for specifying safety properties of a large classes of protocols. The first-order theory of $\mathbb{I}$ (essentially, the Presburger arithmetic) admits quantifier elimination, if we extend the signature of $\mathbb{I}$ by the predicates expressing $m \mid (x - n)$ for concrete natural numbers m, n.

4.1 Integer-Based Transition Systems

We will call the transition systems over $\mathbb{I}$ the *integer transition systems* and the reachability problem over $\mathbb{I}$ the *integer reachability problem*.

We will often use the following simple property of $\mathbb{I}$.

LEMMA 11 *In $\mathbb{I}$ every quantifier-free formula A is equivalent to a disjunction of simple constraints.* ❑

Off-the-shelf tools, such as Omega Library [25] or the Composite Library [32], are available to decide the Presburger arithmetic or its fragments.

The fact that satisfiability- and entailment-checking are decidable does not necessarily imply that the reachability problem is decidable. Neither does it imply that termination of the reachability algorithms is decidable. In this section we note the following undecidability result which will carry over to all classes of transition systems and reachability algorithms studied in this paper.

THEOREM 12 *Consider the instances of the integer reachability problem with three state variables c_1, c_2, s whose transition relation is a disjunction of conjunctions of the following formulas: $c_i > 0$ or $s = n$, where n is a natural number, $c'_i = c_i + 1$, $c'_i = c_i - 1$, $c'_i = c_i$, or $s' = m$, where m is a natural number. There is exactly one initial and one final state. Then*

1. *the reachability problem for this class is undecidable.*
2. *termination of the algorithm* Forward *is undecidable.*
3. *termination of the algorithm* Backward *is undecidable.* ❑

One can easily prove this theorem by encoding two-counter machines by transition systems of this form. The systems used in the theorem are simple guarded assignment systems (the definition is given below).

Some decidability results for integer-based systems are given in e.g., [13, 1]. For example, safety properties of broadcast protocols can be represented as instances of the integer reachability problem, and the backward reachability algorithm terminates on these instances [13].

4.2 Types of Protocols

A large number of protocols can be easily modelled by integer transition systems. Examples are Petri nets, parametric and broadcast protocols. A typical way to change parametric systems into integer transition systems is as follows. Suppose that there is a (parametric) number of processes represented as variables $v_1, v_2, \ldots$ ranging over a finite domain $\{d_1, \ldots, d_n\}$. Then, instead of using formulas $v_i = d_j$ to denote that v_i has the value d_j we introduce new variables n_j to denote the number of variables v_i having the value d_j. The transitions of the original system can then be reformulated using new variables. For example, the transition $v_i = d_1 \wedge v_i' = d_2$ becomes $n_1 > 0 \wedge n_1' = n_1 - 1 \wedge n_2' = n_2 + 1$. Most of the benchmarks described in Section 9 are obtained by such a transformation.

5 Guarded Assignment Transition Systems

In this section we consider an important special case of transition systems in which quantifier-free formulas can be used for backward reachability, even without the use of expensive quantifier elimination algorithms. Guarded assignment systems are in the core of the model-checking system BRAIN [27]

5.1 Guarded Assignments

As usual, we assume that $\mathcal{V}$ the set of state variables of the transition system.

DEFINITION 13 (Guarded Assignment) A *guarded assignment* is any formula (or transition) of the form

$$P \wedge v_1' = t_1 \wedge \ldots \wedge v_n' = t_n \wedge \bigwedge_{v \in \mathcal{V} - \{v_1, \ldots, v_n\}} v' = v,$$

where P is a formula with free variables $\mathcal{V}$, $\{v_1, \ldots, v_n\} \subseteq \mathcal{V}$, and $t_1, \ldots, t_n$ are terms with variables in $\mathcal{V}$. We will write guarded assignments as

$$P \;\Rightarrow\; v_1 \;:=\; t_1, \ldots, v_n \;:=\; t_n. \tag{2}$$

The formula P is called the *guard* of this guarded assignment.

A guarded assignment is *quantifier-free* (respectively, *positive existential*), if so is its guard. A guarded assignment is called *simple existential* if its guard is a conjunctive constraint, and *simple* if its guard is a simple constraint. ❑

Formula (2) represents a transition which applies to states satisfying P and changes the values of variables v_i to the values of the terms t_i. Note that a guarded assignment T is a deterministic transition: for every state s there exists at most one state s' such that $(s, s') \in T$. Moreover, such a state s' exists if and only if the guard of this guarded assignment is true at s, i.e., $\mathbb{M}, s \models P$.

$$\begin{array}{rlll} drinks > 0 \wedge customers > 0 \Rightarrow & drinks & := drinks - 1 & \texttt{(*dispense-drink*)} \\ true \Rightarrow & drinks & := drinks + 64 & \texttt{(*recharge*)} \\ true \Rightarrow & customers & := customers + 1 & \texttt{(*customer-coming*)} \\ customers > 0 \Rightarrow & customers & := customers - 1 & \texttt{(*customer-going*)} \end{array}$$

Fig. 4. Guarded assignment system

DEFINITION 14 (Guarded Assignment System) A transition system is called a *guarded assignment system*, or simply *GAS*, if its transition relation is a union of a finite number of guarded assignments. A GAS is called *quantifier-free* (respectively, *positive existential*, *simple existential*, or *simple*) if every guarded assignment in it is also quantifier-free (respectively positive existential, simple existential, or simple). ❑

An example integer guarded assignment system for representing a drink dispenser is given in Figure 4. This system is a union of transitions shown in this figure. Since all guards are simple constraints, the system is simple.

Note that every guarded assignment represents a deterministic transition, but a guarded assignment system may represent a non-deterministic transition system because several guards may be true in the same state. Not every transition system is a guarded assignment system. Indeed, in every guarded assignment system with a transition relation $\mathcal{T}$ for every state s there exists a finite number of states s' such that $(s, s') \in \mathcal{T}$. The greatest transition *true* over any infinite structure $\mathbb{M}$ does not have this property.

THEOREM 15 *Every integer quantifier-free GAS is also a simple GAS.*

Proof. Since every quantifier-free formula is equivalent to a formula in disjunctive normal form, we can assume that the guards are in disjunctive normal form. Since in $\mathbb{I}$ the negation of an atomic formula is equivalent to a disjunction of atomic formulas (for example $\neg x = y$ is equivalent to $x > y \vee y > x$ and $\neg x > y$ is equivalent to $y \geq x$), we can assume that every guard is a disjunction $C_1 \vee \ldots \vee C_n$ of conjunctions of atomic formulas. But every guarded assignment

$$C_1 \vee \ldots \vee C_n \Rightarrow A \tag{3}$$

is equivalent to the union of the simple guarded assignments $C_1 \Rightarrow A, \ldots, C_n \Rightarrow A$,

$$C_1 \Rightarrow A, \quad \ldots, \quad C_n \Rightarrow A,$$

so we can equivalently replace it by n simple guarded assignments.

One can generalize this theorem to structures different from $\mathbb{I}$. Indeed, the only property of $\mathbb{I}$ used in this proof is that the negation of an atomic formula is equivalent to a positive quantifier-free formula.

The notion of a guarded assignment system is not very restrictive. Indeed, broadcast protocols and Petri nets can be represented as integer simple guarded assignment systems. All transition systems for cache coherence protocols described in [10] are integer simple GAS.

Let us note one interesting property of GAS related to properties other than reachability. Let us call a state s a *deadlock state* if there is no state s' such that $(s, s') \in T$, where T is the transition relation.

THEOREM 16 *Let the transition relation T be a union of guarded assignments $(P_1 \Rightarrow A_1), \ldots, (P_n \Rightarrow A_n)$. Then the set of all deadlock states is represented by the formula $\neg(P_1 \vee \ldots \vee P_n)$.* ❑

This theorem shows that checking *deadlock-freedom* (i.e., non-reachability of a deadlock state) of GAS may be as easy as checking reachability properties, for example, if the GAS is quantifier-free, then the set of deadlock states can also be represented by a quantifier-free formula. Theorem 16 is also used in the system BRAIN to generate the deadlock-freedom conditions automatically.

5.2 Reachability Algorithms for Guarded Assignment Systems

We introduced simple guarded assignment systems because they have a convenient property related to backward reachability algorithms. Essentially, one can use backward reachability algorithms for these systems when the underlying structure has a much weaker property than the decidability of the first-order theory.

Let us see how the formulas FR_i and BR_i look like in the case of simple guarded assignment systems. Let u be a guarded assignment of the form

$$P(v_1, \ldots, v_n) \Rightarrow v_1 \;:=\; t_1, \ldots, v_n \;:=\; t_n.$$

For simplicity we assume that $\mathcal{V} = \{v_1, \ldots, v_n\}$. This can be achieved by adding "dummy" assignments $v \;:=\; v$ for every variable $v \in \mathcal{V} - \{v_1, \ldots, v_n\}$. Let also $A(v_1, \ldots, v_n)$ be a formula whose free variables are in $\mathcal{V}$. For every term t denote by t' the term obtained from t by replacing every occurrence of every state variable v_i by v_i'.

Define the following formulas:

$$A^u(v_1, \ldots, v_n) \stackrel{\text{def}}{=} \exists \mathcal{V}'(A(v_1', \ldots, v_n') \wedge P(v_1', \ldots, v_n') \wedge v_1 = t_1' \wedge \ldots \wedge v_n = t_n');$$
$$A^{-u}(v_1, \ldots, v_n) \stackrel{\text{def}}{=} P(v_1, \ldots, v_n) \wedge A(t_1, \ldots, t_n).$$

LEMMA 17 *Let $A(v_1, \ldots, v_n)$ represent a set of states S. Then*

1. *the formula $A^u(v_1, \ldots, v_n)$ represents the set of states reachable in one step from S using u;*
2. *the formula $A^{-u}(v_1, \ldots, v_n)$ represents the set of states backward reachable in one step from S using u.*

Proof. For forward reachability the statement is straightforward, so let us prove the backward reachability part. The set of states backward-reachable in one step from S is represented by the formula

$$\exists \mathcal{V}'(P(v_1, \ldots, v_n) \wedge A(v_1', \ldots, v_n') \wedge v_1' = t_1 \wedge \ldots \wedge v_n' = t_n). \tag{4}$$

Using the fact that in first-order logic the formula $\exists x(x = t \wedge B(x))$, where t does not occur in x, is equivalent to $B(t)$, it is not hard to argue that formula (4) this formula is equivalent to $A^{-u}(v_1, \ldots, v_n)$.

```
procedure GASForward
input: formulas In, Fin,
       finite set of guarded assignments U
output: "reachable" or "unreachable"
begin
  current := In
  while M ⊭ ∃V(current ∧ Fin)
    next := current ∨ ⋁_{u∈U} current^u
    if M ⊨ ∀V(next → current)
      then return "unreachable"
    current := next
  return "reachable"
end
```

Fig. 5. Forward reachability algorithm for GAS

```
procedure GASBackward
input: formulas In, Fin,
       finite set of guarded assignments U
output: "reachable" or "unreachable"
begin
  current := In
  while M ⊭ ∃V(current ∧ In)
    prev := current ∨ ⋁_{u∈U} current^{-u}
    if M ⊨ ∀V(prev → current)
      then return "unreachable"
    current := prev
  return "reachable"
end
```

Fig. 6. Backward reachability algorithm for GAS

One can deduce from this lemma that the formulas A^u and A^{-u} have a special form, as expressed by the following lemma.

LEMMA 18 *Let u be a guarded assignment. Then the function $A \mapsto A^u$ is stable w.r.t. positive existential formulas and conjunctive constraints. The function $A \mapsto A^{-u}$ is stable w.r.t. positive existential formulas, conjunctive constraints, quantifier-free formulas, and simple constraints.* ❑

So far forward and backward reachability were treated symmetrically. Lemma 18 shows an asymmetry between forward and backward reachability for simple GAS: the predicate transformer corresponding to backward reachability yields simpler formulas. Using Lemma 17, one can modify forward and backward reachability algorithms of Figures 2 and 3 for guarded assignment transition systems.

DEFINITION 19 The *forward reachability algorithm* GASForward and the *backward reachability algorithm* GASBackward for GAS are shown in Figures 5 and 6. When we check reachability for a GAS whose transition relation is a union of guarded assignments $u_1, \ldots, u_n$, we let $U = \{u_1, \ldots, u_n\}$ in the input of the algorithm. ❑

THEOREM 20 (Soundness and Semi-Completeness) *The algorithms* GASForward *and* GASBackward *have the following properties:*

1. *there is a final state reachable from an initial state if and only if the algorithm returns "reachable";*
2. *if the algorithm returns "unreachable", then there is no final state reachable from an initial state.*

Moreover, GASForward *(respectively,* GASBackward*) terminates if and only if* Forward *(respectively* Backward*) terminates.* ❑

One can prove soundness and semi-completeness of the algorithms GASForward and GASBackward similar to that of Theorem 9.

THEOREM 21 *If the input GAS and the formulas In, Fin are quantifier-free, then all of the formulas* current *in the algorithm* GASBackward *are quantifier-free too. If, in addition, the input GAS and In, Fin are negation-free, then* current *is also negation-free.* ❑

This theorem shows that for quantifier-free GAS satisfiability and entailment in the algorithm GASBackward should only be checked for quantifier-free formulas. It does not hold for the forward reachability algorithm GASForward. Checking satisfiability and validity of quantifier-free formulas is possible if the existential theory (i.e., the set of all sentences of the form $\exists X A$, where A is quantifier-free) of the structure $\mathbb{M}$ is decidable. There exist theories whose full first-order theory is undecidable but existential theory is decidable. A simple example is Presburger arithmetic extended by the divisibility predicate [4, 23, 22], though it is not clear how useful is this theory for applications. A more practical example are some theories of queues considered in [6, 5]. This shows that there are infinite domains for which backward reachability may be easier to organize than forward reachability. Of course one can omit the entailment checks from the algorithms, but then they will only be applicable for checking reachability, but not for checking non-reachability.

6 Local Algorithms

In this section we study so-called *local* algorithms, which are different from the algorithms discussed above. They are simpler since checking for satisfiability and entailment for simple GAS is only performed for conjunctive constraints, but not so powerful since they may diverge for instances of reachability for which the previously described algorithms terminate. The idea of these algorithms is to get rid of disjunctions which appear in the formulas current even when the input contains no disjunctions at all. Instead of a disjunction $C_1 \vee \ldots \vee C_n$, one deals with the set of formulas $\{C_1, \ldots, C_n\}$. The entailment check is not performed on the disjunction, but separately on each member C_i of the disjunction, and is therefore called a *local entailment check.*

DEFINITION 22 (Local Reachability Algorithms) The *local forward reachability algorithm* LocalForward and *local backward reachability algorithm* LocalBackward are given in Figures 7 and 8. They are parametrized by a function *select* which selects a formula in a set of formulas. The input set of guarded assignments U is defined as in the algorithms GASForward and GASBackward. The input sets of formulas IS and FS are any sets of formulas such that $In = \bigvee_{A \in IS} A$ and $Fin = \bigvee_{A \in FS} A$. ❑

Local algorithms can be used not only for guarded assignment systems, but for every system in which the transition relation is represented by a finite union of transitions.

THEOREM 23 (Soundness and Semi-Completeness) *The algorithms* LocalForward *and* LocalBackward *have the following properties:*

```
procedure LocalForward
input: sets of formulas IS, FS,
       finite set of guarded assignments U
output: "reachable" or "unreachable"
begin
  if there exist I ∈ IS, F ∈ FS such that
              M ⊨ ∃V(I ∧ F) then
    return "reachable"
  unused := IS
  used := ∅
  while unused ≠ ∅
    S := select(unused)
    used := used ∪ {S}
    unused := unused − {S}
    forall u ∈ U
      N := S^u
      if there exists F ∈ FS such that
                M ⊨ ∃V(N ∧ F) then
        return "reachable"
      if for all C ∈ used ∪ unused
              M ⊭ ∀V(N → C) then
        unused = unused ∪ {N}
        forall C' ∈ used ∪ unused
          if M ⊨ ∀V(C' → N) then
            remove C' from used or unused
  return "unreachable"
end
```

Fig. 7. Local forward reachability algorithm

```
procedure LocalBackward
input: sets of formulas IS, FS,
       finite set of guarded assignments U
output: "reachable" or "unreachable"
begin
  if there exist I ∈ IS, F ∈ FS such that
              M ⊨ ∃V(I ∧ F) then
    return "reachable"
  unused := FS
  used := ∅
  while unused ≠ ∅
    S := select(unused)
    used := used ∪ {S}
    unused := unused − {S}
    forall u ∈ U
      N := S^{-u}
      if there exists I ∈ IS such that
                M ⊨ ∃V(N ∧ I) then
        return "reachable"
      if for all C ∈ used ∪ unused
              M ⊭ ∀V(N → C) then
        unused = unused ∪ {N}
        forall C' ∈ used ∪ unused
          if M ⊨ ∀V(C' → N) then
            remove C' from used or unused
  return "unreachable"
end
```

Fig. 8. Local backward reachability algorithm

1. *there is a final state reachable from an initial state if and only if the algorithm returns "reachable";*
2. *if the algorithm returns "unreachable", then there is no final state reachable from an initial state.* ❑

One can prove soundness and semi-completeness of the algorithms LocalForward and LocalBackward similar to that of Theorem 9. However, we cannot guarantee that the algorithms terminate if and only if their non-local counterparts terminate.

Example 1. Consider an example. Consider an example integer GAS with one variable v whose transition relation is represented by a single guarded assignment

$$true \Rightarrow v \ := \ v - 1.$$

Take $0 > 0$ as the initial condition and $v \neq 0$ as the final condition. The non-local backward reachability algorithm for GAS at the second iteration will generate the formula $v \neq 0 \lor v \neq 1$, and at the third iteration $v \neq 0 \lor v \neq 1 \lor v \neq 2$. Since these formulas entail each other, the algorithm terminates. However, the local backward reachability algorithm for GAS generates formulas $v \neq 0$, $v \neq 1$, $v \neq 2$ which do not entail one another, and hence diverges. One can give a similar example for forward reachability.

It is not difficult to prove the following result.

THEOREM 24 *If* LocalForward *(respectively,* LocalBackward*) terminates, then so does* GASForward *(respectively,* GASBackward*).* ❑

Note that termination of the local algorithms may depend on the selection function *select*. Let us call the selection function *fair* if no formula remains in *unused* forever.

THEOREM 25 *If the local forward (respectively backward) algorithm terminates for some selection function, then it terminates for every fair selection function.* ❑

It is difficult to say if there are problems coming from real applications for which a non-local algorithm terminates but its local counterpart does not. One can prove, for example, that for broadcast protocols LocalBackward always terminates. In fact, this algorithm is implicitly used in the proof of decidability of broadcast protocols in [13].

The main property of the local algorithms is that subsumption and entailment-checking is only performed on simpler classes of formulas.

THEOREM 26 *If the input GAS is conjunctive and the sets IS, FS are sets of conjunctive constraints, then all of the formulas* current *in both algorithms* LocalForward *and* LocalBackward *are conjunctive constraints. If the input GAS is simple and the sets IS, FS are simple constraints then all of the formulas* current *in* LocalBackward *are simple constraints too.* ❑

The last property of local reachability algorithms is heavily used in the system BRAIN. For some structures (for example the reals or the integers) the algorithms for satisfiability-checking of simple constraints are a heavily investigated subject. If the structure is $\mathbb{I}$, then a simple constraint is essentially a system of linear equations and inequations. Checking satisfiability of simple constraints means solving such a system. For solving systems of linear equations and inequations over integers or reals several off-the-shelf tools are available. For example, [11] implement satisfiability- and entailment-checking for conjunctive constraints over real numbers using the corresponding built-in functions of the SICStus Prolog constraint library.

When a backward reachability algorithm is performed on GAS, we get a quantifier elimination effect due to the special form of formulas A^{-u}. Theorems 21 and 26 show that in many cases entailment can be checked only between quantifier-free formulas or even simple constraints. Let us show that in some cases entailment-checking can be performed using satisfiability-checking. It is easy to see that a formula $\forall \mathcal{V}(A \rightarrow B)$ is valid if and only if the formula $\exists \mathcal{V}(A \land \neg B)$ is unsatisfiable. Therefore, for quantifier-free GAS and backward reachability algorithms one can use satisfiability-checking for quantifier-free formulas also for entailment-checking.

In the case of simple GAS the situation is not much more complicated. Suppose that we would like to check an entailment problem of the form $\forall \mathcal{V}(A_1 \land \ldots \land A_n \rightarrow$

$B_1 \wedge \ldots \wedge B_m)$. This problem is equivalent to unsatisfiability of the formula $\exists\mathcal{V}(A_1 \wedge \ldots \wedge A_n \wedge (\neg B_1 \vee \ldots \vee \neg B_m))$. This formula is equivalent to

$$\begin{array}{rl}\exists\mathcal{V}(A_1 \wedge \ldots \wedge A_n \wedge \neg B_1) & \vee \\ \ldots & \vee \\ \exists\mathcal{V}(A_1 \wedge \ldots \wedge A_n \wedge \neg B_m). & \end{array}$$

Therefore, to check the original entailment problem, one has to check m satisfiability problems for the formulas $A_1 \wedge \ldots \wedge A_n \wedge \neg B_i$. These formulas are not simple constraints any more, but can be made into simple constraints or disjunctions of simple constraints if the first-order theory of the structure $\mathbb{M}$ has the following property: the negation of any atomic formula is effectively equivalent to a disjunction of atomic formulas. The first-order theory of $\mathbb{I}$ has this property, for example the formula $\neg x = y$ is equivalent to $x > y \vee y > x$.

This observation shows that in many situations it is enough to implement satisfiability-checking for simple constraints in order to implement backward reachability. For example, for $\mathbb{I}$ it is enough to implement algorithms for solving systems of linear equations and inequations.

One general obstacle for efficiency of reachability algorithms is the problem of *accumulated variables*, i.e., existentially quantified variables introduced by applications of the corresponding predicate transformers. We have shown that for quantifier-free GAS and backward reachability algorithms no new variables are accumulated. For forward reachability algorithms, GAS still have an advantage of (normally) accumulating only a small number of extra variables. Indeed, consider a guarded assignment u of the form

$$P(v_1, \ldots, v_k) \Rightarrow v_1 \ := \ t_1, \ldots, v_n \ := \ t_n$$

and assume that $\mathcal{V} = \{v_1, \ldots, v_n, v_{n+1}, \ldots, v_k\}$. Denote by $t'_1, \ldots, t'_n$ the terms obtained from $t_1, \ldots, t_n$ by replacing each variable v_i for $i \in \{1, \ldots, n\}$ by v'_i (note that the variables $v_{n+1}, \ldots, v_k$ are not replaced). It is not hard to argue that $A^u(v_1, \ldots, v_k)$ is equivalent to

$$\begin{array}{l}\exists v'_1 \ldots \exists v'_n (A(v'_1, \ldots, v'_n, v_{n+1}, \ldots, v_k) \wedge P(v'_1, \ldots, v'_n, v_{n+1}, \ldots, v_k) \wedge \\ \qquad v_1 = t'_1 \wedge \ldots \wedge v_n = t'_n).\end{array}$$

In this formula the new quantifiers bind the variables $v'_1, \ldots, v'_n$, i.e., only those variables whose values are changed by the transition. In transition systems formalizing protocols (e.g., from [10]) the number of variables whose values are changed by the transition is usually small compared to the overall number of state variables.

In some cases accumulated existentially quantified variables are not the only source of complexity. For example, for finite domains the number of variables is usually large, and repeated substitutions of terms t_i for v_i make the formula current grow exponentially. One way to make current shorter is to introduce extra existentially quantified variables to name common subexpressions, in which case the formulas current will not be quantifier-free even for simple GAS.

7 Integer-Based Systems and Hilbert's Basis

To implement the LocalBackward one has to implement procedures for the following problems:

1. *Backward application of guarded assignments*: given a simple constraint S and guarded assignment u, compute S^{-u}.
2. *Satisfiability of simple constraints*: given a simple constraint C, is C satisfiable in $\mathbb{I}$?
3. *Entailment of simple constraints:* given simple constraints N and C, is the formula $N \rightarrow C$ valid in $\mathbb{I}$?

To implement the reachability algorithm efficiently, one has to implement efficiently these three procedures. As our experimental data show, for hard problems the number of entailment-checks is considerably larger than the number of transition applications and satisfiability-checks. Therefore, entailment-checking should be implemented especially efficiently.

A simple constraint over $\mathbb{I}$ can be considered as a system of linear equations and inequations over integers with variables in $\mathcal{V}$. Since ever equation $u\mathcal{V} + l = 0$ can be equivalently replaced by two inequations $u\mathcal{V} + l \leq 0$ and $-u\mathcal{V} - l \leq 0$, in the sequel we will only discuss inequations. Satisfiability of simple constraints is known to be NP-complete and entailment coNP-complete. If one uses relaxation (see [10]) to use real numbers instead of integers, then both problems can be solved in polynomial time.

In this section we present some properties of the set of non-negative solutions to a simple constraint. Proofs can be found in, e.g., Schrijver [28]. In this section we consider $\mathcal{V}$ as a vector of variables rather than a set and restrict ourselves to simple constraints with the variables $\mathcal{V}$. Denote by $\mathbf{0}$ a vector of 0's. Every simple constraint C with variables in $\mathcal{V}$ can be written as a system of linear inequations with integer coefficients:

$$L\mathcal{V} + l \leq \mathbf{0}. \tag{5}$$

where L is a matrix with integer coefficients and l is a integer vector. We call a *solution* to such a system any vector V of non-negative integers which satisfies all inequations in the system. Let us emphasize that in this section we will only consider non-negative solutions. We will show below in Section 8 how to treat arbitrary integer solutions. For every system C of the form (5) denote by C^{hom} the corresponding system of homogeneous linear Diophantine inequations

$$L\mathcal{V} \leq \mathbf{0}. \tag{6}$$

We call a solution v to (5) *non-decomposable* if it cannot be represented in the form $v_1 + v_2$, where v_1 is a solution to (5) and v_2 is a non-zero solution to (6). Likewise, we call a solution to (6) non-decomposable if and only if it cannot be represented as a sum of two non-zero solutions to (6). It is easy to see that for systems of homogeneous linear Diophantine equations a solution is non-decomposable if and only if it is minimal w.r.t. the point-wise partial order on vectors, see e.g., Contejean and Devie [9].

DEFINITION 27 (Basis) A pair of sets of vectors (N, H) is called a *basis* for a simple constraint C if the following conditions hold.

1. Every vector in N is a non-decomposable solution to C.
2. Every vector in H is a non-zero non-decomposable solution to C^{hom}.
3. Every solution v to C can be represented as a sum $v = w + \sum_{i=1...k} m_i w_i$, where $w \in N$, $k \geq 0$ and for all $i = 1 \dots k$ m_i is a non-negative integer and $w_i \in H$.
4. Every solution v to C^{hom} can be represented as a sum $v = \sum_{i=1...k} m_i w_i$, where $k \geq 0$ and for all $i = 1 \dots k$ m_i is a non-negative integer and $w_i \in H$. ❑

This definition is a modification of the standard definition of Hilbert's basis [17] for the case of systems of linear inequations.

THEOREM 28 *Every simple constraint has a basis, and this basis is unique.* ❑

Algorithms for finding the basis of systems of linear Diophantine inequations are described in, e.g., Contejean and Devie [9], Ajili and Contejean [3], and Tomas and Filgueiras [29]. BRAIN uses a novel algorithm [30]. This algorithm, as well as other algorithms for funding Hilbert's basis, is too complex to be described here. In general, it is more difficult to find the basis of a simple constraint than to check its solvability[1]. The solvability problem is NP-complete, but the number of vectors in the basis can be exponential in the size of the system. Nevertheless, we will show that the construction of the basis may speed up reachability-checking.

BRAIN uses an *incremental algorithm* for building the basis. We call an *incremental basis-finding function* any function ibff of two arguments, such that for every pair of simple constraints (C_1, C_2), if B is the basis for C_1, then ibff(B, C_2) is the basis of $C_1 \wedge C_2$. Essentially, an incremental basis-funding function uses a basis computed previously for C_1 to find a basis for $C_1 \wedge C_2$.

8 BRAIN

In this section we explain how BRAIN implements the three important procedures used in the local backward reachability algorithm: backward application of guarded assignments, satisfiability and entailment. All three algorithms are implemented using repeated calls to an incremental basis-finding function ibff. In order to use the basis incrementally, BRAIN stores the basis together with every computed simple constraint. This technique is similar to a technique used by Halbwachs, Proy, and Roumanoff [14] for real-valued systems. We assume that all variables range over non-negative integers and show how to handle arbitrary integers later. We call an *augmented constraint* a pair (C, B) consisting of a simple constraint C and its basis B.

[1] Strangely enough, our experiments have shown that the existing algorithms for building the basis often outperform some well-known algorithms for checking solvability taken from integer programming packages. This could probably be explained by the fact that these packages are mostly intended for optimization and do not cope well with systems having several unbounded variables.

Entailment-checking. The algorithm for entailment-checking in BRAIN is based on the following theorem.

THEOREM 29 *Let* $(C_1, (N_1, H_1))$ *be an augmented constraint and* C_2 *be a simple constraint. Then* $\mathbb{I} \models \forall\mathcal{V}(C_1 \rightarrow C_2)$ *if and only if the following two conditions hold: (i) every vector* $v \in N_1$ *is a solution to* C_2*; (ii) every vector* $w \in H_1$ *is a solution to* C_2^{hom}.
❑

This theorem gives us an algorithm for entailment-checking: to check the entailment problem for augmented constraints, one has to check that the vectors of the basis of C_1 are solutions to C_2 or to the corresponding homogeneous system C_2^{hom}. Checking that a particular vector is a solution to a system can obviously be solved in time polynomial in the size of the vector and the system. As a consequence, we obtain the following theorem.

THEOREM 30 *Entailment of augmented constraints can be solved in polynomial time.*
❑

The algorithm implicit in Theorem 29 is used in BRAIN to check the entailment. To check entailment in polynomial time one can use instead of the basis any pair of sets of vectors (N, H) satisfying conditions (3) and (4) of the definition of basis, that is non-decomposability is not necessary. However, it is easy to prove that every pair of vectors (N, H) with these properties contains the basis, and thus using only non-decomposable vectors saves both space and time.

Satisfiability-checking. Evidently, a simple constraint C is satisfiable if for its basis (N, H) we have $N \neq \emptyset$. So satisfiability-checking for augmented constraints is trivial. Note that the reachability algorithm makes two kinds of satisfiability-checks:

1. checking whether the new formula N (i.e., S^{-u}) is satisfiable;
2. *intersection-checks*, when we check satisfiability of the formula $N \wedge I$.

The latter kind of satisfiability-checking can be performed by any satisfiability-checking procedure. But the first kind of satisfiability checks in BRAIN is combined with the backward applications of transitions for the reasons mentioned below.

Backward application of transitions. Repeated backward applications of transitions in the reachability algorithm may create too large constraints. To explain this, let us consider the formula for computing the set of states backward reachable from the set states presented by a simple constraint $C_1(v_1, \ldots, v_n)$. If the guarded assignment u has the form $C_2 \Rightarrow v_1 := t_1, \ldots, v_n := t_n$, then by Lemma 17 the formula C_1^{-u} is $C_2 \wedge C_1(t_1, \ldots, t_n)$. The number of atomic formulas in this simple constraint is the number of atomic formulas in C_1 plus the number of atomic formulas in C_2. Every iteration of the reachability algorithm yields longer constraints in which, for hard examples described below in Section 9, the number of atoms may be over a hundred. It is often the case that a large number of these atoms are a consequence of the remaining atoms in the constraint and can be safely removed (in our hardest examples the number of

```
procedure Basis
input: sequence of atoms A_1, ..., A_n,
output: pair (C, B), where C is equivalent to A_1 ∧ ... ∧ A_n,
                              and B is the basis for A_1 ∧ ... ∧ A_n
begin
  C := true; B := the basis of C
  for i = 1...n
    B' = ibff(B, A_i)
    if B' contains no solution then return (false, B')
    if B' ≠ B then(C, B) := (C ∧ A_i, B')
  return (C, B)
end
```

Fig. 9. Incremental building of the basis

non-redundant atoms usually does not exceed ten). Redundant atoms in constraints do not change the basis, but they slow down entailment, since our algorithm for checking validity of $(C_1 \rightarrow C_2)$ is, roughly speaking, linear in the number of atoms in C_2.

We can get rid of redundant constraints in $C_2 \wedge C_1(t_1, \ldots, t_n)$ together with checking satisfiability of this constraint and building a basis for it using an incremental basis-finding function. The procedure for this is given in Figure 9. The input to this procedure is the sequence of atoms in $C_2 \wedge C_1(t_1, \ldots, t_n)$ in any order. When a new atom A_i should be added to the constraint, it is first checked whether the addition of this atom changes the basis. If it does not, then the atom is redundant.

The current version of BRAIN only works with variables ranging over non-negative integers. Integers can be implemented using the same technology as follows. If an integer-valued variable is restricted by $v \leq n$ (or respectively by $n \leq v$), then it can be replaced by a variable $w = n - v$ (or respectively by $w = v - n$) ranging over non-negative integers. For every unrestricted integer-valued variable v one can introduce two variables w_1, w_2 ranging over non-negative integers and replace all occurrences of v by $w_1 - w_2$.

9 Experiments

In this section we present the results of experiments carried out on a number of benchmarks taken from several Web pages. The examples can be found on the Web page `www.cs.man.ac.uk/~voronkov/BRAIN/`. We compare the performance of our system BRAIN with that of the following systems: HyTech (Henzinger, Ho, and Wong-Toi [15]), Action Language Verifier (Bultan [7]), and DMC (Delzanno and Podelski [11]).

All benchmarks were carried out on the same computers (Sparc 300 with 2G of RAM memory). These computers are slow (about 8–10 times slower than the modern PCs), but we did not have access to a network of PCs with large RAM. The systems HyTech, Action Language, and BRAIN are implemented in C++ or C, and were compiled using the same version 2.92 of the GNU C/C++ compiler. DMC is implemented in Sicstus Prolog. In several cases we had to interrupt the systems because they consumed over 2G of memory. DMC never consumed more than 14M of memory, but was interrupted after several weeks of running. We were interested in hard benchmarks, but occasionally, for the sake of comparison, included figures for relatively easy benchmarks, because only HyTech and BRAIN could solve some of the hard ones. All runtimes are given in seconds.

Note that HyTech and DMC use relaxation, i.e., they solve real reachability problems instead of integer reachability problems. Therefore, they are correct only when they report non-reachability. Among the systems compared with BRAIN only Action Language Verifier checks for integer reachability.

We took most of the benchmarks presented in this paper from Giorgio Delzanno's Web page `www.disi.unige.it/person/DelzannoG/`. The problems specified in these benchmarks were used to verify cache coherence protocols, properties of Java programs, and some other applications. The results are presented in Table 1. For each we present the runtimes and memory consumption (in megabytes). We write – when the compared system could not solve the problem because of the time or memory limit.

The table shows that BRAIN is normally faster than HyTech, and sometimes considerably faster. It also consumes less memory than HyTech. There are three problems (with the suffix `-inv` in the name) on which HyTech was faster (denoted by negative numbers in the speedup column). We will comment on these problems below. Considering that HyTech's implementation uses a polyhedra library based on [14] we cannot explain a considerable difference in the memory consumption between BRAIN and HyTech.

For non-trivial problems BRAIN is normally several hundred times faster than DMC, except for problems with invariants, where the difference is not so high. On non-trivial problems BRAIN without invariants is also normally at least 500 times faster than Action Language Verifier, on problems with invariants the difference is not so high. BRAIN also uses less memory than Action Language Verifier.

The problems with invariants were obtained from the original problems by adding *invariants*: some simple properties obtained by forward reachability analysis. A typical invariant has the form $v_1 + \ldots + v_k = m$, where m is a natural number. In fact, it bounds the variables $v_1, \ldots, v_k$ to a finite region. Such invariants cause a problem to BRAIN, because the basis for problems with such an invariant usually contains all, or a large portion, of the points in this region explicitly. We believe that this problem is not essential for the approach, but is rather particular to the current implementation of BRAIN in which the basis is stored explicitly, point-wise. A symbolic representation of this finite region, or the use of suitable datastructures for presenting finite-domain variables should solve this problem.

There are several problems which could only be solved by BRAIN, but not by any of the other systems. However, we would like to note that all of these systems are on some benchmarks more powerful than BRAIN since they can use techniques such as widening or transitive closure which the current version of BRAIN does not have. Examples are some versions of the ticket protocol.

To give the reader an idea of the complexity of the problems solved by BRAIN, we present statistics about the number of operations such an entailment-checks performed

Table 1. Statistics

problem	variables	intersections	entailment	transitions	time	memory	time	memory	speedup	time	memory	speedup	time	memory	speedup
		operations			BRAIN		HyTech			ALV			DMC		
csm5	13	3,313	790,119	3,576	9.26s	2	40.32s	18	4.35	87m32s	297	567	13h15m	11	5165
csm10	13	27,308	60,803,697	29,736	457s	11	868s	140	1.9	–			–		
csm15	13	107,503	990,874,884	117,496	119m13s	60	154m32s	509	1.3	–			–		
csm5-inv	13	106	835	152	0.37s	1	0.96s	2	2.6	2.87s	28	7.77	3.2s	10	8.6
csm10-inv	13	106	835	152	0.89s	1	0.96s	2	1.07	2.87s	28	3.22	3.2s	10	3.59
csm15-inv	13	106	835	152	1.88s	1	0.96s	2	-1.95	2.87s	28	1.52	3.2s	10	1.7
consistencyprot	12	813	30,557	880	0.96s	1	17.66s	10	18.4	17.66s	10	18.4	888s	10	925
consistencyprot-inv	12	813	29,803	880	7.25s	1	0.13s	2	-55	164.8s	73	22.7	19m52s	11	164.4
consprod	18	162,817	698,478,060	181,650	173m31s	22	–			–			–		
consprod-inv	18	187	5,126	742	1.19s	1	1.2s	4	1.01	25.87s	68	21.7	4.9	10	4.12
incdec	32	41,971	12,762,257	42,252	170.3s	10	–			–			–		
incdec-inv	32	873	54,824	4,004	22.9s	1	120.9s	80	5.3	10h20m	846	1625	96.3s	10	4.2
bigjava	44	122,516	93,410,447	134,828	171m31s	25	–			–			–		
bigjava1	44	127,185	95,800,396	139,688	189m16s	25	–			–			–		
bigjava-inv	44	7,581	3,378,979	45,103	40m2s	11	32m5s	849	-1.25	–			–		
bigjava1-inv	44	49,531	48,384,856	104,538	744m14s	19	–			–			–		

by BRAIN during each run. This statistics shows why DMC is hopelessly slow on some of these problems: for example, in the case of `csm15` one can hardly check almost 10^9 entailment problems in reasonable time using general-purpose constraint-solving tools. BRAIN solves them in less than 2 hours (on a fast PC with Intel this time would be less than 15 minutes). The table shows that, for most of the benchmarks, entailment seems to be the most important operation. It also demonstrates slowdown of BRAIN on the problems with invariants: indeed the number of operations per second in these problems is much smaller than that for their original formulations without the invariants.

10 Other Issues

In this section we discuss some other issues related to our formalization of transition systems. There are still many topics not covered here, such as fairness constraints, or concurrent applications of transitions.

10.1 Termination of the Local Algorithms

Simple constraints are usually easier to handle, especially in the case of constraints over the reals or integers. To deal only with simple constraints, one should use local backward reachability algorithms (see Theorem 26). As Example 1 shows, the use of local algorithms can make backward reachability analysis not terminate when it terminates for non-local algorithms.

10.2 Widening

When the reachability algorithms diverge, one can apply technique known as *approximation* or *widening*. This technique can be formalized as follows. Instead of using an equivalence-preserving simplification function simp we use a *widening function* wide, such that $\mathbb{M} \models \forall \mathcal{V}(A \rightarrow \mathsf{wide}(A))$. As a result, the set of states computed by the reachability algorithms in the limit, is a superset of the set of reachable states. The algorithms using widening are not semi-decision procedures and correct only if they return "unreachable". Widening can be used to make diverging algorithms converge in some cases, but also to make the formulas current simpler.

10.3 Transitive Closure

Transitive closure is a technique used by [8] to make the reachability analysis converge. The idea is to try to use a transitive closures of some transitions instead of the transitions themselves. For example, the transition

$$v_1 > 0 \Rightarrow v_1 \ := \ v_1 - 1, v_2 \ := \ v_2 + 1$$

over $\mathbb{I}$ can be replaced by its transitive closure

$$v_1 > 0 \wedge v_1' < v_1 \wedge v_1' + v_2' = v_1 + v_2.$$

The use of transitive closure can make otherwise non-terminating reachability analysis terminate. Unlike widening, it does not change the set of computed reachable states, so it is sound and semi-complete in the same sense as the other algorithms presented here. However, transitive closure applied to a GAS gives a system which is not a GAS, and hence our results on GAS are not applicable any more.

10.4 Relaxation

The system DMC [11] uses the technique known in integer programming as *relaxation*. Instead of solving the reachability problem over the integers, one solves "the same" reachability problem over the reals. Similarly, [10] analyzes integer reachability problems using the Hy-Tech model checker in which the state variables are real-valued [15]. Relaxation is attractive since solving satisfiability and entailment problems over the reals is easier than solving these problems over the integers. For example, satisfiability and entailment of simple constraints can be solved in polynomial time over the reals but are respectively NP-complete and coNP-complete over the integers. In this section we formalize relaxation in a general setting.

Let $\mathbb{M}_1$ and $\mathbb{M}_2$ be two structures of the same signature Σ over the domains $\mathcal{D}_1$ and $\mathcal{D}_2$ respectively, and ϵ be an *isomorphic embedding* of $\mathbb{M}_1$ into $\mathbb{M}_2$, i.e., a function from $\mathcal{D}_1$ to $\mathcal{D}_2$ with the following properties

1. For every predicate symbol P of arity n of the signature Σ and valuation s for a set of variables $\{x_1, \ldots, x_n\}$ in $\mathbb{M}_1$ we have $\mathbb{M}_1, s \models P(x_1, \ldots, x_n)$ if and only if $\mathbb{M}_2, \epsilon \circ s \models P(x_1, \ldots, x_n)$.
2. For every function symbol P of arity n of the signature Σ and valuation s for a set of variables $\{x, x_1, \ldots, x_n\}$ in $\mathbb{M}_1$ we have $\mathbb{M}_1, s \models x = f(x_1, \ldots, x_n)$ if and only if $\mathbb{M}_2, \epsilon \circ s \models x = f(x_1, \ldots, x_n)$.

These properties guarantee that for any quantifier-free formula A and valuation s for its variables in $\mathbb{M}_1$ we have $\mathbb{M}_1, s \models A$ if and only if $\mathbb{M}_2, s \models A$.

DEFINITION 31 (Relaxation) Let $R = (Tr, In, Fin)$ be an instance of the reachability problem over a structure $\mathbb{M}_1$ and there exists an isomorphic embedding of $\mathbb{M}_1$ into a structure $\mathbb{M}_2$. The instance (Tr, In, Fin) of the reachability problem over $\mathbb{M}_2$ is called the *relaxation* of R. ❑

That is, the original problem and its relaxation use the same formulas but are considered over different structures.

THEOREM 32 *Let $R_1 = (Tr, In, Fin)$ be an instance of the reachability problem over $\mathbb{M}_1$ and R_2 be its relaxation over $\mathbb{M}_2$. Suppose that all formulas in (Tr, In, Fin) are equivalent to existential formulas in $\mathbb{M}_1$. If R_2 does not hold, then R_1 does not hold too.* ❑

This theorem means that one can solve relaxation instead of the original problem to show non-reachability as soon as the initial and final conditions and the transition formula are existential or equivalent to existential formulas in $\mathbb{M}_1$. All special classes of transition systems (for example, positive existential) introduced in this paper have this property. One should be careful about using relaxation when properties other than reachability are checked. For example, if guards in a GAS are existentially quantified formulas, than one cannot in general check deadlock-freedom using relaxation. As stated in Theorem 16, the deadlock formula for GAS is $\neg(P_1 \lor \ldots \lor P_n)$, which may not be equivalent to an existential formula, as soon as at least one of the P_i is existentially quantified.

10.5 Other Temporal Formulas

As observed in many papers, a possibility to build fixpoints implies that properties more complex than reachability could be verified. It is not hard to argue that the following property, related to the fixpoint construction, is true for all of the algorithms of this paper.

THEOREM 33 *If the forward reachability algorithm terminates with the answer "unreachable" then* current *symbolically represents the set of all states reachable from initial states. If the backward reachability algorithm terminates with the answer "unreachable" then* current *symbolically represents the set of all states backward reachable from the final states.* ❑

This property allows one to extend the reachability algorithms and some results to more general classes of temporal formulas than those expressing reachability, but the analysis of such extended algorithms is beyond the scope of this paper.

11 Related Work

Podelski [24] and Delzanno and Podelski [11] formalize model checking procedures as constraint solving using a rather general framework. In fact, what they call constraints are first-order formulas over structures. They treat a class of properties more general than reachability and use GAS and local algorithms. Our formalization has much in common with their formalization but our results are, in a way, orthogonal, because they do not study special classes of formulas. Among all results proved in this paper, only soundness and semi-completeness are similar to those studied by Delzanno and Podelski. All other results of this paper are new. In addition, Delzanno and Podelski do not discuss selection functions, so their algorithms are less general than ours. The class of systems for which Delzanno and Podelski's algorithms can be applied is also less general than the one studied here. Indeed, they require a constraint solver also to implement variable elimination. If a theory has such a solver, then this theory also has quantifier-elimination (every formula is equivalent to a boolean combination of constraints). The solver is also able to decide satisfiability of constraints, so in fact they deal with theories which are both decidable and have quantifier elimination. They do not notice that in many special cases (as studied in this paper) quantifier elimination is not required. Thus, our results encompass some important theories (such as those of queues in Bjørner [5]) not covered by the results of Delzanno and Podelski.

An approach to formalizing reachability for hybrid systems using first-order logic is presented by Lafferriere, Pappas, and Yovine [21]. They also note that one can use arbitrary first-order formulas (over reals) due to quantifier elimination. Henzinger and Majumdar [16] present a classification of state transition systems, but over the reals.

Lemma 17 which observes that for GAS backward reachability algorithms do not introduce extra variables is similar to the axiom of assignment for the weakest preconditions. This property is heavily used in BRAIN [27]. It is possible that it has been exploited in other backward reachability symbolic search tools but we could not find papers which observe this property.

Non-local reachability procedures for model checking were already formulated in Emerson and Clarke [12], Queille and Sifakis [26]. Delzanno and Podelski [11] studied algorithms based on local entailment in the framework of symbolic model checking. They do not consider the behaviour of local algorithms for different selection functions. Local algorithms could be traced to early works in automated reasoning and later works on logic programming with tabulation (e.g., Warren [31]), constraint logic programming, and constraint databases (e.g., Kanellakis, Kuper, and Revesz [18]).

Kesten, Maler, Marcus, Pnueli, and Shahar [19] present a general approach to symbolic model checking of infinite-state systems, but based on a single language rather than different languages for different structures.

Guarded assignment systems (under various names) are studied in many other papers, too numerous to be mentioned here. There are also many papers on (un)decidability results for infinite-state systems, both over integers and other structures, not mentioned here.

Our technique of using Hilbert's basis is similar to a technique used to deal with real-valued systems described in Halbwachs, Proy and Roumanoff [14]. They represent convex polyhedra using *systems of generators*, i.e., two finite sets of vectors (called *vertices* and *rays*). This representation allows one to perform efficient entailment checks using a property similar to that of Theorem 29.

One can implement satisfiability- and entailment-checking using the decision procedure for the first order theory of $\mathbb{I}$. In some cases, one can even use off-the-shelf decision procedures or libraries. For example, Bultan, Gerber, and Pugh [8] use the Omega Library [25] for deciding Presburger arithmetic, Bultan [7] uses the Composite Library (Yavuz-Kahveci, Tuncer and Bultan [32]), Delzanno and Podelski [11] use the CLP(R) library of Sicstus Prolog. The use of decision procedures for Presburger arithmetic has several advantages, since formulas more general than simple constraints can be handled. As a consequence, one can use non-local reachability algorithms, forward reachability, and apply techniques such as widening and transitive closure which cannot be handled by the current version of BRAIN. However, the use of general-purpose algorithms to decide specific classes of formulas may be inefficient, which is confirmed by our experiments.

In the future we are going to develop BRAIN into an advanced infinite-state model-checker based on the implementation method proposed here, which will be both faster and more flexible. In particular, we will include in BRAIN other reachability algorithms and techniques such as widening. This will, however, require an implementation of quantifier elimination and an extension of our method, and especially Theorem 29, to constraints with existentially quantified variables and divisibility constraints.

To cope with the problem of large finite regions, one has to introduce their convenient symbolic representation, which may require reworking of all algorithms. It would also be interesting to apply our method to real-valued systems, or systems with both integer- and real-valued variables.

Acknowledgments

We thank Howard Barringer, Giorgio Delzanno, and Andreas Podelski for their comments on earlier versions of this paper.

References

1. P.A. Abdulla, K. Cerans, B. Jonsson, and Y.-K. Tsay. Algorithmic analysis of programs with well quasi-ordered domains. *Information and Computation*, 160(1-2):109–127, January 2000.
2. P.A. Abdulla and B. Jonsson. Ensuring completeness of symbolic verification methods for infinite-state systems. *Theoretical Computer Science*, 256:145–167, 2001.
3. F. Ajili and E. Contejean. Avoiding slack variables in the solving of linear Diophantine equations and inequations. *Theoretical Computer Science*, 173(1):183–208, 1997.
4. A.P. Beltyukov. Decidability of the universal theory of natural numbers with addition and divisibility (in Russian). *Zapiski Nauchnyh Seminarov LOMI*, 60:15–28, 1976. English translation in: Journal of Soviet Mathematics.
5. N.S. Bjørner. *Integrating Decision Procedures for Temporal Verification*. PhD thesis, Computer Science Department, Stanford University, 1998.
6. N.S. Bjørner. Reactive verification with queues. In *ARO/ONR/NSF/DARPA Workshop on Engineering Automation for Computer-Based Systems*, pages 1–8, Carmel, CA, 1998.
7. T. Bultan. Action Language: a specification language for model checking reactive systems. In *ICSE 2000, Proceedings of the 22nd International Conference on Software Engineering*, pages 335–344, Limerick, Ireland, 2000. ACM.
8. T. Bultan, R. Gerber, and W. Pugh. Model-checking concurrent systems with unbounded integer variables: symbolic representations, approximations, and experimental results. *ACM Transactions on Programming Languages and Systems*, 21(4):747–789, 1999.
9. E. Contejean and H. Devie. An efficient incremental algorithm for solving of systems of linear Diophantine equations. *Information and Computation*, 113(1):143–172, 1994.
10. G. Delzanno. Automatic verification of parametrized cache coherence protocols. In A.E. Emerson and A.P. Sistla, editors, *Computer Aided Verification, 12th International Conference, CAV 2000*, volume 1855 of *Lecture Notes in Computer Science*, pages 53–68. Springer Verlag, 2000.
11. G. Delzanno and A. Podelski. Constraint-based deductive model checking. *International Journal on Software Tools for Technology Transfer*, 3(3):250–270, 2001.
12. E.A. Emerson and E.M. Clarke. Using branching time temporal logic to synthesize synchronization skeletons. *Science of Computer Programming*, 2(3):241–266, 1982.
13. J. Esparza, A. Finkel, and R. Mayr. On the verification of broadcast protocols. In *14th Annual IEEE Symposium on Logic in Computer Science (LICS'99)*, pages 352–359, Trento, Italy, 1999. IEEE Computer Society.

14. N. Halbwachs, Y.-E. Proy, and P. Roumanoff. Verification of real-time systems using linear relation analysis. *Formal Methods in System Design*, 11(2):157–185, 1997.
15. T.A. Henzinger, P.-H. Ho, and H. Wong-Toi. Hy-Tech: a model checker for hybrid systems. *International Journal on Software Tools for Technology Transfer*, 1(1–2):110–122, 1997.
16. T.A. Henzinger and R. Majumdar. A classification of symbolic state transition systems. In H. Reichel and S. Tison, editors, *STACS 2000*, volume 1770 of *Lecture Notes in Computer Science*, pages 13–34. Springer Verlag, 2000.
17. D. Hilbert. über die Theorie der algebraischen Formen. *Mathematische Annalen*, 36:473–534, 1890.
18. P. Kanellakis, G.M. Kuper, and P.Z. Revesz. Constraint query languages. *Journal of Computer and System Sciences*, 51:26–52, 1995.
19. Y. Kesten, O. Maler, M. Marcus, A. Pnueli, and E. Shahar. Symbolic model checking with rich assertional languages. *Theoretical Computer Science*, 256(1-2):93–112, 2001.
20. O. Kupferman and M. Vardi. Model checking of safety properties. *Formal Methods in System Design*, 19(3):291–314, 2001.
21. G. Lafferriere, G.J. Pappas, and S. Yovine. Symbolic reachability computation for families of linear vector fields. *Journal of Symbolic Computations*, 32(3):231–253, 2001.
22. L. Lipshitz. The Diophantine problem for addition and divisibility. *Transactions of the American Mathematical Society*, 235:271–283, January 1978.
23. V.I. Mart'janov. Universal extended theories of integers. *Algebra i Logika*, 16(5):588–602, 1977.
24. A. Podelski. Model checking as constraint solving. In J. Palsberg, editor, *Static Analysis, 7th International Symposium, SAS 2000*, volume 1924 of *Lecture Notes in Computer Science*, pages 22–37. Springer Verlag, 2000.
25. W. Pugh. Counting solutions to Presburger formulas: how and why. *ACM SIGPLAN Notices*, 29(6):121–134, June 1994. Proceedings of the ACM SIGPLAN'94 Conference on Programming Languages Design and Implementation (PLDI).
26. J.P Queille and J. Sifakis. Specification and verification of concurrent systems in Cesar. In M. Dezani-Ciancaglini and M. Montanari, editors, *International Symposium on Programming*, volume 137 of *Lecture Notes in Computer Science*, pages 337–351. Springer Verlag, 1982.
27. T. Rybina and A. Voronkov. Using canonical representations of solutions to speed up infinite-state model-checking. In *CAV 2002*, 2002. To appear.
28. A. Schrijver. *Theory of Linear and Integer Programming*. John Wiley and Sons, 1998.
29. A.P. Tomás and M. Filgueiras. An algorithm for solving systems of linear diophantine equations in naturals. In E. Costa and A. Cardoso, editors, *Progress in Artificial Intelligence, 8th Portugese Conference on Artificial Intelligence, EPIA'97*, volume 1323 of *Lecture Notes in Artificial Intelligence*, pages 73–84, Coimbra, Portugal, 1997. Springer Verlag.
30. A. Voronkov. An incremental algorithm for finding the basis of solutions to systems of linear Diophantine equations and inequations. unpublished, January 2003.
31. D.S. Warren. Memoing for logic programs. *Communications of the ACM*, 35(3):93–111, 1992.
32. T. Yavuz-Kahveci, M. Tuncer, and T. Bultan. A library for composite symbolic representations. In T. Margaria, editor, *Tools and Algorithms for Construction and Analysis of Systems, 7th International Conference, TACAS 2001*, volume 1384 of *Lecture Notes in Computer Science*, pages 52–66, Genova, Italy, 2001. Springer Verlag.

Winning Strategies and Synthesis of Controllers

Igor Walukiewicz

LaBRI, Université Bordeaux-1

A system consist of a process, of an environment and of possible ways of interaction between them. The synthesis problem is: given a specification ϕ find a program P for the process such that the overall behaviour of the system satisfies no matter what the bahaviour of the environment is.

One of the motivations for development of automata theory was the interest in verifying and synthesizing switching circuits. A circuit can be modeled as a function transforming infinite sequence of inputs into an infinite sequence of outputs. Church in 1963 has posed the problem of synthesizing such a function for a specification given in S1S, the monadic second order logic of one successor. The solution to this problem was given by Buchi and Ladweber in 1969. Nowadays we understand that this problem, and many other variations of the synthesis problem, reduce to finding a winning strategy in a two player game with perfect information and regular winning conditions.

A distributed system consist of some number of processes, each interacting in some way with the environment and other processes. The distributed synthesis problem is to find a program for each of the processes so that the global behaviour satisfies the given specification no matter what the behaviour of the environment is.

There are numerous ways of describing a distributed system and hence there are numerous ways of formalizing the distributed synthesis problem. One way, suggested by Pnueli and Rosner, is to consider a system as an architecture: a graph of communication channels connecting boxes into which we need to put I/O programs. The other way, introduced by Rudie and Wonham, is to specify the sets of controllable and observable actions for each of the components of the system. For these and other settings there are algorithms solving the distributed synthesis problem for some cases.

In the distributed synthesis we do not understand very well what makes the problem decidable or undecidable. Each decidability/undecidability proof depends very strongly on the formalism being used. One of the reasons for this is that, unlike for simple synthesis, we cannot reduce distributed synthesis problems to the problem of solving a two player game for a simple reason that in the distributed case there are more than two parties.

In this talk we will summarize the existing approaches to synthesis and distributed synthesis problems. We will propose a notion of distributed game that can serve as a common framework for distributed synthesis problems. In such a game there is one environment player and there is a coalition of process players, each having only a partial information about the play. We will state a couple of results on solvability of such games and indicate why they imply most of the known results for distributed synthesis problems.

M. Baaz and J.A. Makowsky (Eds.): CSL 2003, LNCS 2803, p. 574, 2003.

Logical Relations for Dynamic Name Creation*

Yu Zhang and David Nowak

LSV, CNRS UMR 8643, ENS de Cachan
94235 Cachan Cedex, France
{zhang,nowak}@lsv.ens-cachan.fr

Abstract. Pitts and Stark's nu-calculus is a typed lambda-calculus which forms a basis for the study of interaction between higher-order functions and dynamically created names. A similar approach has received renewed attention recently through Sumii and Pierce's cryptographic lambda-calculus, which deals with security protocols. *Logical relations* are a powerful tool to prove properties of such a calculus, notably observational equivalence. While Pitts and Stark construct a logical relation for the nu-calculus, it rests heavily on operational aspects of the calculus and is hard to be extended. We propose an alternative Kripke logical relation for the nu-calculus, which is derived naturally from the categorical model of the nu-calculus and the general notion of Kripke logical relation. This is also related to the Kripke logical relation for the name creation monad by Goubault-Larrecq et al. (CSL'2002), which the authors claimed had similarities with Pitts and Stark's logical relation. We show that their Kripke logical relation for names is strictly weaker than Pitts and Stark's. We also show that our Kripke logical relation, which extends the definition of Goubault-Larrecq et al., is equivalent to Pitts and Stark's up to first-order types; our definition rests on purely semantic constituents, and dispenses with the detours through operational semantics that Pitts and Stark use.

Keywords: Kripke logical relation, name creation, nu-calculus, categorical models of lambda calculi

1 Introduction

The *nu-calculus* of Pitts and Stark is a typed lambda-calculus, extended with state in the form of dynamically created names. Through the interaction between names and functions, the language can capture notions of scope, visibility and sharing. It has also connections to local declarations in general; to the mobile processes of the π-calculus; to security protocols in the spi-calculus; and particularly, to the cryptographic lambda-calculus of Sumii and Pierce [15]. When we regard dynamically created names as dynamically created keys, the nu-calculus is indeed a syntactical subset of the cryptographic lambda-calculus. It contains a primitive for key generation but lacks those for encryption and decryption.

* Partially supported by the RNTL project EVA, the ACI jeunes chercheurs "Sécurité informatique, protocoles cryptographiques et détection d'intrusions" and the ACI Cryptologie "PSI-Robuste".

M. Baaz and J.A. Makowsky (Eds.): CSL 2003, LNCS 2803, pp. 575–588, 2003.

Logical relations are a powerful technique to prove properties of typed lambda calculi such like the nu-calculus and the cryptographic lambda-calculus. In [11, 12], Pitts and Stark define an operational logical relation to establish the observational equivalence between nu-calculus expressions. They also prove that this logical relation is complete for first-order types. But the operational logical relation is a quite syntactic one and it is defined based on some operational semantics of the nu-calculus, which makes it hard to be extended.

We propose another logical relation for the nu-calculus in a natural way. This logical relation follows the general notion of Kripke logical relations [5] and is derived from a denotational model of the nu-calculus, which is a functor category with a strong monad. It is also related to the Kripke logical relation for the name creation monad by Goubault-Larrecq, Lasota and Nowak [3]. The latter claims to have similarities with Pitts and Stark's operational logical relation, while we shall show that it is indeed strictly weaker than Pitts and Stark's. We extend Goubault-Larrecq, Lasota and Nowak's definition so as to derive naturally our Kripke logical relation in a categorical style. We also show that, this logical relation is equivalent to Pitts and Stark's operational logical relation up to first-order types.

As preliminaries, Section 2 gives a brief review of the nu-calculus and the operational logical relation, more details of which can be found in [11, 12, 14]. Section 3 introduces a categorical model $\boldsymbol{Set}^{\mathcal{I}}$ for the nu-calculus [12, 13] to provide a basis for our Kripke logical relation. We give in Section 4 the definition of the Kripke logical relation, while before that, we derive another one from Goubault-Larrecq, Lasota and Nowak's definition for name creation monad in [3] and show that this one is strictly weaker than Pitts and Stark's operational logical relation. At the end of Section 4, we show that ours is equivalent to Pitts and Stark's up to first-order types.

2 The nu-Calculus and the Operational Logical Relation

2.1 The nu-Calculus

The nu-calculus is a small language designed to show the interaction between dynamically created names and higher-order functions [11, 12]. It is a typed call-by-value lambda-calculus extended with the notion of a *name*; names have their own type ν, they can be created fresh, passed around and tested for equality. Higher-order functions and booleans are also available, but to ensure termination functions cannot be defined recursively. New names are created in expressions of the form $\nu n.M$, which binds a fresh name to the identifier n and then evaluates M. The use of a call-by-value semantics means that although only the expression M can refer to n explicitly, the new name itself may escape from this scope. For example, $(\lambda x^{\nu}.x = x)(\nu n.n)$ evaluates to *true*, with x bound to a name rather than to $\nu n.n$ itself.

The syntax of the nu-calculus is based on the simply-typed lambda-calculus. Types take the form

$$\sigma ::= o \mid \nu \mid \sigma \to \sigma$$

where o and ν are two ground types denoting *booleans* and *names* respectively. Terms take the form

$$\begin{array}{lll} M ::= & x & \text{variable} \\ & \mid n & \text{name} \\ & \mid \mathit{true} \mid \mathit{false} & \text{boolean values} \\ & \mid \mathit{if}\ M\ \mathit{then}\ M\ \mathit{else}\ M & \text{conditional} \\ & \mid M = M & \text{equality of names} \\ & \mid \nu n.M & \text{local name declaration} \\ & \mid \lambda x^{\sigma}.M & \text{function abstraction} \\ & \mid MM & \text{function application} \end{array}$$

where $x \in Var$, an infinite set whose elements are called *variables*, and $n \in Nme$, an infinite set (disjoint from Var) whose elements are called *names*. Function abstraction is a variable-binding construct (occurrences of x in M are bound in $\lambda x^{\sigma}.M$), whereas local name declaration is a name-binding construct (occurrences of n in M are bound in $\nu n.M$). We implicitly identify expressions which only differ in their choice of bound variables and names (α-conversion).

An expression is *closed* if it has no free variables; a closed expression may still have free names. We denote by $M[M'/x]$ (respectively $M[M'/n]$) the result of substituting the expression M' for free occurrences of the variable x (respectively, the name n) in the expression M. The substitution is *capture avoiding*. This can always be arranged by α-converting M.

Expressions will be assigned types via typing assertions of the form

$$s, \Gamma \vdash M : \sigma$$

where s is a finite subset of Nme, Γ is a finite set of typed variables, σ is a type, and M is an expression (more precisely, an α-equivalence class of expressions) with free names in s and free variables in Γ.

An expression is in *canonical form* if it is either a name, a variable, one of the boolean constants true or false, or a function abstraction. These are the *values* of the nu-calculus. As what Pitts and Stark do in [11, 12], we define the sets

$$\begin{aligned} \mathrm{Exp}_{\sigma}(s, \Gamma) &= \{M \mid s, \Gamma \vdash M : \sigma\} \\ \mathrm{Can}_{\sigma}(s, \Gamma) &= \{C \mid C \in \mathrm{Exp}_{\sigma}(s, \Gamma), C \text{ canonical}\} \\ \mathrm{Exp}_{\sigma}(s) &= \mathrm{Exp}_{\sigma}(s, \emptyset) \\ \mathrm{Can}_{\sigma}(s) &= \mathrm{Can}_{\sigma}(s, \emptyset) \end{aligned}$$

of expressions and canonical expressions at any type σ and for any finite sets s, Γ of names and typed variables.

The operational semantics of the nu-calculus is specified in [11, 12] by an inductively defined evaluation relation. Elements of the relation take the form

$$s \vdash M \Downarrow_{\sigma}^{s'} C$$

where s and s' are disjoint, $M \in \mathrm{Exp}_\sigma(s)$ and $C \in \mathrm{Can}_\sigma(s \oplus s')$. This is intended to mean that in the presence of the names s, expression M of type σ evaluates to canonical form C while creating the fresh names in s' in the process. The sets s and $s \oplus s'$ can be seen as initial and final *states* of the computation. See [11, 12] for details of the evaluation relation.

2.2 The Operational Logical Relation

An operational logical relation between nu-calculus expressions is given by Pitts and Stark in [11, 12]. It rests on the operational semantics of the nu-calculus and a certain kind of relation between sets of names.

Definition 1 (Spans). If s_1 and s_2 are sets of names, then a *span* $R : s_1 \rightleftharpoons s_2$ is an injective partial map from s_1 to s_2. That is, the graph $R \subseteq s_1 \times s_2$ satisfies

$$(n_1 \; R \; n_2) \;\&\; (n_1' \; R \; n_2') \Longrightarrow (n_1 = n_1' \Longleftrightarrow n_2 = n_2').$$

A span can also be represented as a pair of injections $s_1 \leftarrowtail R \rightarrowtail s_2$.

If $R' : s_1' \rightleftharpoons s_2'$ is another span, with s_1' and s_2' disjoint from s_1 and s_2 respectively, then the disjoint union of R and R' is also a span:

$$R \oplus R' : s_1 \oplus s_1' \rightleftharpoons s_2 \oplus s_2'.$$

The domain and co-domain of definition for $R : s_1 \rightleftharpoons s_2$ are defined by :

$$\mathrm{dom}(R) = \{n_1 \in s_1 \mid \exists n_2 \in s_2 \; (n_1 \; R \; n_2)\}$$
$$\mathrm{cod}(R) = \{n_2 \in s_2 \mid \exists n_1 \in s_1 \; (n_1 \; R \; n_2)\}$$

The intuition behind spans is that they should capture the use that nu-calculus expressions make of public and private names. So if $R : s_1 \rightleftharpoons s_2$, then the bijection between $\mathrm{dom}(R) \subseteq s_1$ and $\mathrm{cod}(R) \subseteq s_2$ represents matching use of "visible" names, while the remaining elements not in the graph of R are "unseen" names.

Definition 2 (Operational logical relation). If $R : s_1 \rightleftharpoons s_2$ is a span then the relations

$$R_\sigma^{can} \subseteq \mathrm{Can}_\sigma(s_1) \times \mathrm{Can}_\sigma(s_2)$$
$$R_\sigma^{exp} \subseteq \mathrm{Exp}_\sigma(s_1) \times \mathrm{Exp}_\sigma(s_2)$$

are defined by induction over the structure of the type σ, according to:

$$b_1 \; R_o^{can} \; b_2 \Longleftrightarrow b_1 = b_2$$
$$n_1 \; R_\nu^{can} \; n_2 \Longleftrightarrow n_1 \; R \; n_2$$

$$(\lambda x^\sigma.M_1) \; R_{\sigma \to \sigma'}^{can} \; (\lambda x^\sigma.M_2) \Longleftrightarrow$$
$$\forall R' : s_1' \rightleftharpoons s_2', C_1 \in Can_\sigma(s_1 \oplus s_1'), C_2 \in Can_\sigma(s_2 \oplus s_2')$$

$$(C_1 \ (R \oplus R')^{can}_{\sigma} \ C_2 \Longrightarrow M_1[C_1/x] \ (R \oplus R')^{exp}_{\sigma'} \ M_2[C_2/x])$$

$$M_1 \ R^{exp}_{\sigma} \ M_2 \Longleftrightarrow$$
$$\exists R' : s_1' \rightleftharpoons s_2', C_1 \in Can_{\sigma}(s_1 \oplus s_1'), C_2 \in Can_{\sigma}(s_2 \oplus s_2')$$
$$(s_1 \vdash M_1 \Downarrow^{s_1'}_{\sigma} C_1 \ \& \ s_2 \vdash M_2 \Downarrow^{s_2'}_{\sigma} C_2 \ \& \ C_1 \ (R \oplus R')^{can}_{\sigma} \ C_2)$$

The relation R^{can}_{σ} and R^{exp}_{σ} coincide on canonical forms, and we may write them as R_{σ} indiscriminately.

This operational logical relation can be used to establish observational equivalences in the nu-calculus. Consider the two nu-calculus expressions of type $\nu \to o$: $\nu n.\lambda x^{\nu}.(x = n)$ and $\lambda x^{\nu}.false$. According to the evaluation relation in [11, 12],

$$s, \Gamma \vdash \nu n.\lambda x^{\nu}.(x = n) \Downarrow^{\{n\}}_{\sigma} \lambda x^{\nu}.(x = n)$$

and

$$s, \Gamma \vdash \lambda x^{\nu}.false \Downarrow^{\emptyset}_{\sigma} \lambda x^{\nu}.false.$$

The first term creats a new name, which is definitely private to the function, and then compare it with the argument. Since no external context can supply this name, the function always returns the value $false$, so it is observationally equivalent to the second term. The only possible span for new names is the empty relation $R' : \{n\} \rightleftharpoons \emptyset$. Whatever the span $R : s_1 \rightleftharpoons s_2$ is, we have

$$\lambda x^{\nu}.(x = n) \ (R \oplus R')^{can}_{\nu \to o} \ \lambda x^{\nu}.false$$

and consequently, the above two expressions are related.

3 A Categorical Model for the nu-Calculus

As the definition of R^{exp}_{σ} shows, the operational logical relation relies heavily on the operational semantics of the nu-calculus. Moreover this definition silently assumes that each term has a canonical form — it is indeed the case in nu-calculus. This prevents one from extending this relation to non-terminating calculi with names. This section introduces the categorical model $\boldsymbol{Set}^{\mathcal{I}}$ defined by Stark in [12, 13] to give the denotational semantics for the nu-calculus. This model is a functor category equipped with a strong monad and it provides us not only a denotational model for the behavior of nu-calculus expressions, but also a sound basis for a logical relation to be defined next.

3.1 Categorical Model

Let $\mathcal{I}$ be the category of finite sets and injective functions. Let $\boldsymbol{C}$ be the category $\boldsymbol{Set}^{\mathcal{I}}$ of functors from $\mathcal{I}$ to $\boldsymbol{Set}$ and natural transformations. For short, we write $\boldsymbol{T}As$ for $\boldsymbol{T}(A)(s)$ and similarly for other notations. Let $+$ denote disjoint union in $\mathcal{I}$ [1].

We consider the strong monad $(\boldsymbol{T}, \boldsymbol{\eta}, \boldsymbol{\mu}, \boldsymbol{t})$ on $\boldsymbol{C}$ defined in [12, 13, 3]:

[1] Note that $+$ is not a coproduct in $\mathcal{I}$ which in fact does not have a coproduct. However $+$ is functorial in both components, associative, and has a neutral element.

- $\boldsymbol{T}A = \mathrm{colim}_{s'} A(\_ + s') : \mathcal{I} \to \boldsymbol{Set}$. Explicitly:
  - $\boldsymbol{T}As$ is the set of equivalence classes of pairs (s', a), where $s' \in \mathcal{I}$ and $a \in A(s+s')$, modulo the smallest equivalence relation $\equiv_C$ such that $(s', a) \equiv_C (s'', A(\mathrm{id}_s + i)a)$ for each morphism $i : s' \to s''$ in $\mathcal{I}$. We write $[s, a]$ for the equivalence class of (s, a).
  - On morphism i, $\boldsymbol{T}Ai$ maps the equivalence class of (s', a) to the equivalence class of $(s', A(i + \mathrm{id}_{s'})a)$.
- For any $f : F \to G$ in $\boldsymbol{C}$, $\boldsymbol{T}fs : \boldsymbol{T}Fs \to \boldsymbol{T}Gs$ is defined by $\boldsymbol{T}fs[s', a] = [s', f(s+s')a]$. This is compatible with $\equiv_C$ because f is natural.
- $\boldsymbol{\eta}_A s : As \to \boldsymbol{T}As$ is defined by $\boldsymbol{\eta}_A sa = [\emptyset, a]$.
- $\boldsymbol{\mu}_A s : \boldsymbol{T}^2 As \to \boldsymbol{T}As$ is defined by $\boldsymbol{\mu}_A s[s', [s'', a]] = [s' + s'', a]$
- $\boldsymbol{t}_{A_1,A_2} s : A_1 s \times \boldsymbol{T}A_2 s \to \boldsymbol{T}(A_1 \times A_2)s$ is defined by $\boldsymbol{t}_{A_1,A_2} s(a, [s', b]) = [s', (A_1 i_{s,s'} a, b)]$ where $i_{s,s'} : s \to s + s'$ is the canonical injection.

It is standard that this category is cartesian closed. If $F, G : \mathcal{I} \to \boldsymbol{Set}$ are functors then their exponent is defined by the standard construction in covariant presheaves [4]:

$$G^F s = \boldsymbol{Set}^{\mathcal{I}}(\mathcal{I}(s, -) \times F, G)$$
$$G^F ifs''\langle j, a''\rangle = fs''\langle j \circ i, a''\rangle,$$

where $s, s', s'', i : s \to s', j : s' \to s'' \in \mathcal{I}$ and $f \in G^F s, a'' \in Fs''$.

3.2 The Interpretation of the nu-Calculus

The interpretation of the nu-calculus is based on a computational metalanguage [12, 13], which stems from Moggi's computational lambda calculus [8–10]. In this particular metalanguage, besides the usual boolean type, name type and function types, we have a unary type constructor T: if A is a type, then elements of TA are *computations* of type A. The difference between a value and a computation is that a computation may generate new names before returning a value. A nu-calculus expression is regarded as a computation in the metalanguage which generates a set of names and then returns a value, and an expression in canonical form simply returns a value without generating any name.

The category $\boldsymbol{Set}^{\mathcal{I}}$ is qualified to be a categorical model of this metalanguage [12, 13]. In this category, objects of $\mathcal{I}$ represent *stages of computation*, i.e., what names have been created. For a functor $A : \mathcal{I} \to \boldsymbol{Set}$, the set As is composed of values defined over the names in s. Morphisms in $\mathcal{I}$ and their images in $\boldsymbol{Set}$ correspond to name substitutions.

We begin by defining two functors B and N from $\mathcal{I}$ to $\boldsymbol{Set}$. Functor B is defined by:

$$\forall s \in \mathcal{I}, Bs = Bool \;\;\&\;\; \forall i : s \to s' \in \mathcal{I}, Bi = \mathrm{id}_{Bool},$$

and functor N:

$$\forall s \in \mathcal{I}, Ns = s \;\;\&\;\; \forall i : s \to s' \in \mathcal{I}, Ni = i,$$

where *Bool* is a set containing only two elements — *tt* and *ff*. Types in the nu-calculus are interpreted as objects of $\boldsymbol{Set}^{\mathcal{I}}$:

$$\begin{aligned} [\![o]\!] &= B \\ [\![\nu]\!] &= N \\ [\![\sigma \to \sigma']\!] &= (\boldsymbol{T}[\![\sigma']\!])^{[\![\sigma]\!]}. \end{aligned}$$

According to what Stark does in [12, 13], for each valid type assertion $s, \Gamma \vdash M : \sigma$ in the nu-calculus, the semantics of M in the categorical model is given with an index s:

$$[\![M]\!]_{\Gamma,s} \in (\boldsymbol{T}[\![\sigma]\!])^{[\![\Gamma]\!]}s \quad \text{where} \quad [\![\Gamma]\!] = \prod_{x_i:\sigma_i \in \Gamma} [\![\sigma_i]\!].$$

If M is a closed term, i.e., $M \in \mathrm{Exp}_\sigma(s)$, then $[\![M]\!]_s \in \boldsymbol{T}[\![\sigma]\!]s$, identifying A^1 with A, where 1 is the terminal object in $\boldsymbol{Set}^{\mathcal{I}}$ and $1s$ is a singleton for all s. The semantics of a closed canonical expression $[\![C]\!]_s$ is always of the form $[\emptyset, |C|_s]$, an element of the set $\boldsymbol{T}[\![\sigma]\!]s$. Here $|C|_s \in [\![\sigma]\!]s$ and $|-|_s$ is the canonical interpretation defined for each closed canonical form. If $s \vdash M \Downarrow_\sigma^{s'} C$, where $M \in \mathrm{Exp}_\sigma(s)$ and $C \in \mathrm{Can}_\sigma(s)$, the semantics of the expression M can also be given by the canonical interpretation of the corresponding canonical form C:

$$[\![M]\!]_s = [\![\nu s'.C]\!]_s = [s', |C|_{s+s'}] \in \boldsymbol{T}[\![\sigma]\!]_s. \tag{1}$$

A definability result for the model $\boldsymbol{Set}^{\mathcal{I}}$ is also given in [12].

Lemma 1 (Definability). *Suppose that σ is a ground or first-order type of the nu-calculus and s is some set of names. If $a \in [\![\sigma]\!]s$ and $e \in \boldsymbol{T}[\![\sigma]\!]s$, then there are expressions $C \in \mathrm{Can}_\sigma(s)$ and $M \in \mathrm{Exp}_\sigma(s)$ such that $a = |C|_s$ and $e = [\![M]\!]_s$.* □

4 The Kripke Logical Relation

4.1 A Weak Logical Relation

We try to define a logical relation on the categorical model $\boldsymbol{Set}^{\mathcal{I}}$ so that it does not depend on any syntactical aspect of the nu-calculus. The general Kripke logical relation [5] gives a natural way to construct such a logical relation. A Kripke logical relation is based on a Kripke lambda models [5, 6] which includes a partially-ordered set of *possible worlds* and have a set of elements of each type at each possible world. Informally, the relationship between elements of type σ at different worlds is that every element of σ at w will continue to be an element of σ at every possible $w' \geq w$.

Strictly speaking, a Kripke logical relation is indexed by elements of a poset. It has a relation for each type and a *possible world* instead of just having a relation for each type [5]. This can be naturally extended to be indexed by objects of a category as argued in [1]. The category $\boldsymbol{Set}^{\mathcal{I}}$ in the last section just defines a Kripke model. We take the sets in $\mathcal{I}$ as the possible worlds and set inclusion as the partial order between them.

In [7], a categorical definition of logical relations based on subsconing was proposed. This definition was extended in [3] in order to define a notion of logical relation for monadic types. According to this extended logical relation, we define a Kripke logical relation $\mathbb{R}^s_A \subseteq As \times As$ in the category $\boldsymbol{Set}^{\mathcal{I}}$ by induction over the structure of the corresponding type of A in a metalanguage for $\boldsymbol{Set}^{\mathcal{I}}$ [8–10] (here we do not distinguish on notation between an object of a category and the corresponding type in the metalanguage). Notably, we specify the types corresponding to the objects B and N as two constant types:

$$\begin{aligned}
b_1 \ \mathbb{R}^s_B \ b_2 &\iff b_1 = b_2, \\
n_1 \ \mathbb{R}^s_N \ n_2 &\iff n_1 = n_2, \\
f_1 \ \mathbb{R}^s_{G^F} \ f_2 &\iff \forall s', i : s \to s' \in \mathcal{I}, a_1, a_2 \in Fs' \\
&\qquad (a_1 \ \mathbb{R}^{s'}_F \ a_2 \Longrightarrow f_1 s'\langle i, a_1\rangle \ \mathbb{R}^{s'}_G \ f_2 s'\langle i, a_2\rangle), \\
[s_1, a_1] \ \mathbb{R}^s_{TA} \ [s_2, a_2] &\iff \exists s_0, i_1 : s_1 \to s_0, i_2 : s_2 \to s_0 \in \mathcal{I} \\
&\qquad A(\mathrm{id}_s + i_1)a_1 \ \mathbb{R}^{s+s_0}_A \ A(\mathrm{id}_s + i_2)a_2.
\end{aligned}$$

To define the above logical relation, we do not need any operational semantics for the nu-calculus such as Pitts and Stark define in [11, 12]. And since it is defined on the pure categorical model $\boldsymbol{Set}^{\mathcal{I}}$, it can be easily generalized to study other typed lambda-calculi which $\boldsymbol{Set}^{\mathcal{I}}$ is suitable to model.

However, there is a small flaw in this logical relation. With the *span* in Pitts and Stark's operational logical relation, we can distinguish between public and private names while we cannot in the above Kripke logical relation. In fact, the above logical relation is strictly weaker than the operational one. Consider the two expressions which we have already shown to be related by the operational logical relation:

$$\nu n.\lambda x^\nu.(x = n) \ R^{exp}_{\nu \to o} \ \lambda x^\nu.false.$$

By (1), their semantics are

$$\begin{aligned}
[\![\nu n.\lambda x^\nu.(x = n)]\!]_s &= [\{n\}, |\lambda x^\nu.(x = n)|_s] \in \boldsymbol{T}((\boldsymbol{T}B)^N)s \\
[\![\lambda x^\nu.false]\!]_s &= [\emptyset, |\lambda x^\nu.false|_s] \in \boldsymbol{T}((\boldsymbol{T}B)^N)s.
\end{aligned}$$

If these two terms can be related by the above Kripke logical relation, then by the definition of $\mathbb{R}^s_{\boldsymbol{T}A}$ with $A = (\boldsymbol{T}B)^N$ we get

$$(\boldsymbol{T}B)^N(\mathrm{id}_s + i_1)|\lambda x^\nu.(x = n)|_{s+s_0} \ \mathbb{R}^{s+s_0}_{N \to \boldsymbol{T}B} \ (\boldsymbol{T}B)^N(\mathrm{id}_s + i_2)|\lambda x^\nu.false|_{s+s_0}$$

for some $s_0 \in \mathcal{I}$ and $i_1 : \{n\} \to s_0, i_2 : \emptyset \to s_0 \in \mathcal{I}$. Without losing the universality, we just let $s_0 = \{n\}, i_1 = \mathrm{id}_{\{n\}}$, and then get

$$|\lambda x^\nu.(x = n)|_{s+\{n\}} \ \mathbb{R}^{s+\{n\}}_{N \to \boldsymbol{T}B} \ |\lambda x^\nu.false|_{s+\{n\}}.$$

Now applying n to both functions, we get consequently $tt \ \mathbb{R}^{s+\{n\}}_B \ ff$, which is a contradiction to the definition of the relation for boolean type.

A direct way to represent this identification of public names is to provide the set of public names in the logical relation. We modify the structure of possible worlds by adding a name set w so as to make the pair $\langle w, s\rangle$ instead of s as the possible world. Intuitively, the new set w stands for the set of public names of two nu-calculus expressions, so we require $w \subseteq s$. The partial order on the modified possible worlds is given by

$$\langle w_1, s_1\rangle \leq \langle w_2, s_2\rangle \Longleftrightarrow w_1 \subseteq w_2 \ \ \& \ \ s_1 \subseteq s_2,$$

and we change the relation for name type:

$$n_1 \ \mathbb{R}_N^{\langle w,s\rangle} \ n_2 \Longleftrightarrow n_1 = n_2 \ \ \& \ \ n_1, n_2 \in w.$$

The intuition behind this relation is that two names are related if and only if they are equal and they are both public names.

4.2 Observation Category

Whereas in [3] the chosen observation category was the categorical model $\boldsymbol{Set}^{\mathcal{I}}$ itself, here we refine the observation category in order to be able to derive a stronger logical relation.

Consider the comma category $\mathcal{I}^{\rightarrow}$, whose objects are tuples $\langle w, i, s\rangle$ with $i : w \rightarrow s$ in $\mathcal{I}$, and whose morphisms are pairs $\langle j, k\rangle : \langle w, i, s\rangle \rightarrow \langle w', i', s'\rangle$ where $j : w \rightarrow w'$ in $\mathcal{I}$ and $k : s \rightarrow s'$ in $\mathcal{I}$ such that the following diagram commutes:

$$\begin{array}{ccc} w & \xrightarrow{i} & s \\ {\scriptstyle j}\downarrow & & \downarrow{\scriptstyle k} \\ w' & \xrightarrow[i']{} & s' \end{array}$$

We write i for $\langle w, i, s\rangle$ when the domain w and the codomain s of i are clear from context.

Let $\mathbb{C}$ be the category $\boldsymbol{Set}^{\mathcal{I}^{\rightarrow}}$ of functors from $\mathcal{I}^{\rightarrow}$ to $\boldsymbol{Set}$ and natural transformations. We define a strong monad (T, η, μ, t) on $\mathbb{C}$ by:

- $TA = \mathrm{colim}_{i'} A(\_ + i') : \mathcal{I}^{\rightarrow} \rightarrow \boldsymbol{Set}$. Explicitly:
  - TAi is the set of equivalence classes of pairs (i', a), where $i' : w' \rightarrow s'$ in $\mathcal{I}$ and $a \in A(i + i')$, modulo the smallest equivalence relation $\equiv_{\mathbb{C}}$ such that $(i', a) \equiv_{\mathbb{C}} (i'', A(\mathrm{id}_i + \langle j, k\rangle)a)$ for each morphism $\langle j, k\rangle : \langle w', i', s'\rangle \rightarrow \langle w'', i'', s''\rangle$ in $\mathcal{I}^{\rightarrow}$. We write $[i, a]$ for the equivalence class of (i, a).
  - On morphism $\langle j, k\rangle$, $TA\langle j, k\rangle$ maps the equivalence class of (i', a) to the equivalence class of $(i', A(\langle j, k\rangle + \mathrm{id}_{i'})a)$.
- For any $f : A \rightarrow B$ in $\mathbb{C}$, $Tfi : TAi \rightarrow TBi$ is defined by $Tfi[i', a] = [i', f(i + i')a]$. This is compatible with $\equiv_{\mathbb{C}}$ because f is natural.
- $\eta_A i : Ai \rightarrow TAi$ is defined by $\eta_A ia = [\emptyset, a]$.
- $\mu_A i : T^2Ai \rightarrow TAi$ is defined by $\mu_A i[i', [i'', a]] = [i' + i'', a]$
- $t_{A,B} i : Ai \times TBi \rightarrow T(A \times B)i$ is defined by $t_{A,B}i(a, [i', b]) = [i', (Ai^{\rightarrow}_{i,i'}a, b)]$ where $i^{\rightarrow}_{i,i'} : i \rightarrow i + i'$ is the canonical injection.

4.3 Natural Derivation of the Logical Relation

Let $U : \mathcal{I}^{\rightarrow} \rightarrow \mathcal{I}$ be the forgetful functor which maps an object $\langle w, i, s\rangle$ to s and a morphism $\langle j, k\rangle$ to k. Let $|\_| : \boldsymbol{C} \rightarrow \mathbb{C}$ be the functor $\mathrm{id}_{\boldsymbol{Set}}{}^{U}$ (where $\boldsymbol{C} = \boldsymbol{Set}^{\mathcal{I}}$ and $\mathbb{C} = \boldsymbol{Set}^{\mathcal{I}^{\rightarrow}}$). On an object A, $|A|$ is equal to $A \circ U$. On a morphism f (which is a natural transformation), for each $\langle w, i, s\rangle$ in $\mathcal{I}^{\rightarrow}$, the component $|f|_{\langle w,i,s\rangle}$ is equal to $(f * U)_{\langle w,i,s\rangle}$, that is to say f_s. Let $\sigma : T|\_| \rightarrow |\boldsymbol{T}\_|$ be the monad morphism[2] defined by

$$\sigma_A i[\langle w', i', s'\rangle, a] = [s', a]$$

This is well defined as $|A|(i + i') = A(s + s')$. We can then define $\sigma_{\langle A_1, A_2\rangle} : T|A_1 \times A_2| \rightarrow |\boldsymbol{T}A_1| \times |\boldsymbol{T}A_2|$ by

$$\sigma_{\langle A_1, A_2\rangle} i = \langle \sigma_{A_1} i \circ T\pi_1, \sigma_{A_2} i \circ T\pi_2\rangle$$

That is to say

$$\sigma_{\langle A_1, A_2\rangle} i[\langle w', i', s'\rangle, (a_1, a_2)] = ([s', a_1], [s', a_2])$$

$\mathbb{C}$ has a mono factorization system consisting of point-wise surjections and point-wise injections. And it is clear that T and finite product preserve point-wise surjections. The conditions required in [3] are therefore satisfied. We thus define the monadic type component of our Kripke logical relation. Each injection $f : S\langle w, i, s\rangle \hookrightarrow |A_1 \times A_2|\langle w, i, s\rangle = A_1 s \times A_2 s$ is a representation of a relation $S\langle w, i, s\rangle \subseteq A_1 s \times A_2 s$ up to isomorphism. The logical relation, denoted $\widetilde{S}$, is given by mono factorization of the composition of Tf with $\sigma_{\langle A_1, A_2\rangle}$:

$$\begin{array}{ccc} TS & \xrightarrow{Tf} & T|A_1 \times A_2| \\ \downarrow & & \downarrow \sigma_{\langle A_1, A_2\rangle} \\ \widetilde{S} & \hookrightarrow & |\boldsymbol{T}A_1| \times |\boldsymbol{T}A_2| \end{array}$$

Explicitly, $\widetilde{S} i \hookrightarrow |\boldsymbol{T}A_1| i \times |\boldsymbol{T}A_2| i$ is given by:

$$\begin{array}{l} [i_1, a_1] \;\; \widetilde{S} i \;\; [i_2, a_2] \quad \Longleftrightarrow \\ \quad \exists i_0 \in \mathcal{I}^{\rightarrow} \cdot \exists \langle j_1, k_1\rangle : i_1 \rightarrow i_0 \in \mathcal{I}^{\rightarrow} \cdot \exists \langle j_2, k_2\rangle : i_2 \rightarrow i_0 \in \mathcal{I}^{\rightarrow} \cdot \\ \quad |A_1|(\mathrm{id}_i + \langle j_1, k_1\rangle) \;\; S(i + i_0) \;\; |A_2|(\mathrm{id}_i + \langle j_2, k_2\rangle) \end{array}$$

It can be simplified in

$$\begin{array}{l} [i_1, a_1] \;\; \widetilde{S} i \;\; [i_2, a_2] \quad \Longleftrightarrow \\ \quad \exists i_0 \in \mathcal{I}^{\rightarrow} \cdot \exists \langle j_1, k_1\rangle : i_1 \rightarrow i_0 \in \mathcal{I}^{\rightarrow} \cdot \exists \langle j_2, k_2\rangle : i_2 \rightarrow i_0 \in \mathcal{I}^{\rightarrow} \cdot \\ \quad A_1(\mathrm{id}_s + k_1) \;\; S(i + i_0) \;\; A_2(\mathrm{id}_s + k_2) \end{array}$$

where s is the codomain of i.

[2] More precisely, the pair $(|\_|, \sigma)$ is the monad morphism from $\boldsymbol{T}$ to T where σ is a natural transformation. σ was named 'distributivity law' in [3]. It turned out to be a bad choice as it could be confused with 'distributive law' which denotes a close but different concept by Beck [2]. Also, while 'distributive laws' tend to be rare, monad morphisms abound.

Lemma 2. *The full subcategory $\mathcal{I}^{\subseteq}$ of $\mathcal{I}^{\rightarrow}$ consisting only of inclusions is equivalent to the whole category $\mathcal{I}^{\rightarrow}$.*

Proof. Let $F : \mathcal{I}^{\subseteq} \to \mathcal{I}^{\rightarrow}$ be the inclusion functor. Let $G : \mathcal{I}^{\rightarrow} \to \mathcal{I}^{\subseteq}$ be the functor which maps $\langle w, i, s\rangle$ to $\langle i(w), s\rangle$, and $\langle j, k\rangle : \langle w, i, s\rangle \to \langle w', i', s'\rangle$ to $i' \circ j \circ i^{-1} : \langle i(w), s\rangle \to \langle i'(w'), s'\rangle$. Then clearly, $G \circ F$ is identity, and $F \circ G$ maps $\langle w, i, s\rangle$ to $\langle i(w), s\rangle$ which are isomorphic through $\langle i, \mathrm{id}_s\rangle$ and $\langle i^{-1}\restriction_{i(w)}, \mathrm{id}_s\rangle$. So (F, G) is an equivalence of categories. □

Thus the definition of $\widetilde{S}i$ is equivalent to

$$\begin{aligned}&[\langle w_1, s_1\rangle, a_1] \quad \widetilde{S}\langle w, s\rangle \quad [\langle w_2, s_2\rangle, a_2] \iff \\ &\quad \exists s_0 \in \mathcal{I} \cdot \exists w_0 \subseteq s_0 \cdot \exists k_1 : s_1 \to s_0 \in \mathcal{I} \cdot \exists k_2 : s_2 \to s_0 \in \mathcal{I} \cdot \\ &\quad A_1(\mathrm{id}_s + k_1) \quad S(w + w_0 \subseteq s + s_0) \quad A_2(\mathrm{id}_s + k_2).\end{aligned}$$

Note that a Kripke logical relation indexed by elements of a poset can be naturally extended to be indexed by objects of a category (in this case $\mathcal{I}^{\rightarrow}$) as argued in [1]. We therefore arrive at:

Definition 3 (Kripke logical relation). *The relations $\mathbb{R}_A^{\langle w,s\rangle} \subseteq As \times As$, where $w \subseteq s$ and $A \in \boldsymbol{Set}^{\mathcal{I}}$, are defined by induction over the structure of the corresponding type in the metalanguage, according to:*

$$b_1 \ \mathbb{R}_B^{\langle w,s\rangle} \ b_2 \iff b_1 = b_2$$

$$n_1 \ \mathbb{R}_N^{\langle w,s\rangle} \ n_2 \iff n_1 = n_2 \ \wedge \ n_1, n_2 \in w$$

$$\begin{aligned}&f_1 \ \mathbb{R}_{GF}^{\langle w,s\rangle} \ f_2 \iff \\ &\quad \forall s', \forall k : s \to s' \in \mathcal{I}, \forall w' \ s.t. \ k(w) \subseteq w' \subseteq s', \forall a_1, a_2 \in Fs', \\ &\qquad (a_1 \ \mathbb{R}_F^{\langle w',s'\rangle} \ a_2 \Longrightarrow f_1 s'\langle k, a_1\rangle \ \mathbb{R}_G^{\langle w',s'\rangle} \ f_2 s'\langle k, a_2\rangle)\end{aligned}$$

$$\begin{aligned}&[s_1, a_1] \ \mathbb{R}_{TA}^{\langle w,s\rangle} \ [s_2, a_2] \iff \\ &\quad \exists s_0, w_0 \subseteq s_0, i_1 : s_1 \to s_0, i_2 : s_2 \to s_0 \in \mathcal{I}, \\ &\qquad A(id_s + i_1)a_1 \ \mathbb{R}_A^{\langle w+w_0, s+s_0\rangle} \ A(id_s + i_2)a_2.\end{aligned}$$ □

Lemma 3. *For all canonical forms $C_1, C_2 \in \mathrm{Can}_\sigma(s)$,*

$$[\![C_1]\!]_s \ \mathbb{R}_{T[\![\sigma]\!]}^{\langle w,s\rangle} \ [\![C_2]\!]_s \iff |C_1|_s \ \mathbb{R}_{[\![\sigma]\!]}^{\langle w,s\rangle} \ |C_2|_s$$

Proof. The interpretation of nu-calculus canonical forms into category $\boldsymbol{Set}^{\mathcal{I}}$ provides us $[\![C_k]\!]_s = [\emptyset, |C_k|_s]$ $(k = 1, 2)$. According to the definition of the Kripke logical relation,

$$[\emptyset, |C_1|_s] \ \mathbb{R}_{T[\![\sigma]\!]}^{\langle w,s\rangle} \ [\emptyset, |C_2|_s] \iff [\![\sigma]\!](\mathrm{id}_s + i_1)|C_1|_s \ \mathbb{R}_{[\![\sigma]\!]}^{\langle w+w_0, s+s_0\rangle} \ [\![\sigma]\!](\mathrm{id}_s + i_2)|C_2|_s$$

for some $s_0, w_0 \subseteq s_0$ and $i_1, i_2 : \emptyset \to s_0 \in \mathcal{I}$. Since whatever s_0 and w_0 are, i_1 and i_2 are always the empty injection, we simply let s_0, w_0 be the empty set and i_1, i_2 be $\mathrm{id}_\emptyset$. Then we get $[\![\sigma]\!](\mathrm{id}_s + i_k)|C_k|_s = |C_k|_s$, $(k = 1, 2)$, and consequently,

$$[\![C_1]\!]_s \ \mathbb{R}_{T[\![\sigma]\!]}^{\langle w,s\rangle} \ [\![C_2]\!]_s \iff |C_1|_s \ \mathbb{R}_{[\![\sigma]\!]}^{\langle w,s\rangle} \ |C_2|_s.$$ □

4.4 Equivalence between Operational and Kripke Logical Relations

Through this Kripke logical relation, we can relate the two nu-calculus expressions in Section 4.1 again. To prove that, for $\mathbb{R}_{TA}^{\langle w,s\rangle}$ with $A = (\boldsymbol{T}B)^N$, we let $w_0 = \emptyset$ and by the definition of $\mathbb{R}_{TA}^{\langle w,s\rangle}$, we need to prove

$$|\lambda x^\nu.(x = n)|_{s+\{n\}} \ \mathbb{R}_{(\boldsymbol{T}B)^N}^{\langle w,s+\{n\}\rangle} \ |\lambda x^\nu.false|_{s+\{n\}}.$$

By the definition of $\mathbb{R}_N^{\langle w,s+\{n\}\rangle}$, related names must be in set w and are definitely different from n, so applying any related name to the function $|\lambda x^\nu.(x = n)|_{s+\{n\}}$ must return the value tt (by (1)), which is same as the result of applying it to the second function. Then, by the definition of $\mathbb{R}_{N\to\boldsymbol{T}B}^{\langle w,s+\{n\}\rangle}$ and $\mathbb{R}_B^{\langle w,s+\{n\}\rangle}$, the above two functions can be related.

Theorem 1. *If $R : s_1 \rightleftharpoons s_2$ is a span and $s_1 \vdash M_1 : \sigma$, $s_2 \vdash M_2 : \sigma$ where σ is a ground type or first-order type, then*

$$M_1 \ R_\sigma \ M_2 \iff \exists s, w \subseteq s, i_1 : s_1 \to s, i_2 : s_2 \to s \in \mathcal{I},$$
$$\boldsymbol{T}[\![\sigma]\!]i_1([\![M_1]\!]_{s_1}) \ \mathbb{R}_{\boldsymbol{T}[\![\sigma]\!]}^{\langle w,s\rangle} \ \boldsymbol{T}[\![\sigma]\!]i_2([\![M_2]\!]_{s_2}).$$

Proof. If M_1 and M_2 are two expressions in canonical form, by the definition of the category $\boldsymbol{Set}^{\mathcal{I}}$ in Section 3.1,

$$\boldsymbol{T}[\![\sigma]\!]i_k[\![M_1]\!]_{s_k} = \boldsymbol{T}[\![\sigma]\!]i_k[\emptyset, |M_1|_{s_k}] = [\emptyset, [\![\sigma]\!]i_k|M_1|_{s_k}] \qquad (k = 1, 2).$$

Then by Lemma 3, the theorem is equivalent to

$$M_1 \ R_\sigma^{can} \ M_2 \iff [\![\sigma]\!]i_1(|M_1|_{s_1}) \ \mathbb{R}_{[\![\sigma]\!]}^{\langle w,s\rangle} \ [\![\sigma]\!]i_2(|M_2|_{s_2})$$

for some $w \subseteq s$. So we can made the proof by induction on the structure of types in the metalanguage.

- The case for boolean type is obvious.
- For name type, we should define a proper w and the injections i_1, i_2 to represent the span R in the operational logical relation:

  $$\forall n_1 \in s_1, n_2 \in s_2 \ (i_1(n_1) = i_2(n_2) \iff n_1 \ R \ n_2)$$

  and w is the image of the domain of R under i_1, equivalently the image of the codomain of R under i_2. Because $Ni_k = i_k$, we have

  $$n_1 \ R_\nu^{can} \ n_2 \iff Ni_1(n_1) = Ni_2(n_2) \iff Ni_1(n_1) \ \mathbb{R}_N^{\langle w,s\rangle} \ Ni_2(n_2).$$

- For first-order function types in the metalanguage ($\sigma \equiv \tau \to \tau'$), the only possibility is the nu-calculus canonical expressions of the form $\lambda x^\tau.M$. Suppose the type of M is τ', then the canonical interpretation $|\lambda x^\tau.M|$ is of type $[\![\tau]\!] \to \boldsymbol{T}[\![\tau']\!]$. The definability result of the model $\boldsymbol{Set}^{\mathcal{I}}$ (Lemma 1) ensure that for each $f \in (\boldsymbol{T}[\![\tau']\!])^{[\![\tau]\!]}s$, we can find a canonical function expression

$\lambda x^\tau.M'$, where M' is of type τ', such that $|\lambda x^\tau.M'|_s = f$. Then by the induction hypothesis and the definitions of $R^{can}_{\tau\to\tau'}$ and $\mathbb{R}^{\langle w,s\rangle}_{(\boldsymbol{T}[\![\tau']\!])^{[\![\tau]\!]}}$, we get

$$\lambda x^\tau.M'_1 \; R^{can}_{\tau\to\tau'} \; \lambda x^\tau.M'_2 \iff$$
$$(\boldsymbol{T}[\![\tau']\!])^{[\![\tau]\!]} i_1(|\lambda x^\tau.M'_1|_s) \; \mathbb{R}^{\langle w,s\rangle}_{(\boldsymbol{T}[\![\tau']\!])^{[\![\tau]\!]}} \; (\boldsymbol{T}[\![\tau']\!])^{[\![\tau]\!]} i_2(|\lambda x^\tau.M'_2|_s)$$

for some $w \subseteq s$ and $i_1 : s_1 \to s, i_2 : s_2 \to s \in \mathcal{I}$.

- The case for the monadic types in the metalanguage corresponds to R^{exp}_σ in the operational logical relation. If $s_1 \vdash M_1 \Downarrow^{s'_1}_\sigma C_1$ and $s_2 \vdash M_2 \Downarrow^{s'_2}_\sigma C_2$, then by the definition of R^{exp}_σ we can find some $R' : s'_1 \rightleftharpoons s'_2$ such that $C_1 \; (R \oplus R')^{can}_\sigma \; C_2$, and consequently by the induction hypothesis, we get

$$[\![\sigma]\!] j_1(|C_1|_{s_1+s'_1}) \; \mathbb{R}^{\langle w',s'\rangle}_{[\![\sigma]\!]} \; [\![\sigma]\!] j_2(|C_2|_{s_2+s'_2}),$$

for some $w' \subseteq s'$ and $j_1 : s_1 + s'_1 \to s', j_2 : s_2 + s'_2 \to s' \in \mathcal{I}$. On the other hand, by (1) and the definition of $\boldsymbol{Set}^{\mathcal{I}}$, we have

$$\boldsymbol{T}[\![\sigma]\!] i_k([\![M_k]\!]_{s_k}) = [s'_k, [\![\sigma]\!](i_k + \mathrm{id}_{s'_k})|C_k|_{s_k+s'_k}] \qquad (k = 1, 2),$$

and according to the definition of $\mathbb{R}^{\langle w,s\rangle}_{\boldsymbol{T}[\![\sigma]\!]}$,

$$\boldsymbol{T}[\![\sigma]\!] i_1([\![M_1]\!]_{s_1}) \; \mathbb{R}^{\langle w,s\rangle}_{\boldsymbol{T}[\![\sigma]\!]} \; \boldsymbol{T}[\![\sigma]\!] i_2([\![M_2]\!]_{s_2}) \iff$$
$$\exists s_0, w_0 \subseteq s_0, i'_1 : s'_1 \to s_0, i'_2 : s'_2 \to s_0 \in \mathcal{I},$$
$$[\![\sigma]\!](i_1 + i'_1)|C_1|_{s_1+s'_1} \; \mathbb{R}^{\langle w+w_0,s+s_0\rangle}_{[\![\sigma]\!]} \; [\![\sigma]\!](i_2 + i'_2)|C_2|_{s_2+s'_2}.$$

Let i_k, i'_k $(k = 1, 2)$ be defined as

$$\forall n \in s_k, i_k(n) = j_k(n) \qquad \text{and} \qquad \forall n' \in s'_k, i'_k(n') = j_k(n'),$$

and let $s + s_0 = s'$, $w + w_0 = w'$, then we get

$$M_1 \; R^{exp}_\sigma \; M_2 \iff \boldsymbol{T}[\![\sigma]\!] i_1([\![M_1]\!]_{s_1}) \; \mathbb{R}^{\langle w,s\rangle}_{\boldsymbol{T}[\![\sigma]\!]} \; \boldsymbol{T}[\![\sigma]\!] i_2([\![M_2]\!]_{s_2}). \qquad \square$$

We only claim the equivalence between these two logical relations for ground types and first-order function types. The failure of the equivalence for high-order function types is due to the non-fully abstract categorical model, while a similar Kripke logical relation defined directly on the computational metalanguage has been shown to be equivalent to the operational logical relation for any type [16].

5 Conclusion

We construct a Kripke logical relation for Pitts and Stark's nu-calculus in this paper and we show that it is equivalent to their operational logical relation up to first-order types. Because our Kripke logical relation is defined on a categorical model, it does not rely on any operational aspect of the nu-calculus, hence it is easier to be generalized. Our future work is to use a similar method to derive logical relations for the cryptographic lambda calculus, which is syntactically an extension of the nu-calculus (Sumii and Pierce define a logical relation for this language in [15], but it is still very operational).

Acknowledgements

We are grateful to Jean Goubault-Larrecq, John Power and Ian Stark for useful discussions and to anonymous reviewers for their comments.

References

1. Moez Alimohamed. A characterization of lambda definability in categorical models of implicit polymorphism. *Theoretical Computer Science*, 146(1–2):5–23, 1995.
2. Jon Beck. Distributive laws. In *Seminar on Triples and Categorical Homology Theory*, volume 80 of *Lecture Notes in Mathematics*, pages 119–140. Springer-Verlag, 1969.
3. Jean Goubault-Larrecq, Slawomir Lasota, and David Nowak. Logical relations for monadic types. In *CSL'02*, volume 2471 of *LNCS*, pages 553–568. Springer-Verlag, 2002.
4. Joachim Lambek and Philip J. Scott. *Introduction to Higher Order Categorical Logic.* Cambridge studies in advanced mathematics. Cambridge University Press, 1986.
5. John C. Mitchell. *Foundations of Programming Languages.* MIT Press, 1996.
6. John C. Mitchell and Eugenio Moggi. Kripke-style models for typed lambda calculus. In *LICS'87*, pages 303–314. The Computer Society of the IEEE, 1987.
7. John C. Mitchell and Andre Scedrov. Notes on sconing and relators. In *CSL'92*, volume 702 of *LNCS*, pages 352–378. Springer Verlag, 1993.
8. Eugenio Moggi. Computational lambda-calculus and monads. In *LICS'89*, pages 14–23. IEEE Computer Society Press, 1989.
9. Eugenio Moggi. An abstract view of programming languages. Technical Report ECS-LFCS-90-113, LFCS, Department of Computer Science, University of Edinburgh, 1990.
10. Eugenio Moggi. Notions of computation and monads. *Information and Computation*, 93:55–92, 1991.
11. Andrew Pitts and Ian Stark. Observable properties of higher order functions that dynamically create local names, or: What's *new*? In *MFCS'93*, volume 711 of *LNCS*, pages 122–141. Springer-Verlag, 1993.
12. Ian Stark. *Names and Higher-Order Functions.* PhD thesis, University of Cambridge, 1994.
13. Ian Stark. Categorical models for local names. *Lisp and Symbolic Computation*, 9(1):77–107, 1996.
14. Ian Stark. Names, equations, relations: Practical ways to reason about *new*. *Fundamenta Informaticae*, 33(4):369–396, 1998.
15. Eijiro Sumii and Benjamin Pierce. Logical relations for encryption. In *CSFW'01*, pages 256–272. IEEE, 2001.
16. Yu Zhang. Logical relations for names. Master's thesis, University of Paris 7, 2002. *http://www.lsv.ens-cachan.fr/Publis/PAPERS/ZY-dea02.ps.*

Author Index

Lecture Notes in Computer Science

For information about Vols. 1–2690
please contact your bookseller or Springer-Verlag

Vol. 2564: W. Truszkowski, C. Rouff, M. Hinchey (Eds.), Innovative Concepts for Agent-Based Systems. Proceedings, 2002. X, 476 pages. 2003. (Subseries LNAI).

Vol. 2690: J. Liu, Y. Cheung, H. Yin (Eds.), Intelligent Data Engineering and Automated Learning. Proceedings, 2003. XXI, 1141 pages. 2003.

Vol. 2691: V. Mařík, J. Müller, M. Pěchouček (Eds.), Multi-Agent Systems and Applications III. Proceedings, 2003. XIV, 660 pages. 2003. (Subseries LNAI).

Vol. 2692: P. Nixon, S. Terzis (Eds.), Trust Management. Proceedings, 2003. X, 349 pages. 2003.

Vol. 2693: A. Cechich, M. Piattini, A. Vallecillo (Eds.), Component-Based Software Quality. X, 403 pages. 2003.

Vol. 2694: R. Cousot (Ed.), Static Analysis. Proceedings, 2003. XIV, 505 pages. 2003.

Vol. 2695: L.D. Griffin, M. Lillholm (Eds.), Scale Space Methods in Computer Vision. Proceedings, 2003. XII, 816 pages. 2003.

Vol. 2696: J. Feigenbaum (Ed.), Digital Rights Management. Proceedings, 2002. X, 221 pages. 2003.

Vol. 2697: T. Warnow, B. Zhu (Eds.), Computing and Combinatorics. Proceedings, 2003. XIII, 560 pages. 2003.

Vol. 2698: W. Burakowski, B. Koch, A. Bęben (Eds.), Architectures for Quality of Service in the Internet. Proceedings, 2003. XI, 305 pages. 2003.

Vol. 2699: M.G. Hinchey, J.L. Rash, W.F. Truszkowski, C. Rouff, D. Gordon-Spears (Eds.), Formal Approaches to Agent-Based Systems. Proceedings, 2002. IX, 297 pages. 2003. (Subseries LNAI).

Vol. 2700: M.T. Pazienza (Ed.), Information Extraction in the Web Era. XIII, 163 pages. 2003. (Subseries LNAI).

Vol. 2701: M. Hofmann (Ed.), Typed Lambda Calculi and Applications. Proceedings, 2003. VIII, 317 pages. 2003.

Vol. 2702: P. Brusilovsky, A. Corbett, F. de Rosis (Eds.), User Modeling 2003. Proceedings, 2003. XIV, 436 pages. 2003. (Subseries LNAI).

Vol. 2704: S.-T. Huang, T. Herman (Eds.), Self-Stabilizing Systems. Proceedings, 2003. X, 215 pages. 2003.

Vol. 2705: S. Renals, G. Grefenstette (Eds.), Text- and Speech-Triggered Information Access. Proceedings, 2000. VII, 197 pages. 2003. (Subseries LNAI).

Vol. 2706: R. Nieuwenhuis (Ed.), Rewriting Techniques and Applications. Proceedings, 2003. XI, 515 pages. 2003.

Vol. 2707: K. Jeffay, I. Stoica, K. Wehrle (Eds.), Quality of Service – IWQoS 2003. Proceedings, 2003. XI, 517 pages. 2003.

Vol. 2708: R. Reed, J. Reed (Eds.), SDL 2003: System Design. Proceedings, 2003. XI, 405 pages. 2003.

Vol. 2709: T. Windeatt, F. Roli (Eds.), Multiple Classifier Systems. Proceedings, 2003. X, 406 pages. 2003.

Vol. 2710: Z. Ésik, Z, Fülöp (Eds.), Developments in Language Theory. Proceedings, 2003. XI, 437 pages. 2003.

Vol. 2711: T.D. Nielsen, N.L. Zhang (Eds.), Symbolic and Quantitative Approaches to Reasoning with Uncertainty. Proceedings, 2003. XII, 608 pages. 2003. (Subseries LNAI).

Vol. 2712: A. James, B. Lings, M. Younas (Eds.), New Horizons in Information Management. Proceedings, 2003. XII, 281 pages. 2003.

Vol. 2713: C.-W. Chung, C.-K. Kim, W. Kim, T.-W. Ling, K.-H. Song (Eds.), Web and Communication Technologies and Internet-Related Social Issues – HSI 2003. Proceedings, 2003. XXII, 773 pages. 2003.

Vol. 2714: O. Kaynak, E. Alpaydin, E. Oja, L. Xu (Eds.), Artificial Neural Networks and Neural Information Processing – ICANN/ICONIP 2003. Proceedings, 2003. XXII, 1188 pages. 2003.

Vol. 2715: T. Bilgiç, B. De Baets, O. Kaynak (Eds.), Fuzzy Sets and Systems – IFSA 2003. Proceedings, 2003. XV, 735 pages. 2003. (Subseries LNAI).

Vol. 2716: M.J. Voss (Ed.), OpenMP Shared Memory Parallel Programming. Proceedings, 2003. VIII, 271 pages. 2003.

Vol. 2718: P. W. H. Chung, C. Hinde, M. Ali (Eds.), Developments in Applied Artificial Intelligence. Proceedings, 2003. XIV, 817 pages. 2003. (Subseries LNAI).

Vol. 2719: J.C.M. Baeten, J.K. Lenstra, J. Parrow, G.J. Woeginger (Eds.), Automata, Languages and Programming. Proceedings, 2003. XVIII, 1199 pages. 2003.

Vol. 2720: M. Marques Freire, P. Lorenz, M.M.-O. Lee (Eds.), High-Speed Networks and Multimedia Communications. Proceedings, 2003. XIII, 582 pages. 2003.

Vol. 2721: N.J. Mamede, J. Baptista, I. Trancoso, M. das Graças Volpe Nunes (Eds.), Computational Processing of the Portuguese Language. Proceedings, 2003. XIV, 268 pages. 2003. (Subseries LNAI).

Vol. 2722: J.M. Cueva Lovelle, B.M. González Rodríguez, L. Joyanes Aguilar, J.E. Labra Gayo, M. del Puerto Paule Ruiz (Eds.), Web Engineering. Proceedings, 2003. XIX, 554 pages. 2003.

Vol. 2723: E. Cantú-Paz, J.A. Foster, K. Deb, L.D. Davis, R. Roy, U.-M. O'Reilly, H.-G. Beyer, R. Standish, G. Kendall, S. Wilson, M. Harman, J. Wegener, D. Dasgupta, M.A. Potter, A.C. Schultz, K.A. Dowsland, N. Jonoska, J. Miller (Eds.), Genetic and Evolutionary Computation – GECCO 2003. Proceedings, Part I. 2003. XLVII, 1252 pages. 2003.

Vol. 2724: E. Cantú-Paz, J.A. Foster, K. Deb, L.D. Davis, R. Roy, U.-M. O'Reilly, H.-G. Beyer, R. Standish, G. Kendall, S. Wilson, M. Harman, J. Wegener, D. Dasgupta, M.A. Potter, A.C. Schultz, K.A. Dowsland, N. Jonoska, J. Miller (Eds.), Genetic and Evolutionary Computation – GECCO 2003. Proceedings, Part II. 2003. XLVII, 1274 pages. 2003.

Vol. 2725: W.A. Hunt, Jr., F. Somenzi (Eds.), Computer Aided Verification. Proceedings, 2003. XII, 462 pages. 2003.

Vol. 2726: E. Hancock, M. Vento (Eds.), Graph Based Representations in Pattern Recognition. Proceedings, 2003. VIII, 271 pages. 2003.

Vol. 2727: R. Safavi-Naini, J. Seberry (Eds.), Information Security and Privacy. Proceedings, 2003. XII, 534 pages. 2003.

Vol. 2728: E.M. Bakker, T.S. Huang, M.S. Lew, N. Sebe, X.S. Zhou (Eds.), Image and Video Retrieval. Proceedings, 2003. XIII, 512 pages. 2003.

Vol. 2729: D. Boneh (Ed.), Advances in Cryptology – CRYPTO 2003. Proceedings, 2003. XII, 631 pages. 2003.

Vol. 2730: F. Bai, B. Wegner (Eds.), Electronic Information and Communication in Mathematics. Proceedings, 2002. X, 189 pages. 2003.

Vol. 2731: C.S. Calude, M.J. Dinneen, V. Vajnovszki (Eds.), Discrete Mathematics and Theoretical Computer Science. Proceedings, 2003. VIII, 301 pages. 2003.

Vol. 2732: C. Taylor, J.A. Noble (Eds.), Information Processing in Medical Imaging. Proceedings, 2003. XVI, 698 pages. 2003.

Vol. 2733: A. Butz, A. Krüger, P. Olivier (Eds.), Smart Graphics. Proceedings, 2003. XI, 261 pages. 2003.

Vol. 2734: P. Perner, A. Rosenfeld (Eds.), Machine Learning and Data Mining in Pattern Recognition. Proceedings, 2003. XII, 440 pages. 2003. (Subseries LNAI).

Vol. 2735: F. Kaashoek, I. Stoica (Eds.), Peer-to-Peer Systems II. Proceedings, 2003. XI, 316 pages. 2003.

Vol. 2740: E. Burke, P. De Causmaecker (Eds.), Practice and Theory of Automated Timetabling IV. Proceedings, 2002. XII, 361 pages. 2003.

Vol. 2741: F. Baader (Ed.), Automated Deduction – CADE-19. Proceedings, 2003. XII, 503 pages. 2003. (Subseries LNAI).

Vol. 2742: R. N. Wright (Ed.), Financial Cryptography. Proceedings, 2003. VIII, 321 pages. 2003.

Vol. 2743: L. Cardelli (Ed.), ECOOP 2003 – Object-Oriented Programming. Proceedings, 2003. X, 501 pages. 2003.

Vol. 2744: V. Mařík, D. McFarlane, P. Valckenaers (Eds.), Holonic and Multi-Agent Systems for Manufacturing. Proceedings, 2003. XI, 322 pages. 2003. (Subseries LNAI).

Vol. 2745: M. Guo, L.T. Yang (Eds.), Parallel and Distributed Processing and Applications. Proceedings, 2003. XII, 450 pages. 2003.

Vol. 2746: A. de Moor, W. Lex, B. Ganter (Eds.), Conceptual Structures for Knowledge Creation and Communication. Proceedings, 2003. XI, 405 pages. 2003. (Subseries LNAI).

Vol. 2747: B. Rovan, P. Vojtáš (Eds.), Mathematical Foundations of Computer Science 2003. Proceedings, 2003. XIII, 692 pages. 2003.

Vol. 2748: F. Dehne, J.-R. Sack, M. Smid (Eds.), Algorithms and Data Structures. Proceedings, 2003. XII, 522 pages. 2003.

Vol. 2749: J. Bigun, T. Gustavsson (Eds.), Image Analysis. Proceedings, 2003. XXII, 1174 pages. 2003.

Vol. 2750: T. Hadzilacos, Y. Manolopoulos, J.F. Roddick, Y. Theodoridis (Eds.), Advances in Spatial and Temporal Databases. Proceedings, 2003. XIII, 525 pages. 2003.

Vol. 2751: A. Lingas, B.J. Nilsson (Eds.), Fundamentals of Computation Theory. Proceedings, 2003. XII, 433 pages. 2003.

Vol. 2752: G.A. Kaminka, P.U. Lima, R. Rojas (Eds.), RoboCup 2002: Robot Soccer World Cup VI. XVI, 498 pages. 2003. (Subseries LNAI).

Vol. 2753: F. Maurer, D. Wells (Eds.), Extreme Programming and Agile Methods – XP/Agile Universe 2003. Proceedings, 2003. XI, 215 pages. 2003.

Vol. 2754: M. Schumacher, Security Engineering with Patterns. XIV, 208 pages. 2003.

Vol. 2756: N. Petkov, M.A. Westenberg (Eds.), Computer Analysis of Images and Patterns. Proceedings, 2003. XVIII, 781 pages. 2003.

Vol. 2758: D. Basin, B. Wolff (Eds.), Theorem Proving in Higher Order Logics. Proceedings, 2003. X, 367 pages. 2003.

Vol. 2759: O.H. Ibarra, Z. Dang (Eds.), Implementation and Application of Automata. Proceedings, 2003. XI, 312 pages. 2003.

Vol. 2761: R. Amadio, D. Lugiez (Eds.), CONCUR 2003 - Concurrency Theory. Proceedings, 2003. XI, 524 pages. 2003.

Vol. 2762: G. Dong, C. Tang, W. Wang (Eds.), Advances in Web-Age Information Management. Proceedings, 2003. XIII, 512 pages. 2003.

Vol. 2763: V. Malyshkin (Ed.), Parallel Computing Technologies. Proceedings, 2003. XIII, 570 pages. 2003.

Vol. 2764: S. Arora, K. Jansen, J.D.P. Rolim, A. Sahai (Eds.), Approximation, Randomization, and Combinatorial Optimization. Proceedings, 2003. IX, 409 pages. 2003.

Vol. 2765: R. Conradi, A.I. Wang (Eds.), Empirical Methods and Studies in Software Engineering. VIII, 279 pages. 2003.

Vol. 2766: S. Behnke, Hierarchical Neural Networks for Image Interpretation. XII, 224 pages. 2003.

Vol. 2769: T. Koch, I. T. Sølvberg (Eds.), Research and Advanced Technology for Digital Libraries. Proceedings, 2003. XV, 536 pages. 2003.

Vol. 2777: B. Schölkopf, M.K. Warmuth (Eds.), Learning Theory and Kernel Machines. Proceedings, 2003. XIV, 746 pages. 2003. (Subseries LNAI).

Vol. 2783: W. Zhou, P. Nicholson, B. Corbitt, J. Fong (Eds.), Advances in Web-Based Learning – ICWL 2003. Proceedings, 2003. XV, 552 pages. 2003.

Vol. 2786: F. Oquendo (Ed.), Software Process Technology. Proceedings, 2003. X, 173 pages. 2003.

Vol. 2787: J. Timmis, P. Bentley, E. Hart (Eds.), Artificial Immune Systems. Proceedings, 2003. XI, 299 pages. 2003.

Vol. 2789: L. Böszörményi, P. Schojer (Eds.), Modular Programming Languages. Proceedings, 2003. XIII, 271 pages. 2003.

Vol. 2790: H. Kosch, L. Böszörményi, H. Hellwagner (Eds.), Euro-Par 2003 Parallel Processing. Proceedings, 2003. XXXV, 1320 pages. 2003.

Vol. 2794: P. Kemper, W. H. Sanders (Eds.), Computer Performance Evaluation. Proceedings, 2003. X, 309 pages. 2003.

Vol. 2796: M. Cialdea Mayer, F. Pirri (Eds.), Automated Reasoning with Analytic Tableaux and Related Methods. Proceedings, 2003. X, 271 pages. 2003. (Subseries LNAI).

Vol. 2803: M. Baaz, J.A. Makowsky (Eds.), Computer Science Logic. Proceedings, 2003. XII, 589 pages. 2003.